Andreas Birk   Silvia Coradeschi
Satoshi Tadokoro (Eds.)

RoboCup 2001: Robot Soccer World Cup V

Springer

Author

Andreas Birk
Interantional University Bremen, School of Engineering and Science
Campus Ring 1, 28759 Bremen, Germany
E-mail: a.birk@iu-bremen.de

Silvia Coradeschi
Örebro University, Department of Technology
70182 Örebro, Sweden
E-mail: silvia.coradeschi@tech.oru.se

Satoshi Tadokoro
Kobe University, Department of Computer and Systems Engineering
Rokkodai, Nada, Kobe 657-8501 Japan
E-mail: tadokoro@cs.kobe-u.ac.jp

Cataloging-in-Publication Data applied for

Die Deutsche Bibliothek - CIP-Einheitsaufnahme

RoboCup <5, 2001, Seattle, Wash.>:
Robot Soccer World Cup V / RoboCup 2001. Andreas Birk ... (ed.). - Berlin ;
Heidelberg ; New York ; Barcelona ; Hong Kong ; London ; Milan ; Paris ;
Tokyo : Springer, 2002
(Lecture notes in computer science ; Vol. 2377 : Lecture notes in
artificial intelligence)
ISBN 3-540-43912-9

CR Subject Classification (1998): I.2, C.2.4, D.2.7, H.5, I.5.4, J.4

ISSN 0302-9743
ISBN 3-540-43912-9 Springer-Verlag Berlin Heidelberg New York

Springer-Verlag Berlin Heidelberg New York
a member of BertelsmannSpringer Science+Business Media GmbH

http://www.springer.de

Printed in Germany

Typesetting: Camera-ready by author, data conversion by Boller Mediendesign
Printed on acid-free paper SPIN: 10870431 06/3142 5 4 3 2 1 0

Preface

RoboCup 2001, the Fifth Robot World Cup Soccer Games and Conferences, was held from August 2–10, 2001, at the Washington State Convention and Trade Center in Seattle, USA. Like the previous international RoboCup events – RoboCup 97 in Nagoya, Japan; RoboCup 98 in Paris, France; RoboCup 99 in Stockholm, Sweden; and RoboCup 2000 in Melbourne, Australia – RoboCup 2001 included a symposium as well as several robotic competitions. Both parts, the symposium and the tournaments, are documented in this book.

The symposium received over 80 submissions of which 18 were selected for full presentation, i.e., a talk at the symposium and a 10-page contribution to this book, and 40 were selected as posters, i.e., a poster presentation at the symposium and a 6-page contribution to this book. Among the full presentations, five were selected as finalists for the Scientific and the Engineering Challenge Awards. These five papers are presented separately in the second chapter of this book. The Scientific Challenge Award went to the contribution "A Control Method for Humanoid Biped Walking with Limited Torque" by Fuminori Yamasaki, Ken Endo, Minoru Asada, and Hiroaki Kitano. The Engineering Challenge Award was given to the paper "A Fast Vision System for Middle Size Robots in RoboCup" by M. Jamzad, B.S. Sadjad, V.S. Mirrokni, M. Kazemi, R. Ghorbani, A. Foroughnassiraei, H. Chitsaz, and A. Heydarnoori. The symposium also featured an invited talk by Prof. Takanishi from Waseda University on "Humanoid Robots in the 21st Century".

The RoboCup tournaments were organized in five leagues, namely Soccer Simulation, Small-Size Robot (F180), Mid-Size Robot (F2000), Sony Legged Robot, and Rescue Simulation. The results of all games are listed in an overview article at the very beginning of this book. Information about each team is found in its team description paper. The team descriptions, that are grouped by league, serve to catalog the full range of researchers and approaches within RoboCup.

The next international RoboCup events will be held in Fukuoka, Japan (2002) and in Padua, Italy (2003).

March 2002 Andreas Birk, Silvia Coradeschi, and Satoshi Tadokoro

RoboCup Federation

The RoboCup Federation, the governing body of RoboCup, is an international organization that promotes science and technology using soccer games by robots and software agents.

RoboCup 2001 Organization and Support

General Chair:

Manuela Veloso, Carnegie Mellon University, USA

Associate Chairs:

Tucker Balch, Carnegie Mellon University, USA
Peter Stone, AT&T Labs – Research, USA

Simulator League Committee:

Gal Kaminka (chair), Carnegie Mellon University, USA
Mikhail Prokopenko, CSIRO, Australia
Martin Riedmiller, University of Karlsruhe, Germany
Masayuki Ohta, Tokyo Institute of Technology, Japan

Small-Size Robot League (F180) Committee:

Raul Rojas (chair), University of Berlin, Germany
Brett Browning, Carnegie Mellon University, USA
Ng Beng Kiat, NgeeAnn Polytechnic, Singapore
Gordon Wyeth, University of Queens, Australia

Middle-Size Robot League (F2000) Committee:

Pedro Lima (chair), Technical University of Lisbon, Portugal
Gerhard Kraetzschmar, University of Ulm, Germany
Mark Makies, Royal Melbourne Institute of Technology, Australia
Takeshi Ohashi, Kyushu Institute of Technology, Japan
Wei-Min Shen, University of Southern California, USA

Sony Legged Robot League Committee:

Masahiro Fujita (chair), Sony Inc., Japan
Daniele Nardi, University of Rome "La Sapienza" Italy
Jim Ostrowski, University of Pennsylvania, USA

Junior Committee:

Elizabeth Sklar (chair), Boston College, USA
Amy Emura, Cambridge University, United Kingdom
Maria Hybinette, Georgia Institute of Technology, USA
Henrik Lund, University of Southern Denmark, Denmark
Brian Thomas, Bellarine Secondary College, Australia

Supported by:

American Association for Artificial Intelligence (AAAI)
Carnegie Mellon University, School of Computer Science

RoboCup Worldwide Sponsors 2001:

Sony Corporation
SGI

Symposium Program Committee

Giovanni Adorni, Italy
David Andre, USA
Ron Arkin, USA
Minoru Asada, Japan
Hajime Asama, Japan
Tucker Balch, USA
Jacky Baltes, New Zealand
Tony Belpaeme, Belgium
Ryad Benosman, France
Ansgar Bredenfeld, Germany
Hans-Dieter Burkhard, Germany
Hendrik Van Brussel, Belgium
Thomas Christaller, Germany
Ruediger Dillmann, Germany
Marco Dorigo, Belgium
Dominique Duhaut, France
Edmund H. Durfee, USA
Ian Frank, Japan
Masahiro Fujita, Japan
John Hallam, United Kingdom
Gillian Hayes, United Kingdom
Wiebe van der Hoek, The Netherlands
Andrew Howard, USA
Huosheng Hu, United Kingdom
Mansour Jamzad, Iran
Andrew Jennings, Australia
Holger Kenn, Belgium
Gerhard Kraetzschmar, Germany
Ben Kuipers, USA
Paul Levi, Germany
Pedro Lima, Portugal
Henrik Lund, Denmark
Ramon Lopez de Mantaras, Spain
Hitoshi Matsubara, Japan
Fumitoshi Matsuno, Japan
Robin Murphy, USA
Yuki Nakagawa, Japan
Daniele Nardi, Italy
Itsuki Noda, Japan
Tairo Nomura, Japan
Marnix Nuttin, Belgium
Tsukasa Ogasawara, Japan
Masayuki Ohta, Japan
Yoshikazu Ohtsubo, Japan
Koichi Osuka, Japan
Enrico Pagello, Italy
Daniele Polani, Germany
Mikhail Prokopenko, Australia
Martin Riedmiller, Germany
Raul Rojas, Germany
Josep de la Rosa, Spain
Alessandro Saffiotti, Sweden
Paul Scerri, Sweden
Elisabeth Sklar, USA
Patrick van der Smagt, Germany
Frieder Stolzenburg, Germany
Peter Stone, USA
Shoji Suzuki, Japan
Katia Sycara, USA
Milind Tambe, USA
Tomoichi Takahashi, Japan
Kenichi Tokuda, Japan
Micheal Wooldridge, United Kingdom
Jeremy Wyatt, United Kingdom
Willem H. Zuidema, Belgium

Table of Contents

Technical Papers

Full Papers

Poster Descriptions

Team Descriptions

Simulation League

Small-Size Robot (F180) League

Middle-Size Robot (F2000) League

Sony Legged Robot League

The Tournament Results of the Different Leagues of RoboCup-2001

Andreas Birk[1], Silvia Coradeschi[2], and Satoshi Tadokoro[3]

[1] International University Bremen, Germany,
birk@ieee.org
[2] Orebro University, Sweden,
silvia.coradeschi@tech.oru.se
[3] Kobe University, Japan,
tadokoro@cs.kobe-u.ac.jp

1 Overview

A complete overview of all results from the RoboCup 2001 tournaments in the different leagues is given here. In most leagues, the competition consisted of two parts. In the first part, teams were placed in groups were they played round robin games. After this, the teams were ranked in their groups. Based on the ranking, a second part with elimination rounds took place to determine the places.

In addition to the presentation of all results, some basic information on each league is included here. More detailed information on each team is given in its Team Description Paper. The team description are found in a latter part of this book grouped by leagues.

2 Soccer Simulation League

In the soccer simulation league a full 11 against 11 game is played via the so-called soccer server. The soccer server provides a virtual field to which simulated players can connect as clients via UDP/IP. So, simulated soccer teams can be implemented on any platform in any language that provides support for the basic UDP/IP protocol. The soccer server features a companion program, the soccer monitor, that serves for visualization of the game. The soccer server and monitor are open systems that can be downloaded via the RoboCup website www.robocup.org.

The rest of this section includes all the results from the games in the simulator league. The tables show first the preliminary round games, and then the second level group games including numbers of wins, losses, and draws (W/L/D), and the rank of each team within each group. Figure 1 shows the results of the double-elimination championship tournament.

A. Birk, S. Coradeschi, and S. Tadokoro (Eds.): RoboCup 2001, LNAI 2377, pp. 1–11, 2002.

Results from the Preliminary Round Games

Group A

	A1	A2	A3	A4	A5	W/L/D	Rank
A1: FCPortugal2000		17–0	29–0	12–0	24–0	4/0/0	1
A2: Essex_Wizards	0–17		23–0	5–2	15–0	3/1/0	2
A3: NITStones	0–29	0–23		0–23	3–1	1/3/0	4
A4: UTUtd	0–12	2–5	23–0		4–0	2/2/0	3
A5: Dirty_Dozen	0–24	0–15	1–3	0–4		0/4/0	5

Group B

	B1	B2	B3	B4	B5	B6	W/L/D	Rank
B1: YowAI2001		1–2	3–0	5–0	21–0	21–0	4/1/0	2
B2: Cyberoos2001	2–1		7–1	3–0	19–0	22–0	5/0/0	1
B3: Helli-Respina_2001	0–3	1–7		8–0	23–0	3–0	3/2/0	3
B4: Lucky_Lubeck	0–5	0–3	0–8		14–0	14–0	2/3/0	4
B5: Team_SimCanuck	0–21	0–19	0–23	0–14		0–1	0/5/0	6
B6: Girona_Vie	0–21	0–22	0–3	0–14	1–0		1/4/0	5

Group C

	C1	C2	C3	C4	C5	C6	W/L/D	Rank
C1: Sharif-Arvand		2–0	***	16–0	11–0	0–0	3/0/1	1
C2: Gemini	0–2		***	14–0	4–1	0–1	2/2/0	4
C3: BlutzLuck	***	***		***	***	***	0/0/0	–
C4: Virtual_Werder	0–16	0–14	***		0–4	0–8	0/4/0	5
C5: FuzzyFoo	0–11	1–4	***	4–0		0–6	3/1/0	3
C6: A-Team	0–0	1–0	***	8–0	6–0		3/0/1	1

Group D

	D1	D2	D3	D4	D5	W/L/D	Rank
D1: Living_Systems		3–0	0–6	0–7	0–0	1/2/1	3
D2: Harmony	0–3		0–3	0–15	1–1	0/3/1	5
D3: UvA_Trilearn_2001	6–0	3–0		0–7	2–0	3/1/0	2
D4: Tsinghaeolus	7–0	15–0	7–0		5–0	4/0/0	1
D5: rUNSWift	0–0	1–1	0–2	0–5		0/2/2	4

Group E

	E1	E2	E3	E4	E5	E6	W/L/D	Rank
E1: FC Portugal2001		8–0	***	32–0	9–0	29–0	4/0/0	1
E2: RoboLog2k1	0–8		***	24–0	0–0	11–0	2/1/1	2
E3: Pasargad	***	***		***	***	***	0/0/0	–
E4: RMIT_Goannas	0–32	0–24	***		0–30	0–26	0/4/0	5
E5: TUT-groove	0–9	0–0	***	30–0		4–0	2/1/1	2
E6: 11monkeys3	0–29	0–11	***	26–0	0–4		1/3/0	4

Group F

	F1	F2	F3	F4	F5	W/L/D	Rank
F1: ATTUnited01		4–0	0–0	2–0	11–0	3/0/1	1
F2: DrWeb	0–4		0–4	0–4	1–2	0/4/0	5
F3: Zeng01	0–0	4–0		0–0	7–0	2/0/2	2
F4: MRB	0–2	4–0	0–0		4–0	2/1/1	3
F5: Anderlecht	0–11	2–1	0–7	0–4		1/3/0	4

Group G

	G1	G2	G3	G4	G5	G6	W/L/D	Rank
G1: Brainstormers01		2–2	14–0	8–0	18–0	12–0	4/0/1	1
G2: FCTripletta	2–2		5–0	4–0	9–0	1–0	4/0/1	1
G3: Lazarus	0–14	0–5		1–6	6–0	3–0	2/3/0	4
G4: SBCe	0–8	0–4	6–1		12–0	5–2	3/2/0	3
G5: wwfc	0–18	0–9	0–6	0–12		0–5	0/5/0	6
G6: Oulu_2001	0–12	0–1	0–3	2–5	5–0		2/3/0	4

Group H

	H1	H2	H3	H4	H5	W/L/D	Rank
H1: ChaMeleons		0–19	1–0	0–3	8–0	2/2/0	3
H2: WrightEagle2001	19–0		27–0	4–0	47–0	4/0/0	1
H3: saloo	0–1	0–27		0–13	17–0	1/3/0	4
H4: AT_Humboldt	3–0	0–4	13–0		18–0	3/1/0	2
H5: crocaroos	0–8	0–47	0–17	0–18		0/4/0	5

Results from the Second Level Preliminary Round Games

Group A

	A1	A2	A3	A4	A5	A6	W/L/D	Rank
A1: FCPortugal2000		14–0	0–6	10–0	5–0	5–0	4/1/0	2
A2: Cyberoos2001	0–14		0–8	6–0	4–2	3–0	3/2/0	3
A3: UvA_Trilearn_2001	6–0	8–0		4–0	5–0	3–0	5/0/0	1
A4: A-team	0–10	0–6	0–4		0–1	0–0	0/4/1	5
A5: MRB	0–5	2–4	0–5	1–0		3–0	2/3/0	4
A6: TUT-groove	0–5	0–3	0–3	0–0	0–3		0/4/1	5

Group B

	B1	B2	B3	B4	B5	B6	W/L/D	Rank
B1: Sharif-Arvand		0–17	4–2	0–3	5–1	***	2/2/0	3
B2: Tsinghuaeolus	17–0		9–1	3–0	17–0	***	4/0/0	1
B3: Essex_Wizards	2–4	1–9		0–2	5–0	***	1/3/0	4
B4: YowAI2001	3–0	0–3	2–0		8–0	***	3/1/0	2
B5: ChaMelons	1–5	0–17	0–5	0–8		***	0/4/0	5
B6: SBCe	***	***	***	***	***		0/0/0	–

Group C

	C1	C2	C3	C4	C5	C6	W/L/D	Rank
C1: FCPortugal2001		22–0	4–0	13–0	16–0	5–0	5/0/0	1
C2: ATTUnited01	0–22		1–1	3–0	2–0	5–0	2/2/0	3
C3: FCTripletta	0–4	1–1		4–0	3–0	1–0	3/1/1	2
C4: AT_Humboldt	0–13	0–3	0–4		1–3	0–0	0/3/1	6
C5: UTDtd	0–16	0–2	0–3	3–1		0–4	1/4/0	5
C6: Helli-Respina_2001	0–5	0–5	0–1	0–0	4–0		1/3/1	4

Group D

	D1	D2	D3	D4	D5	D6	W/L/D	Rank
D1: WrightEagle2001		0–2	0–0	6–0	2–0	1–1	2/1/2	2
D2: Brainstormers01	2–0		4–0	1–0	12–0	2–1	5/0/0	1
D3: Zeng01	0–0	0–4		0–0	4–0	2–0	2/1/2	2
D4: RoboLog2k1	0–6	0–1	0–0		4–0	1–3	1/3/1	5
D5: Gemini	0–2	0–12	0–4	0–4		1–6	0/5/0	6
D6: Living_Systems	1–1	1–2	0–2	3–1	6–1		2/2/1	4

The Results of the Final Double Elimination Rounds

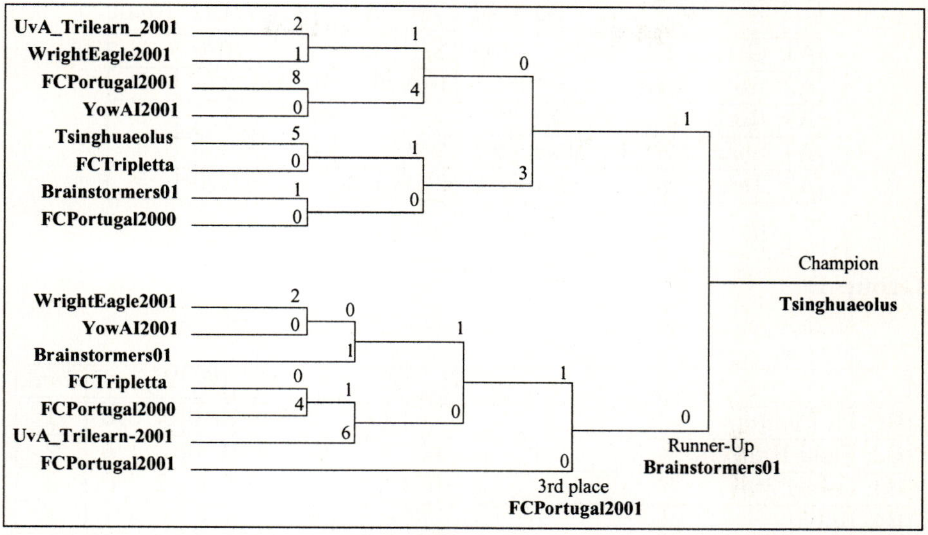

Fig. 1. The double elimination rounds of the soccer simulation league.

3 Small-Size Robot League Results

In this league two teams with five robots each play against each other with an orange golf ball. The field is slightly larger than a ping-pong table and covered with carpet. Each player of a team is equipped with a yellow, respectively blue marker, depending on the team. Also the goals are marked with colors. Additional markers, for example to detect the orientation of a robot, are allowed. Furthermore, global vision, typically in form of a camera perpendicularly hanging over the field, is allowed. But some teams are striving for onboard vision, i.e., cameras and processing on the robots themselves. Robot players can be constructed of anything 18 cm in diameter or less that is not dangerous to other robots, humans, or the playing surface.

The rest of this section includes all the results from the games in the small-size robot league. The tables show the preliminary round games, including numbers of wins, losses, and draws (W/L/D), and the rank of each team within each group. Figure 2 shows the results of the single-elimination championship tournament.

Group A

	A1	A2	A3	A4	A5	W/L/D	Rank
A1: Cornell Big Red		10–0	8–0	10–0	10–0	4/0/0	1
A2: RoGi-Team	0–10		0–0	6–0	10–0	2/1/1	2
A3: Roobots	0–8	0–0		4–2	10–0	2/1/1	2
A4: Sharif CESR	0–10	0–6	2–4		4–1	1/3/0	4
A5: Team Canuck	0–10	0–10	0–10	1–4		0/3/0	5

Group B

	B1	B2	B3	B4	B5	W/L/D	Rank
B1: Fu Fighters		3–0	10–0	5–0	10–0	4/0/0	1
B2: Field Rangers	0–3		3–0	9–1	10–0	3/1/0	2
B3: Owaribito	0–10	0–3		10–0	6–1	2/2/0	3
B4: RobSix	0–5	1–9	0–10		4–4	0/3/1	4
B5: Viper-Roos/ FU-Fighters-Omni	0–10	0–10	1–6	4–4		0/3/1	4

Group C

	C1	C2	C3	C4	C5	W/L/D	Rank
C1: LuckyStar II		5–0	10–0	10–0	***	3/0/0	1
C2: 5dpo	0–5		10–0	10–0	***	2/1/0	2
C3: HWH-Cats	0–10	0–10		0–0	***	0/2/1	3
C4: OMNI	0–10	0–10	0–0	***		0/2/1	3
C5: Huskies	***	***	***	***		0/0/0	–

Group D

	D1	D2	D3	D4	D5	W/L/D	Rank
D1: RoboRoos		5–1	11–1	5–0	10–0	4/0/0	1
D2: KU-Boxes	1–5		3–0	4–1	9–0	3/1/0	2
D3: TPOTS	1–11	0–3		3–1	10–0	2/2/0	3
D4: CM-Dragons	0–5	1–4	1–3		10–0	1/3/0	4
D5: 4 Stooges	0–10	0–9	0–10	0–10		0/0/3	5

The Results of the Final Elimination Rounds

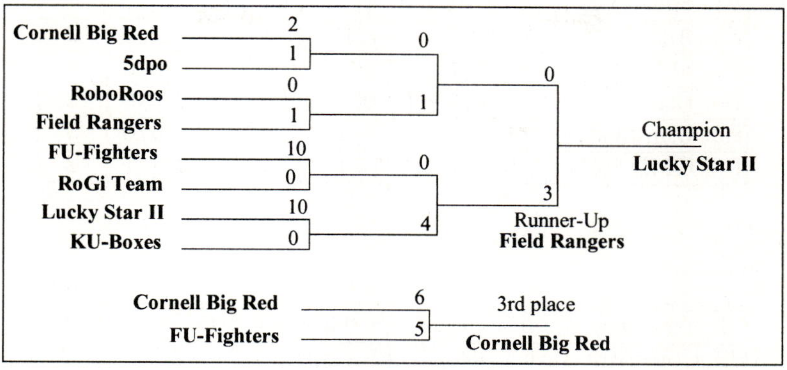

Fig. 2. The final elimination rounds of the small-size robot league.

4 Middle-Size Robot League Results

Teams in the middle-size robot league consist of up to four players. The field is up to 10m by 5m. The ball for middle size league is a FIFA standard size 5 football with orange color. Global sensing is explicitely prohibited in the middle-size league. The players of each team and the goals are uniquely colored. The size of each robot must be such that it fits into an area of 50cm by 50cm and its weight is limited to 80kg.

The rest of this section includes all the results from the games in the middle-size robot league. The tables show the preliminary round games, including numbers of wins, losses, and draws (W/L/D), and the rank of each team within each group. Figure 3 shows the results of the single-elimination championship tournament.

Group A

	A1	A2	A3	A4	A5	A6	W/L/D	Rank
A1: CS Freiburg		2–0	2–0	3–1	9–0	16–0	5/0/0	1
A2: Artisti Veneti	0–2		0–2	3–0	5–0	6–0	3/2/0	3
A3: Fusion	0–2	2–0		5–1	7–0	11–0	4/1/0	2
A4: Minho	1–3	0–6	1–5		0–2	5–0	1/4/0	5
A5: CMU Hammerheads	0–9	0–5	0–7	2–0		5–1	2/3/0	4
A6: Robosix	0–16	0–6	0–11	0–5	1–5		0/5/0	6

Group B

	B1	B2	B3	B4	B5	B6	W/L/D	Rank
B1: Sharif CE		0–7	10–0	1–3	4–1	2–5	2/3/0	4
B2: Agilo	7–0		7–0	0–4	7–0	1–1	3/1/1	2
B3: SPQR	0–10	0–7		0–11	0–6	1–7	0/5/0	6
B4: Eigen	3–1	4–0	11–0		8–1	3–1	5/0/0	1
B5: Ulm Sparrows	1–4	0–7	5–0	1–8		0–9	1/4/0	5
B6: GMD Robots	5–2	1–1	7–1	1–3	9–0		3/1/1	2

Group C

	C1	C2	C3	C4	C5	C6	W/L/D	Rank
C1: Trackies		3–0	8–0	9–0	8–0	2–0	5/0/0	1
C2: Cops Stuttgart	0–3		7–1	4–0	2–2	2–0	3/1/1	2
C3: Fun2maS	0–8	1–7		0–1	0–5	2–0	1/4/0	5
C4: ISocRob	0–9	0–4	1–0		1–3	1–0	2/3/0	4
C5: Clockwork Orange	0–8	2–2	5–0	3–1		2–0	3/1/1	2
C6: JayBots	0–2	0–2	0–2	0–1	0–2		0/5/0	6

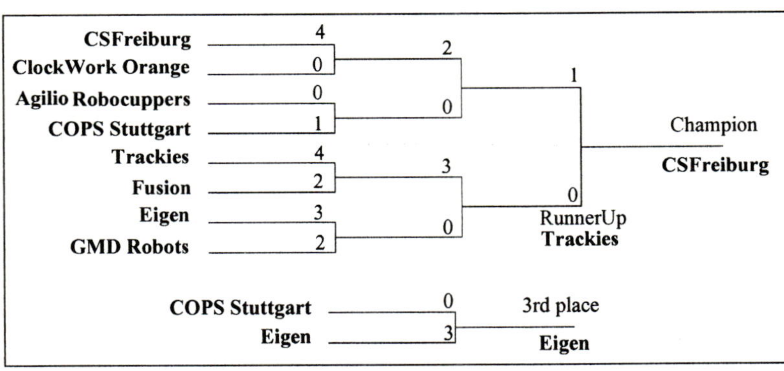

Fig. 3. The final elimination rounds of the middle-size robot league.

5 Sony Legged Robot League Results

In the legged robot league, each team has three Sony Aibo robot dogs as players. The robots are marked with blue, respectively red team markers and the ball is bright orange. It is not allowed to modify the robots, e.g., it is not allowed to install additional sensors. The players hence mainly rely on the build in camera of the Aibo. The goals are marked with yellow, respectively blue. Together with six color coded poles at the four corners and the two centers of the sidelines, the goals can be used by the players to locate themselves on the field.

The rest of this section includes all the results from the games in the Sony legged robot league. The tables show the preliminary round games, including numbers of wins, losses, and draws (W/L/D), and the rank of each team within each group. Figure 4 shows the results of the single-elimination championship tournament.

Group A

	A1	A2	A3	A4	W/L/D	Rank
A1: LPR		3–1	9–0	2–1	3/0/0	1
A2: Essex	1–3		1–0	1–3	1/2/0	3
A3: Melbourne	0–9	0–1		0–1	0/3/0	4
A4: USTC	1–2	3–1	1–0		2/1/0	2

Group B

	B1	B2	B3	B4	W/L/D	Rank
B1: CMU		12–0	5–3	8–0	3/0/0	1
B2: Sweden	0–12		0–6	1–4	0/3/0	4
B3: Tokyo	3–5	6–0		0–5	1/2/0	3
B4: Kyushu	0–8	4–1	5–0		2/1/0	2

Group C

	C1	C2	C3	C4	W/L/D	Rank
C1: Rome		2–3	2–0	5–0	2/1/0	2
C2: UPENN	3–2		1–0	2–0	3/0/0	1
C3: McGill	0–2	0–1		3–0	1/2/0	3
C4: Balkan	0–5	0–2	0–3		0/3/0	4

Group D

	D1	D2	D3	D4	W/L/D	Rank
D1: UNSW		8–1	11–0	12–0	3/0/0	1
D2: Osaka	1–8		4–1	1–0	2/1/0	2
D3: German United	0–11	1–4		3–0	1/2/0	3
D4: Washington	0–12	0–1	0–3		0/3/0	4

5.1 The Results of the Final Elimination Rounds

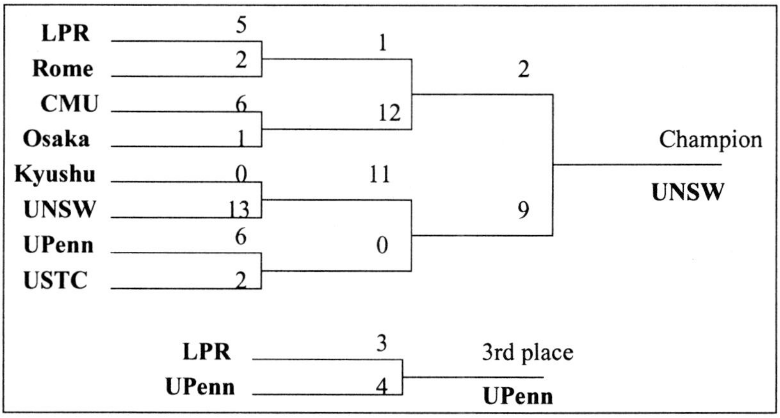

Fig. 4. The final elimination rounds of the Sony legged robot league.

6 Rescue Simulation League

The intention of RoboCup Rescue is to promote research and development for rescue in urban disaster scenarios. This involves work on the coordination of multi-agent teams, physical robotic agents for search and rescue, information infrastructures, personal digital assistants, a standard simulator and decision support systems, and evaluation benchmarks for rescue strategies. Built upon the success of RoboCup Soccer project, RoboCup Rescue will provide forums of technical discussions and competitive evaluations for researchers and practitioners.

RoboCup Rescue Simulation is a new domain of RoboCup. Its main purpose is to provide emergency decision support by the integration of disaster information, prediction, planning, and a human interface. For this purpose, a generic urban disaster simulation environment is constructed on network computers, similar to the soccer server. Heterogeneous intelligent agents such as fire fighters, commanders, victims, volunteers, etc. conduct search and rescue activities in this virtual disaster world. The simulator can be downloaded via the RoboCup website `www.robocup.org`.

The rest of this section includes all the results from the games in the RoboCup Rescue Simulation League. The tables show first the preliminary round games, and then the semifinals and final results.

Results from the preliminary games

Team Name	Total Score	Ranking
YabAI	39	2
Gemini-R	26	5
Rescue-ISI-JAIST	30.5	3
Arian Simulation Team	40	1
JaistR	29	4
NITRescue	19.5	6
RMIT-ON-FIRE	12	7

Semi Final A

Team Name	Score Point (V)		Winning Point
	Map G2	Map N2	
YabAI	17.3	20.2	2
Rescue-ISI-JAIST	17.7	20.4	0

Semi Final B

Team Name	Score Point (V)			Winning Point
	Map G2	Map N2	Map R2	
Arian	13.2	20.3	19.2	1
JaistR	14.7	16.4	17.4	2

3rd Position

Team Name	Final Ranking
Rescue-ISI-JAIST	3
JaistR	default

Final

Team Name	Score Point (V)		Winning Point	Final Ranking
	Map RIJ2	Map C2		
YabAI	19.26	56.82	2	1
Arian	33.40	56.86	0	2

Global Planning from Local Eyeshot: An Implementation of Observation-Based Plan Coordination in RoboCup Simulation Games

Yunpeng Cai, Jiang Chen, Jinyi Yao, and Shi Li

State Key Lab of Intelligent Technology and System,
Deparment of Computer Science and Technology,
Tsinghua University , Beijing, 100084, P.R.China
{Caiyp01, Chenjiang97, yjy01}@mails.tsinghua.edu.cn
lishi@s1000e.cs.tsinghua.edu.cn

Abstract. This paper presents a method of implementing distributed planning in partial-observable, low communication bandwidth environments such as RoboCup simulation games, namely, the method of global planning from local perspective. By concentrating planning on comprehending and maximizing global utility, the method solved the problem that individual effort might diverge against team performance even in cooperative agent groups. The homogeny of the agent decision architecture and the internal goal helps to create privities among agents, thus make the method applicable. This paper also introduces the application of this method in the defense system of Tsinghuaolus, the champion of RoboCup 2001 Robot Soccer World Cup Simulation League.

1 Introduction

Cooperation is an essential topic in the study of multi-agent systems, among which Distributed Planning is a challenging issue. In a multi-agent system, distributed plans can sometimes be formulated in a centralized manner by appointing a single agent as the planner. The planning agent executes the process of task decomposition, selection and assignment, and then distributes subtasks to each given member of the team. But centralized planning requires that at least one agent has the ability to possess accurate information of the global environment, and that the communication bandwidth is sufficient to accomplish task distribution in time. If the above condition is not satisfied, distributed planning is required.

The key to achieving cooperation is to establish agreement among the group of agents. This can be done either through communication or privities. Communication is the usual way to maintain coherence because it is more reliable and interpretable. Many methods such as Contract Net, Blackboard System and KQML [1] have been applied for cooperation protocols. And PGP (Partial Global Planning) [1] has proven useful in dealing with distributed planning. However, in a real-time environment such as RoboCup simulation, sometimes the communication ability of an agent is not adeque to perform conversations before they

A. Birk, S. Coradeschi, and S. Tadokoro (Eds.): RoboCup 2001, LNAI 2377, pp. 12–21, 2002.

reach an agreement. One way to coordinate without explicit communication is to allow agents to infer each others plans based on observations [1]. This is what we do in our simulation team.

Undoubtedly in circumstances like robot soccer where teamwork is so critical, agents must be designed to be altruist. Agents do not have any interest other than those relative to the common goal. But two main obstacles still exist preventing the group from acting as a unified entity. The first is the cognition difference: Agents' different observation on environment leads to different judgment about current mission of the team. The second is the divergence of individual effort against team performance. The maximization of individual contribution to the team does not certainly lead to best team performance.

In the design of the Tsinghuaeolus RoboCup simulation team, we introduced a method called Global Planning from Local Perspective to solve the second problem and to make an attempt to reduce the influence of the first problem to an acceptable degree. In this paper, we chose planning of defense as an example to demonstrate the implementation of this method in our RoboCup simulation team.

Defense plays an indispensable part both in real soccer game and in RoboCup. The defense task in RoboCup simulation game is obviously a process of distributed planning and dynamic coordination. The diversity of opponent attack strategies makes it impossible to set up a fixed defense strategy beforehand. Due to the limitation of communication bandwidth in RoboCup simulation environment, it is impossible to exchange enough information to perform centralized planning. Players cannot even tell the team what he himself is planning to do because the situation might change so fast that communication cannot catch up. In a word, a player has to decide what to do by himself according to his knowledge about the current state of the environment, but the team must keep a high level of consistency to create a solid defense system.

2 Main Points of Global Planning from Local Perspective

The essence of Global Planning from Local Perspective is to abstract the concept of global utility, then concentrate the purpose of planning on maximizing it. Sometimes global utility can be represented by a simple variable that can be noticed by all members, e.g., total output. But generally it is a quantity that exceeds the ability of a single agent to grasp directly. In RoboCup system, we cannot even have an objective description on what global utility is.

In fact, we model global utility instead of trying to describe it as it is. Since the decision of planning will eventually turn into behaviors of team members, global utility can be considered as the integration of influences each individual behavior contributes to the environment, which can be considered as individual utilities. Decompounding global utility into individual ones will attain great benefit because the latter are far easier to estimate.

Obtaining global utility from individual utilities is a process of evaluating a set of single agent behaviors and synthesizing them. During the procedure of synthesis, a set of operators has to be introduced to model the joint effect of behaviors. There are two basic kinds of synthesizers. Mutexs means that two behaviors should not be executed together, which can be expressed in mathematics as leading to a negative infinite compound utility. Interferences means that the compound utility P (AB) of action A with utility P (A) and action B with utility P (B) can be formulated as P (AB) = P (A) + P (B) + I (A, B), where I (A, B), may be either positive or negative, is a term that measure the influences of putting A and B together. I (A, B) is zero if A and B are independent.

Having defined individual and global utilities, we divide the process of planning into five steps:

- Step 1, Task Decomposition [1]: A series of subtasks that can be executed by a single agent are generated. Note that although fulfilling all subtasks means the accomplishment of the overall task, it is not necessary to do this all the time.
- Step 2, Generation of subtask-executor pairs: All subtasks are linked to each agent that is able to carry them out. Each pair is called an arrangement.
- Step 3, Evaluation of arrangement: Some evaluate functions are given to measure the performance of the subtask if a given arrangement is applied (that is, the executor really takes up the task). The value represents the individual utility of an arrangement.
- Step 4, Generation of scheme: A selection of arrangement (the execution set) is made based on the evaluation functions. The execution set includes all key subtasks that are required to achieve the global goal, and is considered the best plan to finish the task. The execution set is the one with the highest global utility among all feasible plans. In order to figure out the execution set, we use the algorithm of branch and bound. Mutexs, as a standard synthesizing operator, are used prior to interference operator so as to prevent redundancy or conflicts in combination of arrangements, thus speeding up searching.
- Step 5, Task assignment and execution: Agents pick out their own share of tasks from the execution set and carry it out.

The above procedure looks similar to centralized planning. The difference is that in centralized planning the procedure is carried out by only one planner, and in distributed planning it is carried out by all agents, each agent having a distinct viewpoint. In centralized planning, any inaccuracy of information may lead to fault in results. But in distributed planning, an agent only needs to guarantee that assignment concerning itself is proper, and the diversity of viewpoint provided more opportunity not to have all members making the same mistake. So, in partially observable environments, distributed planning might even achieve more reliability.

3 Reducing the Influence of Cognition Difference

So far we have not addressed the problem of cognition difference, which exists in every partial-observable environment. The idea of global utility maximization involves assumptions on teammate behaviors, which indicates the risk of misunderstanding. When an agent abandons the action that can bring maximal individual utility, it assumes that its teammates will have corresponding actions to make up this loss, which is doubtful. The entire plan might be broken even if simply one agent acts singularly because of its distinctive viewpoint. This should be prevented. Indeed, we do not propose a radical solution to this problem. But we found that through proper choice of opponent modeling and good construction of evaluation functions, we can make our system less sensitive to slight differences in observation, hence increasing coherency in plan execution.

For convenience, in this section we call the agent executing the procedure of task planning the planning agent. But note that it is different from the one in centralized procedure, for all agents are planning agents.

As mentioned above, we made use of privities among agents in observation-based planning. Privities are knowledge of something secret shared between individuals, which is very important in cooperation among people. For example, in a human soccer team, many times team members may know how a player will pass the ball without any notification. To reach privities, some conditions must be satisfied:

- I, Agents have mutual beliefs [2] about the environment.
- II, All agents know the behavior model of others, so that they can to some degree predict teammates' future actions.
- III, Some public rules are set up as norms [4] to clarify roles of a given member in a given situation.

The latter two conditions can be realized in the structure of decision framework. What needs to be thoroughly considered is the first one.

Mutual beliefs can be represented in the following form:

If A and B mutually believe that p, then:

(a) A and B believe that p,

(b) A and B believe that (a). [2]

From this representation we found that the confirmation of mutual beliefs involving inferring others' internal states. For low bandwidth communication systems, to exchange parameters of internal states is not realistic. So, the ordinary manner of interpreting others' intention is through observation. A set of action styles is defined and characteristic behavior patterns are assigned to them. Through information collection an agent watches the activities of others and captures characteristic behaviors so as to comprehend their purpose. Thus mutual belief can be formed and coherence can be reached.

But the procedure of pattern recognition might be time consuming since agents' behaviors are not always distinct enough. Scarcity of information about teammate's intention might frequently emerge when an agent tries to make a decision. Four choices are offered to deal with this case.

The first one is to exclude all those whose intention is not clearly known from the cooperation scheme. This is advisable if the result of failed collaboration will be much more severe than working alone. But in environments where cooperation is emphasized, the choice is not adaptable.

In RoboCup simulation games, players perform decision-making almost every simulation cycle. Batches of new schemes come out replacing the old ones. So in most situations the fault of a single cycle won't have severe consequences. That means the environment is fault-tolerant. And the nature of soccer game made teamwork – even with poor coordination - more essential than individual practice. So agents must try to collaborate even with communication difficulties.

The second one is to choose the action with the largest probability (according to the agents expressive action) as the intention of that agent. The third one is to describe its intention in the form of a probability distribution, then to make a Bayesian decision. These two ways are popular in behavior modeling and do have great benefits. But both have to face the fact that the action inferred in this way is the one agents performed in the past instead of the one they are going to carry out in the future. In dynamic systems where agent might reconsider at any time, modeling in these two ways will cause bluntness in environment changes.

In global planning from local perspective, we chose the last approach - assumption of mutual knowledge. That is, the planning agent never tries to assume what teammates are intended to do before the entire plan is completed. Instead, it treats its own view of the situation as the global view, and assumes that teammates will perform those actions that the planning agent considers to achieve global utility maximization.

The introduction of assumed mutual knowledge exploits more collaboration ability from agents, at the same time raises more risk of cognition mistakes. But the homogeny of agent decision architecture and internal goal help to reduce it. As is mentioned, "just the fact that the agents share membership in a community provides a wealth of evidence upon which to infer mutual belief and common knowledge" [2]. Although in a distributed system with local information acquisition, agents have diversified partial knowledge of the external world, the fact that they share the same goal and use the same goal-directed information acquisition function makes it easy to have all the team members mutually grasp the key parameters of the current situations.

So, during the phase of planning the planning agent does not collect evidence of collaborative behaviors; instead, it collects those that shows that some agent failed to collaborate. Agents obviously acting inconsistently with the rest of the team will be wiped off from the candidate list of task executors. And new scheme will be brought out as a remedy. This can usually be implemented by adding some factor in the evaluation of subtasks.

Looking at partial knowledge as the picture of the entire environment is of course dangerous. But it is not always infeasible. Notice that partial knowledge acquired by an agent via its local information collecting mechanism is the true (although inaccurate and incomplete) mapping of the real world, so partial views differ from each other but the difference is bounded. If (and only if) the evaluation of individual and global utilities will not magnify these quantitative differences to qualitative ones, team coherency will be maintained.

Plainly speaking, the design of evaluation functions must take into account the bound of quantitative difference a variable may reach when observed by different agents. The mathematical characters of such functions should be continuous and smoothly varied. Further more, often when evaluation can proceed from various starting points, it is important to choose parameters that are easy to observe for all agents, as inputs to the function, instead of choosing hidden or trivial ones.

4 Application to Defense Play in RoboCup Simulation

Defense can be interpreted as a matter of positioning. Tsinghuaeolus uses a kind of Situation Based Strategy Positioning policy [3]. One of the most important positioning is the basic formation, which determines a player's position by standard role, ball position and ball controlling state. In the defense problem, basic formation serves as a standard, namely, a common rule, for role assignment and task evaluation.

Another important factor is the defense position sensitivity, which measure how dangerous it is if an opponent is staying in a given point. Defense position sensitivity for any point in on the field is known beforehand by all players of the team.

Four types of defense actions are defined:

- 1, Block: intersecting an opponent possessing the ball in the outer side of our goal, preventing him from pushing forward
-
- 2, Press: running after an opponent possessing the ball who is nearer to our goal, keeping threat on him
-
- 3, Mark: keeping up with an opponent without ball so that his teammates cannot pass the ball to him
-
- 4, Point Defend: staying at the basic formation position, this will benefit when a nearby teammate fails in 1 vs. 1 defense or when the team regains control of ball.

To simplify the problem, we set the rule that one defender cannot defend against two opponents at the same time and two defenders should not defend against the same opponent. The mutex operator is applied on both of these circumstances. The Interference operator is not used here.

There is an alternative where block and press are set as interference instead of mutex, providing a different defense strategy. Since either is ok, we do not discuss it further.

After decomposing a defense task into defending against each one of the attackers and staying at the basic formation point, arrangements are generated by linking every defender with every subtask. Then, the procedure goes to the evaluation of individual utilities.

We defined one evaluation function for each type of action. Three variables are used as the input of each evaluation function: distance, measuring the interval

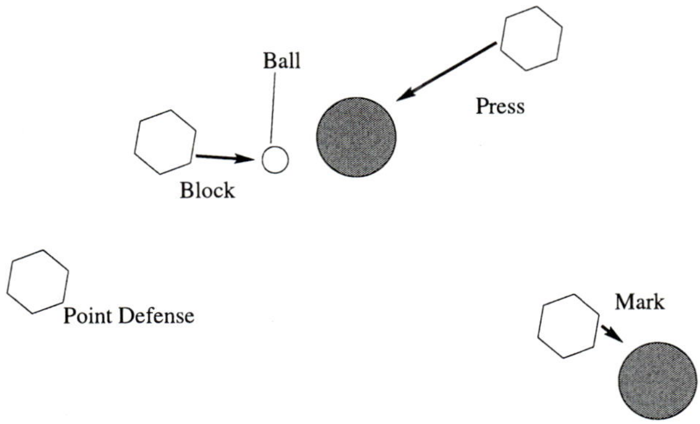

Fig. 1. Defend Actions

from the player's current position to the defend point; deviation, measuring the distance from the defend point to the player's basic formation position; threat, namely the defend position sensitivity of the opponent's current position. The function output increases with threat and decreases with distance and deviation. For point defend, only threat is applied as input.

To attain the actual value of each function, we set some typical scenes, extract input values from them, and endow output values to the functions. The values are carefully tuned to keep mild variation. In this way a list of input and output data is created.

We use BP neural networks for encoding evaluation functions. The list generated above is used as the training set of the neural networks. After training, networks are tested to ensure that they fit requirements mentioned in the previous section, thus evaluation functions are defined.

The rest of the planning goes as the standard procedure presented above and no further discussion is needed.

One thing left is the generation of remedy scheme. Sometimes the team does fail to maintain coherency. For example, when an opponent is about equally far away from two defenders' current position and, at the same time, formation point, two arrangements will have similar evaluation and the two defenders may be in conflict. Our design takes advantage of the positive feedback that exists in the system. Since the ball will keep moving during the game, and opponents must move forward to perform attacking, either the basic formation points or the defend point will change in the next cycle. There is little chance that the two arrangements still keep the same evaluation. Once the difference between the two arrangements is obvious enough for both defenders to recognize, the prior defender will execute the defend action, moving nearer to the opponent, and the other one will move in another direction. So, the contrast is increased and the system is drawn to a status apart from the foregone dilemma, thus coherence is achieved again.

5 Experimental Results

We performed an experiment in order to discover to what extent agents coordinate internally and externally. We played a game against FC Portugal 2000 (December binary) in 3000 cycles. During the game all defense players (7 players altogether) recorded their plans and actions, then the log files were analyzed. The statistical result is below:

Table 1. Statistic data on defend planning

Arrangements after merge (1):	13830
Arrangements executed:	10056
Num of cycles the team is in defend state:	1569
Num of cycles the team have dissidence in arrangement:	1219
Num of executed arrangement per cycle:	6.4
Num of dissidence per cycle:	2.4
Num of cycles the team conflict in executions (2):	70

(1) Arrangements generated at the same time by different planners and with the same decision are considered the same one

(2) Including both defending by too many players and missing a attacker that need being defended against

From the data we can conclude that for most of the time agents have dissidence in planning, but they have reached agreement on the major part of the plan (at least 4 arrangements in 6.4 is agreed by all planners in average). Another point to recognize is that most of time dissidence in planning does not lead to incoherency in execution. Most of the time dissidence is about whether a player should mark an attacker in the position near a turning point or stay at the formation point, which is hard to judge but scarcely affects the performance. For over 95 percent of the time the team acts uniformly despites of dissidence in 77 percent of the time, which shows that our method has perfectly reached its design goal.

In RoboCup 2001, Tsinghuaeolus won the championship, scoring 90 goals, while conceding only 1 goal. The defense system faced the challenge of diversified strong attack styles from various teams and functioned well, leaving few chances for opponents, while not involving too many players in defense.

6 Related Works

In [5], Lesser makes a survey in recent progress of multiagent research and provides some principles that are useful to build a multiagent system. While explaining his viewpoint he proposes to move the focus from the performance of individual agents to the properties and character of the aggregate behavior of a group of agents. The basis of Global Planning from Local Perspective has embodied such an idea.

In RoboCup, one of the hot topics relative to coordination is to achieve cooperation by learning (e.g.[6],[7],[8]).Many other teams tried to coordinate via communication. Paper [9] presented an implementation of this approach in Middle-sized Robot League.

Peter Stone proposed the TPOT-RL algorithm and the Flexible Teamwork architecture [6][7]. In his architecture, teammates are considered as part of dynamic environments, so the process of achieving coordination is also the one that agents learn to adapt to the environment. Our method differs from this approach in terms that we do not design an algorithm to learn a decision structure; instead, we provide a fixed but general framework beforehand. The flexibility of the structure lies in its ability to incorporate varied knowledge (that is, evaluation function of different forms). Both human knowledge and machine learning results can become elements of the decision module.

7 Conclusions

Practice results show that merely by introducing the concept of global utility can we achieve a significant enhancement of system performance. The design of evaluation functions enhances performance mainly by decreasing the probability of cognition fallacies and dead locks, hence making performance more stable. Global Planning from Local Perspective contributes mainly by shortening the planning phase to adapt to rapid changing environments, and exploiting the potential hidden in agents reactions.

Further more, the evaluation functions have to be contrived to fit specific environments. More skills other than those described above must be employed. It might be worthwhile to explore the possibility to obtain a good evaluation function by online learning.

References

1. Gerhard Weiss . Multiagent Systems: A Modern Approach to Distributed Artificial Intelligence. pp.83-111, pp.139-156. The MIT Press,1999
2. Stephen A.Long and Albert C.Esterline . Fuzzy BDI Architecture for social agents. North Carolina A&T Univ. Proceedings of the IEEE in Southeastcon, 2000. pp.68-74
3. Luis Paulo Reis and Nuno Lau. FC Portugal Team Description: RoboCup 2000 Simulation League Champion.Univ. of Porto & Univ. of Aveiro. In Stone P., Balch T., and Kaetzschmar G., editors, RoboCup 2000:Robot Soccer World Cup IV. pp.29-41,Springer Verlag, Berlin, 2001
4. Towards socially sophisticated BDI agents .Dignum, F.; Morley, D.; Sonenberg, E.A.; Cavedon, L. MultiAgent Systems, 2000. Proceedings. Fourth International Conference on , 2000 pp.111-118

5. Lesser, V.R. Cooperative multiagent systems: a personal view of the state of the art. Knowledge and Data Engineering, IEEE Transactions on, Volume: 11 Issue: 1,Jan.-Feb. 1999 pp.133-142
6. Peter Stone. Layered Learning in Multi-Agent Systems. PhD Thesis, Computer Science Dep.,Carnegie Mellon University, December 1998
7. Peter Stone and Manuela Veloso. Layered Learning and Flexible Teamwork in RoboCup Simulation Agents. CMU. In Veloso M., Pagello E., and Kitano H., editors, RoboCup-99: Robot Soccer World Cup III. pp.495-508. Springer Verlag, Berlin, 2000
8. Masayuki Ohta. Learning in Cooperative Behaviors in RoboCup Agents. Tokyo Institute of Technology. In Hiroaki Kitano, editor, RoboCup-97: Robot Soccer World Cup I. pp.412-419. Springer Verlag, Berlin, 1998
9. K. Yokota, K.Ozaki, et al. Cooperative Team Play Based on Communication. Utsunomiya Univ. and Toyo Univ. In Minoru Asada, editor, RoboCup-98: Robot Soccer World Cup II. Proceedings of the second RoboCup Workshop. pp.491-496. Paris, 1998.

LuckyStar II - Team Description Paper

Ng Beng Kiat (nbk@np.edu.sg), Quek Yee Ming, Tay Boon Hock, Yuen Suen Yee, Simon Koh

Ngee Ann Polytechnic, Singapore

1 Introduction

Our first robotic soccer team, LuckyStar, competed in Stockholm 1999 and was lucky enough to win third place. From that experience, we found that our team weaknesses were due to lack of vision reliability, smooth robot movement control, shooting capability and team cooperation. We decided to concentrate on areas with the most immediate impact on the game, i.e. vision, robot movement control and shooting mechanism.

2 System Description

We continue to use a global-vision-based system except that the vision system now resides in the host computer. The host computer computes the next move for each robot. The outputs, which are the robot's translational and rotational speeds, are sent to the robots via a Radiometrix RF transceiver.

3 Robot

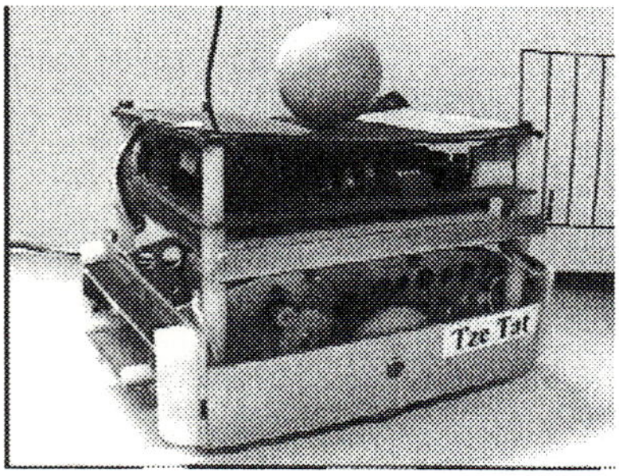

Figure 1. Picture of robot

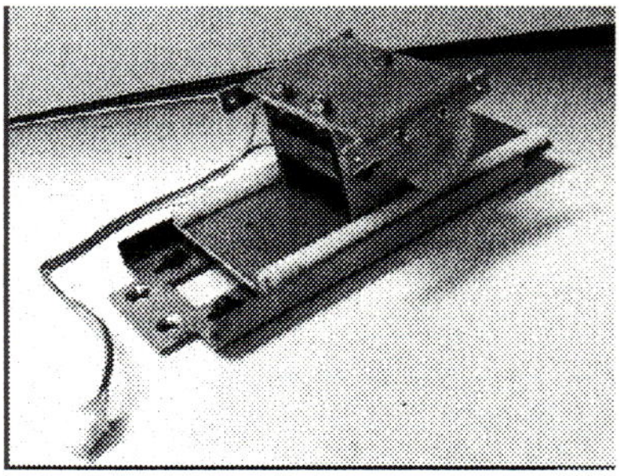

Figure 2. Kicking mechanism exposed

The new robots are similar to the old robot except that they have kicking mechanism. The kicking mechanism is based on a rack and pinion system driven by a DC motor. This allows the robot to kick both way and with varying amount of force. The robot is

A. Birk, S. Coradeschi, and S. Tadokoro (Eds.): RoboCup 2001, LNAI 2377, pp. 22-25, 2002.

able to shoot with full force at opponent goal and with just enough force when clearing the ball towards the sloped wall so as to avoid putting the ball off field.

The host system decides completely when to kick the ball. In general, once the ball is within kicking distance, angular error between the robot orientation and the ball-target-to-robot angle is used to determine shooting criterion. The further the robot is from the ball-target, the smaller the angle error allowance.

4 Vision system

One of the problems faced last year is the detection of opponent color. The ping pong balls are small and not evenly illuminated as they are not flat. In order to be able to use only simple vision algorithm, the camera and the frame grabber must be carefully chosen so as not to lose any ping pong ball color pixel unnecessarily.

4.1 Hardware

We tested some cameras and concluded that 3-ccd cameras are the best in term of resolution but they are extremely expensive. Not being able to afford one, we settled for the second best, a single-ccd camera with RGB output. With composite video, the color information are encoded at the camera end and decoded by the frame grabber board. The result is the degradation of signal-noise ratio, hence poorer color resolution.

It is also important that camera has electronic shutter capability in order to be able to capture high-speed object. The shorter the exposure time, the sharper the high-speed object. Of course, we must make sure the lighting requirement for short exposure can be met.

For the frame grabber, we chose the FlashBus MV Pro from Integral Technologies as it provides RGB input and field capture interrupt capability for fast vision processing response.

4.2 Software

We rely on color patches on top the robot to determine robot orientation. In order to play a game, the vision system must be able to detect the orange ball, the yellow and blue team colors and one other secondary colors. True enough, with proper choice of hardware, we are able to use HSI thresholding to detect 5 good colors without any post-processing. This enables us to process the vision data at 50 field per second on a Pentium 500MHz PC with spare capacity for the host software.

5 Robot Motion Control

As in all sport, speed is a major factor in determining the outcome of the game. But if you have speed without good skill and control, then the game will end up pretty rough and unpredictable. One of our major goals is to achieve high speed and smooth robot control.

5.1 Video lag

In general, by the time the robot receives the movement speed commands, there is a time lag of about 4 video field time between its current position and the processed vision data. The time lag is due to a) camera exposure and capture (1 field), b) frame grabber capture(1 field), c) vision and host processing (slightly less 1 field) and d) RF transmission (slightly less 1 field). To have smooth high-speed control of the robot, the time lag must be compensated. There are two ways to do it.

1) Use filtering (e.g. G-H filter[2]) to estimate the robot position, speed and orientation in 4 field time or

2) Use the last four speed commands sent to the robots to estimate the robot position, speed and orientation in 4 field time.

We use the second method, which gives a better estimate unless the robot is obstructed.

5.2 Moving Target

In order to block or kick a moving ball, the system must be able to move the robot to the predicted ball position. This must take into account the current ball speed and direction and current status of robot. For our system, an estimate of the time to get to the current ball position is calculated based on the robot-ball distance, current-robot-orientation and final-robot-orientation difference, and average robot translational and rotational speed. With the estimated time, we can estimate the new ball position. With the estimated new ball position, the process is performed for another 2 iterations in order to get a better estimate of the ball position. In general, this works well for only slow ball and is pretty useless for ball speed greater than 0.2 m/sec. We will be spending more effort in this area for our next team.

7 Conclusion

Our robotic soccer team is fairly reliable and able to play a match continuously without many problems. But it is still very lacking in the area of team cooperation.

Next year, we hope to develop ball-passing skills for our robots. With good ball-pass skills, team cooperation will be more feasible.

References

[1] Manuela Veloso, Michael Bowling, and Sorin Achim. The CMUnited-99 small robot team.

[2] E. Brookner, Tracking and Kalman Filtering Made Easy. A Wiley-Intersience Publication, 1998

CS Freiburg 2001*

Thilo Weigel[1], Alexander Kleiner[1], Florian Diesch[1], Markus Dietl[1],
Jens-Steffen Gutmann[2], Bernhard Nebel[1], Patrick Stiegeler[1], and Boris Szerbakowski[3]

[1] Institut für Informatik, Universität Freiburg
79110 Freiburg, Germany
{last-name}@informatik.uni-freiburg.de
[2] Digital Creatures Laboratory, Sony Corporation
141-0001 Tokyo, Japan
gutmann@ieee.org
[3] Optik Elektronik, SICK AG
79183 Waldkirch, Germany
boris.szerbakowski@sick.de

Abstract. The CS Freiburg team has become F2000 champion the third time in the history of RoboCup. The success of our team can probably be attributed to its robust sensor interpretation and its team play. In this paper, we will focus on new developments in our vision system, in our path planner, and in the cooperation component.

1 Introduction

Although the general setup, the hardware and software architecture of CS Freiburg at the RoboCup 2000 competition [16] turned out to be quite satisfactory, there appeared to be room for improvements. First of all, the hardware was not as reliable as it used to be when the robots were bought in 1997. Secondly, there were quite a number of software components in the system that could be improved.

For this reason, we invested in new hardware, which led to a significant improvement in robustness. In fact, the final game against the *Osaka Trackies* was partially won because our hardware was very reliable.

On the software side, a number of improvements were made on the system level, such as integrating a new camera system. Furthermore, we worked on components that we considered as critical, such as the vision system, the path planning module, and the module for cooperative play. The improvements of these components, which are described below, did indeed give us the competitive edge. However, it also became clear that the reactiveness by the *Osaka Trackies* is hard to match with our team.

* This work has been partially supported by *Deutsche Forschungsgemeinschaft* (DFG), by *Medien- und Filmgesellschaft Baden-Württemberg mbH* (MFG), by *SICK AG* and *Sony Corporation*

A. Birk, S. Coradeschi, and S. Tadokoro (Eds.): RoboCup 2001, LNAI 2377, pp. 26–38, 2002.

2 Hardware

When we first participated at RoboCup in 1998 we used *Pioneer-1 robots* as manufactured by *ActivMedia*. To recognize the ball ball we employed the *Cognachrome* vision system manufactured by *Newton Labs* which was available installed in the Pioneer-1 robots. Our effort in hardware development was limited to mounting a *SICK laser range finder* for self localization and object recognition and building a simple kicking device with parts from the Märklin Metallbaukasten. For local information processing the robots were equipped with *Libretto 70CT* notebooks and they used the *Wavelan radio ethernet* from *Lucent Technologies* to communicate with each other [6].

Even though we maintained the same general hardware setup, we virtually replaced every part of this setup in the past four years. In 1999 we replaced the laser range finders with the *SICK LMS200* model. They provide depth information for a 180° field of view with an angular resolution of 0.5° and an accuracy of 1 cm [9]. For RoboCup 2000 we started to make major mechanical and electronical modifications to our robots. We installed nickel-cadmium batteries because of their light weight and high speed charging capability. To improve the ball handling and shooting capability of our robots *SICK AG* assisted us in building a new kicking device with movable ball steering flippers, which can be turned to an upright position and back. The kicking device is strained by a wind-screen wiper motor and released by a solenoid [16].

Fig. 1. A CS Freiburg Player

This year we made further hardware modifications in order to increase the performance of our team. Figure 1 shows one of our robots as it competed this year. We installed Pioneer-2 boards, which now allow the robots to move considerably faster with the same *Pittman* motors that we have been using during the past years. For more precise movements we substituted the rear caster wheel by a caster roller. To be able to develop our own vision software we replaced the old vision system by a *Sony DFW-V500 digital camera* and switched to *Sony Vaio PCG-C1VE notebooks* because of their IEEE 1394 interface. We also upgraded the *WaveLan* cards to the new 11 Mbit/s (2.4 GHz) cards. To get our laser range finders to work via the USB port we had to find a circuit diagram for a RS422-USB converter board, capable of a 500Mbaud rate[1], which *SICK AG* then manufactured for us. For an artificial intelligence research group hardware development and maintenance isn't a trivial task at all and would certainly not have been possible without the help of SICK AG. Even only adapting our software to the new hardware consumed a lot of our preparation time for this year's tournament.

[1] http://www.ftdichip.com/Files/usb-422.zip

3 Vision

The implementation of our new vision system consists of two major software parts, one for region segmentation by color and one for the mapping from image coordinates to positions of objects on the field. The region segmentation by color is carried out on 320x240 Pixels in YUV color space using the CMVision library.[2] Due to the fast algorithm [3], our system is capable to work with even two cameras at a rate of 30Hz. From the segmented regions the world coordinates are computed. For this purpose we implemented two alternative solutions, described below.

3.1 Tsai Camera Calibration

Tsai's camera model [15] is based on the pinhole model of perspective projection, and consists of six extrinsic and five intrinsic parameters. The extrinsic parameters describe the translation (T_x, T_y, T_z) and rotation (R_x, R_y, R_z) from the world coordinate frame **W** to the camera coordinate frame **C**. The intrinsic parameters include the effective focal length f, the radial lens distortion κ_1, a scale factor s_x and the image center (C_x, C_y), also known as principal point. Generally, the objective of calibration is to determine optimal values for these parameters from a set of known points in the world coordinate frame (x_w, y_w, z_w) and their corresponding pixels in the sensor plane (x_u, x_v). Once the intrinsic parameters are determined, they can be used for different positions and orientations of the camera. The extrinsic parameters, however, have to be re-calibrated when the camera moves relatively to the world coordinate frame origin.

With the calibrated parameters, one can predict the pixel (x_u, x_v) in the sensor plane from a given point (x_w, y_w, z_w) in **W** by three transformations:

1. **Rigid body transformation** Transformation from the world coordinate frame to the camera coordinate frame: $\left[x_c \; y_c \; z_c\right]^T = R\left[x_w \; y_w \; z_w\right]^T + \left[T_x \; T_y \; T_z\right]^T$, where $R = Rot(R_x)Rot(R_y)Rot(R_z)$
2. **Perspective projection** Due to the pinhole model, the undistorted pixel coordinates can be calculated by the theorem of intersecting lines: $x_u = f\frac{x_c}{z_c}, x_v = f\frac{y_c}{z_c}$
3. **Distortion transformation** Transformation from undistorted pixels to distorted ones with the distortion factor κ_1: $x_u = x_d(1 + \kappa_1\rho^2), y_u = y_d(1 + \kappa_1\rho^2)$, where $\rho = \sqrt{x_d^2 + y_d^2}$

Since the ball is usually located on the field plane, a coplanar calibration was used. As mentioned above, the calibration process can be divided into two parts, which are intrinsic and extrinsic parameter calibration. For the intrinsic parameter calibration a small dot matrix was placed on different positions in front of the camera. Data for extrinsic parameter calibration was generated from a large dot matrix, which was placed on the soccer field plane in front of the robot.

Another possibility to generate calibration data is to take snapshots of the ball on the field directly. This comes with the advantage that the error from calculating the blob center on the image is reflected in the data. However, it turned out that the data was too noisy for the Tsai optimization procedure, which was indicated by a high calibration error.

[2] http://parrotfish.coral.cs.cmu.edu/cmvision/

3.2 Interpolation

Besides the analytical Tsai method, we also used a method from the last years team for linear interpolation [13]. The method interpolates the position of the ball based on triangles generated from *a priori* collected world to image correspondences. The mapping takes place by identifying the triangle which encloses the considered pixel. This triangle can be mapped to a triangle on the field which then is used to interpolate the objects position.

For a uniform distribution of the triangles the *Delaunay* triangulation has been used. Figure 2 shows the generated triangles on the image and the field respectively. The dense occurrence of samples at the bottom of both pictures indicates that more samples have been collected from positions in the vicinity of the robot than from distant positions.

The collection of adequate samples has been carried out by assistance of our accurate self-localization. The ball has been placed on a fixed position on the field, which has been taken as reference. Subsequently, snapshots of the ball from different robot position have been taken and converted to samples in the camera coordinate frame. Finally, a triangulation algorithm produced a list of triangles, such as shown in figure 2.

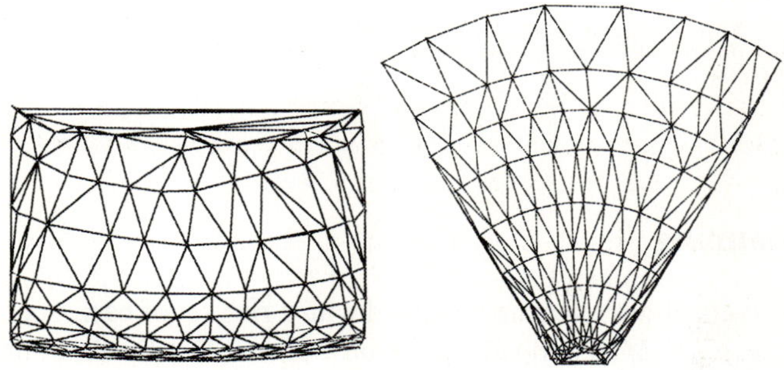

(a) (b)

Fig. 2. Distribution of triangles on the camera image (a) and on the field (b) after the *Delaunay* triangulation

3.3 Evaluation

Figure 3 shows the measured accuracy in estimating the ball distance and ball angle by the two approaches. While both approaches perform well in estimating the direction of the ball, the Tsai method seems to be inaccurate in estimating larger distances. We assume that this is partially caused by a poor coverage of the field by calibration dots during the calibration of the extrinsic parameters. A larger matrix would probably improve the quality of the calibration. However, this is not always practicable during RoboCup competitions.

In contrast, data collection for the interpolation is handled easier. The robot is programmed to take samples at particular positions autonomously. Furthermore, the quality of the collected data can be evaluated directly by considering the triangle distribution as shown in figure 2. Also the effect of noisy measurements is not as critical as it is for the analytical method, which uses non-linear optimization to estimate the model parameters.

For most demands in the RoboCup domain, however, the resulted accuracy of both methods suffice, since in most cases the direction of the ball is more important than an accurate distance estimation beyond three meters.

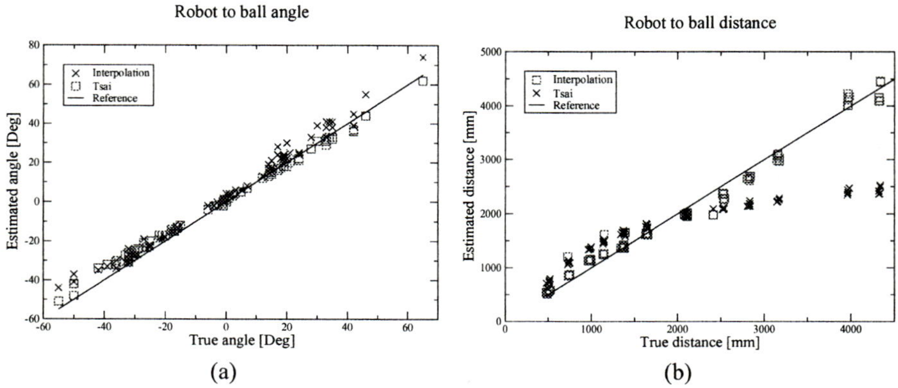

Fig. 3. Accuracy in estimating the direction (a) and distance (b) of the ball by interpolation and the Tsai method

4 Path-Planning

The robotic players of our team are equipped with a rich set of basic soccer skills such as *GoToBall*, *GetBall*, *MoveShootFeint*, *ShootGoal*, etc. Details about these skills can be found in our last year's team report [16]. Some of these skills, e.g. *GoToBall* which enable a player to move around obstacles close to the ball, require the planning of a path from the robot's current position to a target position. In addition, more team oriented skills like moving to a strategic position on the field, or the automatic positioning procedure before game start need the ability to find collision-free paths, too.

A prerequisite for planning paths is the existence of a model of the world. Although, a crude world model can already be used for generating paths, the integrity and accuracy of the world model has a big influence on the quality of the computed paths. Fortunately, our robots have access to a comprehensive and accurate world model built from observations of the LRF and the vision system by the full team of robots [5, 6, 16].

Path planning has been well researched with a lot of interesting results found [8]. There are at least 3 basic categories approaches can be divided into: roadmaps, cell decomposition, and potential fields. Recently, probabilistic approaches became popular, in particular for finding paths in a high-dimensional configuration space [7]. As our robots only move in a 2 dimensional world and are able to turn on the spot, plus all objects can be reasonably represented as circles, we can restrict the configuration space to only two dimensions. Thus, a deterministic approach can be employed.

Roadmap algorithms are a popular way for finding paths in the RoboCup environment, e.g. the *All Botz* RoboCup team in the small size league developed a variant of the *extended visibility graph* for finding paths around other players [2]. We also implemented a version of the extended visibility graph for our first participation in 1998 [6]. However, this approach has the drawback that it depends on the chosen minimum distance the robot should keep to obstacles. If chosen small, the algorithm can generate paths that are too close to the robots. Especially for soccer this can be disadvantageous as obstacles like opponent robots are moving and collisions are likely to happen. On the other hand, a large value for the minimum distance might cause the planner to fail finding a path. Since it is difficult to find a good trade off we developed a new path planning approach based on a combination of potential fields and cell decomposition. We will show a comparison of the two approaches at the end of this section.

The basic algorithm is described in detail in Topor's work [14]. Due to limited space we will only give a brief overview and present extensions that improved the performance of the algorithm. The path planning algorithm uses a potential field for finding its way through the obstacles. This potential field consists of an attracting force towards the target position and repulsive forces arising from other players, the ball (optionally), and the field boundaries. An additional potential field directs the search into the robot's current heading. In our approach we approximate the world into a grid of 10×10 cm cells.

The search algorithm maintains a grid of the world where each cell holds a boolean indicating if the cell has already been visited. Furthermore, a priority queue is used with elements consisting of a cell index and its corresponding potential value. The algorithm starts with the target position and follows the negative gradient of the potential field to the robot. In each iteration the potential and gradient value of the cell referring to the current position are computed by summing up pre-computed potential and gradient fields of the individual forces. Cell index and potential value of cells that are not already visited are inserted into the priority queue. Note, that although we discretize the world into a grid, the algorithm still computes positions with floating point precision.

Figure 4(a) illustrates how this algorithm finds a path around an obstacle. Note that we search from the target towards the robot and not vice versa, since otherwise the robot would first move directly in direction of the obstacle before turning to the side. In each cycle of the robot control program (every 100 ms), a new path is computed and the robot smoothly moves around the obstacle (Figures 4(b) and (c)). The example also shows that in our algorithm re-using a previously computed path like in the approach of Baltes and Hildreth [2] is not possible without major modifications.

Figure 5 displays two features we added to the original algorithm. If a player is in possession of the ball and is heading towards the opponent goal, we surely do not want other robots to interfere with it. This can be achieved by adding additional, virtual potentials in front of the ball-owning robot (Figure 5(a)).

A problem in path planning in general is that of oscillation. A situation where an obstacle is in between robot and target can lead to such a problem (see Figure 4(a)). If the position of the obstacle is noisy, e.g. because the sensors delivers noisy data, the path planning algorithm might find paths around the obstacles on both sides with equal chance. Since paths are re-planned every 100 ms, the robot might continuously

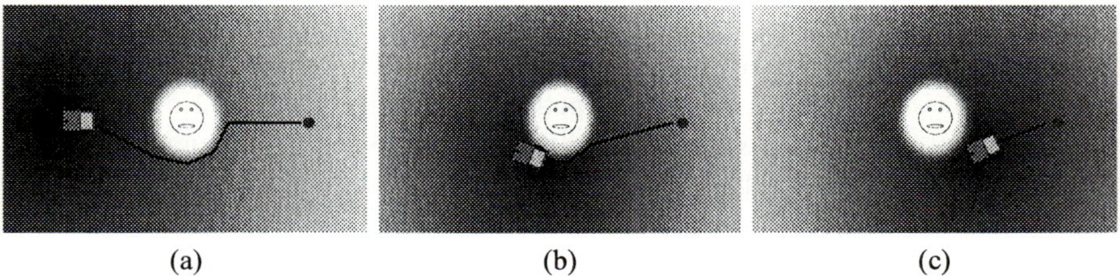

Fig. 4. Path planning around an obstacle using potential fields: initial plan (a), plan after robot moved closer to obstacle (b), and final straight-line path to target (3). Potential values are indicated by grey-scale levels where lower potentials appear darker.

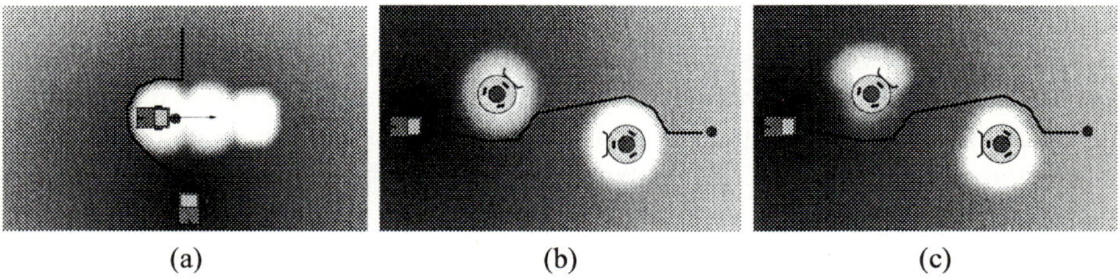

Fig. 5. Extra potential fields are added in front of the active player (a). To reduce oscillation the center of gravity of an obstacle is shifted according to the path computed in the last cycle: without (b) and with hysteresis shift (c).

choose the other way. This could be observed in experiments in our laboratory including the robot bumping into the obstacle because it wasn't able to decide which way to go around.

A directional potential field that prefers the robot's current heading helps avoiding such scenarios. However, if the robot is far from the obstacle, the algorithm still can oscillate. For RoboCup 2001 we added a new technique for directing the search into the right way. Based on the path computed in the previously cycle a hysteresis shift is applied to the individual potential fields of the obstacles. This shift moves the center of gravity of a potential field while still retaining the shape at the border of the obstacle, i.e. the center of the obstacle is shifted while the obstacle still occupies the same space. Due to this shift, the gradient changes and forces the search to go around the obstacle in the same way as in the previous cycle. Figures 5(b) and (c) give an example.

Potential field methods can be trapped in local minima. In our approach this can be detected by examining the visited flag of the cell that refers to the current position. If a cell is visited for the second time, a best first search is started beginning with the cell that has the lowest potential in the priority queue. In best first search, the current cell is removed from the queue and its 4 neighbors are examined and inserted into the priority queue if not already visited. This leads to a wave-front search in the potential field until the negative gradient of a cell points towards a non-visited cell where the algorithm can follow steepest decent again. Figure 6 shows an example.

For evaluating the path planner, we conducted several experiments using random start, target and object positions in a simulated environment of size 81×45. In average,

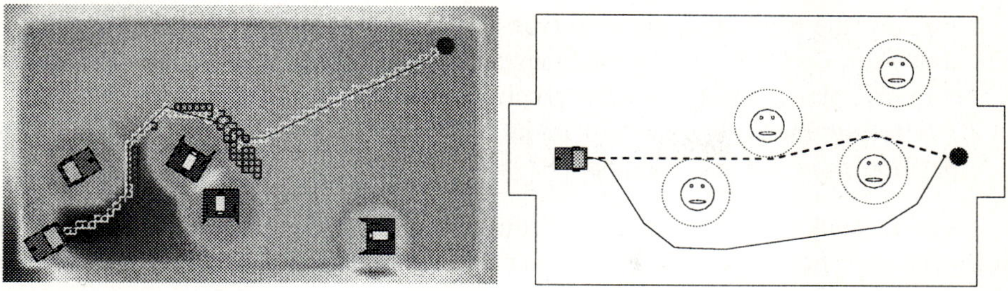

(a) (b)

Fig. 6. Evaluation of the path planner. Best first search in local minima (a). Dark cells indicate steps where best first search has been applied, light cells refer to steepest decent iterations. Comparison of path planning methods (b). Dashed line refers to shortest path as computed by the extended visibility graph, solid line reflects path generated by our new potential field method.

the running time was less than 1.5 msec on a Pentium MMX 233 MHz. In 35 out of 5000 cases, the algorithm needed more than 15 msec and the maximum run time was about 29 msec. Thus, the approach is well suited for continuous path planning.

A last experiment was conducted to compare the performance of the new algorithm to the extended visibility graph approach we employed for our first participation [6]. Figure 6 (b) shows the setup. The extended visibility approach has the property that it computes shortest paths and for the given scenario it returned the path that goes through the middle of the field. In contrast, our new approach leads the robot around the obstacles close to the wall. Although the length of the path at the wall is significantly longer than the shortest one, our robots need much less time following the longer path (13 s) than following the shorter one (24 s) because our motion controller allows for higher speeds if obstacles are far.

5 Team Coordination

Soccer is a complex game where a team usually has to meet several requirements at the same time. To ensure that in any game situation a team is prepared to defend its own goal, but also ready to attack the opponent goal, the various team players have to carry out different tasks and need to position themselves at appropriate strategical positions on the field. In this section we describe our method for determining such positions depending on the current game situation and illustrate how our players decide which position they should occupy.

To express that a player has a task which is related to a position in the team formation we say a player pursues a certain *role* [12]. Distinguishing between different areas of responsibility we established the following roles:

- *active*: the active player is in charge of dealing with the ball. The player with this role has various possible actions to approach the ball or to bring the ball forward towards the opponent goal.
- *strategic*: the task of the strategic player is to secure the defense. It maintains a position well back in our own half.

- *support*: the supporting player serves the team considering the current game situation. In defensive play it complements the teams' defensive formation and in offensive play it presents itself to receive a pass close to the opponents goal.
- *goalkeeper*: the goalkeeper stays on its goal line and moves depending on the balls position, direction and velocity.

As our goalkeeper has a special hardware setup for his task it never changes its role. The three field players, however, are mechanically identical and switch their roles dynamically whenever necessary.

5.1 Positions

Our approach to determine the target positions associated with the field player roles is similar to the SPAR method of the *CMU* team in the small size league [11]. From the current situation as observed by the players, a potential field is constructed which includes repulsive forces arising from undesirable positions and attracting forces from desirable ones.

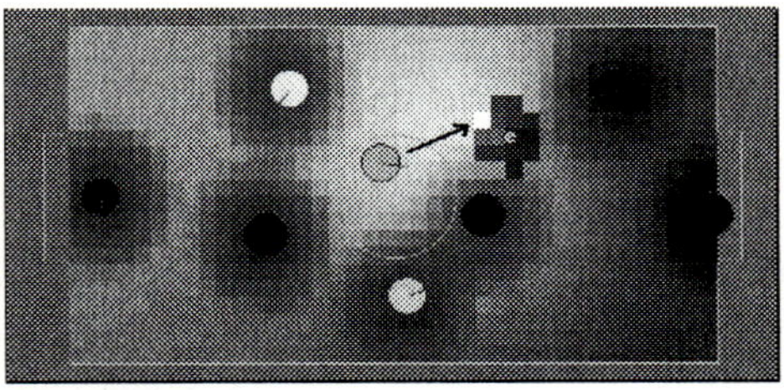

Fig. 7. Potential field for determining the active position

Figure 7 shows an example of a potential field for the desired position of the active player. Dark cells indicate very undesirable positions whereas light positions represent very desirable positions. The resulting position is marked white. We consider the ideal position for the active player to be at least a certain distance away from other players and at an optimum distance and angle to the ball. While the optimum distance is fixed, the optimum angle is determined by interpolating between a defensive and an offensive variant depending on the balls' position. A defending player should be placed between the ball and our own goal, but in offensive play the ball should be between the player and the opponent goal.

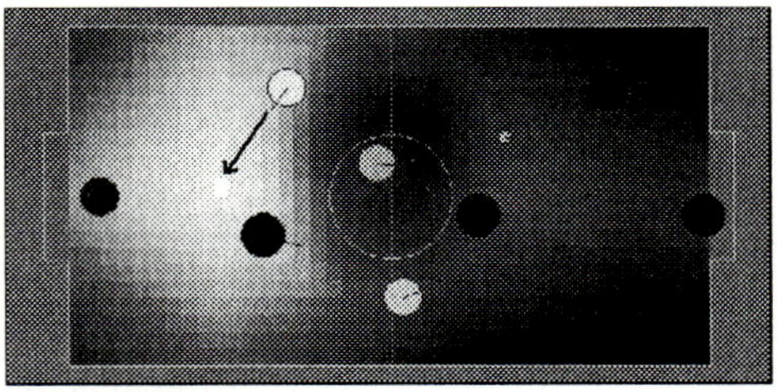

Fig. 8. Potential field for determining the strategic position

Figure 8 shows the potential field for the desired position of the strategic player. It is based on the same game situation and uses the same colors as the example for the active player. We want the strategic player to stay well behind all players and the ball and prefer positions close to the horizontal centerline of the field. Only the active player is assigned a repulsive force explicitly in order to enforce staying out of its way. Other players are avoided implicitly by the path planner which finds an appropriate position close to the desired one.

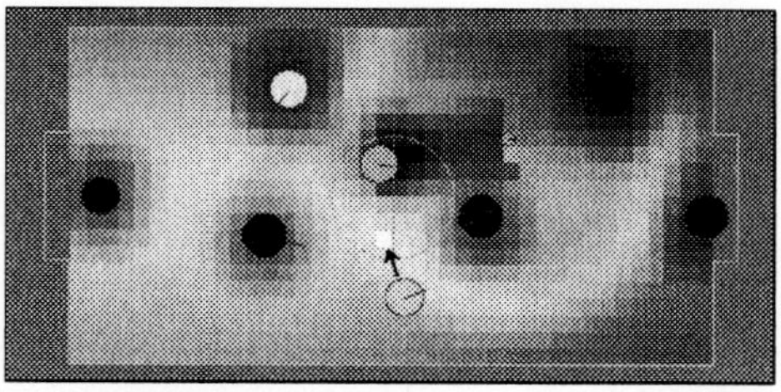

Fig. 9. Potential field for determining the support position

Figure 9 shows how in the same game situation as above the defensive support position is determined. We want the supporter to stay away from all other players and at a certain distance to the ball. As the supporting player should complement the teams' defensive formation, we additionally prefer positions behind and aside the active player.

Even though the resulting positions are in general similar to the ones a human observer would establish, we needed to make some restrictions in order to avoid "hysteric" behavior resulting from "overreacting" to the constantly changing environment. Because our robots are turning rather slowly they need a lot of time for even minor position changes. We therefore favor a players' current position with a persistence value and are quite tolerant regarding how close our defending players actually need to get to their positions. In order not to loose too much of precision either, we adjust this tolerance dynamically depending on the players current angle to its target position. By allowing large tolerances for large angles but requiring small tolerances for small angles, we achieve that a player only starts to update its position if the new position differs considerably from the old one. But once the player is moving towards that position the player intends to approach it as close as possible.

Statistics based on the log files of our Seattle games show that the balance between teams is reflected in the average distances between the target positions of our three field players. Against strong teams our players intended to stay in a rather compact formation with average distances of 2051mm against *Trackies* (1:0) or 2270mm against *Fusion* (2:0). When our team was clearly dominant, the players were spread wider over the field with average distances of 3000mm against *CMU* or 3657mm against *Robosix* (16:0).

The fact that in the Seattle games the active player was on average only 973mm away from its target position indicates that our players managed to maintain a formation where always at least one robot was close to the ball. Also the fact that the area for possible strategic positions is quite restricted is reflected in the respective average distance of only 892mm. As the support position differs considerably in offensive and defensive play, the corresponding player showed the largest offset to its target position (1857mm). Evaluating the log files also showed that there is still room for improvement. As it is a tedious task to test and evaluate team behavior online with real robots, we intend to rely more on the simulator that we are currently implementing.

5.2 Roles

After a field player has determined the best active, strategic and support position from its perspective, it calculates utilities which are basically a heuristic for the time it would take the player to reach the corresponding position. The following criteria are taken into account:

- The (euclidian) distance to the target position
- The angle necessary to turn the robot towards the target position
- The angle necessary for the robot to turn at the target position in order to face in direction of the ball
- The angle between the robot and the ball (it is more desirable, especially for the active player, to approach the target position already facing in direction of the ball)
- Objects between the player and the target position

The total utility is now computed as the weighted sum of all criteria. In order to decide which role to take a player sends these utilities to its teammates and compares them with the received ones.

Following a similar approach taken by the *ART* team [4],[3] each player's objective is to take a role so that the sum of the utilities of all players is maximized. In contrast to the *ART* approach a player doesn't take its desired role right away, but checks first if not another player currently pursues the same role and considers that role best for itself as well. As the world models of our players are not identical, the perspectives of our players can in fact differ. Therefore a player only takes a role if either no other player is currently pursuing that role or the pursuing player signals that it actually wants to change its role. That way we reduce with only little extra communication effort the number of situations where more than one player owns the same role.

A problem for this approach are situations, where different players come up with very similar utilities for a certain role and the roles might oscillate. However, by adding a hysteresis factor to the utility of a player's current role we ensure that a player only gives up a role if its real utility for that role is clearly worse than the one of its teammate.

	Goalkeeper	Active	Strategic	Support
Role kept	986.9 s	5.4 s	5.7 s	8.1 s
Role not unique % of playing time	0 %	2.17 %	1.23 %	1.06 %
Role not unique average time	0 s	0.203 s	0.218 s	0.206 s

Table 1. Evaluation of role assignment method

Table 1 displays a statistics evaluating our role assignment method. All values are averaged over the games played at RoboCup 2001 in Seattle. In the first line the times our players kept a certain role are listed. Interestingly the values for the field players are similar to the average role switch time of 7 seconds stated by the *ART* team [4]. The second line shows how much of the total playing time a role was pursued by more than one player. The fact that for the active player this happened in only 2.17% of the total playing time and in even less cases for the other roles, demonstrates, that our method to avoid such situations works successfully. The average values for the times, a role was not unique (in line three), gives further evidence for this.

When evaluating the log files of our Seattle games we also noted that the roles were usually quite equally distributed between our players. However, in games where we

[3] Note, however, that our approach was developed independently.

scored a lot of goals, the player with the most offensive starting position held the active role considerably longer than its teammates.

6 Conclusion

The development of robotic soccer during the last five years was quite impressive. In 1997 the robots hit the ball only occasionally – and often kicked it in the wrong direction (or even into the own goal). In 2001, the games looked much more interesting. The development of our team followed this general path. In 1998, our main advantage was that our robots knew their own position – which helped to avoid own goals. Over the last four years, we concentrated on the improvement of our hardware, on the sensor interpretation process, on cooperative sensing and on team play. As demonstrated, this gave CS Freiburg the chance to win the championship three times. However, having watched the final game against *Osaka Trackies*, we got the impression that a point is reached where it is hard to improve our robots so that they are able to dominate a game against a fast, reactive team such as the *Trackies*.

References

[1] M. Asada and H. Kitano, editors. *RoboCup-98: Robot Soccer World Cup II.* Lecture Notes in Artificial Intelligence. Springer-Verlag, 1999.

[2] Jacky Baltes and Nicholas Hildreth. Adaptive path planner for highly dynamic environments. In Stone et al. [10], pages 76–85.

[3] J. Bruce, Tucker Balch, and Maria Manuela Veloso. Fast and inexpensive color image segmentation for interactive robots. In *Proceedings of the 2000 IEEE/RSJ International Conference on Intelligent Robots and Systems (IROS '00)*, volume 3, pages 2061 – 2066, October 2000.

[4] Claudio Castelpietra, Luca Iocchi, Maurizio Piaggio, Alessandro Scalzo, and Antonio Sgorbissa. Communication and coordination among heterogeneous mid-size players: ART99. In Stone et al. [10], pages 149–158.

[5] Markus Dietl, Jens-Steffen Gutmann, and Bernhard Nebel. Cooperative sensing in dynamic environments. In *Proceedings of the IEEE/RSJ International Conference on Intelligent Robots and Systems (IROS'01)*, Maui, Hawaii, 2001.

[6] Jens-Steffen Gutmann, Wolfgang Hatzack, Immanuel Herrmann, Bernhard Nebel, Frank Rittinger, Augustinus Topor, Thilo Weigel, and Bruno Welsch. The CS Freiburg robotic soccer team: Reliable self-localization, multirobot sensor integration, and basic soccer skills. In Asada and Kitano [1], pages 93–108.

[7] L.E. Kavraki and J.C. Latombe. Probabilistic roadmaps for robot path planning. In K. Gupta and A. del Pobil, editors, *Practical Motion Planning in Robotics: Current Approaches and Future Directions*, pages 33–53. John Wiley, 1998.

[8] Jean-Claude Latombe. *Robot Motion Planning.* Kluwer Academic Publishers, 1991.

[9] Bernhard Nebel, Jens-Steffen Gutmann, and Wolfgang Hatzack. The CS Freiburg '99 team. In M. Veloso, E. Pagello, and H. Kitano, editors, *RoboCup-99: Robot Soccer World Cup III*, Lecture Notes in Artificial Intelligence, pages 703–706. Springer-Verlag, 2000.

[10] P. Stone, G. Kraetzschmar, T. Balch, and H. Kitano, editors. *RoboCup-2000: Robot Soccer World Cup IV.* Lecture Notes in Artificial Intelligence. Springer-Verlag, Berlin, Heidelberg, New York, 2001.

[11] P. Stone, M. Veloso, and P. Riley. CMUnited-98: Robocup-98 simulator world champion team. In Asada and Kitano [1], pages 61–76.

[12] Peter Stone and Maria Manuela Veloso. Task decomposition, dynamic role assignment, and low-bandwidth communication for real-time strategic teamwork. *Artificial Intelligence*, 1999.

[13] Maximilian Thiel. Zuverlässige Ballerkennung und Positionsschätzung (in German). Diplomarbeit, Albert-Ludwigs-Universität Universität Freiburg, Institut für Informatik, 1999.

[14] Augustinus Topor. Pfadplanung in dynamischen Umgebungen. Diplomarbeit, Albert-Ludwigs-Universität Freiburg, Institut für Informatik, 1999.

[15] Roger Y. Tsai. An efficient and accurate camera calibration technique for 3D machine vision. In *Proceedings of IEEE Conference on Computer Vision and Pattern Recognition*, pages 364–374, 1986.

[16] Thilo Weigel, Willi Auerbach, Markus Dietl, Burhard Dümler, Jens-Steffen Gutmann, Kornel Marko, Klaus Müller, Bernhard Nebel, Boris Szerbakowski, and Maximiliam Thiel. CS Freiburg: Doing the right thing in a group. In Stone et al. [10].

The UNSW RoboCup 2001 Sony Legged Robot League Team

Spencer Chen, Martin Siu, Thomas Vogelgesang, Tak Fai Yik,
Bernhard Hengst, Son Bao Pham, and Claude Sammut

School of Computer Science and Engineering
University of New South Wales
Sydney, Australia

Abstract. In 2001, the UNSW United team in the Sony legged robot league successfully defended its title. While the main effort in last year's competition was to develop sound low-level skills, this year's team focussed primarily on experimenting with new behaviours. An important part of the team's preparation was playing practice matches in which the behaviour of the robots could be studied under actual game-play conditions. In this paper, we describe the evolution of the software from previous years and the new skills displayed by the robots.

1 Introduction

In the RoboCup robot soccer tournament, the Sony legged league differs from the other leagues in that all the teams use identical hardware, the Sony ERS-210 robot. Thus, a team's strength is based on the software that implements low-level skills as well as game-play behaviour. UNSW's team in RoboCup 2000 made many significant advances in programming the low-level skills of the previous model, the ERS-111 robot. These included improved methods for vision, localisation and locomotion [3,2]. The goal of the 2001 team was to build on these skills to develop new behaviours for stronger game play.

The first step in developing the software was to port last year's code to the new robots. The ERS-210 has several advantages over the older model. The leg motors are substantially stronger, permitting the robots to move more quickly. The on board MIPS R4000 series processor is also much faster. However, the shape of the new robot is somewhat different from the previous robot. Since many of the skills to do with pushing the ball are dependent on the geometry of the body, substantial changes were required to the code, including abandoning some skills and inventing new ones.

Once the porting of the infrastructure modules for vision, localisation, locomotion and sound were completed, development of new behaviours could begin. A very important part of our development strategy was to play weekly matches between "last week's champion" and new code written during the week. The winning code retained the trophy for that week. We believe that observing and testing behaviours in realistic game play is vital. One of the major lessons we

A. Birk, S. Coradeschi, and S. Tadokoro (Eds.): RoboCup 2001, LNAI 2377, pp. 39–48, 2002.

learned from our 1999 entry [1], was that understanding the interactions of the robots on the field was paramount. Behaviours tested using a single robot are unlikely to work as expected when the robot is being interfered with by other players. Furthermore, close observation of a game can reveal where subtle adjustments to behaviours in particular situations can result in an advantage for the team.

In the remainder of this paper, we describe the modifications made to the infrastructure and we describe the new skills developed for this year's team. We refer the reader to the report of the 2000 team [3] for detailed descriptions of the previously developed software.

2 Vision

2.1 High Resolution

The vision system of the Sony robots can deliver images at low, medium or high resolution. In previous years, we limited ourselves to using medium resolution so that we could maintain a high camera frame rate. The faster CPU in the new model allowed us to process high resolution images with no loss of speed, almost keeping up with the maximum frame rate of the camera. The benefits of high resolution were that images contained larger and better defined blobs of colour which made object recognition and localisation more reliable and, in particular, detection of the other robots was improved. Furthermore, we were able to reliably detect the ball across the length of the field. Switching to high resolution involved minor changes to the image processing routines and changes to tools for classifying sample images and generating the colour lookup table.

2.2 Polygon Growing and Decision Trees

All of our object recognition relies on finding regions of specific colours. Thus, accurate colour classification is critical. Since colours may appear differently under different lighting conditions, we train the colour classifier on sample images taken on the competition playing field.

Starting in 1999 and until half way through the 2001 competition, we trained the system using a Polygon Growing Algorithm (PGA). The polygon based classifier was used during the pool matches, however, we were unhappy with its behaviour. The lights at the competition venue were very bright and caused some significant misclassifications. Earlier in the year, we had briefly experimented with using Quinlan's C4.5 decision tree learning algorithm [4] to construct the classifier. We decided to try this again during the rest day between the pool matches and the finals and found that the decision tree classifier was more reliable under the competition lighting conditions. We took the chance of using the decision trees in the final rounds and this gamble seems to have paid off.

The input to C4.5 is a set of examples, each of which is characterised by a list of attribute/value pairs and a class label. In this case, the attributes are the Y, U, V values of each pixel and the class label is the colour obtained from manually classifying sample images. The output of the learning algorithm is a decision tree that is easily translated into a large if-statement in C.

The most likely cause of the improved classification accuracy is that in the PGA, we arbitrarily divide the colour space into a fixed set of Y-planes and construct polygons for each plane. So, in a sense, the PGA is not fully 3D. Whereas, the decision tree classifier uses the full range of all the attributes in its classification. One drawback of using C4.5 are that it requires more training examples, particularly negative examples, to avoid over generalisation.

2.3 Blob Formation

Since all object recognition is based on finding blobs of particular colours, the blob formation algorithm is critical. However, it is also the most time consuming section of the code and causes a slow down in the camera frame rate. When the ball is close to the camera, it fills the image,creating a large patch of orange. When large blobs appeared in the image, our previous code would slow down from about 25fps to 19fps. A new blob forming algorithm was written to solve this problem. It requires two scans over an image. The first pass goes through each horizontal lines and creates run length information on continuous colour pixels. The second pass merges the same colour run length segments if they overlap each other. The the new blob forming algorithm can keep up with the maximum frame rate of the camera.

2.4 Extending Occluded Beacons

The playing field of the Sony legged robot league has six coloured landmarks or beacons placed at each corner of the field and at the centres of each side. These beacons are used by the robots to localise. When a beacon is partially obscured, the distance to it may be incorrectly calculated. The distance to the beacon is based on the Cartesian distance between the centroids of the two coloured blobs that make up a beacon. When one half of the beacon is partially occluded, the distance between the two centroids will appear smaller than its actual distance. In this case the distance to the object will be over estimated and result in poor localisation. To circumvent this problem, we extend the partially occluded blob and shift its centre, as in Figure 1.

2.5 Looking at the Ball at Close Range

One of the differences between the ERS-111 and the ERS-210 robots is that the older model's head has a longer "snout". Since the camera is in the tip of the snout, the ERS-111 loses sight of the ball as it comes under the chin and close to the robot's chest. The ERS-210's shorter snout permits the head to look down and track the ball much closer to the robot's body, which improves ball handling skills.

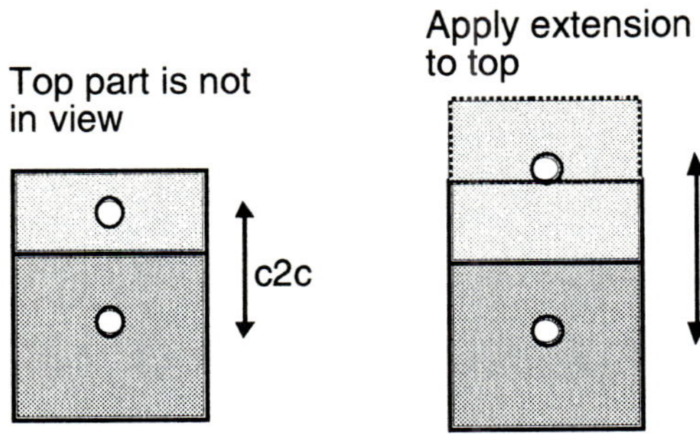

Fig. 1. Extension to Beacons

When the ball is close, the limited field of view of the camera means that the view of the ball is clipped. This affects the calculation of the distance to the ball since it is based on the size of the orange blob. When we recognise that clipping may be occurring, only the top of the ball's bounding box is used to estimate distance. Combining this with tracking the top of the ball rather than the centre of the ball, improves the accuracy of ball distance estimation when the ball is within 20cm of the robot.

This case is a good illustration of the fact that the overall level of performance of the robots results from an accumulation of skills for handling the many different situations that might occur during a game. So far, the only way we have of identifying all of these possible situations is by playing practice matches and carefully observing the behaviour of the robots. These observations also resulted in finding various bugs, including problems with the 3D transforms that take into account the orientation of the head relative to the robot's body.

3 Localisation and World Model

Each robot maintains a world model. Since we know the dimensions of the playing field, the world model is simply a map in which the goals and landmarks are fixed and the mobile objects are placed in the map at their estimated locations. A consequence of using only medium resolution in last year's vision system was the it was difficult to reliably detect the colours of the robots' uniforms. Therefore, the world model did not include the locations of other robots.

With the switch to high resolution, it became possible to add other players into the world model. Each robot now keeps track of the last seen team mate and opponent, allowing more complex team play, as described in section 6. Because the uniform of each robot consists of many patches and the robots are constantly in motion, it remains difficult to separate multiple robots and to accurately judge the distance to the robot. These are topics of on-going investigation.

Several other adjustments were made to the localisation system, including modifications to improve the accuracy of triangulation on two beacons and discarding distances to far away beacons, since these measurements are often unreliable.

4 Locomotion

The walking style developed for the ERS-210 is based on last year's low stance walk originally developed for the ERS-111. However, the shape and dimensions of the robot are somewhat different. The shoulder joints are closer together and the diameters of the limbs are larger. Thus, the locomotion algorithms had to be tuned for the new robots. The limbs were extended further outward, which helped us to develop some new behaviours. For example, by having the front limbs extended further to the front, the ball is more likely to be trapped between the legs. This gives us the ability to turn with the ball (see section 5.1). One of the main features of the low stance walk is its stability. Our robots rarely fall over, while robots with a higher stance frequently fall when pushed. On the ERS-210, with the help of stronger motors, we are able to keep the same level of stability while raising the height of the walk slightly. This has improved sideway movement and consequently the general agility of the robots.

5 Skills

5.1 Paw Dribble and Turn Kick

Through playing many practise games we noticed that if one of the robots hits the ball with its front paw, often a straight powerful pass results. We decided to implement this as a paw dribble. This requires a very accurate approach to position the ball so that it connects with the paw for a straight kick. The walk is dynamically adjusted, based on the ball distance and heading, and in the final approach a downward tilt and rotation of the head is used for finer adjustment. This kick makes the game more dynamic and often allows the robot to push the ball past an opponent on the wing.

The turn kick was developed to allow the robots to kick the ball in directions 90° and greater from the robot's heading when it is able to trap the ball between its front legs. It is especially useful on the edge of the field when a perpendicular approach is often the only one possible. Once the ball is under the head, a quick turning motion is executed resulting in a powerful and reasonably accurate pass. There is a complex interplay between the exact turn parameters, and the position of the head. Thus, a substantial amount of experimentation is required to construct a reliable kick.

The very last goal of the tournament was scored when a UNSW robot approached the ball facing its own goal (a very dangerous thing to do) and then executed an almost perfect 180° turn to take the ball down the field and past the CMU goalie.

5.2 Chest Push and Double-Handed Kick

When the ball is close to the front of the robot's body, the body is thrust forward while the paws remain still on the ground, resulting in the chest pushing the ball forward. This kick is not a powerful one but it is easy to setup, reliable and quick for the robot to return to its original position. It is very effective near the target goal as it is able to push the ball past the goalie into the goal.

The "double-handed kick" was introduced in the 2000 code. This version brought both of the front arms up and pushed them down onto the ball to shoot. Because of changes in the robot's shape, this simple method no longer works. In the current version, the pressing force is changed to a squeezing force. The two arms are moved inward together, at floor level, to grab the ball and then the arms are moved up straight. When the ball is raised, it hits the head and is pushed downward under the arms. Now the ball is underneath the raised arms. The two arms now squeeze the ball out by moving both hands down quickly. This is very powerful and accurate and can score a goal across the full field length. However, to make the grab part reliable requires time to set up and during the game we usually do not have this luxury, so this kick was rarely effective. This is true of most of the kicks developed by all the teams.

6 Interaction with Other Robots

6.1 Team Mate Avoidance

During a game, robots tend to get into scrums with both team mates and opponents. Legs lock together and odometry is confused. Furthermore, other robots at close range block the field of view. To avoid these problems, our team has a "back off from team mate policy". When a team mate is detected within a range of about 20cm and the location of the ball is not known, the robot aborts the current action and takes a few steps backward. If ball position is known and a team mate is detected, the distance to the ball and the distance to the team mate are compared to determine which robot is closer to the ball. If the player itself is closer, than the team mate is ignored, and continues its attack, otherwise, the attack is aborted and the robot begins backing away from the team mate. As the robot backs away, it also turns to get behind the team mate and face towards the target goal. This often blocks an opponent's attack and also gives the player a better view of the situation. Since it tends to stay behind the scrum, the robot is often in a position to pick up a ball that comes loose.

At the same time that the player decides to back away from its team mate, it emits a high pitched sound signal telling the team mate that it is backing off. This signal is used by the other robot to decide in which direction to turn the ball.

6.2 Opponent Avoidance

The turn kick is useful in avoiding opponents and keeping the ball away from them. When a player has the ball in its possession and it detects an opponent at close range, it performs a turn kick. If it has heard a sound signal from its team mate, it knows that there is a supporting robot behind and to one side. so it turns in the direction of the team mate. If no signal was heard, it turns away from the opponent. This behaviour is effective in moving the ball away from an opponent and keeping possession of the ball, especially in a head-on situation.

6.3 Sound

The ERS-210 has on board stereo microphones and a speaker. Using sine waves, we are able to drive the speaker at any frequency within the sampling rate limitations. To detect a tone frequency, we utilize the orthogonality of the complex Fourier series to extract the power coefficient of that particular frequency and are able to detect signals on several channels at the same time without effecting the strategy performance. Since there are stereo microphones, it is possible to roughly determine if the signal originated from the left or right side of the head. However, due to the noise of motors and collision with other robots, we are only able to separate signal from noise within 70cm radius of the head.

6.4 Goalie Avoidance

A new "keeper charging" foul was introduced this year. A player is removed from the field for 30 seconds if it charges the goal keeper while it is holding the ball. We were unsure what effect this rule wold have on the game since losing a player is obviously a serious disadvantage but it was not clear how often such a situation would arise. Furthermore, it is not possible for the robot's vision system to recognise if the goalie has possession of the ball. To make matters worse, judging the distance to a robot is very inaccurate.

Because of our uncertainty about handling this foul, we decided to run two different forwards, one which attempted to avoid the goalie when the ball appeared to be close to it and another that simply carried on as normal. Figure 2 explains how the goalie avoidance is implemented.

6.5 Kick Off

There are several alternatives for the kick off, depending on the position of the team mate and opponents. If the player kicking off sees a team mate to one side, it will always kick the ball in the team mate's direction at an angle of about 40°. This helps the team retain possession of the ball after kick off and is often make it possible to get behind the opponent's forwards.

before kicking off, the robot looks up to check if there are any robots ahead. If it sees an obvious gap on either side of the goal, it attempts to shoot directly using a head butt. If the player is placed on the kick off position facing one side, there are two options. One is to head butt the ball straight ahead to get the ball to the target end. The second option is to kick off at an angle if an opponent is ahead.

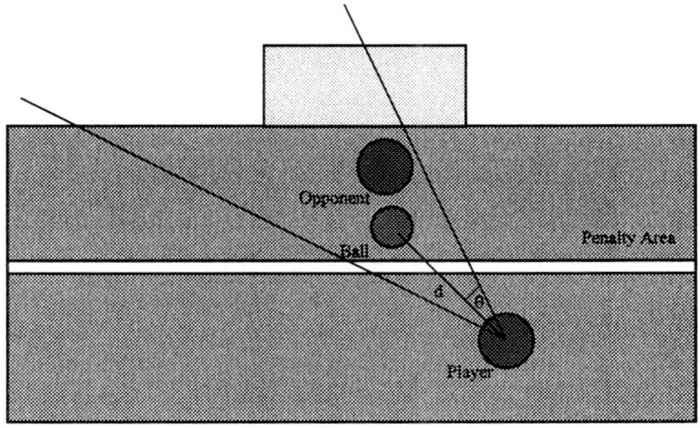

Fig. 2. If the ball is inside opponent's penalty area and the opponent robot is detected within an angle of θ from the direction of the ball, the player assumes the goalie is holding the ball and backs away from the keeper. The value of θ is inversely proportional to *d*, the distance to the ball.

7 The Challenge Competition

In addition to the normal soccer competition, the Sony legged league includes a set of "challenges". These are short tests of skill that may demonstrate abilities that are not necessarily used in the soccer matches.

7.1 Challenge 1

The first challenge in the 2001 competition was the "goalie" challenge. The goalie is placed on the side, at the centre line and has a maximum of 20 seconds to reach its defensive position within the penalty area. Once there, a ball is rolled down a ramp, aimed at the goal. The end of the ramp is placed at the centre of the field. The goalie must block the ball and clear it out of the penalty area.

We found that if the robot moves forward, near the penalty line, when it fully spreads its front legs to block a shot, it can cover almost the entire shooting angle. So the challenge 1 program simply moves the robot to a fixed position slightly behind the penalty line, facing directly forward. When the ball is released and approaches the robot, the spread motion is triggered. The robot does not attempt to move left or right to block the shot, nor is there any need to calculate the trajectory of the ball.

Robots that successfully completed the task were ranked according to the length of time taken for the robot to enter the penalty area. UNSW came second to the University of Pennsylvania.

7.2 Challenge 2

The second challenge tested the robot's ability to localise. Five marks were placed on the field. One robot was given the coordinates of these points and then instructed to move to each mark, in turn. At each mark, it was required to stop for three seconds and indicate that it knew it was over the mark by wagging its tail or beeping. Teams that successfully completed the task were ranked according to the total time taken.

Since we were not sure how accurately the robots would be required to locate the marks, we initially concentrated on making our localisation as accurate as possible. Later, maintaining this level of accuracy, we attempted to make the robot as fast as possible. Our strategy was to always face the robot towards the centre and pan the head slowly, to maximize the number of beacons in a frame. Once we were close to a point the speed was adjusted dynamically to allow for exact positioning. A shortest path algorithm was used to select the order of marks visited.

Once again, the UNSW robot was ranked second, this time to CMU's winning robot. Since the marks placed on the field were 8 cm in diameter, we were too conservative in our localisation and could have sacrificed some accuracy for greater speed.

7.3 Challenge 3

Challenge three was indeed challenging. None of the teams successfully completed this task. However, with some small adjustments to the starting conditions, the UNSW robot did demonstrate that the skill could be achieved. The task was to let two robots walk down the field, one on each side, passing the ball back and forth until one of the robots scores. Two unmoving "opponent" robots were placed on the middle of each half of the field as obstacles.

Our philosophy for Challenge 3 was to do it slowly but surely. This was necessary because it was a complex task, involving many stages, all requiring a high degree of reliability. The strategy involved grabbing the ball and then carrying it with the front paws in a customised walk. Once in position, we used the double-handed kick to ensure the ball made it to the other side and was positioned so that the other robot could pick it up. There was much experimentation to ensure that the robot would behave correctly. For example, the robot would back away from the edge to give it enough room to turn and align its kick. Unfortunately, in the actual challenge, a small error in localisation, as a result of the robot's vision being covered on boot up caused the first challenger to switch into its final state. The challenger did not pass to its team mate, instead going directly to the goal and shooting.

Although we did not earn any points, we demonstrated Challenge 3 again, with the robots' cameras uncovered at the start. On this occasion, both robots performed flawlessly.

On aggregate points the UNSW team also won the challenge competition.

8 Conclusion

Early in the competition, it was evident that two teams, UNSW and CMU, were clearly stronger than the others. Both teams had powerful, agile and stable locomotion. Both teams had good vision and localisation and both teams were recording high scores against their opponents. The UNSW vs CMU final was arguably the best contest seen in this league. Most games are won because of a mismatch in low-level skills. For example, one team may be faster, see the ball better, localise more accurately, etc. However, in the 2001 final, both teams had good low-level skills. The difference seemed to be more in game-play tactics.

One of the reasons for the strength of the UNSW team is due to that fact that we always test new behaviours under game-play conditions. We also run frequent practice matches that enable us to develop tactics that are more likely to work during a game. For example, several teams developed very good kicks. While these are useful to have, they are not game winners against a fast opponent that can block the ball. The UNSW team concentrated on small things like when the robot loses sight of the ball, turning in a direction that is most likely to knock the ball in the right direction. Other small behaviours dealt with minimising head movement when localising so that we do no lose sight of the ball and do not waste time looking around. While the paw-dribble and turn kicks were not "killer" shots, they tended to move the ball around and keep it away from opponents. Finally, the 2001 team improved the robot avoidance behaviours to ensure that team mates were well spread on the field, not interfering with each other and in positions where they could take advantage of loose balls.

It is clear that a great deal of hand-crafting was required to achieve championship winning performance. The real challenge is to devise means by which situation-specific behaviours can be learned. However, we believe there is still much for us humans to learn before we understand the problem well enough to be able to achieve this goal.

Acknowledgements

We thank Mike Lawther for his assistance in setting up the CVS repository, used to control the 2001 source code. Mike and Ken Nguyen also provided valuable support during the Seattle competition.

References

1. J. Dalgliesh and M. Lawther. Playing soccer with quadruped robots. Computer engineering thesis, School of Computer Science and Engineering, University of New South Wales, 1999.
2. B. Hengst, D. Ibbotson, S. B. Pham, J. Dalgliesh, M. Lawther, P. Preston, and C. Sammut. The UNSW Robocup 2000 Sony Legged League Team. In P. Stone, T. Balch, and G. Kraetzschmar, editors, *Robocup 2000: Robot Soccer World Cup IV*, pages 64–75. Springer-Verlag, 2001.
3. B. Hengst, D. Ibbotson, S. B. Pham, and C. Sammut. UNSW RoboCup 2000 Team Report. Technical report, School of Computer Science and Engineering, University of New South Wales, 2000. `http://www.cse.unsw.edu.au/~robocup`.
4. J. R. Quinlan. *C4.5: Programs for Machine Learning.* Morgan Kaufmann, San Mateo, CA, 1993.

YabAI
The First Rescue Simulation League Champion

Takeshi Morimoto, Kenji Kono, and Ikuo Takeuchi

Department of Computer Science, University of Electro-Communications,
1-5-1 Chofugaoka, Chofu, Tokyo 182-8585, Japan,
morimoto@takopen.cs.uec.ac.jp,
http://ne.cs.uec.ac.jp/~morimoto/

Abstract. RoboCupRescue project aims to simulate large urban disasters. In order to minimize damage resulting from disasters, various rescue agents try to accomplish their missions in the disaster space in the simulation system. Ability of an individual agent, however, is utterly insufficient. Agents need to cooperate with other same and different types utilizing as little communication as possible under stringently limited visual sensory information. Our YabAI team, however, successfully implemented effective cooperations under this limitation.

1 Introduction

RoboCupRescue project aims to simulate large urban disasters. Rescue agents such as ambulance teams, police forces, and fire brigades act on the disaster space in the simulation system. Buildings and houses collapse and burn, and roads are blocked in the disaster space. So number of civilian agents are sacrificed and injured. In order to minimize damage resulting from disasters, these agents have to try to accomplish their missions.

This paper, at first, briefly describes the outline of the disaster space, the simulation system, and agents. Then it describes technical problems which researchers and developers encounter, and some of the relevant research issues and innovations that lead YabAI team to the success.

2 Disaster Space

Building collapses, fires and blockage of roads occur in the disaster space by a large earthquake as was recently experienced in the Hanshin-Awaji Earthquake which killed more than 6,500 citizens. Many agents such as civilians, ambulance teams etc. lived in the disaster space. Soon after the large earthquake, buildings collapse, many civilians are buried in the collapsed buildings, fires propagate, and it becomes difficult for agents to move roads because these are blocked by debris of buildings and something else.

A. Birk, S. Coradeschi, and S. Tadokoro (Eds.): RoboCup 2001, LNAI 2377, pp. 49–59, 2002.

The objective of agents is, collectively, to minimize casualties. The behavior and performance of agents are evaluated by the following equation in the RoboCupRescue competition.

$$V = DeathToll + \frac{TotalDamage}{TotalLifePoint} \cdot \frac{BurnedSquare}{TotalBuildingSquare} \tag{1}$$

represents the degree of damage resulting from the disaster. The less is V, the higher is the evaluation. Note that the second term is less than or equal to 1.

The size of the disaster space used in the first competition is 500m × 500m. Within this area there are 740 buildings, 820 roads, 97 agents, etc. The simulation

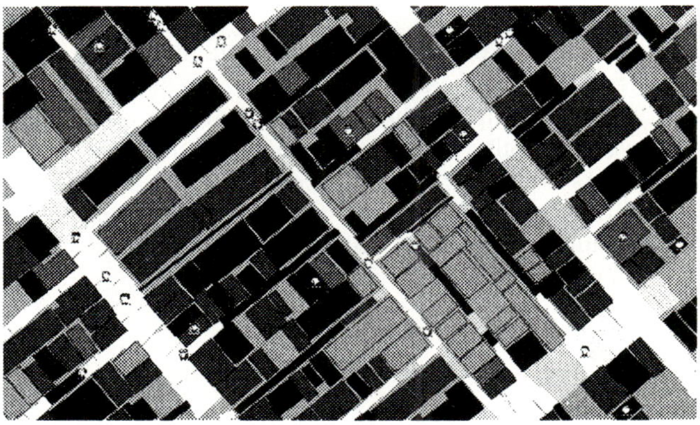

Fig. 1. Disaster space

system simulates 5 hours, or 300 minutes of the disaster space after an earthquake occurs. One minute in the disaster space is referred to as one turn. At each turn, each agent decides its action, and the simulation system reflects these actions if possible, action cannot be always carried out in the disaster space.

2.1 Simulation System

The simulation system consists of several disaster sub-simulators, the GIS (Geographical Information System), the viewer and the kernel. In addition, civilian agents are also regarded as a part of the simulation system in the competition. The simulation system is managed by the kernel. and The kernel proceeds the simulation as follows.

1. The kernel sends individual sensory informations to all the agents in the simulation world.
2. Agents submit their action command to the kernel individually.
3. The kernel sends agents' commands to all relevant disaster sub-simulators.

4. Sub-simulators submit updated states of the disaster space to the kernel.
5. The kernel sends updated states to the viewer.
6. The kernel advances the simulation clock in the disaster space.

The simulation system repeats this loop 300 times. Typically it takes a few seconds to simulate one turn. All the agents must decide an action for a few seconds.

For more details of the simulation system, refer to [2000].

2.2 Types and Abilities of Agents

Types of Agents: In the disaster space, there are seven types of agents; civilian, ambulance team that rescues injured persons and take them to refuges, fire brigade that extinguishes fires and arrests the spread of fires, police force that repairs blocked roads, ambulance center, fire station, and police office. Agents are classified to three kinds; civilians, platoons, and centers.

Some of civilian agents are buried, injured or sacrificed. Other civilian agents probably go to refuges or places safer than their current position. some of them panic, and cause traffic jams perhaps. Because civilian agents are regarded as a part of the simulation system in the competition, agent developers need not develop them.

Ambulance team agents, fire brigade agents and police force agents are platoon agents. A platoon agent consists of several people. They can move by a vehicle and have special skills according to their agent type.

Ambulance center agents, fire station agents and police office agents are center agents. Center agents represent buildings and have the ability of people who work there. They only communicate with same type platoons and other type centers.

In the first RoboCupRescue Simulation League, there are initially 72 civilian agents, 5 ambulance team agents, 10 fire brigade agents, 10 police force agents, 1 ambulance center agent, 1 fire station agent and 1 police office agent in the disaster space.

Abilities of Agents: Each agent has abilities according to their type (Table 1). Platoon agents can sense visually information about buildings, roads, other agents and etc. within a radius of 10m, and fires of any distance. Center agents cannot sense information visually.

Any agent can communicate with other agents. When an agent says a message by voice, the message is immediately heard by agents within a radius of 30m. When an agent tells a message by electrical communications devices, the message is immediately heard by agents whose type is the same as the speaker's. Furthermore, when a center agent tells something, the message is transferred to other type centers, too (Fig.2). The number of Say, Tell and Hear commands that can be issued by an agent in one turn is limited. An agent can say or tell up to four messages in a turn, and can hear four messages in a turn. Information amount of messages is measured by the number of sentences involved. When an

Table 1. Ability of agent

Type	Ability
Civilian	Sense, Hear, Move, Say
Ambulance team	Sense, Hear, Move, Say, Tell, Rescue, Load, Unload
Fire brigade	Sense, Hear, Move, Say, Tell, Extinguish
Police Force	Sense, Hear, Move, Say, Tell, Clear
Ambulance center	Hear, Say, Tell
Fire station	Hear, Say, Tell
Police office	Hear, Say, Tell

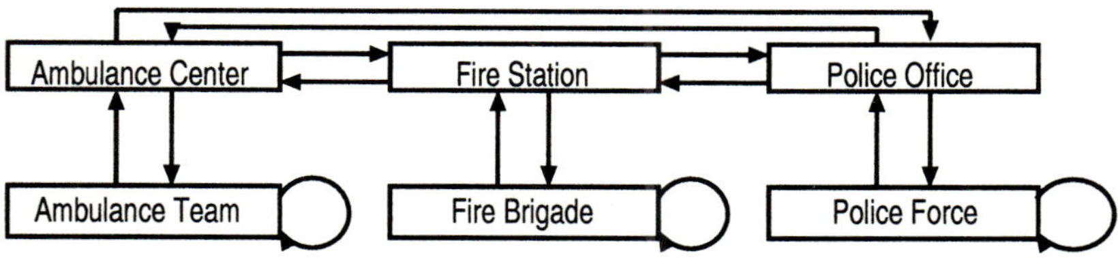

Fig. 2. Tell by Electrical communication

agent receives hear information from the simulation system, the agent may select whether it hear individual messages or not by checking who is the speaker.

3 Problems

This section describes technical problems researchers and developers will encounter when they develop agents.

Causes of casualties are mainly the damage by burial on land and fire. Almost all of civilian agents which are buried in collapsed buildings are injured, and become weaker as time goes by. Furthermore, when they are enveloped in fire spreading, they suffer from a fatal injury. In order to rescue and extinguish, ambulance team and fire brigade agents must go to the scene of the accident. It is, however, difficult for those agents to reach there, because many roads are blocked.

In order to minimize casualties, not only rescue operation by ambulance team agents but also extinguishing operation by fire brigade agents and road repairing operation by police force agents are necessary.

Agents must try to accomplish their missions under stringently limited conditions. Ability of an individual agent is insufficient against the scale of disasters. Agents can only get visual sensory information within a radius of 10m in the large disaster space which has an area of 500m $\times$ 500m. They have not enough ability to cope with disaster alone. Furthermore, it is stringently limited for agents to communicate with each other.

Cooperation among Same Type Agents: Single agent can hardly be useful. It takes a long time to rescue a buried civilian by single ambulance team agent. Single fire brigade agent can hardly extinguish a fire. Single police force agent lag demand. Consequently, platoon agents need to cooperate with other agents of the same type.

Cooperation with Different Type Agents: The ability of each agent is complementary to each other, so agents need to cooperate with agents of different types. Police force agents clear blocked roads for other types, or agents cannot go through these roads. Fire brigade agents extinguish fires and arrests the spread of fires for buried agents, or they are enveloped in fire. Ambulance team agents are not possible to rescue buried agents without these cooperations by other types.

Cooperation under Stringently Limited Communications: Rescue agents need to communicate with each other in order to accomplish their missions efficiently. However it is difficult to communicate, because they can speak only four messages and hear only four messages in a turn. There is a dilemma. In order to communicate enough, agents want to speak many messages. However, if agents speak many messages, it increases the likelihood that other agents miss the messages. For example, when ten fire brigade agents and one fire station agents speak four messages respectively in a turn, they receive 40 messages except for four messages of their own and then they must select four messages from 40 messages by only relying on the information who is the speaker. In this case, each message can be heard by others with only 10% probability. Agents need to hold speaking messages in bare minimum. Agents are required cooperating strategies with minimum communication.

Above problems are summarized as follows. In order to minimize casualties, rescue agents need to cooperate with same and different types with as little communication as possible under stringently limited visual sensory information.

4 Modularization of Fundamental Functions

We developed YabAI in incremental steps. First, we modularized several fundamental functions. Second, we developed agents which act without cooperation, and improved fundamental abilities such as an advanced world modeling and a reliable routing ability. Finally, we introduced several cooperation strategies; a fire brigade distribution strategy, a blocked road reporting strategy etc.

4.1 Modules

In order to develop agents efficiently, we modularized several fundamental functions;

- communication function between the simulation system and agents,
- world modeling function,
- world condition describing function,
- object extracting function,
- routing function,
- communication function among agents in the simulation system, and
- visual debugger.

Agent developers can concentrate all of their energies on the development of agent intelligence by using these modules. We chose JAVA as description language. Almost all intelligent parts of YabAI are described as a production system. For example, an agent intelligence that finds fires and goes there is described as follows:

$$\begin{aligned}
&isBurning := FIERYNESS_F \geq 1 \;\wedge\; FIERYNESS_F \leq 3\\
&fireSet := \{b \in buildingSet\ in\ world \mid isBurning(b)\}\\
&route := route\ from\ here\ to\ fireSet\ minimizing\ costFunction\\
&move\ along\ route
\end{aligned}$$

we can easily describe a program by using the modules as follows.

```
Predicate isBurning = FIERYNESS_F.gte(1).and(FIERYNESS_F.lte(3));
Set fireSet = world.buildingSet().get(isBurning);
Route route = router.get(self.position(), fireSet, costFunction);
move(route);
```

4.2 Visual Debugging Module

It is impossible to develop agents without viewing various information about agents;

- an internal world model of individual agent,
- action which an agent did,
- a route which an agent wanted to go along,
- messages which agents spoke, etc.

We introduced a visual debugger into YabAI. Visual information is much more easily handled than text information. Agent developers can get information visually as they wish by using the visual debugger (Fig.3).

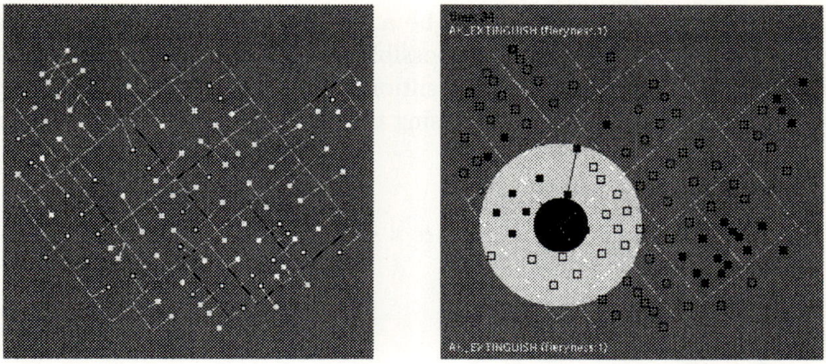

Fig. 3. Default Viewer (left) & Visual debugger (right)

5 Fundamental Abilities

Agents need to have well-tuned fundamental abilities in order to act efficiently. Mechanisms for multi-agent cooperation are of no use unless the individual agents can do their low level tasks properly. There are inseparable two fundamental abilities which intelligent agents must have; one is the ability of an appropriate perception of the current situation, and the other is the ability to decide an appropriate action according to the current situation. If agents does not have an appropriate perception, they cannot decide an appropriate action, and vice versa.

In RoboCupRescue, agents must have two fundamental abilities in particular; the advanced world modeling ability and the reliable routing ability.

5.1 Advanced World Modeling Ability

The disaster space changes every moment. Injured persons becomes weaker. Fires spread. Other agents move from space to space. Agents can not pass through even the roads which they have passed before, because of a traffic jam. On the contrary, agents could pass through the roads which they have never passed before as a result of clearance by police force agents. Agents have only a narrow range of view, so they can hardly grasp perception of the whole disaster space. However, agents can get more information within their range of view by observing others. Agents can get informations from visible others; their type, position, life point, damage, route along which they came from other place etc. By taking advantage of such route information, YabAI infers that these route are passable. This technique is highly effective, and is a base for all actions.

5.2 Reliable Routing Ability

The ability by which agents reach destination in a short time is very important for any agents. Especially when there are mortally wounded persons and early fires, even slight delay turned out to be a fatal mistake. Agents must route so as to move surely. YabAI becomes possible to move fast and surely by using certainty factor based on the above-mentioned advanced world modeling ability. To be specific, YabAI gets routes by using the routing module with the following cost function:

$$costFunction = actualDistance \times \begin{cases} 1 & movable \\ 10 & not\ confirmed \\ 100 & unmovable \end{cases}$$

6 Cooperation Strategies

6.1 Ambulance Team

Ambulance team agents search for injured persons in a distributed manner, and concentrate to rescue them together with others by using communication. Ambulance team agents can hardly grasp states of all injured persons, because there are many injured persons and they are often appear suddenly, for example, by the spread of a fire. So it is nearly impossible to optimize a rescue plan. Ambulance team agents put a premium on the certainty, and rescue injured persons soon after they decide it is necessary to rescue just now. In order to restrict rescue operation from which no pay-back can be expected, they divide injured persons into four rescuing priorities; "need immediately", "need", "no need" and "abandon".

6.2 Fire Brigade

Fire rapidly changes its condition, so it must be especially rapidly coped with. It may not be difficult to extinguish an early fire for even a few fire brigade agents. On the contrary, it is very difficult to extinguish a late and big fire for even many. Time is valuable very much. It is better that individual fire brigade agent selects fire on its own arbitrarily, than spending time to communicate with others.

There are four tactical points about fire fighting;

- give early fire top priority,
- barricade by extinguishing edges of spread of fires,
- concentrate flashing power as possible, and
- distribute power as far as possible finally.

It is wasteful that many fire brigade agents extinguish a small-scale fire, and it is ineffective that a small number of agents extinguish a large-scale fire. However, under a situation where fires are scattered, it is a formidable task to balance concentration and distribution of flashing powers.

YabAI becomes possible to distribute fire brigade agents efficiently by estimating the priority of fires based on three criteria; (Fig.4)

– state of the fire, especially burning duration,
– surrounding circumstances of the fire, and
– distance from the agent to the fire.

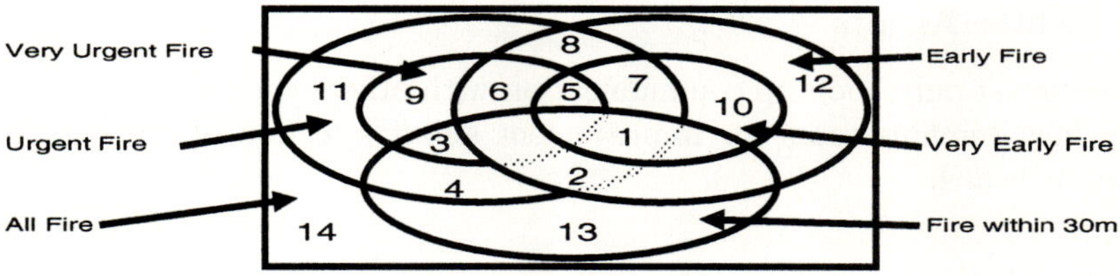

Fig. 4. Priority of fire

Fire brigade agents look at the surrounding circumstance of the fire to estimate whether fire spreading can be blocked by extinguishing it. If extinguishing the fire is useful for blocking of fire spreading, then it is valuable. The value of fire extinguishment is calculated with the numbers of unburned buildings, extinguished buildings and fires around it. To be specific, the fire around where has many unburned buildings and few fires is valuable to extinguish, that is, urgent. Fire brigade agents should extinguish such urgent fires rapidly, or the fires probably propagate to the surroundings.

Fig.4 shows the priority of fires which is described by numeric character. For example, the first priority the smallest the integer is given to a fire which is very early and located within a radius of 30m.

6.3 Police Force

Police force agents repair blocked roads through which others need to pass. Other types are often put in a situation that they cannot reach any destinations because routes are cut off. Though they are able to calculate optimal routes, they cannot reach any destinations in such a situation. As previously mentioned, agents must reach their destinations as soon as possible when they need to be there. Activity of police force agents especially affects others'. It is, however, difficult to get a criterion which roads they had better to repair first.

YabAI mainly repairs roads requested by others, at this point. This strategy is effective, but it takes time for others to be able to pass. In order to save others' time, it is necessary to repair important roads in advance. After a number of trials and errors, we introduced two strategies without using communication. One strategy is to make a route from the passable police force agent to the nearest fire brigade agent. This strategy has two merits. First, the police force agent can go to the fire brigade agent soon after being requested to repair a road. Second, this produces an optional route for the fire brigade agent. The other strategy is to make a route passable from the police force agent to fires. This also produces optional routes for fire brigade agents who want to go to those fires. These two strategies are implemented in a trivial short-range planning, but these are effective them expected.

6.4 Center Agents

Center agents only hook up communication with other type center agents, now. There may remains room for improvement about selection of messages to be spoken or heard.

7 Results

At the RoboCup-2001 competition, YabAI won the champion.

In the preliminaries there were seven games. All participants contested in the same disaster space at each game. At every game, they were placed in order, and got a point according to their order. Top four placings in total point passed the preliminaries. YabAI became 1st in 3 games, 2nd in 2 games, 4th in 1 game and 6th in 1 game. YabAI became 4th and 6th unfortunately, because of the simulation system's malfunction. YabAI was 2nd place in the preliminaries.

At the semifinal YabAI won all 2 games over Rescue-ISI-JAIST (3rd place).

At the final YabAI won all 2 games over Arian (2nd place). YabAI got overwhelming victory because of the remarkable activity of ambulance team agents at the first game, and won the second game which was very close by virtue of our early extinguishment.

It can be said that YabAI overwhelmed other teams except Arian.

8 Conclusions

In RoboCupRescue, agents try to accomplish their missions in order to minimize casualties. Ability of an individual agent is limited compared with the scale of disasters. Agents need to cooperate with same and different type agents without communication as far as possible under stringently limited visual sensory information.

YabAI shows effective cooperation based on well-tuned fundamental abilities such as the advanced world modeling ability and the reliable routing ability. Ambulance team cooperates by using communication with other ambulance teams in order to search and rescue injured persons efficiently. Fire brigade cooperates without communication with other in order fire brigades to extinguish fires surely and efficiently. Police force cooperates with agents of other types in two ways; one uses explicit communication with others to facilitate their requests, the other does not use communication, but implicitly cooperates by clearing important routes in advance.

The success of YabAI due to the effective cooperations with minimum communication.

For a more thorough understanding of the implementation details, the reader is encouraged to scrutinize the algorithms described here by looking at the YabAI source code [2001]

References

[2000] Satoshi Tadokoro, Hiroaki Kitano: "RoboCupRescue", 2000. Kyoritsu publisher.

[2001] Takeshi Morimoto: "YabAI source code", 2001. Accessible from http://ne.cs.uec.ac.jp/~morimoto/.

A Control Method for Humanoid Biped Walking with Limited Torque

Fuminori Yamasaki[1,2], Ken Endo[3], Minoru Asada[2], and Hiroaki Kitano[1,4]

[1] Kitano Symbiotic Systems Project, ERATO, JST,
yamasaki@symbio.jst.go.jp
[2] Graduate School of Engineering, Osaka University
[3] Graduate School of Science and Technology, Keio University
[4] Sony Computer Science Laboratories, Inc.

Abstract. This paper presents an energy-efficient biped walking method that has been implemented in a low-cost humanoid platform, PINO, for various research purposes. For biped walking robots with low torque actuators, a control method that enables biped walking with low torque is one of the most important problems. While many humanoids use high-performance motor systems to attain stable walking, such motor systems tend to be very expensive. Motors that are affordable for many researchers have only limited torque and accuracy. Development of a method that allows biped walking using low-cost components would have a major impact on the research community as well as industry. From the view point of high energy-efficiency, many researchers have studied a simple planar walker without any control torque. Their walking motions, however, are decided by the relationship between a gravity potential effect and structural parameters of their robots. Thus, there is no control of walking behaviors such as speed and dynamic change in step size. In this paper, we propose a control method using the moment of inertia of the swing leg at the hip joint, and confirm that a robot controlled by this method can change its walking speed when the moment of inertia of the swing leg at the hip joint is changed without changing the step length. Finally, we apply this control method to the PINO model with torso in computational simulations, and confirm that the method enables stable walking with limited torque.

1 Introduction

The RoboCup Humanoid League [1], which is scheduled to start in 2002, is one of the most attractive research targets. We believe that the success of the humanoid league is critical for the future of RoboCup, and will have major implications in robotics research and industry. Thus, PINO[2][3] has been developed as a humanoid platform that can be widely used for many RoboCup researchers in the world. Figure 1 shows the whole view and the mechanical architecture of PINO.

In this paper, we present a first step toward energy-efficient biped walking methods that can be attained through low-cost, off-the-shelf components. At

A. Birk, S. Coradeschi, and S. Tadokoro (Eds.): RoboCup 2001, LNAI 2377, pp. 60–70, 2002.

(a) Whole view (b) Mechanism

Fig. 1. Picture and mechanism of PINO

the same time, energy-efficient walking may reflect actual human biped walking. Stable biped walking generally requires highly precise control and powerful actuator systems. It forces the use of expensive actuators and gear systems, so the total cost of the robot is very high. Due to the cost constraints, the motors that we can afford to use tend to have limited torque. Thus, we cannot apply conventional control methods of biped walking to PINO because of limited torque. It should be noted that most of the current control methods for humanoid walking are designed independent of the structural properties of the robot hardware. In general, these control methods require extremely large torque to realize desired walking patterns. The use of structural properties of a mechanical platform may significantly reduce the need for high torque motors by attaining energy-efficient behavioral patterns. Actuators can realize various kinds of behavior.

Recently, many researchers have reported studies on biped walking[4][5][6]. McGeer, especially, has studied PDW (Passive Dynamic Walk), in which simplest walkers can walk down a gentle slope without any actuators for controlling them[7], their walking motions are decided by the relationship between gravity potential effect and their structural parameters. Asano et al. realized the virtual PDW on a horizontal floor with a torque which generates a virtual gravity effect[8]. These control methods for PDW make much use of the structural properties to achieve highly energy-efficient walking. However, passive dynamic walkers cannot change their walking motions by themselves, thus such a control method is not adequate for humanoid walking for flexible of walking patterns.

On the other hand, Ono et al. regarded biped locomotion as the combined motion of an inverted pendulum and a 2 DOF pendulum. As a result, they achieved human-like planar biped walking using simple self-excitation control[9][10].

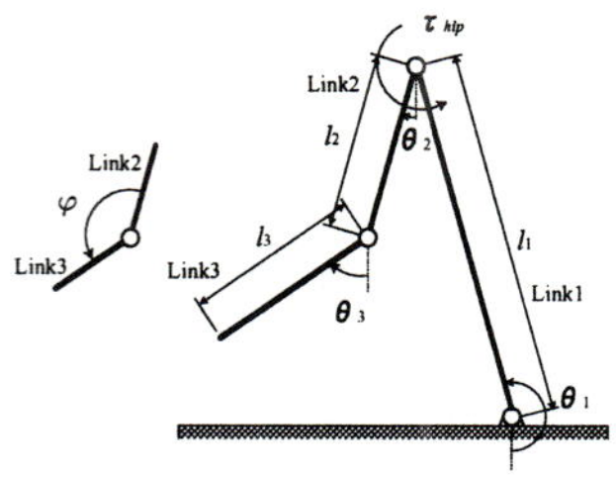

Fig. 2. Planar three link model of the robot

In the biped walking method, we started from the hypothesis that the walker can change the walking speed without changing the step length, if the moment of inertia of the swing leg at the hip joint has changed. Then, we proposed a simple control method based on our hypothesis, and confirmed that this control method is valid for biped walking on a planar model. Finally, we applied this control method to a planar four-link model with torso, and showed that appropriate swings of the torso enable the robot to walk with lower energy consumption. Through this study, we aim to achieve stable walking for PINO.

2 The Three-Link Model

2.1 The Model

We consider the legs of PINO as a combination of an inverted pendulum model and a 2 DOF pendulum model, assuming the structure of PINO to be a planar walker. In this case, the inverted pendulum represents the supporting leg and the 2 DOF pendulum represents the swing leg. The inverted pendulum model is the most energy-efficient model at the supporting leg. Figure 2 shows the planar three-link model of the robot we assume. This model consists of link1, link2 and link3, and every link is connected in series. Link1 has a joint with the ground. Link1 and link2 are connected by the hip joint, and link2 and link3 are connected by the knee joint. We assume that every joint has a viscosity coefficient of 0.01 [N·m·s/rad], and that the knee joint also has a knee stopper. We define every joint angle θ_1, θ_2 and θ_3 as the absolute angle of link1, link2 and link3 respectively. Also, φ represents the angle between link2 and link3. Each link has uniformly distributed mass m_1, m_2 and m_3, respectively.

2.2 The Control Method

It is desirable that the control method is represented simply by structural properties. In this study, we choose the control method with the property of moment of inertia on the hip joint I_{hip} in order to verify our hypothesis. The moment of inertia of the swing leg on the hip joint I_{hip} can be represented as follows,

$$I_{hip} = \frac{1}{3}m_2 l_2^2 + \frac{1}{12}m_3 l_3^2 + m_3(x'^2 + y'^2) \tag{1}$$

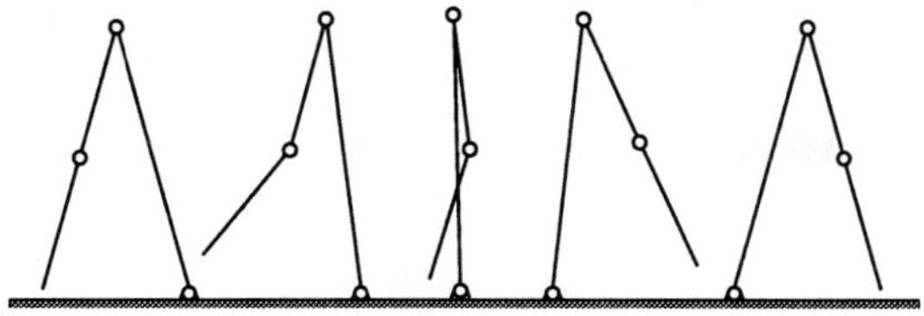

Fig. 3. The motion phases of one cycle

where x' and y' denote the center of gravity position of link3 as follows.

$$x' = l_2 sin\theta_2 + \tfrac{1}{2} l_3 sin\theta_3$$
$$y' = l_2 cos\theta_2 + \tfrac{1}{2} l_3 cos\theta_3 \qquad (2)$$

From Eq. 1 and 2, φ can be represented as follows using I_{hip},

$$\varphi = f(Arccos(I_{hip})) \qquad (3)$$

From Eq. 3, we define the control method as $k_{leg}\varphi$. Here, k_{leg} denotes the feedback gain. Also, we assume the walking motion consists of two phases as shown in Fig. 3.

In the first phase $(0 \le t \le t_1)$, the feedback torque τ_{leg} is added at the hip joint. This feedback torque τ_{leg} makes the swing leg bend at the knee joint and swing forward without other torque. From the relationship between τ_{leg} and the time until the swing leg touches the ground t_2, the time to cut off the torque t_1 is decided. In the second phase $(t_1 < t \le t_2)$, the feedback torque τ_{leg} does not add to the hip joint. The swing leg moves forward freely until it collides with the ground. Equation 4 shows the feedback torque at the hip joint.

$$\tau_{leg} = \begin{cases} -k_{leg}\varphi & (0 \le t \le t_1) \\ 0 & (t_1 < t \le t_2) \end{cases} \qquad (4)$$

where t_1 and t_2 are decided uniquely satisfying the law of conservation of angular momentum between immediately before and after foot collision as shown in Eq. 5.

$$\left[I(\theta)^-\right]\left\{\dot{\theta}^-\right\} = \left[I(\theta)^+\right]\left\{\dot{\theta}^+\right\} \qquad (5)$$

where $[I(\theta)^-]$ and $[I(\theta)^+]$ denote inertia matrices. Also $\left\{\dot{\theta}^-\right\}$ and $\left\{\dot{\theta}^+\right\}$ denote the angular velocity vectors of immediately before and after foot collision. We assume that the collision between the swing leg and the ground, and the collision caused by the knee lock are perfectly non-elastic collisions. From Eq. 5, the final angular velocities of each link of one step are decided uniquely when we define the initial angular velocities of each link.

Table 1. Link parameters

m_1	[kg]	1.500	l_1	[m]	0.5000
m_2	[kg]	0.750	l_2	[m]	0.2500
m_3	[kg]	0.750	l_3	[m]	0.2500

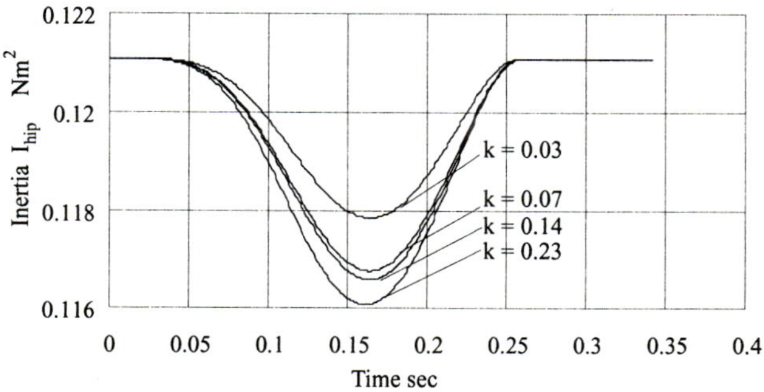

Fig. 4. Time series of I_{hip}

Also, the dynamic equation of this three-link model can be represented as,

$$[M(\theta)]\left\{\ddot{\theta}\right\}+[C(\theta)]\left\{\dot{\theta}^2\right\}+[K(\theta)]=[\tau] \tag{6}$$

where $[M(\theta)]$, $[C(\theta)]$ and $[K(\theta)]$ denotes the parameter matrices of the mechanism and angular positions of the links, and $[\tau]$ denote feedback torque vector. Feedback torque vector $[\tau]$ can be represented as,

$$[\tau]=\left\{-\tau_{hip},\,\tau_{hip},\,0\right\}^T \tag{7}$$

2.3 Verification of Effect of the Control Method on Three-Link Model

We conducted preliminary computational simulations to verify that this control method is valid for the biped walking.

First, we define the link parameters of the robot as given in Table 1. We set the initial angular velocity $\dot{\theta}_i$ and we conduct computational simulations to change the feedback gain k_{leg} in order to change φ.

As a result, we obtain time series data of the inertia I_{hip} of the swing leg at the hip joint shown in Fig. 4. From the comparison of I_{hip} in each k_{leg}, we can use the minimum moment of inertia I_{min} as a performance index of I_{hip}. Figure 5 shows the effect of k_{leg} on t_2, and Fig. 6 shows the effect of k_{leg} on minimum moment of inertia I_{min} during walking. From these graphs, it is clear that the increase of k_{leg} causes a linear reduction of t_2 and I_{min}.

The time series of torque values of every k_{leg} and clearance to the ground of every k_{leg} are shown in Figs. 7 and 8 respectively. The maximum torque that we

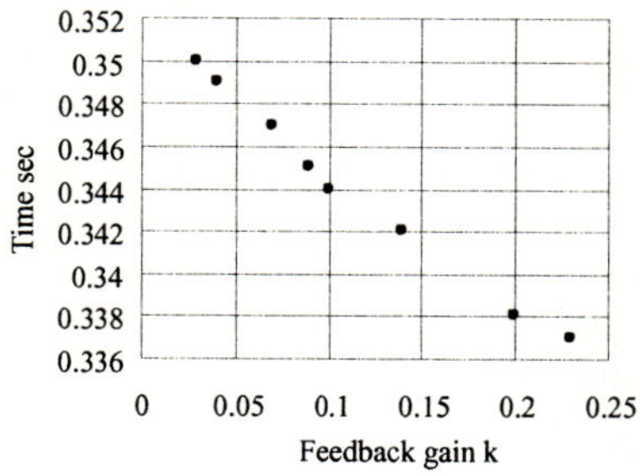

Fig. 5. Effect of k_{leg} on t_2

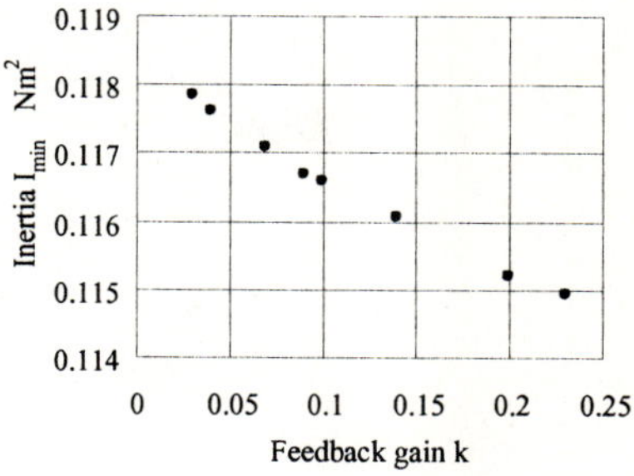

Fig. 6. Effect of k_{leg} on I_{min}

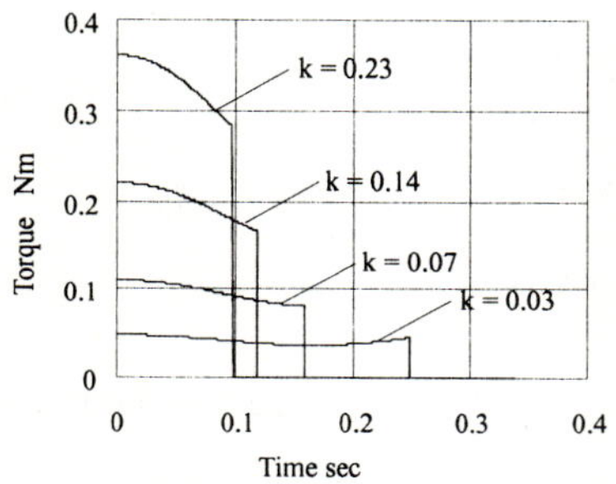

Fig. 7. Time series of torque value

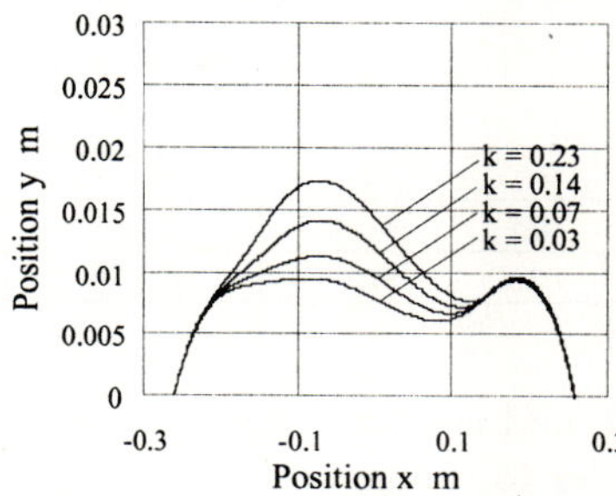

Fig. 8. Clearance to the ground

obtain from every k_{leg} ranges from 0.05 [N·m] (at $k_{leg} = 0.03$) to 0.35 [N·m] (at $k_{leg} = 0.23$). Considering this two-link swing leg model as a pendulum model consisting of a mass (M = 1.5 [kg]) and a link (L = 0.2795 [m]), the desired torque for swinging this pendulum is approximately 4.11 [N·m]. This control method, therefore, required the torque below one tenth than this torque, and it is clear that it can achieve the walking motion with high energy-efficiency. From Fig. 8, the clearance of the swing leg to the ground becomes larger as k_{leg} increases.

From the relationship among k_{leg}, t_2 and I_{min}, we can obtain the relationship between I_{min} and t_2.

From Fig. 9, it is clear that the reducing I_{min} causes fast walking, and increasing I_{min} causes slow walking. According to this relationship, we confirmed our hypothesis that the walker can change the walking speed when we change the moment of inertia of the swing leg at the hip joint, and this control method is valid for biped walking.

3 The Lower Limb Model of PINO

We apply this control method to the lower limb model of PINO. First, we define the link parameters of the robot as given in Table 2. We set the initial angular

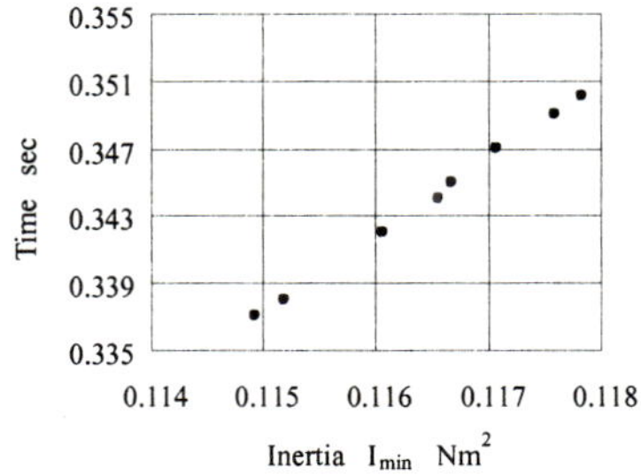

Fig. 9. Relationship between I_{min} and t_2

Table 2. Link parameters

m_1	[kg]	0.718	l_1	[m]	0.2785
m_2	[kg]	0.274	l_2	[m]	0.1060
m_3	[kg]	0.444	l_3	[m]	0.1725

velocity $\dot{\theta}_i$ and conduct computational simulations to change the feedback gain k_{leg} in order to change φ.

As a result, we obtain the time series of torque values of every k_{leg} and clearance to the ground of every k_{leg} as shown in Figs. 10 and 11.

The maximum torque that we obtain from every k_{leg} ranges from 0.2 [N·m] (at $k_{leg} = 0.13$) to 0.35 [N·m] (at $k_{leg} = 0.22$). Considering this two-link swing leg model as a pendulum model consisting of a mass (M = 0.718 [kg]) and a link (L = 0.155 [m]), the desired torque for swinging this pendulum is approximately 1.09 [N·m]. This control method, therefore, required the torque below one third than this torque, and it is clear that it can achieve the walking motion with high energy-efficiency at lower limb model of PINO, too. From Fig. 11, the clearance of the swing leg to the ground becomes larger as k_{leg} increases. The maximum clearance of the swing leg to the ground ranges from 0.025 [m] (at $k_{leg} = 0.13$) to 0.035 [m] (at $k_{leg} = 0.22$). The minimum clearance ranges from 0.012 [m] (at $k_{leg} = 0.13$) to 0.017 [m] (at $k_{leg} = 0.22$).

4 Application to the Four-Link Model with Torso

4.1 The Model and Control Method

We have verified that our proposed control method of the swing leg is valid for biped walking. Thus, as an application of the three-link model, we apply this control method to the four-link model with torso, and verify the influence of the torso on the biped walking motion. Figure 12 shows the four-link model with torso. This model consists of link1, link2, link3 and link4, and link1 has a joint with the ground. We define every joint angle θ_1, θ_2, θ_3 and θ_4 as an absolute angle of link1, link2, link3 and link4 respectively. Each link has uniformly distributed

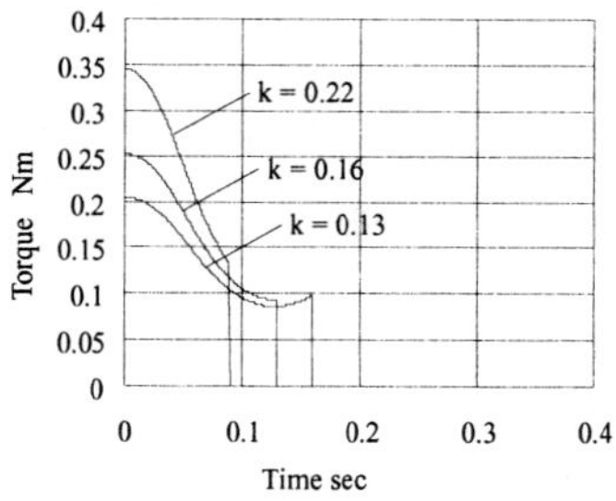

Fig. 10. Time series of torque value

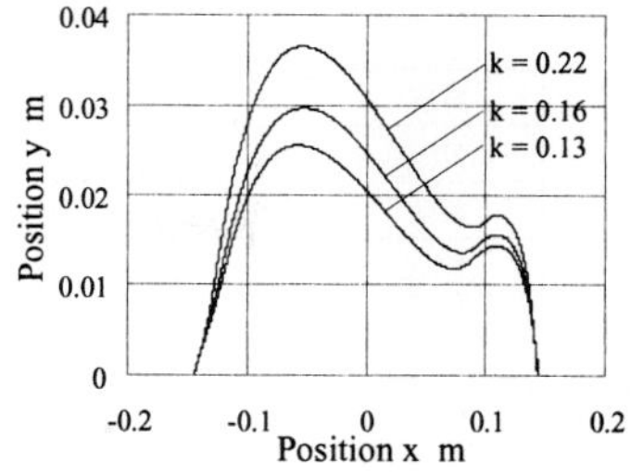

Fig. 11. Clearance to the ground

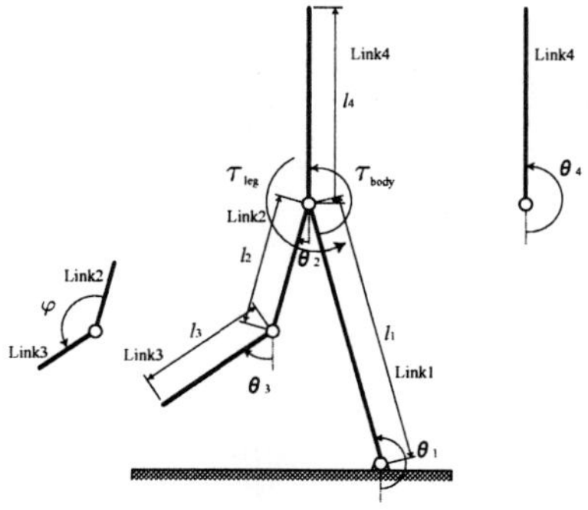

Fig. 12. Planar four-link model of the robot

mass m_1, m_2, m_3 and m_4, respectively. Table 3 shows the link parameters of the four-link model. We define the control method for the torso as $-k_{body}\theta_4$. Here, k_{body} denotes the feedback gain. We examined the effect of the feedback gain k_{body} on the walking motion. Feedback torque vector $[\tau]$ can be represented as follows,

$$[\tau] = \left\{ -\tau_{leg} + \tau_{body},\ \tau_{leg},\ 0,\ -\tau_{body} \right\}^T \tag{8}$$

4.2 Results and Discussion of the Model

We examine the three cases of the torso posture.

Case 1: We set the body feedback gain as $k_{body} = 14.0$, and the four-link model with torso moves using τ_{leg} and τ_{body}.

Case 2: We fix the torso vertically, and the four-link model with torso moves using only τ_{leg}.

Case 3: We examine the three-link model without torso to compare the three-link model without torso and the four-link mode with torso.

Also, we define the leg feedback gain as $k_{leg} = 0.5$ in every case in order to verify the effect of the torso on the biped walking motions.

Table 3. Link parameters

m_1	[kg]	0.718	l_1	[m]	0.2785
m_2	[kg]	0.274	l_2	[m]	0.1060
m_3	[kg]	0.444	l_3	[m]	0.1725
m_4	[kg]	3.100	l_4	[m]	0.4515

Table 4. Results of every case

Value		Case 1	Case 2	Case 3
$\dot{\theta}_1$	[rad/sec]	1.736	0.962	3.374
$\dot{\theta}_2$	[rad/sec]	1.692	0.223	1.384
$\dot{\theta}_3$	[rad/sec]	0.000	0.000	0.000
$\dot{\theta}_4$	[rad/sec]	1.309	—	—
t_2	[sec]	0.319	0.406	0.296
Energy consumption	[J]	0.064	0.109	0.025

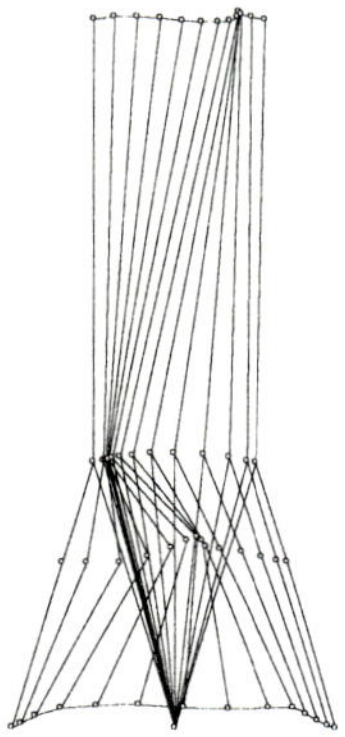

Fig. 13. Result of the foot gait of case 1

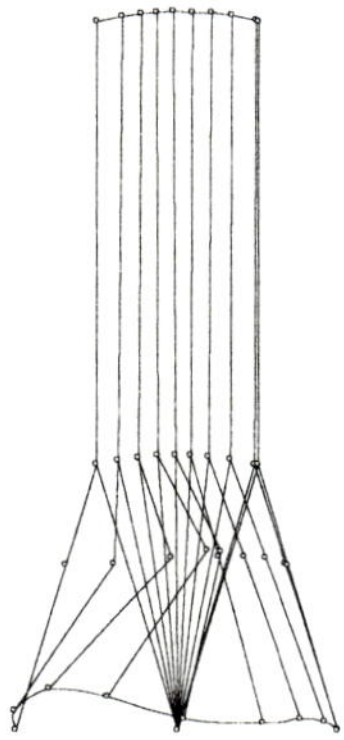

Fig. 14. Result of the foot gait of case 2

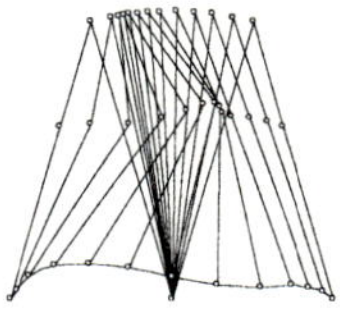

Fig. 15. Result of the foot gait of case 3

Figures 13, 14 and 15 show the foot trajectory of every case. Table 4 shows the initial angular velocity $\dot{\theta}_1$, $\dot{\theta}_2$, $\dot{\theta}_3$, $\dot{\theta}_4$, time to touch down t_2 and energy consumption. From Table 4, t_2 of the four-link model with torso is longer than that of the three-link model without torso t_2, and energy consumption of case 1 is smaller than that of case 2 though every angular speed is larger. From these results, it is verified that the walking motion with appropriate swings of the torso enables the robot to walk with lower energy consumption.

5 Conclusion

In this paper, we presented a method for energy-efficient biped walking, and applied the method to PINO, a low-cost humanoid robot. The proposed method aims to achieve stable biped walking with limited torque, so that it can be applied to humanoids that are built from low-cost, off-the-shelf components.

The conventional control method can not be used for such a robot, because it does not have sufficient torque to accommodate conventional control. Instead of using the conventional control method, we proposed and verified an energy-efficient biped walking method that exploits the structural properties of the mechanical platform. In this paper, therefore, we proposed a simple control method which is highly energy-efficiency using the structural properties. We chose the moment of inertia of the swing leg at the hip joint, and we applied feedback torque $\tau_{leg} = -k_{leg}\varphi$ to the hip joint.

As a result, in the lower limb model of PINO, the maximum torque required was reduced to the range of approximately 0.2 [N·m] (at $k = 0.13$) to 0.35 [N·m] (at $k = 0.22$), and it was confirmed that this enables the low-cost humanoid PINO to perform reasonably stable biped walking.

Also, the maximum clearance of the swing leg to the ground was adjusted to the range of approximately 0.025 [m] (at $k = 0.13$) to 0.035 [m] (at $k = 0.22$), and the minimum clearance to 0.012 [m] (at $k = 0.13$) to 0.017 [m] (at $k = 0.22$) which are sufficient for humanoid walking on a flat floor. Further, in the four-link model with torso, it was verified that appropriate swings of the torso enable the robot to walk with lower energy consumption low as 0.064 [J]. In the future, we intend to expand this control method to 3D space, and to apply this control method to the real humanoid PINO.

6 Acknowledgments

The dynamic simulation was supported by Mr. Masaki Ogino. The authors thank him and members of Asada Laboratory of Osaka University.

References

1. Kitano, H., Asada, M., "RoboCup Humanoid Challenge: That's One Small Step for a Robot, One Giant Leap for Mankind", *Proc. of International Conference on Intelligent Robots and Systems*, 1998.
2. Yamasaki, F., Matsui, T., Miyashita, T. and Kitano, H., "PINO The Humanoid that Walk", *Proc. of The First IEEE-RAS International Conference on Humanoid Robots*, CD-ROM, 2000.
3. Yamasaki, F., Miyashita, T., Matsui, T. and Kitano, H., "PINO the Humanoid: A basic architecture", *Proc. of The Fourth International Workshop on RoboCup*, CD-ROM, 2000.
4. Hashimoto, S., Narita, S., Kasahara, K., Shirai, K., Kobayashi, T., Takanishi, A., Sugano, S., et al, "Humanoid Robots in Waseda University – Hadaly-2 and WABIAN", *Proc. of The First IEEE-RAS International Conference on Humanoid Robots*, CD-ROM, 2000.

5. Inaba, M., Kanehiro, F., Kagami, S. and Inoue, H., "Two-armed Bipedal Robot that can Walk, Roll Over and Stand up", *Proc. of International Conference on Intelligent Robots and Systems*, 1995.
6. Kajita, S. and Tani, K., "Experimental study of biped dynamics walking in the linear inverted pendulum mode", *Proc. of IEEE International Conference on Robotics & Automation*, pp.2885–2891, 1995.
7. T. McGeer, "Passive dynamic walking", *The International Journal of Robotics Research*, Vol.18, No.6, pp.62 – 82, 1990.
8. Asano, F., Yamakita, M. and Furuta, K., "Virtual Passive Dynamic Walking and Energy-based Control Laws", *Proc. of International Conference on Intelligent Robots and Systems*, 2000.
9. Ono, K., Takahashi, R., Imadu, A. and Shimada, T., "Self-Excitation Control for Biped Walking Mechanism", *Proc. of International Conference on Intelligent Robots and Systems*, CD-ROM, 2000.
10. Rongqiang Liu and Ono, K., "Energy Optimal Trajectory Planning of Biped Walking Motion", *Proc. of the International Symposium on Adaptive Motion of Animals and Machines*, CD-ROM, 2000.

A Fast Vision System for Middle Size Robots in RoboCup

M. Jamzad, B.S. Sadjad, V.S. Mirrokni, M. Kazemi, H. Chitsaz, A. Heydarnoori, M.T. Hajiaghai, and E. Chiniforooshan

Computer Engineering Department
Sharif University of Technology, Tehran, Iran
jamzad@sina.sharif.ac.ir
{sadjad,mirrokni,kazemi,chitsaz,noori,hajiagha,chinif}@ce.sharif.ac.ir
http://sina.sharif.ac.ir/~,ceinfo

Abstract. A mobile robot should be able to analyze what it is seeing in real time rate and decide accordingly. Fast and reliable analysis of image data is one of the key points in soccer robot performance. In this paper we suggest a very fast method for object finding which uses the concept of perspective view. In our method, we introduce a set of jump points in perspective on which we search for objects. An object is estimated by a rectangle surrounding it. A vector based calculation is introduced to find the distance and angle of a robot from objects in the field. In addition we present a new color model which takes its components from different color models. The proposed method can detect all objects in each frame and their distance and angle in one scan on the jump points in that frame. This process takes about $\frac{1}{50}$ of a second. Our vision system uses a commercially available frame grabber and is implemented only in software. It has shown a very good performance in RoboCup competitions.

1 Introduction

In a soccer robot, there are several key components such as good mechanical design and its stability, reliable hardware and fast vision system. In this paper we focus on our experiments in developing a fast image processing and vision system for our robots. The basic mechanics and hardware of our robots are fully described in [1]. The hardware of our vision system consists of a Genius grabber which uses BT878 IC. This grabber is mounted on a main board with Pentium, 233 MHz processor. The camera used is a Sony handycam. The software is written in C++ under DJGPP compiler (DJ Gnu Plus Plus) in MS-DOS.

The fast changing environment in RoboCup prevents us from using standard image processing methods for segmentation and edge detection. Although there exist many reliable such methods, but with our processor power, they can not be implemented in real time (25 frames per second). Therefore, we developed a very fast algorithm for region estimation and object detection which can be used for finding mobile targets as well.

A. Birk, S. Coradeschi, and S. Tadokoro (Eds.): RoboCup 2001, LNAI 2377, pp. 71–80, 2002.

2 Proposing a New Color Model

We used to use the HSI color model for segmentation, but in practice we found some limitation especially in I and H parameters. As a result of a search for a more suitable color model, we concluded that we should combine the parameters of different color models and introduce a new color model. In this way we propose a color model named $\hat{H}SY$. Where the $\hat{H}$ is taken from CIELab, S from HSI and Y from YIQ color models [2]. We have obtained satisfactory results from this new color model. The reason for this selection is that, the component Y in YIQ represents the conversion of a color image into a monochrome image. Therefore, comparing with I in HSI which is simply the average of R,G and B, the component Y is a better mean for measuring the pixel intensity in an image. The component S in HSI is a good measure for color saturation. And finally the parameter $\hat{H}$ in CIELab is defined as follows:

$$\hat{H} = tan^{-1}\frac{b^*}{a^*}$$

Where a^* denotes relative redness-greenness and b^* shows yellowness-blueness [2]. The reason for selecting $\hat{H}$ is that, it has been reported that $\hat{H}$ is a good measure for detecting regions matching a given color [3]. This is exactly the case in RoboCup where we have large regions covered with a single color.

3 Segmentation

Any class of objects in RoboCup environment is supposed to have a predefined color. We have eight such colors according to the rules in practice in year 2001. These colors are green, white, red, yellow, blue, black, light blue, and purple. Therefore, we have to segment the image frames taken by the robot into one of the above colors. In practice we accept one don't care color for objects which can not be assigned to any of the eight standard colors.

In RoboCup, we think it is worth getting fast and almost correct results than slow but exact. Therefore, unlike the traditional image processing routines which define a segment to have a well defined border with its neighboring segments [4], the segments that we define always have rectangular shapes. This shape is a rough estimation for a bounding box around the object that should be segmented. An object is defined to be a connected component with a unique color.

3.1 Segmentation Training Stage

The output of most commercially available CCD cameras are in RGB color model. In training stage, we put a robot in the field and run programs which convert RGB values of each pixel into $\hat{H}SY$ components as described above. However, to obtain a relatively good segmentation, by try and error, we find the minimum and maximum range values for each parameter $\hat{H}$, S and Y such that the interval in between covers for example almost all tones of red colors seen in the ball. Doing this, we put the ball and robot in different locations on the field and update the interval ranges for red color. The same procedure is done for the remaining 7 colors. At the end of this training stage, 8 (colors) $\times$ $3(\hat{H}, S, Y) =$ 24 intervals are determined. All RGB values in one interval will be assigned to one of the 8 standard colors. However, there might be a few pixels which can not be assigned to any of the 8 colors. These pixels will be given a don't care color code and are considered as noise.

At the end of training stage an LUT (look up table) is constructed which can assign any RGB to a color code 1 to 9 representing one of the eight standard colors and the don't care color.

4 Search in a Perspective View

In real world we see everything in perspective, in which objects far away from us are seen small and those closer are seen larger. But, to recognize the objects in real world, we do not examine the scene pixel by pixel and do not check for certain conditions, but, somehow we just recognize the object. It will be a great job if we could define an algorithm for the term "just recognize the object".

The following method is an effort to make the robot detect objects in a quick way which is different from traditional image processing methods.

We use the fact that the robot camera sees everything in perspective. Figure 1 shows an image of RoboCup middle size soccer field with certain points on it. These points that are displayed in perspective with respect to robot front camera are called "Jump points". Actually the jump points are set at equal physical distance. They have equal spacing on each horizontal perspective line. The vertical spacing is relative to the RoboCup soccer field prespective view. The find the actual spacing between jump points on an image, we consider the facts that, the smallest object in RoboCup soccer field is the ball, and the longest distance between a robot and the ball is when they are located in two opposite corners diagonally. If the ball is located in its farthest position and we imagine a square bounding box around it, we want to make sure that robot can detect the ball. Therefore, this square should be large enough to house approporate number of jump points in itself.

The size of this square is very critical, because it determines the maximum distance between adjacent jump points in the perspective view of figure 1. In practice we found it safe to take the distance of each two horizontal jump points on the farthest distance to be about $\frac{1}{3}$ of the side of a bounding box drawn around the ball at that position. In this way, at least 5 jump points (3 horizontal and 2 vertical), will be located on the ball. Therefore, there is a high probability to find the ball, even in the farthest area, by checking only the jump points. In this method, all objects can be found by examining the jump points only, and there is no need to process the image pixel by pixel. In our system we obtained satisfactory results with 1200 jump points.

Fig. 1. Position of jump points in a perspective view of robot.

4.1 Search for the Ball

We scan the jump points from lower right point towards upper left corner. The reason for this order of search is that we assume a higher priority for finding objects closer to robot.

At each jump point, the RGB values are passed to a routine which searches in a look up table as described in section 3.1, and returns the color code corresponding to this RGB. If this color is red, then we conclude that this jump point can be located on the ball. Since this jump point can be any point on the ball, therefore, from this jump point, we move toward right, left, up and down, checking each pixel for its color. As long as the color of pixel being checked is red, we are within the ball area. This search stops in each direction, when we reach a border or limit point which is a non red color pixel. However, in checking the pixels on the ball, we perform a simple noise reduction method as follows. A non red pixel is determined to be red if at least 6 of its 10 connected neighbor pixels on the same direction are red.

In figure 2-a, point A_1 is the initial jump point, and points T_1,T_2,T_3 and T_4 are the 4 border points on the ball detected by the above algorithm. Rectangle R_1 passes through these 4 border points and is the initial guess for the rectangle surrounding the ball. O_1 is the center of rectangle R_1.

However, rectangle R_1 is considered to be the final answer if the distance of A_1O_1 is less than a certain threshold. This threshold guarantees that point A_1 is close enough to the rectangle center. If the above condition does not satisfy for rectangle R_1, we repeat the procedure of finding another rectangle from point O_1 which is assumed to be a jump point. This procedure is stopped when the above mentioned threshold condition is satisfied. An example of a satisfactory result is shown in figure 2-b, where the distance A_2O_2 is less than the threshold. Rectangle R_2 is the acceptable fit rectangle which surrounds the ball. Figure 3. shows a real illustration of this situation.

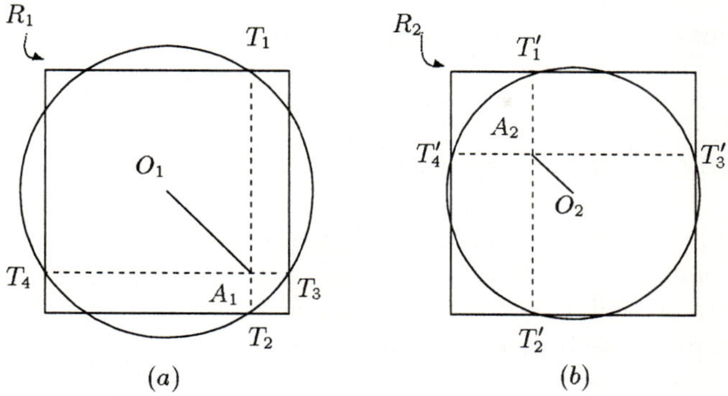

Fig. 2. (a) R_1 is the initial guess rectangle to surround the ball. (b) R_2 is the acceptable fit rectangle surrounding the ball.

Finally, we perform a size filter to eliminate the very small objects, which is, if the rectangle size is less than a certain threshold (this threshold is different for each object) it is considered to be a noise. Then the procedure of examining jump points is continued from the next jump point.

Fig. 3. An illustration of ball segmentation by a surrounding rectangle.

4.2 Search for Other Objects

In one scan of all jump points, we can find all objects such as robots (our team or opponent team robots), yellow goal, blue goal, etc., in the image.

For each object such as ball, robot, yellow goal, etc. we return its descriptive data such as, its color, size and the coordinate of its lower left and upper right corner of its surrounding rectangle and a point Q on the middle of its lower side. Point Q is used to find the distance and angle of the robot from that object. However, there is an exception when the ball is too close to our robot. In this case point Q will be on the lower side of the image and is handled differently. The procedure for calculating the distance and angle from an object is given in section 5. Our software can provide this information for an image in about $\frac{1}{50}$ of a second.

5 Finding Object Distance and Angle from Robot

In RoboCup, we have fixed position objects such as walls, goals, lines on the field and dynamically changing position objects such as ball and robots. But, as a robot moves around, the distance of fixed and moving objects from that robot continuously changes. To calculate distance and angle of the robot from objects, we need to define a dynamic coordinate system of which origin is located on the middle front side of robot body. Figure 4-a shows the projection of robot body on the field and point O is the origin of the robot dynamic coordinate system (we selected the reversed direction for the x axis).

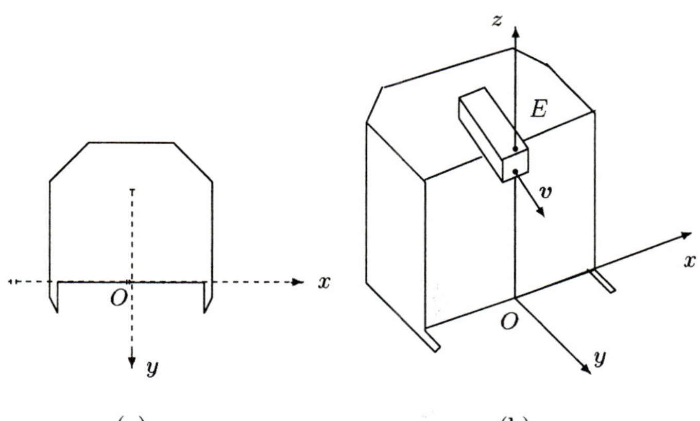

Fig. 4. (a) Projection of a robot body on the field. (b) A 3D view of robot, and the 3D coordinate system.

Figure 4-b shows a 3D view of the CCD camera in a 3D coordinate system on the robot body. For simplicity of calculations, we assume, point E which is the camera gimbal center to be located on the z axis which passes through origin O and is perpendicular to xy plane of figure 4-b.

The idea behind the following mathematical description is that we assume to calculate the distance of robot from an object as the distance from point O to a point Q. If we assume a rectangle surrounding an object such that its lower side touches the soccer field, then point Q is considered to be in the middle of the lower side of this rectangle. By this assumption we do not need to consider the z coordinate value of objects in the 3D world.

Let α be the angle of CCD camera with respect to the xy plane (pan angle: the angle between the line passing through image plane center and lens center with the xy plane or the field). And β be the angle of CCD camera with zy plane (the tilt angle, which is usually zero or has a small amount).

Figure 5 shows the geometry according to which the mathematical relations are constructed. The point Q on the field is projected on point T on the image plane in such a way that the line EQ passes through image plane at point T.

Now we can define the problem as "Finding the length of line OQ which is the distance from an object to our robot". In addition, the angle θ is the relative angle of robot front side with the object represented by point Q.

A unit vector $\hat{v}$ with origin of E and along the camera axis (along the line passing through image plane center and lens center) is calculated as $\hat{v}_x = \cos\alpha \sin\beta$, $\hat{v}_y = \cos\alpha \cos\beta$, and $\hat{v}_z = E_z - \sin\alpha$. Thus,

$$\hat{v} = (\cos\alpha \sin\beta, \cos\alpha \cos\beta, E_z - \sin a)$$

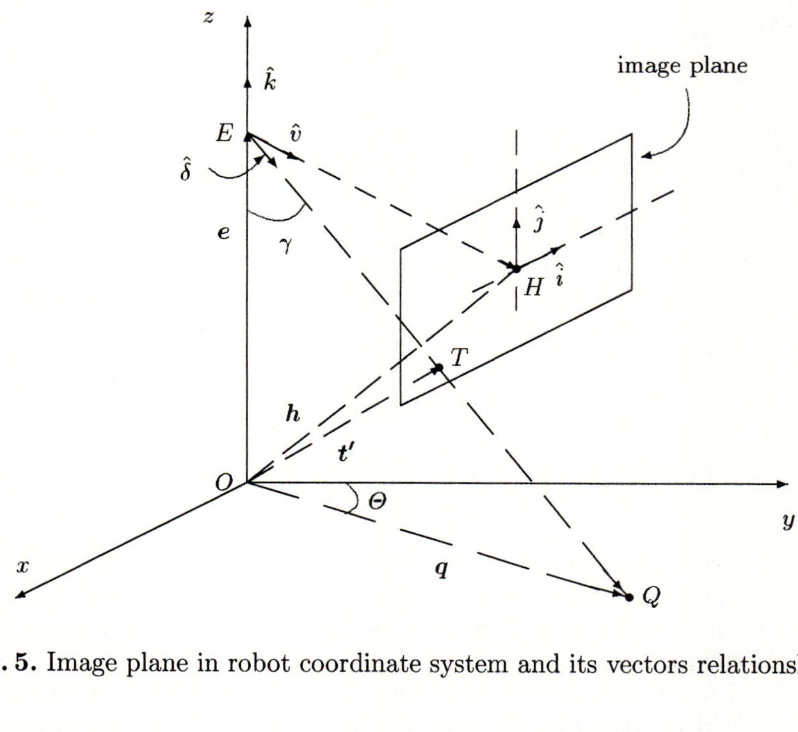

Fig. 5. Image plane in robot coordinate system and its vectors relationships.

We assume the image plane is perpendicular to the camera axis and is formed in a distance f from point E. Let H be the the center of the image plane. So the image plane is perpendicular to vector $\hat{v}$ and f is the distance between E and H. Therefore, we can define the following vector equation.

$$\boldsymbol{h} = \boldsymbol{e} + f\hat{v}$$

Let $\hat{k}$ be the unit vector along the z axis in figure 5. Unit vectors $\hat{\imath}$ and $\hat{\jmath}$ are along the x and y axis of the image plane. $\hat{\imath}$ is perpendicular to $\hat{k}$ and $\hat{v}$, thus unit vector $\hat{\imath}$ is the cross product of unit vector $\hat{k}$ by unit vector $\hat{v}$ and unit vector $\hat{\jmath}$ is the cross product of unit vectors $\hat{v}$ and $\hat{\imath}$.

$$\hat{\imath} = \hat{k} \times \hat{v}$$

$$\hat{\jmath} = \hat{v} \times \hat{\imath}$$

But the coordinate of point T on image plane is given by (T_x, T_y) and vector $\boldsymbol{t'}$ is calculated as follows:

$$\boldsymbol{t'} = \boldsymbol{h} + T_x\hat{\imath} + T_y\hat{\jmath}$$

and

$$\boldsymbol{\delta} = \boldsymbol{t'} - \boldsymbol{e}$$

Where $\boldsymbol{t'}$ is the vector between points O and T as seen in figure 5, and $\boldsymbol{\delta}$ is the ET vector and $\hat{\delta}$ is the unit vector on $\boldsymbol{\delta}$.

We know that

$$\boldsymbol{q} = \boldsymbol{e} + EQ \Rightarrow \boldsymbol{q} = \boldsymbol{e} + (\frac{E_z}{\cos\gamma})\hat{\delta}$$

Where E_z is the z component of vector OE (i.e. the height of camera).

Referring to figure 5, we see that $\cos\gamma$ is the dot product of unit vectors $\hat{k}$ and $\hat{\delta}$, because $\hat{k}$ and $\hat{\delta}$ are unit vectors and γ is the angle between these two lines.

Therefore, the x, y components of vector $\boldsymbol{q}$ represent the (x, y) coordinates of point Q on the field. The angle θ which represents the angle of object with respect to the robot front is defined as

$$\theta = \arctan(\frac{Q_y}{Q_x})$$

and

$$OQ = \sqrt{Q_x^2 + Q_y^2}$$

To find the parameters f, α and β we start from an initial estimates and update them by minimizing the error measured for a set of sample points on the field. This optimization process is done by fine tuning f, α and β around their estimated values.

However, after finding these parameters, we introduce a set of new sample points on field for which the distance OQ is calculated. By comparison of the real distance and the calculated one, we find an interpolation function which are used in practice.

6 Straight Line Detection

During the matches, there are many cases when the robot needs to find its distance from walls. In addition, the goal keeper in all times need to know its distance from walls and also from white straight lines in front of the goal. If we have one such line, and we know to what wall it belongs to, then we can find the angle of our robot with respect to that wall. However, if two such lines can be detected, the robot can detect its position on the field. This method was used to position the goal keeper in the middle of goal area.

Although there are many edge detection routines which use different filters [4], but all filtering methods are very time consuming and are not appropriate in a real time environment. To solve this speed problem, we propose a fast method for straight line detection. Since in RoboCup soccer field, the border line between walls and field, and also the 5cm wide white lines in front of goal keeper, all are straight lines, in our method, we detect a few points on each straight line and then by using Hough transform [4] we find the equation of the straight line passing through these points.

As it is seen in Figure 6, to locate points on the border of wall and field, we select a few points on top of the image (these points are on the wall), and assume a drop of water is released at each point. If no object is on its way, the water will drop on the field, right on the border with wall. To implement this idea, from a start point w_i we move downward, until reaching a green point f_i.

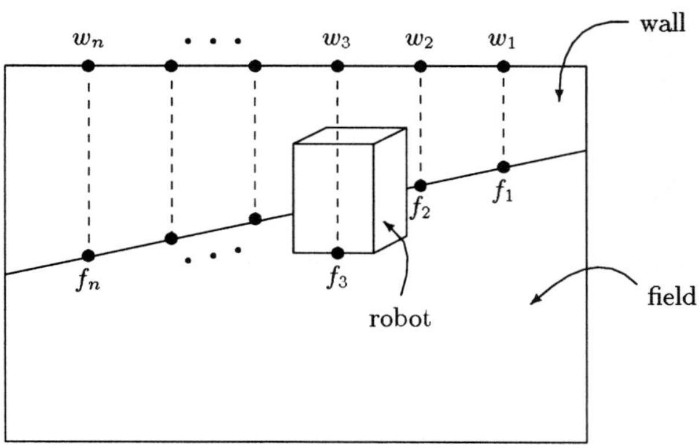

Fig. 6. An illustration of a robot view. Straight lines $w_i f_i$ shows the pass of water dropped from top of wall.

A green point will be accepted as a border point between the wall and field, if it is below a white point (wall), for example points such as f_1 and f_2. By this criteria, point f_3 is considered to be noise. One of the advantage of using Hough transform is that it can tolerate many noise points and also the selected non noise points need not be in close neighborhood of each other.

7 Discussion

There are many aspects in constructing a reliable mobile robot. But even if we have good mechanical and hardware control on our robots, still there are many situations we wonder why the robot can not do things it must do. Here, we would like to bring the issue of sensing devices and the software control capability. The more sophisticated the sensing system and the software, the more time consuming it will be. Therefore, to make a soccer robot act fast enough in a dynamically changing environment we shall reduce the number of sensors as much as possible and design fast software for object finding and good decision making. We believe in long term, when the soccer robot becomes like a humanoid robot to play in the real soccer field, most sensing devices such as infrared, laser, ultrasonic, etc. might not have the same good performance as at present they have in a relatively small field (4x9 meter with 50cm high wall around the field). Therefore, we think it is more appropriate to focus more on vision sensors, and develop methods using images taken from different camera systems. We believe the combination of a CCD camera in front and an omnidirectional viewing system on the top of robot will give a reliable performance [5]. The omnidirectional camera system can be used for localization and also sensing objects close to or touching the robot.

Although, the fast object finding method proposed in this paper has shown a high performance in previous years in RoboCup competitions, we are currently working on improving the system by using several cameras which enable us to have more reliable localization and sensing methods. Because, when we use only one camera, and detect the straight border lines between walls and field, it is difficult to know to what wall does this line belong. In these situations, an omnidirectional view can provide additional data for correct detection of walls.

8 Acknowledgments

We would like to thank a lot, the mechanical engineering students of Sharif CE team, A.Forough Nassiraei and R.Ghornabi who played important role in design and construction of the robots. Also, we are very grateful to all our sponsors.

References

1. M. Jamzad, et al, *Middle sized Robots: ARVAND*, RoboCup-99: Robot Soccer world Cup II, Springer (2000), pp 74-84 .
2. S.J Sangwine and R.E.N Horne, *The colour image processing handbook*, Chapman and Hall (1998).
3. Y. Gong, and M. Sakauchi, *Detection of regions matching specified chromatic features* , Computer vision and Image Understanding, 61(2), pp163-269, 1995.
4. R.C. Gonzalez, and R.E. Woods, *Digital Image Processing*, Addison-Wesley, 1993.
5. A. Bonarini, P. Aliiverti, M. Lucioni, *An Omnidiretional vision sensor for fast tracking for mobile robot.*, IEEE Instrumentation and Measurement technology Conference, IMTC 99, Piscataway, NJ, 1999, IEEE Computer press.

Designing an Omnidirectional Vision System for a Goalkeeper Robot

Emanuele Menegatti[1], Francesco Nori[1], Enrico Pagello[1,2], Carlo Pellizzari[1], and Davide Spagnoli[1]

[1] Intelligent Autonomous Systems Laboratory
Department of Informatics and Electronics
University of Padua, Italy
{emg,epv}@dei.unipd.it
[2] *also with* Institute LADSEB of CNR
Padua, Italy

Abstract. The aim of this paper is to provide a guideline for the design of an omnidirectional vision system for the Robocup domain. We report the design steps undertaken, with a detailed description of the design of an omnidirectional mirror with a custom profile. In the second part, we present the software written to exploit the properties of the designed mirror. The application for which the vision system is designed is the Middle-Size Robocup Championship. The task performed by the vision system is to localise the robot and to locate the ball and other robots in the field of play. Several practical tricks and suggestions are provided.

1 Introduction

In the Robocup competitions several teams use omnidirectional vision sensors and their number is increasing year after year. Asada uses a goal-keeper fitted with an omnidirectional vision system with a learning capability [8]. To reduce the learning time, the omnidirectional sensor is fitted with an attention control provided by an active zoom mechanism that permits to select a restrict area of the omnidirectional image. Lima uses an omnidirectional sensor for the self-localization of the robot in the field of play [6]. The omnidirectional mirror is designed to give a bird's eye view of the field of play. This permits to exploit the natural landmarks of the soccer field (goals and fields lines) for a reliable self-localization. Despite this second example, most of the teams use simple commercial mirrors, often used for other task than playing Robocup competitions. Yagi, one of the pioneers of omnidirectional vision with his COPIS system [10], in [9] proposed to look at the mirror of a catadioptric systems as a design variable. The insight is: the robot task determines the design of the mirror. This idea was stressed by Marchese and Sorrenti in [5]. Their paper has been fundamental for our work as well as the work of Bonarini [2]. In [5] they presented the design of a mirror composed by three parts:

- the *isometric part*, that permits to determine the position of the objects in the Robocup field;

A. Birk, S. Coradeschi, and S. Tadokoro (Eds.): RoboCup 2001, LNAI 2377, pp. 81–91, 2002.

- the *markers mirror*, that is devoted to the identification of the robots' markers;
- the *proximity mirror*, that is used to perform high resolution measures in an area close to the body of the robot;

In this paper we follow the approach of [5], but we change totally the characteristics of the mirror.

Fig. 1. (Left) Our goalkeeper robot. (Right) Our multi-part mirror

In Figure 2, we sketch the image formation mechanism in an omnidirectional vision sensor. Consider the point P laying on the field of play. Using the pinhole camera model and the optical laws, for every point at distance d_{OP} from the sensor, it is possible to calculate the coordinates (x, y) of the corresponding image point P' on the CCD. Vice versa, knowing the coordinate (x, y) of a point in the image plane, we can calculate the distance d_{OP} of the corresponding point in the world (for a detailed theory of catadioptric image formation, refer to the homonym paper of Nayar [1]). Because of the finite size of the sensitive element of the CCD, we do not have access to the actual coordinates (x, y) of the image point, but to the discrete corresponding pair (x_d, y_d) (i.e. the location of the corresponding pixel of the CCD). So, if we calculate the distance d_{OP} from (x_d, y_d), it will be discrete. The discrete distance deviates from the actual distance with an error that depends on the geometry of the mirror.

2 How to Design a Mirror

In this section we delineate an algorithm for the design of a custom mirror profile. The algorithm is a modification of the one presented in [5]. The main point is to identify the function that maps point of the world into points of the CCD. This is a function $f : R^3 \rightarrow R^2$ that transform world point (X',Y',Z') into image points (x,y). Actually, what we want it is a simpler function. We want a function that maps points laying on the plane Y=0 around the sensor from a distance

D_{MIN} up to a distance D_{MAX}. Therefore, exploiting the rotational symmetry of the system, the problem can be reduced to finding a one dimensional function $f^* : [D_{MIN}, D_{MAX}] \rightarrow [0, d_{MAX}]$ where d_{MAX} is the maximum distance on the CCD. The exact solution of this problem can be found with a quite complex differential equation. In [4] a solution of this equation is reported for the simple case $d = f^*(D) = KD$. In [5] is reported an approximated solution. We used this same solution, consisting in a local linear approximation of the mirror profile with its tangent. Let us see in the detail the algorithm exploited.

1. discretise the interval $[D_{MIN}, D_{MAX}]$ in a set $[D_{MIN} = D0, D1, D2..., D_N = D_{MAX}]$ that will be mapped by f^* in the set $[0, d_1, d_2, ..., d_N = d_{MAX}]$;
2. the tip of the mirror is in $P_0 = (0, Y_0)$ and the tangent to the mirror is $tan(arctan(D_{MIN}/y_0)/2)$. With this choice the point at distance $D = D_{MIN}$ is mapped into d=0. Let us call r_2 the line passing by P_0 whose derivative is $tan(arctan(D_{MIN}/y0)/2)$;
3. r_1 is the line passing by the focal point (0, f) and the point $(-d_1, h)$ on the CCD, where h is the height of the CCD on the ground. The intersection between r_1 and r_2 determines the point P_1. The line r_3 will be created as the line passing by the point P_1 and the point $(D_1, 0)$. Now the line r_3 and r_1 constitute the path the light has to follows if we want to map the world point $(D_1, 0)$ into CCD point $(-d_1, h)$. Therefore the tangent to the profile of the mirror in the point P_1 must be perpendicular to the bisector formed by r_3 and r_1. And so on, until all the point in the set $[D_{MIN} = D_0, D_1, D_2..., D_N = D_{MAX}]$ are mapped in the set $[0, d_1, d_2..., d_N = d_{MAX}]$;
4. The mirror profile is obtained by the envelope of all the previously calculated tangents in the points P_i, i=0, 1, 2, ..., N;

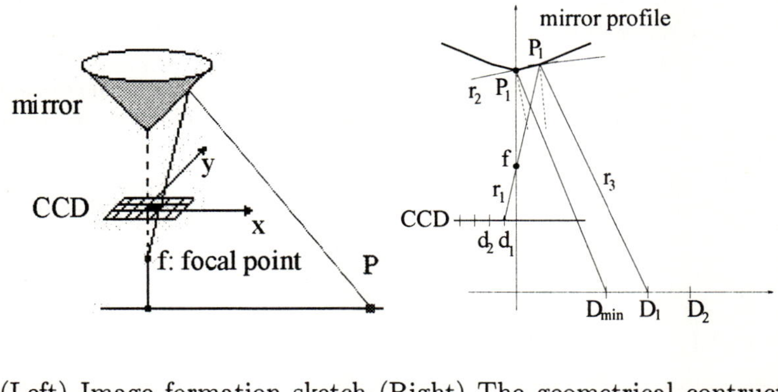

Fig. 2. (Left) Image formation sketch (Right) The geometrical contruction of the custom mirror

Now, we know how to design a mirror that realises a custom mapping of world points into points of the CCD sensor. In the following we will present the choices we made for the mapping function.

3 Comparison between Different Type of Mirrors

To better understand the advantages and the disadvantages of the different mirror profiles, we summarised them in Table 1.

Table 1. Comparison between different mirrors type

Mirror	Advantages	Disadvantages
Conic	– With an opportune choice of the geometric parameters it is possible to eliminate from the image the body of the robot. – Small *relative* error on the calculation of distances from the sensor.	– Maximum measurable distance is low. – Bad resolution close to the sensor.
Conic with spherical vertex	– Good resolution close to the sensor. – Small *relative* error on the calculation of the distances from the sensor.	– The inner part of the image is not utilizable because occupied by the body of the robot. – Maximum measurable distance is low.
Isometric	– Apparent size of the objects does not depend on the distance from the sensor. – Constant *absolute* error on the calculation of the distances from the sensor. – It is possible to perform the conversion from distances on the image to distances on the playground via a simple multiplication.	– The inner part of the image is not utilisable because occupied by the body of the robot. – Big *relative* error on the calculation of the distances for objects close to the sensor.

4 The Mirror We Designed

We designed a multi-part mirror composed by a inner part and two circular rings. As in [5] the two circular rings are devoted to the observation of the robot markers and of the area surrounding the robot. Let us consider first the inner part of the mirror, what we called the *measurement mirror*. This part is the one that observes the wider area of the field of play. For the *measurement mirror* we had two possibility: a mirror with a continuous curvature or a mirror with a discontinuity in the vertex, like conical mirrors. The latter has been our choice. In fact, in a multi-part mirror is possible to eliminate the disadvantages presented by a conic mirror, exploiting other sections of the mirror. Consider Table 1 and the disadvantages of conic mirrors there presented. The first disadvantage is eliminated by the middle circular ring of the mirror, i.e the *marker mirror*, that permits a long range field of view. The second is overcome by the outer part of the mirror: the *proximity mirror*, that permits to have a very good resolution near the body of the robot. Therefore, considering the whole structure of the mirror we have no drawback to the advantages introduced by the conic mirror. Nevertheless, are we ready to renounce to the important features of the isometric mirror proposed in [5]? In an isometric mirror, the image of the field of play is not distorted. Usually, this implies that the apparent dimensions of an object does not change with the distance from the sensor. Moreover, it is possible to calculate the distance of the object from the sensor via a simple multiplication. This is not possible with a distorting mirror, like a conic one. We can overcome the problem using a look-up table (LUT). The LUT should contain an association between the coordinates of points in the image and the coordinates of the corresponding points in the real world. Thus, calculating the position of a point in the world is reduced to a memory access. The second advantage of an *isometric mirror*, i.e. constant absolute error, is not such an advantage in our minds. In fact, a constant absolute error implies a big relative error at small distances. In the Robocup environment we need a good precision determining the position of objects near the robots, while we can allow a certain amount of absolute error for distant objects.

4.1 The Measurement Mirror

We chose to design a mirror that maps world points on the CCD with a constant *relative error* α, i.e. an error not depending on the distance from the sensor. So, the position of an object close to the sensor will be determined with good precision while an object far away from the robot will have a sensible absolute error. Let us see how we can design such a mirror. We identify a set of discrete positions in the world and the corresponding discrete positions on the CCD sensor. We fix the minimum distance from the sensor for a visible object, called D_{MIN}. All the points at a distance D_{MIN} are mapped into the central pixel of the CCD (distance d_0). In the pixels contiguous to the central one (distance d_1) we will map the points at a distance $D_1 = D_{MIN} + \alpha D_{MIN}$, in the next pixels (distance d_2) we will map pixels at distance $D_2 = D_1 + \alpha D_1$, and so on. In such a way we map the set of point in the world $[D_0 = D_{MIN}, D_1, D_2 ..., D_N = D_{MAX}]$ into the set $[0, d_1, d_2 ..., d_N = d_{MAX}]$ in the image. Therefore if a point has a distance D from the sensor such as $D_{i-1} < D < D_i$ it will be mapped at a distance d from the centre of the CCD such as $d_{i-1} \leq d \leq d_i$. The worst case is that the point P at distance D_{i-1} will be mapped into the pixel at distance d_i. If we reconstruct the position of P from the image we have an error $e = D^* - D_{i-1} = D_i - D_{i-1} = \alpha D_{i-1} = \alpha D$ and then a relative error α. The resulting profile is the first part of the mirror profile sketched in Figure 3 and it maps the world points into the area I_1 of the CCD.

4.2 The Marker Mirror

In the Robocup competitions every robot has to wear a coloured marker that identifies the team the robot belongs to. The marker must be positioned at a maximum height of 60cm. Therefore, we need a part of the mirror pointing to objects over the field of play. This implies to see objects out of the game arena. The vision of object out of the field of play causes troubles to the vision systems of the robots that are designed for the highly structured Robocup environment. In order to reduce the bad influence caused by seeing many unknown objects, we dedicated just a small portion of the image to the vision of the markers. The light reflected by the *marker mirror* will be mapped in a low resolution annular ring: area I_2 in Figure 3 (left). In this area we do not care about precision of the measurements, we want only to be able to detect the markers and to associate them to the robots localised with the *measurement mirror*.

4.3 The Proximity Mirror

As we stated before, the *measurement mirror* produces low resolution images close to the sensor. To clearly see objects close to the body of the robot (one of the most important part of the pitch) we designed the *proximity mirror*. The proximity mirror is the outer part of the multi-part mirror, so it has the most convenient position to observe objects close to the body of the robot. This part is designed with a low curvature and a quite large portion of the image is dedicated to the light it gathers. Thus, objects close to the robot have quite big apparent dimensions. To enhance this effect we made this latter part concave, so points closer to the robot are mapped in outer points of the image, Figure 3 (left). Because the body of our robot does not have a circular symmetry (it is a rectangle $37cm \times 27cm \times 30cm$), some portion of the robot's body (the corners) will appear in the image, Figure 4. We can not avoid this if we want to see objects as close as possible to the body of the robot. Figure 3 (left) presents a sketch showing how the different areas of the world are mapped on the image. Note that there is an overlapping region that is seen by both the measurement mirror and the proximity mirror. This is not a problem, on the contrary, it can be used to perform more precise measures.

5 The Omnidirectional Software

The designed omnidirectional mirror has been mounted on a goalkeeper robot, called Lisa. Lisa has been the reserve goalkeeper of the Azzurra Robot Team (ART), the Robocup Italian National Team. Lisa has played successfully some of the games in Robocup Euro 2000 in Amsterdam, where ART won the second place cup. As a goal keeper, Lisa is required to present high reactivity and high

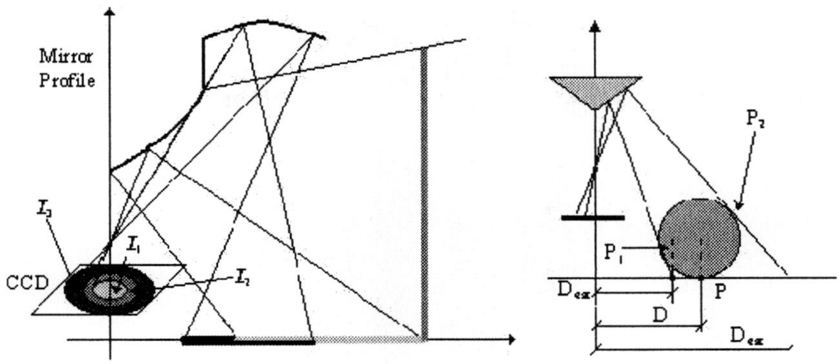

Fig. 3. (Left) A sketch of the multi-part mirror profile. (Right) The closest and farthest pixels of the ball

precision in the calculation of trajectories. To achieve a high reactivity, the vision software should be able to process at least 10 frames per second. Because the hardware of this robot has limited capabilities (the CPU is a K6 200MHz with 64 MB of RAM memory), we decided not to process the whole image, but to extract only the information needed by a goalkeeper: its position with respect to the goal and the position of the ball. In particular, we did not process pixels recognized as belonging to other robots and pixels in the area of the image containing the markers information. This was possible because we did not need to build a behavior of collision avoidance for the goalkeeper. The rules of the Robocup competitions consider every contact between a player and a goalkeeper as a foul move against the goalie.

5.1 Processing the Image

The first operation performed on the image acquired by the omnidirectional sensor is to identify the pixels belonging to the ball and to the goal. Because the colours in the Robocup field of play are coded, we simply need to locate red pixels to find the ball and, to find the goal, yellow pixel (or blue pixels depending on the goal the robot is defending). The search for these pixels is not performed on the whole image but just on *samples* of it. There is a subset of pixels of the image, called *receptors* [3] disposed in a spiral starting from the centre of the image. During the first scan of the image, just these receptors are analysed. As soon as the colour of one of these receptors corresponds to the colour of an object of interest (the ball or the defended goal) the scan is interrupted. A `CreateBlob Algorithm` starts. This algorithm builds the blob corresponding to all the connected pixels of the same colour. Once the blob is completed the scan of the receptors continues. This procedure is chosen in order to avoid scanning every single pixel of the image, that would be too time consuming.

Fig. 4. An image taken with our multipart mirror. Note the ball both in the measurement and in the proximity mirror and the corners of the robot body on the outer annular ring

5.2 Calculating the Ball Position

Once the blob corresponding to the ball is built, the ball location is calculated from the position of its closest and farthest pixel, Figure 3 (Right). The azimuthal coordinate of the ball corresponds to the azimuth of the closest and farthest pixel. The radial coordinate is more complicated. As we stated before, the distance of points seen in the image is calculated as if the point were on the field of play. As depicted in Figure 3 (Right), neither the closest nor the farthest pixel corresponds to the actual point of support of the ball. There is an error $e = D - D_{est}$. Knowing the true dimensions of the ball, it is possible to correct this error. In the calculation of the actual support point of the ball, we used both the closest and farthest pixel. Using two estimates of the ball distance,permits to reduce the position error. Moreover, it is possible to calculate the position of the ball even in situation in which only one of the two points is visible (for instance, if the ball is too close to the robot's body, the closest point is not visible to the sensor). To speed up this calculation, we created two look-up tables. The first contains the association between every image pixel and the corrected ball distance calculated from the closest pixel of the ball. The second contains the same data but calculated using the farthest pixel of the ball. The precision attainable with this method depends on the precision of the vision system. We designed our multipart mirror in order to have a *relative* maximum error of 3%. The described algorithm with the actual mirror is estimating the ball position with a maximum error smaller than 2%.

5.3 Localising the Robot Using the Goalposts

A goal keeper should always know his position with respect to the goal he is defending. Our robot uses the odometric information to calculate its position.

The odometers suffer of cumulative errors, so time by time they need to be reset. To precisely determine the robot location we used the information on the apparent position of the goalposts. The goalposts are easily detected with an omnidirectional vision sensor. They are mapped, as all vertical lines, into radial lines in the image. We choose to extract from the image the azimuths of two goalposts and the radial distance of one of the two. Resetting the position of the robot during the game is dangerous, even steady objects appear to move. If the goalkeeper re-locates its-self during a shoot, the ball appears to experience a sharp deviation from its direction, invalidating the reliability of the trajectory predictor module. Therefore, we decided to perform a re-location process just every 10 seconds, even if our robot could re-locate twice a second.

6 The Goal-Keeper Behavior

The designed omnidirectional mirror permitted to create an innovative moving for the robot. Most of the Robocup Teams have their goalkeepers moving along the goal line. Our robot moves on the semi-circumference, as shown in Figure 4a). This permits to design peculiar behavior for a more effective moving. Up to now, The deliberative level of our robot handles four cases triggering different behaviours.

Case 1: ball not seen. If the robot cannot see the ball or if the ball is farther than the midfield line, a behaviour called `CentreGoal` is activated. This positions the robot at 60 cm from the goal line, parallel to it. This is a *waiting behaviour*, i.e. the robot positions in the best location to reach the ball whenever the ball will be seen again.

Case 2: inactive ball. The ball is steady out of the penalty area or it aims out of the goal. In this case the ball is not *dangerous.* The behaviour called `ProtectGoal` is activated. The robot moves to cover the goal mouth. It moves to the point where the line joining the ball and the centre of the goal intersects the semi-circumference of 60 cm radius centred in the goal centre, Figure 5 (left). As it is easy to understand from Figure 5, this disposition, compared with the one adopted by most Robocup team, permits to protect a bigger position of the goal mouth and to intercept faster a shot.

Case 3: shot in goal. If the ball is moving, the *Arbiter* module calculates the line joining the current and the previous position of the ball. If this line intersects the goal line, the ball is heading the goal. The `CatchBall` behaviour is activated. The robot moves in a straight line toward the point of intersection of the shot line and the line on which is the robot. This is a simple linear motion that combines high reactivity with good stability, avoiding problems that occur when the robot maneuvers or makes fast turns, Figure 6 (left)

Case 4: dangerous ball. This is one of the most delicate situations. The ball was not catch with one of the previous behaviour and now it is steady within the penalty area. In this situation, a wrong move of the robot, trying to sweep the ball out of the area, could result in pushing the ball in its own goal. Up to now we do not have a behaviour dedicated to this event. So, we find a trick that resulted to work properly on the field. We take advantage of a Robocup rule stating that if the ball is steady for more than 10 seconds, it will be repositioned by the referee. So, we try to shield the ball from the other robots with the behaviour used in *Case 2*, i.e. `ProtectGoal` behaviour, Figure 6 (middle). This solution could be considered a kind of *emerging behaviour*, because it was not coded for this case but it resulted to be effective. On the other hand, there are some situation in which this solution does not work. For instance, if the ball is close to the centre of the semi-circle, the robots will move close to a goalpost, making it easier for the opponents to score, Figure 5 (right).

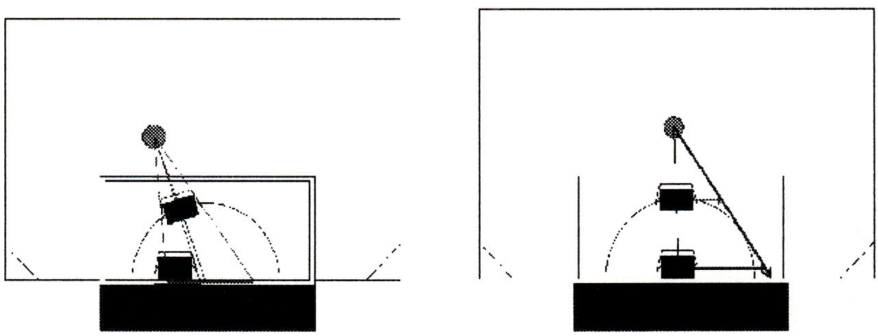

Fig. 5. Comparison between the proposed position of the goalkeeper (upper robot) and the standard position (lower robot).

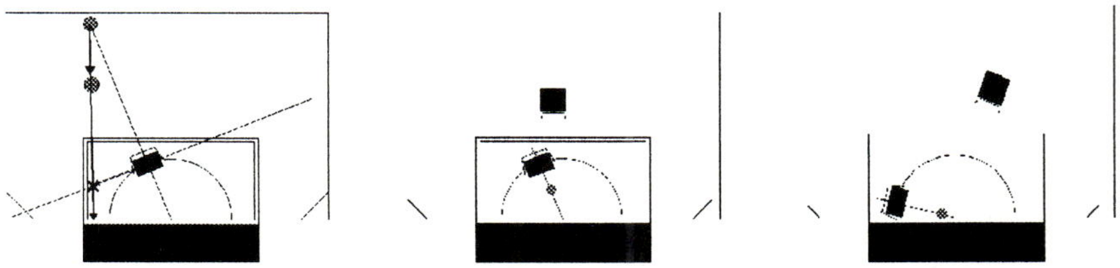

Fig. 6. (Left) Intercepting a shot. (Middle) Shielding the ball. (Right) Bad shielding of the ball.

7 Conclusion

In the first part of this paper we showed how to design a custom mirror for an omnidirectional sensor. The application we chose is the Robocup competition.

Condidered the task we decided to design a mirror that allows measures with a relative maximum error of 3%. The mirror produced with the profile outlined in this paper, showed to be reliable and effective in the Robocup domain. Using the algorithm described in Section 5.2 the position of the ball is calculated with a relative error smaller than 2%. In the second part, we presented the behaviour that have been implemented to exploit the custom design of the mirror. The synergy of the custom mirror and the dedicated behaviour proved to be effective in field of play. The Robocup competition is not the only domain in which we are testing these ideas. In fact, we are working with mirrors designed for other tasks, like navigation and mapping in unknown environments [7] with good results.

References

[1] S. Baker and S. K. Nayar. A theory of catadioptric image formation. In *Proceeding of the 6th Intern. Conf. on Computer Vision*, January 1998.

[2] A. Bonarini. The body, the mind or the eye, first? In M. Veloso, E. Pagello, and H. Kitano, editors, *RoboCup99: Robot Soccer World Cup III*, volume 1856 pp. 210-221 of *LNCS*. Springer, 2000.

[3] A. Bonarini, P. Aliverti, and M. Lucioni. An omnidirectional vision sensor for fast tracking for mobile robot. *Proceedings of the IEEE IMTC99*, 1999.

[4] A. Hicks and R. Bajcsy. Reflective surfaces as computational sensors. In *PProc. of the Second Workshop on Perception for Mobile Agents, Fort Collins*, pages pp. 82–86, 1999.

[5] F. Marchese and D. G. Sorrenti. Omni-directional vision with a multi-part mirror. *The fourth international workshop on robocup*, pages pp. 289–298, 2000.

[6] C. F. Marques and P. U. Lima. Vision-based self-localization for soccer robots. In *Proceedings of the 2000 IEEE/RSJ International Conference on Intelligent robots and systems*, 2000.

[7] E. Menegatti, E. Pagello, and M. Wright. A new omnidirectional vision sensor for the spatial semantic hierarchy. In *International Conference on Advanced Intelligent Mechatronics (to appear)*, July 2001.

[8] S. Suzuki and M. Asada. An application of vision-based learing in robocup for a real robot with an omnidirectional vision system and the team description of osaka university "trackies". In M. Asada and H. Kitano, editors, *RoboCup98: Robot Soccer World Cup II*, volume 1604 pp. 316-325 of *LNCS*. Springer, 1999.

[9] Y. Yagi. Omni directional sensing and its applications. *IEICE TRANS. INF. & SYST.*, VOL. E82-D(NO. 3):pp. 568–579, MARCH 1999.

[10] Y. Yagi, Y. Nishizawa, and M. Yachida. Map-based navigation for a mobile robot with omnidirectionalimage sensor copis. *IEEE Transaction on Robotics and automation*, VOL. 11(NO. 5):pp. 634–648, October 1995.

RoboBase: An Extensible Framework Supporting Immediate Remote Access to Logfiles

John A. Sear and Rupert W. Ford

Centre for Novel Computing
Department of Computer Science
The University, Manchester
M13 9PL, United Kingdom
{jas,rupert}@cs.man.ac.uk

Abstract. This paper describes RoboBase, a system that provides immediate access to entire libraries of RoboCup logfiles. A centralised database stores the logfiles allowing them to be viewed remotely. Instead of downloading a 2MB uncompressed logfile, the match is transferred and displayed in real-time.
The system has been designed specifically to perform well in low bandwidth situations by using a domain specific compression method. Dynamic frame-rates are also employed, providing uninterrupted viewing in fluctuating network conditions.
The system conforms to an Object Oriented methodology and is implemented in Java allowing extension of the software by the user.

1 Introduction

The RoboCup Soccer Simulation League (SSL) provides an interesting environment for research in domains such as artificial intelligence and multi-agent collaboration. The SSL community evaluate their research through periodically held football competitions, the most prestigious being the annual Robo World Cup. The relative success of teams in these football competitions is typically taken as a measure of the advancement in the associated field of research.

A logfile is produced for every match held in these competitions and made publicly accessible. Logfiles record all the information required for visual replays of games (such as player and ball positions during the game and meta-information such as goals and team names). Logfiles provide a very useful way for team owners and their competitors to evaluate the strengths and weaknesses of teams. As such it is important that these logfiles are easily accessible.

Currently there is no official repository for these logfiles and they are typically spread across the World-Wide-Web (WWW); this makes locating a logfile a potentially laborious process. After a logfile has been located, its download can be relatively time consuming. These files are typically greater than 2MB uncompressed and are required in their entirety before viewing can commence.

A. Birk, S. Coradeschi, and S. Tadokoro (Eds.): RoboCup 2001, LNAI 2377, pp. 92–101, 2002.

This paper describes a central repository of logfiles, a logfile compression technique and a viewing technique, which allow near instant remote viewing of these files on low bandwidth connections, such as modems. These components have been developed using an Open-Architecture (OA) approach, facilitating modification or addition to any component. To illustrate the potential of this approach, a full working implementation has been developed and is freely available from `http://www2.cs.man.ac.uk/~searj6`. When evaluating teams there are many metrics (statistics) that may be used and different users will potentially have very different requirements. An OA approach is expected to be particularly useful here as provides a means to add new metrics.

This paper makes the following contributions: it develops logfile specific compression techniques, describes a new open architecture framework which can be easily extended, and introduces a Dynamic Frame-Rate Algorithm (DFRA) for graceful degradation of viewing in poor network conditions.

2 Related Work

There are many ways of viewing existing logfiles. LogPlayer is included as part of the SoccerServer system and replays the logfiles through the standard viewer (SoccerMonitor). It is also possible to connect other viewers to the SoccerServer, e.g. a 3D viewer: Virtual RoboCup [6]. However, stand-alone players are simpler alternatives and several have been developed.

RoboMon [3] is a java applet which has both a 2D and 3D (VRML) version. This is a commonly used viewer which has graphics that are simple but effective. LogMonitor [8] is a simple 2D Java applet/application viewer with relatively advanced game analysis (statistics) features. The LogMonitor website [10] suggests that new versions of the software may support some form of database, however the intention of this is unknown to the author. Windows SoccerMonitor 1.3[4] combines the SoccerMonitor and LogPlayer into a single program for Windows operating systems. It is currently the easiest option for viewing files within a Windows environment.

The above logplayers are able to play the logfiles at the standard frame-rate of 10fps and provide basic features, such as fast-forward and rewind. However, these logplayers are all limited in that the user must first trawl the internet to locate the file and then wait to download the 2-3MB logfiles[1] before play may begin. Furthermore all facilities are hard coded and the user is therefore limited to features implemented by the programmer at creation time.

Our approach solves the above problems by employing a central database, allowing easy access to logfiles, providing near immediate random access to games, avoiding logfile download times and being developed with an OA approach allowing new features to be added.

[1] These reduce to approximately 500k when compressed with gzip

3 Design Issues

Figure 1 shows the overall structure of the proposed system. Data may be stored in any database which supports the Java DataBase Connectivity (JDBC) protocol. An SQL command file generates a database capable of storing individual matches and information regarding the competition structure. The system enables the inclusion of individual games into the database but is also designed to allow entire competitions to be added at once.

The test data for the database was generated from all of the matches played in the Japan Open 2000 [1], a total of 7 groups and 53 games.

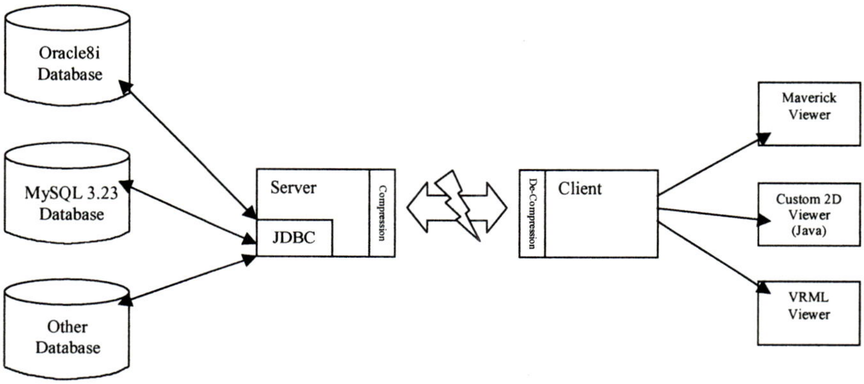

Fig. 1. Overall System Structure

The client application communicates with the RoboBase server via a compressed message format, which then expands these requests into full SQL statements. For example, when the server receives an 'M', this indicates that the next timestep block is required, and since the server has kept a record of the last transmitted block, it will be able to generate an SQL statement such as:

"select * from playertimestep where time>=400 and time<425;"

The data may be compressed before being sent through the communication medium to the client. The data is sent using a send and acknowledge protocol (rather than a constant byte stream) as it allows random access to matches and instant match changes.

At the client side, the first step is to decompress the data (if necessary) and store it in a buffer. Once the data is available, the client software analyses the data and visualises it using the chosen viewer.

The system (Client, Server and StoreFile software) are implemented entirely in Java. Java was chosen for portability as the developers of RoboCup clients use a variety of platforms. Its Object Oriented (OO) nature also means that it is particularly suited to the OA approach, through the extensibility of classes.

4 Implementation Issues

The system is built from a number of components. This section discusses those most relevant to this paper.

4.1 Client/Server

Once the decision was taken to store the data in a database, it was soon clear that the overhead associated with accessing the database directly, meant that real-time playback across low-bandwidth connections would not be possible.

This problem is solved by incorporating a server in between the database and the client. Its function is to remove the JDBC overhead and compress the data into a more concise format. In addition, the JDBC client code (150-500K) can now be excluded resulting in a substantial saving in the program size. Furthermore, this approach solves another problem. There is an applet security issue that needs to be considered, since applets are prohibited from connecting to an IP address other than that from which they were downloaded. Therefore, the server can act as the bridge between the client and database. An additional security benefit is that the the database can restrict access solely to a particular IP address, the server, thus removing the potential problem of unwanted direct connections.

4.2 Compression

In order to achieve the required viewing rate of 10fps through low bandwidth connections some form of efficient data compression is required.

Gzip has already been shown to reduce logfiles from above 3MB to below 1MB and is therefore a potential option. Java includes support to zip and unzip streams of data, hence, implementation of such a system would be relatively simple. However, for low latency of playback, RoboBase requires small blocks of data to be transferred. Under these conditions the compression ability of Gzip is reduced. The processing requirement also places a high burden at the client end, suggesting it might be too slow for real-time decompression.

Another solution is to send every nth frame and simply interpolate for unknown values. Again, this is relatively easy to implement and has the advantage that it requires little computational effort. However, it would introduce inaccuracies into the data at an early stage. This could cause problems in RoboCup as players are not bound by normal physical limits and are able to teleport around the field. Any generated statistics could become inaccurate, e.g. a goal may not be detected if the timestep is absent where the ball has crossed the goal-line.

An alternative technique is to transmit only the differences between frames in the data. Player positional prediction is an example of this. The current direction and heading of a player are used to calculate the next timestep position. Since the time between frames is only 0.1s, the differences are typically small and therefore the prediction will be relatively accurate.

Player prediction was chosen as the final compression routine because it provides the highest compression ratio, allows parameters to alter data size, can provide 100% accurate data and requires little computational effort.

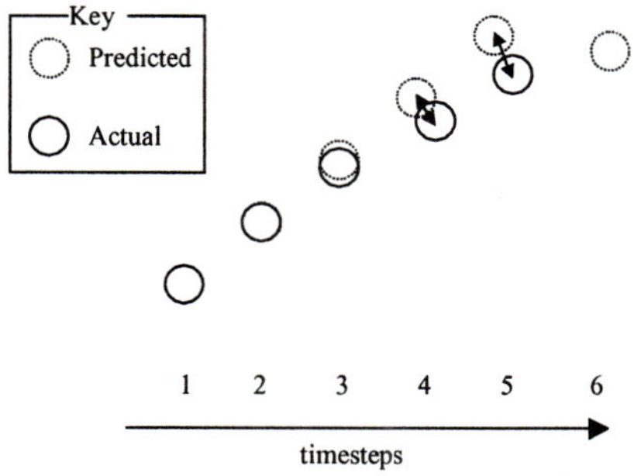

Fig. 2. Prediction Concept

Figure 2 illustrates the prediction technique. Timesteps 1 and 2 are used to calculate a heading and a velocity. The third timestep has only deviated very slightly. Timestep 4 shows a reasonably large deviation. In a real game this would be greater than the allowable tolerance, however, for illustrative purposes, the threshold has been increased. Timestep 5 shows a deviation greater than the threshold, therefore the real position must be stored. The predicted timestep 4 and the actual timestep 5 have been used to generate the predicted location in timestep 6.

It is possible to compress the data further by only transferring relative positions. If a player only moves a small distance between timesteps then this can be represented using a small number of bits. If a player has moved a considerable distance then their absolute location must be conveyed. The compression algorithm is depicted in Figure 3. The first bit defines whether the players are within the tolerance of the bitsize, i.e. if the data sent is relative or absolute. (In the example the bitsize is 4, therefore this tolerance is from -8 to +7). The next 23 bits correspond to which of the ball and players have been updated during this timestep. Each bit set corresponds to a player exceeding the prediction tolerance. After this the player co-ordinate data is listed, ordered by player number. The result is then separated into bytes which are sent across the network as a byte stream.

The compression parameters are the bit size of relative co-ordinates, the maximum player distance tolerance and the number of timesteps per message. These parameters were tuned by experimental examination of file size and accuracy. Future work will examine the possibility of dynamically tuning these parameters depending on network performance.

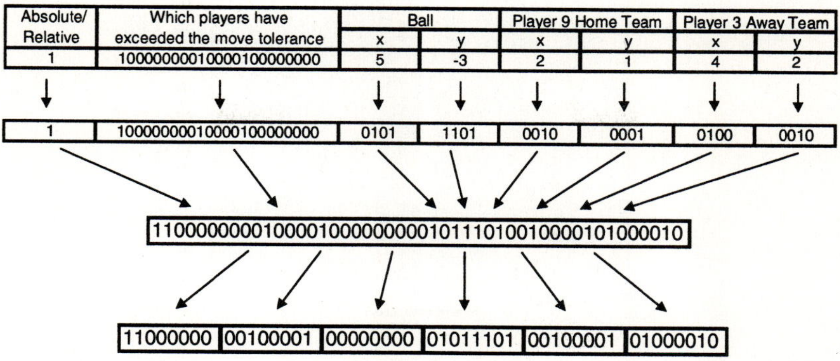

Fig. 3. Compression Algorithm

As the compression class is implemented in the OA framework, new compression algorithms can be added as appropriate.

4.3 Dynamic Frame-Rate Algorithm (DFRA)

If there is a delay in receiving data across the network, then the viewer may not be able to sustain the desired frame-rate of 10fps. A buffer is used to reduce the effect of network performance fluctuations, ensuring that saved data is available when the network is slow and filling up when the network improves. In poor conditions, using a buffer may not be enough to sustain the frame-rate.

Below are three possible approaches to the problem of poor network conditions:

1. Play at a constant frame-rate of 10fps. Once the buffer empties, the viewer pauses until more data arrives.
2. Reduce the frame-rate as the buffer empties. A potential problem is that when the next message arrives, the frame-rate suddenly jumps back to full speed.
3. Use a smoothed variable frame-rate. The frame-rate is reduced as the buffer empties, as above, however once the new data arrives, the frame-rate is gradually increased.

The third option was implemented as it provides the best results in terms of maximising the minimum frame-rate, and provides the smoothest (most pleasing) view to the user. The variables used to configure the DFRA include the buffer threshold, desired, maximum, and minimum frame-rates, which have been set as 50,10,10 and 2fps respectively.

Figure 4 shows how the frame-rate is reduced, once the buffer has less than 50 timesteps left. It is clear that the client has 18 seconds to receive the next message, not the 10 if it were displayed at a constant (full speed) frame-rate.

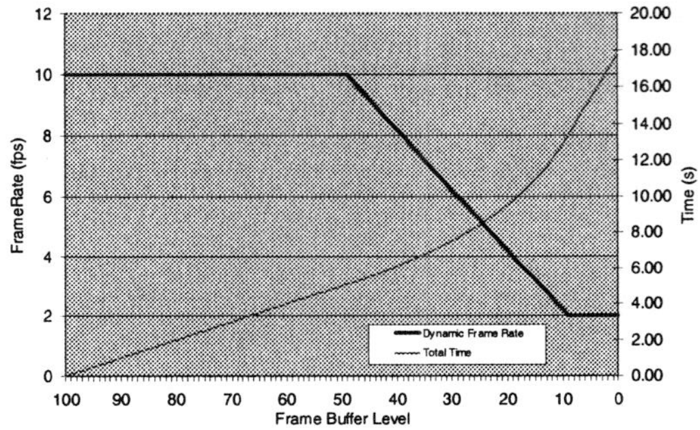

Fig. 4. Dynamic Frame-Rate - Buffer initialised with 100 timesteps

4.4 Statistics

The statistical analysis component is also completely configurable. As the RoboCup simulation community is large and diverse, it is impossible to target analysis for every users requirement. Instead, an extendible class is provided to allow users to implement their own statistics. It is also possible to use this class to interface with other software, such as commentators.

As an illustration the standard statistics given in Sky-Sports English Premiership televised matches are recreated and may be overlaid at anytime throughout the game. A separately windowed statistic, ManMarking, can be seen in the bottom right corner of the screenshot in Figure 5. Alternative metrics can also be graphed using this approach.

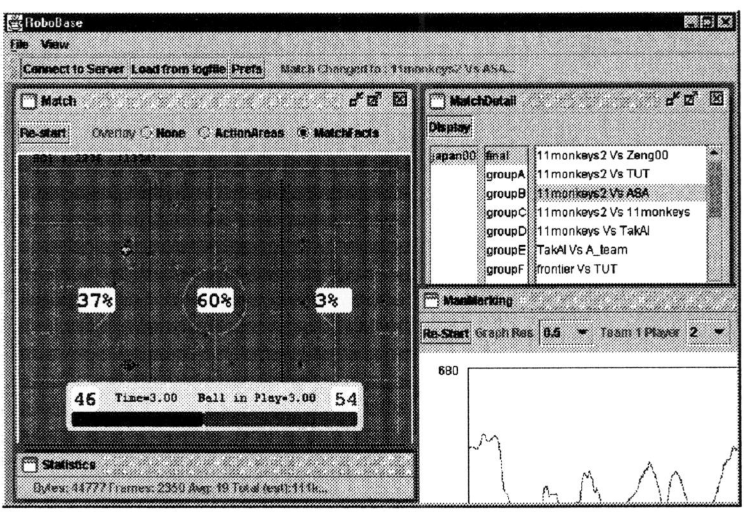

Fig. 5. Screenshot of RoboBase Client

4.5 Visualisation

The visualisation component is another completely configurable unit. At present three versions have been implemented. The first is a basic display using simple polygons to represent the match. The second is depicted in the Match Window of Figure 5 and displays a more realistic 2D view, using images to represent pitch, ball and players. The third employs Java3D to create a 3D view of the stadium and players. It is possible to use this class to interface to other viewing systems, such as VRML [9] using the External Authoring Interface (EAI) [5], or Maverick [7].

5 Experiments and Results

5.1 Dynamic Frame-Rate

In order to test RoboBase's playback performance, three client test conditions were created: a) a 450Mhz Pentium 3 connected to the Internet via a 44000bps modem connection, b) a 450Mhz Pentium 3 connected to the Internet via a 10Mbs ethernet connection and c) local playback of a file. The server is a MySQL database running on a Pentium 3 800Mhz system.

For each test the logfile was played in its entirety using three slightly varying compilations of RoboBase: a) visualisation enabled, b) visualisation disabled and c) visualisation and statistical analysis disabled. All these experiments were performed with the frame-rate limiter disabled. This combination of experiments allowed the frame-rate to be attained for user playback and the effect of the compression and analysis components to be assessed.

The timings for the locally stored logfile are included only as a guideline. This is because when playing back a local logfile, stored in the standard format, the prediction and compression routines are not required.

	Visualisation Enabled	Visualisation Disabled	Visualisation and Analysis Disabled
Modem (Ping time = 250ms)	33.84	34.09	34.20
Ethernet (Ping time<10ms)	57.51	63.42	65.79
Local logfile	59.04	74.06	74.93

Fig. 6. Frame-rate in test conditions

The results of these experiments are illustrated in Figure 6. For all the experiments attempted the minimum frame-rate attained was 34fps; this lower limit was demonstrated when using a modem with the visualisation enabled. This is significantly in excess of the desired 10fps. The limiting factor in these experiments was not the bandwidth of the modem but rather the time it takes to acknowledge a message (ping time). This suggests further experimentation should be performed by increasing the block size (timesteps per message).

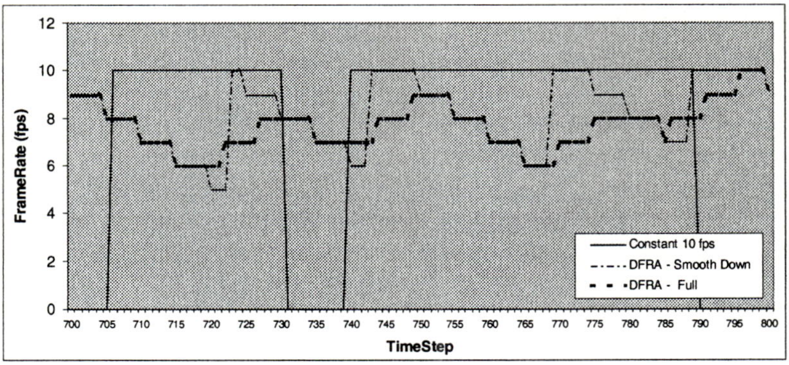

Fig. 7. Frame-rate in poor conditions

Further experiments were performed with numerous clients attached. A group of 10 400Mhz Pentium 3 machines running a mixture of Windows NT and Linux were all connected to the 800Mhz server. Whilst the frame-rate did decrease slightly, from the desired 10fps, with all 10 machines connected, the DFRA ensured a 'pleasing' playback, with no machines dropping below 5fps and average frame-rates above 7fps. This was due to the heavy load at the server end. This may be improved by placing the database and server on different machines.

In our final network test we manufactured poor network conditions, by adding a significant random delay to the response time of the server. Running the viewer at a constant 10fps resulted in many periods of display inactivity, see Figure 7. It can be seen that the DFRA ensures a constant playback and higher minimum frame-rate.

5.2 Compression

The matches from group F of the Japan 2000 RoboCup were played using the different compression ideas discussed earlier and the quantity of data transferred was recorded, see Figure 8. The benefits of the logfile specific techniques described in this paper are clear. In this case the file size is approximately 25 smaller than the original uncompressed version and 5 times smaller than the gzipped version. When downloading the Japan00 competition in its entirety, the logfile specific compression technique reduces the uncompressed 104MB of logfiles to to an estimated 5.5MB compared to 23MB with gzip.

6 Conclusions and Future Work

This paper has described Robobase, a system designed for near instant, remote viewing of RoboCup logfiles. It comprises of a database, storage utility, server and client which allows access to a library of games. Using compression and

Match	File Size (bytes)		
	Uncompressed	GZ	RoboBase
11monkeys Vs PSI	2,014,534	453,158	93,707
PSI Vs Revolvers	2,118,202	399,728	75,183
Revolvers Vs 11monkeys	2,130,638	415,562	75,837
ThinkingAnts Vs 11monkeys	2,030,326	521,122	103,742
ThinkingAnts Vs PSI	2,060,962	505,609	102,370
ThinkingAnts Vs Revlovers	2,137,918	355,270	69,337
Total	**12,492,580**	**2,650,449**	**520,176**

Fig. 8. Compression Results for Japan00 Group F

DFRA it is able to achieve frame-rates in excess of the desired, even over a low-bandwidth modem connection and in fluctuating network conditions. Examples have also illustrated how the open architecture structure may be extended if required.

The server and database support random access allowing retrieval of any timestep without incurring additional time penalty. This aspect is planned to be used to playback match highlights. Matches would be separated into obvious sections, e.g. when the ball goes out of play. Each section is then rated according to an interest value, such as a high rating for goals. Highlight programmes may then be produced by specifying either a time limit or a minimum interest setting.

The work described in this paper also forms the start of prototyping work for a new application area, 'Interactive Sports Entertainment', which provides both video and data to the home viewer [2]. The results shown in Section 5.1 demonstrate that it is possible to watch a representation of a football match over a low bandwidth connection. 3D rendering can then be used at the client end to provide realistic generated match views, allowing virtual cameras to be placed anywhere in the scene.

References

[1] Japan Open 00. `http://ci.etl.go.jp/~noda/soccer/JapanOpen00/`.

[2] Guy Blair, Rajeeb Hazra, and Richard Qian. *White Paper: Intel's vision of sports entertainment and marketing.* Intel Architecture Labs, 2000.

[3] Antonio Cisternino. Robomon, 2000.

[4] SoccerMonitor Klaus Dorer. `http://www.iig.uni-freiburg.de/cognition/team/members/dorer/robocup/`.

[5] External Authoring Interface. `http://hiwaay.net/~crispen/vrmlworks/`.

[6] B. Jung, M. Oesker, and H. Hecht. Virtual robocup: Real-time 3d visualization of 2d soccer games, 1999.

[7] MAVERICK. `http://aig.cs.man.ac.uk/systems/Maverik/index.html`.

[8] Tomoichi Takahasi. *LogMonitor.* 2000.

[9] VRML. `http://www.vrml.org/technicalinfo/specifications/vrml97/`.

[10] LogMonitor WebSite. `http://157.110.40.100/robocup/LogMonitor/`.

Agent Based Approach in Disaster Rescue Simulation - From Test-Bed of Multiagent System to Practical Application -

Tomoichi Takahashi[1], Satoshi Tadokoro[2], Masayuki Ohta[3], and Nobuhiro Ito[4]

[1] Chubu University, 1200 Matsumoto, Kasugai, AICHI, 487-8501 Japan
ttaka@isc.chubu.ac.jp
[2] Kobe University, Rokkodai, Nada, Kobe 657-8501 Japan
tadokoro@cs.kobe-u.ac.jp
[3] Tokyo Inst. of Tech., Dept. of Math.and CS., 2-12-1 Ookayama, Meguro-ku, Tokyo 152-8550 Japan. m-oota@is.titech.ac.jp
[4] Nagoya Inst. of Tech., Gokisochou, Showa-ku, Nagoya, 466-8555 Japan
bobson@phaser.elcom.nitech.ac.jp

Abstract. We apply multi-agent approach to search and rescue in a large-scale domain. The simulator is designed to simulate various domain specific simulation and human behaviors. Kobe-Awaji earthquake data is used as disaster scenarios and a prototype system was made open at RoboCup 2000. A rescue team composed of heterogeneous agents, – fire brigades, ambulances, and polices –, takes active part in the disastrous situation where about 100 civilian agents move autonomously. By comparing with rescue operations of two teams, we showed that the search and rescue in disasters can be used as a test-bed for multi-agent systems. The comparing experiments made clear that rescue task is not well defined in spite of its practical importance, and planning based on multi-perspectives on disaster losses is necessary. It points that the rescue problem is not only a test-bed for multi-agent system but also for laboratory work for practical system.

1 Introduction

Society consists of multi/heterogeneous entities. Simulating activates in the society requires handling social structures that is difficult to handle systematically. Multi-agent system is one of methods to simulate such human activities. Casti presented that behaviors of individual change traffic condition of a city [1].

Management of environmental emergency is concerned as one of application fields of multi-agent system [7]. Cuena and Ossowski handled a flood management with four types of agent - Local Emergency Management Agent, Dam Management, Fire Brigade Management Agent, and Transport & Ambulance Management Agent.

The agents decide their actions to save victims according to disaster situations. Escape problem from a theater has been studied as application field of multi-agent learning [9]. The problem consists of two phases. The first is that agents search for a button to open doors and one agent pushes it. Other agents can escape from doors after that. Their field is 8×8 mesh world with one exit.

A. Birk, S. Coradeschi, and S. Tadokoro (Eds.): RoboCup 2001, LNAI 2377, pp. 102–111, 2002.

Earthquake at urban areas causes various disasters, such as, fire, building damages, disruption of roads and life lines - electricity, water supply, gas - etc. Fire brigades extinguish fires, ambulance teams bring victims to hospitals, while civilians seek refuge and ask for help.

RoboCup - Rescue is proposed to develop a series of technologies that can actually save people in case of large-scale disasters [2,4,3]. The simulation Project that applies multi-agent research to search and rescue domain, is composed of disaster simulations, rescue people, and residents in that area.

In this paper, we first describe rescue operations in disaster simulation. Secondly, what urban disasters are is described, and the prototype system of *RoboCup - Rescue* is outlined. Simulations of rescue agent behavior and the simulated world is demonstrated in the third section. Problems that were made clear in applying multi-agent system to rescue domains are discussed in fourth section. Finally, points to be studied in future are presented.

2 Disaster & Rescue Simulation Task

At January 17, 1995, a large earthquake of a moment magnitude 6.9 hit Kobe City, Japan. Over 6,000 people were dead, at least 300,000 were injured, and more than 100,000 buildings were collapsed. Similar tragedies also took place in California, Turkey, Taiwan, and other places.

2.1 Rescue Domain Characteristics as Multi-agent Task

Domain characteristics of soccer, and rescue are illustrated in Table 1. In rescue domain, there are heterogeneous agents - civilian, fire fighter, ambulance, police, city offices - to act autonomously. Some of them have hierarchical structure in themselves, for example, fire fighters are composed of fire brigades and fire office. Fire office agents can order fire brigade agents.

	Rescue (Prototype)	Soccer
Number of Agents	100 or more	11 per a team
Agents in the team	Heterogeneous, Hierarchical	Homogeneous
Logistics	Major Issue	No
Information Access	Very Bad	Reasonably good
Representation	Symbolic & Non-Symbolic	Non-Symbolic
Control	Distributed/Semi-Central	Distributed
Filed structure	dynamic (GIS)	static (goal, line)
motion	only move along roads	can move to any point in xy plane

Table 1. Features of Rescue and Soccer

2.2 Fire Department Decisions in Disaster

When a fire occurs, a fire station receives calls from residents. The fire office deploys fire brigades to put off the fire effectively. One of the deployment policies is that extinguish power of fire engines suppress the fire, $\sum_i extinguish\_power(i) > fire$.

On the other hand, an earthquake occurs at urban area, various disasters - fire, collapses of buildings and roads, damages of electric powers and telephone lines - occurs simultaneously over the area. The power of fires all over the area is more than extinguish powers of fire engines. $\sum_i extinguish\_power(i) < \sum_j fire(j)$. And roads covered with debris may disturb fire engines' move and the flow of civilians who refuge from damaged areas makes situations worse.

According to Tokyo Fire Department, their main office controls their fire engines. In a case of big fires, they change their strategies from extinguishing fires respectively to not spreading the fires. Fire engines are deployed around the fires to make water wall. And in a case of worse disaster where there may be troubles in communication, they suppose the center-controlled system will not work well. At that time, local fire stations are permitted to control their fire engines at their decisions.

2.3 *RoboCup - Rescue* Simulation Prototype System

RoboCup - Rescue has been undertaken to put large-scale simulations in use in the domain of search and rescue for large-scale disasters [5].

specification The prototype simulator was designed based on Kobe-Awaji earthquake's case.

- **target area:** Nagata Ward was one of the most damaged area. The area was 11.47 km^2 and 130,466 people (53,284 households) lived there.
- **simulation period:** The purposes of rescue activities change as time passes. The rescue activities are classified into five stages: chaos stage, initial operation stage, recovery stage, reconstruction stage, and normal stage. At the first chaos stage, there is no aid from outside, and the main purpose is saving the victims using local facilities. Considering that survival rate of buried people decreases rapidly after a few days, the period to be simulated is set to first 72 hours.
- **rescue agents:** When earthquakes occur, there are many calls asking for fire fighters. So local rescue agents will do the first rescue actions. There were a total of 7 rescue agents at Nagata fire office.
- **space resolution:** Representing disaster situations or rescue activities requires displaying items at the size of cars. GIS (Geographic Information System) data is maintained with a resolution of 5 m.

architecture The simulator is designed to combine various disaster simulations and human behaviors, and to present them as coherent and comprehensive scene. Disaster simulators are fire, building and road collapse. Human behaviors are rescue activities and evacuation to safe places, etc. Traffic simulator manages the movement of human agents.

A kernel of the simulator combine all information and up-date the status of the simulated world. Several domain specific simulators are connected to the kernel. Information on an entire disaster field is stored in geographic information system (GIS).

Numbers of agents are deployed in this simulation environment to test strength and weakness of the search and rescue strategies.

3 Rescue Scenarios

3.1 Model of Agent Behaviors

In simulation, statistics data will help to model human activities in disasters. Disaster simulators such as fire or collapse have been designed so that results of simulations match the real disasters data. Statistics data on human, such as White paper of National Land Agency, Japan, says that 88% of the cause of death was crushed to death, 10% people were burned to death, and others by fall of things, etc. The data are used to set initial conditions, however, it is very difficult to model human behavior during the simulation.

The behavior model how a civilian acts in emergency or how people act collectively in evacuation will be needed in simulation of rescue operations and their estimation.

3.2 Agents and Their Hierarchy

The word *agent* is used to refer to autonomous entity in the simulated world. Figure 1 shows agents prepared in the prototype system.

Agents – Civilian, Car, Fire Brigade, Ambulance, Police – can move in the area. Civilian represents a resident in that area, and Car represents a resident in a car and can move faster than Civilian. Fire Brigade is a fire engine, Ambulance can bring civilians to a refuge, and Police can repair collapsed roads. All agents have properties such as damage of body, stamina. The properties affected by disaster simulations determine agents' life and death.

Fire Station, Ambulance Center, Police Office, Refuge are building objects and represent functions of people in the building provide. Fire Station, Ambulance Center, and Police Office can instruct corresponding agents. When buried civilians are brought to a refuge, they can receive medical treatments. Otherwise their health point decreases with time, and they will be dead as time passes.

Agents take actions against disastrous situations. The situation changes with time and is changed by the agent rescue operations. Using predefined protocols, the agents

see situations around them with `sensory information` sent from kernel. The information is objects within a certain radius around the agent,
hear voices that other agents ask for help,
plan next actions according to their objectives such as, to search for victims, to rescue them, to evacuate them to safe places, etc.,
send their actions to the kernel as `act` command. It consists of `extinguish` *Target*, `rescue` *Target*, `load` *Target*, `unload`, `open` *Target*.

The kernel gathers all messages sent from agents, and broadcasts them to the component simulators. The component simulators individually compute how the world changes its internal status. These results are sent back to the kernel.

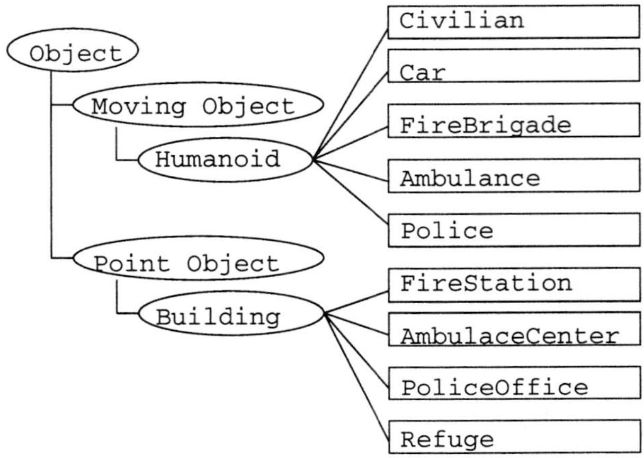

Fig. 1. Hierarchy of rescue agents

3.3 Collaboration among Rescue Agents

The above agent's communication must be done within one simulation step, otherwise the agent stays with doing nothing,

And agents communicate each other using the communication to collaborate in rescue operations. The following are types of collaboration that should be appeared in rescue domain.

type I: collaboration among homogeneous agents.
The first task is n Fire Brigade agents extinguish fires that break out at m places. Using 1/10 scale mode with $m = 3, n = 1 \sim 9$, Ohta showed that extinguishing task has a feature that collaboration among n homogeneous agents obtains more than n times results [8]. And his simulation showed it better to deploy fire brigades intensively to an ignition point than to deploy them widely to ignition points. This corresponds to one of fire departments' methodologies of extinguishing.

type II: collaboration among homogeneous agents with hierarchy.
Fire departments consist of fire brigades and stations. Fire brigades extinguish fires based on local information that they can get within the field of their vision, while fire stations may grab global situation from civilians reports. Fire stations instruct their fire brigades to minimize loss caused from fires. For example, fire stations decide to change their strategies from extinguishing fires to preventing their spread. This hierarchical structure exists in ambulance agents, police agents, etc.

type III: collaboration among heterogeneous agents.
It is very important for fire brigades to go quickly to fires. Earthquakes cause traffic troubles as collapses of roads and buildings, civilian's evacuate-flow. Police agents get rid of debris from roads and make fire brigades or ambulance moves swiftly. Collaboration among police agents and fire brigades is necessary to rescue effectively.

Fig. 2. snapshot of simulation at time=5

4 Simulation Experiments

4.1 Simulation of Rescue Operations

Table 2 shows GIS data that was used in our experiment. 1/1 scale model is a square area 2,217 m on a side. The numbers of agents and objects are real ones except civilian's number. The real number of civilians, which is 100 times this number, is too large to simulate at this time.

Figure 2 shows a snapshot of a 2D viewer. The legible part in the center is the area selected as 1/10 scale model. The blocks are buildings and the lines are roads. The color of buildings is green, and buildings turn red when they burn.

Table 3 shows the results of two teams rescue activities. Both teams have 10 fire brigades, 5 ambulances, and 10 polices. Team A's fire agents are programmed to look for fires individually by walking randomly. On the other hand, the agents of team B are programmed to rush fires after looking around them. Ambulance and police agents provided as sample agents were used for both teams [6].

An earthquake occurred at time=1, more than 740 houses were collapsed. The collapsed houses buried citizens who were accidentally near them. The buried citizens are alive at first, however their physical strength is decreased with time. If rescue agents save the buried citizens, they will lose their lives. Figure 3 shows the number of dead people during the first 50 steps. After that period, the numbers remain the same.

Table 2. Numbers of agents and objects used in test

	autonomous agents				GIS objects		
scale	1/100	1/10	1/1		1/100	1/10	1/1
Civilian	8	76	934	road	125	818	9,776
Ambulance Team	2	5	5	node	119	765	9,143
Fire Brigade	2	10	10	building	99	778	9,357
Police Force	2	10	10	area size (m)	160	521	2,217

Table 3. Statistics on rescue operations

time	dead /living	buildings on fire	burnt	ext. †	road collapsed
5	1/98	1	0	0	28
team A					
50	16/83	29	2	2	27
100	17/82	48	12	5	25
150	17/82	61	47	10	24
200	17/82	81	84	12	24
250	17/82	102	136	13	24
300	17/82	77	216	15	23
team B					
50	18/81	17	0	8	28
100	18/81	18	4	13	26
150	18/81	3	15	22	26
‡ 200	18/81	1	15	23	24

† extinguished and saved buildings
‡ the data is the same from 200 to 300

The number of collapsed roads at time=5 shows that road collapse simulator worked and the number of burning houses increases as fire spreads. Team A could not extinguish fires, while team B put off at time=200. The number of burnt houses by team A is more 10 times than by team B. However team A saved more people than team B. Figure 4, 5 show the simulation results of team A and B respectively.

4.2 Estimation of Rescue Operations

The rescue operations are estimated from how many lives or houses are saved, not form how well their simulations fit the real one. It is important to validate whether the simulation results can be applied to the real situations. Different from physical phenomena, we cannot experience earthquakes repeatedly to collect data that show some relationship between rescue operations and damages. The disaster data we can get are the total number not as a form of time sequence [10]. The fire and road collapse simulators

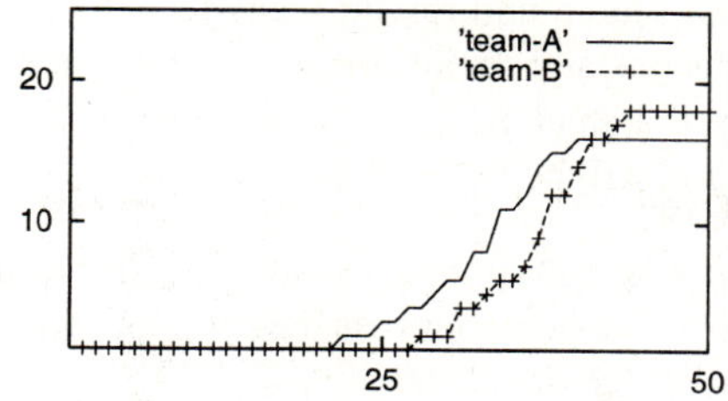

Fig. 3. time sequence of the number of dead people

are adjusted to produce the reported numbers, while other simulators have no reliable models.

There are many indexes that show how large the disaster damaged the society. The indexes are loss of human life, loss of lost buildings, or the money required to restore , etc. It is difficult to say which index is the best or how to combine the indexes to estimate rescue operations. Moreover the rescue agents cannot calculate their values to optimize their behavior when they are in action.

The right window in Figure 4, 5 indicates the losses over the area that is total of all rescue operations at each simulation step. The indexes are the same ones in Table 3 and the values at the final step are used to estimate rescue operations. Actions of agents can be rephrased as teamwork. Although we implemented only type I collaboration at present, it is important to evaluate teamwork as rescue activities. From this point, team-B is better than team-A. However, their evaluations change places form saving human lives.

5 Discussion and Conclusion

Applying multi-agent system to rescue operation in large disaster made clear the following problems, in addition to themes of information engineering that makes simulators stable to move over 10,000 agents;

- Rescue activities are so closely related with human activities, that are easy to express with task-level language but are hard to describe them as command-level operation.
- Objectives of rescue operations come from different fields. And they change with time. Different form other fields, such as soccer that can be scored as goals, rescue operations include something mental aspect like victims' feeling. Attentions from sociological aspect must be paid to refer to disaster victims.

Even though the rescue domain is one of the most serious social issues, goals of rescue simulation are not well defined as in Shakey's world, wumpus world, etc. Besides representing state space of rescue, our first simulation presents new themes;

- In order to represent a civilian as an individual agent, ontology is necessary to communication among agents and representing their feel of the place [11]

- There are numbers of search and rescue scenario in city offices. Human planners will do political decisions and social decision that are not easily incorporated into automated planning of agents. As a way of interactions with real world, planning with human interaction will be necessary.

Search and rescue problem in disasters is an ideal platform for practical application of multi-agent system. We hope our research will not only provide many research themes but also help real rescue operations.

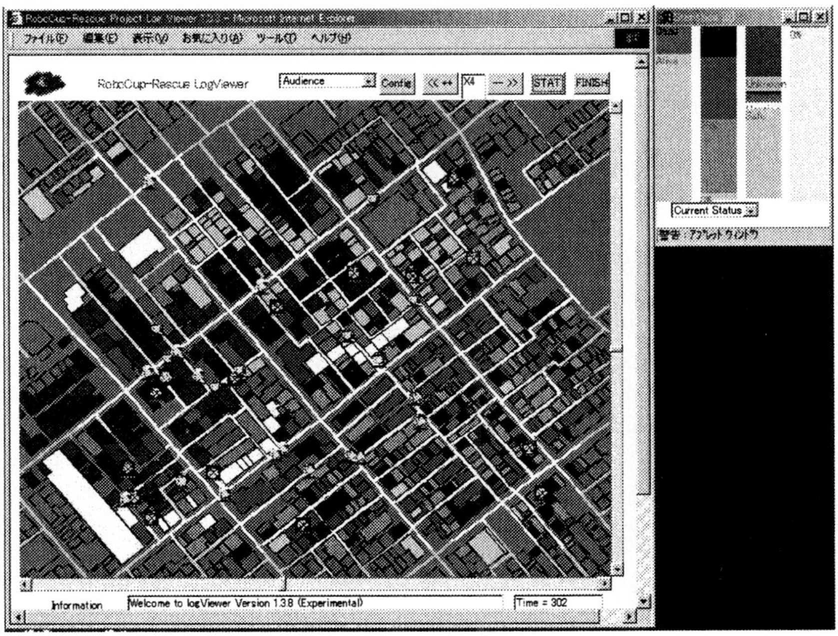

Fig. 4. snapshot of Team A at time=300

The author would like to thank other members of *RoboCup - Rescue* Technical Committee.

References

1. J. L. Casti, "Would-Be Worlds," John Wiley & Sons, Inc. 1996.
2. Kitano, H., *et al.*, "RoboCup Rescue: Search and Rescue in Large-Scale Disasters as a Domain for Autonomous Agents Research," *IEEE International Conference on System, Man, and cybernetics (SMC-99),* Tokyo, 1999.
3. Kitano, H., "RoboCup Rescue: A Grand Challenge for Multi-Agent Systems," *International Conference on Multi-Agent Systems (ICMAS-2000),* Boston, 2000.

Fig. 5. snapshot of Team B at time=200

4. Tadokoro, S., *et al.*, "The RoboCup Rescue Project: a multi-agent approach to the disaster mitigation problem," *IEEE International Conference on Robotics and Automation (ICRA-00),* San Francisco, 2000.
5. Takahashi, T., *et al.*, "Disaster Simulator Architecture," *International Conference on Multi-Agent Systems (ICMAS-2000),* Boston, 2000.
6. http://robomec.cs.kobe-u.ac.jp/robocup-rescue/
7. Cuena, J. & Ossowski, S. *Distributed Models for Decision Support.* in Multiagent Systems edited by G. Weiss, MIT Press, 1999.
8. Ohta, M., *et al.*, "RoboCup-Rescue Simulation: in case of Fire Fighting Planning," *Proc. of ICMAS RoboCup Rescue Workshop,* , 2000.
9. Mita, H., *et al.*, "Solving an Escape Problem with Generic Programming," (Japanese) *Proc. of MACC2000 ,* , 2000.
10. http://www.fdma.go.jp/html/infor/hansinawaji105.htm
11. Noda, I., *et al.*, "Language Design for Rescue Agents" *Proc. of RoboCup Symposium,*, 2001.

Planning and Executing Joint Navigation Tasks in Autonomous Robot Soccer

Sebastian Buck, Michael Beetz, and Thorsten Schmitt

Munich, University of Technology, Germany
{buck,beetzm,schmittt}@in.tum.de

Abstract. In this paper we propose a hybrid navigation planning and execution system for performing joint navigation tasks in autonomous robot soccer. The proposed system consists of three components: an artificial neural network controller, a library of software tools for planning and plan merging, and a decision module that selects the appropriate planning and execution methods in a situation-specific way. The system learns by experimentation predictive models for the performance of different navigation planning methods. The decision module uses the learned predictive models to select the most promising planning method for the given navigation task.
In extensive experiments using a realistic and accurate robot simulator that has learned the dynamic model of the real robots we show that our navigation system is capable to (1) generate fast and smooth navigation trajectories and (2) outperform the state of the art planning methods.

1 Introduction

In order to perform plays competently, teams of autonomous robots playing robot soccer must be capable to perform joint navigation tasks given by a target state for each robot. In this paper we describe how the navigation system of the AGILO RoboCup team [3] solves the joint navigation problem. The distinctive characteristics of the AGILO navigation system are the following ones. First, the system uses a recursive neural network controller as its basic execution component. This neural network controller automatically learns how to best transform the robot's current dynamic state into a given target state. Second, the navigation system employs a library of single robot navigation planning and plan merging methods that it can select, combine, and apply to a given navigation task. Third, the navigation system automatically learns to predict which methods are best for which kinds of navigation tasks. Exploiting this knowledge it can apply the most promising planning method of its toolbox to the navigation tasks it is to perform. This way the navigation system performs better than any of the individual planning methods that it is using.
One of the key problems in the design of a multi robot navigation system is that different multi robot navigation planning methods make different assumptions about the kind of navigation problems they are to solve and the capabilities of the robots they are to control. Because these assumptions are implicit and not

A. Birk, S. Coradeschi, and S. Tadokoro (Eds.): RoboCup 2001, LNAI 2377, pp. 112–122, 2002.

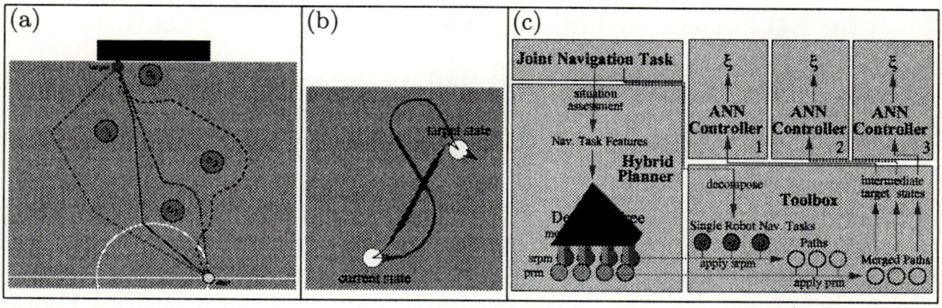

Fig. 1. Subfigure (a) shows navigation plans for one robot proposed by different path planning methods. Subfigure (b) shows two possible trajectories a controller can guide the robot. The orientation of start and target state is indicated by an arrow. The width of the trajectory indicates the robot's translational velocity. Subfigure (c): The planning and execution system: The *hybrid planner* computes the situation dependent navigation task features and uses a decision tree to determine an appropriate single robot planning method (srpm) and plan repair method (prm). In the *toolbox* first the srpm is applied to the decomposed single robot navigation tasks. Thereafter the obtained paths are merged by the prm. The intermediate target states of the merged paths are passed to the *artificial neural network (ANN) controller* of each robot. The controller then computes the low level command ξ to be executed.

well understood, selecting the **right** planning method for a given application domain and parameterize it optimally is very difficult.

Let us illustrate this point using a practical example. Figure 1(a) depicts a single robot navigation task in a typical game situation and the navigation plans proposed by different navigation planning algorithms. The figure illustrates that the paths computed by the different methods are qualitatively very different. While one path is longer and keeps larger distances to the next obstacles another one is shorter but requires more abrupt directional changes. The performance that the paths accomplish depends on many factors that the planning algorithms have not taken into account. These factors include whether the robot is holonomic or not, the dynamic properties of the robot, the characteristics of change in the environment, and so on.

The conclusion that we draw from this example is that the choice of problem-adequate navigation planning methods should be based on empirical investigations. For our investigations we develop a feature language which allows us to classify navigation tasks along dimensions that challenge planning methods. The remainder of this paper is organized as follows. An overview of the software architecture is given in section 2. Section 3 introduces our basic robot controller. Section 4 shortly describes the path planning algorithms used in our empirical investigation. Section 5 introduces a set of features that can be used to measure the characteristics of multi robot navigation problems along certain interesting dimensions. Section 6 then summarizes the results of our empirical investigation. We conclude with a review of related work and a discussion of the results.

2 The Structure of the Hybrid Navigation System

To plan *and* execute a joint navigation task we use a hybrid system containing a software robot controller, a toolbox for path planning, and a hybrid planner to select the appropriate algorithms for execution.
The hybrid navigation planning and execution system works as follows (see figure 1(c)). The system is given a joint navigation task that specifies a target state (position, orientation and velocity) for each robot of the team. The objective of the navigation system is to achieve a state where each robot is at its target state as fast as possible. The first step in the performance of the navigation task is the selection of the most appropriate planning tools. This is done by first assessing the given navigation task and the situation in which it is to be executed. The assessment assigns a navigation task a set of characteristics described in section 5.1. Thus features are then tested by a learned decision tree that assigns to each feature vector the most promising single robot path planning and plan merging method provided by the toolbox. In the second step, after the planning and plan repair methods have been chosen, the joint navigation task is decomposed into single robot navigation problems. The individual problems are then solved using the selected planning methods. Then, the individual plans are repaired in order to avoid negative interferences between the different plans. Finally, the toolbox extracts sequences of target states from the repaired plans and sends those sequences to the neural network controllers of the respective robots.
The *robot controllers* are used for basic navigation to reach a given target state as fast as possible not considering any constraints. All constraints including moving obstacles, walls etc. are considered in the path planning algorithms of the *toolbox* which determine the intermediate target states.

3 A Recursive Neural Robot Controller

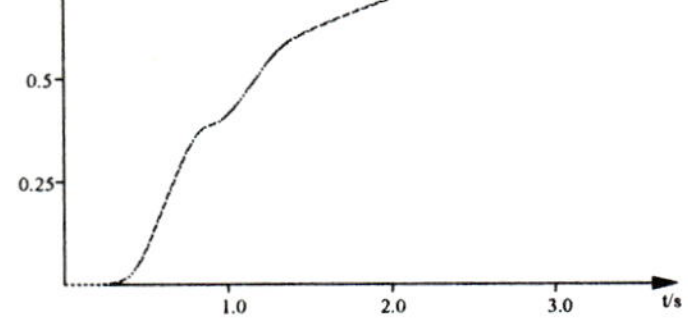

Fig. 2. The acceleration curve of a Pioneer I robot resulting from $\xi = (0, 0)$ for any $t < 0$ and $\xi = (0.75, 0)$ for $t \geq 0$. The dead time delay is about 300 ms.

The basic component of a navigation system is a controller that enables the robot to achieve specified dynamic states quickly.
Such a controller receives the target state (for example, the next state on a planned path) of a robot and returns low level commands that transform the current state into the target state as fast as possible. As shown in figure 1(b) there are different ways to solve this problem. To arrive at the target state different trajectories are possible. The dynamic state of a Pioneer I robot [11] can be summarized as a quintuple $\zeta = \langle x, y, \varphi, V_{tr}, V_{rot} \rangle$, where x and y are coordinates in a global system, φ is the orientation of the robot and V_{tr} (V_{rot}) are the translational (rotational) velocities. Using the Saphira software [7] one can set a command $\xi = \langle V_{tr}, V_{rot} \rangle$ at the frequency of 10 Hz where $V_{tr}(V_{rot})$ denote the target velocities in meters per second (degrees per second). But how to set them to quickly

reach the target state? We have measured a time delay of around 300 ms from the setting of ξ until the robot executes the command (see fig. 2). In order to take that delay into account we map commands ξ_i to the state change $\zeta_{i+3} \rightarrow \zeta_{i+4}$ as depicted in figure 3. In the remainder of the paper ζ_i and ξ_i denote a state and the respective command causing it.

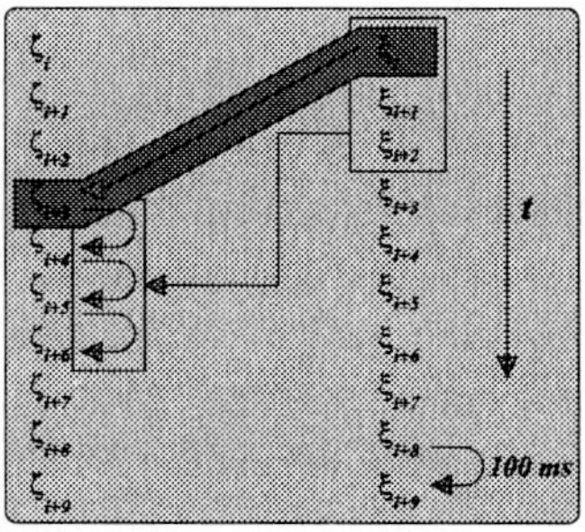

Fig. 3. Overcoming the dead time delay by assigning ξ to the state change 300 ms ahead.

A common approach (as proposed by [10] and described more thoroughly in [16]) is the use of fuzzy functions. This means to turn as long as the robot does not head towards its target state and to drive forward dependent on the orientation with respect to the target state.

In contrast, we learn a direct mapping from the robot's current state (ζ_0) and the robot's target state (ζ_{target}) to the next command to be executed (ξ_0) using multi layer artificial neural networks [4] and the RPROP [13] algorithm: *Net*: $\langle\zeta_0, \zeta_{target}\rangle \mapsto \xi_0$. Considering the start state at $x = 0, y = 0, \varphi = 0$ in a local system we can reduce the input dimension to 7 by converting the target state's x, y, φ into that local system (that means we regard $\Delta x, \Delta y, \Delta\varphi$).

To collect training data we do several runs with ξ set to certain constant values[1] and recording the state changes resulting. Out of it we get a huge number of patterns $\langle\langle\zeta_i, \xi_i\rangle \mapsto \zeta_{i+1}\rangle$. These patterns are inverted to $\langle\langle\zeta_i, \zeta_{i+1}\rangle \mapsto \xi_i\rangle$. Successive patterns $\langle\langle\zeta_i, \zeta_{i+1}\rangle \mapsto \xi_i\rangle$ and $\langle\langle\zeta_{i+1}, \zeta_{i+2}\rangle \mapsto \xi_{i+1}\rangle$ can recursively be combined to $\langle\langle\zeta_i, \zeta_{i+2}\rangle \mapsto \xi_i\rangle$ as illustrated in figure 4(a). From one simple trajectory we get lots of useful training patterns as we can see in figure 4(b). Patterns are created not only from start and target state but from any two consecutive states or patterns of the tra-

(a)

(b)

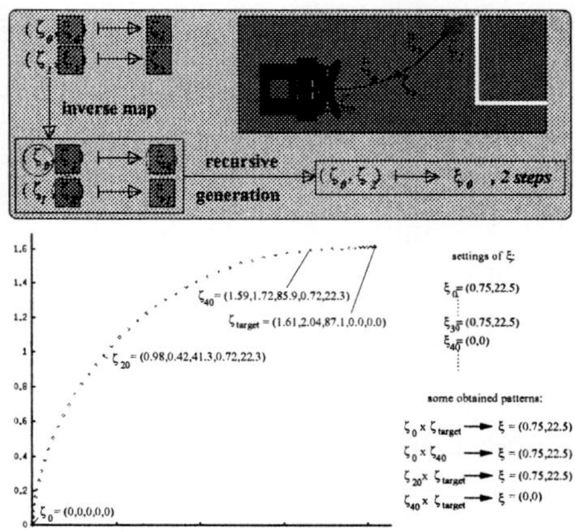

Fig. 4. Subfigure (a): Recursive generation of training patterns for the neural controller: Two successive state changes observed are inverted and combined to one new pattern. Subfigure (b): Driving 4 seconds with $\xi = (0.75, 22.5)$ and thereafter with $\xi = (0, 0)$ leads to numerous different patterns. The states from $\zeta_0 = (0, 0, 0, 0, 0)$ to ζ_{target} and 4 example patterns are plotted in the graph.

---

[1] one should always try to drive *fast* and not to increase, decrease and increase velocity again. Driving with only half speed on a straight line for example will teach the controller to do so even if it could reach the target state faster!

jectory. Recapitulatory we learn the dynamic driving behavior of Pioneer I robots and exploit it for navigation.

A Robot Simulator that Learns Models of Dynamic Behavior. An important means for developing competent navigation systems for robot soccer is a robot simulator that allows for controllable and repeatable experiments. For this reason we have developed a robot simulator that simulates how the dynamic state of the robot changes as the robot's control system issues new driving commands such as setting the target translational and rotational velocities.

The dynamic state of the robot is given as defined above. The robot control system issues driving commands $\langle V_{tr}, V_{rot} \rangle$. The dynamic model used by the simulator is acquired by learning the mapping $\Delta : \zeta_i \times \xi_i \mapsto \zeta_{i+1}$ from experience, that is recorded data from real robot runs. Our simulator learns this mapping using a simple multi layer neural network [4] and supervised learning with the RPROP algorithm [13]. Using this learning simulator we have learned the dynamics of Pioneer I robots. During data acquisition we have executed a wide variety of navigation scenarios that correspond to soccer plays. We have collected a total of more than 10000 training patterns from runs with a real Pioneer I robot. The accuracy for navigation tasks (including acceleration) is around 99% The accuracy decreases to about 92% in situations where both velocities, the translational and rotational one, are changed abruptly at the same time. These inaccuracies are caused by the lack of representative training patterns as well as the high variance in navigation behavior with maximal acceleration.

4 A Toolbox for Multi Robot Navigation Planning

The second component of our navigation system is a library of software tools for planning and repairing multi robot navigation plans. So far the planning methods library contains single robot navigation planning methods and methods for integrating the single robot plans through plan merging and plan repair afterwards.

4.1 Single Robot Path Planning

The path planning methods we will use in our toolbox include the *Potential Field Method* [5,1,15], the *Shortest Path Method* [9,6], *Circumnavigating Obstacles* [14], and *Maximizing the Clearance* [8]. Buck et al. [2] discusses in a more detailed overview the employed algorithms, their features, and abilities.

4.2 Plan Merging and Repair

The algorithms for single robot path planning can be coupled with plan merging and repair methods. We will now address the question of how to combine the individual plans in order to obtain a good performance on the joint navigation tasks.

Waiting Simply combining the paths computed by each robot without taking further precautions entails the danger that two robots might collide if their paths intersect. The simplest fix is to let one robot wait when two robots are to reach an intersection at about the same time until the other robot has crossed the intersection.

Path Replanning The remaining methods try to revise the individual plans such that no negative interferences will occur. Again we assign priorities to the robots according to the importance of their navigation tasks and ask the robots with lower priority to revise their plans to avoid conflicts with the paths of the higher priority robots. We have considered three different methods for path revision called *Defining Temporary Targets*, *Hallucinating Obstacles at Critical Sections*, and *Inserting New Obstacles.* The first one modifies the path by introducing additional intermediate target points. The second one hallucinates additional obstacles at the positions where collisions might occur. The third one simply considers the other robot at its respective position as a static obstacle. Advantages and drawbacks of these methods are discussed in [2].

5 A Hybrid Navigation Planner

The third component of our navigation system is the hybrid navigation planner. The hybrid navigation planner selects based on the particular navigation task and the situation the task is to be performed in the most appropriate single robot planning and plan repair method. The working horse of the hybrid navigation planner is a decision tree where pairs consisting of a single robot planning and a plan repair method are stored in the leaves. The branches of the tree are labelled with tests on the characteristics of the navigation problem. The semantics of the decision tree is the following. For any navigation task that satisfies all the tests on a branch b is the planning method proposed by the leaf of the branch b the most appropriate one. In our case the main reason for using a decision tree for classification is that it leads to understandable rules while a neural network for example does not (by the advantage of quantifying the planning methods).

In the remainder of this section we describe the feature language that we use for the characterization of navigation tasks and the automatic learning of decision trees for the selection of the most appropriate planning methods.

5.1 A Feature Language for Multi Robot Navigation Tasks

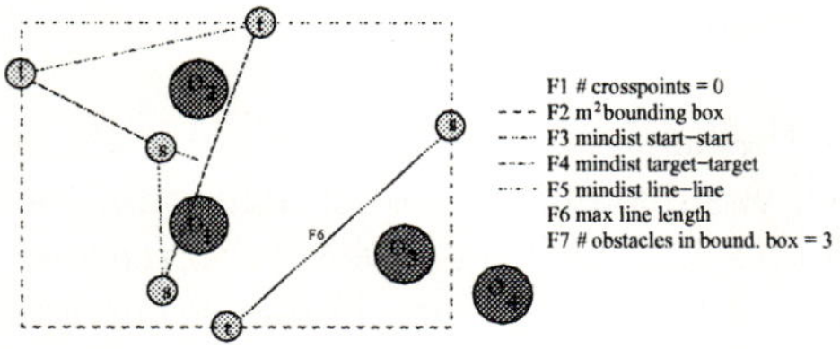

Fig. 5. Visualization of navigation task features that are used for classifying navigation tasks.

As mentioned in section 4 a large variety of different single robot navigation planning and plan merging methods exist, that have different strengths and weaknesses, and make different assumptions about the navigation problems at hand. This observation suggests that we need a language in which we can describe the characteristics of navigation problems in a given robot control application and use these characteristics to select the appropriate new planning method.

In our investigations we will use 7 different features (see fig. 5). These features are: (1) the number of intersections between the line segments that represent the navigation tasks, (2) the size of the bounding box of the navigation tasks, (3) the minimal linear distance between different starting positions, (4) the minimal linear distance between different target positions, (5) the minimal distance between the line segments that represent the navigation tasks, (6) the maximum length of the linear distances of the individual navigation tasks, and (7) the number of obstacles in the bounding box of the joint navigation task.

5.2 Predicting the Performance of Planning Methods

A natural way for encoding a predictive model of the expected performance of different navigation planning methods in a given application domain is the specification of rules that have the following form:

if $c_1 \wedge ... \wedge c_n$ **then** fastest-method($\langle$ *srpm,prm*$\rangle$)

In this rule pattern the c_is represent conditions over the features that we use to classify navigation problems. The **then**-part of the rule asserts that for navigation problems that satisfy the conditions c_i the combination of the single robot planning method *srpm* together with the plan repair method *prm* can be expected to accomplish the navigation task faster then any other combination of single robot planning and plan repair method.

We have collected a training set of 1000 data records and used it for decision tree learning. From this training set the C4.5 algorithm [12][2] with standard parameterization and subsequent rule extraction has learned a set of 10 rules including the following one:

if there is one intersection of the navigation problems
$\wedge$ the navigation problems cover a small area ($\leq 10.7m^2$)
$\wedge$ the target points are close to each others ($\leq 1.1m$)
$\wedge$ the starting/target point distances are small ($\leq 5m$)
then fastest-method($\langle$*potential field,temp. targets* $\rangle$)

This rule essentially says that the potential field method is appropriate if there is only one intersection and the joint navigation problem covers at most one fourth of the field, and the target points are close to each others. This is because the potential field algorithm tends to generate smooth paths even for cluttered neighborhoods.

The accuracy of the ruleset for predicting the fastest navigation method is about 50% both for the training and the test set. A substantially slower algorithm was chosen only in very few cases. The inaccuracies of the rules have several reasons. First, the feature language as it has been introduced is not yet expressive enough. We expect that the accuracy of the rules can be substantially increased by adding

[2] For our experiments, we have used the public domain version of Quinlan's C4.5 algorithm

additional features such as the angle between the robot's current orientation and the direction to the target position. Second, in many navigation problems different methods achieved almost the same performance. In those cases we only selected the best one even when the margins where very narrow. Obviously, these data records are very noise sensitive. Third, in many runs collisions were caused by dynamic obstacles and have caused robots to get stuck. These runs resulted in outlier results that are not caused by the planning methods. Thus, for higher accuracy those runs have to be handled differently.
The conclusions that we draw from this experiment are that even with crude feature languages, without sophisticated data transformations and outlier handling a robot can learn useful predictive models for the performance of different navigation methods in a given application domain.

6 Experimental Results

In order to empirically evaluate our hybrid navigation planning and execution system we have performed extensive experiments using the simulator introduced in section 3. This simulator has enabled us to collect the necessary amount of realistic data, to make a simple quantitative comparison with respect to the average performance of the individual methods, and to find clusters of navigation tasks within the navigation tasks in the application that the individual methods solve well or poorly.
In the remainder of this section we describe two experiments. The first one is a quantitative comparison of the performance achieved by the different planning and plan repair methods provided by the toolbox. The results of this experiment suggest that it can, in general, not be expected that a single planning method can achieve heterogeneous navigation tasks equally well. In other words, we cannot expect navigation methods that dominate other ones in such complex application domains as robot soccer.
The second experiment shows that the hybrid navigation planner outperforms the individual planning methods on large collections of randomly sampled joint navigation tasks. This result suggests that it is indeed possible — even with such a simple feature language as ours — that a robot team can automatically learn predictive models of their planning methods that enable them to improve their performance both in a substantial and a statistically significant way.

6.1 Comparative Experiments

We set the number of obstacles to 4 (which is the team size of the mid size league). We carry out experiments with 3 robots (resulting from 3 field players in RoboCup). All our experiments underlie the same randomly generated situations: A robot starts at a randomly defined state in its configuration space and needs to get to a randomly defined target state. The obstacles move linearly from a randomly generated start point to a randomly defined target. If an obstacle reaches its target a new target is defined immediately. Obstacles move with a

Algorithm	Mean time values of 1000 train and 1000 test problems			
	TRAIN		TEST	
	μ/sec	significance level $P(\mu_{tree} < \mu)$	μ/sec	significance level $P(\mu_{tree} < \mu)$
Simple Potential Field	15.49	99.99 %	15.92	99.99 %
Shortest Path	13.36	99.99 %	13.14	99.99 %
Maximum Clearance	12.35	99.71 %	12.31	99.84 %
Viapoint	12.14	94.62 %	11.95	96.25 %
Decision Tree	11.64	50.00 %	11.44	50.00 %

Fig. 6. Results (mean time needed to solve a navigation task) of four evaluated algorithms and the trained decision tree. The significance level is based on a t-test.

random but constant velocity. Defining these preconditions we consider different dynamic behaviors of the obstacles without building a complex behavior model which would mean another laborious task.

Figure 6 pictures the mean value of the time resources required to complete a joint navigation task using the different planning methods introduced in section 4. The statistical data was acquired by performing 1000 randomly chosen navigation problems and performing the planning methods at a frequency of 10 Hz. The results show that based on the empirical data we cannot determine a single method that outperforms the other ones in a statistically significant way. This suggests that we should try to identify specializations of the navigation problems for which one planning method outperforms the other ones.

6.2 A Hybrid Navigation Planner

We have performed a bootstrapping t-test based on 1000 different joint navigation tasks (fig. 6) in order to empirically validate that the hybrid navigation planner performs better than the individual planning methods that we are using. Based on these experiments we obtained a 99.9% confidence in the test set (99.9% in the training set) that the hybrid method outperforms the potential field method (with its respective parameterization). The respective probabilities for the shortest path method are 99.9% (99.9%), for the maximum clearance method 99.84% (99.71%), and for the viapoint method 96.25% (94.62%). This means that our hypothesis that the hybrid planner dominates the other planning methods could be validated with statistical significance ($\geq 95\%$).

7 Conclusions

In this paper we have shown that in complex multi robot navigation planning domains it is extremely difficult to predict the performance of different kinds of planning methods. This is because the different planning methods make different kinds of assumptions. The maximum clearance method, for example, assumes that it is better to keep larger distances to the objects and follow a longer path with higher speed, on the other hand, the circumnavigation methods assume that it is better to pass obstacles at a closer distance. This, however, requires the robot to have more accurate control over its dynamics, for example to be equipped with an omnidirectional drive. Therefore, it is unclear for most application domains to which degree the navigation tasks match the assumptions of the different methods. Even worse, yet within a single application domain we can identify classes of navigation problems for which the algorithms' suitability varies.

In this paper, we have therefore proposed to select problem-adequate navigation planning methods based on empirical investigations, that is the robots should learn by experimentation to use the best methods. To support this development strategy we have provided a feature language for classifying navigation problems and software tools that enable the robots to automatically learn predictive models for the performance of different navigation planning methods in a given application domain. We have shown, in the context of robot soccer, that a hybrid planning method that selects planning methods based on a learned predictive model outperforms the individual planning methods. The results were validated in extensive experiments using a realistic and accurate robot simulator that has learned the dynamic model of the real robots.

Even though these results are conclusive we expect that the performance can be substantially improved by using a more sophisticated feature language for the classification of navigation problems and by devising more sophisticated preprocessing methods for the data elements used for learning. In particular, we would expect that a competent module for rejecting outliers would reduce the deviations in the data substantially and improve the predictability drastically. These extensions of our basic framework are subject of our future investigations.

References

1. J. Barraquand, B. Langlois, and J. Latombe: *Numerical potential field techniques for robot path planning.* IEEE Transactions on Systems, Man, Cybernetics, 22(2):224–241, March/April 1992.
2. S. Buck, U. Weber, M. Beetz, and T. Schmitt: *Multi Robot Path Planning for Dynamic Environments: A Case Study.* Accepted for publication at IEEE/RSJ IROS, 2001.
3. S. Buck, R. Hanek, M. Klupsch, and T. Schmitt: *Agilo RoboCuppers: RoboCup Team Description.* RoboCup 2000: Robot Soccer World Cup IV, Springer, 2000.
4. J. Hertz, A. Krogh, R.G. Palmer: *Introduction to the Theory of Neural Computation.* Addison-Wesley, 1991.
5. Y. K. Hwang and H. Ahuja: *A Potential Field Approach to Path Planning.* IEEE Transactions on Robotics and Automation, vol. 8, 1, 23-32, 1992.
6. K. Konolige: *A Gradient Method for Realtime Robot Control.* Proceedings of the IEEE/RSJ IROS, 2000.
7. K. Konolige, K. Myers, E. Ruspini, and A. Saffiotti: *The Saphira Architecture: A Design for Autonomy.* Journal of Experimental and Theoretical Artificial Intelligence, 9:215-235, 1997.

8. J.-C. Latombe: *Robot Motion Planning.* Kluwer Academic Publishers, 1991.
9. J. Lengyel, M. Reichert, B. Donald, and D. Greenberg: *Real–time robot motion planning using rasterizing computer graphics hardware.* Proceedings of SIGGRAPH, pages 327–335, August 1990.
10. B. Nebel and T. Weigel: *The CS Freiburg 2000 Team.* Fourth International Workshop on RoboCup, Melbourne, Australia, 2000.
11. Pioneer Mobile Robots, Operation Manual, 2nd edition, Active Media, 1998.
12. R. Quinlan: *Induction of decision trees*, Machine Learning 1 (1), 1986
13. M. Riedmiller and H. Braun: *A direct adaptive method for faster backpropagation learning: the Rprop algorithm.* Proceedings of the ICNN, 1993.
14. A. Schweikard: *A simple path search strategy based on calculation of free sections of motions.* Engineering Applications of Artificial Intelligence, 5, 1, 1 - 10, 1992.
15. P. Tournassoud: *A strategy for obstacle avoidance and its application to multi-robot systems.* Proceedings of the IEEE ICRA, pp. 1224-1229, 1986.
16. T. Weigel: *Roboter-Fuball: Selbstlokalisierung, Weltmodellierung, Pfadplanung und verhaltensbasierte Kontrolle.* Master Thesis, University of Freiburg, Germany, 1998

A Case Study of How Mobile Robot Competitions Promote Future Research

Jennifer Casper, Mark Micire, Jeff Hyams, and Robin Murphy

Computer Science and Engineering, University of South Florida
Tampa, Florida 33620
{jcasper, mmicire, hyams, murphy}@csee.usf.edu
http://www.csee.usf.edu/robotics

Abstract. The purpose of mobile robot competitions is to push the field of research and inspire future work. Our involvement in the Urban Search and Rescue event of the 2000 AAAI Mobile Robot Competition allowed us the opportunity to test our Urban Search and Rescue robot team in a standardized testbed. The result was ideas for future enhancements and research work. This article presents our experience as a case study on how mobile robot competitions promote research.

1 Introduction

One objective of mobile robot competitions is to push the field and inspire future research. The Urban Search and Rescue event of the 2000 AAAI Mobile Robot Competition was the first of its kind. It encouraged involvement from many different types of robotic systems. The competition platform was provided by the National Institute of Standards and Technology (NIST) and contained three different levels to allow for robots of varying capabilities. It is now being used as the basis for the RoboCup/AAAI Rescue Robot League.

The purpose of this paper is to present our experience as a case study on how competitions promote research. Section 2 presents the approach taken for the competition. Next, a description of the competition platform is discussed in Section 3. Section 4 discusses the results obtained from the competition. Enhancements to the current USAR Team and future research topics inspired by the competition are presented in Section 5. This paper does not address the utility of the NIST testbed and our recommendations for scoring USAR competitions; the reader is directed to [6] for such a discussion.

2 Approach

The approach taken for the USAR event of the 2000 AAAI Mobile Robot Competition was novel in two ways. First, the autonomy was implemented in the perception rather than the navigation. The second novel aspect was the idea of collaborative teleoperation. Implementing the autonomy in the perception was done in part by the recommendation of the Hillsborough County Fire Rescue

A. Birk, S. Coradeschi, and S. Tadokoro (Eds.): RoboCup 2001, LNAI 2377, pp. 123–132, 2002.

teams. The result was an Intelligent Assistant Agent which searched for victims based on cues (heat, color, motion, etc.) and informed the operator when a victim was located (or the high potential of a victim). This allowed the operator to concentrate on navigating the robot through the complex environment. The agent was adjustable in that the operator could specify which cues to look for and the ability to notify the operator (beeping in this case). The adjustable autonomy system is discussed in the next section, followed by the description of the Intelligent Assistant Agent. Collaborative teleoperation is presented in 2.3.

2.1 USAR Team

Fig. 1. USF USAR team: Fontana and Taylor.

The USF team used two outdoor robots: 1) a RWI ATRV with sonar, video, and a miniature uncooled FLIR (Fontana) and 2) a customized RWI Urban with a black and white camera, color camera, and sonars (Taylor). This was intended to be a marsupial pair, but the transport mechanism for the team was still under construction at the time of the competition. The USF team used a mixed-initiative or adjustable autonomy approach: each platform was teleoperated for purposes of navigation but the Intelligent Assistant Agent was utilized for autonomous victim detection.

Each operator controlled one robot through the use of dedicated operator control units (OCU), see Figure 2. There were two distinct OCUs. The OCU used for the Urban, Taylor, was designed and implemented by RWI with a proprietary interface using Mobility. It consists of a laptop, battery pack, and a wireless Ethernet in a carrying case. The OCU provided the user with a camera view, battery power levels, speed, inclinometer, compass, pose, and sonar. Taylor was

Fig. 2. The operator station setup during the AAAI conference.

controlled by moving the cross-hairs on the camera view using the touch pad. Her flippers were controlled similarly by cross-hairs on the power level display.

The OCU used for the ATRV, Fontana, was an in-house design using X-windows display widgets. It attempted to simplify the teleoperation control challenges imposed by the RWI OCU as well as permit the display of results from the Intelligent Assistant Agent. It consists of a laptop and wireless Ethernet. The operator was provided with sonar, an intelligent operator system, and keyboard control. The arrow keys were used to drive Fontana.

The robots communicated with the OCUs for the operators through a pair of Breezecom wireless Ethernet nodes and a base unit connected to a hub. The Intelligent Assistant Agent processing resided on Fontana, not on the OCU.

2.2 Intelligent Assistant Agent

The Intelligent Assistant Agent for this work consisted of one sub-agent, a vision agent to aid with victim detection. The vision agent performed cueing and behavioral fusion [8] based on the output of four concurrent detection algorithms: *motion, skin color, color different, and thermal region.* The user interface displayed the extraction results from each applicable algorithm, highlighted in color. Thus, the operator had visual cues for victims. In addition, the operator was notified by beeping when the agent had sufficient confidence in the presence of a victim based on fusion of multiple detectors. All active detectors were tiled side by side along with an un-augmented view from the video camera.

The vision agent can support any number of vision algorithms. The choice of these four algorithms was based on our understanding of the cues used by USAR technical rescue personnel and on the use of low computational affordances. A living victim will always radiate heat, which can be measured by a thermal imager. Heat is a primary affordance of victims in USAR. The only heat sources in the rubble should be people, animals, or indicators of a fire. The region properties of the heat source (e.g., does it look like the profile of a human) are not as critical. The rubble may insulate large portions of the victim, causing a more sophisticated (and computationally expensive) segmentation and matching algorithms to report false negatives (nothing is present).

Another affordance of a victim is color. In a collapsed structure, there is a grey dust from sheet rock or ceiling tiles which covers everything, and as such, most color is washed from the image. In a structure that has experienced a fire, the debris is similarly covered in black soot. Therefore rescuers look for anything that is distinguishable from that grey dust. For instance, a victim may move and shake off some of the dust, or there may be distinguishable colors such as blood.

Notice that the two previous victim detection affordances apply regardless if the victim were conscious or not. If the victim is conscious, an easily detectable affordance is motion, e.g., moving arms and waving for attention as the robot comes near.

Each of the four detection algorithms extracted a single feature that afforded the presence of a victim. The detection algorithms were written in C. Sufficient size of a feature or co-location of cues was interpreted by the vision agent as a possible victim. The operator was notified of the results of each algorithm through a continuously updating display, written in QT. The operator was also given the ability to switch on and off any of the algorithms. For further details of the Intelligent Assistant Agent, see [9].

2.3 Collaborative Teleoperation

The second novel aspect of the approach concerned the robot control used during the competition: *collaborative teleoperation.* Collaborative teleoperation is the use of a secondary robot to provide the operator with an external view of the primary robot. The external view is intended to help performance (e.g., navigate through a tight spot), diagnose and recover from a problem (e.g., high-centering of the robot, loss of a track, etc.), or cooperate (e.g., make an opening for the target robot to go through).

This task was performed by the operators who communicated with one another when help was needed. For instance, Fontana needed Taylor's point of view to assist in navigating around corners of the confined spaces (See Figure 3). Taylor had to be turned 180 degrees to face Fontana. Taylor's view could then be used by the operator to navigate the turn.

3 2000 AAAI Mobile Robot Competition

The USF USAR Team tested the NIST standard test bed for USAR as part of the 2000 AAAI Mobile Robot Competition [6]. The test bed consisted of three sections, each providing a different level of difficulty in order to accommodate most competitors. The easiest section, Yellow, contained mainly hallways, blinds,

Fig. 3. To navigate, Fontana's view and Taylor's view were needed in certain situations (Collaborative Teleoperation).

and covey holes to search through. The intermediate Orange Section provided more challenge with the addition of a second level that was reachable by stairs or ramp. Other challenges included those found in the yellow as well as some added doors. The Red Section was intended to be the most difficult. It contained piles of rubble and dropped floorboards that represented a pancake-like structure. The Orange and Red sections clearly required hardware that was capable of traveling such spaces. The USF Team concentrated on the more realistic Orange and Red sections.

The victims were mannequins and dolls. "Live" victims were adjacent to a heating pad to generate heat. Parts of mannequins were used to represent dismembered body parts and served as a source of false positives. A metronome type of device created a motion signature and a tape recorder playing simulated victims' cries for help were also utilized. The victims placed within the Yellow and Orange sections were surface victims [2]. The Red section victims were either in voids or entombed [2].

The USF Team ran through five events total; preliminary qualification, final competition, and three exhibitions. The ATRV found an average of 3.75 victims per each of the four runs recorded, while the Urban found an average of 4.67 victims. A fifth run was not recorded and no data is available. The data reported here is based on analysis of tape recordings from four of the five events made by USF, since the competition officials failed to record any data (number of identified victims, number of victims in each section, etc.).

4 Data and Results

Table 1 shows the performance of the USF heterogeneous team for victim detection. The data shows that in the best run (was the Final round, respectively),

Run	Time (min)	Total Victims Found	Robot	Victims Found Solo	Victims Found Jointly
Preliminary	32	4	Fontana	1	2
			Taylor	1	2
Final	30	7	Fontana	1	3
			Taylor	3	3
Exhibition #3	not recorded	6	Fontana	0	3
			Taylor	3	3
Exhibition #4	28	6	Fontana	1	4
			Taylor	1	4

Table 1. Results of victim detection in the four recorded runs.

3 of the 7 victims were detected jointly, or cooperatively. In the Exhibition #4 round, 4 of the 6 victims were detected jointly. This was due to two reasons. First, the more agile Urban robot, Taylor, could quickly enter an area and find a sign of survivors, but often could not confirm that the sign was a victim using only a black and white camera with a narrow field of view. Second, Taylor could not confirm whether a victim was "living" or "dead" if the victim was not moving; this had to be disambiguated with the FLIR thermal imager on Fontana. Fontana found at most 1 victim by herself, or solo, because she was larger, less agile, and had to slowly navigate the more restricted spaces of the Orange and Red sections.

Run	Time (min)	Robot	Seconds Spotting
Preliminary	32	Fontana	25
		Taylor	222
Final	30	Fontana	45
		Taylor	160
Exhibition #4	28	Fontana	10
		Taylor	250

Table 2. Results of collaborative teleoperation during the four recorded runs.

Table 2 shows that Fontana spent between 10 and 45 seconds and Taylor spent between 160 and 250 seconds acting as the secondary robot, providing an external view for the primary robot. The table indicates that both robots needed collaborative views at some point during each of the recorded runs (each robot spent some time spotting). As expected, the agile Urban, Taylor, needed much less help navigating, while the larger ATRV, Fontana, had to frequently rely on viewpoints from the Urban to assist the operator in guiding it through confined space. This is apparent by Fontana's low average spotting time of 26.7 seconds and Taylor's higher 210.7 second average spotting time.

5 Research Inspired by the Competition

Our experience with the 2000 AAAI Mobile Robot Competition sparked ideas for future enhancements on the current USAR Team and four research topics pertaining to USAR.

5.1 Enhancements for Robot Control

The collaborative teleoperation results in Table 2 show that between 11 and 16% of the total time of the competition run was spent cooperating (Taylor spotting Fontana or Fontana spotting Taylor). It was quickly realized what an inconvenience it was to have to move the robot to a viewing angle for the sake of assistance. Collaborative teleoperation shouldn't cost valuable search time. However it is necessary, as collaborating resulted in keeping the team progressing through the environment and at least 50% (on average) of the total victims found due to joint effort, see Table 2. Multiple lead robots may be helpful so that one specialized search robot may continue to explore while a designated and less equipped robot assists a larger robot, such as Fontana.

Looking up is commonly overlooked when searching. In the case of the competition, a victim on the second floor of the Orange section was clearly visible but slightly above camera range of the robots searching the first floor. Had the robots had the ability to tilt up, the team may have had a chance to find the victim. Stairwells and open balconies are common in apartment complexes and hotel buildings (urban areas where looking up and down are required).

While controlling the robot, the side sonars were providing displays to the user which were somewhat helpful in determining how close obstacles (e.g., walls) were. Yet between 11 and 16% of the total run time on average was still spent with both of the robots spotting and recovering from wall hand-ups, maneuvering mistakes, and course corrections (see Table 2). A logical addition to the teleoperation control would be an adjustable avoidance system, or "guarded motion", for navigation that would assist in keeping the robot from bumping into wall and other obstacles that might not be very easily judged by the operator.

The average 11-16% of the total run time spent by both robot spotting was needed in situations such as navigating a corner. The side sonars were utilized by the operators to help provide proximity information of the environment for navigating. Collisions, such as running into the wall or door frame, continued to occur. This leads to questioning the performance of the sensors, the sonars in this case, in confined space environments.

5.2 Enhancements for the OCU

The operators' feedback (collected from informal interviewing) pertaining to the OCU interface was negative in the effect that the OCUs were cumbersome. The keyboard and touch pad drive control interfaces for the ATRV and Urban were difficult to use in an environment like the mock-collapse NIST structure. The touch pad on the Urban's OCU disabled the ability to rotate the flippers and the drive forward at the same time. The touch pad pointer could only be on the drive display or the flipper display at any one time. The ability to rotate the flippers and drive forward is very useful in recovering the Urban when it is stuck on top of something (e.g., lodged on a pile of rocks). Another method of interface to the multiple functions of robots for the purpose of controlling is needed.

Another problem with the OCUs was the number of tasks the operator was required to perform at one time. The operator for Fontana was required to drive the robot, pay attention to the Intelligent Assistant Agent, and collaborate with the other operator. The OCU for the ATRV included a display panel for the vision detectors, each of which could be turned on or off by using the mouse to click a button in the X-window display. This diverted the operator's attention from controlling the robot, thus leaving the robot at a standstill, to alter what detectors are used. This results in a loss of time spent searching for victim, not to mention a break in the operator's focus.

The current implementation requires the operator to lean over and view the video display on the other robot's OCU in order to gain a second perspective (e.g., two views were needed to navigate the ATRV around tight corners). This was extremely cumbersome due to the physical separation of the two OCU displays. A method by which an operator may switch between views on the designated OCU is desirable.

5.3 Four Main Research Issues

Four main research issues emerged during the competition and exhibition. These contribute to the development of an adjustable autonomous heterogeneous robot team for urban search and rescue.

Collaborative Teleoperation As previously mentioned, collaborative teleoperation is the use of a secondary robot to provide the operator with an external view of the primary robot. Our experience with fielding USAR robots at the Hillsborough County Fire Rescue (Tampa, FL), SRDR (Miami, FL), Montgomery County Fire Training Academy (Rockville, MD), and George Air Force Base (Victorville, CA) training facilities is that it is very easy to flip or high center a robot in rubble. While it is easy to get the robot into trouble, it is difficult to diagnose the problem or construct a solution using only limited proprioceptive sensors (tilt, pose) or an exteroceptive view of the environment. What is needed is an exproprioceptive view, where the robot's pose is known relative to the external environment. Since the robot cannot see itself, the viewpoint of another agent is helpful. In the case of the competition, a situation occurred in which the view from the ATRV was able to prevent the Urban falling from an elevated position. The operator needed the exproprioceptive view from the ATRV in order to save Taylor as Taylor's viewpoint was worthless in this case. Collaborative teleoperation is a function that could be supported by an Intelligent Assistant Agent.

Persistent Pursuit The main idea of persistent pursuit is to obtain and maintain an ideal vantage point of the daughter or team members in order to provide the operator with multiple usable views. For instance, in the red section of the USAR mock up, the team entered the opening on the east side of the structure. Without much maneuverability, Fontana remained stationary until Taylor began to traverse alongside the back of the rubble pile. It was worth the risk to abandon Taylor momentarily in order to follow the outer wall of the structure in the direction Taylor was heading in hope of acquiring a better vantage point. Plexiglas windows outlining the back of the course provided an opportunity to view Taylor, who continued to rummage around the backside of the pile. A certain amount of persistence was needed in order to find a better point of view. If the Plexiglas hadn't been there, the operator would've retreated to the last known position the Urban was seen at. This concept is useful not only for teleoperation or navigation, but for line of sight communications as well.

Semantic, Systematic, and Opportunistic Search Current search methods include semantic, systematic, and opportunistic [7]. Any one used on its own is insufficient to complete this task. A semantic search involves searching the interesting areas for victims, but does not guarantee a thorough search. The systematic search is a way to more thoroughly investigate a site. An example of a simple systematic search is the right wall follow which is often utilized by the Hillsborough County Rescue teams in low visibility situations. Just utilizing a wall follow might be considered insufficient because it will search the interesting as well as uninteresting areas, thus costing precious time. An opportunistic search is being aware of opportunities that may present themselves (i.e., a victim is discovered on the way to a goal). An optimized search of an USAR site would require the ideal combination of the three search types.

In the case of this competition, the teleoperators were utilizing a right hand wall search and tended to any victims found along the way. The teams hadn't been given any semantic information, such as what kind of building this is, or who was in it. The best solution was to systematically search and hope to opportunistically find victims.

Conflict between Navigation and Sensing There existed moments during the competition in which the navigation and sensing conflicted. For example, Taylor sensed a victim and wanted to get within a certain distance from the victim in order to provide communications but couldn't due to being obstructed. She would have to abandon her view in hopes of finding an alternate route to the victim. In a purely reactive system, her victim find behavior and avoid behavior would conflict causing her to stand still and do nothing. This type of conflict needs to be resolved in order to properly automate victim searching in USAR.

6 Conclusions and Current Work

The purposes of robot competitions are to promote research and push the field forward. The first Urban Search and Rescue Competition was held at the AAAI 2000 Conference in Austin, Texas and intended to do just that. It was successful in promoting ideas and research for future work in USAR in our experience. The competition inspired us to implement and test the ideas we had developed from our work within the USAR field. USAR training (trench rescue, confined space, and structural collapse), collaboration with rescue professionals, and research has provided information as to what is required of potential USAR robots.

The competition was a chance to test and compare USAR systems in a standard and reproducible platform. This is an important step up from testing in a sterile lab, but not as effective as an authentic USAR environment in our experience. The competitions will give the opportunity for the comparison of different USAR systems on a common accepted platform. It will continue to be advantageous to strive to make more authentic USAR platforms for competitions.

Acknowledgments

Portions of this work were funded by DARPA and a travel scholarship from AAAI. The authors would like to thank Mike Larson of IR Cameras for the loan of the Indigo Alpha FLIR, and Chief Ron Rogers and Jeff Hewitt from the Hillsborough County Fire Department for their support. Without them, our work wouldn't be possible.

References

1. National Fire Academy, *Rescue Systems I,* 1993.
2. United States Fire Administration, *Technical Rescue Program Development Manual,* 1996.
3. National Fire Protection Association, *Standard on Operations and Training for Technical Rescue Incidents,* 1999.
4. J. Hyams, M. W. Powell, R. R. Murphy, "Cooperative Navigation of Micro-Rovers Using Color Segmentation," *Autonomous Robots*, Vol. 9, pp. 17-16, 2000.
5. J. L. Casper, M. J. Micire, R. R. Murphy, "Issues in Intelligent Robots for Search and Rescue," *SPIE Ground Vehicle Technology II*, 2000.
6. R. R. Murphy, J. Casper, M. Micire, J. Hyams, "Assessment of the NIST Standard Test Bed for Urban Search and Rescue," *NIST Workshop on Performance Metrics for Intelligent Systems*, 2000.
7. R. R. Murphy, "Biomimetic Search for Urban Search and Rescue," *IROS*, 2000.
8. R. R. Murphy, *Introduction to AI Robotics*, MIT Press, Cambridge, MA, 2000.
9. R. R. Murphy, J. L. Casper, M. J. Micire, J. Hyams, "Mixed-Initiative Control of Multiple Heterogeneous Robots for Urban Search and Rescue," *In review, IEEE Transactions on Robotics and Automation.*
10. K. Osuka, "Rescue Robot Contest in RoboFesta-KANSAI 2001," *Proceedings of 2000 IEEE International Conference on Industrial Electronics, Control and Instrumentation*, pp. 132-137, 2000.

CS Freiburg: Global View by Cooperative Sensing*

Markus Dietl, Jens-Steffen Gutmann**, and Bernhard Nebel

Institut für Informatik
Universität Freiburg
79110 Freiburg, Germany
{dietl,gutmann,nebel}@informatik.uni-freiburg.de

Abstract. Global vision systems as found in the small size league are prohibited in the middle size league. This paper presents methods for creating a global view of the world by cooperative sensing of a team of robots. We develop a multi-object tracking algorithm based on Kalman filtering and a single-object tracking method involving a combination of Kalman filtering and Markov localization for outlier detection. We apply these methods for robots participating in the middle-size league and compare them to a simple averaging method. Results including situations from real competition games are presented.

1 Introduction

Mobile robots usually can only perceive a limited area of their environment at a time. In general, sensors such as laser range finders (LRFs), ultrasonic sensors or cameras have a limited field of view, so an agent cannot sense all objects around him from one sensor frame. Furthermore, a robot does not know about objects that are occluded by other objects, e.g. by walls in an office environment.

There are two ways to overcome these limitations. For one, an agent can keep a history of sensor frames (or interpretations thereof) to reason about possible objects and their locations in the environment. For example, a robot can map an environment by maintaining an occupancy grid [9] where the grid cells represent possible object locations. However, if the environment is dynamic, the grid cells only reflect snapshots of situations when the data was recorded.

Another possibility is to deploy multiple robots in the environment, each with its own sensors, and to communicate their sensor informations to a module for multi-agent sensor fusion. This approach is especially useful for dynamic environments where moving objects are present that cannot be reliably tracked by a single robot.

This work addresses the second class of methods. We develop methods for two different kinds of scenarios in dynamic environments. One, where an unknown number of multiple objects have to be tracked by a group of robots under the assumption that the sensed data is noisy but reliable, and second, the tracking of a single object by a group

* This work has been partially supported by *Deutsche Forschungsgemeinschaft* (DFG), by *Medien- und Filmgesellschaft Baden-Württemberg mbH* (MFG), and by *SICK AG*.

** New address: Sony Corporation, Digital Creatures Laboratory, Tokyo 141-0001, Japan.

A. Birk, S. Coradeschi, and S. Tadokoro (Eds.): RoboCup 2001, LNAI 2377, pp. 133–143, 2002.

of robots where sensor readings are noisy and unreliable.[1] In both cases we assume that each robot in the team knows its own position with high accuracy and that all sensor information is acquired by on-board sensors only.

We apply these methods to the world of RoboCup for the middle size league where in our approach a team of robots maintains a consistent and accurate world model of team mates, opponents and the ball. Note, however, that the methods are general enough to apply to other domains, too, under the assumptions given above.

2 Multi Sensor Fusion in the RoboCup Environment

When developing an object tracking method, one usually has to deal with track initiation, track update including prediction and data association, and track deletion [1, 2]. This section describes the steps in our system necessary for tracking objects in the RoboCup environment.

Each robot first localizes itself by matching a scan from the LRF with an *a priori* line model of the soccer field. Experiments show that this localization technique is very accurate (usually < 2 cm and $< 1°$) and that it is faster and more robust than competing methods [5].

After estimating its own position, the robot extracts other players that are present in the field by discarding all scan points belonging to field walls and clustering the remaining ones. For each cluster the center of gravity is assumed to correspond to the center of another robot. Inherent in this approach is the systematic error due to the different robot shapes. However, it can be noted that measurements are usually reliable (no false positives).

Since our lasers are mounted at a level that prohibits the detection of the ball, we use a camera for obtaining information about where the ball is. Unfortunately our vision system has a limited quality and colors are difficult to train. Thus, ball observations are noisy and unreliable, e.g. we had problems with white field markings because when training the ball color, shiny reflections on the ball appeared to have a similar color.

All players send their own position, the position of observed other players and the position of the ball (if detected) to a global sensor integration unit which in our case runs on a central computer outside the soccer field. Here the measurements are fused in the following way.

For a player's own position, no further actions have to be carried out since the robots already determined their own position with high accuracy. However, for each player a track is initiated for fusing player observations coming from its team mates.

For each player detected by a robot, a new track is initiated or, if the player can be associated with an already existing track, it is fused with this track. We use a geometric method for data association and Kalman filtering for data fusion. Tracks containing measurements from an own player are marked as *team mate*, all others as *opponent*.

The ball position is determined by a probabilistic integration of all ball measurements coming from the players. Here we use a combination of Kalman filtering and Markov localization for achieving maximum accuracy and robustness.

---

[1] By reliable we mean that there are usually no incorrect measurements, that is, no false positives, whereas unreliable means that a robot might sense objects that are actually not present.

If no measurements can be associated with a track for a certain amount of time (e.g. 5 secs in our implementation), the track is deleted.

The fused positions of players and ball are sent back to all robots on a regular basis where they are integrated into each robot's world model. This enables a player to extend its own view with objects it does not currently perceive and also to know about which player is friend or foe. As a result, our players have a much greater knowledge of the world than when using local perception only. Especially knowing where the ball is, appears to be useful in almost all situations and we use the sharing of this information for developing sophisticated skills [14].

Details about our probabilistic sensor fusion methods for tracking multiple and single objects are described in the next section.

2.1 Multi-object Tracking from Reliable Data

Consider the tracking of an *a priori* unknown number of objects by a team of robots under the assumption that the measurements are reliable but might be corrupted by Gaussian noise. This scenario occurs in the RoboCup environment when players extracted from the robot's range finders are fused in the global sensor integration module.

Each of our robots detects other players in the field from data recorded by the LRF and computes heading and velocity information based on the last few observations belonging to this player using differentiation. Position, heading and velocity of each object are then communicated to the multi-sensor integration module. Thus, the observation model is a random variable $\mathbf{x}_s = (x_s, y_s, \theta_s, v_s, \omega_s)^T$ with mean $\hat{\mathbf{x}}_s$ and covariance Σ_s where (x_s, y_s) is the position, θ_s the heading and v_s and ω_s are the translational and rotational velocities of the object respectively.

As our robots know their own position with high accuracy and the LRF provides accurate data, we assume that for player observations, Σ_s is a constant diagonal matrix

$$\Sigma_s = diag(\sigma_{x_s}^2, \sigma_{y_s}^2, \sigma_{\theta_s}^2, \sigma_{v_s}^2, \sigma_{\omega_s}^2) \tag{1}$$

where $diag(\ldots)$ is a square matrix with its diagonal elements set to the given arguments and all other elements set to 0, and σ_{x_s}, σ_{y_s}, σ_{θ_s}, σ_{v_s} and σ_{θ_s} are constant standard deviations for position, heading and velocity which we manually adjusted through experiments.

Whenever a robot sends information about a player for which no already existing track can be found, i.e. if the distance to all existing tracks exceeds a certain threshold, a new track is initiated. Tracks are modeled as Gaussian variables $\mathbf{x}_r$ with mean $\hat{\mathbf{x}}_r$ and covariance Σ_r. Thus, when initiating a new track, it is set to

$$\hat{\mathbf{x}}_r = \hat{\mathbf{x}}_s, \;\; \Sigma_r = \Sigma_s \tag{2}$$

For predicting the state of a track over time, we use a simple motion model where we assume that the object moves and rotates with constant speed. Given a certain time interval t, the track is projected according to

$$\hat{\mathbf{x}}_r \leftarrow F_s(\hat{\mathbf{x}}_r, t) = \begin{pmatrix} \hat{x}_r + \cos(\hat{\theta}_r)\hat{v}_r t \\ \hat{y}_r + \sin(\hat{\theta}_r)\hat{v}_r t \\ \hat{\theta}_r + \hat{\omega}_r t \\ \hat{v}_r \\ \hat{\omega}_r \end{pmatrix} \tag{3}$$

$$\Sigma_r \leftarrow \nabla F_s \Sigma_r \nabla F_s^T + \Sigma_a(t) \tag{4}$$

where ∇F_s is the Jacobian of F_s and $\Sigma_a(t)$ is the covariance of some additive Gaussian noise with zero mean:

$$\Sigma_a(t) = diag(\sigma_{x_a}^2 t, \sigma_{y_a}^2 t, \sigma_{\theta_a}^2 t, \sigma_{v_a}^2 t, \sigma_{\omega_a}^2 t) \tag{5}$$

with σ_{x_a}, σ_{y_a}, σ_{θ_a}, σ_{v_a} and σ_{ω_a} being some constant standard deviations which we estimated through experiments.

Now, when a new measurement $\hat{\mathbf{x}}_s$ arrives from one of our robots which corresponds to a track $\mathbf{x}_r$, we fuse observation and track according to:

$$\hat{\mathbf{x}}_r \leftarrow (\Sigma_r^{-1} + \Sigma_s^{-1})^{-1}(\Sigma_r^{-1}\hat{\mathbf{x}}_r + \Sigma_s^{-1}\hat{\mathbf{x}}_s) \tag{6}$$

$$\Sigma_r \leftarrow (\Sigma_r^{-1} + \Sigma_s^{-1})^{-1} \tag{7}$$

Note, that since our sensor model does directly observe the system state, we can utilize the simplified Kalman filter equations found in Maybeck [8].

The success of a Kalman filter depends on a reliable data association method. In our implementation we use a geometric method that assigns measurements to tracks by minimizing the sum of squared error distances between observations and tracks [13]. Although this already yields reasonable results in practice, we want to note that the application of a probabilistic method such as joint probabilistic data association filters (JPDAF) [2, 11] might still improve our results.

2.2 Single-object Tracking from Noisy Data

Often the task is not to track multiple objects but a single object where observations are both noisy and unreliable. Consider the case of keeping track of the ball in the RoboCup scenario. There can only be one ball in the field during a game. However, since for ball recognition one usually employs vision, accuracy and robustness are in general low compared to LRFs which – in our case – are mounted above a height where they can detect the ball.

Again we use a Kalman filter as described in section 2.1 for accurately tracking the ball. Each of our agents regularly sends ball observations to the global sensor integration module. However, the vision sensor is only able to determine the heading to the ball with good accuracy but fails to provide accurate range data, especially if the ball is far away from the robot. For a ball observation $\hat{\mathbf{x}}_b$ we cannot assume a constant covariance Σ_b due to this characteristics.

Given the range $\hat{r}_b = \sqrt{(\hat{x}_b - \hat{x}_{rob})^2 + (\hat{y}_b - \hat{y}_{rob})^2}$ and heading $\hat{\phi}_b = \tan^{-1}(\hat{y}_b - \hat{y}_{rob})/(\hat{x}_b - \hat{x}_{rob})$ of the ball with respect to the robot located at position $(\hat{x}_{rob}, \hat{y}_{rob})$, we model the uncertainty $\Sigma_{r\phi}$ of ball position as

$$\Sigma_{r\phi} = diag(\hat{r}_b \sigma_{r_b}^2, \sigma_{\phi_b}^2) \tag{8}$$

where σ_{r_b} and σ_{ϕ_b} are some constant standard deviations we adjusted by hand. From this, we can compute the observation error as

$$\Sigma_b = \nabla P \Sigma_p \nabla P^T \tag{9}$$

where

$$P(\hat{r}_b, \hat{\phi}_b, \hat{\theta}_b, \hat{v}_b, \hat{\omega}_b) = \begin{pmatrix} \hat{x}_{rob} + \hat{r}_b \cos(\hat{\theta}_b + \hat{\phi}_b) \\ \hat{y}_{rob} + \hat{r}_b \sin(\hat{\theta}_b + \hat{\phi}_b) \\ \hat{\theta}_b \\ \hat{v}_b \\ \hat{\omega}_b \end{pmatrix}$$

$$\Sigma_p = diag(\hat{r}_b \sigma_{r_b}^2, \sigma_{\phi_b}^2, \sigma_{\theta_b}^2, \sigma_{v_b}^2, \sigma_{\omega_b}^2)$$

and σ_{θ_b}, σ_{v_b} and σ_{ω_b} are further constant standard deviations we estimated by hand.

For track initiation we create a new track $\mathbf{x}_r$ and set it to the last ball observation:

$$\hat{\mathbf{x}}_r = \hat{\mathbf{x}}_b, \;\; \Sigma_r = \Sigma_b \tag{10}$$

For predicting the ball state over time we use a similar function as for player movements but assume that the ball rolls in a straight line and slows down with deceleration a_b,

$$\hat{\mathbf{x}}_r \leftarrow F_b(\hat{\mathbf{x}}_r, t) = \begin{pmatrix} \hat{x}_r + \cos(\hat{\theta}_r)(\hat{v}_r - a_b t')t' \\ \hat{y}_r + \sin(\hat{\theta}_r)(\hat{v}_r - a_b t')t' \\ \hat{\theta}_r \\ \hat{v}_r - a_b t' \\ \hat{\omega}_r \end{pmatrix} \tag{11}$$

$$\Sigma_r \leftarrow \nabla F_b \Sigma_r \nabla F_b^T + \Sigma_a'(t) \tag{12}$$

where $t' = \min(t, \hat{v}/a_b)$ and $\Sigma_a'(t)$ is a similar constant covariance matrix as $\Sigma_a(t)$ to flatten the Gaussian distribution over time. Finally, fusing new observations to this track is analogous to equations (6) and (7).

It should be noted that a Kalman filter with this sensor model produces more accurate results when fusing observations from different viewpoints than e.g. a simple averaging method. This can be seen from Fig. 1(a) where the ball estimates of two players (indicated by grey circles) are integrated. Even though there is a large range error, the Kalman filter can effectively triangulate measurements from two separate viewpoints to localize the object much more precisely, because the angular error is small.

The Kalman filter for ball tracking presented in this section assumes noisy but reliable data, that is, no outliers are integrated. However, in our system we sometimes observed completely wrong ball measurements by one of our robots due to reflections on walls or poorly trained colors. One possibility to filter out such outliers is to use a validation gate that discards measurements whose Mahalanobis distance is larger than a certain threshold d, where d is chosen from a χ^2 distribution.

Such a validation gate, however, has the problem that when one robot is constantly sending out wrong observations and the Kalman filter for some reasons is tracking these

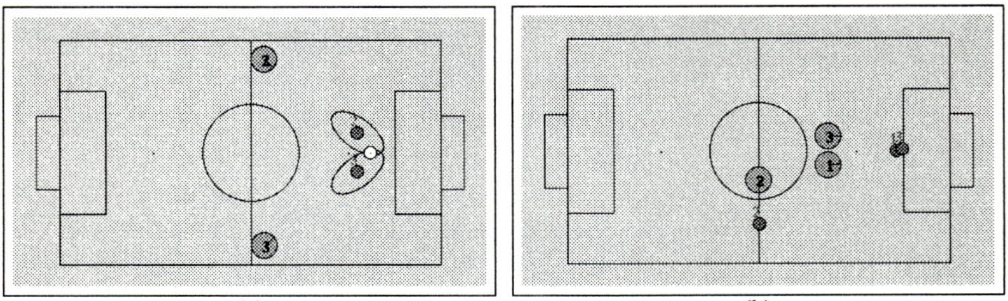

(a) (b)

Fig. 1. Fusing ball observations. Triangulation (a) and false positive observations by Player 2 (b).

wrong observations and filters out all others, the global ball position becomes unusable. Furthermore, when the robot stops sending wrong observations it takes several cycles until other observations are taken into account again. For these reasons we decided to develop a more sophisticated filter method described in the next section.

2.3 Markov Localization as Observation Filter

In localization experiments carried out on a mobile robot, Gutmann, Fox, Burgard and Konolige found out that Markov localization is more robust, while Kalman filtering is more accurate when compared to each other [3]. They propose to use a combination of both methods to get a maximum robust and accurate system.

In this paper, we follow the same idea and show how to use a Markov process as an observation filter for our Kalman filter. We use a grid-based approach with a 2-dimensional (x, y) grid where each cell z reflects the probability $p(z)$ that the ball is in this cell. We initialize this grid with a uniform distribution before any observation is processed. The integration of new ball measurements is then done in two steps: prediction and update.

In the prediction step, ball motion is modeled by a conditional probability $p(z \mid z')$ which denotes the probability that the ball is at position z given that it was at position z'. Upon ball motion, the new ball position is calculated as:

$$p(z) \leftarrow \sum_{z'} p(z \mid z')p(z') \tag{13}$$

Grid-based Markov localization can be computational expensive if the size and especially the dimension of the grid is large. For efficiency, we only use a 2-dimensional grid that does not store any heading or velocity information of the ball, which means that we cannot accurately estimate the position when the ball moves. For the motion model $p(z \mid z')$ we assume that all directions are equally possible and velocities are normally distributed with zero mean and covariance σ_v^2. Therefore, $p(z \mid z')$ can be expressed as a Gaussian distribution around z':

$$p(z \mid z') \sim N(z', diag(\sigma_v^2 t, \sigma_v^2 t)) \tag{14}$$

where t is the time passed during ball motion.

In the update step, a new ball observation z_b is fused into the probability distribution according to Bayes' law:

$$p(z) \leftarrow \frac{p(z_b \mid z)p(z)}{\sum_{z'} p(z_b \mid z')p(z')} \tag{15}$$

The sensor model $p(z_b \mid z)$ determines the likelihood of observing z_b given the ball is at position z. We model it according to:

$$p(z_b \mid z) \sim N(\hat{z}_b, \Sigma'_b) \tag{16}$$

where $\hat{z}_b$ are the (x, y) components of ball observation $\hat{\mathbf{x}}_b$ as defined in section 2.2 and Σ'_b is the upper left 2×2 sub matrix of Σ_b as calculated in equation (9).

Maintaining the multi-modal probability grid makes it very easy to distinguish which ball measurement should be integrated by the Kalman filter and which not. After updating the grid with a new measurement we determine the most likely ball position, that is, the cell with the highest probability. Only measurements that are close to the most likely position are fused into the Kalman filter and all others are considered as outliers. Furthermore, if the current state of the Kalman filter does not correspond to the most likely ball position in the grid, we re-initialize the Kalman filter using this position. By using this dual probabilistic localization method we achieve high accuracy through Kalman filtering together with high robustness through Markov localization. Experimental results using this technique are presented in the next section.

3 Results

The methods presented in the previous section have been implemented on our mobile robot soccer team (see Fig. 2) and have been successfully used since our participation in the RoboCup 1999 world competition. For our first participation in 1998 we developed a similar but much simpler approach for fusing measurements from different robots [4]. In this approach, a greedy nearest-neighborhood method handled data association. For computing object positions, a weighted averaging of observations from the different players was employed. In this section we compare this simple averaging method to our new probabilistic approach from Section 2.

In the case of multi-object tracking, we generally observed a more consistent world model for the new tracking method with a slightly higher run time due to the more sophisticated data association method. However, the differences in the world model were only marginal which we attribute to the fact that after solving the correspondence, the methods are very similar.

On the other hand, we got much better results for our new single-object tracking algorithm compared to the simple algorithm. Therefore, our main focus in this section is on results obtained by the single-object tracker.

3.1 An Ambiguous Situation

We now show how the algorithm for tracking single objects performs in an ambiguous situation. Fig. 1(b) displays a setup where two robots, Player 1 and 3, see the ball at the

Fig. 2. CS Freiburg players. Each one is a *Pioneer I* robot equipped with *SICK* LRF, *Cognachrome* vision system, *Libretto* notebook, *WaveLan* wireless ethernet, and semi-professional kicking device developed by *SICK AG*. For an overview, see [4, 14].

true location in front of the goal but one robot, Player 2, gets it all wrong and thinks the ball is somewhere on the center line. If we assume that all three players send their ball observations to the global sensor integration module on a regular basis, we get the probability distributions as shown in Fig. 3.

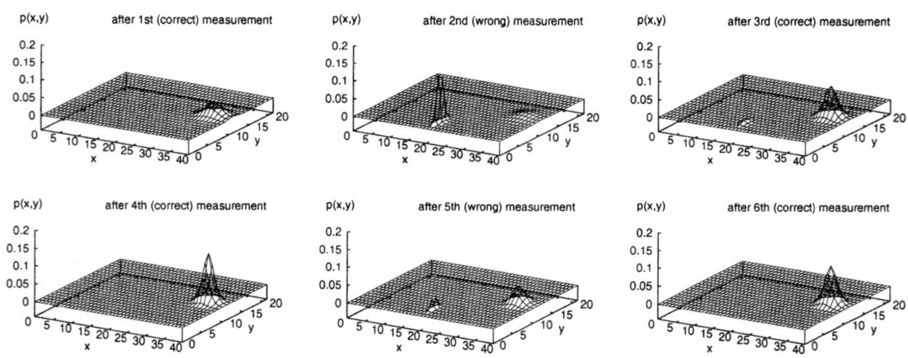

Fig. 3. Evolution of the position probability grid.

When integrating the first three measurements, all of them are fused by the Kalman filter since none of them has been detected to be an outlier yet. Note that after updating the grid with the 2nd measurement, the probability distribution has a sharp peak at the center line caused by the low uncertain measurement of Player 2 which thinks the ball is close. After integrating more measurements, the probability distribution concentrates more and more on the true location of the ball and further measurements from Player 2 (graph in the center of bottom row in Fig. 3) cannot out-weigh the true location of the ball anymore. Thus, after the first integration of observations from all players, subsequent readings from Player 2 are filtered out and the Kalman filter tracks the ball based on observations from Player 1 and 3 only.

3.2 Real Game Scenarios

Another example is shown in Fig. 4(a). This situation has been recorded in the final game against the CoPS Stuttgart team during the German domestic *VisionCup* competition held in October 1999.

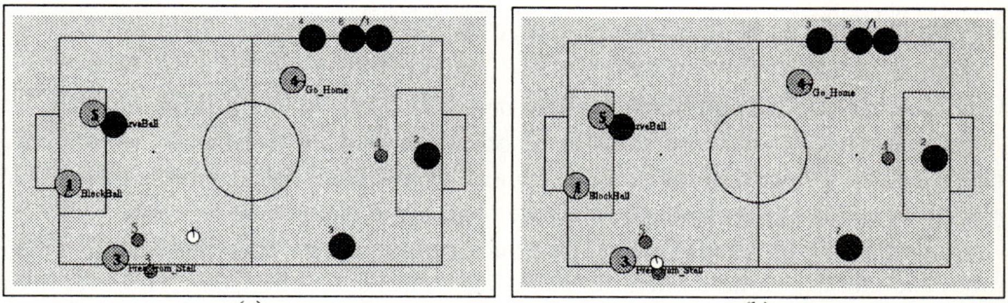

(a) (b)

Fig. 4. Ball tracking using simple averaging (a) and combination of Markov localization and Kalman filtering.

Depicted is a situation where the ball is at the wall close to Player 3, Player 3 and 5 observe the ball approximately at the right location but Player 4 thinks the ball is right in front of the opponent goal. The simple averaging method computes a global ball position somewhere between all these measurements which is shown as a white circle in Fig. 4(a). It is obvious that this global ball position is unusable and – if taken seriously by Player 3 – could lead to a disadvantageous situation where, after Player 3 moving towards the global ball position, the opponent player close to Player 5 could take over ball control.

Using our single-object tracking method, the measurements from Player 5 are reliably filtered out and the global ball position actually represents the true location of the ball (see Fig. 4(b)). Taking this information into account, Player 3 can safely approach the ball and clear the situation by pushing the ball towards the opponent's half.

In a further investigation, we examined data from the RoboCup 2000 competition held in Melbourne, Australia. During this competition, our robots recorded a total of about 120,000 ball observations in 10 games with a total playing time of more than 2 hours[2]. It turns out that in only about 0.7% of all ball observations (about 50 secs) our Markov localization approach detected an outlier. This low number can be explained by the fact that false positive measurements are rare and simultaneous observations by 2 or more robots do not occur all the time. If we compare our dual approach to one where only a Kalman filter integrates all measurements then in 72% of the filtered cases the ball position changes by more than 30 cm. Of course it is hard to tell which method is closer to the real world as there is no ground truth information from the games.

A last experiment was carried out to find out about the accuracy of our approach compared to the simple averaging one. In our laboratory, the ball was fixed at the center

[2] See http::/www.informatik.uni-freiburg.de/~robocup for watching log files of CS Freiburg's competition games in a Java applet.

of the field (the origin of our global world model) with three robots around observing it. From 2000 measurements the simple averaging method reported a mean of (18 cm, 0 cm) with standard deviation (47 cm, 19 cm) whereas our method delivered a mean of (-2 cm, 4 cm) with standard deviation (8 cm, 9 cm). Thus, our new method is significantly more accurate than the averaging one.

4 Related Work

Our work is related to two research areas: object tracking and multi-robot systems.

Kluge *et al.* [7] track multiple moving objects around a wheelchair by employing a maximum-flow algorithm for data association. We use a similar geometric method but also probabilistically integrate the measurements using motion and sensor models.

Schulz *et al.* [11] use JPDAFs [2] for tracking multiple moving targets around their robot and employ particle filters for achieving robust state estimation. However, they report a running time of 2 scans per second which is infeasible on our system where each robot sends all of its observations every 100ms.

Multi-robot systems gained significant attraction in recent years. For all different aspects of mobile robot navigation, multi-robot solutions have been developed that make use of the exchange of information about the robots' beliefs and their intentions.

Probably, the most related work to ours is that of other teams in the RoboCup middle size league. The CMU Hammerheads RoboCup team also uses Kalman filters for object state estimation and reports promising results on the accuracy of this method [12]. However, they do not apply a motion model to the states, thus observations have to be taken at the same time. Furthermore, they do not consider data association and use validation gates for detecting outliers. We use a more sophisticated outlier detection method which is superior to simple validation gates.

A similar probabilistic approach to ours is that of the Agilo team [10]. They use an iterative optimization technique for estimating object positions [6] and employ JPDAFs for data association [10]. However, they do not deal with outliers as we do.

5 Conclusion

We developed new methods for tracking multiple and single objects from noisy and unreliable sensor data and compared them to a simple averaging method. Experiments show that the methods are more robust and accurate than the simple averaging one due to better probabilistic modeling of motion and sensing of objects, and a dual sensor integration approach involving Markov localization and Kalman filtering.

Our approach presented in this paper is very similar to a decentralized filter with feedback to local filters [1, pp. 371-377]. As decentralized filters can be sub-optimal, it would be interesting to compare our results to a central filter that reads all sensor data from our robots.[3] This is part of future work.

[3] We did not consider this approach due to the amount of data that would have to be transmitted through wireless communication during a game.

References

[1] Robert Grover Brown and Patrick Y.C. Hwang. *Introduction to Random Signals and Applied Kalman Filtering*. Wiley & Sons, 1996.

[2] I.J. Cox. A review of statistical data association techniques for motion correspondence. *Int. Journal of Computer Vision*, 10(1):53–66, 1993.

[3] Jens-Steffen Gutmann, Wolfram Burgard, Dieter Fox, and Kurt Konolige. An experimental comparison of localization methods. In *Int. Conf. on Intelligent Robots and Systems*, 1998.

[4] Jens-Steffen Gutmann, Wolfgang Hatzack, Immanuel Herrmann, Bernhard Nebel, Frank Rittinger, Augustinus Topor, and Thilo Weigel. The CS Freiburg team: Playing robotic soccer on an explicit world model. *AI Magazine*, 21(1):37–46, 2000.

[5] Jens-Steffen Gutmann, Thilo Weigel, and Bernhard Nebel. Fast, accurate, and robust self-localization in polygonal environments. In *Proc. International Conference on Intelligent Robots and Systems (IROS'99)*, Kyongju, October 1999.

[6] Robert Hanek and Thorsten Schmitt. Vision-based localization and data fusion in a system of cooperating mobile robots. In *Int. Conf. on Intelligent Robots and Systems*, 2000.

[7] B. Kluge, C. Köhler, and E. Prassler. Fast and robust tracking of multiple moving objects with a laser range finder. In *Int. Conf. on Robotics and Automation*, Seoul, Korea, 2001.

[8] Peter S. Maybeck. The Kalman filter: An introduction to concepts. In I.J. Cox and G.T. Wilfong, editors, *Autonomous Robot Vehicles*, pages 194–204. Springer-Verlag, 1990.

[9] H.P. Moravec and A. Elfes. High resolution maps from wide-angle sonar. In *Proc. International Conference on Robotics and Automation (ICRA'85)*, March 1985.

[10] Thorsten Schmitt, Sebastian Buck, and Michael Beetz. Agilo RoboCuppers 2001: Utility- and plan-based action selection based on probabilistically estimated game situations. In *RoboCup 2001: Robot Soccer World Cup V*. Springer-Verlag, to appear.

[11] D. Schulz, W. Burgard, D. Fox, and A.B. Cremers. Tracking multiple moving targets with a mobile robot using particle filters and statistical data association. In *Int. Conf. on Robotics and Automation (ICRA'01)*, pages 1665–1670, Seoul, Korea, 2001.

[12] A.W. Stroupe, M.C. Matrin, and T. Balch. Distributed sensor fusion for object position estimation by multi-robot systems. In *Int. Conf. on Robotics and Automation (ICRA'01)*, pages 1092–1098, Seoul, Korea, 2001.

[13] M. Veloso, M. Bowling, S. Achim, K. Han, and P. Stone. The CMUnited-98 champion small-robot team. In M. Asada and H. Kitano, editors, *RoboCup-98: Robot Soccer World Cup II*, Lecture Notes in Artificial Intelligence. Springer-Verlag, 1999.

[14] Thilo Weigel, Willi Auerbach, Markus Dietl, Burhard Dümler, Jens-Steffen Gutmann, Kornel Marko, Klaus Müller, Bernhard Nebel, Boris Szerbakowski, and Maximiliam Thiel. CS Freiburg: Doing the right thing in a group. In *RoboCup 2000: Robot Soccer World Cup IV*. Springer-Verlag, 2001.

Multi-sensor Navigation for Soccer Robots

Carlos F. Marques and Pedro U. Lima

Instituto de Sistemas e Robótica
Instituto Superior Técnico, Av. Rovisco Pais, 1 — 1049-001 Lisboa, PORTUGAL
{cmarques,pal}@isr.ist.utl.pt
http://socrob.isr.ist.utl.pt/

Abstract. *This work introduces a method for robot navigation in structured indoors environments, based on the information of multiple sensors. Guidance control is based on odometry, reset at some time instants by a vision-based self-localization algorithm introduced in previous work. Sonar data is used to avoid and go around obstacles. Results from the application of the complete navigation system to a real robot moving on a RoboCup soccer field are presented.*

1 Introduction

Navigation is the robot subsystem that allows the robot to determine its current posture ξ_i=(x_i, y_i, θ_i) and move autonomously from an initial posture ξ_i to another target posture ξ_f=(x_f, y_f, θ_f), without colliding with obstacles. Figure 1.a) shows a block diagram of a Navigation system. The robot vehicle can be characterized by its Dynamics and Kinematics. The wheel angular velocities are controlled in closed loop, by the closed loop actuator controllers. The Guidance controller adjusts the vehicle trajectory, by generating references for the closed loop actuator controllers, so as to take the robot from the initial posture to the target posture, avoiding the obstacles in between. The Self-localization algorithm plus the odometry system estimate the robot posture.

The ability to navigate at relatively high speeds through an environment cluttered with static and dynamic obstacles is a crucial issue for a mobile robot. Most robotic tasks require a robot to move to target postures adequate to carry out its planned activities. In robotic soccer, this includes facing the opponent goal with the ball in between or covering the team goal by positioning itself between the ball and the goal, while avoiding the field walls and the other (stopped and moving) robots.

The most usual approach to mobile robot navigation consists of planning a path between the current and the target postures [1,10]. Path planning is typically based on a priori or sensor-based full knowledge of the surrounding environment. Furthermore, due to the robot kinematic constraints, the planned path must be converted into feasible motion. The whole procedure is thus time-consuming, especially when frequent re-planning is needed to handle dynamic environments [1]. Therefore, this is not the most convenient method for high speed motion within environments cluttered with dynamic obstacles. An elegant

A. Birk, S. Coradeschi, and S. Tadokoro (Eds.): RoboCup 2001, LNAI 2377, pp. 144–153, 2002.

solution consists of formulating the guidance problem for non-holonomic vehicles as one of reaching a desired final posture through time-varying non-linear state feedback [9]. This method allows on-line changes to the desired final posture, and ensures stabilization in the final posture, but the incorporation of (even static) obstacles is not considered. It may also lead to large and/or oscillating trajectories. The Potential Field Method [4,7,8] attempts to handle these problems by adding a repulsive acceleration originated by sensed obstacles and an attractive acceleration vector that pulls the robot towards the target posture. Unfortunately, in its original formulation, this method has some shortcomings, such as the local minima problem, when the robot gets inside an U-shaped obstacle, oscillations in the presence of obstacles, no passage between closely spaced obstacles and oscillations in narrow passages [2]. An alternative is the Vector Field Histogram (VFH) [3], method, where the robot finds the open spaces between the obstacles and chooses the heading closest to the target as the direction for safe motion. Minguez and Montano developed the Nearness Diagram Navigation (NDN) method [6] where the VFH is used, but in this case two polar diagrams are introduced: the Nearness Diagram from the Central Point (PND), to find the safe passages in between the obstacles, called valleys, and the Nearness Diagram from the Robot (RND), used to verify the robot safety conditions. The work described in [2,6] is applied to holonomic robots only.

In this paper we concentrate on guidance control and introduce a guidance control method for non-holonomic (differential drive) vehicles, based on odometry reset by a vision-based self-localization algorithm described in a previous paper [5], endowed with sonar-based obstacle avoidance. The guidance controller is used in the field robots of the RoboCup middle-size league ISocRob team, fully integrated in the state machine that coordinates task execution. The odometry is reset at specific states, such as at restart time or before returning to home position. The algorithm can be generally applied to structured indoors environments, provided that visual features can be observed by the self-localization method [5].

2 The Freezone Method

Borenstein et al.[3], introduced the concept of Vector Field Histogram (VFH), based on a two-stage data reduction technique, with three levels of data representation. The first level is the description of the robot environment, where a two-dimensional Cartesian histogram grid is continuously updated. In the second level the Cartesian histogram is mapped onto an one-dimensional Polar Histogram that describes the density of obstacles around the robot. In the last level, the method determines the open spaces among the obstacles (valleys), and uses them as candidates to possible robot trajectories. The method selects the valley closer to the direction to the target.

The Nearness Diagram Navigation, introduced by Minguez and Montano[6], can be described as follows. First, two nearness diagrams are introduced: the Nearness Diagram from the Central Point (PND), used to find the valleys and one

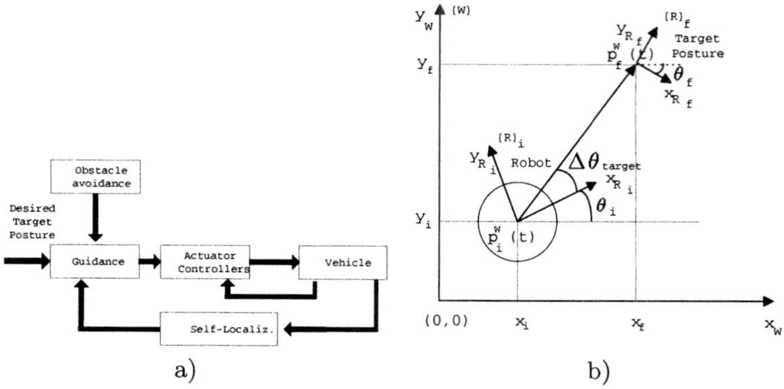

Fig. 1. a) Navigation System. b) Coordinate Systems .

of the regions is selected for the robot to pass, and the Nearness Diagram from the Robot (RND), used to evaluate the robot safety situation. After selecting the free region, and analyzing the safety situation of the robot, the algorithm chooses one navigation strategy based on 5 different heuristics that cover all safety cases. The navigation heuristic chooses the steering angle for the safe path of the robot and, based on this angle, determines the rotational and translational velocities for the robot.

We have developed a guidance control method, named Freezone, that merges the VFH and NDN algorithms, adapting them for non-holonomic (differential drive) robots. Some modifications were introduced, namely the use of a spatial mask that allows finding the valley candidates, and the introduction of a heuristic that allows to avoid oscillations when a small obstacle is in between the robot and the target position.

The coordinate systems used in this work are presented in Fig. 1.b). In the figure, $p_i^W(t) = (x_i^W(t), y_i^W(t))$, is the position of the robot center in the world frame $\{W\}$ at each time instant, $\theta_i(t)$ is its heading in the same frame and $\Delta\theta_{target}$ is the angular error of the robot heading at the initial posture with respect to (w.r.t) the heading towards the target frame origin $p_f^W(t)$:

$$\Delta\theta_{target} = \arctan\left(\frac{y_f^W - y_i^W}{x_f^W - x_i^W}\right) - \theta_i. \tag{1}$$

Here and henceforth, we will use, to simplify the notation, $p(t) = p_i^W(t)$. The Freezone algorithm steps are:

1. At each time t create a polar diagram centered with the origin $p(t)$ of $\{R\}$, using the information of the distance to obstacles obtained from the sensors.
2. Find in this diagram all the possible candidate angles that will let the robot move safely. The candidates are the "valleys" of the diagram that allow the safe passage of the robot without bumping into obstacles. To find "valleys"

in the diagram a mask was created. This mask represents an angular interval with a minimum distance to the obstacles that allow the robot to pass between them.

3. Choose, among the candidates, the one that minimizes the initial relative heading $\Delta\theta_{target}$ w.r.t. the line connecting p_i^W to p_f^W subtarget, should the robot head towards that valley.
4. The safety condition of the robot is analyzed, and described as Low Safety or High Safety. Low Safety occurs if there is an obstacle inside a circular area, centered on the robot, with radius d_{max}. High Safety occurs otherwise.
5. During the robot motion, the desired rotation $\Delta\theta(t)$ suggests the reference for the rotational velocity to be:

$$\omega(t) = \begin{cases} -\omega_{max} & \Delta\theta(t) < -\frac{\pi}{2} \\ \omega_{max}\frac{\Delta\theta(t)}{\pi/2} & -\frac{\pi}{2} \leq \Delta\theta(t) \leq \frac{\pi}{2} \\ \omega_{max} & \Delta\theta(t) > \frac{\pi}{2} \end{cases} \quad (2)$$

where ω_{max} is the maximum rotational speed of the robot;

6. The safety condition of the robot is used to choose between two control laws for the translational velocity of the robot:

$$v(t) = \begin{cases} v_{max}\left(\frac{d_{obs}(t)}{d_{max}}\right)\left(1 - \left|\frac{\Delta\theta(t)}{\pi/2}\right|\right) & \text{If Low Safety} \\ \\ v_{max}\left(1 - \left|\frac{\Delta\theta(t)}{\pi/2}\right|\right) & \text{If High Safety} \end{cases} \quad (3)$$

where:

v_{max} is the maximum speed of the robot;
d_{obs} is the distance to the closest obstacle;
$\Delta\theta$ is the desired rotation angle of the robot.

The first equation has two terms that decrease the velocity in the presence of an obstacle (linear velocity decreases with the distance to the obstacle) and when turning. A saturation was imposed to the rotation of the robot, $\Delta\theta \in [-\pi/2, \pi/2]$. At the saturation limits, the robot must turn with maximum angular velocity, and no translational velocity, thus leading to quick turns towards the desired heading.

3 Freezone Applied to Soccer Robots

This algorithm was implemented in all the Nomadic SuperScout II robots of the ISocRob team. The sensors used for obstacle avoidance were the robot sonars, while odometry was used for guidance, reset after the occurrence of some critical situations (e.g., after bumping) by the vision-based self-localization method described in [5]. The implementation of the Freezone algorithm in the robots was as follows.

Model Identification All the model parameters were determined, from the safety region limits to the maximum rotational and translational speeds, as well as the maximum robot acceleration, and the conditions that ensure a safe path. The resolution of the sectors is 22.5 degrees centered with each sonar, being the geometric configuration of the sonars used as sectors $S = \{s_1, s_2, ..., s_{17}\}$. For example s_9 is the sector of the sonar located at the front of the robot. We also denote S_{ri} as the reading of the sector (sonar) i.

One of the restrictions considered in the sector configuration used was the robot geometry. The robot is a cylinder with 20.7 cm of radius and a kicker device 14 cm long (l_k) and 24 cm wide (w_k). The robot must be allowed to rotate near a wall. Another restriction was that the robot must keep its distance to a wall without steering. This restriction is very important to solve one of the problems of the Potential Fields method. Without this restriction, the robot, while following a wall, starts to oscillate. Imposing the restriction it is possible to maintain the velocity of the robot when near to a wall or obstacles, even when the angle between the target and the robot is more than 90 degrees.

For the first restriction the maximum distance a from the center of the robot to the outmost point of the robot was measured, and this distance was considered to be the minimum distance (d_{min}) from a wall, so that the robot is allowed to rotate near a wall. For the robots used the distance was $d_{min} = 35cm$.

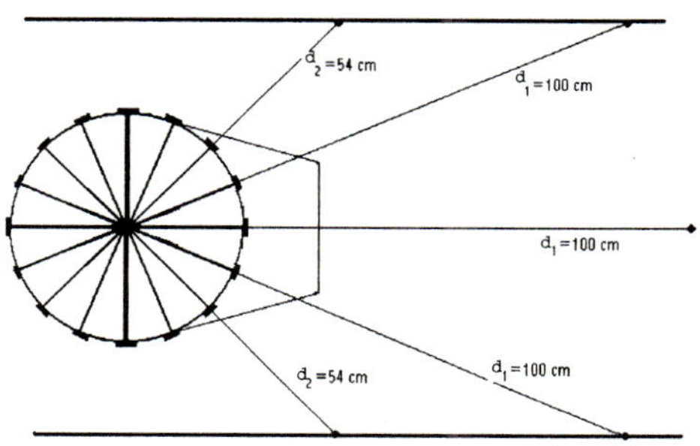

Fig. 2. Expected sonar measurements when traveling near a wall.

The expected sonar measurements when traveling near a wall were computed. These are shown in Fig. 2.

To compute the safety regions, some parameters must be determined first: the maximum acceleration, velocity and sample time of one sonar cycle. From the manufacturer's data, the robot maximum acceleration is a_{max} =0.8 m/s^2, the maximum velocity used is v_{max} =0.6 m/s and the sonar sample time is $T_{sample} = 0.032$ s.

At v_{max}, the robot needs to break until it stops, within a distance $d_{vmaxstop}$ given by:

$$d_{vmaxstop} = \frac{v_{max}^2}{2a_{max}}. \qquad (4)$$

The distance $d_{samplesonars}$ traversed by robot with velocity v_{max} during one sonar cycle is given by

$$d_{samplesonars} = v_{max} \times T_{sample}. \tag{5}$$

The Minimum Safety Distance $d_{Safemin}$ when the robot moves at v_{max} is thus given by

$$d_{Safemin} = d_{min} + d_{vmaxstop} + d_{samplesonars}. \tag{6}$$

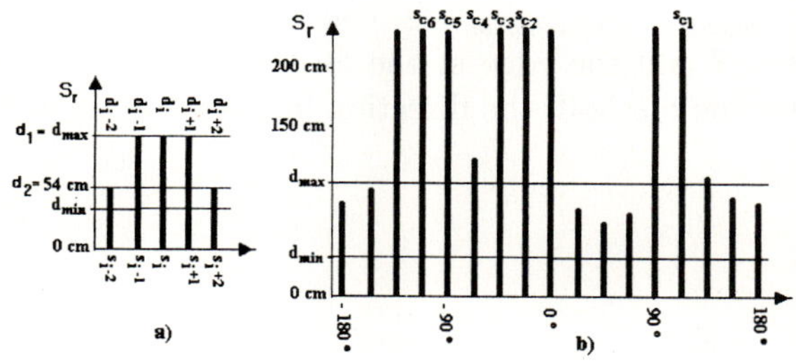

Fig. 3. a) Mask with the geometric configuration of the sectors; b) Typical Polar Histogram, showing the set S_c of sector candidates.

In this implementation, d_{min}=35 cm, $d_{vmaxstop}$=22.5 cm , $d_{samplesonars}$=2 cm hence $d_{Safemin}$=59.5 cm.

The $d_{Safemin}$ was used to evaluate the geometric sector configuration of the sectors used and to find the Maximum Safe Distance d_{max} (which must be greater than or equal to $d_{Safemin}$) inside which the obstacles are considered. In this implementation d_{max}=100 cm. In Fig. 2, d_1 and d_2 are the calculated distances from the robot center to the walls, read by the sonars s_{i-2}, s_{i-1}, s_i, s_{i+1} and s_{i+2}.

After identifying the model, the algorithm proceeds as explained in the following subsections.

Polar Diagram A polar diagram is created with the information given from the robot sonar sensors, as described in step 1 of section 2. In this implementation, the polar diagram is composed of the information given by the sonars only, with an accuracy of 22,5°, from -180° up to 180°, where 0° is the angle corresponding to the velocity axis of the robot. Figure 3.b) shows a typical robot polar histogram.

Find the Possible Candidates For model identification purposes, the configuration used for the sectors was computed (see Fig. 2), and a mask with those sectors was created, as shown in Fig. 3.a).

This mask was used to determine in which directions of S the robot can travel without endangering itself, as described in step 2 of Section 2. The algorithm uses the mask in the polar diagram, to find all the safe directions for the robot passage (step 2 of Section 2). The safety directions are collected into a vector $S_c = (s_{c1}, .., s_{cN}) \subseteq S$, where s_{ci} is a sector given by the mask. The pseudo code to use this mask is shown in Table 1.

Notice that Minguez et al.[6] and Borenstein[3], collect the valley extreme sectors (the right and left sectors). The near extreme is found and a new far extreme computed as the sum of the near border plus S_{max} (the angular interval that allows the passage of the robot), finding the desired steering direction $\Delta\theta$ from $\Delta\theta = (K_{nearborder} + K_{farborder})/2$, where $K_{farborder}$ and $K_{nearborder}$ are the angles between θ and the edge of the farthest and nearest borders of the valley. In the Freezone method, the direction is given by one of the orientation candidates.

```
for (i=2; i<15; i++)
{
  if ((s_r(i-2) ≥ d_i-2) and (s_r(i-1) ≥ d_i-1) and (s_r(i) ≥ d_i) and
   (s_r(i+1) ≥ d_i+1) and (s_r(i+2) ≥ d_i+2)) then
   {
    The direction s_i is a Safe Direction in S_c ;
   }
}
```

Table 1. Mask Pseudo code.

Choose the Orientation The orientation of safe travel selected is the sector from S_c that minimizes the direction to the target, with two exceptions that are presented at the end of this section, under the Special Case Conditions subsection.

The required rotation is designated as $\Delta\theta$ and can be expressed as:

$$\Delta\theta = \arg \min_{s_{ci} \in S_c} \{|s_{ci} - \Delta\theta_{target}|, i = 1, 2, ..., N\}. \tag{7}$$

where $\Delta\theta_{target}$ is computed by (1). In this case, f refers to the goal target. This rotation provides a reference for the rotational velocity (see (2)).

Safety Conditions The safety condition (step 4 of Section 2) of the robot is analyzed. The robot is considered in High Safety if no obstacle is inside d_{max}=100 cm around the robot, and Low Safety if one or more obstacles are inside the safety region.

Translation Velocity As referred in step 6 of Section 2, the safety condition of the robot is used to choose between two alternative control laws of Section 2 for the translational velocity of the robot.

Special Case Conditions There are some cases where the algorithm leads to oscillations. Two of the most common are the following:

When the robot has a wall on its way to the goal posture and the wall is perpendicular to the trajectory of the robot to the goal. In this case the robot will rotate to one of the sides, and the angle between the goal and the direction of traveling will become > 90 degrees. At some point the robot will stop and rotate back 180 degrees, returning to the same point where it first arrived and the process will be repeated.

When a narrow obstacle is in between the target and the robot. In this case the robot will try to turn towards one of the obstacle sides. However, while turning, the angle to the target will increase and the the robot will eventually choose to go to the other side of the obstacle. This process will create an oscillatory motion that will make the robot collide with the obstacle.

To solve these two problems a new condition was introduced. If there is an obstacle in the direction of the target, in a distance less than d_{minobs}, the safe travel orientation selected is the one that minimizes the robot steering, aligning the robot with the direction $\Delta\theta(t-1)$, corresponding to the last reference for the steering angle.

$$\Delta\theta(t) = \arg\min_{s_{ci} \in S_c} \{|s_{ci} - \Delta\theta(t-1)|, i = 1, 2, ..., N\}. \quad (8)$$

4 Experimental Tests

The tests presented with the Freezone method illustrate the resolution of the Potential Fields algorithm problems. The path shown was obtained from the robot odometry and the initial position of each run starts by determining the robot posture using the self-localization method. Notice that robot posture was determined with obstacles surrounding him. The robot's maximum speed was set to 0.6 m/s, $d_{max} = 100$ cm and $d_{min} = 35$ cm .

4.1 Test of Trap Situations Due to Local Minima

In these tests, some obstacles were placed inside the field (the four rectangles in the figures are boxes and the circle is the representation of a robot), and form a V-shaped obstacle between the robot and the goal target. This test allows to find the distance d_{minobs} that must be used to determine if there is an obstacle between the robot and the direct path to the target posture. This distance can be used to solve the local minima of a U or V-shaped obstacle.

When using a minimum distance of only d_{minobs} =10 cm from the robot to an obstacle, the robot oscillates inside the V-shaped obstacle. In the test depicted in Fig. 4.a), the robot made a smooth trajectory from inside the V obstacle to the target posture, using d_{minobs} =40 cm.

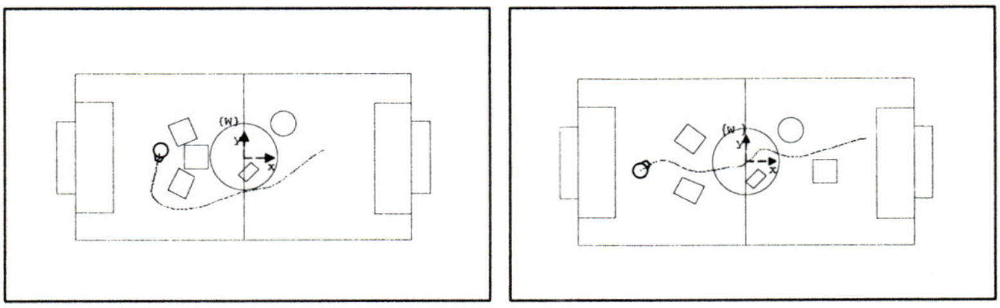

a) b)

Fig. 4. a) Test of trap situations using a distance of d_{minobs} =40 cm. b)Smooth passage between closely spaced obstacles. $\xi_i = (-3, 0.5, 45°)$ and $\xi_f = (3, 1, 0°)$.

4.2 The Passage between Closely Spaced Obstacles

In these tests some of the obstacles were placed near each other and the robot had to choose either to pass through or to go around the obstacles. Notice that the two boxes in the left of the Figure 4.b), are 70 cm apart (closer point), while the robot and the box of the right field are separated by only 60 cm (closer point). In the following tests, the initial robot posture was changed in order to analyze different cases. It is possible to notice that the robot made a smooth path between the obstacles.

4.3 Oscillations in the Presence of Obstacles

The robot always tries to keep a certain distance from the obstacle, as seen in previous tests, and, because it is always looking for valleys, the path trajectory will be smooth, except if an obstacle is placed in front of the robot, in which case it turns towards one of the directions. However, in the presence of an obstacle and increasing the steering angle, the translational velocity is reduced to almost zero, making the robot stop to decide where to go.

5 Conclusions

A novel method for guidance with obstacle avoidance was implemented in the team robots : the Freezone. The Freezone algorithm was chosen to be used in the ISocRob robots state-machine, due to the superior results obtained when compared to the Potential Fields methods, namely higher velocity and smoother trajectories while avoiding the obstacles in between the initial posture and the target posture. This method was used to place the robots in different postures, such as facing the goal with the ball in between or going to its home position, avoiding obstacles in between. Future work to improve the guidance with obstacle avoidance of the robots include: introduction of a certainty grid or histogram grid to include more information or uncertainty of the sonar data, adaptation of the Freezone method for situations when the robot has the ball to allow control of the ball by controlling the path geometry and velocity, sensor fusion, using both the sonars and the catadioptric system to identify obstacles.

References

1. Baltes, J., and Hildreth, N., "Adaptive Path Planner for Highly Dynamic Environments", RoboCup-2000: Robot Soccer World Cup IV, Springer Verlag, pp. 76-85, Berlin, 2001
2. Borenstein, J., and Koren, Y., "Potential Fields Methods and Their Inherent Limitactions for Mobile Navigation", Proceedings of the IEEE International Conference on Robotics and Automation, Sacramento, California, pp. 1398-1404, April 7-12, 1991
3. Borenstein, J., "Real-time Obstacle Avoidance for Fast Mobile Robots", IEEE Transactions on Systems, Man and Cybernetics, Vol. 19, No. 5, pp. 1179-1187, Sept./Oct. 1989
4. Krogh, B. H., "A Generalized Potencial Field Approach to Obstacle Avoindance Control", Proceedings of International Robotics Research Conference, Bethlehem, Pennsylvania, August, 1984
5. Marques, C. and Lima, P., "A Localization Method for a soccer Robot using a Vision-Based Omni-directional sensor", RoboCup-2000: Robot Soccer World Cup IV, Springer Verlag, pp. 96-107, Berlin, 2001
6. Minguez, J. and Montano, L., "Nearness Diagram Navigation(ND): A New Real Time Collision Avoidance Approach.",Proceedings IEEE/IROS2000, Takamatsu, Japan, Oct.30-Nov.5, 2000
7. Nagasaka, Y., Murakami, K., Naruse, T., Takahashi, T., Mori, Y., "Potential Field Approach to Short Term Action Planning in RoboCup F180 League", RoboCup-2000: Robot Soccer World Cup IV, Springer Verlag, pp. 345-350, Berlin, 2001
8. Khatib, O., "Real-Time Obstacle Avoidance for Manipulators and Mobile Robots." 1985 IEEE International Conference on Robotics and Automation, St. Louis, Missouri, pp. 500-505, March 25-28, 1985
9. Canudas de Wit, C., Siciliano, B., Bastin, G. (Eds), "Theory of Robot Control", CCE Series, Kluwer, 1996
10. Minoru Asada, Tucker Balch, Raffaelo D'Andrea, Masahiro Fujita, Bernhard Hengst, Gerhald Kraetzschmar, Pedro Lima, Nuno Lau, Henrik Lund, Daniel Polani, Paul Scerri, Satoshi Tadakoro, Thilo Weigel, Gordon Wyeth, "RoboCup-2000: The Fourth Robotic Soccer World Championships", AI-Magazine, volume 22, NO.1, Spring 2001

Visual Attention Control by Sensor Space Segmentation for a Small Quadruped Robot Based on Information Criterion

Noriaki Mitsunaga and Minoru Asada

Dept. of Adaptive Machine Systems, Osaka University, Suita, Osaka, 565-0871, Japan

Abstract. Since the vision sensors bring a huge amount of data, visual attention is one of the most important issues for a mobile robot to accomplish a given task in complicated environments. This paper proposes a method of sensor space segmentation for visual attention control that enables efficient observation by taking the time for observation into account. The efficiency is considered from a viewpoint of not geometrical reconstruction but unique action selection based on information criterion regardless of localization uncertainty. The method is applied to four legged robot that tries to shoot a ball into the goal. To build a decision tree, a training data set is given by the designer, and a kind of off-line learning is performed on the given data set. The visual attention control in the method and the future work are discussed.

1 Introduction

Mobile robots often have visual sensors that bring a huge amount of data for decision making. Therefore, attention control which extracts necessary and sufficient information for the given task is demanded for efficient decision making. We have proposed a method of efficient observation for decision making [1], which assumed that the sensor values were quantized in advance and observation cost (the time needed for the camera head motion and image acquisition) did not vary. For more adaptive and efficient observation, self-segmentation of the sensor space for attention control is necessary.

In the reinforcement learning area, the number of states should be minimum because the learning time is exponential to the size of the state space [2]. Then, several sensor space segmentation methods for state space construction have been proposed (Takahashi et al.[3], Yairi et al.[4], Kamiharako et al.[5], and Noda et al.[6]). However, they did not consider the actual time for observation nor used an active vision system. Kamiharako et al.[5] showed some results with the coarse to fine attention control but they adopted the omni-directional vision system by which the robot capture the image whole around of itself.

In this paper, we propose a method of sensor space segmentation for visual attention control that enables efficient observation taking the time needed for observation into account. The efficiency is considered from a viewpoint of not

A. Birk, S. Coradeschi, and S. Tadokoro (Eds.): RoboCup 2001, LNAI 2377, pp. 154–163, 2002.

geometrical reconstruction but unique action selection based on information criterion regardless of localization uncertainty. The method is applied to four legged robot that tries to shoot a ball into the goal. To build a decision tree, a training data set is given by the designer, and a kind of off-line learning is performed on the given data set.

The rest of the paper is organized as follows. First, the method is introduced along with basic ideas related to information criterion, efficient observation, prediction model, and decision making. Then, the experimental results using RoboCup four-legged robot league platform (almost same as Sony AIBO) are shown. Finally, discussion on the visual attention control in the method is given and the future works are shown.

2 The Method

2.1 Assumptions

We assume, 1) the robot needs to pan and tilt its camera to acquire the necessary information for action selection, since the visual angle of the camera is limited. 2) The environment includes several landmarks, which provide the robot sufficient information to uniquely determine the action from their views. 3) Sufficient training data for decision making are given. We used a teaching method to collect such data. A training datum consists of a set of the appearance of the landmarks and the action to accomplish the task at the current position. During the training period, the robot pans its camera head from the left-most angle to the right most one, and observes as many landmarks as possible.

There are methods which construct a classifier in the form of decision tree with information gain, such as ID3 and C4.5 [7]. To construct a decision tree, we need a training data set. Each datum consists of a class which it belongs to and attribute values by which we classify it to the class. If we apply these methods, a class and an attribute correspond to an action and a sensor, respectively. When we use ID3 which only handles quantized attribute values, 1) we calculate each information gain I_i in terms of action after observing sensor i, 2) divide data set according to the sensor values with the largest information gain. We iterate these process until the information gains for all sensors become zero or the action in the divided data becomes unique. In an action decision tree, a node, an arc, and a leaf indicate the sensor to divide data set, the sensor value, and the action to take, respectively. C4.5 handles continuous attribute values by dividing data set with a threshold. The threshold to divide is determined so that the information gain can be the largest after the division. Due to the limited view angle of its camera, the robot needs to change its gazes in order to know whether a landmark is observed on the left side of the threshold or not. However, it needs only one gaze to know whether a landmark is observed inside a limited area (attention window) or not. Therefore we use an attention window which maximize the information gain for dividing the training set into two sub-sets.

2.2 Information Gain by Observation

Suppose we have r kinds of actions and n training data. First, calculate the occurrence probabilities of actions p_j ($j = 1, ..., r$) as $p_j = n_j/n$, where n_j denotes the number of taken action j. Therefore, the entropy H_0 for the action probability is given by

$$H_0 = -\sum_{j=1}^{r} p_j \log_2 p_j. \tag{1}$$

Next, calculate the occurrence probabilities of actions after observation. After the observation, it knows whether the landmark is inside the attention window $(\theta_{Lk}, \theta_{Uk}]$ or not. The lower and upper limits of a window θ_{Lk}, θ_{Uk} are a pair of middle points of adjacent sensor values of the training data. We denote the number of times action j was taken as n^I_{ijk} when the landmark i was observed in $(\theta_{Lk}, \theta_{Uk}]$ and n^O_{ijk} when not observed. Then, the occurrence probability becomes, $p^I_{ikj} = n^I_{ikj}/n^I_{ik}$, $p^O_{ikj} = n^O_{ikj}/n^O_{ik}$. Where $n^I_{ik} = \sum_j^r n^I_{ikj}$, and $n^O_{ik} = \sum_j^r n^O_{ikj}$. Next, calculate the entropy after the observation, as follows:

$$H_{ik} = - \sum_{x=\{I,O\}} \frac{n^x_{ik}}{n_{ik}} \sum_{j=1}^{r} (p^x_{ikj} \log_2 p^x_{ikj}). \tag{2}$$

The information gain by this observation is $I_{ik} = H_0 - H_{ik}$. The larger I_i is, the smaller the uncertainty is after the observation.

2.3 Actual Time for Observation

When the time for observation is constant, we can use information gain for making action decision tree. The tree becomes compact and the robot can determine its action at shortest observation time by following the tree. However, if the time for observations changes depending on the gaze directions, the time for action decision can be longer using the tree. Therefore, we use the information gain per time, in other words the velocity, rather than information gain.

We denote T as the time to get the observation after previous observation, and information gain per time i_{ik} as,

$$i_{ik} = \frac{I_{ik}}{T + a}. \tag{3}$$

Here a is a positive constant. When the direction is already observed $T = 0$.

2.4 Making an Action Decision Tree

We put the attention windows into the tree in decreasing order of uncertainty after its observation. Based on i_{ik} we divide training data into two subsets until the action in the subset becomes unique. For the training data which take different actions for the same situation, we add a leaf for each action and record the probability that it was taken.

For example, suppose we have training data as shown in the left of Table 1. The numbers in the table indicates the direction in which the landmark was observed. The view angle is limited and it can gaze and observe three areas $[0, 15)$, $[15, 30)$, $[30, 45)$. It gazes in $[15, 30]$ at the beginning of action decision, and needs one period of time to change the direction to observe. Since $p_x = 2/4$, $p_y = 1/4$, and $p_z = 1/4$, H_0 is 1.5. The information gain by observation I_{ik} and the information gain per time i_{ik} are shown in the right of Table 1. Since i_{ik} of the observation which checks whether the landmark A is in $[27, 30)$ or not is the largest, we put this to the root of the tree. If the landmark is in $[27, 30)$ the action is unique and it is y. Else, the subset has three training data and the actions are not unique. The information gain per time of observation whether landmark B is in $[0, 15)$ or not and observation whether landmark A is in $[30, 40)$ or not is 0.05. We prefer left $[0, 15)$ to observe and the action decision tree is shown in Fig.1.

Table 1. Example training data (left) and calculated information and information per time (right). Lm means landmark.

Data #	Lm A	Lm B	action
1	5	5	x
2	25	15	x
3	30	10	y
4	40	30	z

Observation	Info. I_{ik}	Info. / time i_{ik}
$0 \leq (LmA) < 15$	.31	.15
$15 \leq (LmA) < 27$	.31	.31
$15 \leq (LmA) < 30$	.50	.50
$27 \leq (LmA) < 30$	1.4	1.4
$30 \leq (LmA) < 45$	1.4	.70
$0 \leq (LmB) < 7$	.31	.15
$0 \leq (LmB) < 12$	.5	.25
$0 \leq (LmB) < 15$	1.4	.70
$7 \leq (LmB) < 12$	1.4	.70
$7 \leq (LmB) < 15$	.5	.25
$30 \leq (LmB) < 40$	1.4	.70

```
Observe [15, 30), if 27<=(Landmark A)<30
then
        take action y
else
        Observe [0, 15) and if 0<=(Landmark B)<15
        then
                take action x
        else
                take action z
```

Fig. 1. Action decision tree of the example data

2.5 Making a Decision

In order to make a decision on which action to take, first, the robot calculates the observation probability $\boldsymbol{x}(t)$ from previous $\boldsymbol{x}(t-1)$ and action $a(t-1)$ if possible. If no prediction model is applicable, use the probability 1 or 0 if the direction of the window has been observed, otherwise the probability is 0.5. Then it updates $\boldsymbol{x}(t)$ by the observation and calculate action probabilities. An action probability is the sum of the probability to reach leaves to take that action in the action decision tree. If one of the action probabilities is very high, it takes that action. Otherwise, until one of them becomes high enough, it continues to check attention windows from the root of the action decision tree, update the observation probability, and the action probabilities.

3 Experiments

3.1 Task and Environment

The task is to push a ball into a goal based on the visual information. We used a legged robot for the RoboCup SONY legged robot league (Fig.2). The robot is equipped with a limited view angle camera. In the field, there are 8 landmarks, that is, target goal (TG), own goal (OG), north west pole (NW), north east pole (NE), center west pole (CW), center east pole (CE), south west pole (SW), and south east pole (SE). All the landmarks and the ball are distinguished by their colors.

The view angle / number of image pixels of the robot's camera are about 53 degrees / 88 pixels in width, and about 41 degrees / 59 pixels in height. Each leg and the neck have three degrees of freedom. We fixed the joint angles of the legs and the role of the neck joint when it observes the landmarks and the ball to make its decision. The robot can rotate the pan joint from -88 to 88 degrees and the tilt joint from -80 to 43 degrees. We prepared five directions (every 44 degrees) in the pan joint and four directions (every 40 degrees) in the tilt joint to observe. The maximum angular velocity of the pan joint is $5.9[rad/s]$ and $3.9[rad/s]$ in the tilt joint. The robot waits at least for $0.16[s]$ after rotating pan or tilt joint. Since it needs at least 0.29[s] before action decision after changing observing directions, we prepared $a = 0.29[s]$.

As vision sensors, we used the coordinates of the image centers of the landmarks and the ball, the minimum x/y and maximum x/y (totally four) of the goals. We did not directly use the values in the images, but converted to the pan and tilt angles when the targets are viewed at center of an image. And, we used pair of the pan (x) and tilt (y) angles as a sensor value rather than one of them, or we divided training data-set by the observation to check whether a sensor value is in the rectangle of $(x_{\min}, y_{\min}) - (x_{\max}, y_{\max})$ (attention window). This is because if we divide the data-set by the observation to check whether the x is in $[10, 20)$ or not, we have to rotate the tilt angle to check whole y variation.

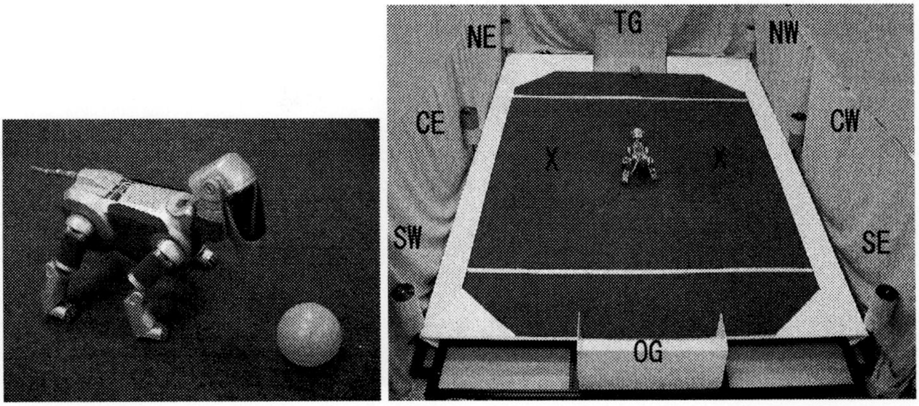

Fig. 2. The legged robot for RoboCup SONY legged robot league (left) and the field (right).

3.2 Experimental Results

We trained the robot starting from one of three positions in the middle of the field. We prepared seven actions, try to reach the ball, move forward, move left / right forward, turn left / right, and turn leftward / rightward. For each starting position, we trained five times and obtained 99 data points to construct trees. We show the part of the action decision tree in Fig.3. The unit of the figures are degrees. When the robot has no prediction and not observed yet, first attention window is rectangle of $(-19,-8)-(-8,18)$ for the ball and observe the direction $(3,2)$ (Fig.4(a)). Direction (X,Y) indicates the direction of observation. X is the panning direction $(1,...,5)$ and 3 if it is center. Y is the tilt direction $(1,...,4)$ and 2 if it is horizontal. Since the robot is facing $(3,2)$ at the beginning of action decision time, the direction $(3,2)$ is preferred by nodes near the root of the tree. If the center of the ball image is in the window, the next window is $(-19,-10)-(-18,-13)$ for minimum y of TG and the direction is same $(3,2)$ (Fig.4(b)). If it is in the window, try to reach the ball otherwise move forward. If the center of the ball image is not in the first window, the next windows to check is $(-19,-18)-(18,18)$ for minimum x of TG and the direction is $(3,2)$ (Fig.4(c)). If it is in the window check the next window $(-19,-10)-(4,4)$ (Fig.4(d)) and so on.

We show and compare the attention windows generated by three methods in Fig.5-7. Fig.5 shows the attention windows generated with pre-quantized sensor values in every 20 degrees without time consideration (we directly used information gain rather than velocity). Fig.6 shows the one generated by the proposed segmentation method without time consideration. Fig.7 shows the one generated with proposed segmentation method with time consideration. With these figures we can see that pre-quantized segmentation is not efficient for decision making

```
[ball] -19<x<11, -8<y<18 (Fig.5(a)) then
  [TG ymin] -19<x<-10, -18<y<-13 (Fig.5(b)) then
    Do TryReachBall
  else
    Do Forward
  [TG xmin] -19<x<18, -18<y<18 (Fig.5(c)) then
    [TG xmin] -19<x<4, -10<y<4 (Fig.5(d)) then
      Do Forward
    else
      [TG xmin] -19<x<11, -15<y<-3 then
        Do RightTurn
      [TG xmin] -6<x<18, 15<y<18 then
        [TG xmin] -12<x<11, -1<y<18 then
          [ball] 27<x<46, 26<y<32 then
            Do RightFoward
          else
            Do RightTurn
        else
          Do LeftRotate
          [TG xmin] 15<x<18, -8<y<4 then
            [CW] -52<x<-30, -15<y<5 then
              Do Forward
            else
              Do RightTurn
          else
            Do Forward
                    :
```

Fig. 3. Part of action decision tree generated by proposed method

and with time consideration the windows are preferred which bring information gain faster rather than the ones which bring higher information gain.

We show the comparison of quantization and consideration of time in Table 2. We compare the number of nodes (windows) in a tree (# of nodes), depth of the tree, the number of expected observing directions, the expected time for observation (time). The expected number of observing directions and time are the one when it does not use predictions. The proposed quantization shows smaller size of the tree and the number of expected observing directions. And it can reduce the time for observation to half by using information gain per time.

The mean time for observation without predictions in experiments with real robots were 6.6[s] with pre-quantization method, 5.2[s] with proposed quantization without time consideration and 2.5[s] with proposed method. Though the time is longer than expected, the time of proposed method is below half of other methods.

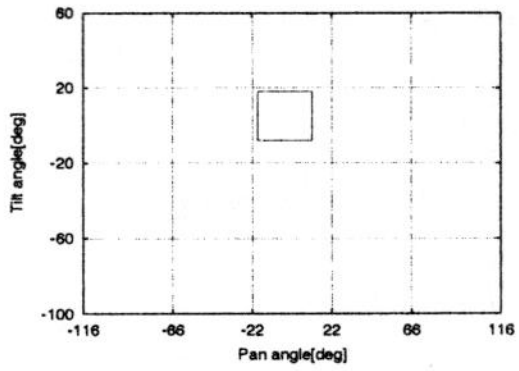

(a) Window for the ball

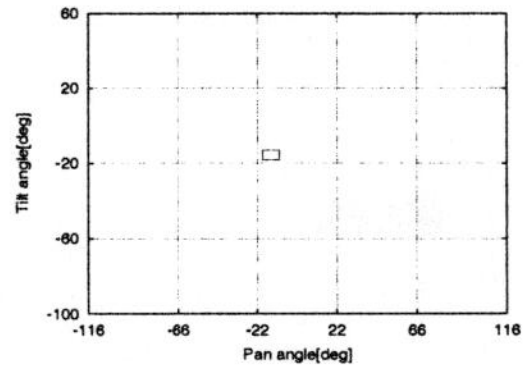

(b) Window for min. y of TG

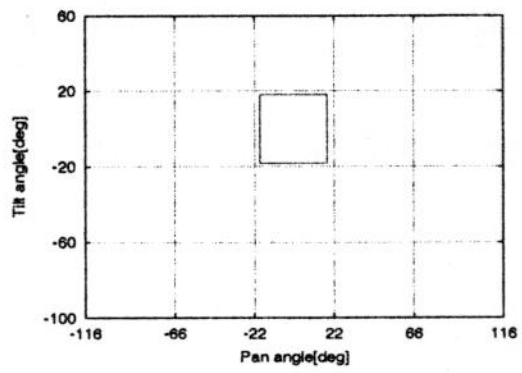

(c) Window for min. x of TG

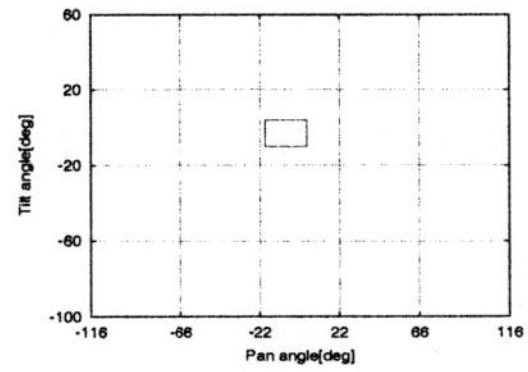

(d) Another window for min. x of TG

Fig. 4. Attention windows

Table 2. Comparison of sizes of the tree, expected number of observing directions, and expected time for observation

	# of nodes,	depth,	# of leaves,	directions,	time[s]
pre-quant.	61	18	31	7.6	2.8
quant. only	39	11	20	5.5	2.0
proposed	59	15	30	3.4	0.83

4 Discussions and Conclusions

We showed that a decision tree which is constructed with greedy for information gain or information gain per time. Efficient observation for decision making was achieved by greedy approach. However, decision making with tree constructed with greedy approach may prone to sensor noise, occlusions, and so on [4]. Occlusions are ignored if an action is determined without occluded windows, otherwise they may lead to wrong action selection. Currently, training data should contain the some variations with sensor values which cover noises or occlusions, so that the reliability is reflected to the information gain. The sufficiency of training data is measured with the actions determined by the action decision tree. If you can use some prediction model it will help to solve the problem of temporal occlusions.

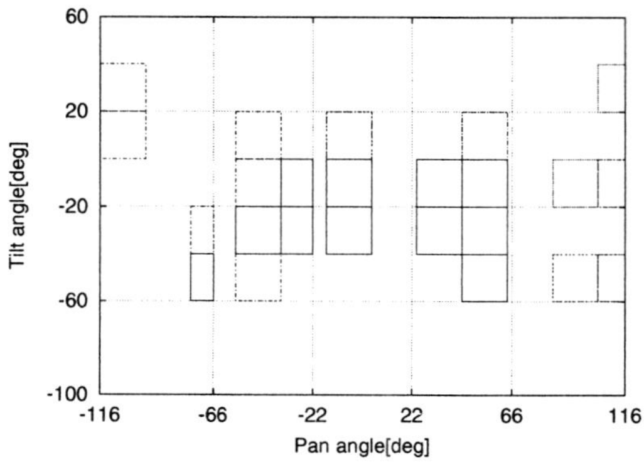

Fig. 5. Created attention windows by pre-quantization method without time consideration.

We showed only the results without prediction model. Since the data were not enough for simple least square method. A prediction model which can be consists from small data robustly is expected.

To conclude, we proposed a method to make a decision tree with an autonomous sensor value segmentation with consideration for variance in time interval to acquire observation. Attention control is done by observation following a decision tree which is constructed based on information criterion with sensor space segmentation. The validity of the method was shown with a four legged robot.

Acknowledgement

This research was supported by the Japan Science and Technology Corporation, in Research for the Core Research for the Evolutional Science and Technology Program (CREST) title Robot Brain Project in the research area "Creating a brain".

References

[1] Noriaki Mitsunaga and Minoru Asada. Observation strategy for decision making based on information criterion. In *Proceedings of the 2000 IEEE/RSJ International Conference on Intelligent Robots and Systems*, pages 1038–1043. 2000.

[2] S. D. Whitehead. A complexity analysis of cooperative mechanisms in reinforcement learning. In *Proceedings of AAAI-91*, pages 607–613, 1991.

[3] Yasutake Takahashi, Minoru Asada, and Koh Hosoda. Reasonable performance in less learning time by real robot based on incremental state space segmentation. In *Proceedings of the 1996 IEEE/RSJ International Conference on Intelligent Robots and Systems*, pages 1518–1524, 1996.

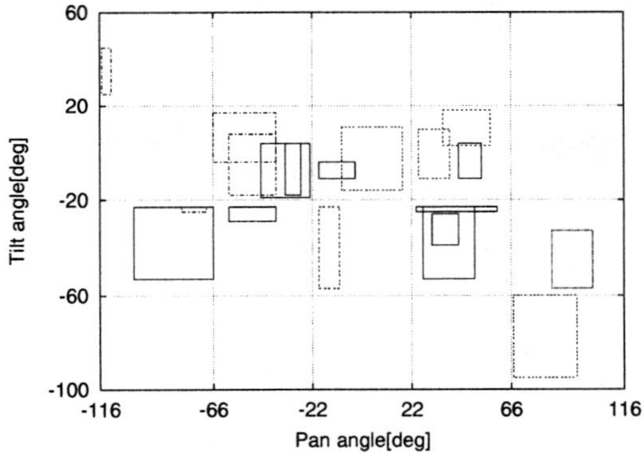

Fig. 6. Created attention windows by proposed segmentation without time consideration.

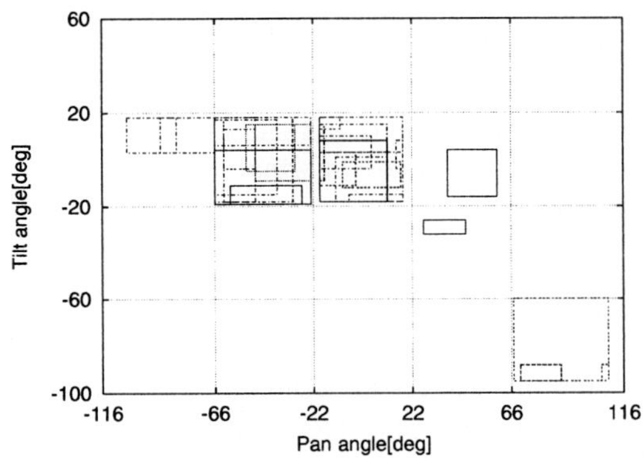

Fig. 7. Created attention windows by proposed method.

[4] Takehisa Yairi, Shinichi Nakasuka, and Koichi Hori. State abstraction from heterogeneous and redundant sensor information. In Y. Kakazu, M. Wada, and T. Sato, editors, *In Proc. of the Intelligent Autonomous Systems 5*, pages 234–241, 1998.

[5] Masatoshi KAMIHARAKO, Hiroshi ISHIGURO, and Toru ISHIDA. Attention control for state space construction. In Y. Kakazu, M. Wada, and T. Sato, editors, *In Proc. of the Intelligent Autonomous Systems 5*, pages 258–265, 1998.

[6] Minoru Asada, Shoichi Noda, and Koh Hosoda. Action based sensor space segmentation for soccer robot learning. *Applied Artificial Intelligence*, 12(2-3):149–164, 1998.

[7] J. Ross Quinlan. *C4.5: PROGRAMS FOR MACHINE LEARNING.* Morgan Kaufmann Publishers, 1993.

Language Design for Rescue Agents

Itsuki Noda[1], Tomoichi Takahashi[2], Shuji Morita[3], Tetsuhiko Koto[4], and Satoshi Tadokoro[3]

[1] AIST & PRESTO
[2] Chubu Univ.
[3] Kobe Univ.
[4] Univ. of Electro-Communications

Abstract. We are proposing a model of communication and a specification of language for civilian agents in RoboCup Rescue Simulation. Robust information systems are critical infrastructure for rescue in huge disasters. The main issue of the information system in the rescue domain is that the most of information sources are civilians who may report incomplete information. Proposed model and specification is designed to reflect natural features of human communication behaviors. The model is also designed in order to provide for researchers to implement new design of information devices in the simulation flexibly. Using the model and specification, we can evaluate information systems and behaviors of rescue specialist agents.

1 Introduction

Communication of agents in the rescue domain is an important factor to affect the performance of rescue activities. Especially communication among civilians, who are the majority of agents in damaged area, will be the primary information source for rescue activities.

We are designing agent simulator modules for the RoboCup rescue simulator. The main purpose of the design is to provide a standard of communication among agents, especially civilians.

In this article, we describe a brief overview of the rescue simulator system (Section 2), and propose a layered model of agent communication (Section 3) and specification of a communication language for civilian agents (Section 4).

2 RoboCup Rescue Simulation System

Here, we like to provide brief overview of the rescue simulator (version 1) that we are designing.

Similar to the version 0 simulator[TTK+00] and FUSS[NOD00], the system consists of the kernel and plug-in modules (Figure 1). The kernel provides functions to synchronize distributed expert modules and to manage shared data among modules. The plug-in modules are classified into three types: expert simulation modules, human-interface modules, and agent simulation/proxy modules.

A. Birk, S. Coradeschi, and S. Tadokoro (Eds.): RoboCup 2001, LNAI 2377, pp. 164–172, 2002.

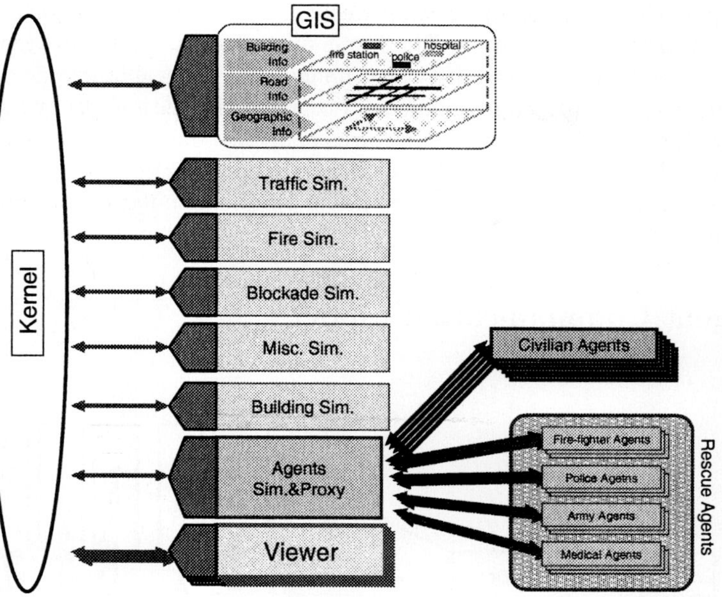

Fig. 1. Rescue Simulation System

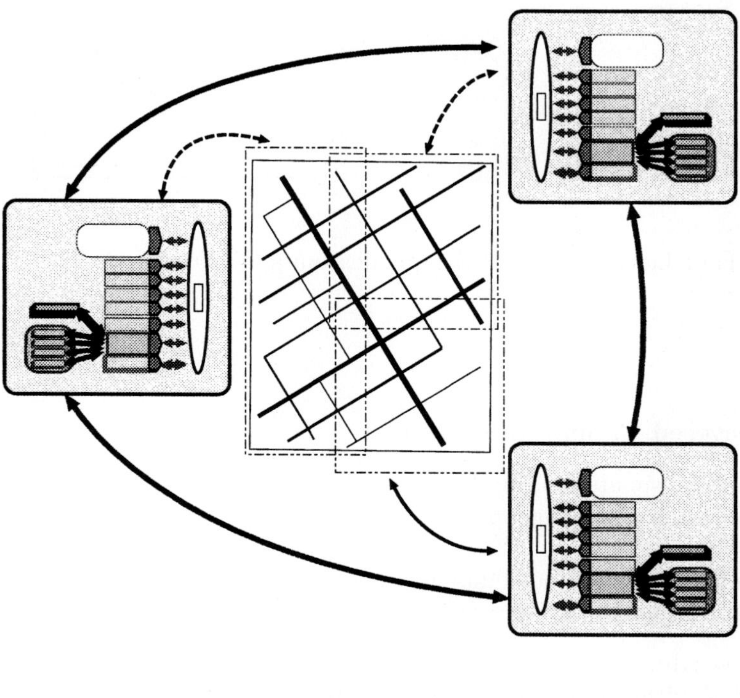

Fig. 2. Parallel Simulation with Overlapping

Each expert simulation module calculates changes of a phenomenon from an aspect like file, traffic, and so on. The human-interface modules provide a facility to generate real-time 2D and 3D visual image, and natural language description of the process of simulated phenomena. The agent simulation/proxy module performs two functions: (1) it simulates status of agents' bodies. (2) it communicates with agent-clients that become brains of the agents.

The system also has a facility to enable parallel simulation of divided areas as shown in Figure 2.

3 Agent Communication Model

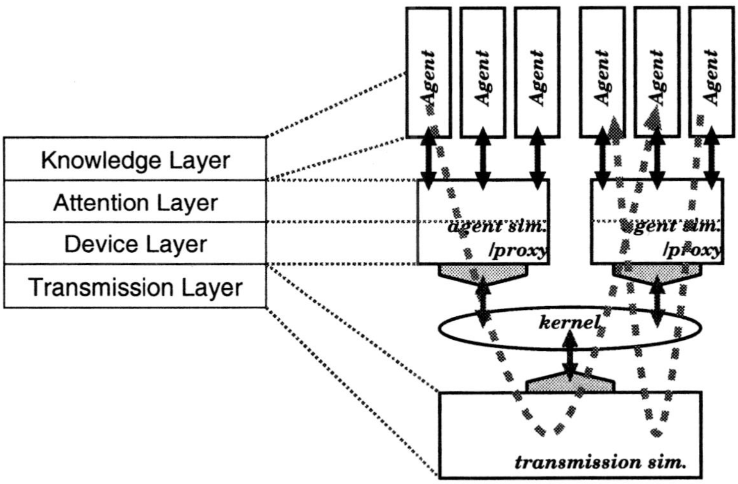

Fig. 3. Four Layers Model of Communication and Mapping to The Simulation System

3.1 Layered Commutation Model

In the rescue simulation, modeling agents' activities is a key issue. Especially, communication among agents is an important factor to decide the agents' physical performance.

In order to model the communication, we design *four layers model of communication.* The four layers are:

- **knowledge layer**
  to deal with syntax and semantics of communication. This layer is open for agent programmers.

- **attention layer**
  to simulate responsibility of agents' communication. For example, reading bulletin boards require agents' active attention. Even if information is on a bulletin board, an agent can not know the information without giving his/her attention to the bulletin board. On the other hand, a siren attracts attention of most of agents' even if they do not pay their attentions to the sound. This layer also controls the capacities of agents' communication.
- **device layer**
  to simulate availability and functionality of devices like telephone, PDA, PC, and so on.
- **transmission layer**
  to simulate transmission of physical signals like sound, electric wave, telephone line, and so on.

Figure 3 shows a mapping from the four layers to the simulation systems. The knowledge layer correspond to agent clients, in which original messages are generated and interpreted. The attention layer is in the agent simulator/proxy modules, which filter received information according to status of agents' attention and resources of agents' body. For example, when an agent is not paying attention to a bulletin board, it can not get information from the memo on the board. When an agent is listening a message, it can not listen another messages in the same time.

The device layer is also in the agent simulator/proxy modules, which filter information according to devices required in a specified communication method. For example, an agent need to have a mobile phone or find a wired phone to call a fire department office. The transmission layer corresponds to an expert simulator for the transmission. In the case of mobile phone, the simulator checks resource for the connection. When agents communicate by direct conversation, the simulator calculates the range the voice can reach.

The separation to the four layers helps for researchers to introduce a new method of communication into the simulation system. When researchers want to introduce their new PDA device designed for rescue into the simulator to test their facility for rescue domain, they should only develop their own device layer and combine it into the agent simulator/proxy modules. If they want to introduce new transmission method (like bluetooth) with PDA, it is possible by implementing both of device and transmission layers.

4 Agent Language for Civilian

Basically, specification of language used in the knowledge layer is open in the rescue simulation system. Agents can transfer information by any kind of formats with each other. However, it will be required to provides interoperability among heterogeneous types of agents that are developed separately. Therefore, we need a specification of the rescue agent language. For the first step, we are designing a language for civilian.

FIPA already specify languages for several applications. For example, FIPA KIF [FIP00c] and FIPA CCL [FIP00a, WCF+00]are designed respectively for interchange of knowledge and for information gathering. But no proposed specifications in FIPA fits for our purpose in the rescue simulation, because the language needs to reflect natural language features. Especially, civilian agents should behave as standard human civilians in the virtual world, so that civilian's conversation will include incomplete and ambiguous expression.

The natural language itself, on the other hand, is too wide and complex to be handled in the rescue simulation. There are many unsolved issues in the researches on natural language processing. Fortunately, we can limit variation of utterances for the rescue domain, where there are few complicated language phenomena.

Based on these consideration, we design a formal language.

4.1 Basic Syntax

Basically we use a similar syntax as S-expression of FIPA's communication language, that is:

```
(SpeachAct :sender Agent
           :receiver Agent
           :content Content
           ...)
```

Currently, the variations of *SpeachAct* are `inform`, `query-if`, `query-ref`, `request`, `commit`, `acknowledge`, and `not-understood`. An `inform` clause tells that the sender believe the *Content* is true. A `query-if` clause asks the receiver whether the *Content* is true or not. A `query-ref` clause asks the receiver to fill the information denoted by '`:wh`' in the *Content*. A `request` clause requests the receiver to do the *Content*. A `commit` clause tells the sender will do the *Content* in future. An `acknowledge` clause tells that the sender receives the *Content* (or a message if no contents are specified). In contrast to it, an `not-understood` clause informs that the sender can not understand the *Content*. We also permit to use `say` speech act for the general purpose, because sometimes it may be difficult to determine speech acts of uttered sentences. A `say` clause can have an optional `:acttype` slot to specify the speech act. For example, `(query-if ...)` is equivalent to `(say :acttype query-if ...)`.

4.2 Syntax for Content

In order to reflect flexibility of natural language, we use an S-expression of frame representation for the content. The syntax of the representation is

```
(FrameName :SlotName1 SlotValue1
           :SlotName2 SlotValue2
           ...)
```

```
          Clause ::= '(' ActType Role_Vale_Pair* ')'
         ActType ::= 'inform' | 'query-if' | 'query-ref' | 'request' | 'commit'
                     | 'acknowledge' | 'not-understood' | 'say'
 Role_Value_Pair ::= Role Value
            Role ::= Keyword
         Keyword ::= ':'RoleName        ; symbols that begin with ':'.
        RoleName ::= Symbol
           Value ::= Data
            Data ::= AtomData | Frame | SpecialForm
        AtomData ::= Symbol | Number
           Frame ::= '(' Tag Role_Value_Pair* ')'
     SpecialForm ::= Collection | LogicalForm
      Collection ::= Set | List
             Tag ::= Symbol        ; symbols that begin with ":" are reserved.
             Set ::= '(' ':set' Data* ')'        ; non-ordered list
            List ::= '(' ':list' Data* )         ; ordered list
     LogicalForm ::= Negation
        Negation ::= '(' ':not' Data )
```

Table 1. Definition of Civilian's Communication Language

SlotName should be a *Symbol*, and *SlotValue* should be an S-expression of a *Data*, where *Data* is one of *Frame*, *Symbol* or *SpecialForm*. For example, "building (id=300) is in fire" and "agent (id=100) carries agent (id=200) to the building (id=300)" are respectively represented as:

```
(is :subject (building :id 300) :state in-fire)
(do :action carry
    :subject (agent :id 100)
    :object (agent :id 200)
    :destination (building :id 300))
```

SpecialForm includes the following expressions:

```
(:set Data*)
(:list Data*)
(:not Data)
(:wh [Data])
```

`:set` and `:list` forms denote collections of *Data*. `:not` form denotes negation of *Data*. `:wh` is used in `query-ref` clause to specify which information is queried.

Table 1 shows whole definition of the civilian language.

4.3 Examples of Communication

Here, we show several typical examples of communication that civilian agents should be able to understand.

(1) Agent(ID=100) say to agent(ID=200) "person(ID=150) is buried at a building(ID=250)".

```
(inform :sender (agent :id 100)
        :receiver (agent :id 200)
        :content (is :subject (agent :id 150)
                     :state (buried)
                     :place (building :id 250)))
```

We also can denote "`:state buried`" instead of "`:state (buried)`". A symbol is equivalent to a frame its tag is the same symbol and has no slots.

(2) Agent(ID=100) say to agent(ID=200) "person(ID is unknown) is buried at place(100,200)".

```
(inform :sender (agent :id 100)
        :receiver (agent :id 200)
        :content (is :subject (agent)
                 :state (buried)
                 :place (position :x 100 :y 200)))
```

We can omit `:id` slot in the agent frame when it is unknown. Generally, any kind of slots can be omitted.

(3) Agent(ID=100) ask to agent(ID=200) "which building is broken, how much, and when?". Then Agent(ID=200) answer to agent(ID=100) "at time 15, building(ID=150) is broken in degree 100."

```
(query-ref :sender (agent :id 100)
           :receiver (agent :id 200)
           :content (is :subject (building :id (:wh))
                        :time (:wh)
                        :state (broken :degree (:wh))))
(inform :sender (agent :id 200)
        :receiver (agent :id 100)
        :content (is :subject (building :id 150)
                 :time 15
                 :state (broken :degree 50)))
```

(4) Agent(ID=100) say to someone "help!".

```
(request :sender (agent :id 100)
         :receiver (agent :tid T1)
         :content (do :subject (agent :tid T1)
                      :action (help)
                      :object (agent :id 100)))
```

When the same agent (or building) occurs twice in the expression, we can assign a temporal id (`:tid`) to denote the identity in the expression.

(5) Agent(ID=100) say to agent(ID=200) and agent(ID=300) "move away!".

```
(request :sender (agent :id 100)
         :receiver (:set (agent :id 200) (agent :id 300))
         :content (do :action (move)))
```

(6) Agent(ID=100) say to agent(ID=200) "please say to person(ID=300) to move away".

```
(request :sender (agent :id 100)
         :receiver (agent :id 200)
         :content (do: action (request :receiver(agent :id 300)
                                       :content(do:action(move)))))
```

We can embed a clause in any part of content.

5 Discussion for Future Extensions

5.1 Situated Communication

A benefit of using the frame representation is that it is easy to add or omit slots from a frame. The benefit enables to reflect an important feature of natural language conversation, *situated communication*. In human communication, we tend to omit words when they are well known for both of sender and receiver, or when they are not so important. As shown in example 4.3, we can omit slots when they can be filled by receiver. In this example, `:subject` slots in the `do` frames and `:sender` slot in the embedded `request` frame are omitted.

One of open issues in the situated communication is how to treat references like pronouns. We can avoid this by assigning temporal id for referenced information as shown in example 4.3. But this method is not applicable easily in the case the reference points outside of a clause, because we need to define a scope of the temporal id over discourses.

Another issue of situated communication is how to control contexts. Contexts of conversation is an important information source to fill omitted information in utterances, so agents need to manipulate contexts properly. However, the context control is still an open issue on natural language processing. Fortunately, flows of contexts seem to be straight-forward in conversation in the rescue domain. In this case, we can control contexts by introducing `:in-reply-to` slots into the *Clause*.

5.2 Complex Logical Forms

We suppose that all civilian's utterances are simple sentences. Therefore, we introduce only negation (`:not` form) as logical forms. This specification has the following merit:

- It is easy to control the amount of communication transfered by an agent in a time unit. We can treat a simple sentence as a unit of communication, so that we may be able to limit the number of sentences for an agent to say or hear.
- It is easy to define the civilian's behavior as a response to listened messages.

In future, however, we will need to introduce complex or compound sentences, which require more complicated logical forms like '`and`', '`or`', '`if`' and so on.

5.3 Languages for Rescue Agents

While we define a civilian's language to reflect features of natural language, we are thinking that rescue agents will use more formal and well formed language between themselves. Compared with civilians, rescue specialists will be trained to use unambiguous and clear expressions. It is also required that the language can be specify complex combination of conditions and actions, which is similar to a kind of programming language. For this purpose, we can use more formal languages specification like FIPA's Content Language Library [FIP00b].

6 Summary

In this article, we proposed a framework of agents' communication in the RoboCup Rescue simulation, *four layers model of communication*, in which factors of establishment of the communication are classified into knowledge, attention device and transmission layers. Using this framework, researchers can introduce easily new ideas of various level of communication into the simulation.

We also propose a specification of a language for civilian agents. Because civilian agents will be designed to behave as usual people, their utterances will include incomplete and ambiguous expression. The specification is designed to reflect such kinds of natural language features.

Both propositions are prototypes for future extensions. Especially, language specification need to modified to reflect more natural language features.

References

[FIP00a] FIPA, Geneva, Switzerland. *FIPA CCL Content Language Specification*, Aug. 2000. Document number XC00009A (`http://www.fipa.org/`).

[FIP00b] FIPA, Geneva, Switzerland. *FIPA Content Language Library Specification*, Jul. 2000. Document number XC00007A (`http://www.fipa.org/`).

[FIP00c] FIPA, Geneva, Switzerland. *FIPA KIF Content Language Specification*, Aug. 2000. Document number XC00010A (`http://www.fipa.org/`).

[NOD00] Itsuki NODA. Framework of distributed simulation system for multi-agent environment. In *Proc. of The Fourth International Workshop on RoboCup*, pages 12–21, Aug. 2000.

[TTK+00] Tomoichi Takahashi, Ikuo Takeuchi, Tetsuhiko Koto, Satoshi Tadokoro, and Itsuki Noda. Robocup-rescue disaster simulator architecture. In *Proc. of thr fourth International Workshop on RoboCup*, pages 100–105, Aug. 2000.

[WCF+00] Steven Willmott, Monique Calisti, Boi Faltings, Santiago Macho-Gonzalez, Omar Belakhdar, and Marc Torrens. Ccl: Expressions of choice in agent commnication. In *The Fourth International Conference on MultiAgent Systems (ICMAS-2000)*. IEEE, July 2000.

Specifying Rational Agents with Statecharts and Utility Functions

Oliver Obst*

Universität Koblenz-Landau, AI research group
Rheinau 1, D–56075 Koblenz, GERMANY
fruit@uni-koblenz.de

Abstract. To aid the development of the robotic soccer simulation league team RoboLog-2000, a method for the specification of multi-agent teams by statecharts has been introduced. The results in the last years competitions showed that though the team was competitive, it did not behave adaptive in unknown situations. The design of adaptive agents with this method is possible, but not in a straightforward manner. The purpose of this paper is to extend the approach by a more adaptive action selection mechanism and to facilitate a more explicit representation of goals of an agent.

1 Introduction

Creating a team of autonomous agents for dynamic environments like simulated robotic soccer is a difficult task, because agents can encounter a number of different situations the designer might not have been thinking of at the time of creation. Statecharts [3] are a visual approach to formalize complex systems. They can be used to describe the behavior of a system on different levels of granularity with one representation. The specification can be transformed in a straightforward manner into running Prolog code [7, 5]. The method facilitates a high degree of reactivity, since only a few rules have to be checked each cycle. As a proof of concept, we developed the RoboLog 2000 simulator team using our approach [6]. During the competitions in Amsterdam (EuRoboCup 2000) and Melbourne (RoboCup 2000), it turned out that our team performed quite well in many situations. On the other hand, in new situations our team did not behave adaptive enough. This became clear against new teams *crossing* the ball in front of our goal and in matches with a high loss or a high win. Designing adaptive teams with our approach would have been possible by intensively using hierarchically structured statecharts [12]. On the other hand, this was not straightforward, and the goals of an agent were determined implicitly by its actions. The purpose of this paper is to extend our approach by a more adaptive action selection mechanism and to facilitate a more explicit representation of goals of an agent.

* This research is supported by the grant *Fu 263/6-1* from the German research foundation *DFG*.

A. Birk, S. Coradeschi, and S. Tadokoro (Eds.): RoboCup 2001, LNAI 2377, pp. 173–182, 2002.

The idea of our new approach is as follows: We want to specify multi-agent scripts with statecharts, so agents are using scripts as parameterized, predefined plans. An agent should select which of the applicable scripts is most useful for its current goals. The goals of an agent are in its set of beliefs and can be changed by a script. The action selection mechanism becomes more adaptive, because the predefined plans are chosen more carefully for a new situation. Though a second step towards more adaptiveness is the alteration of predefined plans, we focus on the sooner aspect in this paper.

Additionally, we want to provide a mechanism to evaluate the utility of a script that takes the commitment to selected options into account. Like in our former approach, there should be a possibility to generate the agent program from its specification.

1.1 A First Approach to Multi-agent Systems with Statecharts

In [7], we introduced a method for specifying multi-agent behavior by statecharts and the translation of statecharts into code. For the design of soccer agents, the statecharts we are using are a natural way of describing a sequence of actions for a certain situation: While designing agents, the designer might think of typical situations for a certain player and some appropriate actions during this situation. Situations and appropriate actions can be formalized with the statecharts. Once the agent design is finished, it consists of a number of scripts.

At the time of execution, the agent looks for an applicable script and executes a step of the first one it can find. Because of the way actions are specified in this approach, representation of goals of our agents is implicit in the agent's behavior specification: if the preconditions for an action hold, the action is executed. The conditions are regarded in the order of specification, that is, most of the time, more specific rules are regarded first. If no specific rule is applicable, as a last resort a default rule is applied. As long as a specific rule holds, this mechanism ensures that the script the agent is going to execute is applicable. What it does not guarantee is that the script which is going to be used is the best or most useful one.

1.2 Action Selection

A rational agent is one that tries to maximize its performance measure [10], that is in our case the expected success of its actions. Until now, the measure for expected success were boolean conditions on edges of a statechart. Actions at an edge with applicable conditions were expected to be successful. If an agent had more than one option to execute actions, the first applicable one was selected. A more adaptive action selection mechanism should evaluate the usefulness of applicable options and execute the most useful one. As an intended side effect of an improved option evaluation and explicit goal representation, our agent model is closer to the BDI model [9].

2 Utilities in Scripts

In [13], an architecture for action selection based on preferences and the probability of success is introduced. In this architecture, options can be scored and executed. The score of an option depends on a preference (or reward) and on the probability for success of that option. Both the reward and the probability values can be hand written or are subject to reinforcement learning. The option evaluation described in their paper is used for the player with the ball only. It is a single decision what to do with the ball: the player decides what to do, executes it and during the next simulation step, the situation is evaluated anew.

Their approach supports the construction of reactive agents that behave locally optimal with respect to their information. However, the disadvantage is that building complex plans consisting of several steps or including some team mates can only happen implicitly by intelligently selecting preference values when building the agent.

For an option selection mechanism that can be used for all players in a team throughout the whole match, it is important to consider how long an agent remains committed to an option. There are situations where it is an advantage to continue a locally non-optimal script, since we are specifying the behavior of a multi-agent system. If agents do not switch behavior due to only small changes in the environment, agents can rely on each other. As Kinny [4] found out, agents with strong commitment to their plans behave superior to cautious agents that always reconsider their plans, even in rapidly changing environments.

2.1 Preferences and Probabilities in Statecharts

Basically, a statechart without preferences consists of the four finite and pairwise disjoint sets S, E, G and A, where S is a set of states, E is a set of events, G is a set of guard conditions and A is a set of actions. Usually, a transition from a state s to state s' is labeled with an annotation of the form $e[g]/a$, where e is from the set of events E, g from the set of guards G and a from the set of actions A. The (informal) meaning of an annotation like this is that if the agent is in state s and an event e occurs, the agent should execute action a and find itself in state s' afterwards, provided that condition g holds.

Additionally, it is possible to structure statecharts hierarchically. Within hierarchically structured statecharts, a transition from one state to the next is possible, if both states belong to the same state(chart). A configuration within such a statechart is a rooted tree of states [12]. A step from one configuration to the next can happen exactly once every simulation step. After a transition, the agent saves the current script and configuration. In the simulated soccer environment, the agent returns control to the soccer server and waits until the next simulation cycle starts. After the belief update for the next cycle, the agent tries to continue the script with the saved configuration.

For our new approach, we need to change the statecharts so that the agent can estimate the preference value for an action annotated at a transition. The preference for a script is clearly dependent on the current belief (in the BDI sense) of the state of the environment. The second important parameter for the preference for a script is the current state of the statechart that should be executed. Usually, this is going to be the initial state of a statechart, since each script should be progressed from its beginning. Only the current executed script might be in a different state. So for each script, the preference function maps the agent's belief and the state of the statechart to a value, the preference value of a script. If we want to regard the uncertainty whether a script will or will not be successful, the preference for a script and the probability for success make up the utility of a script.

Instead of events and actions, transitions are now labeled with annotations $p[g]/a$, where p is a preference function returning a value evaluating the preference for the annotated action, given the agents' belief of the environment. A preference function consists of one or more *attribute*, the conditions in the former approach.

Definition 1 (Attribute function). *Let S be a set of states of a statechart, A a set of actions, and Bel the current belief of the agent about the state of the world. An attribute α is a function $\alpha : S \times A \times Bel \mapsto [0..1]$.*

The preference of an agent for a certain action is given by the set of attributes at a transition from one state to the next. Like in [13], the difficulty lies in the selection of appropriate values for different options. Preferences have to be comparable, so that options with a higher score turn out to be a better decision for the agent.

For most actions, it will be useful to combine some attributes to a preference, for instance the preference value of a pass could depend on the possibility to kick the ball, the distance to the team mate and the positions of the opponents. For our purposes, we make the simplifying assumption that all attributes are independent of each other. The advantage of this approach is that we can use an multiplicative value function for conjunctions.

Definition 2 (Preference Function).

1. *If α is an attribute function and $w \in \mathbb{R}$, $w \cdot \alpha$ is a preference function. w is the weight of the preference function. w_{max} is the maximum preference for all preference functions.*
2. *If p_1 and p_2 are preference functions, $p_1 \wedge p_2$ is a preference function.*
3. *If p_1 and p_2 are preference functions, $p_1 \vee p_2$ is a preference function.*
4. *If p is a preference function, $\neg p$ is a preference function.*

For a conjunction, the value of a preference of an action is the product of values of all weighted attributes at the transition: $P = \prod_{i=0}^{n-1} w_i \cdot \alpha_i$, where α_i are the attributes and w_i are the weights. For disjunctions of attributes, we use the maximum value of all attributes in the disjunction: $P = \max w_i \cdot \alpha_i|_0^{n-1}$. The preference value of a negated preference function $\neg p$ is w_{max} minus the value of p.

Additionally, it is possible to use probabilities whether an action will succeed. Besides the preferences, the probabilities are a measure for the utility of an action. The probability whether an action will succeed can be estimated by statistics collected over a series of matches against other teams. The probability of an action is a value $k \in [0..1]$ that is multiplied with the preference value for the respective action.

2.2 Commitment

As said earlier, most of the the time it is an advantage if agents continue their plans once started instead of switching behaviors all the time. However, there are still situations where an agent should stop its current plan and restart the script evaluation. For instance, a statechart for double passing could have a global outgoing edge that should be selected if the play mode changes. In this case, the selection of another statechart was intended at the time the agent was developed. So the agent has to evaluate all alternatives emitting from its current state, those continuing the statechart and those leading to a terminal state. In the latter case, the configuration of the statechart is reset. Additionally, at every step the agent has to evaluate the utilities of some other actions, namely those annotated at edges emitting from initial states of statecharts different from the current one. If the agent finds an action from another statechart to be more useful, it resets the configuration of the current statechart to the initial state and starts to progress the transitions of the new statechart beginning from its initial state.

In order to reflect the stronger commitment to an already selected script, utilities of scripts with a state unequal to the initial state are multiplied by a commitment factor c ($c \geq 1$). The commitment factor is a global value, so it is easy to adjust the tendency of the agent to switch behaviors.

There is yet another case to look at, namely circular edges. Circular edges might be useful for specifying repetition, for instance successive dashes or waiting for a pass. However, the utility of a script could decrease if a certain step in a script is executed a number of times. In case of the pass, it is certainly going to be the case that after some steps of unsuccessful waiting to receive the ball the usefulness of a longer wait is low. In other cases, the number of repetitions does not have any influence on the utility of a script, for instance when dashing to a certain position. To provide means to distinguish different passes of execution of a statechart, both the state s and the number of the current pass p are used to estimate the utility of a script. For each state s in a statechart $\mathcal{S}$ there is a value $d_s \in [0..1]$ which is used to decrease the utility of an action exponentially by the number of passes through a state. s_0 denotes the initial state of a statechart, while p_0 stands for the first pass through a state. The commitment of an agent to an action is defined by the following equation.

$$\mathrm{C}_{\mathcal{S}}(s,p) = \begin{cases} 1, & \text{if } s = s_0 \text{ and } p = p_0 \\ c \cdot {d_s}^p, & \text{otherwise} \end{cases} \tag{1}$$

Altogether, the following parameters are used to estimate the utility of an action: k the probability of success for an action, the according preference function as the product of all its attributes and $\mathrm{C}_{\mathcal{S}}(s,p)$, the commitment to selected options given the current state s and the current pass p. For the utility function, see the equation below.

$$\mathrm{U}_{\mathcal{S}}(s,p) = k \cdot \mathrm{C}_{\mathcal{S}}(s,p) \cdot \prod_{i=0}^{n-1} w_i \cdot \alpha_i \tag{2}$$

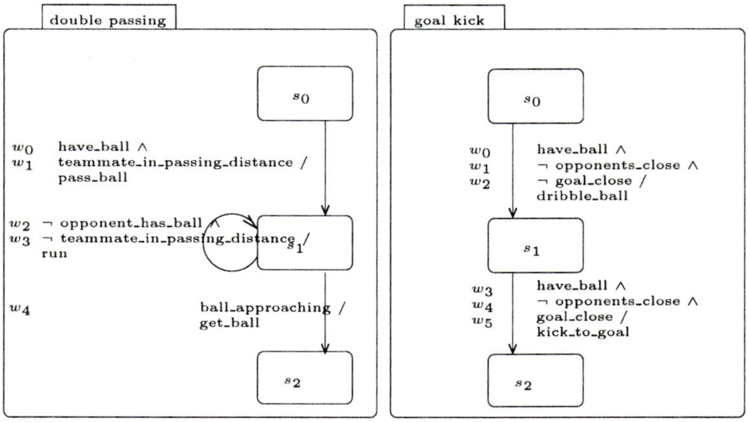

Fig. 1. Statecharts with preferences

If the utility of an action in the current statechart is lower than the utility of some applicable action in another statechart, state s and pass p are reset to s_0 and p_0 respectively.

In Fig. 1, there is a (simplified) example for statecharts specifying two kinds of behavior of the agent. The designer annotates the transitions with some attributes and actions. Additionally, weights have to be specified. If both statecharts are in state s_0, the agent program evaluates the preference for each transition from state s_0 to state s_1 and executes the annotated action, i.e. `pass_ball` or `dribble_ball`, in this case.

Specification of team behavior is possible in the same manner: the transitions between states of a script performed by some agents are annotated with the preference functions for the roles of the respective agent. A description of a situation where some agents work together is given by the configuration of the collaborating agents and the state of the environment (following the description in [12]). Transitions within the same statechart describe the behavior of participating agents, where each agent selects actions according to its preferences. In situations where teamwork is useful, the preference functions assign a higher value to the behaviors in a multiagent script. To resolve possible mismatches concerning the current state of environment, communication can be used.

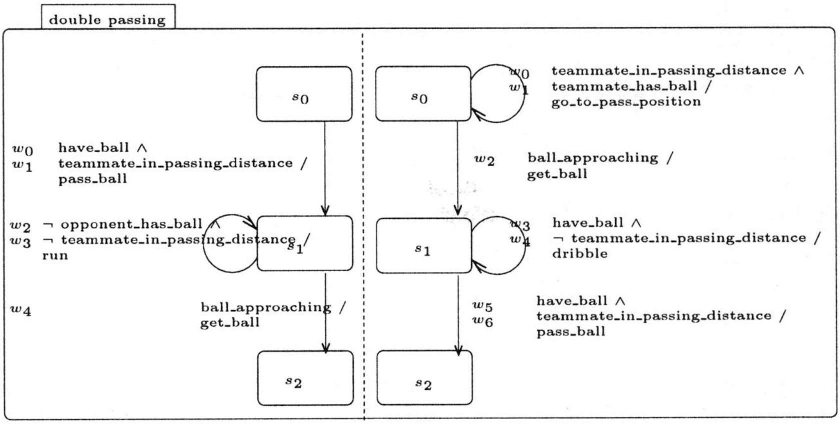

Fig. 2. Statechart for two agents

3 Translating Statecharts with Preferences into Code

One of the advantages of our former approach was, that the statecharts could be translated easily into Prolog code. But what has to be done to translate statecharts with utility functions into running code? Assuming the system provides the attribute functions defined in the previous section and an interface to the actions of an agent, the agent designer specifies statecharts for typical situations. Additionally, the designer combines attribute functions and weights to utility functions and annotates transitions with appropriate utility/action pairs. It is clearly more complicated to find the right conditions *plus* the appropriate values for the weights. As in [13], it is possible to leave this calibration to some learning procedure. On the other hand, simulation of our former approach is possible when using only preference values of 0 and 1 as truth values. From this point of view, it is possible to adjust the values "by hand" and to do a refinement later on.

To translate the statecharts into running code, the system has to be equipped with some control loop processing the resulting scripts. For an outline of the agent control loop, have a look at Fig. 3.

To actually map the relevant part of belief for an action to some real value, we were assuming that the system provides attribute functions doing this. Dependent on the type of information, continuous or binary functions have to be used. An attribute encoding the fact, that the player has the ball could be binary (0 or 1), whereas an attribute saying that the team mate is in passing distance could employ the sigmoid function, as described in [13].

For each statechart, the translation procedure takes the transitions from one state to the next and produces code like in Fig. 4. Unlike in the former approach,

```
while true do
   update belief
   if the guard condition at the current transition does not hold then
      reset state and pass of current script
   end if
   calculate utility for all scripts
   make script with the highest utility the current script
   execute a step of current script
end while
```

Fig. 3. Agent Control Loop

now there are two types of rules for each transition: the first one for estimating the utility of an action and the second one for actually executing the action.

```
transition(pass_ball, s0, s1, Utility) :-
     have_ball(U0),
     closest_teammate(No),
     in_passing_distance(No,U1),
     Utility is w0 · U0 · w1 · U1.

transition(pass_ball, s0, s1) :-
     closest_teammate(No),
     teammate_dist(No,Dist),
     teammate_dir(No,Dir),
     power_for_dist(Dist, Power),
     kick_ball(Power, Dir).
```

Fig. 4. Prolog code for a transition in a statechart

4 Related Work

Scerri and Ydrén [11] work on an approach providing means to create complex agent systems for the end user. End users are domain experts, but non-programmers. The high level behavior of agents in their approach can be edited using a graphical tool. Diagrams the user creates are translated into a behavior based system, where different behaviors run in parallel according to their activation function. End users cannot change the lower levels of the behavior, in contrast to our approach, where the specification method can be used for low level and high level behavior specification. The user interface in [11] is tailored for the soccer domain and the user can specify the behavior directly on the soccer field.

Dorer [2] tackles the problem of action selection by extending behavior networks to continuous domains. Goals of an agent activate competence modules assigned to the goals and, if the value of activation and executability for the most activated module lies above some threshold, the associated behavior is executed.

In Dix et al. [1], the authors present a general framework for dealing with uncertainty in agents. They provide different semantics for probabilistic agent programs based on deontic logic of actions. They define an agent as a set of probabilistic rules under which an agent is obliged, permitted, or forbidden to take some actions. However, they do not address the problem of selecting the best or most useful action.

Poole [8] introduces the independent choice logic (ICL), a framework for specifying decision problems in multi-agent systems. The approach is very expressive and tries to combine advantages of Markov decision process, influence diagrams and logic programming.

5 Conclusions

In this paper we introduced statecharts with utilities as a method to specify autonomous agents. The specification of an agent can be translated into a running program automatically. The resulting agent program is both reactive and committed to actions it once started. Compared to our previous approach, we improved the adaptiveness of our agents.

For our future work, we want to investigate the use of learning methods to estimate preference values and weights. Offline learning could aid the developer during the design process. Especially online learning seems to be challenging, as this would – again – contribute to adaptive behavior.

References

[1] Jürgen Dix, Mirco Nanni, and V. S. Subrahmanian. Probabilistic Agent Programs. *ACM Transactions of Computational Logic*, 1(2):207–245, 2000.

[2] Klaus Dorer. Behavior networks for continuous domains using situation-dependent motivations. In *International Joint Conference on Articial Intelligence (IJCAI'99)*, pages 1233–1238, 1999.

[3] David Harel. Statecharts: A visual formalism for complex systems. *Science of Computer Programming*, 8(3):231–274, June 1987.

[4] David Kinny and Michael P. Georgeff. Commitment and effectiveness of situated agents. In *Proceedings of the Twelfth International Joint Conference on Artificial Intelligence (IJCAI-91)*, Sydney, Australia, 1991.

[5] Jan Murray. Soccer agents think in UML. Master's thesis, Universität Koblenz-Landau, Fachbereich Informatik, 2001.

[6] Jan Murray, Oliver Obst, and Frieder Stolzenburg. RoboLog Koblenz 2000. In Peter Stone, Tucker Balch, and Gerhard Kraetzschmar, editors, *RoboCup-2000: Robot Soccer World Cup IV*, Lecture Notes in Artificial Intelligence, pages 469–472. Springer, Berlin, Heidelberg, New York, 2001. Team description.

[7] Jan Murray, Oliver Obst, and Frieder Stolzenburg. Towards a logical approach for soccer agents engineering. In Peter Stone, Tucker Balch, and Gerhard Kraetzschmar, editors, *RoboCup-2000: Robot Soccer World Cup IV*, Lecture Notes in

Artificial Intelligence, pages 199–208. Springer, Berlin, Heidelberg, New York, 2001.

[8] David Poole. The independent choice logic for modelling multiple agents under uncertainty. *Artificial Intelligence*, 94(1-2):5–56, 1997. Special Issue on Economic Principles of Multi-agent Systems.

[9] Anand S. Rao and Michael P. Georgeff. Modeling rational agents within a BDI-architecture. Technical Note 14, Australian Artificial Intelligence Institute, February 1991.

[10] Stuart Russell and Peter Norvig. *Artificial Intelligence. A Modern Approach.* Prentice Hall, Englewood Cliffs, NJ, 1995.

[11] Paul Scerri and Johan Ydrén. End user specification of RoboCup teams. In *RoboCup-99: Robot Soccer World Cup III*, number 1856 in LNAI, pages 450–459. Springer, 2000.

[12] Frieder Stolzenburg. Reasoning about cognitive robotics systems. In Reinhard Moratz and Bernhard Nebel, editors, *Themenkolloquium Kognitive Robotik und Raumrepräsentation des DFG-Schwerpunktprogramms Raumkognition*, Hamburg, 2001.

[13] Peter Stone and David McAllester. An architecture for action selection in robotic soccer. In *Submitted to the Fifth International Conference on Autonomous Agents*, 2001.

COACH UNILANG - A Standard Language for Coaching a (Robo)Soccer Team

Luis Paulo Reis[1] and Nuno Lau[2]

[1] LIACC - Artificial Intelligence and Comp. Sci. Lab. - Oporto University, Portugal
lpreis@fe.up.pt, http://www.ncc.up.pt/liacc/

[2] DET - Electronics and Telecommunications Dep. - Aveiro University, Portugal
lau@det.ua.pt, http://www.det.ua.pt/

Abstract. This document introduces COACH UNILANG, a standard language for coaching (Robo)Soccer teams. This language was developed with two main objectives: to coach FC Portugal 2001 team and as a proposal to be used in Fukuoka 2002 RoboCup coach competition. This language enables high-level and low-level coaching through coach instructions. High-level coaching includes changing tactics, formations used in each situation and changing player behavior. Low-level coaching includes defining formations, situations, player behavior and positioning with high detail. The language also enables the coach (functioning like an assistant coach) to send opponent modeling information and game statistical information to the players.

1 Introduction

A standard language for coaching must be prepared considering the state of the art in simulated RoboCup [1] and its predictable evolution [7]. Games in RoboCup2001 will be, most likely, similar to the FC Portugal vs Brainstormers games in Melbourne (similar to real soccer games). So, a standard language definition must be sufficiently powerful to enable coaching teams like FC Portugal [2,3], Brainstormers [4], CMUnited [6] or other good teams, which are able to play well without using the coach. To be useful, the information and advice given by the coach to the players must be high-level consisting of things that the players are not able individually (or through communication) to spot by themselves.

This paper introduces COACH UNILANG, a general language for coaching robosoccer teams. In Section 2 some possible coaching architectures are discussed along with usual coaching concepts. Section 3 fully describes the requirements of our language. Section 4 introduces COACH UNILANG and describes its message types. The syntax and semantics of the language are described in Section 5. Finally we give some conclusions and give an outlook to future research.

2 Coaching Architectures, Levels, and Concepts

COACH UNILANG enables coaching at several different levels. We introduce the assistant coach agent that is able to analyze a game (directly or through

A. Birk, S. Coradeschi, and S. Tadokoro (Eds.): RoboCup 2001, LNAI 2377, pp. 183–192, 2002.

a log file) and gather opponent modeling and game statistics information at several levels. This agent may be used to assist the coach providing him with high-level information that may be used more efficiently. The assistant coach may be used as a single agent (Fig. 1a) or integrated into the principal coach (functioning both as a single agent). This architecture (Fig. 1b) will be FC Portugal coach architecture for RoboCup 2001. The principal coach may also be included inside the player agents (Fig. 1c). If the principal coach only needs the opponent modeling and game statistical information provided by the assistant coach then he might be included inside the players. In this architecture, players are responsible for changing the team tactic in use (using the same algorithm). In the last architecture (Fig. 1d) the coach is not an autonomous agent. This was FC Portugal 2000 coaching architecture, in which players used simple game statistics and opponent modeling information to change team tactic.

COACH UNILANG enables coaching at several different abstraction levels:

- **Level 1 - Instructions.** Level 1 coaching is intended to train teams with intelligent players that are used to play together. It enables high-level coaching but it is not very flexible because it uses generic (and fixed) soccer concepts.
- **Level 2 - Statistics + Opponent Modeling.** Level 2 coaching is intended to help a coach to train a team of players. It is the language used by our assistant coach to talk with the coach (although the rules do not allow two coaches, they are implemented as a single agent)! The assistant coach is concerned with calculating game statistics (like ball possession by regions, action result, etc.) and trying to get opponent modeling information.
- **Level 3 - Definitions + Instructions.** This is a very flexible coaching level. It enables the coach to redefine the standard soccer concepts at a much more low-level. It enables both high-level coaching and low-level coaching.

COACH UNILANG is based on several real soccer concepts. These include:

- **Field Regions** - corresponding to areas of the field (mostly rectangles but also points and other type of regions are allowed)
- **Time Periods** - correspond to periods of the game (with a given beginning time and duration);
- **Tactics** - enabling high-level configuration of team behavior by defining offensive or defensive characteristics, game pace, pressure placed on the opponent's team, formations used for different situations, etc.

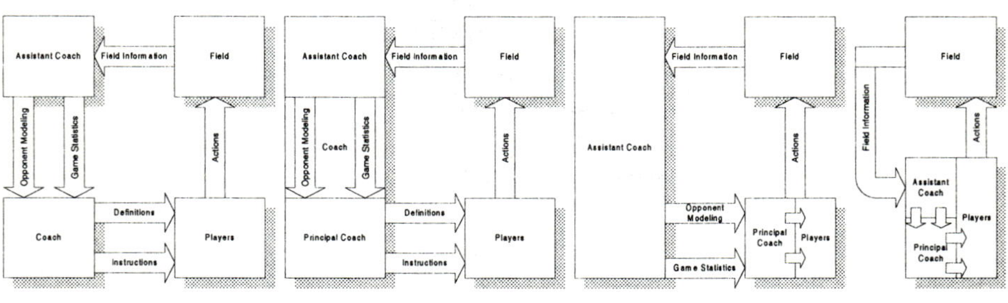

Fig. 1. Coaching architectures possible using COACH UNILANG

- **Formations** - spatial distributions of the players in the field.
- **Situations** - high-level analysis of the game state. Typical situations include offensive and defensive situations, from defense to attack, corners, etc.
- **Player Types** - defining players' behaviors while in possession of the ball and without the ball.

3 Requirements for Our Coaching Language

Our format tries to compromise between generality and simplicity. It is sufficiently general to allow representing the most common variants of coaching strategies but has also a simple syntax and enables coaching at different abstraction levels. The main requirements in the elaboration of this format were:

- It should be independent of implementation details and coaching strategies;
- It should be small but easy to extend including new concepts and definitions;
- Existing languages and coaching strategies should be easily expressed;
- It should be easily readable by the agents but also by the human user;
- Information not directly concerned with coaching should not be included;
- The format should be general enough and multi-resolution to enable the representation of coaching strategies at different abstraction levels;
- Semantics should be completely clear. All concepts must be clear to people with basic soccer and robosoccer knowledge;
- It should enable to represent regions, time periods, tactics, formations, player types and situations;
- Normal soccer regions, tactics, formations, player types and situations should be included as default but new ones must be easily definable;
- It should be robust with simple data validation and override of usual errors.

4 Message Types

COACH UNILANG has four main message types:

- **Definition.** Enables the coach to define a given concept (region, time period, tactic, formation, situation or player type);
- **Statistics.** Enables the coach to broadcast some game statistics;
- **Opponent Modeling.** This message type is used to inform the players about opponent (or own) team characteristics (at several levels).
- **Instruction.** Instruction messages enable the coach to change the different parts of the team tactic (formations used, team pressure, game pace, playing style, player behaviors, etc.)

Each message is composed of a time, identification and one or several parts. Messages may have one of four types: Definition, statistics, opponent modeling or instructions. The language uses several types that include integers, sets, qualitative scales (very high, high, medium, low and very low), etc.

```
<MESSAGE>::= (<TIME> <ID> <MESSAGE PART> {<MESSAGE PART>})
<MESSAGE PART>::= <DEFINITION_MESS> | <STATISTICS_MESS> | <OPP_MOD_MESS>| <INSTRUCTION_MESS>
<DEFINITION_MESS> ::= (definition <ID> <DEFINITION >)
<STATISTICS_MESS> ::= (statistics <ID> <STATISTICS>)
<OPP_MOD_MESS> ::= (opp_mod <ID> <OPP_MOD>)
<INSTRUCTION_MESS> ::= (instruction <ID> <INSTRUCTION>)
<ID>::=[integer] <TIME>::=[integer] <VALUE>::=vhi|hi|mid|low|vlow
<TEAM>::=our|oppon|any  <PLAYER>::=[1..14]|any <POS_NUMB>::=[1..11]|any
```

5 Coach Language

5.1 Definition Messages

Definition messages allow the coach to define concepts used during the game. These concepts include regions, periods, tactics, formations, situations and player types (active and recovery behavior). Concepts are identified by a name so that they may be referenced in future coach messages.

```
<DEFINITION>::= (region <REGION_NAME> <REGION>) | (period <PERIOD_NAME> <PERIOD>) |
  (tactic <TACTIC_NAME> <TACTIC>) | (formation <FORMATION_NAME> <FORMATION>) |
  (situation <SITUATION_NAME> <SITUATION>)|(active_type <ACTIVE_TYPE_NAME> <ACTIVE_TYPE>)|
  (recovery_type <RECOVERY_TYPE_NAME> <RECOVERY_TYPE>)
```

5.2 Regions

One of the most common soccer concepts is the region concept. Usually, coaches, players and journalists communicate speaking about known regions like the penalty box, the left wing, etc. COACH UNILANG provides a predefined usual soccer field regions and a simple mechanism to define new regions. Predefined regions include a simple multi-resolution field description.

Regions may be defined as simple regions or as a list of simple regions (interpreted as a union of those simple regions). Simple regions may be the predefined regions included in the language but also user defined regions and regions defined directly. Regions may be defined using points, rectangles, quadrangles, circles or arcs. In the standard region definition, trying to let it remain simple, we included only rectangles:

```
<REGION>::= <SIMPLE_REGION> | (<SIMPLE_REGION> {<SIMPLE_REGION>})
<SIMPLE_REGION>::= <PREDEFINED_REGION>|<REGION_NAME>| <REGION_DEFINITION>
<PREDEFINED_REGION>::= field|outside|our_middle_field|their_middle_field| our_leftwing|
  our_middle|our_rightwing|our_penalty_box|our_leftwing_back|our_leftwing_middle|
  our_leftwing_front|our_rightwing_back|our_rightwing_middle|our_rightwing_front|
  our_middle_back|our_middle_front| our_middle_back_right|our_middle_back_left|...
<REGION_DEFINITION> ::= <POINT> | (rect <POINT> <POINT>) |
    (quad <POINT> <POINT> <POINT> <POINT>) | (circle <POINT>[real])|
    (arc <POINT> [real][real][real][real])
<POINT> ::= (point [real] [real])
```

5.3 Time Periods

Time is very important in simulated robosoccer and most of the analysis that a coach must perform, are connected with time periods. COACH UNILANG provides a simple mechanism for defining time periods. They may be predefined time periods (defined as general concepts in the language), periods defined by the coach during the game (using the define directive) or periods defined directly when used. Predefined periods include the whole game, first half, second half, extra time, last 1000 cycles, etc. To define other periods the coach may define them in terms of the begin and end times or in terms of length:

```
<PERIOD>::= <PREDEFINED_PERIOD> | <PERIOD_NAME> | <PERIOD_DEFINITION>
<PREDEFINED_PERIOD>::= game | first_half | second_half | extra_time | last_1000 |
   last_500 | last_300 | last_100 | last_50 | last_20
<PERIOD_DEFINITION>::= (<TIME> <TIME>) | <LENGTH>
<TIME>::= [integer]|now         <LENGTH>::= [integer]
```

5.4 Tactics

Tactics are probable the most complex soccer concept. In a given tactic, other complex concepts are included, like: formations, game pace, team pressure, game type, playing style, offside trap use, etc. At each moment in a game, a team uses a given tactic. However, the coach according to the game course and opponent team behavior may change this tactic. Our definition of what is a tactic is based on common soccer concepts like team mentality, game pace, team pressure, field use to develop collective moves, offside trap, risk taken, etc. The following tactic characteristics are used:

- **Team mentality** (very offensive, offensive, normal, defensive, very defensive). A team with an offensive mentality is thinking about attacking while it is defending. Some players may stay in the middle field waiting to attack instead of defending. In a team with a defensive mentality players are mainly concerned with defending. Even while the team is attacking players remain in useful defending positions trying to prevent opponent's counter attacks.
- **Game pace** is concerned with the speed imposed to the offensive and defensive game of the team. Teams with high game pace (for example usual English teams) use high speed for all moves. Teams with low game pace will move the ball slowly. Players will first stop the ball and take their time to decide what to do with the ball. Also, players will only move fast in the field if their active participation may be important for the team, otherwise they will prefer to keep their stamina for more important moves.
- **Team pressure** defines the pressure made over the ball when the team is not in possession. When high team pressures are used, the team players tend to be much more concentrated in the region where the ball is located in the field and more players challenge the opponent in possession. More than one player tries to pressure the attacker with the ball and also several players must give support covering his possible passes and dribbles. Defense may get somewhat unbalanced but ball recovery is usually much faster if the opponent team is not a very good one.
- **Field use** concerns the use of the different regions of the field to develop offensive moves. Teams may use a balanced system or use systems that give priority to moves developed using the wings of the field. Also teams may be more enthusiastic to use the middle of the field to develop offensive moves (for example if the opponent covers well both wings).
- **Playing style** is concerned with the type of attack and defense used. Attacking style is mostly concerned with the type of actions used by the players to attack. These actions may be mainly short passes and dribbles (passing_game also known as possession play) or they may be mainly long passes, forwards and runs by the players (long_ball also known as direct play). Defense type is mainly concerned with the type of marking of the players (individual marking or zonal defense).
- **Risk taken** defines the risk players are willing to take while conducting defensive and offensive moves. If a high-risk tactic is used, then players will try risky interceptions and risky passes. When the team is losing a given game, risk is needed to recover quickly the ball and to construct fast offensive moves. If a low degree of risk is used, then players will only try safe moves.
- **Offside tactic** concerns the measures taken by the team to catch opponents in offside and to prevent team members from being caught in offside. The offside-trap pushes defenders up in a line and they will be attempting to push out of defense

quickly after an opposition attack in the hope that they can catch opposition attackers offside.

– **Positioning exchange** use is concerned with the freedom of the players to exchange positions with their teammates. If positioning exchange is high, players will try to exchange positioning to confuse the opponent's defense.
– **Formations**. One of the more important characteristics of the tactic is concerned with the formations used for different situations. For example, a team may use a narrow 433 formation to defend and a wide 235 formation to attack. Our language lets the coach define the formations that may be used in each situation and their shape (in terms of horizontal and vertical opening).

```
<TACTIC>::= <PREDEFINED_TACTIC>|<TACTIC_NAME>|<TACTIC_DEFINITION>
<PREDEFINED_TACTIC>::= normal|score_lots|counter_attack|keep_possession|.
<TACTIC_DEFINITION>::= <MENTALITY> <GAME_PACE> <PRESSURE> <FIELD_USE> <PLAYING_STYLE>
    <RISK_TAKEN> <OFFSIDE_TACTIC> <POS_EXCHANGE_USE> <FORMATIONS_USED>
<MENTALITY>::= very_offensive|offensive|normal|defensive|very_def
<GAME_PACE>::= sleep | ease_up | normal | speed_up | give_evrthng
<PRESSURE>::= no_pressure| standoff| normal| pressure| high_press
<FIELD_USE>::= wings | middle | balanced | right_wing | left_wing
<PLAYING_STYLE>::= (<ATTACKING_STYLE> <DEFENDING_STYLE>)
<ATTACKING_STYLE>::= short_passing|passing|normal|long_ball|verylong_ball
<DEFENDING_STYLE>::= zonal_defense | normal | individual_marking
<RISK_TAKEN>::= very_safe | safe | normal | risky | very_risky
<OFFSIDE_TACTIC>::= (<OFFSIDE_TRAP_USE> <OFFSIDE_TRAP_CAREFUL>)
<OFFSIDE_TRAP_USE>::= <VALUE>  <OFFSIDE_TRAP_CAREFUL>::= <VALUE>
<FORMATIONS_USED>::= ({(<SITUATION> <FORMATION> <FORMATION_SHAPE>)})
<FORMATION_SHAPE>::= <FORMATION_WIDTH> <FORMATION_HEIGHT>
<POS_EXCHANGE_USE> ::= <VALUE>  <FORMATION_WIDTH>::= <VALUE>  <FORMATION HEIGHT>::= <VALUE>
```

Although we have several predefined tactics in FC Portugal, the definition of common soccer tactics is not an easy task and there are no globally accepted definitions for standard soccer tactics.

5.5 Formations

Formations describe the spatial distribution of the players in each situation. COACH UNILANG includes predefined common soccer formations (433, 442, 343, etc.) It also includes a high-level formation definition mechanism based on an imaginary 7*5 rectangular board. Figure 2 shows the use of this board to define some common soccer formations.

Formation definition also includes the definition of player types for each of the player in the formation. Usually these player types are something like a striker, a sweeper, a central defender, a left-winger, etc. In COACH UNILANG the definition of player types (section 5.7) is performed at two different levels: behavior with the ball and behavior without the ball. This way we may state that a player is an aggressive central defender or a positional striker, etc:

```
<FORMATION>::= <PREDEF_FORMATION> | <FORMATION_NAME> | <FORMATION_DEFINITION>
<PREDEF_FORMATION>::= 433 | 433att | 442 | 442att | 442diamond | 343 | 4123 | 352 | 532 |
   41212 | 4132 | 424 | 451 | 541 | 3223 | 3232
<FORMATION_DEFINITION>::={(<PLAYER> <POSITIONING_NUMBER> <PLAYER_POSITIONING> <PLAYER_TYPE>)}
<PLAYER_POSITIONING>::= <VERTICAL_POSITIONING> <HORIZ_POSITIONING>
<VERTICAL_POSITIONING>::=gk|sweeper|defense|defense_middle|middle|middle_forward|forward
<HORIZ_POSITIONING>::= left | left_center | center | right_center | right
```

5.6 Situations

Situations refer to a high-level analysis of the world state. Basic situations include attack, defense, passing from attack to defense, scoring opportunities, goalie free kicks, goal kicks, kickoffs, corners, kickins and free kicks.

```
<SITUATION>::= <PREDEF_SITUATION | <SITUATION_NAME> | <SITUATION_DEFINITION>
<PREDEF_SITUATION>::= attack | defend | defense_to_attack | attack_to_defense |
   our_opportunity | their_opportunity | our_goalie_kick | their_goalie_kick
<SITUATION_DEFINITION>::= <CONDITION>
<CONDITION>::= (true) | (false) | (team_in_possession <TEAM><PERIOD>) |
   (ball_pos <REGION>) | (ball_velocity <VALUE>) | (ball_possession <TEAM> <PLAYER>) |
   (player_position <TEAM> <PLAYER> <REGION>) | (not <CONDITION>) |
   (and <CONDITION> <CONDITION>) | (or <CONDITION> <CONDITION>)
```

Using our language we may define easily situations like "attack": (situation attack (and (ball_position their_middle_field) (ball_possession our any))).

5.7 Player Types and Actions

Player types refer to player behaviors with and without ball according to what is happening in the field. Active type refers to the player behavior with ball and recovery type refers to the player behavior without ball. We define seven possible actions with ball: shoot, pass, forward (pass not directly to a player but to a point where the player is supposed to collect the ball almost stopped), dribble, run with the ball, hold the ball and clear (through the ball deliberately outside the field). We also define seven possible actions without the ball (recovery actions). These include: interception (intercept a moving ball in the field), tackle (steal the ball that an opponent has controlled), mark a given passing line, mark individually a player, go to the ball position (although not being able to intercept it) and move strategically in the field (respecting the formation in use).

We also defined six different action results (with ball): success (if the action is a shoot then a goal was achieved, etc.), opp_goalie_catch (opponent goalie catch the ball), our_goalie_catch, ball went out of field bounds, interception by the opponent, stolen by the opponent (by a tackle).

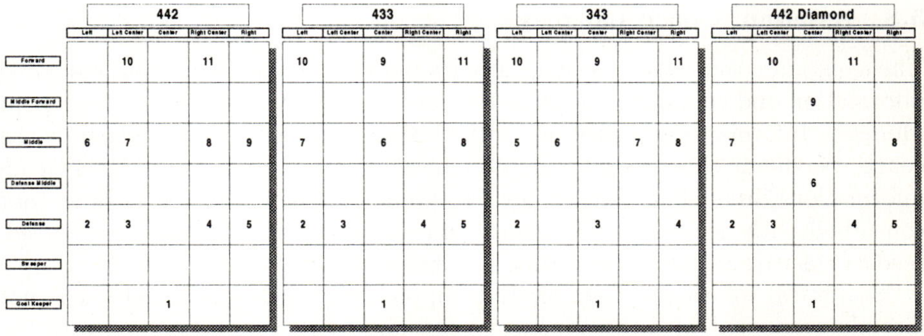

Fig. 2. Definition of some common soccer formations

For recovery actions, besides normal action results, the following action results are also defined: stolen_other (other player took the ball from the opponents while this player was conducting the given recovery action) and intercepted_other (other player intercepted the ball while the player was conducting the recovery action):

```
<ACTION>::= shoot|pass|forward|dribble|run|hold|clear|any
<RECOVERY>::= interception|tackle|mark_pass_line|mark_player|goto_ball|strategical_move|any
<ACTION_RESULT>::=success|opp_goalie_catch|our_goalie_catch|out|interception|stolen|fail|any
<REC_RESULT>::= <ACTION_RESULT>|stolen_other|intercepted_other
<ACTION_PARAMETERS>::= (<DISTANCE> <POWER> <SAFETY>)
<RECOVERY_PARAMETERS>::= (<DISTANCE> <POWER> <SAFETY>)
```

To define a player type, the coach must define the player behavior with ball (active type) and without ball (recovery type). Some simple predefined behaviors are available for active types and recovery types. Some qualitative parameters are available for the different ball possession actions and recovery actions available. These include the distance, power and safety of the action to perform. For example, a coach may state that in the team's defensive middle field, the passing action shall be safe. He may also state that in the attacking third, safety of actions is not a concern.

```
<PLAYER_TYPE>::= <ACTIVE_TYPE> <RECOVERY_TYPE>
<ACTIVE_TYPE>::= <PREDEF_ACTIVETYPE>|<ACTIVE_TYPE_NAME>|<ACTIVETYPE_DEF>
<PREDEF_ACTIVETYPE>::= basic|normal|shooter|dribbler|passer
<ACTIVETYPE_DEF>::= ({(<ACTION> <REGION_FROM> <REGION_TO> <ACTION_PARAMETERS> <VALUE>)})
<RECOVERY_TYPE>::= <PREDEF_RECTYPE>|<RECOVERY_TYPE_NAME>|<RECTYPE_DEF>
<PREDEF_RECTYPE>::= normal | positional | marker | aggressive
<RECTYPE_DEF>::= ({(<RECOVERY> <REGION> <RECOVERY_PARAMETERS> <VALUE>)})
```

The definition of ball possession behaviors in simulated robosoccer is rather complex. This way, our predefined active types definition is quite basic relatively to what is needed to construct good soccer players. It is completely impossible to define or configure the behavior of a decent soccer player using only 2 or 3 conditions. Definition of good ball recovery behaviors is simpler than the definition of good ball possession behaviors. The following statement shows a simple definition of a recovery player type: (recovery_type aggressive ((interception field (any any any) vhi) (tackle field (any any any) vhi) (goto_ball field (any any any) hi) (mark_player field (any any any) vlow) (mark_passline field (any any any) vlow) (strategical_move field (any any any) low)).

5.8 Assistant Coach Language

The assistant coach gathers game statistical information and opponent modeling information and communicates it to the coach and/or to the players. Opponent Modeling Information is concerned with opponent characteristics spotted by the coach. We divide this information into four abstraction levels: social level (team information), individual decision level (players and goalie), low-level skills (not concerned with decision mechanisms) and physical characteristics (like speed, recovery, decay, inertia, acceleration, size, kickable margin and kicking accuracy).

Statistical information contrarily to opponent modeling information is not directly concerned with the opponent behavior but is more concerned with the game course at several levels. The definition of the statistical information included in COACH UNILANG is based on the information used by real soccer coaches and information available in computerized video analysis soccer systems like Match Analysis [5]. Statistical information includes game occurrences, action number and results (by region, team, player and period), ball possession (of each team and player in each region and period), player positions in the game, assistances to shoots on goal, ball losses, ball recoveries and ball circulation information. Due to space limitation we cannot describe here the syntax and semantics of our assistant coach language.

5.9 Coach Instructions

One of the innovations of COACH UNILANG is the fact that it enables very high-level coaching (although not limiting in any way, low level coaching). To change the team strategy, the coach may use the instruction directive.

```
<INSTRUCTION>::= (change_tactic <TACTIC>)|(change_mentality <MENTALITY>)|
  (change_game_pace <GAME_PACE>)|(change_risk_taken <RISK_TAKEN>)|
  (change_field_use <FIELD_USE>)|(change_playing_style <PLAYING_STYLE>)|
  (change_pressure <PRESSURE>)|(change_offside_tactic<OFFSIDE_TACTIC>)|
  (change_pos_exchange_use <POS_EXCHANGE_USE>)|
  (change_formation <SITUATION_NAME> <FORMATION> <FORMATION_SHAPE>)|
  (change_formation_pos <FORMATION_NAME> <POS_NUMBER> <PLAYER_POS> <PLAYER_TYPE>)|
  (change_player_type <PLAYER_TYPE_NAME> <ACTIVETYPE_DEF> <RECTYPE_DEF>)
```

The coach may change the tactic in use completely (change_tactic) or he may change the different parts of that tactic. He may change the team mentality, game pace, team pressure, field use, playing style, offside tactic and, positioning exchange use. Also he may change the formation used (on the tactic in use) for a given situation and that formation's shape.

The coach may also change a single positioning and player behavior of one of the formations. This operation is performed using the change_formation_pos directive. Using this directive, for example, the coach may say that (change_-formation_pos 433 10 (left_center middle_forward) (striker aggressive)). This means that for the formation 433, the positioning number 10 that had the value (left forward) will now be (left_center middle_forward). Also this instructs the player occupying that positioning to behave like an aggressive (ball recovery) striker (ball possession). The final directive enables the change of a given player type, changing some of its parameters.

6 Conclusions and Future Work

This is, to our knowledge, the first attempt to create a language that enables high-level coaching of robo(soccer) teams. Another attempt to create a standard language for coaching robosoccer teams was made by Patrick Riley, Gal Kaminka and Timo Steffens [8]. This later language does not include high-level concepts like tactics, formations, player types, time periods or situations (essential to define the behaviour of a soccer team). It only enables the definition of player's home positions and several directives concerning individual action behaviours (like which player to pass the ball or which player to mark). Also, this language includes some concepts derived from COACH UNILANG (whose draft proposal was made public on January 2001) like definitions for conditions, actions and directives, marking passing lanes and play mode based conditions. This "standard language" was used in a "coaching competition" in Seattle. As expected, no visible tactical changes were seen in any of the teams that did use this "standard language" and the competition was considered by most of the participants as "somewhat random". Opposite, FC Portugal 2001 did change its tactic and even played, when losing near the end of a game, in a very agressive 433 formation (without a goalie)!

Some people may argue that the definition of tactics and formations is too high-level. However, we argue that COACH UNILANG enables the definition of tactics and formations at any level. The coach may use the high-level instruction mechanisms or he may define a tactic, defining all used formations and the behavior of each player used in each formation at low level (defining for each region what are the player preferred actions and configuring them). This enables virtually the definition of any soccer tactic. Other critique that can be made to this language is that it is too complex. However, soccer is a complex game. So, there is no interest in developing a language that is able to train only teams that use simple decision mechanisms and is not able to play like real soccer teams (capturing and using the real soccer complexity).

Future work is related with the discussion of this language with soccer experts: professional coaches, players and journalists enabling us to refine the language. Future work will also be related with the adaptation of FC Portugal team enabling it to be fully trained using COACH UNILANG. We also plan to extend COACH UNILANG with information regarding free kick takers, studied offensive moves, goal keeper instruction and player vision advice. Also we plan to develop player-coach and player-player languages. These languages will effectively enable the construction of teams with coach, coach assistant and players developed by different people that are able to play together as a real soccer team.

References

1. Itsuki Noda et al.: Soccer server: A Tool for Research on Multiagent Systems. Applied Artificial Intelligence, Vol. 12, pp.233-250, 1998
2. Luis Paulo Reis and Nuno Lau, FC Portugal Team Description: RoboCup 2000 Simulation League Champion, pp.29-40, in [7]
3. Luis Paulo Reis, Nuno Lau and Eugenio Oliveira. Situation Based Strategic Positioning for Coordinating a Simulated RoboSoccer Team, in M. Hannebauer et al eds, Balancing Reactivity and Social Deliberation in MAS, Springer LNAI 2103, pp.175-197, 2001
4. Martin Riedmiller et al. Karlsruhe Brainstormers 2000 - A Reinforcement Learning Approach to Robotic Soccer. pp.367-372, in [7]
5. Match Analysis, On Line, available as URL: http://www.matchanalysis.com/
6. Peter Stone, Patrick Riley and Manuela Veloso, The CMUnited99 Champion Simulator Team, RoboCup99: Robot Soccer World Cup III, LNAI 1856, Springer, 2000
7. Peter Stone, Tucker Balch and Gerhard Kraetzschmar, editors, RoboCup-2000: Robot Soccer World Cup IV, Springer, LNCS 2019, 2001
8. Noda et al, RoboCup Soccer Server Users Manual, RoboCup Federation, June 2001

Cooperative Probabilistic State Estimation for Vision-Based Autonomous Soccer Robots

Thorsten Schmitt, Robert Hanek, Sebastian Buck, and Michael Beetz

TU München, Institut für Informatik, Arcisstrasse 21, 80290 München, Germany
{schmittt,hanek,buck,beetzm}@in.tum.de, http://www9.in.tum.de/agilo

Abstract. With the services that autonomous robots are to provide becoming more demanding, the states that the robots have to estimate become more complex. In this paper, we develop and analyze a probabilistic, vision-based state estimation method for individual, autonomous robots. This method enables a team of mobile robots to estimate their joint positions in a known environment and track the positions of autonomously moving objects. The state estimators of different robots cooperate to increase the accuracy and reliability of the estimation process. This cooperation between the robots enables them to track temporarily occluded objects and to faster recover their position after they have lost track of it. The method is empirically validated based on experiments with a team of physical robots.

1 Introduction

Autonomous robots must have information about themselves and their environments that is sufficient and accurate enough for the robots to complete their tasks competently. Contrary to these needs, the information that robots receive through their sensors is inherently uncertain: typically the robots' sensors can only access parts of their environments and their sensor measurements are inaccurate and noisy. In addition, control over their actuators is also inaccurate and unreliable. Finally, the dynamics of many environments cannot be accurately modeled and sometimes environments change nondeterministically.

Recent longterm experiments with autonomous robots [13] have shown that an impressively high level of reliability and autonomy can be reached by explicitly representing and maintaining the uncertainty inherent in the available information. One particularly promising method for accomplishing this is *probabilistic state estimation*. Probabilistic state estimation modules maintain the probability densities for the states of objects over time. The probability density of an object's state conditioned on the sensor measurements received so far contains all the information which is available about an object that is available to a robot. Based on these densities, robots are not only able to determine the most likely state of the objects, but can also derive even more meaningful statistics such as the variance of the current estimate.

Successful state estimation systems have been implemented for a variety of tasks including the estimation of the robot's position in a known environment,

A. Birk, S. Coradeschi, and S. Tadokoro (Eds.): RoboCup 2001, LNAI 2377, pp. 193–203, 2002.

the automatic learning of environment maps, the state estimation for objects with dynamic states (such as doors), for the tracking of people locations, and gesture recognition [12]. With the services that autonomous robots are to provide becoming more demanding, the states that the robots have to estimate become more complex. Robotic soccer provides a good case in point. In robot soccer (mid-size league) two teams of four autonomous robots play soccer against each other. A probabilistic state estimator for competent robotic soccer players should provide the action selection routines with estimates of the positions and may be even the dynamic states of each player and the ball.

This estimation problem confronts probabilistic state estimation methods with a unique combination of difficult challenges. The state is to be estimated by multiple mobile sensors with uncertain positions, the soccer field is only partly accessible for each sensor due to occlusion caused by other robots, the robots change their direction and speed very abruptly, and the models of the dynamic states of the robots of the other team are very crude and uncertain.

In this paper, we describe a state estimation module for individual, autonomous robots that enables a team of robots to estimate their joint positions in a known environment and track the positions of autonomously moving objects. The state estimation modules of different robots cooperate to increase the accuracy and reliability of the estimation process. In particular, the cooperation between the robots enables them to track temporarily occluded objects and to faster recover their position after they have lost track of it.

The state estimation module of a single robot is decomposed into subcomponents for self-localization and for tracking different kinds of objects. This decomposition reduces the overall complexity of the state estimation process and enables the robots to exploit the structures and assumptions underlying the different subtasks of the complete estimation task. Accuracy and reliability is further increased through the cooperation of these subcomponents. In this cooperation the estimated state of one subcomponent is used as evidence by the other subcomponents.

The main contributions of this paper are the following ones. First, we show that image-based probabilistic estimation of complex environment states is feasible in real time even in complex and fast changing environments. Second, we show that maintaining trees of possible tracks is particularly useful for estimating a global state based on multiple mobile sensors with position uncertainty. Third, we show how the state estimation modules of individual robots can cooperate in order to produce more accurate and reliable state estimation.

In the remainder of the paper we proceed as follows. Section 2 describes the software architecture of the state estimation module and sketches the interactions among its components. Section 3 provides a detailed description of the individual state estimators. We conclude with our experimental results and a discussion of related work.

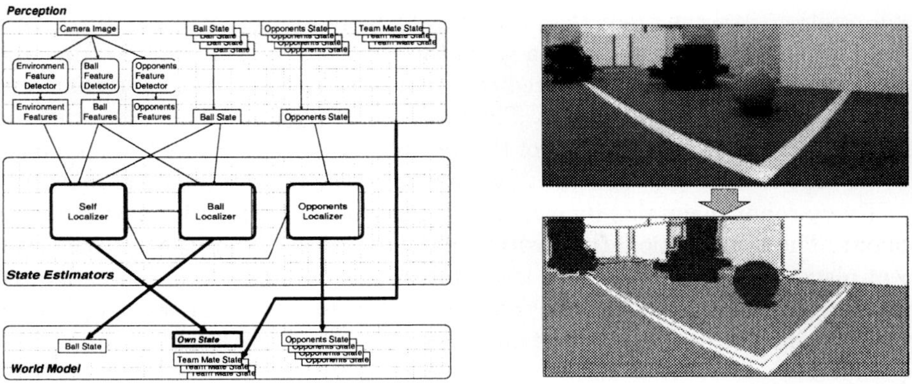

Fig. 1. (a) Architecture of the state estimator. (b) The figure shows an image captured by the robot and the feature maps that are computed for self, ball, and opponent localization.

2 Overview of the State Estimator

Fig. 1a shows the components of the state estimator and its embedding into the control system. The subsystem consists of the perception subsystem, the state estimator itself, and the world model. The perception subsystem itself consists of a camera system with several feature detectors and a communication link that enables the robot to receive information from other robots. The world model contains a position estimate for each dynamic task-relevant object. In this paper the notion of position refers to the x- and y-coordinates of the objects and includes for the robots of the own team the robot's orientation. The estimated positions are also associated with a measure of accuracy, a covariance matrix.

The perception subsystem provides the following kinds of information: (1) partial state estimates broadcasted by other robots, (2) feature maps extracted from captured images, and (3) odometric information. The estimates broadcasted by the robots of the own team comprise the estimate of the ball's location. In addition, each robot of the own team provides an estimate of its own position. Finally, each robot provides an estimate for the position of every opponent. From the captured camera images the feature detectors extract problem-specific feature maps that correspond to (1) static objects in the environment including the goal, the borders of the field, and the lines on the field, (2) a color blob corresponding to the ball, and (3) the visual features of the opponents.

The state estimation subsystem consists of three interacting estimators: the self localization system, the ball estimator, and the opponents estimator. State estimation is an iterative process where each iteration is triggered by the arrival of a new piece of evidence, a captured image or a state estimate broadcasted by another robot. The self localization estimates the probability density of the robot's own position based on extracted environment features, the estimated ball position, and the predicted position. The ball localizer estimates the probability density for the ball position given the robot's own estimated position and its perception of the ball, the predicted ball position, and the ball estimations broadcasted by the other robots. The positions of the opponents are estimated based on the estimated position of the observing robot, the robots' appearances in the captured images, and their positions as estimated by the team mates.

Every robot maintains its own global world model, which is constructed as follows. The own position, the position of the ball, and the positions of the opponent players are produced by the local state estimation processes. The estimated positions of the team mates are the broadcasted results of the self localization processes of the corresponding team mates

3 The Individual State Estimators

3.1 Self- and Ball-Localization

The self- and ball-localization module iteratively estimates, given the observations taken by the robot and a model of the environment and the ball, the probability density over the possible robot and ball positions. A detailed description and analysis of the applied alogrithms can be found in [6,7].

3.2 Opponents Localization

The objective of the opponents localization module is to track the positions of the other team's robots. The estimated position of one opponent is represented by one or more alternative object hypotheses. Thus the task of the state estimator is to (1) detect feature blobs in the captured image that correspond to an opponent, (2) estimate the position and uncertainties of the opponent in world coordinates, and (3) associate them with the correct object hypothesis. In our state estimator we use Reid's Multiple Hypotheses Tracking (MHT) algorithm [10] as the basic method for realizing the state estimation task. In this section we demonstrate how this framework can be applied to model dynamic environments in multirobot systems. We extend the general framework in that we provide mechanisms to handle multiple mobile sensors with uncertain positions.

Multi Hypotheses Tracking We will describe the Multiple Hypotheses Tracking method by first detailing the underlying opponents model, then explaining the representation of tracked opponents position estimates, and finally presenting the computational steps of the algorithm.

The Opponents Model. The model considers opponent robots to be moving objects of unknown shape with associated information describing their temporal dynamics, such as their velocity. The number of the opponent robots may vary. The opponent robots have visual features that can be detected as feature blobs by the perception system.

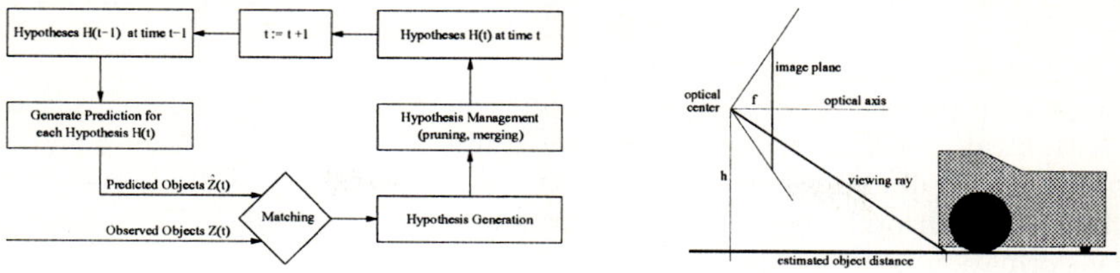

Fig. 2. (a) The multiple hypotheses framework for dynamic environment modeling. (b) An estimate of the robot's distance is given through the intersection of the viewing ray with the ground plane of the field.

The Representation of Opponent Tracks. When tracking the positions of a set of opponent robots there are two kinds of uncertainties that the state estimator has to deal with. The first one is the inaccuracy of the robot's sensors. We represent this kind of uncertainty using a Gaussian probability density. The second kind of uncertainty is introduced by the data association problem, i.e. assigning feature blobs to object hypotheses. This uncertainty is represented by a hypotheses tree where nodes represent the association of a feature blob with an object hypothesis. A node $H_j(t)$ is a son of the node $H_i(t-1)$ if $H_j(t)$ results from the assignment of an observed feature blob with a predicted state of the hypothesis $H_i(t-1)$. In order to constrain the growth of the hypotheses tree, it is pruned to eliminate improbable branches with every iteration of the MHT.

The MHT Algorithm. Fig. 2a outlines the computational structure of the MHT algorithm. An iteration begins with the set of hypotheses $H(t-1)$ from the previous iteration $t-1$. Each hypothesis represents a different assignment of measurements to objects, which was performed in the past. The algorithm maintains a Kalman filter for each hypothesis. For each hypothesis a position of the dynamic objects is predicted $\widehat{Z}_i(t)$ and compared with the next observed opponent performed by an arbitrary robot of the team. Assignments of measurements to objects are accomplished on the basis of a statistical distance measurement. Each subsequent child hypothesis represents one possible interpretation of the set of observed objects and, together with its parent hypothesis, represents one possible interpretation of all past observations. With every iteration of the MHT probabilities describing the validity of an hypothesis are calculated [1]. In order to constrain the growth of the tree the algorithm prunes improbable branches. Pruning is based on a combination of ratio pruning, i.e. a simple lower limit on the ratio of the probabilities of the current and best hypotheses, and the N-scan-back algorithm [10]. The algorithm assumes that any ambiguity at time t is resolved by time $t+N$. Consequently if at time t hypothesis $H_i(t-1)$ has n children, the sum of the probabilities of the leaf notes of each branch is calculated. The branch with the greatest probability is retained and the others are discarded.

The Unscented Transformation

The general problem is as follows. Given an n-dimensional vector random variable x with mean $\bar{x}$ and covariance C_x we would like to estimate the mean $\bar{y}$ and the covariance C_y of an m-dimensional vector random variable y. Both variables are related to each other by a non-linear transformation $y = g(x)$. The unscented transformation is defined as follows:

1. Compute the set Z of $2n$ points from the rows or columns of the matrices $\pm\sqrt{nC_x}$. This set is zero mean with covariance C_x. The matrix square root can efficiently be calculated by the Cholesky decomposition.
2. Compute a set of points X with the same covariance, but with mean $\bar{x}$, by translating each of the points as $x_i = z_i + \bar{x}$.
3. Transform each point $x_i \in X$ to the set Y with $y_i = g(x_i)$.
4. Compute $\bar{y}$ and C_y by computing the mean and covariance of the $2n$ points in the set Y.

Fig. 3. Outline of the unscented transformation.

Feature Extraction and Uncertainty Estimation This section outlines the feature extraction process which is performed in order to estimate the positions and the covariances of the opponent team's robots. Each opponent robots is modeled in world coordinates by a bi-variate Gaussian density with mean Ψ and a covariance matrix C_ψ.

At present it is assumed that the opponent robots are constructed in the same way and have approximately circular shape. All robots are colored black. Friend foe discrimination is enabled through predefined color markers (cyan and magenta, see Fig. 1b) on the robots. Each marker color may be assigned to any of the two competing teams. Consequently it is important that the following algorithms can be parameterized accordingly. Furthermore we assume that (see Fig. 2b) the tracked object almost touches the ground. The predefined robot colors allow a relatively simple feature extraction process.

Step 1: Extraction of Blobs Containing Opponent Robots From a captured image the black color-regions are extracted through color classification and morphological operators. In order to be recognized as an opponent robot a black blob has to obey several constraints, e.g. a minimum size and a red or green color-region adjacent to the bottom region row. Through this we are able to distinguish robots from black logos and adverts affixed on the wall surrounding the field. Furthermore blobs that contain or have a color-region of the own team color in the immediate neighborhood are discarded.
For all remaining regions three features are extracted: The bottom most pixel *row* which exceeds a predefined length, the column *col* representing the center of gravity and a mean blob *width* in pixels. For the latter two features only the three bottom most rows which exceed a certain length are used. In order to determine these rows, we allow also for occlusion through the ball. If the length of these rows exceeds an upper length, we assume that we have detected

$$\bar{\omega} = [\mathbf{\Phi}, row, col, width]$$

$$C_\omega = \begin{pmatrix} C_\phi & 0 & 0 & 0 \\ 0 & 1 & 0 & 0 \\ 0 & 0 & 1 & 0 \\ 0 & 0 & 0 & 1 \end{pmatrix}$$

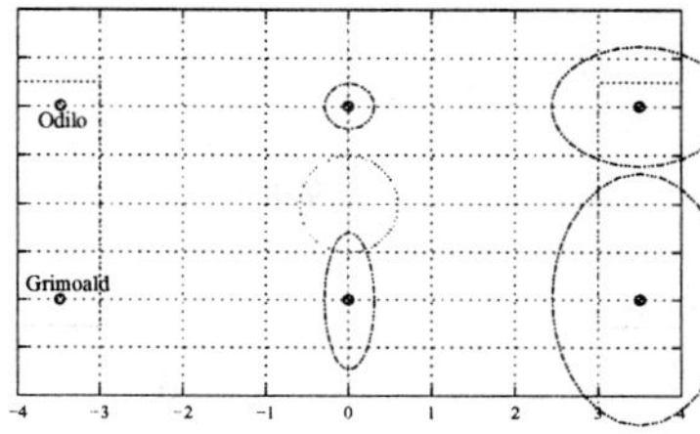

Fig. 4. (a) Intermediate $\bar{\omega}$ mean and covariance C_ω (b) Propagation of uncertainties.

two opponents which are directly next to each other. In this case two centers of gravity are computed and the width is halfed.

In order to detect cascaded robots, i.e. opponent robots that are partially occluded by other robots, our algorithm also examines the upper rows of the blobs. As soon as the length of a blob row differs significantly from the length of its lower predecessor and the respective world coordinates indicate a height of more than 10 cm above the ground we assume that we have detected cascaded robots. In this case we split the blob into two and apply the above procedure to both blobs. Empirically we have found that this feature extraction procedure is sufficient to determine accurate positions of opponent robots. Mistakenly extracted objects are generally resolved in a fast manner by the MHT algorithm.

Step 2: Estimation of Opponent Position and Uncertainty In the following we will estimate the position and covariance of an observed robot. For this the pose and the covariance of the observing robots as well as position of the detected feature blob in the image and the associated measurement uncertainties are taken into account.

We define a function *opp* that determines the world coordinates of an opponent robot based on the pose $\mathbf{\Phi}$ of the observing robot, the pixel coordinates *row*,*col* of the center of gravity and the width *width* of the opponent robot's blob. Due to rotations and radial distortions of the lenses *opp* is non-linear. First the function *opp* converts the blob's pixel coordinates to relative polar coordinates. On this basis and the width of the observed blob the radius of the observed robot is estimated. Since the polar coordinates only describe the distance to the opponent robot but not the distance to its center, the radius is added to the distance. Finally the polar coordinates are transformed into world coordinates taking the observing robot's pose $\mathbf{\Phi}$ into account.

In order to estimate the position ψ and the covariance C_ψ of an opponent robot, we will use a technique similar to the unscented transformation [8] (see Fig. 3). First an intermediate mean $\bar{\omega}$ and covariance C_ω describing jointly the observing robot's pose and the observed robot is set up (see Fig. 4a). $\mathbf{\Phi}$, *row*, *col* and *width* are assumed to be uncorrelated with a variance of one pixel. To this mean and covariance the unscented transformation using the non-linear mapping *opp* is applied. This yields the opponent robot's position ψ and covariance C_ψ. In Fig. 4b the uncertainties of objects depending on the uncertainty of the

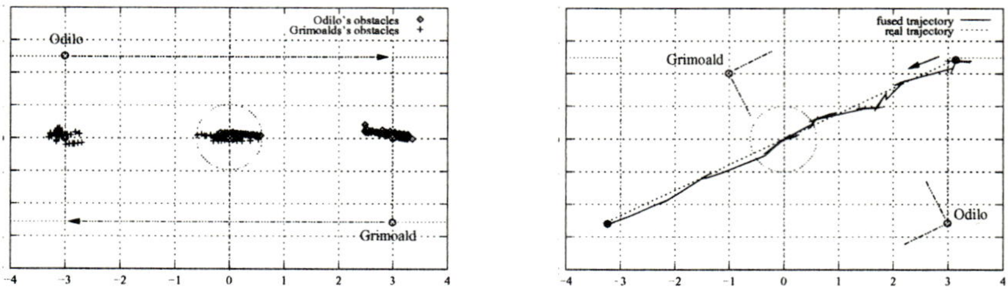

Fig. 5. (a) Two robots are traveling across the field, while they observe three stationary robots of the opponent team. The diamonds and crosses indicate the different measurements performed by the observing robots. (b) The resolved trajectory (continuous line) of an opponent robot, observed by two robots. The real trajectory is displayed as dotted line. The dashed lines indicate the robot's 90 degrees field of view.

observing robot and their relative distances are displayed using 1σ-contours. For illustrative purposes the uncertainty ellipses are scaled by an order of five. Each robot observes two obstacles in 3.5 and 7 meters distance. Robot Odilo is very certain about its pose and thus the covariance of the observed robot depends mainly on its distance. Robot Grimoald has a high uncertainty in its orientation ($\approx$ 7 degrees). Consequently the position estimate of the observed obstacle is less precise and is highly influenced by the orientation uncertainty.

Step 3: Association of Opponents to Object Hypotheses The association of an opponent robot's position with a predicted object position is currently performed on the basis of the Mahalanobis distance. In future we intent to use the Bhattacharyya distance, which is a more accurate distance measure for probability distributions.

4 Experimental Results

The presented algorithms are applied in our middle-size RoboCup team, the AGILO[1] RoboCuppers. At present, the RoboCup scenario defines a fixed world model with field-boundaries, lines and circles (see Fig. 4). Our approach was successfully applied in 1999 and 2000 during the RoboCup World Soccer Championship in Stockholm and Melbourne and the German Vision Cup. During a RoboCup match, every robot is able to process 15 to 18 frames per second with its on-board Pentium 200 MHz computer. When the robots planning algorithms' are turned off the vision system is easily able to cope with the maximum frame rate of our camera (25 fps). The localization algorithm runs with a mean processing time of 18 msec for a 16-Bit RGB ($384 * 288$) image. Only for 4% of the images the processing time exceeds 25 msec. A detailed analysis of the self- and ball-localization algorithm can be found in [6].

[1] The name is an homage to the oldest ruling dynasty in Bavaria, the Agilolfinger. The dynasties most famous representatives are Grimoald, Hugibert, Odilo and Tassilo

In the following we will present experiments that investigate the capability of tracking multiple opponent robots by our system. In the first experiment we have examined the capability of our algorithms to detect and estimate the opponent robots positions. The robots Odilo and Grimoald are simultaneously traveling in opposite directions from one side of the playing field to the other (see Fig. 5a). While they are in motion they are observing three stationary robots which are set up at different positions in the middle of the field. Diamonds and crosses indicate the observed opponents by Odilo and Grimoald, respectively. The variance in the observation is due to positions estimations over long distances (4 to 7 meters) and minor inaccuracies in the robots self-localization. Furthermore it is noteworthy that the vision system of both robots never mistook their teammate as opponent.

The second experiment examines the tracking and data fusion capability of our system. Odilo and Grimoald are set up at different positions on the field. An opponent robot crosses the field diagonally from the corner of the penalty area at the top right to the corner of the penalty area at the lower left (see Fig. 5b). The first part of the journey is only observed by Grimoald, the middle part of both robots and the final part only by Odilo. The 90 degrees field of view of both robots is indicated through lines.

The opponent robot was tracked using the MHT algorithm with a simple linear Kalman filter. The state transition matrix, described a constant velocity model and the measurement vector provided positional information only. The positions of the opponent robots and their uncertainties were computed according to the algorithm described in section 3.2. Positional variance for the pixel coordinates of the region's center of gravity and region's width was assumed to be one pixel. The process noise was assumed to be white noise acceleration with a variance of 0.1 meters. The Mahalanobis distance was chosen such that $P\{X \leq \chi_2^2\} = 0.95$. N-scan-back pruning was performed from a depth of $N = 3$. In general the update time for one MHT iteration including N-scan-back and hypo pruning was found to be less than 10 msec. This short update time is due to the limited number of observers and observed objects in our experiment. We expect this time to grow drastically with an increasing number of observing and observed robots. However within a RoboCup scenario a natural upper bound is imposed through the limited number of robots per team. A detailed analysis of the hypothesis trees revealed that only at very few occasions new tracks were initiated. All of them were pruned away within two iterations of the MHT. Overall the observed track (see Fig. 5b, continuous line) diverges relatively little from the real trajectory (dotted line).

5 Related Work

Related work comprises work done on object tracking and probabilistic localization in the robot soccer domain and probabilistic and vision-based tracking of moving targets. In the research area of autonomous robot soccer algorithms for probabilistic self-localization have been proposed. Gutmann et al. [5] have proposed a self localization method based on a Kalman filter approach by matching observed laser scan lines into the environment model. We differ from this approach mainly by using vision data instead of laser data. The advantage of using vision sensors is that the method can be applied more easily to other kinds of environments, for example outdoor environments. Enderle et al. [3] have developed a vision-based self-localization module using a sample-based Markov localization method. The advantage of Markov localization is that no assumption about the form of the probability distribution is made. However, in order to achieve high accuracy, usually a high number of samples is required. A good match between the observation and a sample pose leads to new randomly generated sample poses in the vicinity of the good sample. Hence, Markov localization leads to limited accuracy and/or relatively high computational cost. The self localization method that is proposed here has the advantage that it is faster and can be more easily integrated with the other state estimation modules. Our approach extends the Kalman filter approach [4] to self localization in that we are able to deal with nonlinearities because our approach performs a iterative optimization instead of a linear prediction.

To the best of our knowledge no probabilistic state estimation method has been proposed for tracking the opponent robots in robot soccer or similar application domains. Gutmann et al. [5] estimate the positions of the opponents and store them in the team world model but they do not probabilistically integrate the different pieces of information. This results in a less accurate and reliable estimate of the opponents positions. Probabilistic tracking of multiple moving has been proposed by Schulz et al. [11]. They apply sample-based Markov estimation to the tracking of moving people with a moving robot using laser range data. The required computational power for the particle filters is opposed by the heuristic based pruning strategies of the MHT algorithm.

Our approach to multiple hypothesis tracking is most closely related to the one proposed by Cox and Miller [2]. Indeed our algorithm is based on their implementation. We extend their work on multiple hypothesis tracking in that we apply the method to a much more challenging application domain where we have multiple moving observers with uncertain positions. In addition, we perform object tracking at an object rather than on a feature level.

6 Conclusions

In this paper, we have developed and analyzed a cooperative probabilistic, vision-based state estimation method for individual, autonomous robots. This method enables a team of mobile robots to estimate their joint positions in a known environment and track the positions of autonomously moving objects.

References

1. I. Cox and J. Leonard. Modeling a dynamic environment using a bayesian multiple hypothesis approach. Artificial Intelligence, 66:311–344, 1994.

2. I.J. Cox and S.L. Hingorani. An Efficient Implementation of Reid's Multiple Hypothesis Tracking Algorithm and Its Evaluation for the Purpose of Visual Tracking. *IEEE Trans. on PAMI*, 18(2):138–150, February 1996.
3. S. Enderle, M. Ritter, D. Fox, S. Sablatng, G. Kraetzschmar, and G. Palm. Soccer-robot locatization using sporadic visual features. In *IAS-6*, Venice, Italy, 2000.
4. O.D. Faugeras. Three-dimensional computer vision: A geometric viewpoint. *MIT Press*, page 302, 1993.
5. J.-S. Gutmann, W. Hatzack, I. Herrmann, B. Nebel, F. Rittinger, A. Topor, T. Weigel, and B. Welsch. The CS Freiburg Robotic Soccer Team: Reliable Self-Localization, Multirobot Sensor Integration, and Basic Soccer Skills. In *2nd Int. Workshop on RoboCup, LNCS*. Springer-Verlag, 1999.
6. R. Hanek and T. Schmitt. Vision-based localization and data fusion in a system of cooperating mobile robots. In *IROS*, 2000.
7. R. Hanek, T. Schmitt, M. Klupsch, and S. Buck. From multiple images to a consistent view. In *4th Int. Workshop on RoboCup, LNCS*. Springer-Verlag, 2000.
8. S. Julier and J. Uhlmann. A new extension of the kalman filter to nonlinear systems. The 11th Int. Symp. on Aerospace/Defence Sensing, Simulation and Controls., 1997.
9. C. Marques and P. Lima. Vision-Based Self-Localization for Soccer Robots. In *IROS*, 2000.
10. D. Reid. An algorithm for tracking multiple targets. IEEE Transactions on Automatic Control, 24(6):843–854, 1979.
11. D. Schulz, W. Burgard, D. Fox, and A.B. Cremers. Tracking Multiple Moving Targets with a Mobile Robot using Particle Filters and Statistical Data Association. In *ICRA*, 2001.
12. S. Thrun. Probabilistic algorithms in robotics. *AI Magazine*, 2000.
13. S. Thrun, M. Beetz, M. Bennewitz, W. Burgard, A.B. Cremers, F. Dellaert, D. Fox, D. Haehnel, C. Rosenberg, N. Roy, J. Schulte, and D. Schulz. Probabilistic algorithms and the interactive museum tour-guide robot minerva. *International Journal of Robotics Research*, 2000.

High-Speed Obstacle Avoidance and Self-Localization for Mobile Robots Based on Omni-directional Imaging of Floor Region

Daisuke Sekimori[1], Tomoya Usui[2], Yasuhiro Masutani[2], and Fumio Miyazaki[2]

[1] Akashi College of Technology, Akashi, Hyougo Japan
sekimori@akashi.ac.jp
[2] Osaka University, Toyonaka, Osaka Japan
{usui,masutani,miyazaki}@robotics.me.es.osaka-u.ac.jp
http://robotics.me.es.osaka-u.ac.jp/OMNI/

Abstract. In this paper, we propose a method of obstacle avoidance and a method of self-localization based on floor region provided by omni-directional imaging. With our methods, omni-directional imaging is used not for recognition of the three-dimensional environment but for detecting obstacles and landmarks in a wide area at high speed. Several experiments with a real robot according to the rules of the RoboCup Small-Size League was demonstrated, and proved the effectiveness of these methods.

1 Introduction

In obstacle avoidance and self-localization for autonomous mobile robots, information about the environment provided by an external sensor is indispensable.

Conventional studies on navigation, obstacle avoidance and self-localization of mobile robots using the omni-directional visual sensor propose methods to utilize vertical edges in the environment that appears radially from the image center[1][2]. However, these are based on the premise that vertical edges in the environment are stably extracted. Accordingly, these methods do not apply when the edges are blocked by moving objects or do not exist in the environment.

In this paper, we propose a method of avoiding obstacles around a robot and a method of self-localization based on omni-directional imaging of the floor region. We use an omni-directional visual sensor to detect obstacles and landmarks quickly and widely instead of a method to recognize the three-dimensions of the environment, in order to utilize the special features of an omni-directional visual sensor. Furthermore, we consider obstacle avoidance and self-localization in an environment where other obstacles exist.

In section 2, we explain omni-directional imaging of the floor region. In sections 3 and 4, we describe algorithms of obstacle avoidance and self-localization based on omni-directional imaging in detail. Finally, in section 5, we verify the effectiveness of these algorithms through several experiments with a real robot.

A. Birk, S. Coradeschi, and S. Tadokoro (Eds.): RoboCup 2001, LNAI 2377, pp. 204–213, 2002.

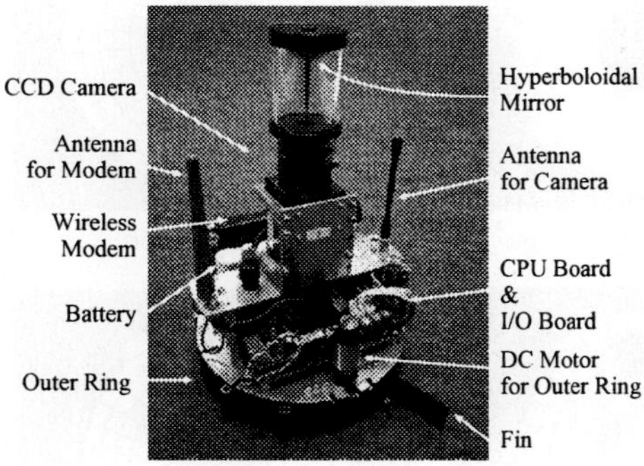

Fig. 1. Overview of the robot

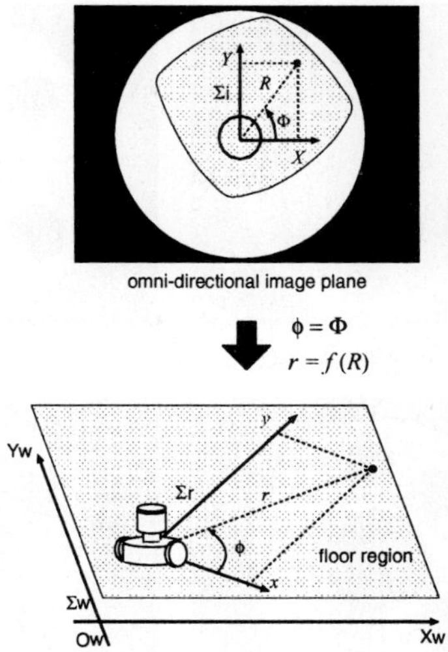

Fig. 2. Distance and direction of object

2 Omni-directional Imaging of Floor Regions

In a lot of indoor environments, it is possible to distinguish between the floor region and other objects by color. Accordingly, in this paper, we focus on an omni-directional, vertical-axis visual sensor model that is mounted on a mobile robot. The mobile robot that we have developed for the RoboCup Small-Size League is shown in **Fig.1**.

Here we consider a conversion from a point of the image coordinate system Σ_i to the correspoinding point on the floor of the robot coordinate system Σ_r, as shown in **Fig.2**. We assume that the conversion from (X, Y) to (x, y) is expressed as follows:

$$R = \sqrt{X^2 + Y^2}, \quad \Phi = \tan^{-1} Y/X \tag{1}$$

$$\phi = \Phi \tag{2}$$

$$r = f(R) \tag{3}$$

$$x = r\cos\phi, \quad y = r\sin\phi, \tag{4}$$

where $f(R)$ is a function determined from the shape of the omni-directional mirror.

Moreover, we make the following assumptions about the omni-directional image. (i) An omni-directional image with color is used, (ii) The floor region has a single color and other regions have other colors, (iii) White lines of constant width exist in the floor region.

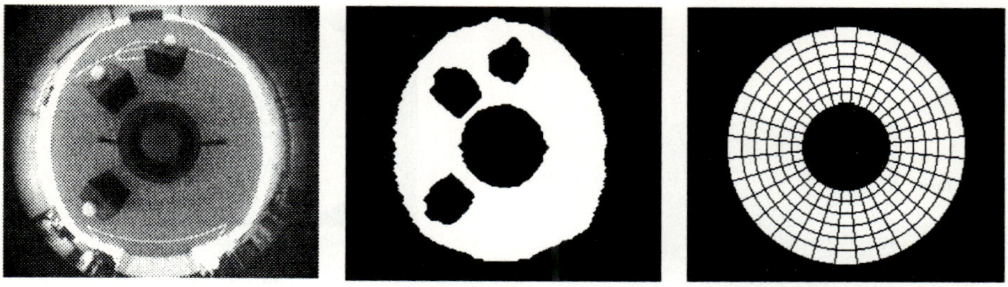

(a)Camera image (b)Binarized image (c)Template

(d)After AND operation of (b) and (c) (e)After labeling and integrating (f)Detected free space

Fig. 3. Example of the method of detecting free space

3 Obstacle Avoidance Based on Omni-directional Imaging of Floor Regions

If the assumptions in the previous section are satisfied, the free space around the robot can be determined by observing the floor color in the robot's neighborhood. Therefore, in this section, we propose a method of high-speed obstacle avoidance based on omni-directional imaging of the floor region. In this method, the omni-directional image is first divided into many cells by superimposing a template image. Then, the free space around the robot is detected at high speed by using the labeling function, which is one of basic image processing functions. It is possible to modify resolution of the floor region totally or partially by changing the number of partitions in the template. In the following part of this section, first, we describe the algorithm for detecting free space around the robot. Second, as an example of obstacle avoidance based on the detected free space, we explain an obstacle avoidance method considering the mobile velocity of the robot.

3.1 Detection of Free Space

Step 1 (Binarizing)
The binarized image is taken from the omni-directional image (**Fig.3** (a)) by extracting floor color. Next, noise in the floor region is removed by using dilation and erosion processes a few times (**Fig.3** (b))

Step 2 (Dividing the floor region)
A template which is divided equally into I individual cells in a circumferential direction and J individual cells in a radial direction is prepared (**Fig.3** (c)). Let S_{ij} and A_{ij} ($i = 1, \cdots, I; j = 1, \cdots, J$) denote a closed domain and the area of each cell, respectively. The floor region is divided into small areas of polar coordinates by using the AND operation between the binarized image and the template(**Fig.3** (d)).

Step 3 (Labeling and determination of the presence of obstacles)
N labeled areas are derived from the divided image. We define area A_k and the center of gravity $\boldsymbol{p}_k = (x_k, y_k)^T$ of each area ($k = 1, \cdots, N$). Then the area ratio r_{ij} of the floor color is computed for every cell, using the following equation:

$$r_{ij} = \frac{1}{A_{ij}} \sum_{\substack{k \\ \boldsymbol{p}_k \in S_{ij}}} A_k \tag{5}$$

Next, the presence of an obstacle is judged for each cell by using the following equation (**Fig.3** (e)):

$$q_{ij} = \begin{cases} 0 \ (r_{ij} < r_{th}) \\ 1 \ (r_{ij} \geq r_{th}) \end{cases}, \tag{6}$$

where $q_{ij} = 0/1$ signifies the presence/absence of obstacles, and r_{th} denotes a threshold which determines the presence of an obstacle.

Step 4 (Detecting free space)
The distance to the closest obstacle in each direction around the robot is derived as discrete one-dimensional data a_j by using the following equation:

$$a_j = \min_{q_{nj}=0} (n-1) \qquad n = 1, \cdots, I+1, \tag{7}$$

where $q_{(I+1)j} = 0$ for convenience (**Fig.3** (f)).

3.2 Obstacle Avoidance Based on Mobile Velocity

If free space around the robot is detected, we can propose various algorithms for obstacle avoidance for various purposes. We have developed an omni-directional mobile robot equipped with an omni-directional visual sensor as shown in **Fig.1**[3]. For this robot, we proposed a simple obstacle avoidance utilizing the free space detection mentioned above which considers only a goal direction[4]. In this section, we propose a method also considering the mobile velocity of the robot.

Although the robot we developed has an omni-directional mobile mechanism, the magnitude of the acceleration vector is limited in order to avoid slip and shock during sudden changes in direction. Actually, the modified command velocity $\boldsymbol{v}_r(t)$ is determined in the following equation:

$$\boldsymbol{v}_r(t) = \begin{cases} \boldsymbol{v}_a + \frac{\Delta \boldsymbol{v}}{|\Delta \boldsymbol{v}|} \alpha \cdot t & (t \leq t_a) \\ \boldsymbol{v}_b & (t > t_a) \end{cases} \quad t_a = \frac{|\Delta \boldsymbol{v}|}{\alpha}, \ \Delta \boldsymbol{v} = \boldsymbol{v}_b - \boldsymbol{v}_a, \tag{8}$$

where the original command velocity(desired velocity) is $\boldsymbol{v}_b$. The time is 0 when the original command velocity is received, and the velocity of the robot at that time is $\boldsymbol{v}_a$. α is the upper limit value of acceleration.

Because of the acceleration limit, the robot can not turn immediately in a mobile direction at the same rate as the original command velocity in case that the mobile velocity is high. Therefore, it is not enough to detect obstacles only in the command direction. In this method, first, the position of an obstacle is extracted from the detected free space. Next, the trajectories of the robot are predicted based on current mobile velocity and projected mobile velocity in every direction. Finally, the mobile direction for obstacle avoidance can be determined beforehand. If this is impossible, the robot stops immediately and keeps searching until it finds a free space. This process is demonstrated in **Fig.4**. Details are omitted due to lack of space.

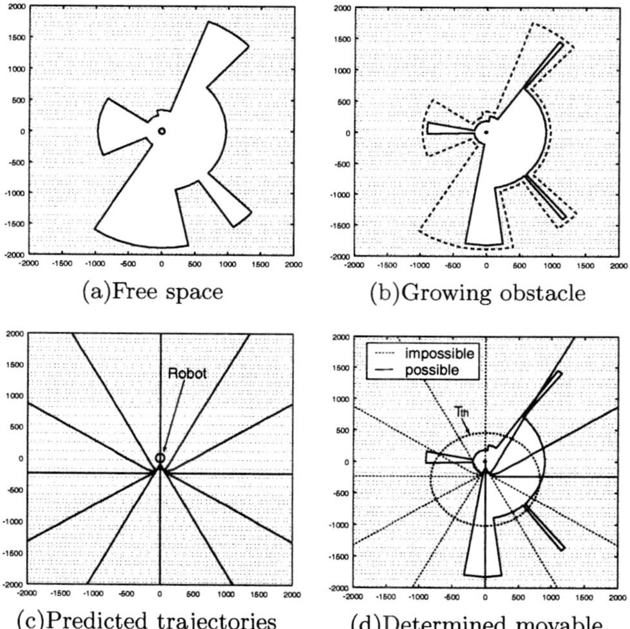

(a)Free space (b)Growing obstacle

(c)Predicted trajectories (d)Determined movable direction

Fig. 4. Example of the method for determining movable direction

4 Self-Localization Based on Omni-directional Imaging of Floor Region

It is important for mobile robots to estimate self-location based on landmarks in the environment. However, if a specified point is regarded as a landmark, self-localization can not be estimated when the landmark is hidden.

On the other hand, a whole image of a known floor region can be acquired at one time with omni-directional imaging. Therefore, we propose a method of self-localization that regards the whole floor region as a landmark. In this method, even if part of the floor region is hidden by obstacles, the hidden part can be compensated by computing the convex hull of the set of boundary points of the floor region in the omni-directional imaging. After that, using the least squares method, considering the property of omni-directional imaging, a set of boundary points after restoration is applied to the already-known shape of the floor region. Although the problem is non-linear, by utilizing good approximate values provided from the geometric features (e.g. center of gravity, principle axis of second moment) of the region surrounded by the set of points, estimated values are calculated by solving the linearized equation. The algorithm of self-localization using the field in the RoboCup Small-Size League is explained as follows:

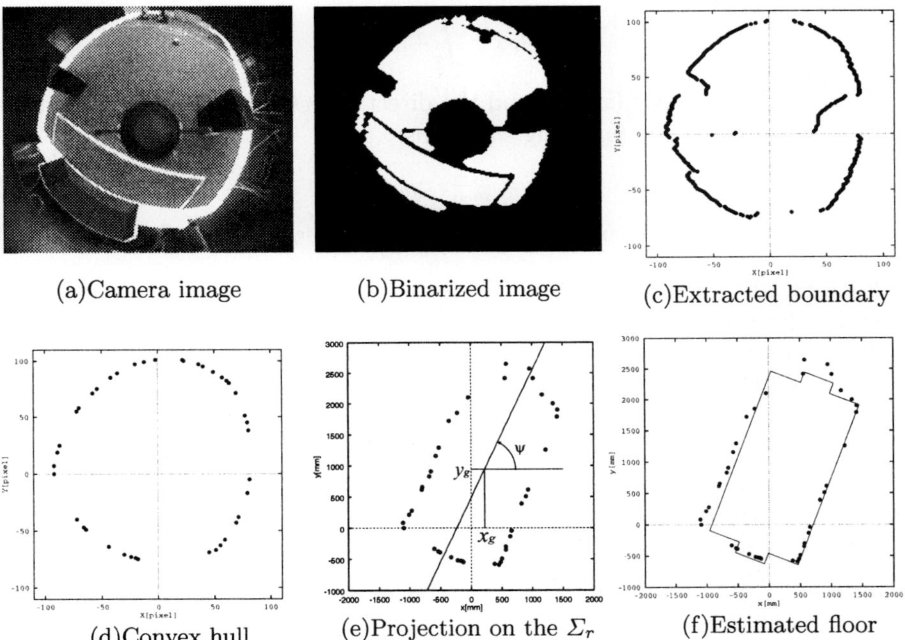

(a)Camera image (b)Binarized image (c)Extracted boundary

(d)Convex hull (e)Projection on the Σ_r (f)Estimated floor region

Fig. 5. Example of the method of self-localization

Step 1 (Extracting the floor region)
The binarized image is provided by extracting floor color (green in the case of the RoboCup) from the omni-directional image(**Fig.5** (a)). Noise is removed by using dilation and erosion processes several times (**Fig.5** (b))

Step 2 (Extracting the boundaries of the floor region)
Minimum $X_{min,i}$ and maximum $X_{max,i}$ of the X coordinates of the region

corresponding to each Y coordinate Y_i are detected by extracting the boundaries of the floor region.

Step 3 (Computing the convex hull)
In order to compensate for the boundaries, the convex hull [5] of a set M of points $\{(X_1, Y_1), \cdots, (X_M, Y_M)\}$ is derived from the set of boundary points $\{(X_{min,1}, Y_1), \cdots, (X_{min,N}, Y_N), (X_{max,1}, Y_1), \cdots, (X_{max,N}, Y_N)\}$(**Fig.5** (d)).

Step 4 (Converting)
The convex hull of a set of points is converted into a floor coordinate system from the image coordinate system, and a set of points $\{(x_1, y_1), \cdots, (x_M, y_M)\}$ is provided by Eq. (1)~Eq. (4).

Step 5 (Geometric features)
The closed domain surrounded by the set of points and the infinitesimal area of the region are S and da, respectively. And we derive the following equation:

$$m_{pq} = \int_S x^p y^q da \tag{9}$$

The center of gravity (x_g, y_g) and the directions ψ_1, ψ_2 of the principal axes are calculated by the following equations:

$$x_g = \frac{m_{10}}{m_{00}}, \quad y_g = \frac{m_{01}}{m_{00}} \tag{10}$$

$$\psi_1 = \frac{1}{2} \tan^{-1} \frac{2(m_{11} - x_g \cdot y_g \cdot m_{00})}{m_{20} - m_{02} - (x_g^2 - y_g^2) m_{00}}, \quad \psi_2 = \psi_1 + \frac{\pi}{2} \tag{11}$$

Each m_{pq} is derived easily from the set of points by dividing the region into triangular segments. Moreover, we calculate the second moments I_1 and I_2 about the principle axes ψ_1 and ψ_2 respectively, and select the direction ψ as follows (**Fig.5** (e)):

$$\psi = \begin{cases} \psi_1 \ (I_1 \leq I_2) \\ \psi_2 \ (I_1 > I_2) \end{cases} \tag{12}$$

Step 6 (Matching with the model)
The set of points $\{(x_1, y_1), \cdots, (x_M, y_M)\}$ is converted into the set of points $\{(\tilde{x}_1, \tilde{y}_1), \cdots, (\tilde{x}_M, \tilde{y}_M)\}$ by the following equations:

$$\tilde{x}_i = (x_i - x_g) \cos \psi + (y_i - y_g) \sin \psi \tag{13}$$

$$\tilde{y}_i = -(x_i - x_g) \sin \psi + (y_i - y_g) \cos \psi \tag{14}$$

It is assumed that the boundary of the floor region consists of some segments of a line. Here, we assume the field of the RoboCup, which consists of six straight lines, as shown in **Fig.6**. Based on relational position, the line each point $(\tilde{x}_i, \tilde{y}_i)$ belongs to is determined, and the distance d_i between the point and the corresponding line is calculated. Next, we calculate parameters a, b, ζ that minimize the following evaluation function J:

$$J(a, b, \zeta) = \sum_{i=1}^{M} w_i {d_i}^2, \tag{15}$$

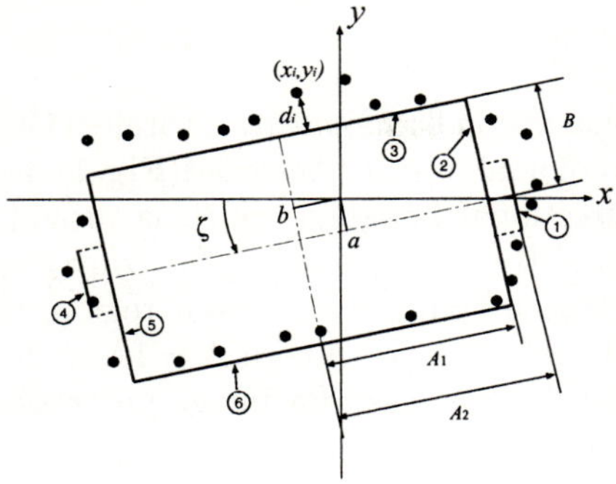

$$x \cos\zeta + y \sin\zeta - (A_2 - a) = 0 \cdots ①$$
$$x \cos\zeta + y \sin\zeta - (A_1 - a) = 0 \cdots ②$$
$$-x \sin\zeta + y \cos\zeta - (B - b) = 0 \cdots ③$$
$$-x \cos\zeta - y \sin\zeta - (A_2 + a) = 0 \cdots ④$$
$$-x \cos\zeta - y \sin\zeta - (A_1 + a) = 0 \cdots ⑤$$
$$x \sin\zeta - y \cos\zeta - (B + b) = 0 \cdots ⑥$$

$A_1 = 1370[\text{mm}], A_2 = 1550[\text{mm}],$
$B = 762.5[\text{mm}]$

Fig. 6. Boundary lines of the floor region (RoboCup field)

where w_i is the weighting factor and set in inverse proportion to its estimated variance. Since the error in the image is considered uniformly distributed, the weighting factor is given as $w_i = 1/\{\frac{df}{dR}(R_i)\}^2$. Although this minimization problem is non-linear, because the set of points is converted into a good approximation from the geometric features, the following approximation can be introduced, $\zeta \approx 0, \sin\zeta \approx \zeta, \cos\zeta \approx 1$. Then a,b and ζ which satisfies $\partial J/\partial a = 0, \partial J/\partial b = 0, \partial J/\partial \zeta = 0$ are solved. Using the solution, new x_g, y_g and ψ values are given in the following equations:

$$x_g^{'} = x_g - a\cos\zeta + b\sin\zeta \tag{16}$$
$$y_g^{'} = y_g - a\sin\zeta - b\cos\zeta \tag{17}$$
$$\psi^{'} = \psi + \zeta \tag{18}$$

In order to decrease the error caused by the approximation, if conditions: $|a| > a_{th}, |b| > b_{th}, |\zeta| > \zeta_{th}$ ($*_{th}$ means each threshold) are satisfied, x_g, y_g, ψ is replaced by $x_g^{'}, y_g^{'}, \psi^{'}$ and we return to the beginning of this step.

Step 7 (Converting to self-location)

The position and orientation of the robot in the field coordinate system ${}^w x, {}^w y, {}^w \theta$ are computed in the following equations:

$${}^w\theta = \psi^{'}, \quad \pi - \psi^{'} \tag{19}$$
$${}^w x = -x_g^{'}\cos({}^w\theta) + y_g^{'}\sin({}^w\theta) \tag{20}$$
$${}^w y = -x_g^{'}\sin({}^w\theta) - y_g^{'}\cos({}^w\theta) \tag{21}$$

For the two solutions of ${}^w\theta$, either is selected according to other landmarks(e.g. goal position) or the time series. The estimated floor region is finally shown in **Fig.5** (f). By weighting various factors, an estimation that attaches great importance to nearby points is realized.

5 Experiments

We equipped the robot with omni-directional vision by mounting a wireless CCD camera (RF System, PRO-5) with a hyperboloidal mirror (Accowle)(**Fig.1**), and applied the proposed method of omni-directional imaging, which is provided at the lens center 165[mm] above the floor. By using model[6] for conversion from distance R[pixel] in the image to distance r[mm] on the floor region, the following relation is derived: $f(R) = R/(A - B\sqrt{C + R^2})$ where $A = 10.12, B = 2.054 \times 10^{-2}, C = 2.298 \times 10^5$. IP5005(Hitachi) was used for image processing, with a resolution of 256×220[pixel^2].

5.1 Obstacle Avoidance

As an example of obstacle avoidance, we undertook an experiment in which the robot turned at an acute angle corner of about 55 deg. and aimed at a goal point.

We experimented on two types of obstacle avoidance. One took no account of the robot's velocity and the other took the robot's velocity into account under the following conditions: number of cells of template $I = 6, J = 32$, maximum velocity of the robot v=400mm/s, upper limit of acceleration α=300mm/s^2.

As a result, both turns were executed without colliding with the wall and the robot reached the goal point. However, because the method that takes the robot's velocity into acount allows for early directional changes, it facilitated turning corners smoothly. And, the mean cycle of the total process was about 40ms when the robot's velocity was taken into account.

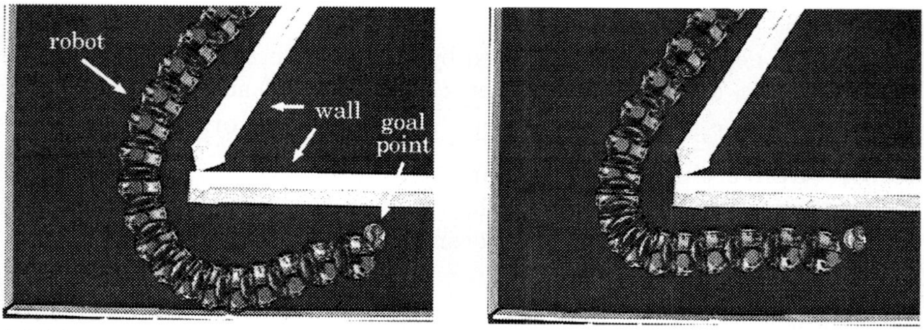

(a)Taking no account of the robot's velocity (b)Taking account of the robot's velocity

Fig. 7. Experimental results of obstacle avoidance (top view)

5.2 Self-Localization

In the field of the RoboCup Small-Size League, cubic obstacles with sides of 130[mm] were put as shown in **Fig.8**. The robot stopped at six places and estimated self-position 10 times at each spot. The RMS of error is shown in **Fig.8**. Results showed errors at each place amounted to dozens of mm, precise enough for the RoboCup. And, the mean cycle of the total process for the robot, including estimation of self-position, was about 45ms.

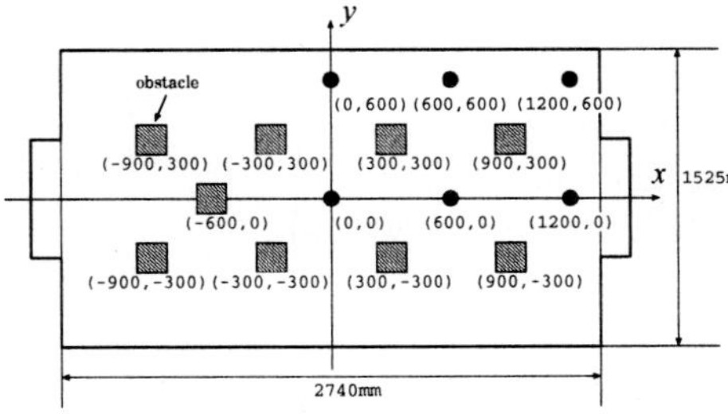

Robot position		Error(RMS)		
x[mm]	y[mm]	Δx[mm]	Δy[mm]	$\Delta\theta$[deg]
0	0	84	19	1.6
600	0	41	16	2.6
1200	0	11	30	2.2
0	600	37	7	0.5
600	600	73	20	3.3
1200	600	21	10	5.2

Fig. 8. Observing points and errors in estimation

6 Conclusion

In this paper, we propose a method of obstacle avoidance and a method of self-localization based on floor region provided by omni-directional imaging by the robot. In addition, high-speed processing is realized by observing a large area of the floor region quickly instead of observing a large area of the floor region in detail. We are planning to demonstrate the effectiveness of our methods in an actual competition of the RoboCup.

References

1. Y.Yagi, S. Kawato and S. Tsuji : "Collision Avoidance for Mobile Robot with Omni-directional Image Sensor (COPIS)", Proceedings IEEE the International Conference on Robotics and Automation, Vol.1, pp. 910-915, 1991
2. A.Clerentin, L.Delahoche, E.Brassart: "Cooperation between two omnidirectional perception systems for mobile robot localization", Proceedings of the 2000 IEEE/RSJ Intemational Conference on Intelligent Robots and Systems, pp. 1499-1504, 2000
3. N.Mori et al.: "Realization of Soccer Game with Small Size Robots which have Omni-directional Mobile Mechanism and Omni-directional Vision — Strategies of Team OMNI —"(in Japanese), Proceedings of the 6th SIG-Challenge of Japanese Society for Artificial Intelligence, pp. 42-47, 2000
4. D.Sekimori et al.: "High-speed Obstacle Avoidance of Mobile Robot Based on Labeling of Omni-directional Image"(in Japanese), Proceedings of the 18th Annual Conference of the Robotics Society of Japan, Vol.3,pp. 995-996,2000
5. Mark de Berg et al.: "Computational Geometry: Algorithms and Applications", Springer-Verlag, 2000
6. Y.Yagi: "Real-time Omnidirectional Image Sensor"(in Japanese), Journal of Robotics Society of Japan, Vol.13, No.3, pp. 347-350, 1995

Keepaway Soccer: A Machine Learning Testbed

Peter Stone and Richard S. Sutton

AT&T Labs — Research
180 Park Ave.
Florham Park, NJ 07932
{pstone,sutton}@research.att.com
http://www.research.att.com/{~pstone,~sutton}

Abstract. RoboCup simulated soccer presents many challenges to machine learning (ML) methods, including a large state space, hidden and uncertain state, multiple agents, and long and variable delays in the effects of actions. While there have been many successful ML applications to portions of the robotic soccer task, it appears to be still beyond the capabilities of modern machine learning techniques to enable a team of 11 agents to successfully learn the full robotic soccer task from sensors to actuators. Because the successful applications to portions of the task have been embedded in different teams and have often addressed different subtasks, they have been difficult to compare. We put forth keepaway soccer as a domain suitable for directly comparing different machine learning approaches to robotic soccer. It is complex enough that it can't be solved trivially, yet simple enough that complete machine learning approaches are feasible. In keepaway, one team, "the keepers," tries to keep control of the ball for as long as possible despite the efforts of "the takers." The keepers learn individually when to hold the ball and when to pass to a teammate, while the takers learn when to charge the ball-holder and when to cover possible passing lanes. We fully specify the domain and summarize some initial, successful learning results.

1 Introduction

RoboCup simulated soccer has been used as the basis for successful international competitions and research challenges [4]. It is a fully *distributed*, *multiagent* domain with both *teammates* and *adversaries*. There is *hidden state*, meaning that each agent has only a partial world view at any given moment. The agents also have *noisy sensors and actuators*, meaning that they do not perceive the world exactly as it is, nor can they affect the world exactly as intended. In addition, the perception and action cycles are *asynchronous*, prohibiting the traditional AI paradigm of using perceptual input to trigger actions. *Communication* opportunities are limited, and the agents must make their decisions in *real-time*. These italicized domain characteristics combine to make simulated robotic soccer a realistic and challenging domain [10].

In principle, modern machine learning methods are reasonably well suited to meeting the challenges of RoboCup simulated soccer. For example, reinforcement

A. Birk, S. Coradeschi, and S. Tadokoro (Eds.): RoboCup 2001, LNAI 2377, pp. 214–223, 2002.

learning is all about sequential decision making, achieving delayed goals, and handling noise and stochasticity. However, RoboCup soccer is a large and difficult instance of many of the issues which have been addressed in small, isolated cases in previous reinforcement learning research. Despite substantial previous work (e.g., [1, 13, 15, 9]), the extent to which modern reinforcement learning methods can meet these challenges remains an open question.

Similarly, evolutionary learning methods are promising candidates for learning complex behaviors. Indeed, there have been two previous attempts to learn the entire simulated RoboCup task via genetic programming (GP) [5, 2]. While both efforts were initially intended to test the ability of GP to scale to the full, cooperative robotic soccer task, the first system ended up evolving over hand-coded low-level behaviors, and the second achieved some successful individual behaviors but was unable to generate many collaborative team behaviors. Whether this approach can be scaled up to produce more successful teams remains to be seen.

One frustration with these and other machine learning approaches to RoboCup is that they are all embedded within disparate systems, and often address different subtasks of the full soccer problem. Therefore, they are difficult to compare in any meaningful way.

In this paper, we put forth keepaway soccer as a subtask of robotic soccer that is suitable for directly comparing different machine learning methods. The remainder of the paper is organized as follows. In Section 2 we describe keepaway and specify the sensors and actuators for this task. In Section 3 we summarize the successful results of our reinforcement learning approach to this task [12]. Section 4 concludes.

2 Keepaway Soccer

We consider a subtask of RoboCup soccer, *keepaway*, in which one team, the *keepers*, is trying to maintain possession of the ball within a limited region, while another team, the *takers*, is trying to gain possession. Whenever the takers take possession or the ball leaves the region, the *episode* ends and the players are reset for another episode (with the keepers being given possession of the ball again).

Parameters of the task include the size of the region, the number of keepers, and the number of takers. Figure 1 shows a screen shot of 3 keepers and 2 takers (called 3 vs. 2, or 3v2 for short) playing in a 20m x 20m region.

All of our work uses version 5.25 of the standard RoboCup soccer simulator [6] but is directly applicable to more recent versions. An omniscient coach agent manages the play, ending episodes when a taker gains possession of the ball for a set period of time or when the ball goes outside of the region. At the beginning of each episode, the coach resets the location of the ball and of the players semi-randomly within the region of play. The takers all start in one corner (bottom left). Three randomly chosen keepers are placed one in each of the three

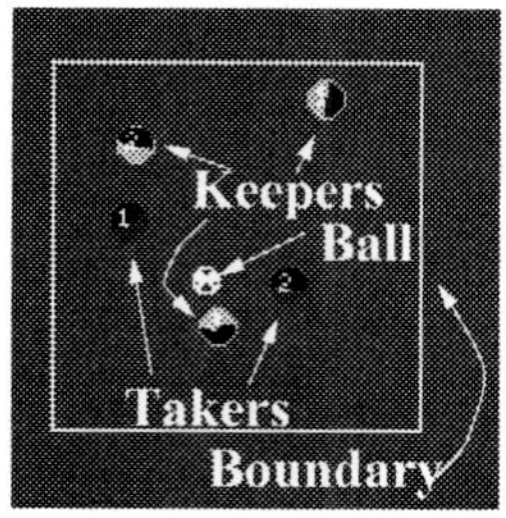

Fig. 1. A screen shot from a 3 vs. 2 keepaway episode in a 20m x 20m region.

remaining corners, and any keepers beyond three are placed in the center of the region. The ball is initially placed next to the keeper in the top left corner.

In the RoboCup soccer simulator, agents typically have limited and noisy sensors: each player can see objects within a 90° view cone, and the precision of an object's sensed location degrades with distance. To simplify the problem, we gave our agents (both keepers and takers) a full 360° of vision. We have not yet determined whether this simplification was necessary or helpful.

Keepaway is a subproblem of robotic soccer. The principal simplifications are that there are fewer players involved; they are playing in a smaller area; and the players are always focused on the same high-level goal—they don't need to balance offensive and defensive considerations. Nevertheless, the skills needed to play keepaway well are also very useful in the full problem of robotic soccer. Indeed, ATT-CMUnited-2000—the 3rd-place finishing team in the RoboCup-2000 simulator league—incorporated a successful hand-coded solution to an 11 vs. 11 keepaway task [11].

Keepaway is a complex enough task that simple, hand-coded solutions are not likely to yield the best behaviors. However, it is also simple enough that complete machine learning approaches are feasible. These complete approaches can then be directly compared. In order to do so, it is necessary that the different systems work with similar action and feature spaces. In the remainder of this section, we specify the actions and state variables in our formulation of the task.

2.1 State Variables and Actions

In this learning problem, the learners choose from among mid-level actions constructed from skills. Those skills include

HoldBall(): Remain stationary while keeping possession of the ball in a position that is as far away from the opponents as possible.

PassBall(k): Kick the ball directly towards keeper k.

GetOpen(): Move to a position that is free from opponents and open for a pass from the ball's current position (using SPAR [16]).

GoToBall(): Intercept a moving ball or move directly towards a stationary ball.

BlockPass(k): Move to a position between the keeper with the ball and keeper k.

All of these skills except PassBall(k) are simple functions from state to a corresponding action; an invocation of one of these normally controls behavior for a single time step. PassBall(k), however, requires an extended sequence of actions—getting the ball into position for kicking, and then a sequence of kicks in the desired direction [10]; a single invocation of PassBall(k) influences behavior for several time steps. Moreover, even the simpler skills may last more than one time step because the player occasionally misses the step following them; the simulator occasionally misses commands; or the player may find itself in a situation requiring it to take a specific action, for instance to self-localize. In these cases there is no new opportunity for decision making until two or more steps after invoking the skill.

Such possibilities may present problems for machine learning approaches. One candidate for dealing with them is to treat the problem as a *semi-Markov* decision process, or SMDP [7, 3]. An SMDP evolves in a sequence of jumps from the initiation of each SMDP action to its termination one or more time steps later, at which time the next SMDP action is initiated. SMDP actions that consist of a subpolicy and termination condition over an underlying decision process, as here, have been termed *options* [14].

2.2 Keeper State Variables

Our task specification involves learning by the keepers when in possession[1] of the ball. Keepers not in possession of the ball are required to select the Receive option:

Receive: If a teammate possesses the ball, or can get to the ball faster than this keeper can, invoke GetOpen() for one step; otherwise, invoke GoToBall() for one step. Repeat until a keeper has possession of the ball or the episode ends.

A keeper in possession, on the other hand, is faced with a genuine choice. It may hold the ball, or it may pass to one of its teammates. That is, it chooses an option from {Holdball, PassK_2ThenReceive, PassK_3ThenReceive, ..., PassK_nThenReceive} where the Holdball option simply executes HoldBall() for one step (or more if, for example, the server misses the next step) and the Pass$k$ThenReceive options involve passes to the other keepers. The keepers are numbered by their closeness to the keeper with the ball: K_1 is the keeper with the ball, K_2 is the closest keeper to it, K_3 the next closest, and so on up to K_n, where n is the number of keepers. Each Pass$k$ThenReceive is defined as

Pass$k$ThenReceive: Invoke PassBall(k) to kick the ball toward teammate k. Then behave and terminate as in the Receive option.

---

[1] "Possession" in the soccer simulator is not well-defined because the ball never occupies the same location as a player. One of our agents considers that it has possession of the ball if the ball is close enough to kick it.

The keepers' learning process thus searches a constrained policy space characterized only by the choice of option when in possession of the ball. Examples of (non-learned) policies within this space include:

Random: Choose randomly among the n options, each with probability $\frac{1}{n}$.

Hold: Always choose HoldBall()

Hand-coded:

If no taker is within 10m, choose HoldBall();

Else If teammate k is in a better location than the keeper with the ball and the other teammates, and the pass is likely to succeed (using the CMUnited-99 pass-evaluation function, which is trained off-line using the C4.5 decision tree training algorithm [8]), then choose PassBall(k);

Else choose HoldBall().

We turn now to the representation of state used by the keepers. Note that in our formulation the state variables are only needed when one of the keepers is in possession of the ball (were keepers to learn where to move when they don't have the ball, additional state variables would be needed). On these steps the keeper determines a set of state variables, computed based on the positions of: the keepers K_1–K_n, ordered as above; the takers T_1–T_m (m is the number of takers), ordered by increasing distance from K_1; and C, the center of the playing region (see Figure 2 for an example with 3 keepers and 2 takers). Let $dist(a, b)$ be the distance between a and b and $ang(a, b, c)$ be the angle between a and c with vertex at b. For example, $ang(K_3, K_1, T_1)$ is shown in Figure 2. With 3 keepers and 2 takers, we used the following 13 state variables:

- $dist(K_1, C)$, $dist(K_2, C)$, $dist(K_3, C)$
- $dist(T_1, C)$, $dist(T_2, C)$
- $dist(K_1, K_2)$, $dist(K_1, K_3)$
- $dist(K_1, T_1)$, $dist(K_1, T_2)$
- $\text{Min}(dist(K_2, T_1), dist(K_2, T_2))$
- $\text{Min}(dist(K_3, T_1), dist(K_3, T_2))$
- $\text{Min}(ang(K_2, K_1, T_1), ang(K_2, K_1, T_2))$
- $\text{Min}(ang(K_3, K_1, T_1), ang(K_3, K_1, T_2))$

This list generalizes naturally to additional keepers and takers, leading to a linear growth in the number of state variables.

2.3 Takers

The takers are relatively simple, choosing only options of minimum duration (one step, or as few as possible given server misses) that exactly mirror the available skills. When a taker has the ball, it tries to maintain possession by invoking HoldBall() for a step. Otherwise, it chooses an option that invokes one of {GoToBall(), BlockPass(K_2), BlockPass(K_3), ..., BlockPass(K_n)} for one step or as few steps as permitted by the server. In case no keeper has the ball (e.g., during a pass), K_1 is defined here as the keeper predicted to next gain possession. Examples of (non-learned) policies within the taker policy space include:

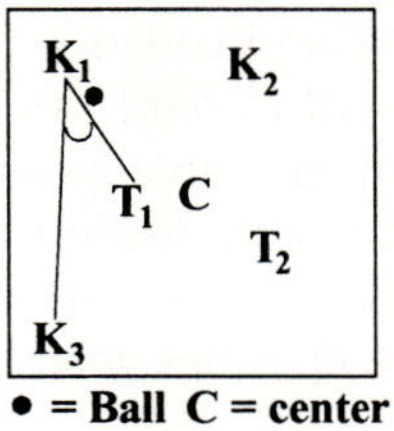

Fig. 2. The state variables used for learning with 3 keepers and 2 takers. Keepers and takers are numbered by increasing distance from K_1, the keeper with the ball.

Random-T: Choose randomly from the n options, each with probability $\frac{1}{n}$.
All-to-ball: Always choose the GoToBall() option.
Hand-coded-T:
If fastest taker to the ball, or closest or second closest taker to the ball: choose the GoToBall() option;
Else let k be the keeper with the largest angle with vertex at the ball that is clear of takers: choose the BlockPass(k) option.

The takers' state variables are similar to those of the keepers. As before, C is the center of the region. T_1 is the taker that is computing the state variables, and T_2–T_m are the other takers ordered by increasing distance from K_1. $K_i$mid is the midpoint of the line segment connecting K_1 and K_i for $i \in [2, n]$ and where the K_i are ordered based on increasing distance of $K_i$mid from T_1. That is, $\forall i, j$ s.t. $2 \leq i < j$, $dist(T_1, K_i\text{mid}) \leq dist(T_1, K_j\text{mid})$. With 3 keepers and 3 takers, we used the following 18 state variables:

- $dist(K_1, C)$, $dist(K_2, C)$, $dist(K_3, C)$
- $dist(T_1, C)$, $dist(T_2, C)$, $dist(T_3, C)$
- $dist(K_1, K_2)$, $dist(K_1, K_3)$
- $dist(K_1, T_1)$, $dist(K_1, T_2)$, $dist(K_1, T_3)$
- $dist(T_1, K_2\text{mid})$, $dist(T_1, K_3\text{mid})$
- $\text{Min}(dist(K_2\text{mid}, T_2), dist(K_2\text{mid}, T_3))$
- $\text{Min}(dist(K_3\text{mid}, T_2), dist(K_3\text{mid}, T_3))$
- $\text{Min}(ang(K_2, K_1, T_2), ang(K_2, K_1, T_3))$
- $\text{Min}(ang(K_3, K_1, T_2), ang(K_3, K_1, T_3))$
- number of takers closer to the ball than T_1

Once again, this list generalizes naturally to different numbers of keepers and takers.

3 Sample Experiments

We have conducted initial experiments in this keepaway domain using the SARSA(λ) reinforcement learning algorithm [12]. Our main results to date have focused on learning by the keepers in 3v2 keepaway in a 20x20 region. For the opponents (takers) we used the Hand-coded-T policy (note that with just 2 takers, this policy is identical to All-to-ball). To benchmark the performance of the learned keepers, we first ran the three benchmark keeper policies, Random, Hold, and Hand-coded, as laid out in Section 2.2. Average episode lengths for these three policies were 5.5, 4.8, and 5.6 seconds respectively. Figure 3 shows histograms of the lengths of the episodes generated by these policies.

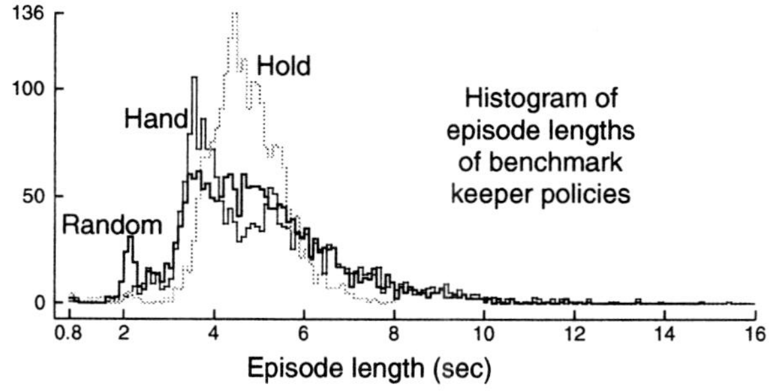

Fig. 3. Histograms of episode lengths for the 3 benchmark keeper policies in 3v2 keepaway in a 20x20 region.

We then ran a series of eleven runs with learning by the keepers against the Hand-coded-T takers. Figure 4 shows learning curves for these runs. The y-axis is the average time that the keepers are able to keep the ball from the takers (average episode length); the x-axis is training time (simulated time $\approx$ real time). The performance levels of the benchmark keeper policies are shown as horizontal lines.

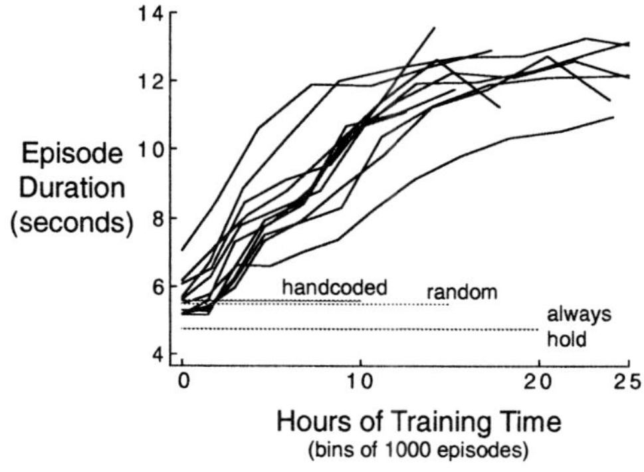

Fig. 4. Multiple successful runs under identical characteristics: 3v2 keepaway in a 20x20 region against hand-coded takers.

This data shows that we were able to learn policies that were much better than any of the benchmarks. All learning runs quickly found a much better policy than any of the benchmark policies, including the hand-coded policy. A better policy was often found even in the first data point, representing the first 1000 episodes of learning. Qualitatively, the keepers appear to quickly learn roughly how long to hold the ball, and then gradually learn fine distinctions regarding when and to which teammate to pass.

Next, we applied the same algorithm to learn policies specialized for different region sizes. In particular, we trained the takers in 15x15 and 25x25 regions without changing any of the other parameters, including the representation used for learning. The results shown in Table 3 indicate that the learned policies again outperformed all of the benchmarks. In addition, it is apparent that policies trained in the same size region as they are tested (shown in boldface), performed better than policies trained in other region sizes. In general it is easier for the keepers to keep the ball in a larger field since the takers have further to run. Thus, we observe a general increase in possession time from left to right in the table. These results indicate that different policies are best on different field sizes. Thus, were we to take the time to laboriously fine-tune a hand-coded behavior, we would need to repeat the process on each field size. On the other hand, the same learning mechanism is able to find successful policies on all three field sizes without any additional human effort.

		Testing Field Size		
Keepers		15x15	20x20	25x25
Trained on field of size	15x15	**11.0**	9.8	7.2
	20x20	10.7	**15.0**	12.2
	25x25	6.3	10.4	**15.0**
Benchmarks	Hand	4.3	5.6	8.0
	Hold	3.9	4.8	5.2
	Random	4.2	5.5	6.4

Table 1. Performance (possession times) of keepers trained with various field sizes (and benchmark keepers) when tested on fields of various sizes.

Finally, we applied our learning algorithm to learn policies for a slightly larger task, 4 vs. 3 keepaway. Figure 5 shows that the keepers learned a policy that outperformed all of our benchmarks in 4v3 keepaway in a 30x30 region. In this case, the learning curves still appear to be rising after 40 hours: more time may be needed to realize the full potential of learning.

We have also begun exploring taker learning. As noted in Section 2.3, the representation used for the keepers may not be suited to taker learning. Nonetheless, using this representation, takers were able to learn policies that outperformed the Random and All-to-ball taker policies. The best learned policies discovered so far perform roughly equivalently to the Hand-coded-T policy.

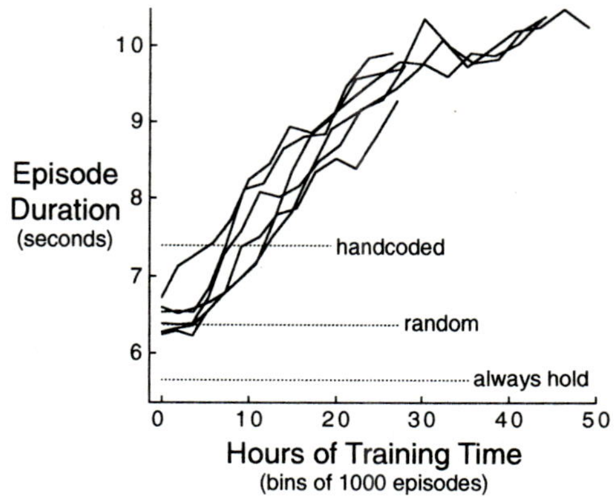

Fig. 5. Multiple successful runs under identical characteristics: 4v3 keepaway in a 30x30 region against hand-coded takers.

Our on-going research is aimed at improving the ability of the takers to learn by altering their representation and/or learning parameters. One promising line of inquiry is into the efficacy of alternately training the takers and the keepers against each other so as to improve both types of policies.

4 Conclusion

Keepaway soccer is a simplification of the full RoboCup soccer task that is suitable for solution with machine learning methods. Our previous research in the domain has demonstrated that a machine learning approach can outperform a reasonable hand-coded solution as well as other benchmark policies.

We put this domain forth as a candidate for use for direct comparisons of different machine learning techniques on a concrete task. It is complex enough that it can't be solved trivially, yet simple enough that complete machine learning approaches are feasible.

In our formulation, learning is done over a very large state space but with only a few possible actions at any given time. The domain could be incrementally expanded by introducing more actions. For example, as a first step, the agents could learn precisely where to move rather than calling functions such as GetOpen() or BlockPass(). Eventually, the agents could learn directly from the simulator low-level, parameterized action space.

References

[1] Tomohito Andou. Refinement of soccer agents' positions using reinforcement learning. In Hiroaki Kitano, editor, *RoboCup-97: Robot Soccer World Cup I*, pages 373–388. Springer Verlag, Berlin, 1998.

[2] David Andre and Astro Teller. Evolving team Darwin United. In Minoru Asada and Hiroaki Kitano, editors, *RoboCup-98: Robot Soccer World Cup II*. Springer Verlag, Berlin, 1999.

[3] S. J. Bradtke and M. O. Duff. Reinforcement learning methods for continuous-time Markov decision problems. pages 393–400, San Mateo, CA, 1995. Morgan Kaufmann.

[4] Hiroaki Kitano, Milind Tambe, Peter Stone, Manuela Veloso, Silvia Coradeschi, Eiichi Osawa, Hitoshi Matsubara, Itsuki Noda, and Minoru Asada. The RoboCup synthetic agent challenge 97. In *Proceedings of the Fifteenth International Joint Conference on Artificial Intelligence*, pages 24–29, San Francisco, CA, 1997. Morgan Kaufmann.

[5] Sean Luke, Charles Hohn, Jonathan Farris, Gary Jackson, and James Hendler. Co-evolving soccer softbot team coordination with genetic programming. In Hiroaki Kitano, editor, *RoboCup-97: Robot Soccer World Cup I*, pages 398–411, Berlin, 1998. Springer Verlag.

[6] Itsuki Noda, Hitoshi Matsubara, Kazuo Hiraki, and Ian Frank. Soccer server: A tool for research on multiagent systems. *Applied Artificial Intelligence*, 12:233–250, 1998.

[7] M. L. Puterman. *Markov Decision Problems*. Wiley, NY, 1994.

[8] J. Ross Quinlan. *C4.5: Programs for Machine Learning*. Morgan Kaufmann, San Mateo, CA, 1993.

[9] M. Riedmiller, A. Merke, D. Meier, A. Hoffman, A. Sinner, O. Thate, and R. Ehrmann. Karlsruhe brainstormers—a reinforcement learning approach to robotic soccer. In Peter Stone, Tucker Balch, and Gerhard Kraetszchmar, editors, *RoboCup-2000: Robot Soccer World Cup IV*. Springer Verlag, Berlin, 2001.

[10] Peter Stone. *Layered Learning in Multiagent Systems: A Winning Approach to Robotic Soccer*. MIT Press, 2000.

[11] Peter Stone and David McAllester. An architecture for action selection in robotic soccer. In *Proceedings of the Fifth International Conference on Autonomous Agents*, 2001.

[12] Peter Stone and Richard S. Sutton. Scaling reinforcement learning toward robocup soccer. In *Proceedings of the Eighteenth International Conference on Machine Learning*, 2001.

[13] Peter Stone and Manuela Veloso. Team-partitioned, opaque-transition reinforcement learning. In Minoru Asada and Hiroaki Kitano, editors, *RoboCup-98: Robot Soccer World Cup II*. Springer Verlag, Berlin, 1999. Also in *Proceedings of the Third International Conference on Autonomous Agents*, 1999.

[14] R. S. Sutton, D. McAllester, S. Singh, and Y. Mansour. Policy gradient methods for reinforcement learning with function approximation. In *Advances in Neural Information Processing Systems*, volume 12, pages 1057–1063. The MIT Press, 2000.

[15] Eiji Uchibe. *Cooperative Behavior Acquisition by Learning and Evolution in a Multi-Agent Environment for Mobile Robots*. PhD thesis, Osaka University, January 1999.

[16] Manuela Veloso, Peter Stone, and Michael Bowling. Anticipation as a key for collaboration in a team of agents: A case study in robotic soccer. In *Proceedings of SPIE Sensor Fusion and Decentralized Control in Robotic Systems II*, volume 3839, Boston, September 1999.

Strategy Learning for a Team in Adversary Environments

Yasutake Takahashi, Takashi Tamura, and Minoru Asada

Dept. of Adaptive Machine Systems,
Graduate School of Engineering Osaka University, Suita, Osaka 565-0871, Japan

Abstract. Team strategy acquisition is one of the most important issues of multiagent systems, especially in an adversary environment. RoboCup has been providing such an environment for AI and robotics researchers. A deliberative approach to the team strategy acquisition seems useless in such a dynamic and hostile environment. This paper presents a learning method to acquire team strategy from a viewpoint of coach who can change a combination of players each of which has a fixed policy. Assuming that the opponent has the same choice for the team strategy but keeps the fixed strategy during one match, the coach estimates the opponent team strategy (player's combination) based on game progress (obtained and lost goals) and notification of the opponent strategy just after each match. The trade-off between exploration and exploitation is handled by considering how correct the expectation in each mode is. A case of 2 to 2 match was simulated and the final result (a class of the strongest combinations) was applied to RoboCup-2000 competition.

1 Introduction

Team strategy acquisition is one of the most important issues of multiagent systems, especially in an adversary environment. RoboCup has been providing such an environment for AI and robotics researchers as one of the most challenging issues in the field since 1997 with increasing number of various leagues.

In the simulation league, a number of methods for team strategy acquisition have been applied. Wunstel et al.[1] proposed a method to acquire an opponent team model in order to build the own team strategy to beat the opponent. However, this method requires all kinds of information and observations on every player on the field during the game. Stone and Veloso [2] proposed a locker room agreement to decide which to take, that is, a defensive formation or an offensive one considering the score just before the match. Dynamic role exchange during the game is also attempted through the broadcasting line.

Not so many teams in the real robot league have tried to apply the method for team strategy acquisition because they cannot afford to pay any attention to such an issue but need to care about more hardware stuffs. Castelpietra et al. [3] proposed a method of cooperative behavior generation by exchanging information and roles through the wireless network. Uchibe et al. [4] proposed a method of dynamic role assignment with shared memory of task achievements.

A. Birk, S. Coradeschi, and S. Tadokoro (Eds.): RoboCup 2001, LNAI 2377, pp. 224–233, 2002.

These methods need explicit communication lines to realize cooperative behaviors. Therefore, the realized cooperation might be damaged by the perception and/or communication noises which affect the opponent model estimation and communication itself.

More deliberative approaches to the team strategy acquisition seem less useful due to its ineffectiveness in such a dynamic and hostile environment. One alternative is to use the dynamics of environment itself with behavior-based approach [5]. Depending on the environment dynamics consisting of not only the teammate actions but also the opponent ones, the cooperative behaviors are expected to happen much more often than by deliberative approaches. However, the design principle is rather conceptual and has not revealed which combination of players can emerge more cooperative behaviors.

This paper presents a learning method to acquire team strategy from a viewpoint of coach who can change a combination of players each of which has a fixed policy. Assuming that the opponent team has the same choice for the team strategy but keeps the fixed strategy during one match, the coach estimates the opponent team strategy (player's combination) by changing the own team strategy based on game progress (obtained and lost goals) and notification of the opponent strategy just after each match. The trade-off between exploration and exploitation is handled by considering how correct the prediction of the opponent team strategy is. A case of 2 to 2 match was simulated and a part of the final result (a class of the strongest combinations) was applied to RoboCup-2000 competition.

The rest of the paper is organized as follows. First, the definitions of the task and the team strategy are given along with a basic idea of the use of environmental dynamics in the context of emergence of cooperative behaviors. Next, a learning method for team strategy estimation is introduced based on the information such as game progress. Then, the results of the computer simulation are given and the real robot experiments are shown.

2 Basic Ideas and Assumptions

As the environment becomes more complex, the system usually tends to be more complicated to cope with the complexity of the environment. Especially, in a multiagent system, adding a capability of communication seems useful to realize cooperative behaviors. However, such a system becomes useless when the environment is much more dynamic and therefore the prediction of the environment changes is hard.

Simon pointed out the use of the environment itself [6]. The walking of ants on the sands is not so complicated but just simple to adapt it to the environment without any predictions or plans. A behavior-based approach is one of such ways to utilize the dynamic environments. However, it seems difficult to design each robot in a multiagent hostile environment because the environment dynamics seriously depends on the actions of other agents who might have their own policies to take actions.

Then, we propose a learning method of team strategy acquisition assuming the following settings.

- Each team has various kinds of players that are designed by behavior based approach without communication lines to each other.
- A team strategy is defined as a combination of players.
- A coach may change its strategy by exchanging players.
- The opponent team has the same choice of the player's combination, that is, it can take one of the team strategies.

The problem here is to find the strategy to beat the opponent from a viewpoint of coach based on the game progress and notification of the opponent strategy after the match. Since the opponent has the same choice, the task here is to identify the unknown opponent strategy and to know the relationship (stronger or weaker) between the opponent's and its own. Then, we add one more assumption.

- The opponent team strategy changes randomly every match but is fixed during one match while the learning team's coach may change its team strategy to estimate the opponent's team strategy during one match. Hereafter, the coach means the learning team's coach.

3 Team Strategy Learning

First of all, we prepare two modes of the coach policy.

Exploration Mode Sampling data in order to fill the game scores in the obtained and lost goals table.

Exploitation Mode Estimating the better team policy against the predicted opponent team.

The coach decides the team strategy based on a weighted sum of the two modes. The weight w_{er} is the ratio between the exploration mode and the exploitation one, and updated gradually from exploitation to exploration as the coach becomes able to estimate the own team strategy to beat the opponent. We define an probability to select the j-th strategy $P(j)$ as following:

$$P(j) = (1 - w_{er})P_r(j) + w_{er}P_e(j) \tag{1}$$

where $P_r(j)$ and $P_e(j)$ are the occurrence probabilities to select the strategy j defined by exploration mode and exploitation one, respectively.

3.1 Exploration Mode

The exploration mode is simple and straightforward; select the most inexperienced strategy to obtain more data. One match consists of several periods between which the coach may change its team strategy while the opponent one

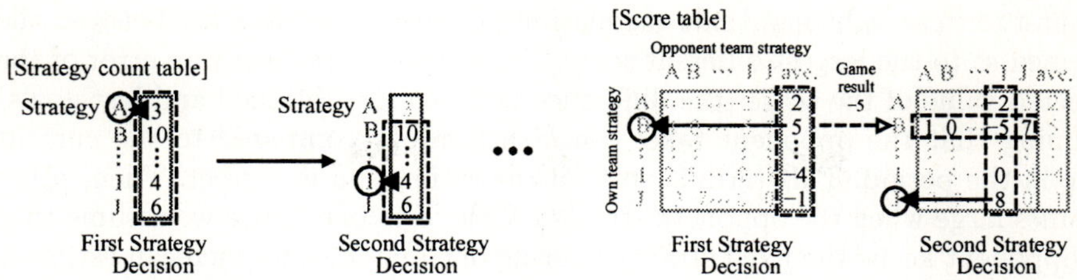

(a) Exploration Mode (b) Exploitation Mode

Fig. 1. Strategy selection

is fixed during one match. The coach has a table $T_r(j)$ which counts number of periods in which the j-th one among n strategies is applied. We define the probability $P_r(j)$ to select the j-th strategy in the next period as following:

$$P_r(j) = \frac{p_r(j)}{\sum_{l=1}^{n} p_r(l)}, \tag{2}$$

where

$$p_r(j) = \begin{cases} 0 & \text{(if coach has selected the } j \text{ th strategy in the same match already)} \\ 1 & \text{(if } T_r(j) = 0 \text{)} \\ \dfrac{1}{T_r(j)} & \text{(else)} \end{cases} \tag{3}$$

Fig.1 (a) shows the idea of the exploration mode. The coach selects the most inexperienced strategy according to the strategy count table at the first period. At the second and the following periods, the coach skips the previously taken strategies in the current match and selects the most inexperienced strategy among the rest.

3.2 Exploitation Mode

The exploitation mode can be decomposed into two parts. One is the estimation process of the opponent team's strategy, and the other is the decision of an own team's strategy to beat the opponent. The coach has a score table $T_s(k, j)$ which stores the difference between obtained and lost goals in a period in which the opponent team takes the k-th strategy and own team takes the j-th strategy. When the coach had score (difference between the obtained and lost goals) s_j^i using the j-th strategy at the i-th period, the coach estimates the opponent team's strategy at the i-th period by:

$$P_s^i(k) = \frac{p_s^i(k)}{\sum_{l=0}^{n} p_s^i(l)}, \tag{4}$$

where

$$p_s^i(k) = \max_l |s_j^i - T_s(l, j)| - |s_j^i - T_s(k, j)|. \tag{5}$$

The first term at right-hand side of equation (5) means the difference between the obtained s_j^i to the largest different score. The second term means an error of the expected value of the score (the difference between the obtained and lost goals) assuming that the opponent takes the k-th strategy compared to the current score in the period. This term is small if the estimation is correct. Then, $p_s^i(k)$ becomes large when the opponent strategy is the k-th one. Since we assume that the opponent keeps the fixed strategy during one match, the coach can estimate the opponent team's strategy after m periods in the match as following:

$$P_s(k) = \frac{1}{m} \sum_{i=0}^{m} P_s^i(k) \tag{6}$$

At the second step, the coach predicts an expected score (difference between the obtained and lost goals) $x_s(j)$ in the next period when the coach takes the j-th strategy by:

$$x_s(j) = \sum_{k=1}^{n} P_s(k) T_s(k, j). \tag{7}$$

We define a occurrence probability $P_e(j)$ to select the the j-th strategy in the next period using $x_s(j)$ by:

$$P_e(j) = \frac{p_e(j)}{\sum_{l=0}^{n} p_e(l)}, \tag{8}$$

where

$$p_e(j) = \begin{cases} 0 & (\text{if } x_s(j) \leq 0) \\ x_s(j) & (\text{else}) \end{cases}. \tag{9}$$

Fig.1 (b) shows the idea of the exploitation mode. The coach selects the strongest strategy according to the average in score table $T_s(k, j)$ at the first period (taking the strategy B which has the largest average score 5). At the second and the following periods, the coach estimates the opponent team's strategy using the score at the previous periods (according the match result, the opponent strategy is predicted as I), and estimates the best strategy against the estimated opponent team's strategy (taking the best strategy J against the regarded opponent as I).

3.3 Update of w_{er}

The coach updates the weight w_{er} which is the ratio between the exploration mode and the exploitation one after a match consisting of n_p periods, and the coach took strategy a^i at the i-th period as following:

$$w_{er} \leftarrow w_{er} + \Delta w_{er}, \tag{10}$$

where

$$\Delta w_{er} = \alpha \sum_{i=2}^{n_p} s^i \frac{i}{n_p} x_s(a^i). \tag{11}$$

α is an update rate. If the weight w_{er} becomes more than 1 or less than 0, it is set to 1 and 0, respectively. These equations intend to update the weight w_{er} so that the coach takes the exploitation mode if the estimated score is correct, or it takes the exploration mode. $\frac{i}{n_p}$ means that the coach makes the weight of estimation at early periods small because it doesn't have data enough to estimate the opponent team's strategy at early periods.

4 Experiments

4.1 Match Setting

We assume the following settings.

- Players and field obey the regulations of the RoboCup2000 middle size league.
- Each team has two players.
- Players have its own fixed policy.
- A player has no communication devices with other teammate.
- One match consists of 500 trials.
- One period consists of 100 trials. That means the coach has five opportunities to select the team strategy during one match.
- The opponent strategy changes randomly every match.
- Players' initial positions are random in their own areas and ball is set at the center of the field.

4.2 Team Strategy

Team strategy is defined from a viewpoint of coach who can change a combination of players each of which has a fixed policy. We prepare 4 types of players' policies from view points of ball handling and social behaviors.

Ball Handling

ROUGH a quick approach and rough ball handling
CAREFUL a slow approach and careful ball handling

Social Behaviors

SELFISH regarding other all players as obstacle
SOCIAL yielding a way to the teammate if the teammate is on its own way

Totally there are 4 types of players' policies because of the combination of the ball handling controllers and the social behaviors. Then, there are 10 kinds of team strategies because one team has two individual players: ${}_4C_2$ strategies of heterogeneous controllers and ${}_4C_1$ strategies of homogeneous controllers.

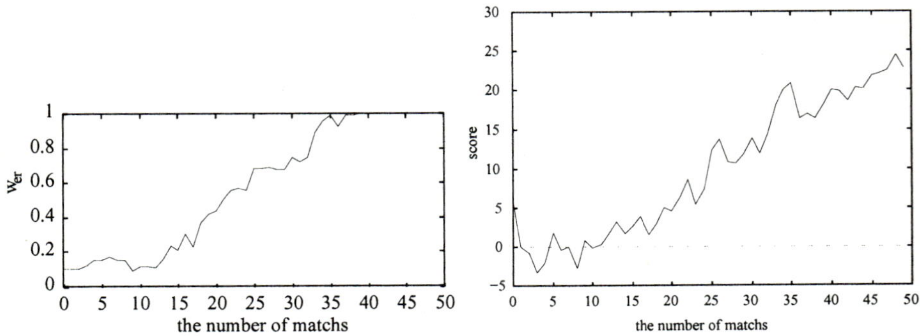

(a) w_{er} (b) score

Fig. 2. The change of w_{er} and score (difference of obtained and lost goals)

4.3 Results

Fig.2 (a) shows the change of w_{er} and that the coach switches mode from exploration to exploitation over the matches. Fig.2 (b) shows the sequence of average score (difference of obtained and lost goals) in the last 10 matches. The team becomes able to beat the opponent team while the coach switch mode to exploitation one. The team obtained more points against the opponent when the coach takes exploitation mode. These figures show that the coach appropriately controls the switch of two modes.

Table 1. The result of score table by learning

own	opponent										average
	1	2	3	4	5	6	7	8	9	10	
1	-	2.3	-4.2	5.5	5.7	6.3	3.0	-	3.2	11.6	4.2
2	-2.3	-	1.2	1.1	6.4	4.1	11.3	-	-	8.3	4.4
3	4.2	-1.2	-	-	-	-3.4	-	-	-3.8	-	-0.6
4	-5.5	-1.1	-	-	-	-	-	-	-	-	-2.8
5	-5.7	-6.4	-	-	-	-	-	-	-	-	-4.1
6	-6.3	-4.1	3.4	-	-	-	-	-	-	-	-2.9
7	-3.0	-11.3	-	-	-	-	-	-	-	-	-4.3
8	-	-	-	-	-	-	-	-	1.1	-	-1.3
9	-3.2	-	3.8	-	-	-	-	-1.1	-	-	-1.0
10	-11.6	-8.3	-	-	-	-	-	-	-	-	-4.3
average	-4.2	-4.4	0.6	2.8	4.1	2.9	4.3	1.3	1.0	4.3	-

1:(RO-SE,CA-SO) , 2:(RO-SE,CA-SE) , 3:(RO-SE,RO-SO) , 4:(CA-SE,CA-SO) , 5:(RO-SO,CA-SE) , 6(CA-SE,CA-SE): , 7:(RO-SO,CA-SO) , 8:(RO-SE,RO-SE) , 9:(RO-SO,RO-SO) , 10:(CA-SO,CA-SO)
RO:ROUGH , CA:CAREFUL , SE:SELFISH, SO:SOCIAL

Table 1 shows the score table after the learning (after 50 matches). The values in the table are the the differences of obtained and lost points in one period. The table has data of stronger strategies and no data of weaker strategies. This means the coach eliminates the experience of week strategies and the learning process is efficient. The learning time needs 1/3 compared to the exhaustive search.

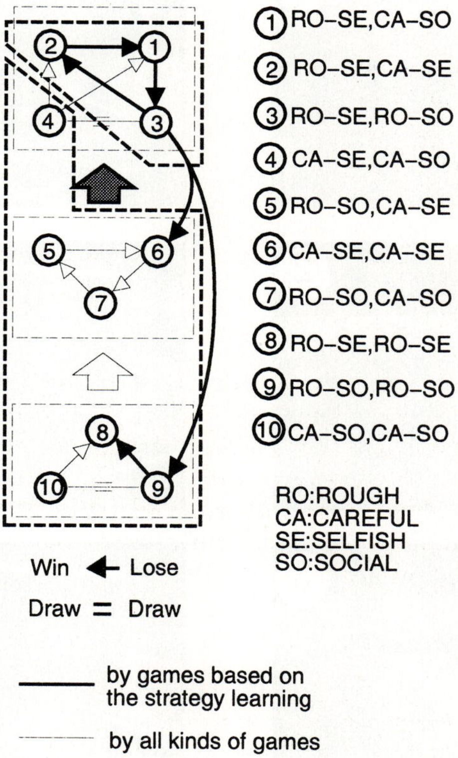

Fig. 3. The learned relationship among strategies

Fig.3 shows the learned relationship among strategies. The strategy pointed by a head of arrow is stronger than strategy pointed by a tail of arrow. The system has a cycle at the top 3 strategies. It seems impossible to select the best team among the all teams, and the coach has to learn the relationship between strategies to beat the opponents.

We applied the part of the final result to the match in the RoboCup2000 middle size league, that is two forward players take a strategy of CAREFUL-SELFISH(Type A) and ROUGH-SOCIAL(Type B) and obtained many goals. Fig.4 shows an example applying the strategy. In this situation, the two robots recover each others' failures quickly. ① indicates that the two different robots follow a ball. The type B robot tries to shoot a ball to the opponent goal at ②. But it failed at ③ because the ball handling skill of type B is not so good, and type A robot recovers the failure soon. The type A robot tries to shoot the ball,

Fig. 4. A sequence of a failure recovery behavior among two robots

but the opponent goalie defends it at ④. The type A robot tries to shoot the ball from left side of the goal at ⑤ and ⑥, but unfortunately fails again while the type B robot moves its position behind the type A robot. The type B robot tries to recover the failure of type A robot's shooting at ⑦, and it shoots the ball successfully after all at ⑧.

5 Conclusion

This paper presented a learning method to acquire team strategy from a viewpoint of coach who can change a combination of players each of which has a fixed policy. Through the learning process the coach gradually obtained the knowledge about the opponent strategy and which strategy to take in order to beat the opponent. The part of the results is applied to the RoboCup2000 middle size league matches and obtained many goals (Top of the preliminary games). As future works, we will make the match settings for the experiments be a real RoboCup middle size scenario, and investigate the theoretical formulation of our approach.

References

[1] Michael Wunstel, Daniel Polani, Thomas Uthmann, and Jurgen Perl. Behavior classification with self-organization maps. In *The Fourth International Workshop on RoboCup*, pages 179–188, 2000.

[2] Peter Stone and Manuela Veloso. Layerd learning and flexible teamworks in robocup simulation agents. In *Robocup-99: Robot Soccer World Cup III*, pages 495–508, 1999.

[3] Claudio Castelpietra, Luca Iocchi, Daniele Nardi, Maurizio Piaggio, Alessandro Scalso, and Antonio Sgorbissa. Communication and coordination among heterogeneous mid-size players: Art99. In *The Fourth International Workshop on RoboCup*, pages 149–158, 2000.

[4] Eiji Uchibe, Tatsunori Kato, Minoru Asada, and Koh Hosoda. Dynamic Task Assignment in a Multiagent/Multitask Environment based on Module Conflict Resolution. In *Proc. of IEEE International Conference on Robotics and Automation*, 2001 (to appear).

[5] Barry Werger. Cooperation without deliberation: Minimalism, stateless, and tolerance in multi-robot teams. *Artificial Intelligence*, 77:293–320, 1999.

[6] H. A. Simon. *The sciences of the artificial.* MIT Press, 1969.

Evolutionary Behavior Selection with Activation/Termination Constraints

Eiji Uchibe[1], Masakazu Yanase[2], and Minoru Asada[3]

[1] Kawato Dynamic Brain Project, ERATO, JST, 2-2-2 Hikaridai, Seika-cho, Soraku-gun, Kyoto 619-0288, Japan
[2] Production Technology Development Center, Production Technology Development Group, Sharp Corporation, 2613-1, Ichinomoto-cho, Tenri, Nara, 632-8567, Japan
[3] Dept. of Adaptive Machine Systems, Graduate School of Eng., Osaka University, Suita, Osaka 565-0871, Japan

Abstract. In order to obtain the feasible solution in the realistic learning time, a layer architecture is often introduced. This paper proposes a behavior selection mechanism with activation/termination constraints. In our method, each behavior has three components: policy, activation constraints, and termination constraints. A policy is a function mapping the sensor information to motor commands. Activation constraints reduce the number of situations where corresponding policy is executable, and termination constraints contribute to extract meaningful behavior sequences, each of which can be regarded as one action regardless of its duration. We apply the genetic algorithm to obtain the switching function to select the appropriate behavior according to the situation. As an example, a simplified soccer game is given to show the validity of the proposed method. Simulation results and real robots experiments are shown, and a discussion is given.

1 Introduction

Machine learning techniques such as reinforcement learning [8] and genetic algorithm [4] are promising to obtain purposive behaviors for autonomous robots in a complicated environment. Many learning and evolutionary techniques can obtain purposive behaviors such as wall-following [2,6], shooting a ball into the goal [1], and so on. However, if the robot has no *a priori* knowledge about the given tasks, it takes enormous time to obtain the complicated behaviors. Consequently, the resultant behavior seems trivial in spite of the long learning time.

In order to obtain the feasible solution in the realistic learning time, a layer architecture is often introduced to cope with large scaled problems [7]. In this approach, the upper layer learns the switching function to select the suitable behavior already designed or learned. Because the designed behaviors can generate purposive action under the limited situations, they can help the evolutionary computation to search the feasible solutions.

A. Birk, S. Coradeschi, and S. Tadokoro (Eds.): RoboCup 2001, LNAI 2377, pp. 234–243, 2002.

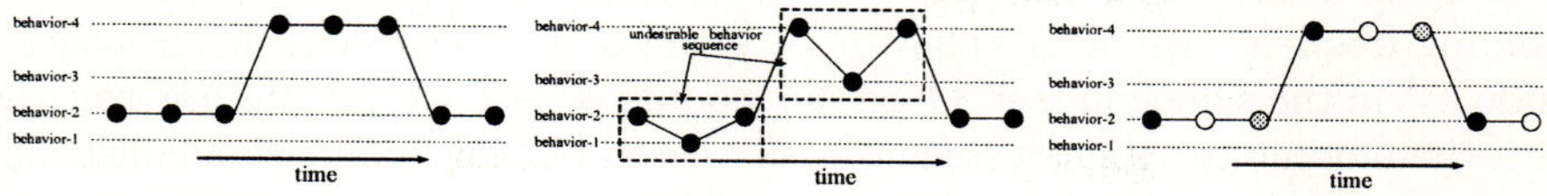

(a) ideal behavior sequence

(b) resultant behavior sequence

(c) using a termination constraint

Fig. 1. A timing to select the next behavior

In this approach, we are faced with the following problems:

1. how to coordinate and switch the behaviors,
2. when to select the behaviors, and
3. when to terminate the currently executing behavior.

Since behavior selection is one of the fundamental issues in robotics, a lot of methods have been proposed. In our case, we have applied genetic programming (GP) to behavior selection problem in the robotic soccer domain [9]. However, the resultant GP trees were not represented in a compact style. For example, the robot selected the `collision avoidance` behavior to search the ball. In addition, the robot did not use the hand-coded `shooting` behavior when it is suitable to be activated. There are two major reasons why a layered approach could not obtain the appropriate behavior sequences: (1) GP does not take account of the precondition of the given behavior explicitly, and (2) the behavior is often switched although the goal of the behavior is not achieved. The first reason prevents GP from reducing the learning time, and the second causes the scraps of behavior sequence.

An example sequence of the selected behaviors is shown in Fig.1, where the black circle indicates the timing to change the behavior to another one. Fig.1 (a) indicates the ideal behavior sequence, whereas Fig. 1 (b) the resultant behavior sequence often obtained by the learning or evolutionary approaches. In the case (a), the robot selects the same behavior for a while. On the other hand, in the case (b), the robot often switches the behaviors according to the situation. Although this can be regarded as an acquisition of the new behavior sequences instead of the given behavior, it causes enormous learning time because the robot does not make good use of the behavior given by the designer.

In order to take advantage of the behavior given by the designer, we have to consider the the precondition and the goal of the behavior. Therefore, we propose a behavior selection mechanism with *activation/termination constraints.* The former constraints reduce the number of situations where each behavior is executable, and the latter contribute to extract meaningful behavior sequences. We call behavior with activation/termination constraints *module.* These constraints enable the robot to modify the timing to select the next behavior as shown in Fig.1 (c), where the gray circle indicate a vague state whether a new behavior is changed or not. From the case (c), the termination constraint contributes to avoiding frequent switching behavior. It enables us to deal with heterogeneous modules[1] in the same manner. Once the module is selected, actions are executed until the module terminates stochastically based on the termination constraint. Thus, we can obtain the behavior sequence like Fig.1 (c).

The lower layer consists of multiple modules, while the upper layer selects the appropriate module according to the situation. Genetic algorithm is applied to obtain the switching function to select the appropriate module and the timing to terminate it according to the situation. Activation/termination constraints affect not the genetic operations such as crossover and mutation but the individual representation. Although we utilize standard genetic operations, we can obtain purposive behaviors owing to the activation/termination constraints within a reasonable computational time. The results of computer simulation are shown, and a discussion is given.

2 Behavior Selection with Activation/Termination Constraints

2.1 Lower Layer

Suppose that the robot has L modules M_i $(i = 1, \cdots, L)$. A module M_i consists of three components: a *behavior* $\pi_i : \boldsymbol{X}$ (state space) $\to \boldsymbol{U}$ (action space), an *activation constraint* and I_i a *termination constraint* T_i. There are several ways to implement the behaviors, but they must be given to the robot in advance.

The activation constraint I_i gives a set of states where the module should be executable.

$$I_i(\boldsymbol{x}) = \begin{cases} 1 & M_i \text{ is executable at state } \boldsymbol{x}, \\ 0 & \text{otherwise.} \end{cases} \tag{1}$$

If the designer gives the behavior π to the robot, it is not difficult to give the precondition of the behavior. For example, the collision avoidance behavior is implemented for the mobile robot with sonars, where the designed behavior is activated only when the obstacles are detected by sonar.

In addition, each module has one termination constraint T_i consisting of a probability to sustain the module or not. In other words, this function gives the time to continue to execute the selected module.

$$T_i(\boldsymbol{x}, t) = \begin{cases} 0 & \text{the goal of the module } m_i \text{ is achieved,} \\ 0 & t > t_{p,i}, \\ p_i & \text{otherwise,} \end{cases} \tag{2}$$

where t and $t_{p,i}$ denote the elapsed steps and the pre-specified time interval, respectively. If the robot continues to execute the same module $t_{p,i}$ steps, it is forced to stop. The robot judges whether the selected module should be terminated or sustained with probability p_i.

---

[1] In this paper, "heterogeneous" means the differences of time to achieve the goal of the module.

2.2 Upper Layer

In the upper layer, the modules are switched according to the current situation. Let the value of the module m_i at the state $\boldsymbol{x}$ be $V_i(\boldsymbol{x})$. The robot selects the module of which value is the highest:

$$i^* = \arg\max_{i=1,\cdots,n} V_i(\boldsymbol{x})I_i(\boldsymbol{x}). \tag{3}$$

Once the module is selected, then actions are executed according to the current behavior π_i until the module terminates stochastically based on T_i.

In order to approximate $V_i(\boldsymbol{x})$, we use a RBF (Gaussian) network expressed by

$$V_i(\boldsymbol{x}) = \sum_{j=1}^{N} w_j g_j(\boldsymbol{x}), \quad g_j(\boldsymbol{x}) = \exp\left(-\sum_{k=1}^{n} (m_{j,k}(x_j - c_{j,k}))^2\right), \tag{4}$$

where n: the number of the input space, N: the number of RBF units, $(c_{j,1}, \cdots, c_{j,n}) = \boldsymbol{c}_j$: the center of the j-th RBF unit, $(m_{j,1}, \cdots, m_{j,n}) = \boldsymbol{m}_j$: the inverse of the size of the j-th RBF unit, and w_j: the connection weight from the j-th RBF unit. Consequently, the number of parameters is

$$1(p_i) + N \times n(\boldsymbol{c}) + N \times n(\boldsymbol{m}) + N(w) = N(2n+1) + 1,$$

for each function V_i.

From Eq.(3), there seems to be no need for the activation constraints I_i since setting V_i to zero whenever I_i would be zero should be equivalent. However, the function V_i is often inappropriate during the evolution processes. The activation constraints I_i suppress the unexpected behavior selection.

3 Genetic Operations

In order to obtain the appropriate behaviors, the robot has to learn $N(2n+1)+1$ parameters. We use the genetic algorithms. In GA, it is an important role to design the genetic operations such as crossover and mutation. We perform two types of crossover.

Global crossover : For each module m_i, a pair of parameters is selected randomly from each parent. Then, we swap two parameters.

Local crossover : At the beginning, we find the RBF units of which distance between them is minimum.

$$(j^*, k^*) = \arg\min_{j,k=1,\cdots N} ||\boldsymbol{c}_{ij}^1 - \boldsymbol{c}_{ik}^2||.$$

For the selected two RBF units j^* and k^*, we utilize BLX-α [3] based on real-coded GA. The BLX-α uniformly picks parameter values from points that lie on an interval that extends αI on either side of the interval I between parents. In other words, it randomly generates two children around their two parents by using uniform distribution in the hyper rectangular region whose sides are parallel to axes of the coordinate system.

Mutation : One of the elements of parameters is replaced to a new random value.

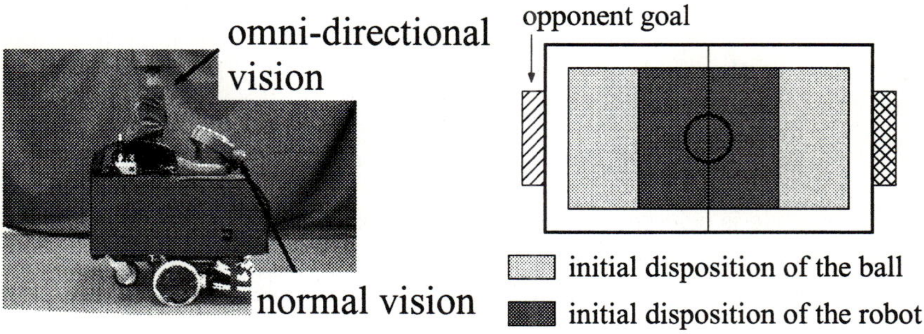

(a) our mobile robot (b) simulated environment

Fig. 2. robot and environment

4 Task and Assumptions

4.1 Robot and Environment

We have selected a simplified soccer game as a test-bed. The task for the learner is to shoot a ball into the opponent goal. The environment consists of a ball and two goals, and a wall is placed around the field except the goals. The sizes of the ball, the goals and the field are the same as those of the middle-size real robot league of RoboCup Initiative [5] that many AI and robotics researchers have been involved in.

The robot moves around the field based on the power wheeled steering system. As motor commands, our mobile robot has two degrees of freedom. The input $\boldsymbol{u}$ is defined as a two dimensional vector:

$$\boldsymbol{u}^T = \begin{bmatrix} u_l \; u_r \end{bmatrix} \quad u_l, u_r \in \{-1, \; 1\},$$

where u_l and u_r are the velocities of the left and right wheels, respectively. In addition, the robot has a kicking device to kick the ball.

The robot has two vision systems; one is a normal vision system to capture the front view, and the other is an omni-directional one to capture the visual information whole around the robot. The omni-directional vision has a good feature of higher resolution in direction from the robot to the object although the distance resolution is poor. The robot observes the center positions of the ball and two goals in the image plane using two vision systems. Therefore, the maximum number of image features is 12. A simple color image processing is applied to detect the ball and the goal areas in the image plane in real-time (every 33 [msec]).

4.2 Module Design

Basic Modules: We prepare five basic modules of which behavior just generates a simple action regardless of sensory information. That is, the motor command generated by each basic module is described as follows:

- m_1 : go forward : $\boldsymbol{u}^T = [1.0,\ 1.0]$
- m_2 : go backward : $\boldsymbol{u}^T = [-1.0,\ -1.0]$
- m_3 : stop : $\boldsymbol{u}^T = [0.0,\ 0.0]$
- m_4 : turn left : $\boldsymbol{u}^T = [-1.0,\ 1.0]$
- m_5 : turn right : $\boldsymbol{u}^T = [1.0,\ -1.0]$

These modules are always executable, in other words, for all $\boldsymbol{x}$ we set $I_i(\boldsymbol{x}) = 1$ $(i = 1, \cdots, 5)$. Then, the termination constraints only depend on the elapsed steps. In this experiment, we set the termination parameters in Eq.(2) as $p_i = 0.4$ and $t_{p,i} = 150$ steps $(i = 1, \cdots 5)$. 150 steps are equal to 5 sec.

Reactive Modules: We prepare four reactive modules:

- m_6 : search the ball
  The purpose of this module is to capture the ball image using the normal vision. Therefore, the robot searches the ball by turning to left or right. T_6 is set to zero when the ball is observed.
- m_7 : avoid collisions
  The purpose is to avoid collisions with the wall. If the wall is not detected, $\boldsymbol{u}^T = [1.0,\ 1.0]$. This module is activated when the robot moves near the wall.
- m_8 : kick the ball
  In a case where the ball is in front of the robot and this module is selected, the robot succeeds in kicking the ball. Of course, this module has no effects when the ball is not in front of the robot. This module is activated when the ball image is captured by the normal vision.
- m_9 : shoot the ball
  The purpose of this module is to push the ball into the opponent goal. This module is activated when both the ball and the opponent goal images are captured by the normal vision. The resultant behavior is the same as that of m_1, that is, $\boldsymbol{u}^T = [1.0,\ 1.0]$. This module does not always succeed in shooting behaviors, especially when the ball position is shifted from the goal direction.

In this experiment, we set the termination parameters in Eq.(2) as $p_i = 0.8$ and $t_{p,i} = 150$ for $i = 6, \cdots 9$.

Complex Modules: This module generates a motor command based on the feedback controller using a part of the image features. For further details, see [11]. Using the simple feedback controller, we prepare the following two modules:

- m_{10} : move to the defensive position
  The purpose of this module is to move to the place between the ball and the own goal.
- m_{11} : move to the offensive position
  The purpose of this module is to move to the opposite side of the opponent goal to shoot.
- m_{12} : pass the ball to the teammate
  The robot kicks the ball toward a teammate using a kicking device.
- m_{13} : receive the ball
  The robot just stops and waits for the ball to come.

These two modules can be executed when the desired state is not achieved, that is,

$$I_i(\boldsymbol{x}) = \begin{cases} 1 & \|\boldsymbol{x} - \boldsymbol{x}_d\| \leq \varepsilon \\ 0 & \text{otherwise} \end{cases},$$

where $\boldsymbol{x}_d$, ε, and $||\cdot||$ denote the desired state, a small threshold, and the norm of $\cdot$, respectively. In this experiment, we set the termination parameters in Eq.(2) as $p_i = 0.8$ and $t_{p,i} = 150$ for $i = 10$ and 11.

5 Experimental Results

5.1 Efficiency of the Constraints

In order to show the validity of our method, we apply our method to a simplified soccer game. We perform two experiments. One is to shoot a ball into the opponent goal. In this experiment, the robot is placed between the ball and the opponent goal and the robot can not see the ball using the normal vision. Therefore, the robot must round the ball until the robot can see both ball and goal. The other is cooperation between two robots. In this experiment, there are two learners and one opponent. Two learners improve their behaviors simultaneously in computer simulation.

One of the most important issues is to design the fitness measures. In this experiment, we set up four fitness measures as follows:

- f_1 : the total number of obtained goals,
- f_2 : the total number of lost goals,
- f_3 : the total number of elapsed steps until all trials end
- f_4 : the total number of ball-kicking,

In order to cope with multiple fitness measures, we create the new scalar function based on the weighted summation of multiple fitness measures by

$$f_c = \sum_{i=1}^{n} w_i f_i, \tag{5}$$

where w_i denotes the weight for i-th evaluation. The problem is to design the value of w_i since we must consider the tradeoff among all the fitness measures. In

case	activation	termination	probability
(1)	*	*	*
(2)	*	o	fixed
(3)	o	o	fixed
(4)	o	o	learned

*: this parameter is ignored.
o: this parameter is taken into account.

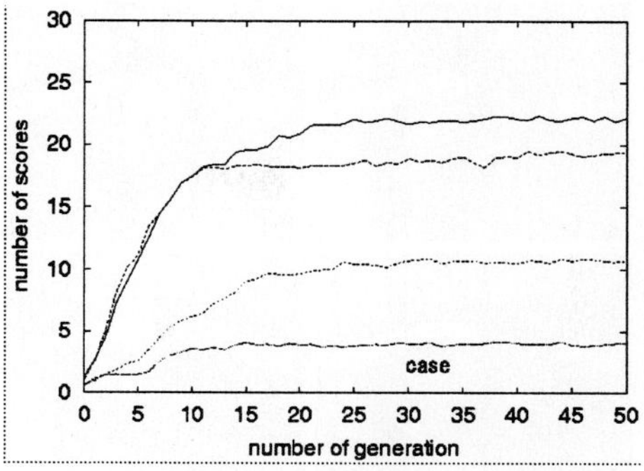

Fig. 3. Average of scores

this experiment, we use the adaptive fitness function [10] to decide the weights. Based on this method, the weights are modified considering the relationships among the changes of the four evaluations through the evolution process.

In addition, other parameters in GA are used as follows: the population size is 50, and we perform 30 trials to evaluate each individual. One trial is terminated if the robot shoots a ball into the goal or the pre-specified time interval expires. In order to select parents for crossover, we use tournament selection with size 10.

At first, we compared three modifications of our proposed method (case (2), (3), and (4)) with the standard method (case (1)). In other words, the case (1) did not take account of both constraints. Fig.3 (left) shows the setting

Fig.3 shows that a case (1) did not fulfill the goal of shooting behavior. Since the robot selects the new module in real time (every 33 [msec]), the robot changed the module frequently. As a result, the robot failed to shoot the ball into the goal until the pre-specified time interval expired. A case (2) caused the successful shooting behavior, and took shorter learning time than the case (1). However, this approach took longer time to evolve than the case with both constraints. Fig.3 shows that a case (4) took the shortest learning time to obtain the shooting behavior, and got best scores. In this method, the basic module, for example, m_5 (turn right) was terminated quickly. The learning processes of the cases (3) and (4) are almost same as shown in Fig.3. For detailed arguments, see [11].

5.2 Real Robots Experiments

In this section, we present the experimental results in the real environment. We transfer the result of computer simulation to the real robot. A simple color image processor (Hitachi IP5000) is applied to detect the ball and the goal area in the image in real-time (33 [msec]).

Fig. 4. Obtained behavior in a real environment

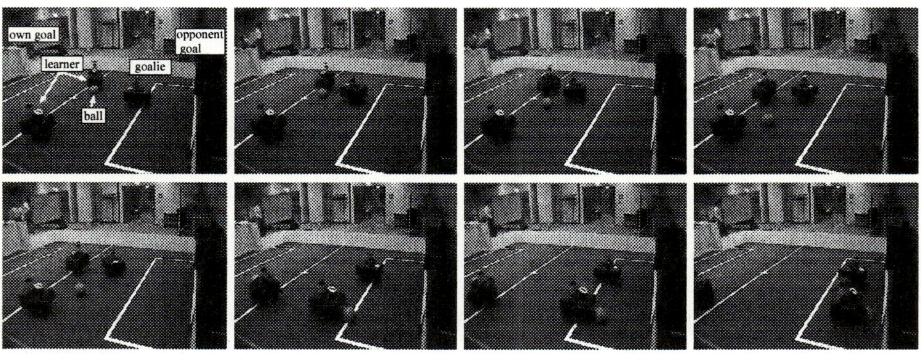

Fig. 5. Obtained behavior in a real environment using three mobile robots

Fig.4 shows an example sequence of obtained behavior in the real environment. Because of the low image resolution of the omni-directional vision system, the robot sometimes failed to detect the objects at a long distance. In this case, the module could not be performed appropriately. In spite of those troubles, our robot could accomplish the given task.

Next, we show the experimental results using three mobile robots. In this case, two offensive teammates obtained their behaviors based on the proposed method simultaneously, while a policy of the defensive robot is hand-coded. This defensive robot is a goalkeeper of our team. Fig.5 shows an example sequence of obtained behaviors. At the beginning, the goalkeeper approached the learner who possessed the ball (hereafter, learner A), while another learner (hereafter, learner B) waited and saw at the initial placement. Then, the learner A kicked the ball toward the learner B, and the leaner B pushed the ball to the opponent goal. Although our proposed method does not consider cooperation explicitly, two learners can achieve the task.

6 Conclusion

This paper presented an architecture for behavior selection with activation/termination constraints. We applied the proposed method to a soccer situation and demonstrated the experiments. Owing to the both constraints, the standard GA can search the feasible solutions for purposive behaviors much faster than the case without constraints.

Acknowledgments This research was supported by the Japan Society for the Promotion of Science, in Research for the Future Program titled Cooperative Distributed Vision for Dynamic Three Dimensional Scene Understanding (JSPS-RFTF96P00501).

References

1. M. Asada, S. Noda, S. Tawaratsumida and K. Hosoda. Purposive Behavior Acquisition for a Real Robot by Vision-Based Reinforcement Learning. *Machine Learning*, 23:279–303, 1996.
2. M. Ebner and A. Zell. Evolving a behavior-based control architecture – From simulations to the real world. In *Proc. of the Genetic and Evolutionary Computation Conference*, pages 1009–1014, 1999.
3. L. J. Eshleman and J. D. Schaffer. Real-Coded Genetic Algorithms and Interval-Schemata. In *Foundations of Genetic Algorithms 2*, pages 187–202. 1993.
4. D. E. Goldberg. *Genetic Algorithms in Search, Optimization, and Machine Learning.* Addison–Wesley, 1989.
5. H. Kitano, ed. *RoboCup-97 : Robot Soccer World Cup I.* Springer Verlag, 1997.
6. J. R. Koza. *Genetic Programming I : On the Programming of Computers by Means of Natural Selection.* MIT Press, 1992.
7. P. Stone. *Layered Learning in Multiagent Systems.* The MIT Press, 2000.
8. R. S. Sutton and A. G. Barto. *Reinforcement Learning.* MIT Press/Bradford Books, March 1998.
9. E. Uchibe, M. Nakamura, and M. Asada. Cooperative and Competitive Behavior Acquisition for Mobile Robots through Co-evolution. In *Proc. of the Genetic and Evolutionary Computation Conference*, pages 1406–1413, 1999.
10. E. Uchibe, M. Yanase, and M. Asada. Behavior Generation for a Mobile Robot Based on the Adaptive Fitness Function. In *Proc. of Intelligent Autonomous Systems (IAS-6)*, pages 3–10, 2000.
11. E. Uchibe, M. Yanase, and M. Asada. Evolution for Behavior Selection Accelerated by Activation/Termination Constraints. In *Proc. of the Genetic and Evolutionary Computation Conference*, pages 1122–1129, 2001.

Stereo Obstacle Detection Method for a Hybrid Omni-directional/Pin-Hole Vision System

Giovanni Adorni[2], Luca Bolognini[1], Stefano Cagnoni[1], and Monica Mordonini[1]

[1] Department of Computer Engineering, University of Parma
[2] Department of Communication, Computer and System Sciences, University of Genoa

Abstract. This paper describes an application of a vision system based on the use of both an omnidirectional vision sensor and a standard CCD camera. Such a hybrid sensor permits implementation of peripheral/foveal vision strategies, that can be very useful for navigation tasks. This paper focuses particularly on the use of the device as a stereo vision system to locate obstacles in a semi-structured environment through a perspective removal algorithm.

1 Introduction

Omni-directional vision systems are usually based on a catadioptric sensor, consisting of an upwards-oriented camera that acquires the image reflected on a convex mirror hanging above it [1, 2]. Such systems are very efficient as concerns target detection, but critical from the point of view of the accuracy with which the target is detected. For these reasons, in several cases, the omni-directional vision sensor has been integrated with different kinds of sensors (see, for example [3, 4, 5]). In this paper we present results obtained using a vision system prototype (called HOPS, Hybrid Omnidirectional/Pin-hole Sensor), that consists of an omni-directional sensor coupled with a standard CCD camera. HOPS was designed with the main intent to assist navigation tasks. This paper focuses on an obstacle detection method that was implemented for use with the hybrid sensor.

HOPS (see figure 1) integrates omnidirectional vision with traditional pin-hole vision, to overcome the limitations of the two approaches. It can be best described as a color CCD camera mounted on top of an omnidirectional catadioptric sensor. Its lower part consists of an omnidirectional sensor that consists of a color CCD camera pointing upwards at a convex mirror (see [6] for details). The omnidirectional sensor provides a base for the more traditional CCD-camera based sensor that leans on it. The CCD camera is fixed, looks down with a tilt angle of 60° with respect to the ground plane and has a field of view of about 80°. For the application described in this paper, to allow for a higher stereo disparity between the two images, it is positioned slightly off the center of the device. An example of the images that can be obtained by the two sensors is provided in fig. 1.

2 Hybrid Stereo Vision for Obstacle Detection

Inverse perspective mapping for obstacle detection was first introduced in [7]. If one applies the inverse perspective mapping transform with respect to the same plane to a

A. Birk, S. Coradeschi, and S. Tadokoro (Eds.): RoboCup 2001, LNAI 2377, pp. 244–250, 2002.

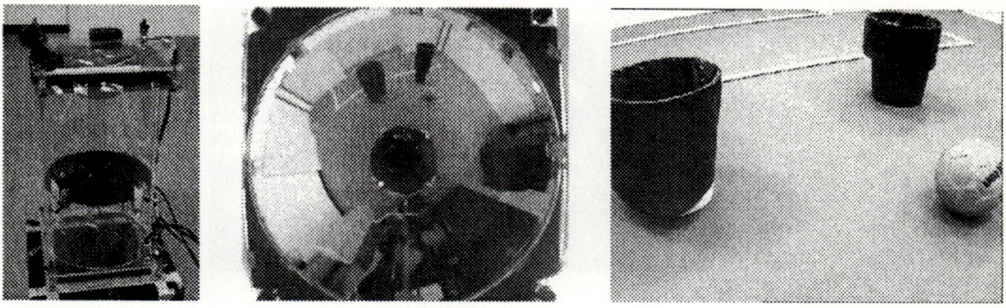

Fig. 1. The HOPS prototype (left) and an example of images that can be acquired through the omni-directional sensor (center) and the top CCD camera (right).

pair of stereo images, everything that lies on that plane looks the same in both views, while everything that does not is distorted differently. This property is particularly useful for tasks in which a relevant reference plane can be easily found. The steps required to detect obstacles based on stereo vision include: i) applying the inverse perspective transform to the two images; ii) subtracting one transformed image from the other one to detect differences; iii) labelling pixels of at least one of the acquired images as free space or obstacles.

Stereo vision is usually obtained by two cameras slightly displaced from each other or by a single moving sensor that can move to obtain the same effect. In the case of HOPS such a displacement is not constant, since the angle between the optical axis of the camera and the plane that is tangent to the mirror surface is different for each point of the mirror. If the traditional camera were positioned on the axis of the omnidirectional sensor, disparity would be close to zero in a few points. Positioning the pin-hole sensor off the omnidirectional sensor axis ensures that a different perspective be obtained from the two sensors for all points.

The problem dealt with by the inverse perspective transform (IPT) consists of computing a correspondence function $P_{x,y} = C(I_{i,j})$ that maps the pixels belonging to image I onto points of a new image P that shows a bird's view of a reference plane. From the view obtained by applying $C(I)$ information about the relative positions of objects (e.g., distances) that lie on the plane can be extracted.

If all parameters related to the geometry of the acquisition systems and to the distortions introduced by the camera were known, the derivation of C could be straightforward [7, 8]. However, this is not always the case, most often because of the lack of a model of the camera distortion. However, assuming an ellipsoidal model for the camera lens usually allows for an easy empirical derivation of the related parameters anyway.

In computing C_o, the IPT for the catadioptric omnidirectional sensor, while, on one side, the problem is complicated by the non-planar profile of the mirror, on the other side the simmetry of the device is such that it is enough to compute the restriction of C_o along a radius of the mirror projection on the image plane to be able to compute the whole function. However, in doing so, one must take into account possible manufacturing flaws, that affect both the shape of the mirror and the smoothness of the mirror surface. In addition to "regular" flaws, that can be found all over the surface and can be included in the radial model of the mirror, a few other minor "random" flaws can be found. These

Fig. 2. Results of perspective removal (right) from the omnidirectional image (left).

require that the IPT be derived empirically, or at least that the ideal model be corrected in flawed points.

A first empirical derivation of C_o can be made by considering a set of equally-spaced radii, on each of which values of C_o are computed for a set of uniformly-sampled points. To obtain the values of the function for a generic point P_i located anywhere in the field of view of the sensor, a bi-linear interpolation is made between the four points among which P_i is located, within a uniformly-sampled polar grid. This makes reconstruction accuracy better close to the robot, as the area of the cells used for interpolation increases with radial distance. Reconstruction quality increases with the number of radii, as the angular sampling step of the grid decreases. However, distortions can still be noticed. To limit the problem, a sixth-grade approximating polynomial is used to make the restriction of C_o along each radius smoother. The approximating models are "manually corrected" in the neighborhood of interpolating points in which the approximation error is above a tolerance threshold, taking care not to affect smoothness. Results obtained using eight radii are shown in figure 2.

After applying the IPT, two "bird's view" images are obtained that can be compared to detect obstacles. However, differences in the two sensors require that the two IPT images be pre-processed. In particular, the two images differ in their pictorial features. Since the region of interest is the same for both sensors, after identifying the region $\mathcal{R}$ of the omnidirectional image O that matches the field of view of the image P acquired by the upper camera, a punctual transformation is applied to O, such that

$\sum_0^n H_{o,col}(l, \mathcal{R}) = \sum_0^n H_{p,col}(l, P) \quad \forall n, col,$

$H_{o,col}(l, \mathcal{R})$ and $H_{p,col}(l, P)$ being the discrete histograms of the *col* (*col* $\in R, G, B$) component for region $\mathcal{R}$ of O and for P, respectively. The variable l represents the intensity level of the color component. Since the input images are color images, the result of the difference operator must combine information from all three color components. We use a max operator followed by a thresholding to do so. Each pixel value of the resulting binary image D is computed as:

$$D_{i,j} = \begin{cases} 0 \text{ if } \max(|R_{i,j}(\mathcal{R}) - R_{i,j}(P)|, \\ \qquad |G_{i,j}(\mathcal{R}) - G_{i,j}(P)|, \\ \qquad |B_{i,j}(\mathcal{R}) - B_{i,j}(P)|) \geq T \\ 1 \text{ otherwise} \end{cases}$$

To make the algorithm less sensitive to light changes, the threshold T is a linear function of the average lighting intensity of the acquired images. The difference image

D computed as described is still affected by the presence of noise, either in the form of isolated white pixels and thin lines, or of small gaps in the significant regions. These artifacts are corrected using a majority-vote morphologic filter.

3 Obstacle Detection with HOPS

The white regions in D derive from two kinds of discrepancies. If they derive from different pan angles or from a lateral displacement of the two cameras, they are located to the left or right of obstacle projections in the IPT transformed images. With different tilt angles or vertical displacements of the cameras, they are above and/or below the obstacle projections. For HOPS aplications the reference plane is the ground plane on which a robot that mounts the sensor is moving.

Thus: i) lateral differences are almost symmetrical and not very large; ii) no differences can be detected at the base of obstacles, as obstacle bases lie on the reference plane and their transforms are identical there; iii) vertical differences are more relevant near obstacle top, and somehow proportional to obstacle height.

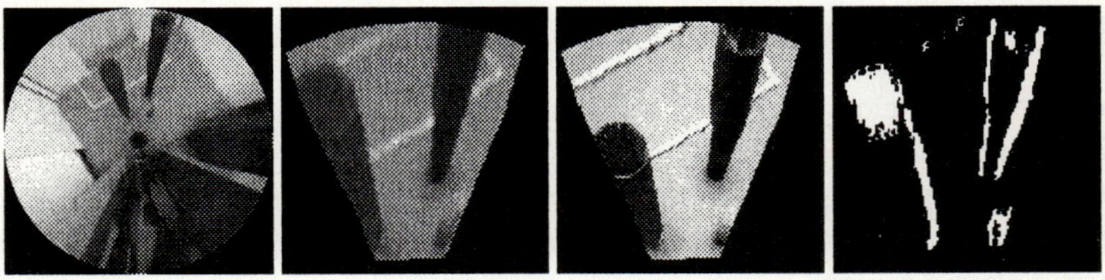

a b c d

Fig. 3. IPT images acquired by the omni-directional sensor (a) and the standard CCD camera (c). The region of the IPT omnidirectional image corresponding to the field of view of the standard camera (b) and the thresholded difference image (d) are also shown.

Therefore, in the difference image, obstacle candidates are represented by regions bounded by two thin roughly-vertical regions laterally and by a rather large blob vertically. Sometimes one or more of the three region delimiters may be absent. Typical exceptions are small roundish obstacles, for which a single blob is usually detected, or cases in which the distortion of the matter that is located above the ground is such that the obstacle (pseudo)projection extends beyond the common field of view of the two cameras. This is the case shown in figure 1. The results of perspective effect removal and the difference image D are shown in figure 3.

The difference image D is the input of the obstacle-detection method, along with color information from the input images. Segmentation is performed to detect image regions having the features listed above. The analysis of D is a bottom-up process in which segments of regions of interest are matched and merged together to obtain larger regions corresponding to obstacles. First, low-level segmentation is performed

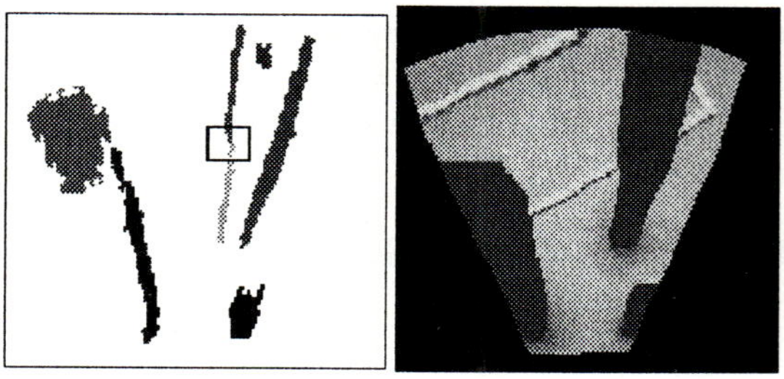

Fig. 4. Left: the results of the blob-coloring algorithm. The two regions evidentiated with a square belong to class 3. The two elongated ones belong to class 2; all other regions belong to class 4. Right: the obstacle regions detected by "closing" the regions surrounded by, or corresponding to, the obstacle segments detected in the previous steps.

to detect connected regions in *D* and label them using a blob-coloring technique. The regions whose area is below a pre-set threshold are discarded. For all other regions, a set of geometrical features is computed, to classify them into four possible categories of interest, namely:

1. long vertically-elongated thick regions with two apices in their lower part, that directly represent obstacles;
2. long vertically-elongated thin regions with a single lower apex, that may represent one of the segments of the typical obstacle "footprint" in *D* and need merging to form a class-1 region;
3. short vertically-elongated regions, a sub-class of class 2, to which lower significance is attributed in the matching phase;
4. small roundish regions (i.e., characterized by a height/width ratio close to 1) that may represent small obstacles, part of them or the upper blob of typical obstacle representations in *D*.

Classification is based on simple empirically-derived rules, implemented as boolean functions of the geometrical features. Although in most cases they can be interpreted as traditional crisp rules, they also allow for fuzzy interpretation in more complex cases. Figure 4 (left) shows the result of the blob-coloring algorithm.

The next step is aimed at merging regions that do not belong to class 1 into larger ones. The matching between regions is based on orientation and color of the neighboring background pixels in the IPT images. More precisely, regions are merged if they are close enough, their orientation differ by less than .2 *rad*, and have regions of the same colors as neighbors. Color similarity is evaluated according to a set of fuzzy rules.

The final step can be divided in two phases: in the first one a preliminary "closure" of the obstacle regions is made. Since the typical obstacle footprint in image *D* consists of two vertically elongated regions connected to a roundish one on top, the closure is obtained by connecting the two lower apices that can be detected in such a configuration. When some parts of the typical configuration lack, detecting two vertically-elongated regions or one vertically-elongated region and the boundary of the region of interest common to both cameras can be enough for the obstacle to be circumscribed. In the second phase, segmentation is refined, based on uniformity of color or, more generally, of pictorial features in the input images. A horizontal scan is made along the right and left boundaries of the closed regions, that are made coincident with the pixel in which the highest gradient between obstacles and background is found. If the maximum gradient falls below a preset threshold in one of the two IPT images the line is discarded: that is why the large round region on the left (the top of the trash can on the left in the IPT image of the upper camera) has almost completely disappeared in figure 4 (right). Results obtained in indoor environments are described in [9].

4 Final Remarks

In the application described in the paper, HOPS has been used as a stereo sensor. This was certainly not among the primary goals for which HOPS was designed. Actually, omni-directional vision is a complicating factor for inverse-perspective based obstacle-detection. However, results show that, besides the immediate advantages provided by HOPS (the joint availability of omni-directional vision and high-resolution information about a region of interest), HOPS provides also the capabilities of a stereo system by which navigation sub-tasks can be accomplished in real-time: our prototype application can process about 10 frames/s on recent high-end PCs.

Acknowledgements

This work has been partially supported by CNR under the ART, Azzurra Robot Team (http://robocup.ce.unipr.it) grant, ASI under the "Hybrid Vision System for Long Range Rovering" grant, and by ENEA under the "Intelligent Sensors" grant.

References

[1] S. K. Nayar. Omnidirectional vision. In *Robotics Research. 8th International Symposium*, pages 195–202, 1998.

[2] T. Svoboda and T. Pajdla. Panoramic cameras for 3D computation. In *Proc. Czech Pattern Recognition Workshop*, pages 63–70, 2000.

[3] L. Delahoche, B. Maric, C. Pégard, and P. Vasseur. A navigation system based on an omnidirectional sensor. In *Proc. IEEE/RSJ Int. Conf. on Intelligent Robots and Systems*, pages 718–724, 1997.

[4] J. S. Gutmann, T. Weigel, and B. Nebel. Fast, accurate, and robust selflocalization in polygonal environments. In *Proc. 1999 IEEE/RSJ Int. Conf. on Intelligent Robots and Systems*, pages 1412–1419, 1999.

[5] A. Clérentin, L. Delahoche, C. Pégard, and E. Brassart-Gracsy. A localization method based on two omnidirectional perception systems cooperation. In *Proc. 2000 ICRA. Millennium Conference*, volume 2, pages 1219–1224, 2000.

[6] A. Bonarini, P. Aliverti, and M. Lucioni. An omnidirectional vision sensor for fast tracking for mobile robots. *IEEE Trans. on Instrumentation and Measurement*, 49(3):509–512, 2000.

[7] H. A. Mallot, H. H. Bülthoff, J. J. Little, and S. Bohrer. Inverse perspective mapping simplifies optical flow computation and obstacle detection. *Biological Cybernetics*, 64:177–185, 1991.

[8] G. Adorni, S. Cagnoni, and M. Mordonini. An efficient perspective effect removal technique for scene interpretation. In *Proc. Asian Conf. on Computer Vision*, pages 601–605, 2000.

[9] G. Adorni, L. Bolognini, S. Cagnoni, and M. Mordonini. A non -traditional omnidirectional vision system with stereo capabilities for autonomous robots. In *Proc. 7th AI*IA Conference*, 2001. in press.

Self-Localisation Revisited

Joscha Bach and Michael Gollin

Humboldt University of Berlin, Department for Computer Science
{bach, gollin}@informatik.hu-berlin.de

Abstract. We discuss different approaches of self-localisation in the Simulation League. We found that the properties of the soccer server's quantization function tend to produce artifacts with the common approaches, which we try to deal with using a new method: We simulate the player's position, and dynamically correct this estimate with a gradient descent function by the minimal amount necessary to make it consistent with the perceived flag positions (where we allow for error margins according to the quantization function). It can be shown that self-localisation using the new method is not only relatively exact (between 6.6% and 35% smaller average error than the 'Nearest Flag' algorithm) but also yields much more consistency than the other approaches. ...

1 Introduction

The soccer agents in the simulation league of RoboCup [5] do not have the ability to perceive their environment (the field with players and ball) as a whole. Rather, the soccer server sends 'sense data' in regular intervals, relative to the agent's position and restricted to a certain viewing angle. Frequency of sense data and angle are inversely related: for the available viewing angles of 45, 90 and 180 degrees, data arrive at intervals of 75 ms, 150 ms and 300 ms, respectively [4]. (It is possible to receive data without distance information at a faster rate, but we do not take this into account here.) Whenever the agent gets no sense data during a simulation step, its position should be simulated with respect to its movement actions. To maintain a world model, the agent has to analyze this data, update the positions of 'seen' objects and perhaps simulate the positions of hidden objects. While there are no restrictions on the internal format of the data, most teams opted for an absolute representation in Cartesian coordinates. Self-localisation is crucial with this approach. Usually, the position is derived from the fixed objects in the field: there are 55 flags and four lines (the latter ones are helpful in determining the direction the agent is facing). Objects within the visible sector are determined by an angle relative to the agent's viewing direction (±0.5 degrees) and the distance. The soccer server adds noise on the latter using a quantization function; closer objects are more accurately seen than distant ones.

Several ways of computing the position of the agent lend themselves to the task, for example:

A. Birk, S. Coradeschi, and S. Tadokoro (Eds.): RoboCup 2001, LNAI 2377, pp. 251–256, 2002.

- With known viewing direction, the position can be calculated from any seen flag. Since closer flags are more accurate, the nearest one should be used. This method (called "Nearest Flag") is both fast and easy to implement and is used for instance by the CMU-98 team. [6]
- Use a weighted sum of the positions derived from all visible flags. This method has been used by our team, AT Humboldt 98. [3]
- Triangulate from two or more visible flags. To reduce the computational complexity, not all flags should be combined. Rather, one might choose to look only at the two closest flags. It is also possible to use heuristics to determine reliable pairs of flags and weight the result. [2]
- The visible flags may be fed into a neural network that delivers the agent's position and viewing angle as output. We have tried this successfully, but we do not consider it an improvement over the algorithmic methods.
- Other learning methods, like Case Based Reasoning, can do the job as well. Again, while it works, the implementation overhead does not appear to be justified by the result. [7]
- It is possible to compute the maximum positional error for each seen flag using the quantization function, and therefore an area where the agent could be possibly located. By intersecting these areas, an 'error figure' can be determined. Whenever this intersected space is small enough, the position of the agent can be calculated with high accuracy. We use a simplified approach that makes use of this property.

2 Quantization and Consistency

To add noise on the sense data, the soccer server makes use of a special function, which we call Noise here (in the soccer server, this takes place in two steps):

$$\text{Noise}(d) = \frac{1}{10}\text{rint}\left(10e^{\text{rint}\left(\frac{\log(d+EPS)}{qstep}\right)qstep}\right) \qquad (1)$$

where d is the distance of the object, *qstep* determines the quantization level (0.01 for fixed objects) and *EPS* is an offset of almost zero (to prevent undefined values from the log function, when d is zero. *Noise* returns the original distance with an added (quantization) noise component, with one decimal place.

The noise produced by this quantization function is very different from common kinds of noise (like, for instance, from sensor input). The values are discrete and not Gaussian distributed, that is, the likelyhood for a value to be close to the exact value is just as high as the likelyhood of it being located at the fringes of the error limit. *Noise* tends to produce artifacts, especially for distant, slowly moving objects, which seem to move in small, erratic jumps or even appear to change their direction temporarily (though within the error limit, of course). This may lead to difficulties, when localisation algorithms do not take the resolution of *Noise* into account; a weighting of values might not be enough.

The positional errors in subsequent self-localisations do not remain constant or, which is worse, add up surprisingly often. The quantization function actively supports agents in beliefs about a jerking, wobbling world around them. This can render any attempt to determine the speed of objects impossible, if positional differences from one step to the next are used, which is often necessary. [1]

These pitfalls of a wobbling world model could be completely avoided by the introduction of a localized world model, where data is maintained relatively to the agent's position. But since the positions of fixed objects are known, and for reasons of joint planning and communication, a global world model relative to the field is immensely practical. The prospect of maintaining two world model components seemed not very elegant to us, and we dropped the idea. Instead, we thought of ways of increasing the accuracy of self-localisation and thereby reducing the possible cumulation of errors, and about steadying the algorithm.

3 The Gradient Descent Algorithm

The maximum positional error of each flag f_i is linearly dependent on its distance to the player and can be approximated as

$$\delta_{max} = 0.0045d_i + 0.05 \tag{2}$$

where d_i is the seen distance of f_i. Thus, the visible distance determines a circular ring of possible positions around the flag. (If angles were taken into account, it would be a ring segment. However, we found it unnecessary to consider angles at all.) If several flags are seen, the intersections of their respective rings define an area of possible residence (the 'error figure'), such that for all points p within the area can be stated

$$\forall f_i : \left|\overrightarrow{p, f_i}\right| \leq \delta_{max} \tag{3}$$

where is the distance between p and f_i. Furthermore, we can guarantee that

$$\exists p : \forall f_i : \left|\overrightarrow{p, f_i}\right| \leq \delta_{max} \tag{4}$$

Theoretically, this area might be disconnected (imagine the case of two flags), but this almost never happens. Nevertheless, we will have to keep this in mind. The size of the error figure determines the maximum accuracy with that the agent's position can be determined from the available sense data. It might be a good idea to localize the agent in the middle of this figure (for instance, by finding the smallest δ to keep condition (4) satisfied and use the corresponding p as position of the agent, provided there is only one). Actually, this yields an average positional error of less than 5 cm at a visible angle of 180 degrees. However, for smaller visible sectors, the error figure tends to be much bigger and the result becomes inaccurate.

If position of the agent in the previous simulation step were known, we could estimate the current position more precisely than we could extract it from the sense data. But since errors from simulated movements accumulate, we have to integrate sense data into the calculation. Ideally, what we want is a p that satisfies [4] and is as close as possible to our estimate. (This constraint helps also in the case of a disconnected or cavernous error figure.) Thus, quantization noise will only affect the resolution, but not the consistency of the localisation.

A suitable p might be found by generating a 'force field' that pulls our estimate towards the error figure. Imagine, all the flags were pulling or pushing the estimated position in the direction of their respective error limit. Note that there is no 'force' whenever the estimate is located within the error limits of the flag. Whenever there is a resultant force different from zero, there is a point in the direction of this force that is closer to the desired p than the current estimate. Of course, we can not simply add the individual forces to get the length of the vector towards this point. We have to make sure that three flags pulling into one direction do not produce three times the necessary force! Instead, we may use the according directional component of the strongest force pulling into the desired direction. To produce faster convergence, it is also acceptable to move in the direction and by the amount of the strongest individual force. This leads to the following simple algorithm:

1. Estimate the current position from previous position and movement actions. If previous positions are not available, use any other self-localisation method.
2. For all visible flags, compute the error limit using (2).
3. For the current estimate, calculate the distances to all visible flags (the real flag positions are known). Calculate the 'force vectors', which are given by the difference between seen distance and calculated distance. The error limit has to be subtracted from each; only forces over this threshold are considered.
4. If there is no force left, or the maximum number of iterations is reached, return the current estimate.
5. Otherwise, move the estimate by the longest force vector, continue with 3.

Since the algorithm sometimes shows asymptotic behaviour, the number of iterations must be limited (for instance, to 10 or 20). One could also increase the respective forces by a small amount to speed up the convergence and escape the asymptote. There is little danger in using only a small number of iterations, because the process will continue in the next simulation step at the position of the current estimate.

4 Results

When comparing the gradient descent algorithm with other methods, we have to bear in mind that data does not arrive in every simulation step (except when the visible angle is restricted to 45 degrees), so all methods have to use a certain amount of simulation. If sense data was to arrive in every simulation step at a visible angle of 180 degrees, the accuracy of our method would vastly increase.

We have tested the algorithms under what we hope are realistic conditions by repeatedly randomly placing an agent on the field and performing straight dashes until the border of the field is reached.

4.1 Accuracy

To determine the accuracy of the algorithm, we compare it to the 'nearest flag' method (CMU-98), the previous AT Humboldt algorithm and a heuristic triangulation algorithm that has been implemented as part of a Diploma Thesis at our department. [2]

Table 1. Average positional errors (in m), standard deviations and maximum errors at different visibility angles (10^7 samples)

	Algorithm:	Gradient	Nearest Flag	ATH 98	Triangulation
180 degrees	Av. Error	0.086	0.123	0.139	0.104
	Std. Dev.	0.101	0.144	0.162	0.122
	Maximum	0.569	0.744	0.743	2.179
90 degrees	Av. Error	0.083	0.127	0.146	0.130
	Std. Dev.	0.097	0.155	0.179	0.177
	Maximum	0.711	0.852	0.885	3.643
45 degrees	Av. Error	0.142	0.152	0.155	0.272
	Std. Dev.	0.182	0.192	0.195	0.353
	Maximum	2.218	1.247	1.027	5.173

At a visible angle of 180 degrees and a sense data frequency of 300 ms, our gradient descent algorithm is 30% better than the 'nearest flag' method. We found maximum values not very informative, since they often occur only rarely (for the triangulation method, only two out of one million values were in the 2 m range). We list the standard deviations to give an idea of the distribution.

At a visible angle of 90 degrees, data arrive in 2 out of 3 simulation steps. The average error is 35% better than for 'nearest flag'. At a visible angle of 45 degrees (sense data in every simulation step), few flags are visible at a time. Because our prototypical implementation does not check whether the number of visible flags allows reliable self-localisation with the gradient descent, the maximum positional error grows more than proportionally.

At an average error of 14.2 cm, the gradient descent algorithm is still slightly more (6.6%) accurate than the 'nearest flag' algorithm.

4.2 Consistency

As mentioned before, we consider consistency (i.e. only small differences between subsequent simulation steps between simulation and perception) as even more important than accuracy. The gradient descent algorithm yields indeed better consistency than the competitors.

While the other tested algorithms tend to produce bigger positional differences with growing unreliability of the input values, the gradient descent algorithm is relatively robust. Consistency appears to be almost independent from

Table 2. Average positional differences between simulation steps in m, standard deviations and maximum differences, at different visibility angles (10^7 samples)

	Algorithm:	Gradient	Nearest Flag	ATH 98	Triangulation
180 degrees	Pos. Diff.	0.050	0.064	0.056	0.061
	Std. Dev.	0.099	0.125	0.106	0.119
	Maximum	0.636	0.980	0.620	2.246
90 degrees	Pos. Diff.	0.058	0.113	0.179	0.130
	Std. Dev.	0.089	0.160	0.117	0.207
	Maximum	0.629	0.962	0.761	5.546
45 degrees	Pos. Diff.	0.051	0.189	0.131	0.322
	Std. Dev.	0.079	0.226	0.148	0.460
	Maximum	1.300	1.307	0.961	7.384

the number of visible fixed objects, and the values for the average positional differences are between 22% (at 180 degrees) and 73% (at 45 degrees) better than for 'nearest flag'.

References

1. Badjonski, M., Schröter, K., Wendler, J., Burkhard, H.-D.: Learning of Kick in Artificial Soccer. Pre-Proceedings of the fourth RoboCup Workshop, Melbourne (2000)
2. Meinert, T., Sander, G.: Der Intention-Agent. Diplomarbeit, Humboldt-Universität zu Berlin (2001)
3. Müller-Gugenberger, P., Wendler, J.: AT Humboldt 98: Design, Implementierung und Evaluierung eines Multiagentensystems fr den RoboCup-98 mittels einer BDI-Architektur. Diplomarbeit, Humboldt-Universität zu Berlin (1998)
4. Noda, I: Web Site: RoboCup Soccer Server. http://ci.etl.go.jp/~noda/soccer/server/ (March 2001)
5. Website: RoboCup Federation. http://www.robocup.org (March 2001)
6. Stone, P.: Layered Learning in Multi-Agent Systems. PhD Thesis, Carnegie Mellon University (1998)
7. Wendler, J., Brüggert, S., Burkhard, H.-D., Myritz, H.: Fault-tolerant self localization by case-based reasoning. In T. Balch, P. Stone, G. Kraetzschmar (eds.): Proceedings of the fourth RoboCup Workshop Melbourne (2000) 82–89

Yue-Fei: Object Orientation and Id without Additional Markers

Jacky Baltes

Centre for Image Technology and Robotics
University of Auckland,Auckland
New Zealand
j.baltes@auckland.ac.nz
http://www.citr.auckland.ac.nz/~jacky

Abstract. This paper describes a novel approach to detecting orientation and identity of robots using a global vision system. Instead of additional markers, the shape of the robot is used to determine an orientation using a general Hough transform. In addition the movement history as well as the command history are used to calculate the quadrant of the orientation. The identity of the robot is determined by correlating the motion of the robot with the command history. An empirical evaluation shows that the performance of the new video server is at least as good as that of a traditional approach using additional coloured markers.

1 Introduction

This paper describes a new approach to image processing in the RoboCup domain, that has been implemented in the video server of the All Botz, the University of Auckland F180 team. The videoserver has been used with good success last year, but problems remained with correctly identifying robots. These shortcomings have been resolved in our new videoserver, called YUE-FEI.

In the F180 league, most teams use a global vision system to control up to five robots per team in a game of robotic soccer. In order to be able to control the robots, coloured markers or bar codes are put on top of the robots to simplify the vision task.

Coloured markers or bar codes are an easy and robust method, but have two big disadvantages. Firstly, the calibration of sets of colours, so that they can be detected over the entire playing field and do not interfere with each other is a very time consuming task. The resulting calibration is not robust. Even small changes in the lighting conditions require a re-calibration. Secondly, these methods do not scale to large teams of robots with eleven players or more.

Therefore, the All Botz developed a new flexible and scalable approach, that uses the original shape of the robot as its sole source of vision information. In other words, the robots are unaltered except for the addition of a coloured marker, which is required by the F180 RoboCup rules. A generalized Hough transform is used to infer the orientation of the robot from a sub-image. The image processing determines an exact orientation of one side of the robot (an angle

A. Birk, S. Coradeschi, and S. Tadokoro (Eds.): RoboCup 2001, LNAI 2377, pp. 257–262, 2002.

between 0 and 90 degrees), but there is not sufficient information to determine the quadrant of the angle. Thus, the quadrant is determined by correlating the movement history (e.g., dx, dy) and current command (e.g., move forward) to the motion of the robot.

The most difficult vision problems in the RoboCup domain is to determine the identity of a robot. All other teams use unique features of the robots to determine their id. As the number of robots increases it becomes more difficult to find unique features that can be recognized efficiently and robustly. In our system, the identity of the robot is determined through correlating the command stream from the individual controllers to the observed behavior of the robot.

Section 2 describes the vision problems associated with the F180 league and how these problems were addressed by other teams previously. Section 3 describes the design of YUE-FEI, the new video server of the All Botz. The results of an empirical evaluation comparing the performance of YUE-FEI against that of a traditional video server are shown in section 5. Directions for future research and further improvements are shown in section 6.

2 Global Vision in the RoboCup

Most teams in the RoboCup competition use additional coloured markers to create feature points on the robot. In the simplest case, the two points have a distinct colour, which makes it easy to determine the orientation of the robot by relating it to the orientation of the line between the two points.

One of the most difficult aspects of the vision processing is the visual identification of robots. To be able to control a robot, the control system needs to know the identity of the robot, so that the commands are sent to the correct robot (e.g., Robot 3 move forward).

So far, the only solution to this problem suggested by teams competing in the RoboCup are to use individual colour markers, "bar codes" , manual tagging, or placing robots into predefined positions.

Most teams identify their robots through different colours. Although this approach is conceptually simple, it is non-trivial to find a parameters for a colour model that allows the detection of a colour over the entire playing field.

Another possibility is to identify a robot using some easily distinguishable geometrical pattern. Popular methods are to identify different patterns based on their size or their aspect ratio.

A third possibility is to manually identify (tag) each robot before play starts or play continues after a stoppage. For example, the user may click on robot one through five in turn. The vision server then continues to track the robot.

3 The Yue-Fei Videoserver

The solutions described in the previous section have severe disadvantages since they do not scale up to larger teams and to more flexible camera positions.

It is difficult to distinguish between more than six different colours reliably over the whole playing field. The selection of the additional colours becomes important and often leads to conflicts with other team, since they also try to find a suitable set of colours. At RoboCup-99, the All Botz as well as most other teams spend a lot of time calibrating their colours and there were some discussions about what colours are permissible.

Based on these observations and our experiences at RoboCup-99, we started work on the design of a new video server YUE-FEI. The design goals for the new video server were to provide a scalable, robust, flexible solution to the orientation and identification problem. For the reasons mentioned above, we ruled out additional colours and additional patterns on the robot.

If we do not want to use additional patterns, then what else is there? The only information left is the image of the robot itself. So the goal was to design a videoserver that uses only a single marker and no other patterns on the robot.

3.1 Orientation Information Using the Generalized Hough Transform

Figure 1 contains three zoomed views of our robots from our video camera. The views correspond to the worst case (i.e., the robot is at the far end of the playing field) for our vision system. As can be seen, the most prominent features are the edges along the top of the robot. Other features (e.g., the wheels are not always visible and hard to distinguish). Therefore, we decided to use these edges as features and to infer the orientation of the robot from them.

Fig. 1. Some sample images of our robots taken at the far side of the field.

This idea faces an immediate problem, since the robots are almost square. This means that it is *impossible* to determine the orientation of the robot from a single image. Given the angle of the edge, there are four possible orientations for the robot, which can not be distinguished without further information.

Furthermore, since all robots have exactly the same shape, it is impossible to identify the robot. Therefore, we decided to use additional information (e.g., history of the cars, current commands, motion of the robot) available to the video server to disambiguate the orientation and to identify the robot. This part of the system is described in section 4.

Given the real world coordinates of the robot, the surrounding image corresponding to a square area of the diameter of the robot is extracted. All further processing is limited to this local neighborhood.

Figure 2 shows the output for the three sample images given in Fig. 1. The contour of the robot is shown in black. As can be seen, using even a very coarse colour calibration, the edges of the robot can be traced accurately even under bad conditions.

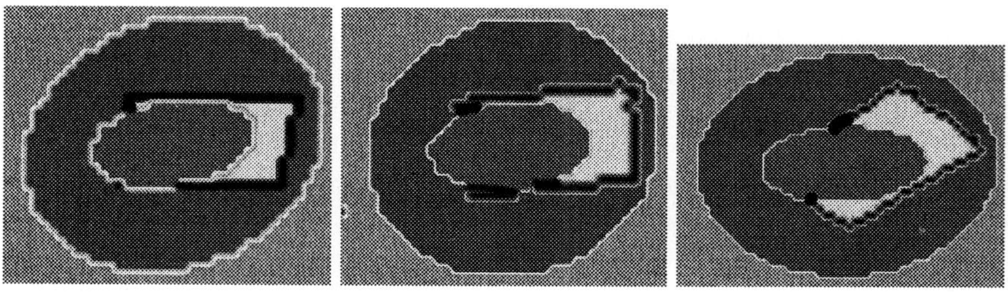

Fig. 2. The image of the robot after preprocessing. Pixels that are too far or too close are ignored. Pixels matching the colour of the top of the robot and pixels on the contour.

Given the position of the edge pixels, a general Hough transform is used to compute the possible orientation of the robot in the first quadrant [1].

The Hough transform is a popular method in computer vision to find lines and other shapes. The basic idea for the Hough transform is to collect evidence to support different hypothesis. The evidence for different hypotheses is accumulated and the hypothesis with the strongest support is returned as the solution.

Figure 3 shows an example of the geometry in our problem. Each edge pixel can be at most on four possible edges (E_1, E_2, E_3, E_4 in the figure). It is easy to see that

$$\alpha = sin^{-1}(w/d)$$
$$\beta = sin^{-1}(l/d)$$

Therefore, the corresponding angles for the edges can be computed as:

$$E_1 = \theta + \beta$$
$$E_2 = \theta - \beta$$
$$E_3 = \theta + \alpha$$
$$E_4 = \theta - \alpha$$

Edges E_1 and E_2 are only possible solutions if the distance d between the center point (X_C, Y_C) and the edge pixel (X_E, Y_E) is greater than the length of the robot l. Similarly, edges E_3 and E_4 are only solutions if d is greater than the width w of the robot.

The algorithm works by allowing each pixel to "vote" for the orientation supported by this pixel. Once all pixels have been processed, the orientation with the most number of votes, that is supported by the most pixels is chosen as the correct orientation. The hough space (vote space) consists of a one-dimensional array with 18 entries, which gives us a resolution of 5 degree.

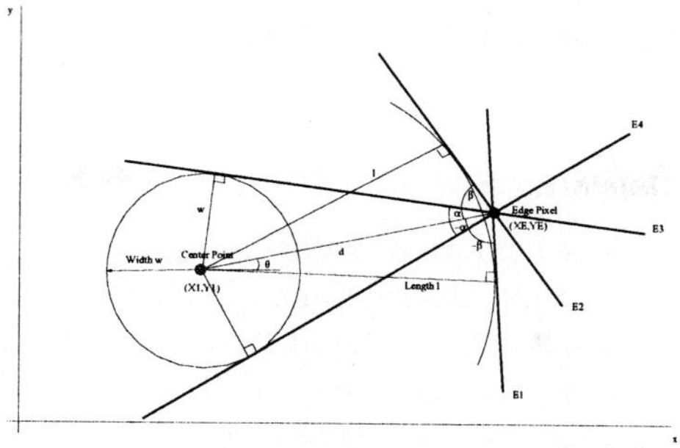

Fig. 3. The four possible edges E_1, E_2, E_3, and E_4 (and therefore, the possible orientations) for an edge pixel (X_E, Y_E).

4 Identification Using Bayesian Probability

As mentioned previously, since all robots in our team look identical, the vision information is insufficient to identify them. YUE-FEI uses two additional sources of information to determine the identity.

Firstly, YUE-FEI predicts the motion of the robot and tracks it. Should the robot be found in the predicted position, its identity and its associated probability is not changed. If the robot is found in the neighborhood of the predicted position, its identity is not changed, but the probability in the identity is reduced by a factor of 0.9 or 0.7, dependent on how far the robot was found from the predicted position.

Secondly, YUE-FEI observes the motion of the robot over a number of frames and assigns it one of seven states: not moving, accelerating, decelerating, turning clockwise, not turning, and turning anti-clockwise. The actual steering angle is not used, so there is no difference between, for example, full left and gently left.

YUE-FEI listens to the communication between the clients and the executors. When an executor receives a command (e.g., Robot 3 forward left!), YUE-FEI stores the last command sent to the robot and observes the motion of the robots. Should the motion of the robot agree with the command, the probability of the identity is slightly increased (a factor of 1.1). On the other hand, there are many reasons why a robot does not follow the command: malfunction, an unknown obstacle in the path, noise in video processing, the robot is being pushed by another robot, occlusion, etc. In this case, YUE-FEI slowly decreases the probability of the identity assignment by a factor of 0.9. When the probability of an identity drops below a certain threshold, the identity of the robot is marked as unknown.

5 Evaluation

The performance of YUE-FEI was compared against the performance of our original video server, both with respect to speed and accuracy. The evaluation shows that the performance of the new videoserver is at least as good as that of our original video server.

5.1 Yue-Fei's Processing Speed

The Hough transform is a computationally intensive method, which as a rule should be avoided in real time applications. However, since the position of the robot is known, only a small neighborhood (64x64 pixels) of the image needs to be considered. Also, the number of edge pixels in that neighborhood is small. In the worst case, there are 256 edge pixels. Also, the Hough space is reduced, since we are not interested in the location of the line and since the possible orientations are only between 0 and 90 degrees. These factors explain why there is no noticeable difference in the processing speed of the two videoservers. Both videoservers are able to maintain a 50 fields/second frame rate in the RoboCup domain with eleven objects.

5.2 Yue-Fei's Accuracy

Knowing the orientation of static objects is rarely useful. We are interested in moving our robots to their targets, so the accuracy of the orientation information for a dynamic object is much more important. A dynamic evaluation is more difficult than a static one, since we have no way of knowing the correct orientation for a moving object. We tested YUE-FEI by driving a simple pattern and by observing the orientation information. The correct information was inferred from a kinematic model of the robot. This test showed that the average error of YUE-FEI was slightly less (less than approx. 5 degrees) than that of our original videoserver (less than approx. 10 degrees).

6 Conclusion

This paper presents a new approach to vision in the RoboCup domain. Instead of coloured markers, the system uses geometrical features of the robots to determine their orientation. This means, that the only coloured marker on the robots are markers are used to determine the position of the robot. The orientation is determined by the projection of the robot in the image. The system uses a generalized Hough transform to find edges in the neighborhood of the position of the robot. These edges are used to determine four possible angles (offset by 90 degrees) for the orientation of the robot.

The videoserver correlates the movement of the different robots with the observed behavior to disambiguate the quadrant of the orientation and the identity of the individual robots.

References

1. D. H. Ballard. Generalizing the hough transform to detect arbitrary shapes. *Pattern Recognition*, 13(2):111–122, 1981.

Efficient Image Processing for Increased Resolution and Color Correctness of CMOS Image Sensors

Jacky Baltes

Centre for Image Technology and Robotics
University of Auckland,Auckland
New Zealand
j.baltes@auckland.ac.nz
http://www.citr.auckland.ac.nz/~jacky

Abstract. This paper describes fast demosaicing methods to quadruple the resolution of a CMOS camera. The resulting increase in accuracy in camera calibration and object detection is important for local vision robots, especially those that use computer vision as their only source of information about the state of the world. The paper describes two methods for demosaicing: interpolation and variance demosaicing. A comparison of three sample views is shown to demonstrate the increased resolution and the difference between the interpolation and variance demosaicing methods. Both demosaicing methods work well. Variance demosaicing performs better around edges in the image, but is computationally more expensive.

1 Introduction

This paper presents demoasicing, a well known technique in digital cameras and investigates its application to problem of computer vision in real-time, real world robotics. Demoasicing quadruples the resolution of CMOS sensors based on the Bayer pattern [1]. This increase in resolution leads to more accuracy in determining the camera calibration and feature locations. For example, the distance to the ball and the angle to the goal can be computed more accurately.

Given the limited amount of processing power available on most robotic platforms and the harsh constraints of the real world, the extra amount of memory required and the additional computational overhead is of interest. Both of these issues are investigated for the two described algorithms.

Section 2 gives a brief introduction into the geometry of a CMOS camera and the Bayer pattern that is most commonly used. Section 3 introduces demosaicing and presents two demosaicing methods: interpolated demosaicing and variance demosaicing. Conclusions are given in section 4.

A. Birk, S. Coradeschi, and S. Tadokoro (Eds.): RoboCup 2001, LNAI 2377, pp. 263–268, 2002.

2 CMOS Camera Hardware

This section describes the actual hardware that is used in most CMOS image sensors. The CMOS sensor is able to measure the itensity of incoming light only. Therefore, a color CMOS sensor uses at least three different image sensors to measure the brightness of the color signals: one red filter, one green filter, and one blue filter.

However, since multiples of three are awkward to manufacture, a common way of manufacturing a CMOS sensor is the so-called "Bayer" pattern. This pattern uses four image sensors as shown in Fig. 1.

R	G	R	G	R	G	R	G
G	B	G	B	G	B	G	B
R	G	R	G	R	G	R	G
G	B	G	B	G	B	G	B
R	G	R	G	R	G	R	G
G	B	G	B	G	B	G	B

Fig. 1. Bayer Pattern as used in CMOS Image Sensors. R, G, and B indicate a red, green, and blue filter sensor element.

It can easily be seen that for a single position only, a single color channel is available. That is pixel (0,0) can only measure the red channel of the signal. The sensor data needs to be further processed to create a RGB color image.

A standard method is to create a single color pixel from four adjacent image sensors. One problem is that there are two image sensors with green information. To speed up the processing, most systems simply discard one channel of green information. Another possibility is to use the average of the two green channels as green value for the pixel. Figure 2 is a graphical illustration of the algorithm used. In this example, a 164 by 124 image sensor produces a 82 by 62 image.

This is the standard method used by the CMOS cameras used by the Eyebot controllers, which are used in the 4 Stooges RoboCup team. A sample images obtained with this method is shown as original image in Fig. 4.

The obvious disadvantage of this method is that information is lost in the conversion process. The position of the pixel segments is not taken into consideration, and the resolution of the resulting image is only a quarter of that of the resolution of the CMOS sensor.

3 Demosaicing

The idea is to infer the missing values for the color information of all pixels by using information about adjacent or near-by pixels. This process is called demoasicing. Figure 3 is a graphical illustration of the main idea.

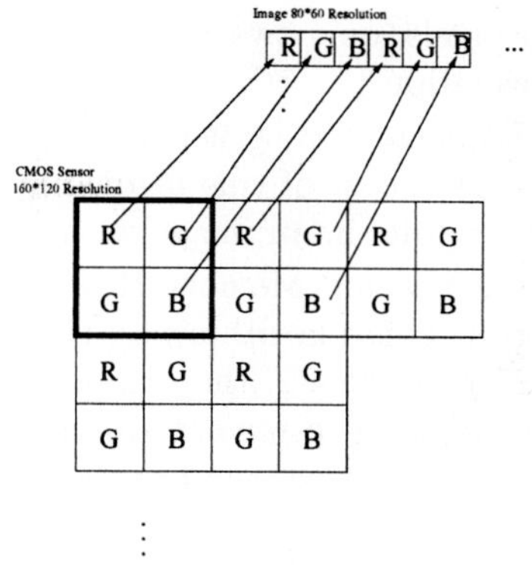

Fig. 2. Standard Processing of CMOS Sensor Data

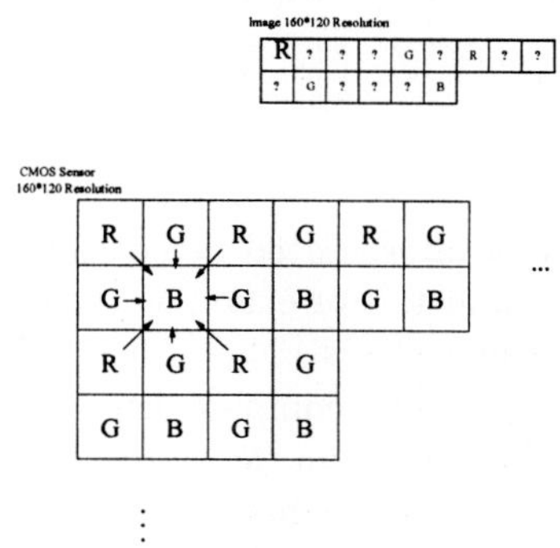

Fig. 3. Demosaicing of CMOS Sensor Data. Green and red information for the pixel (1,1) is inferred from neighboring pixels.

If we take the blue-filter image sensor at location (1,1) as an example, then we can see that the blue information at this pixel is known, but we do not know the values for the red and green channel. However, we do know the values of the red channel for four adjacent pixels and the values of the green channel for four other adjacent pixels.

Similarly the values for the green and blue channel are missing for red-filter sensor elements, and the green filter elements are missing the red and blue color information.

The process of inferring the missing color information is a topic often studied and well known in digital cameras [2,3].

3.1 Averaging

The simplest and fastest algorithm is a bilinear interpolation of the missing pixel values by averaging the values of adjacent pixels. It can be seen that for red and blue pixels four neighboring pixels are available to infer the missing color information. For the green pixels, only two neighboring pixels are available for the red and blue color information.

This algorithm can be implemented by using three rows of pixels as temporary storage. The computational cost of this algorithm is not much more than the standard algorithm, but results in an image with four times the resolution.

The total cost of demosaicing is 8 Memory Read/Write + 6 Add/Sub + 2 Div/Mult operations per pixel. This algorithm was implemented an tested on the Eyebot controllers. On a 25 MHz 68332, the demosaicing routine took 170ms. This is about half the time required to read in the image. This means that the maximum frame rate of the video processing dropped from 15 to 10 frames per second.

A sample image is shown in Fig. 4, The images after interpolated demosaicing look quite natural and provide more information. More pixels are available to determine the center of the ball for example.

3.2 Variance Interpolation

A closer look at the output of interpolated demosaicing shows that it leads to blurring around the edges of objects. For example, the edges of the ball, the goal, or the track are blurred by this process.

To overcome this blurring, a more complex demosaicing algorithm, so called variance demoasicing is used to maintain the sharpness of edges. The main idea is to compute the amount of variance (that is the difference between this pixel and its neighboring pixels) and to assume that the variance is similar in all three color bands. So, if the current red pixel is much brighter than its neighboring pixels, then its green and blue channels are also increased. The algorithm is shown in Alg. 1. Algorithm 2 is used to compute the average brightness in the current color channel. In line 1, the difference between the current pixel and this average is computed. The missing color information (green and blue for a red pixel) is computed similarly to the interpolation algorithm. However, in lines 1 and 1, the computed difference is used to adjust these color values.

This algorithm requires more storage and more computation than the interpolation algorithm described in section. In fact, it requires temporary storage to hold five rows. Furthermore, the cost of computing the calcAverage is 4/1 Memory Read/Write + 4 Add/Sub + 1 Div/Mult.

The average cost of computation for a pixel is 17 Memory Read/Write + 14 Add/Sub + 3 Div/Mult operations. In other words, this algorithm uses more than twice the number of memory accesses and add operations than the previously shown one. This result was verified by timing the implementation of both algorithms, where the variance demoasicing algorithm took sufficiently more time than interpolated demosaicing.

Some sample images with the output of the variance demosaicing algorithm are shown in Fig. 4. The images after variance demosaicing show that the edges around the ball, the cup, and the track are much clearer and not blurred.

Algorithm 1 Variance Algorithm for Demoasicing

```
for i=0 to Height do
  for j=0 to Width do
    if (i mod 2 = 0) AND (j mod 2=0) then
      red = pixel[i,j] {// Red Pixel}
      redAvg = calcAverage(i,j,red)
      redDiff = red - redAvg
      green = (pixel[i − 1,j] + pixel[i,j − 1] + pixel[i + 1,j]
      + pixel[i,j + 1])/4
      green = green + redDiff
      blue = (pixel[i − 1,j + 1] + pixel[i − 1,j − 1] + pixel[i + 1,j − 1]
      + pixel[i + 1,j + 1])/4
      blue = blue + blueDiff
    else if (i mod 2 = 0) AND (j mod 2=1) then
      green = pixel[i,j] {// Green Pixel 1}
      greenAvg = calcAverage(i,j,green)
      greenDiff = green - greenAvg
      red = (pixel[i,j − 1] + pixel[i,j + 1])/2
      red = red + greenDiff
      blue = (pixel[i − 1,j] + pixel[i + 1,j])/2
      blue = blue + greenDiff
    else if (i mod 2 = 1) AND j mod 2=0) then
      {Similar to processing of Green Pixel 1}
      {and omitted here}
    else if (i mod 2 = 1) AND j mod 2=1) then
      {Similar to processing of Red Pixel}
      {and omitted here}
    end if
    image[i,j] = (red, green, blue);
  end for
end for
```

Algorithm 2 calcAverage(x,y,c)

```
avg = (pixel[i − 2,j] + pixel[i + 2,j] + pixel[i,j − 2]
+ pixel[i,j + 2])/4;
return avg
```

We also created difference images to highlight the differences between the interpolated and variance algorithm. As can be seen, the result of the algorithms are very similar and only around the edges are the outputs different.

4 Conclusion

This paper describes two algorithms for the demoasicing of color images: the interpolation and variance demosaicing algorithms. Both algorithms are easy to implement and perform well in practice.

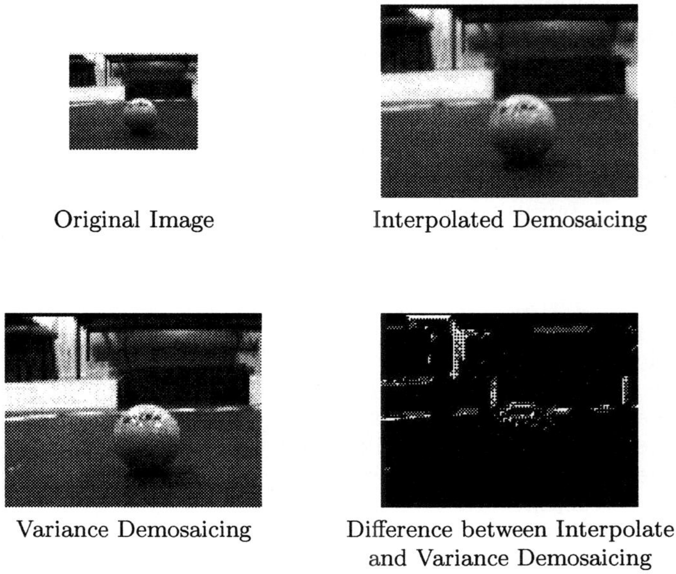

Fig. 4. Ball Scene: Comparison of the original image, the interpolate-demoasiced image, and the variance-demosaiced image. The difference between these two methods is also shown.

An empirical evaluation showed that the interpolation algorithm results in blurring around the edges and that the variance algorithm can overcome this problem.

However, the paper also showed that the computational cost of the variance algorithm is significantly more than that of interpolated demosaicing.

This trade-off depends then on the required accuracy of image features as well as the available processing power. We currently use only interpolated demosaicing in our Eyebot controllers, which allows us to maintain a framerate of 10 frames per second.

References

1. Bryce E. Bayer. Color imaging array. U.S. Patent 3,971,065.
2. Ting Chen. A study of spatial color interpolation algorithms for single-detector digital cameras. WWW:http://www-ise.Stanford.EDU/ tingchen/, Dec. 1999.
3. Tadashi Sakamoto. Software pixel interpolation for digital still cameras suitable for a 32-bit mcu. *IEEE Transactions on Consumer Electronics*, 44(4):1342–1352, November 1998.

Comparison of Several Machine Learning Techniques in Pursuit-Evasion Games

Jacky Baltes and Yongjoo Park

Centre for Image Technology and Robotics
University of Auckland,Auckland
New Zealand
j.baltes@auckland.ac.nz
http://www.citr.auckland.ac.nz/~jacky

Abstract. This paper describes the results of an empirical evaluation comparing the performance of five different algorithms in a pursuit and evasion game. The pursuit and evasion game was played using two robots. The task of the pursuer was to catch the other robot (the evader). The algorithms tested were a random player, the optimal player, a genetic algorithm learner, a k-nearest neighbor learner, and a reinforcement learner. The k-nearest neighbor learner performed best overall, but a closer analysis of the results showed that the genetic algorithm suffered from an exploration-exploitation problem.

1 Introduction

This paper describes the results of an empirical evaluation comparing the performance of three different machine learning algorithms in a pursuit and evasion game.

Because of the inherent difficulty of writing control programs for robots, many researchers have used machine learning techniques to find efficient methods for controlling a robot in the real world. Many of these approaches use machine learning techniques as function optimization. In [1], we describe the use of reinforcement learning to learn the control function of a mobile robot for path tracking. Most of this and similar other approaches focus on acquisition of efficient implementation for low level reasoning.

More recently, there have also been attempts in the RoboCup community to use machine learning to acquire higher level reasoning methods. For example, Stone describes the use of reinforcement learning to learn a subset of three against two keep away soccer [6].

Miller and Cliff argue that pursuit and evasion games are the next logical step in the investigation of robust, intelligent, adaptive behavior. The first step is the avoidance of static obstacles and the achievement of static goals [5].

The quality of the learned solution can often not easily be judged, since the optimal solution for a given problem is often not known. One reason for choosing a pursuit and evasion game in this work is the fact that the optimal strategy is

A. Birk, S. Coradeschi, and S. Tadokoro (Eds.): RoboCup 2001, LNAI 2377, pp. 269–274, 2002.

known for games in an infinitely sized playing field. The quality of the machine learning methods can thus be compared to the optimal performance possible.

The sizes and kinematics of the game were based on the physical robots available in our lab. Those robots are based on remote controlled cars with a length of 18 cm and a width of 10cm. The maximum turn rate of the pursuer was limited to 15cm and its maximum speed to 20cm/s. The maximum turn rate of the evader was limited to 25cm and its maximum speed to 10cm/s.

Figure 1 shows a small example of a game in progress.

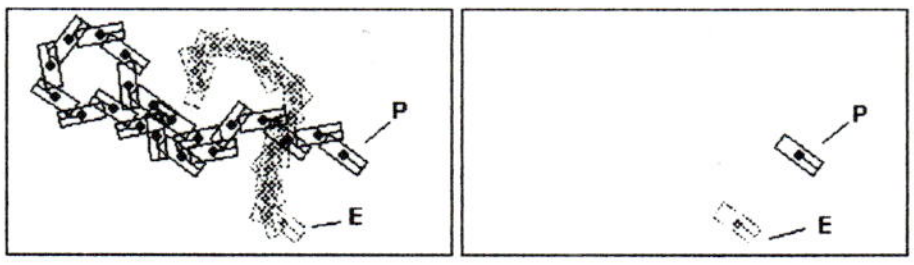

Fig. 1. Example of the Pursuit Evasion Game. The time history is shown on the left, the final state on the right. In this game, the pursuer was successful, since it managed to move within lethal range of the evader.

Section 2 gives a brief introduction into the theory of differential games and leads to the description of the optimal strategy for infinite playing fields in section 3. The machine learning techniques used (Genetic algorithms, k-nearest neighbor, and reinforcement learning) are described briefly in section 4. The empirical evalution of the learned strategies is shown in section 5. The paper concludes with looking at future work (Section 6).

2 Related Work

In this work, we focus our attention on a pursuit and evasion game, the so-called "homicidal chauffeur game" [4].This game takes place on a parking lot. In this game, both pursuer and evader are seen as non-holonomic cars. The pursuer has a higher speed, but a larger turn radius. The evader is more maneuverable, that is, it has a smaller turn radius, but is slower.

The kinematic equations for the pursuer and the evader are shown in Fig. 2. The equations model a unicycle with a limit on the maximum turn cycle.

The control inputs ϕ_p and ϕ_e are normalized from $-1 \ldots + 1$. For example, when $\theta_p = -1$, the pursuer executes a full left turn and for $\theta_e = +1$, the evader executes a full right turn.

3 Optimal Strategy

The kinematic equations of the pursuer and evader need to be simplified to develop optimal strategies for the pursuer and evader. We establish a relative frame of reference, which puts the pursuer at the origin and the evader's position relative to the pursuer is determined. Furthermore, the velocity of the evader and

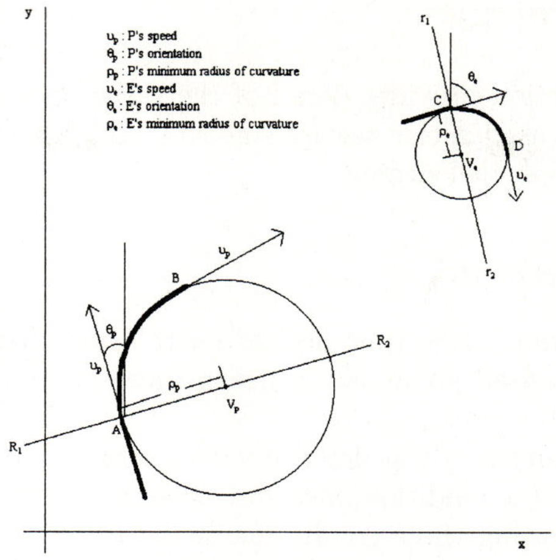

Fig. 2. Kinematic Equations in the Pursuit Evasion Game

the minimum radius of curvature are expressed relative to that of the pursuer. This results in the following coordinate system:

$$\begin{aligned} x &= (x_e - x_p)cos\theta_p - (y_e - y_p)sin\theta_p \\ y &= (x_e - x_p)sin\theta_p - (y_e - y_p)cos\theta_p \\ \dot{x} &= v_e sin\theta_e - y\phi_p \\ \dot{y} &= v_e cos\theta_e - 1 + x\phi_p \end{aligned}$$

The game terminates when the evader is within the *lethal range* of the pursuer, where the lethal range is defined as $x^2 + y^2 < \beta$.

The optimal strategy for the pursuer can be derived mathematically, by realizing that it is the action that will center the evader in the UP. The pursuer should turn towards the evader and move randomly if it is already lined up with the evader.

$$\phi_P = \begin{cases} sgns & \text{if s} \neq 0 \\ \text{else move in a random direction} & \end{cases}$$

The optimal control for the evader when E is on the BUP (the circle $x^2+y^2 = \beta$, where β is the lethal range) is to steer towards the NUP and normal to the BUP. If such a move is not possible because it would exceed the maximum turn radius, it should steer as close to the desired direction as possible.

$$\phi_E = cos^{-1}\sqrt{1 - v_E^2}$$

If the evader is either in UP or NUP, the closest point to the BUP needs to be found. Then E should move towards the direction normal to the BUP and closer to the NUP.

4 Learned Strategies

The following subsection provides detail of the three selected machine learning techniques that were used in this research: genetic algorithms, k-nearest neighbor learner, and reinforcement learning.

4.1 Genetic Algorithm

To reduce the implementation time needed for this, we used the Samuel system [3]. Samuel has been used previously to learn successfully the rules for evading a missile with a fighter jet.

The behavior of an agent was determined by a set of a maximum of 40 rules. Each rule consisted of a condition part and an action part.

We used the following three conditions in the rules:*range*: the distance between the pursuer and evader, *bearing*: the direction from the pursuer to the evader, and *heading*: the direction of the evader relative to the pursuer. Each conditional part of a rule includes values for upper and lower bounds for these three conditions.

The action part of the rules contain one of nine control values (−1, −0.75, −0.50,. . . ,0.00, 0.25,. . . ,+1) corresponding to different steering angles ϕ.

Given a set of rules, a policy needs to be established to select which rules to execute if no rules satisfy the conditions or if more than one rule satisfies all conditions. This problem is solved by associating a rule strength to each rule and by executing the action suggested by the strongest rule. The rule strength is determined by how well the current state matches the conditions of the rule and how successful the rule has been in the past.

If a rule has been selected, all three condition parts are updated by moving the bounds closer to the current conditions. The strength of a rule is updated using the profit sharing plan as described in [2].

We used the following four genetic operators to "stir the gene pool:" rule mutation, rule crossover, rule merging, and plan crossover.

4.2 k-nearest Neighbor Learner

In this work, we used a simple case-based reasoning approach, the 1-nearest neighbor algorithm. The basic idea is that a database of previous situations (so called cases) is maintained by the agent. Anytime the agent is asked to make a decision, the entry in the database that is most similar to the current situation is retrieved and the action that is suggested by this case is executed.

Each state is defined by the same conditions as were used in the genetic algorithm learner: range, bearing, and heading. The similarity measure is the sum over the normalized distances between features as shown below:

$$\mathrm{Sim}(C_i, C_j) = \sum_{k=1}^{3} \frac{|C_{ik} - C_{jk}|}{\mathrm{Max.\ Dist}_k}$$

During learning the agent would execute random actions. All situations that were encountered during this exploratory phase were recorded. If the game was successful, that is the evader was not caught, or the pursuer was able to catch the evader, it is considered as a possible case candidate to be added to the database.

Case-based maintenance is a problem in most case-based reasoning systems. There are processing and memory constraints which limit the size of the database. In this work, we used a simple scheme to control the size of the database. The maximum number of cases for each suggested action was limited to 100 cases. This limited the maximum size of the database to 900 cases.

After a learning episode, a new case would be added to the database, if it was not already subsumed by an existing case in the database. If the maximum size of the case base for a specific action was exceeded by the addition of this case, a randomly selected case was removed from the case base.

4.3 Reinforcement Learner

We also implemented a standard Q learner. Both learners used the standard Q learning update rule shown below. However, since the goal of the pursuer and evader are different, different reward formulas were used for the two players.

Update:$R(x) = R(x) + \rho * (R(i) + \delta(Q(y) - R(x)))$
Reward Evader:$10 * t$ if E is caught at time t
Reward Pursuer:$R(E) = 10 * (T - t)$ if P catches E at time t

5 Comparison of Different Strategies

This section discusses the results of evaluating and comparing the performance of different strategies. In our evaluation, we used a playing field 540cm by 450cm. Each game consisted of a maximum of 100 steps. We run 1000 games and computed the average evasion success rate as well as the standard deviation in 10 blocks of 100 games. In each game, the pursuer started in the centre of the playing field and the evader started in a random position, at least some distance away from the pursuer.

Table 1 shows the result of these experiments. The entries to a random mover (R), the optimized strategy (O), the genetic learner (G), the k-nearest neighbor learner (K), and the reinforcement learner (Q). $\text{Sum}_{P/E}$ summarize the score for a particular pursuer and evader respectively. As should be expected, the optimal evader E_O is the best evader and so is the optimal pursuer P_O. Remember that the entries in the table correspond to the evasion success rate and thus lower numbers show a better performance of the pursuer.

6 Conclusion

The evaluation shows that the k-NN learner performed well compared to the other strategies. Its performance (253) was similar to that of the optimal strategy

	P_R	P_O	P_G	P_K	P_Q	SUM$_E$
E_R	62(4.49)	0(0)	61(6.70)	31(5.21)	60(4.84)	214
E_O	71(4.25)	25(3.22)	78(3.78)	21(1.22)	76(5.43)	271
E_G	62(3.86)	8(2.19)	79(2.26)	12(4.02)	74(2.41)	235
E_K	67(4.28)	4(0.63)	67(4.64)	37(3.36)	78(4.39)	253
E_Q	62(4.20)	0/(0.89)	63(3.79)	42(4.23)	65(4.25)	232
SUM$_P$	324	37	348	143	353	

Table 1. Evasion Success for 1000 Games with 100 Steps/Game. The entry in each cell is the evasion success rate and the standard deviation is given in brackets.

(271). The performance of the k-NN pursuer was worse than that of the optimal strategy, but it was significantly better than that of the other learned strategies.

It is also interesting to note that the genetic algorithm did learn the best evasion strategy against the optimal pursuer of all the learned strategies. Since the all learners were trained against the optimal strategy, it shows that the genetic algorithm did not use enough exploration during the training phase. Although it outperforms the k-NN by 100% against the optimal pursuer, it performs much worse than the k-NN learner against other strategies. The k-NN learner on the other hand uses only randomized moves during training and therefore is able to more general strategies.

Although the genetic learner and the reinforcement learner learned reasonable good strategies for the evasion part of the game, they did perform worse than the randomized mover as pursuers.

References

1. Jacky Baltes and Yuming Lin. Path-tracking control of non-holonomic car-like robots using reinforcement learning. In Manuela Veloso, Enrico Pagello, and Hiroaki Kitano, editors, *RoboCup-99: Robot Soccer World Cup III*, pages 162–173, New York, 2000. Springer.
2. John F. Grefenstette. Credit assignment in rule discovery systems based on genetic algorithms. *Machine Learning*, 3(2/3):225–245, 1988.
3. John F. Grefenstette and Helen G. Cobb. Learning sequential decision rules using simulation models and competition. *Machine Learning*, 5:355–381, 1990.
4. Rufus Isaacs. Differential games: A mathematical theory with applications to warfare and other topics. Technical Report 1, Center of Naval Analysis, Washington, D.C., 1963.
5. G. F. Miller and D. Cliff. Co-evolution of pursuit and evasion 2: Simulation methods and results. Technical report, University of Sussex, Brighton BN1 9QH, U.K., 1994.
6. Peter Stone, Richard Sutton, and Satinder Singh. Towards reinforcement learning for 3 vs. 2 keepaway. In *Proceedings of the Fourth Workshop on RoboCup*, Melbourne, August 2000.

A Simple and Accurate Camera Calibration for the F180 RoboCup League

Ryad Benosman, Jerome Douret, and Jean Devars

Laboratoire des Instruments et Systèmes d'Ile de France,
Université Pierre et Marie Curie,
T 22, coul 22-12, 2eme,
4 place Jussieu,
75252 Paris cedex 05,
Boite 164, France
rbo@lis.jussieu.fr, jdouret@citilog.com, devars@ccr.jussieu.fr

Abstract. Camera calibration is a very important issue in computer vision each time extracting metrics from images is needed. The F180 camera league offers an interesting problem to solve. Camera calibration is needed to locate robots on the field with a very high precision. This paper presents a method specially created to easely calibrate a camera for the F180 league. The method is easy to use and implement, even for people not familiar with computer vision. It gives very acurate and efficient results.

1 Introduction

The robocup F180 league relies on computer vision to locate its robots. Camera calibration is then a very important task to achieve as any slight error in the positionnning will introduce uncertainities in the localization of robots leading to wrong decisions of the behaviour.

Most camera calibrations methods that can be found in the litterature are dedicated to specific tasks, mainly to retreive depth from a binocular vision system. These methods are in most cases too complicated and give too much information that are not necessary in the case of the F180 league.

Camera calibration is usually applied, to find a model of a camera. Calibration patterns are usually used. Since few years in special cases where the camera describes a motion, it is possible to determine the unknown parameters, but these techniques remain uncertain due to degenerencies of the mathematical models[1].

Camera calibration tries to determine the intrinsics and extrinsics parameters of the camera. Intrinsics parameters are the pixels' size, the focal length and the image coordinates of the camera viewpoint[2]. The extrinsics parameters concern the rotation and translation to be applied to pass from the camera coordinates system to the world coordinates system, represented by the calibration pattern.

Retreiving the intrinsics and extrinsics parameters is not needed in the F180 camera configuration. In fact the field and the top of the robots are flat. As we

A. Birk, S. Coradeschi, and S. Tadokoro (Eds.): RoboCup 2001, LNAI 2377, pp. 275–280, 2002.

will see, a simple use of colineations [6]would easely solve the problem. Camera parameters are not even needed.

Many problems in computer vision require a determination of the mapping between points of a scene and their corresponding image locations.

We can then decompose the projection problem. Knowing the position of a point in the field we can compute its location in the image. Applying the opposite transformation, we can, given a pixel in the image compute its line of sight vector [3][4][5].

The simpliest model to describe camera geometry is the pinhole model. Light rays coming from in front the camera converge at a point called the focal point and are projected onto the image plane at the back of the camera. To avoid having a reversed image the image plane is usually positionned in front of the camera. The distance between the focal point and image plane is called the focal length.

A lens can be modelled by a pinhole model, but as a lens is rarely perfect, we usually have to correct the lens distorsions. The paper is organized as follows: the first section introduces the linear transformation which are usefull to map the projection between the image and the field. Section three deals with the problem of correcting lens distorsions. Section four gives a solution to handle the mapping between the image and the field for points that are at different heights. Finally section five gives practical information on how to implement the method.

2 Determining Linear Transformation

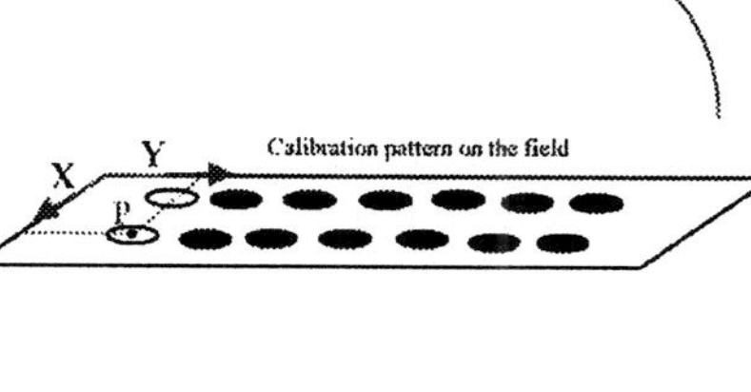

Fig. 1. Linear interpolation between image coordinates and field's points

Let q be an image point which projective coordinates are $[u, v, 1]$ (see Fig.1) its cooresponding point on the field is the point $P = [x, y, 1]$. The relation that connects these points is linear. The linear transformation between q and P is given by :

$$q = HP$$

Where H is 3×3 matrix, given n measurments on each plane we can form the over-determined system :

$$[p_1, p_2 ... p_n] = H [P_1, P_2, ..., P_n]$$

or

$$p_i = HP_i$$

The system can be solved using least squares methods by computing the pseudo-inverse matrix

$$H = [P_i^t P_i]^{-1} P_i^t p_i$$

3 Correcting Lens Distorsions

Lens distorsions lead to a wrong repartition of the incoming rays on the CCD matrix.

The distorsion is radial, which means that a point is translated according to a coeficient $k : d = d \rightarrow \Delta d$.
We have the following relation that links image coordinates to the radial translation :

$$\frac{\Delta u}{\Delta d} = \frac{u}{d}$$
$$\frac{\Delta v}{\Delta d} = \frac{v}{d}$$

We have the following relations where u_d and v_d are the distorted image coordinates and u_{nd} and v_{nd} the non distorted ones :

$$u_{nd} = u_d + \Delta u_d = u_d * \left(1 + k_1 * d^2 + ... + k_2 * d^{2n}\right)$$
$$v_{nd} = v_d + \Delta v_d = v_d * \left(1 + k_1 * d^2 + ... + k_2 * d^{2n}\right)$$

The calibration pattern provides a high number of points, the system is then over-determined we can then find an estimation of the k_i as explained in the previous section. For a better accuracy we used a pattern composed of circles where the gravity center of each circle is extracted with a subpixel accuracy.

4 Taking into Account the Height

The Linear interpolation gives a mapping between field points and image points, if we just apply the transformation on the image points we will not take into acount the error introduced by the height of the robot as shown by Fig.2. We need then to compute a second linear transformation between the image and another plane at a different height L as shown by Fig.3

For each calibration pattern point q appearing on the image using the linear tranformations estimated, we compute its corresponding points in the field (plane P_1) and in P_2 the second calibration plane. Knowing the height L separating the field and the second calibration plane, points P_1 and P_2 can be rewritten as follows :

$$P_1 = (x_1, y_1, 0)$$
$$P_2 = (y_1, y_2, L)$$

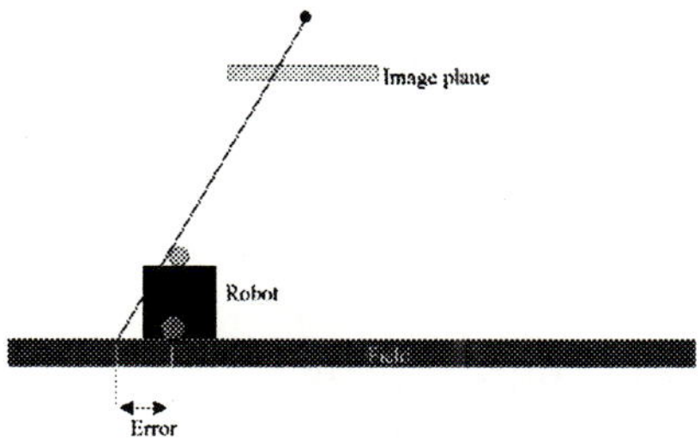

Fig. 2. Localization's errors

We can then compute vector P_1P_2, vector $P_1P_2^n$ is the normalized version of it. We then calculate a scalar $\lambda = \frac{P_1P_2}{P_1P_2^n}$. Finding the position of the pixel point q on a virtual plane located at a height $L^{'}$ is then given by : $\lambda^{'} = \frac{L^{'}\lambda}{L}$. The coordinates of point P_3 on the virtual plane at height $L^{'}$ can be then be retreived as follows :

$$P_3 = P_1 + \lambda^{'} P_1P_2^n.$$

Once the position of point P_3 is known and supposing we apply this to all image points we can then estimate a linear transformation between the image coordinates and the coordinates of their corresponding points on the virtual plane located at a height $L^{'}$.

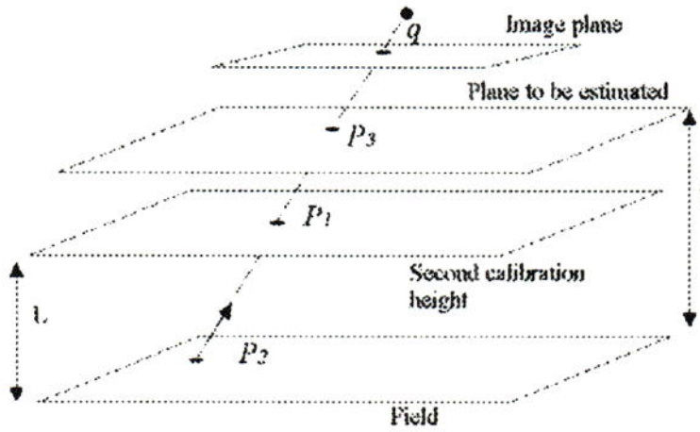

Fig. 3. Computing pixels mapping on a virtual plane placed at a height L'

Fig.4 illustrates the experimental results, we can clearly see the error in the positioning according to the height.

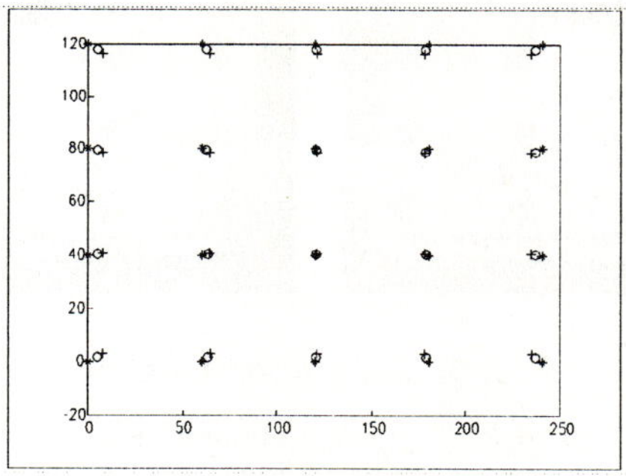

Fig. 4. Experimental results of the estimation of a virtual plane $*$ represents the position of points on the field, o points on the second calibration plane, $+$ the estimated position of points on the virtual plane

5 Practical Use

UPMC-CFA team used this calibration method to locate its robots during the RoboCup European championships and RoboCup 2000. The method is very simple to implement, the first advantage is that computing a robot's position is not time consuming as it deals with four multiplications and two additions. We computed a linear transformation for each present height. One for the ball, one for our robots and one for the other teams. The use of the ping pong ball was also taken into account which means two other transformations. The method was very precise specially if subpixel information are extracted from the image. Most of the errors in the localisation were due to the fact that it is hard to know the angle of view of the ping pong ball, inducing wrong height estimations. We solved this problem which also concerns the ball by computing two transformations one on the height of the ball and another on the three quater of its height, we then calculated a mean matrix. We used a calibration pattern made of paper and for the second height we used plants pots which top was painted green. Fig.5 gives an idea of what the calibration scheme. The code will be soon available for download on the url : www.robosix.org.

6 Conclusion

This paper presented a calibration method that we beleive presents all the requirements for a very simple calibration technique for teams willing to participate to the F180 league and not familiar with the computer vision field. The method

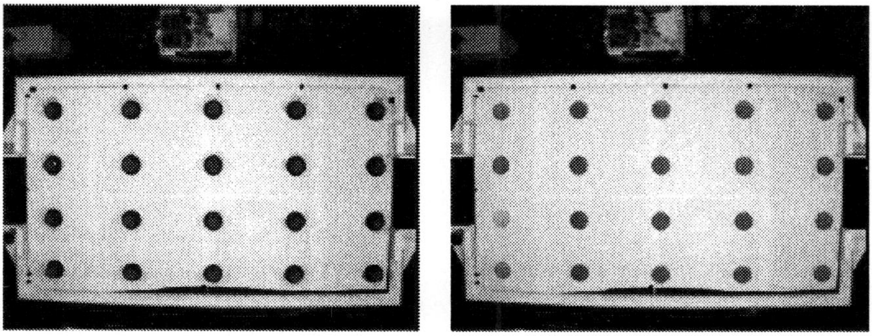

Fig. 5. Calibration pattern

presents many advantages which are mainly simplicity, accuracy and efficiency. Compared to other calibration methods, it does not need to retreive any camera model and deals directly with linear equations very easy to solve. The method is also very easy to implement as it deals with a small amount of data. The accuracy can be increased using more points. The method is not time consuming allowing quick computations. The results and images presented in this paper are those used during the last robocup.

References

1. Quang-Tuan Luong and Olivier Faugeras, Camera Calibration, Scene Motion and Structure recovery from point correspondences and Fundamental matrices. The International Journal of Computer Vision . 22(3): 261-289, 1997
2. Olivier Faugeras Three Dimensional Computer Vision, a Geometric Viewpoint MIT Press, 1993
3. K D Gremban and C H Thorpe and T Kanade Geometric camera calibration using systems of linear equations Proc IEEE Robotics and Automation, 1:562-567, 1988
4. H A Martins and J R Birk and R B Kelley Camera Models Based on Data from two uncalibrated Planes Computer Graphics and Image Processing, 17:173-180 (1981)
5. Y Yakimovsky and R Cunningham A system for extracting three dimensional measurments from a stereo pair of TV camera Computer graphics and image processing
6. J.G Semple and G.T Kneebone Algebraic Projective Geometry, Oxford University Press 1952

Implementing Computer Vision Algorithms in Hardware: An FPGA/VHDL-Based Vision System for a Mobile Robot

Reinaldo A.C. Bianchi[1,2] and Anna H. Reali-Costa[2]

[1] Faculdade de Engenharia Industrial, Departamento de Engenharia Elétrica
Av. Humberto de A. C. Branco, 3972 - 09850-901
São Bernardo do Campo, SP - Brazil
rbianchi@cci.fei.br

[2] Laboratório de Técnicas Inteligentes - LTI/PCS - Escola Politécnica da Universidade de São Paulo
Av. Prof. Luciano Gualberto, trav. 3, 158 - 05508-900
São Paulo - SP, Brazil.
{rbianchi, anna}@pcs.usp.br
http://www.lti.pcs.usp.br/

Abstract. A time critical process in a real-time mobile robot application such as RoboCup is the determination of the robot position in the game field. Aiming at low-cost and efficiency, this paper proposes the use of field-programmable gate array device (FPGA) in the vision system of a robotic team. We describe the translation of well-known computer vision algorithms to VHDL and detail the design of a working prototype that includes image acquisition and processing. The CV algorithms used in the system includes thresholding, edge detection and chain-code segmentation. Finally, we present results showing that an FPGA device provides hardware speed to user applications, delivering real-time speeds for image segmentation at an affordable cost. An efficiency comparison is made among the hardware-implemented and a software-implemented (C language) system using the same algorithms.

1 Introduction

One of the time critical processes of real-time mobile robot applications, such as the RoboCup [1] or any other dynamic and interactive domain is the determination of the robot position. Several approaches have been traditionally used to solve this problem, including custom hardware accelerators and software systems.

On one hand, custom hardware accelerators for image processing are usually high priced, closed systems that implements tracking or threshold algorithms, which can deliver real-time performance but do not allow hardware reconfiguration to adapt to new situations. On the other hand, software systems using low-cost image acquisition boards present lower performance than hardware approaches and hinder robot miniaturization.

A. Birk, S. Coradeschi, and S. Tadokoro (Eds.): RoboCup 2001, LNAI 2377, pp. 281–286, 2002.

Aiming at a low-cost, small-size, hardware-based vision system we propose the use of a Field Programmable Gate Array (FPGA) device for the image processing in the F-180 and the F-2000 leagues.

This paper presents the translation of well-known Computer Vision algorithms to VHDL, a programming language used to define the function of the FPGA circuit and the design of a working prototype that includes image acquisition and processing. The CV algorithms used to define the position of the objects in an image thresholding, Edge Detection and Chain-Code Segmentation.

These algorithms were tested on an ALTERA Laboratory Package and on individual components. Testing results shows that FPGA device provides hardware speed to user applications, delivering real-time speeds for image segmentation at an affordable cost (lower than a custom PC frame grabber).

The reminder of this paper is organized as follows. Section 2 introduces the FPGA and the VHDL language used to write programs that define the FPGA circuit. Section 3 describes the complete vision system, including the image acquisition process and the CV algorithms used. Section 4 presents the translation of the algorithms to VHDL. Section 5 presents the experimental setup and the results obtained. Finally, Section 6 summarizes some important points learned from this research and outlines future work.

2 FPGA and VHDL

Field Programmable Gate Arrays - FPGA [2] are integrated circuit that can have their hardware designed by the user. An FPGA contains a large number of identical logic cells that can be viewed as pre-defined components, which combines a few inputs to one or two outputs according to a Boolean logic function specified by a user defined program. In its turn, individual cells are interconnected by a matrix of wires and programmable switches. The advantage of using a FPGA is that it gives hardware speed to user applications.

A user's design is implemented by specifying the logic function for each cell and selectively closing the switches in the interconnect matrix. This is done by means of a user program that defines the function of the circuit, usually written in VHDL.

VHDL [3] stands for VHSIC Hardware Description Language, where VHSIC means Very High Speed Integrated Circuits. VHDL is a language used to model a digital system, from a simple gate to a complete system.

3 Description of the System

Several approaches have been used to segment the images in the RoboCup domain. The following modules (see Figure 1) compose the vision system implemented: image acquisition, thresholding, edge extraction, and chain-code segmentation.

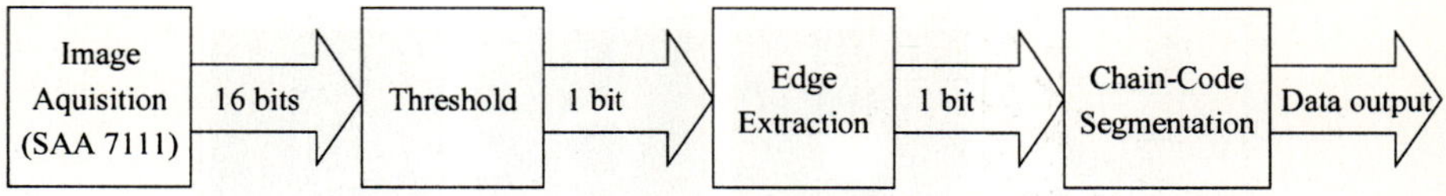

Fig. 1. Block description of the vision system.

3.1 Image Acquisition

The image acquisition in the developed system is made through the Phillips' SAA7111 Video Input Processor. It is a low cost CMOS that receives analogical video signal and transforms it in a digital output.

Its architecture combines a two-channel analog preprocessing circuit (with an anti-aliasing filter, an automatic clamp and gain control), a digital multi-standard decoder and a brightness/contrast/saturation control circuit. The decoder is based on the principle of line-locked clock decoding and is able to decode TV signals (PAL BGHI, PAL M, PAL N, NTSC M and NTSC N) into CCIR-601 compatible color component values. In the vision system, the analog input is an NTSC N CVBS signal and the digital output is an RGB 16-bit signal.

3.2 Image Processing Algorithms

The implementation of edge following algorithms based on Freeman chain coding [4] was chosen because they provide area, perimeter, center of area, minimum enclosing rectangle, orientation, shape factor and other valuable classification information. In this section, algorithms to perform binary image edge following algorithms are presented.

3.2.1 Image Thresholding. The digital RGB 16 bit signal that composes the color image is converted to a 1 bit signal by a simple color detection criteria: if the R, G and B values are in the desired range, the output is true. Else, the output is false. This criteria is used by several researchers [5] and generates a one bit image that will be used by the edge detection algorithm.

3.2.2 Edge Detection. This algorithm apply an edge extracting operator in the image, replacing all the white pixels in the binary image by black ones, unless they are in a black/white border. The operator verifies four elements (2x2 mask) to determine if a point is in a border.

3.2.3 Edge Following. This algorithm, through a procedure that follows the borders in the image, creates a current of connected vectors (called chain) that envelops the objects in the image. The perimeter can be mesured as the sum of the constituent vectors, while the area is mesuring by summing the areas between each vector and a reference line (a procedure similar to area integration).

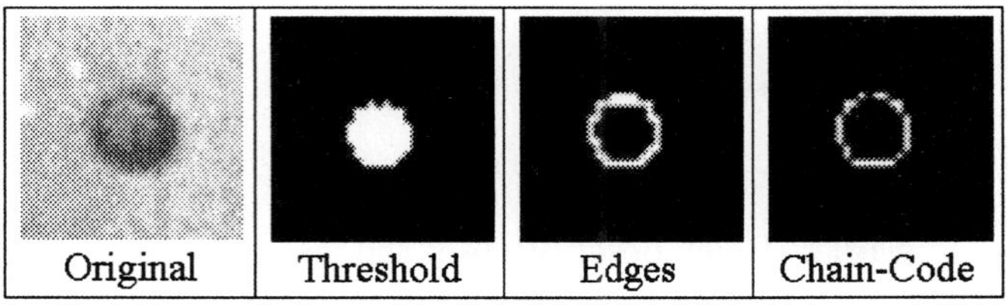

Fig. 2. The result of the image processing steps.

Moment of area and centroids can be found by similar calculations. Finally, a shape factor, defined as (area)/(perimeter) are computed. Figure 2 presents the result of the three steps aplied to the image of a ball.

3.3 Data Output

The information about the objects in the image is stored on elements of an array with 32 bits elements, where 18 bits represents the position where area center of the object is located (9 bits for the line and 9 for the column) and 14 bits register the size of the minimum enveloping chain. This information can be sent to another FPGA module such as the strategy module or to a RS-232 transmission module.

4 Translation to VHDL

While implementing the project the three modes to write a VHDL specification described in section 2 were combined in the so-called mixed-mode design, where the designer can implement part of the system just like a digital circuit and another part just like an algorithm. As the chain-code algorithms have a considerable size, we present here the simplification of the thresholding device.

The first part of the code defines the input and output signals, where there are 5 bits for the red and blue signals and 6 bits for the green signal, and only one bit for the output signal. The second part defines the function: a threshold that will accept only colors with the red, green or blue with half the maximum intensity. If any of the most significant bits is on, the output signal will be off.

Example of a VHDL program that implements a simplified thresholding device using the dataflow method.

```
-- This part defines the input and output ports
Entity Threshold is
   PORT (   R, B:   IN BIT_VECTOR(0 TO 4);
    G   :       IN BIT_VECTOR(0 TO 5);
    I   :       OUT BIT
    );
End Threshold;
```

```
-- This part defines the device function
Architecture Threshold of threshold Is
BEGIN
   I <= not (R(4) or B(4) or G(5));
End Threshold;
```

5 Experimental Results

The algorithms were tested using the simulation tool of the ALTERA MAX+ PLUSII software and on an Educational Board (Figure 3).

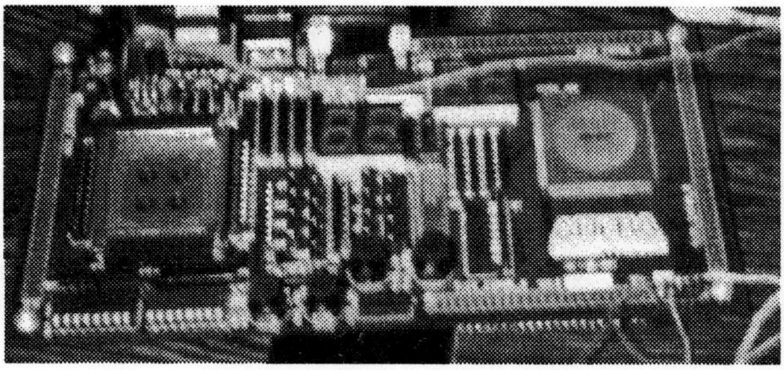

Fig. 3. The ALTERA Education Board.

Table 1 presents simulation results of a 100 MHz FPGA device implementing the vision algorithms, for a 320 x 240 pixels color image. It shows that the edge extraction is the fastest module. This happens because its implementation processes a line at each cycle time. As the other modules takes one cycle time to process each pixel their time are the same. Also, these results were obtained assuming that modules are working sequentially: the edge extraction starts when the image is binarized and the chain-code begins when the edge extraction finishes.

Table 1. Time needed by each module to process an image with 320 x 240 pixels.

Module	Time
Thresholding	770 (μs)
Edge Detection	2,41 (μs)
Chain-Code Segmentation	770 (μs)
Total time	1,54 (ms)

Finally, this table shows that the system can process 649 images (with 320 x 240 pixels) per second. Comparing with a straight forward implementation of the same algorithms in C Language, working on a Pentium 200 computer with Linux OS, the software took 33ms to process each image (21,5 times the FPGA time).

As for the quality of the results, the resulting images from the FPGA were similar to the ones obtained using the software implementation.

6 Conclusion and Future Work

This paper presented the translation of well-known Computer Vision algorithms to VHDL and a working prototype that includes image acquisition and processing.

An efficiency comparison among the hardware-implemented and a software-implemented (C language) system using the same algorithms showed that the first system achieved a superior performance with the same quality. Associated with the low cost of FPGA components (lower than a Video for Windows PC frame grabber), the resulting system is a small-size device suited for the image processing in the F-180 and the F-2000 leagues.

Future works include the implementation of other Computer Vision algorithms such as Blob Coloring in VHDL and final tests of the system on board a F180 Robot Prototypes and Activemedia Pioneer 2 robots.

7 Acknowledgements

This project is supported by Fundaćão de Ciências Aplicadas - FCA-FEI under grant number OS-5886 and by NSF/CNPq-ProTeM CC, project MAPPEL, grant number 68003399-8. The authors would like to thank the undergraduate students who took part in this project: Luis Maia Jr., Paulo V. de Souza, Victor Antônio, and Marcel Gomes.

References

1. Kitano, H. et al.: RoboCup: A challenge Problem for AI. AI Magazine, Vol. 18, No. 1, Spring (1997) 73–85
2. Brown, S.; Rose, J.: Architecture of FPGAs and CPLDs: A Tutorial. IEEE Design and Test of Computers, Vol. 13, No. 2 (1996) 42–57
3. Bhasker, J.: A VHDL Primer. Prentice Hall, Englewood Cliffs (1995)
4. Kitchin, P. W., Pugh, A.: Processing of Binary Images. In: Pugh, A. (ed.): Robot Vision. Springer-Verlag, Berlin Heidelberg New York (1983) 21–42
5. Brusey, J.; Padgham, L.: Techniques for Obtaining Robust, Real-Time, Colour-Based Vision for Robotics. In: Veloso, M., Pagello, E, Kitano, H. (eds.): RoboCup-99: Robot Soccer World Cup III. Lecture Notes in Artificial Intelligence, Vol. 1856. Springer-Verlag, Berlin Heidelberg New York (2000) 243–253

A Framework for Robust Sensing in Multi-agent Systems

Andrea Bonarini[1], Matteo Matteucci[1], and Marcello Restelli[1]

Department of Electronics and Information
Politecnico di Milano, Milan, Italy
{bonarini,matteucc,restelli}@elet.polimi.it

Abstract. We present the framework we have adopted to implement robust sensing in the Milan Robocup F–2000 Team. The main issue concerns the definition of symbolic models both for the single agent and for the whole multi–agent system. Information about physical objects is collected by intelligent sensors and anchored to symbolic concepts. These are used both to control the robots through a behavior–based system, and to improve the global knowledge about the environment.

1 Introduction

We present a framework for robust sensing in a multi–agent system (MAS), based on the cognitive reference model presented in [1]. We have implemented it in our F–2000 Robocup team, obtaining a versatile MAS, able to adapt to different perceptual and communication situations; the team exploit the effectiveness of knowledge sharing, when possible, and shows graceful degradation when each agent should operate on its own. We adopt a behavior–based control architecture, but we feed behaviors with symbolic concepts (implemented by fuzzy predicates) represented in the environment model we are presenting . In this paper, we present the reference framework we have defined to implement the *world modeler* MAP (*Map Anchors Percepts*), used to maintain a consistent instance of the model by anchoring [7] symbolic concepts to physical objects. We have adopted MAP not only in Robocup, but also in other embodied agents applications.

2 Sensorial Integration in Local Maps

The world model we use in MAP considers not only geometrical features, but an integrated representation of objects considering their geometrical, dynamical and perceivable information (e.g., color and odor). The knowledge representation model we propose is based on the notion of *concept* and its *properties*. A property is a tuple

$$p \triangleq < label, \mathbb{D}, \rho >, \tag{1}$$

where *label* denotes the property, $\mathbb{D}$ is the set of all the possible values for that property given a specific representation code (e.g., for the colors we can use the

A. Birk, S. Coradeschi, and S. Tadokoro (Eds.): RoboCup 2001, LNAI 2377, pp. 287–292, 2002.

set $\{red, green, blue, \ldots\}$ or the RGB space $\mathbb{N}^3_{[0,255]}$) and ρ represents a restriction of the domain $\mathbb{D}$ for that property in the specific concept.

A set of properties describes a *concept* C, which is used in our model to represent the knowledge about perceptual images of physical objects. Depending on the concept and on the specific domain, a property can be classified as *substantial* or *accidental*, respectively S and A in equation 2.

$$C \triangleq \{< p, \mathtt{x} >\} : \ \mathtt{x} \in \{\mathtt{S}, \mathtt{A}\}. \tag{2}$$

Substantial properties characterize the immutable part of a concept; for a given object, their values do not change over time, and they can be used for object recognition since they define the essence of the object they represent. Accidental properties are those properties that do not characterize a concept; their values for the specific instance can vary over time, they cannot be used for object recognition, but are the basis of instance formation, tracking, and model validation.

It is possible to describe the agent knowledge by using concepts. We introduce the notion of *model*: given the set of known domains $\mathcal{D}$, a model $\mathcal{M}_d$ is the set of all the concepts known by the agent referring to the specific domain $d \in \mathcal{D}$, linked by (structural and domain specific) relationships. A relationship between concepts may represent:

1. a *constraint* that must be satisfied by concept instances in order to belong to the model
2. a *function* that generates property values for a concept from property values of another (inference function)
3. a *structural constraint* to be used when reasoning about classification and uncertainty

The environment is perceived by a situated agent as a collection of concept instances. The property values for these instances are sensed by means of *intelligent sensors*, which analyze percepts and give them an interpretation at a higher level of abstraction, with the aid of basic domain knowledge.

From the instances of concepts $\overline{C}_i$ and a model $\mathcal{M}_d$ it is possible to infer new instances using relationships between concepts representing specific knowledge for the application domain. An instance of the environment model $\overline{\mathcal{M}}_E$ is the set of all concept instances either derived from the classification process or from inference on concept instances that are compatible with the relationships contained in the model itself:

$$\overline{\mathcal{M}}_E \equiv \{\overline{C} : C \in \mathcal{M}_E\}. \tag{3}$$

The *state* of the system represented by the model instance $\overline{\mathcal{M}}_E$ is the set of all values of accidental properties – time variant and not – of concept instances belonging to the model itself. The *tracking* phase of anchoring consists of maintaining in time a coherent state of $\overline{\mathcal{M}}_E$ and a correct classification of instances. In doing this, accidental properties have to be monitored during time, using state prediction techniques such as linear regression or Kalman filtering.

The model we are presenting in this section can be considered as the logical basis for anchoring, but it is also suitable for classical activities that an embodied agent has to accomplish:

- *sensor fusion*: features perceived by different sensors can be aggregated if they refer to the same object in the environment; this is done to collect as much information as possible about objects before classifying them, to avoid perceptual aliasing [8], and to reduce noise using redundancy in sensorial perception
- *self–localization*: we consider *self–localization* as the process of instantiating the environment model, thus obtaining $\overline{\mathcal{M}}_E$. This definition is a generalization of the common notion of self–localization [3] since it enables reasoning about the own position not only in terms of a geometrical model, but also in terms of more general knowledge (features)
- *virtual sensing*: the instantiation of a model of the environment can be used to produce new information applying *state estimation* techniques to knowledge about the model. This new information can be seen by the agent as new *virtual* features produced by sensors looking at the model of the environment instead than considering the environment itself.

3 Extension to MAS

So far, we have dealt with world modelling processes in a single–agent architecture. It is expected that in a multi–agent context each agent could take advantage of data perceived by its teammates. Having the opportunity to combine different local representations, it is possible to build a shared viewpoint of the common environment, that we call *global representation*. In doing this, we consider that each agent shares the same ontology containing *global concepts* (*GC*).

The global representation builder receives as input the instances of models produced by the local processes. Each model instance contains a set of instances of concepts (e.g., wall, robot, person, etc.). The agent having those instances in its $\overline{\mathcal{M}}_E$ is the *owner* and specifies a reliability value associated to the anchoring process, considering reliability of sensors in the operating conditions, pattern matching, and so on.

The global model building process achieves fusion of concept instances by a clustering process. We define *cluster* a set of concept instances related to concepts whose extensions have a non-null intersection and "similar" values for the accidental properties, where the meaning of "similar" changes according to the property.

Two concept instances $\overline{C}_1$ and $\overline{C}_2$ can belong to the same cluster if:

1. their accidental properties are similar
2. they have a different owner
3. the respective concepts are not mutually exclusive.

For instance, in RoboCup, *robot* and *opponent* are two compatible concepts, while *opponent* and *teammate* cannot belong to the same cluster; moreover, instances of concepts like *opponent* and *goalkeeper* can belong to the same cluster since among opponents there is a goalkeeper.

A new global concept instance ($\overline{GC}$) is generated for each cluster, and its accidental properties are deduced from the accidental properties of the cluster elements by a fusion process that takes into consideration also their reliability values.

A global representation gives to the MAS some interesting qualities (that justify the flourishing of several recent works about this topic [4][6][5]):

- *robustness*: the information coming from several agents that are working together in a given environment can be referred to the same physical objects
- *extensive sensing*: a MAS is comparable to a *super-agent* able to sense and act at the same time in different places, and the agents of a MAS can be considered as virtual sensors
- *fault tolerance*: the global representation can be used to identify and correct sensorial faults
- *cooperation*: it is easier to achieve coordination by reasoning on a global model shared by the agents

It is important to point out that the global representation is not built to substitute the local representation, but to supplement this by providing for lacking information and by recovering possible errors. Each agent weights its own way of integrating the information coming from the global representation, just like it does with information coming from any sensor; for this reason we refer also to the global representation as a *virtual sensor*. In this way, we exploit both the accuracy and the autonomy supplied by the local representation and the completeness and robustness of the global one.

4 The Robocup Application

We have applied the described framework to our agents operating in the Robocup F-2000 league, where a team of four robots play soccer against another four. This domain can be classified as *loosely connected* since robots cannot always perceive everything is happening on the field, because of image resolution, and partial occlusions due to other robots. Each robot maintains its local map, which provides enough information for reactive behaviors; moreover, it exchanges with teammates information aimed at the maintenance of a distributed global map. The behaviors, implemented in our behavior manager BRIAN [1], are activated by evaluating sets of fuzzy predicates which represent concept instances. When information is not available for complex activity, possibly because of poor communication, the agents can still operate with the local map. If a global map can be instantiated, our MAS architecture [2] can assign *jobs* within *schemata* to perform coordinated actions.

In this environment each robot should recognize at least the ball, other robots (possibly distinguishing teammates from opponents), goals and walls. The mentioned objects may be enough for reactive robots, with limited coordination ability, to perform selfish behaviors such as most of those seen till year 2000 in the F-2000 Robocup league: a robot is able to go to the ball, eventually bring it towards the opponent's goal, finally possibly kicking the ball to the goal. This is an important achievement, but it is possible to improve this behavior. Deliberative robots, able to plan complex cooperative behaviors, need to self-localize in order to share their position and their local maps. We use a global map also to overcome problems of limited perception. For instance, if a robot cannot see a relevant object (such as the ball) because it is covered by another robot, it may be informed about the ball position by a teammate, and act consequently. This requires that both the robots are self-localized, and that both share the concept of ball, and its instance. This virtual sensing ability gives to our robots the possibility to perform interesting operating schemata, such as ball passing or coordinated blocking. In the first, a robot can go to a suitable position to receive a passage even if it cannot see the ball; in the second, a robot may intercept an opponent, even if it cannot see the ball, while another teammate is covering the defense area, waiting for the incoming opponent.

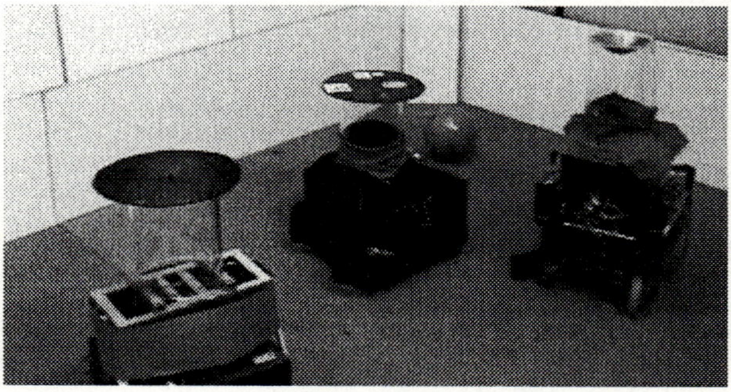

Fig. 1. The robot in front cannot see the ball that its teammate can see.

Another situation when global anchoring is crucial is when a robot has lost its sensing capability. We have made experiments where a robot has been purposely blinded and it is localized by other robots, which inform it about its position and that of ball, goal and obstacles. Of course, the performance of the blind robot could not be optimal, but it was able to intercept the ball when playing as a goal keeper, and to kick the ball in goal when playing as an attacker. This has been implemented as a failure recovery mechanism, to maintain on the field a robot with perception problems, but also to face problems temporarily occurring in specific situations. Our more recent robots can exploit a rotating movement that in certain conditions is fast enough to make the ball disappearing for some instants from the omnidirectional image due to the low frame acquisition rate (PAL standard: 25 frames/sec). They can decide to continue the rotation relying on information present in their local map, but also to consider relevant the information possibly provided by teammates, integrated by global anchoring. As their speed reduces, they directly anchor the ball again, and may consider as less relevant the information coming from other, more distant – and thus, less reliable – virtual sensors of the MAS architecture. Some of the mentioned applications of our approach to world modelling provide information to agents that cannot access to it. Most of the functionalities rely on self-localization, and its improvement is one of the main achievement of our approach. We provide here some data showing the precision obtained in two characteristic situations: self localization of one of the robots of the team by considering information coming from all the others, and ball positioning when the ball is covered by an opponent (Figure 1).

5 Conclusion

We have discussed as anchoring symbolic concepts to physical objects can give robustness to the model of the environment where a robot operates. Reasoning with anchors and established knowledge makes it possible to obtain complex behaviors which cannot be implemented by reactive modules only. Finally, sharing anchors gives the possibility to implement a global map maintained by the MAS as a system; this map may be considered as a virtual sensor which can support local deficiencies, rising the fault tolerance capabilities of the single agent and of the whole system.

References

1. A. Bonarini, G. Invernizzi, T. H. Labella, and M. Matteucci. A fuzzy architecture to coordinate robot behaviors. *Fuzzy Sets and Systems*, Submitted.
2. A. Bonarini and M. Restelli. An architecture for multi-agent systems. *IEEE Transactions on Robotics and Automation*, Submitted.
3. J. Borenstein, H. R. Everett, and L. Feng. *Navigating Mobile Robots: Systems and Techniques.* A. K. Peters, Ltd., Wellesley, MA, 1996.
4. L. Hugues. Griunded representations for a robots team. In *in Proceedings of the 2000 IEEE/RSJ International Conference on Intelligent Robots and Systems*, pages 2248–2253, 2000.
5. D. Jung and A. Zelinsky. Grounded symbolic communication between heterogeneous cooperating robots. *Autonomous Robots*, 8(3):269–292, 2000.
6. T. Nakamura, A. Ebina, M. Imai, T. Ogasawara, and H. Ishiguro. Real-time estimating spatial configuration between multiple robots by triangle and enumeration costraints. In *in Proceedings of 2000 IEEE/RSJ International Conference on Intelligent Robots and Systems*, pages 2048–2054, 2000.
7. A. Saffiotti and K. LeBlanc. Active perceptual anchoring of robot behavior in a dynamic environment. In *Proc. of the IEEE Int. Conf. on Robotics and Automation (ICRA)*, pages 3796–3802, San Francisco, CA, 2000.
8. S. D. Whitehead and D. H. Ballard. Learning to perceive and act by trial and error. *Machine Learning*, 7:45–83, 1991.

Fast Parametric Transitions for Smooth Quadrupedal Motion*

James Bruce, Scott Lenser, and Manuela Veloso

School of Computer Science
Carnegie Mellon University
Pittsburgh, PA 15213
{jbruce,slenser,mmv}@cs.cmu.edu

Abstract. This paper describes a motion system for a quadruped robot that performs smooth transitions over requested body trajectories. It extends the generality of path based approaches by introducing geometric primitives that guarantee smoothness while decreasing (and in some cases entirely removing) constraints on when and what types of parameter transitions can be made. The overall motion system for the autonomous Sony legged robot that served as our test-bed is also described. This motion system served as a component in our entry in the RoboCup-2000 world robotic soccer championship, in which we placed third, losing only a single game.

1 Introduction

The motion system for a legged robot has to balance requests made by an action selection mechanism with the constraints of the robot's capabilities and requirement for fast, stable motions. The desirable qualities of a system are to provide stable and fast locomotion, which requires smooth body and leg trajectories, and to allow smooth, unrestrictive transitions between different types of locomotion and other motions (such as object manipulation).

The platform we used in our work was the a quadruped robot provided by Sony for competition in the RoboCup robotic soccer competitions [6]. In this domain, each team creates the software for a team of three Sony legged robots that compete with another team of three in a game of soccer. The system as a whole is described in [7], and the legged competition in [9]. Although there is an existing system provided by Sony for locomotion, it does not offer the flexibility and low latency required helpful in playing soccer, which motivated our system.

The field of legged locomotion has a long history, with problem formalizations as early as the late 1960s [8]. Recent work on quadrupeds has focused on pattern generator based methods [5] and parameter learning in fixed gaits [2]. Pattern

* This research was sponsored by Grants Nos. DABT63-99-1-0013, F30602-98-2-0135 and F30602-97-2-0250. The information in this publication does not necessarily reflect the position of the funding agencies and no official endorsement should be inferred.

A. Birk, S. Coradeschi, and S. Tadokoro (Eds.): RoboCup 2001, LNAI 2377, pp. 293–298, 2002.

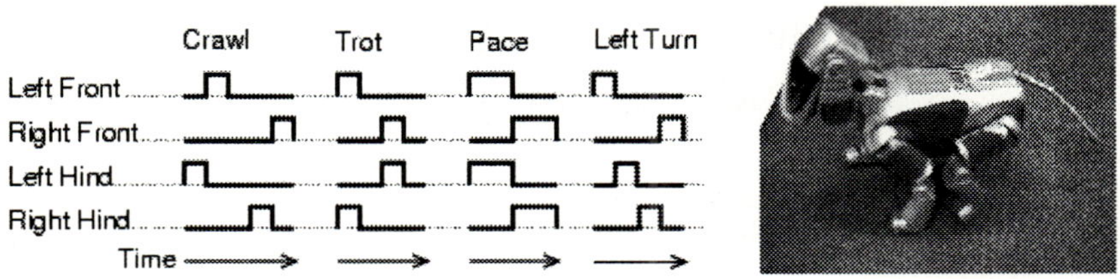

Fig. 1. A timing diagram of various walking gaits. the function is zero when the foot is on the ground and nonzero when it is in the air. To the right is a picture of a robot executing a crawl gait.

generators use internal oscillation methods to drive force based actuators that generate the walking motion through an interaction of motor torques and natural oscillation. While promising, there is still much work to be done in terms of satisfying specific requests on such systems, such as walking or turning on a requested arc. There are also many robots, such as the Sony quadruped, that use more traditional servo actuators which make force-based actuation difficult. State of the art systems for actuated walking include the autonomously learned walks from Sony on their prototype quadruped [2], and later on the release model of the AIBO robot [3]. This system used evolutionary algorithms (EA) to set parameters and test them autonomously for various walking styles. The different gaits focused on are shown in figure 1. The quasi-static crawl moves one foot at a time, maintaining each foot on the ground for at least 75% of the time, this fractional duty cycle is represented as $\beta \geq 0.75$. The supporting basis of feet for a crawl is a triangle. A trotting gait moves diagonally opposite feet at the same time with $\beta \geq 0.50$. Finally the pace gait moves the right and left feet in synchronization and has a $\beta \approx 0.50$ to maintain a side to side oscillation. For our work we focused on the quasi-static crawl for its stability.

While much research has been done in fixed gaits for continuous motion, relatively little work has gone into the area of transitioning among different path targets or from one type of gait to another. One exception is the work of Hugel, which addresses the problem of transitioning among walking and turning gait types as well as arced paths [4]. However a shortcoming of existing transition systems is that the transition points occur at single phase locations in the walk cycle, specifically where leg positions overlap among two gaits or parameter sets. This is quite restrictive on the types of transitions that are then allowed by the system. In the remainder of this paper, we will present a system for transitioning smoothly across more general parameter sets for quasi-static walking and turning gaits. The high level system architecture of our motion system will also be described.

2 Approach

The overall approach taken in implementing the walk is to approach the problem from the point of view of the body rather than the legs. Each leg is represented by a state machine; it is either down or in the air. When it is down, it moves relative to the body to satisfy the kinematic constraint of body motion without slippage on the ground plane. When the leg is in the air, it moves to a calculated positional target. The remaining variables left to be defined are the path of the body, the timing of the leg state transitions, and the air path and target of the legs when in the air. Using this approach smooth parametric transitions can be made in path arcs and walk parameters without very restrictive limitations on those transitions. The leg kinematics was implemented in closed form despite offsets at each joint from the rotation axes, but the derivation is omitted for brevity.

2.1 Spline Path Stitching

We chose to represent the body path of the robot using a piecewise polynomial, specifically a Hermite cubic spline [1]. A Hermite spline is specified by two points (p_0, p_1), and two derivative vectors, $(\delta p_0, \delta p_1)$.

$$\begin{aligned} H(t) &= [x(t)\ y(t)\ z(t)] \\ &= \begin{bmatrix} t^3 \\ t^2 \\ t \\ 1 \end{bmatrix}^T \begin{bmatrix} 2 & -2 & 1 & 1 \\ -3 & 3 & -2 & -1 \\ 0 & 0 & 1 & 0 \\ 1 & 0 & 0 & 0 \end{bmatrix} \begin{bmatrix} p_0 \\ p_1 \\ \delta p_0 \\ \delta p_1 \end{bmatrix} \end{aligned}$$

The body path is modeled as a three component spline $B(t) = [x(t)\ y(t)\ \theta(t)]$, which covers the target configuration space. Using a spline as the body path allows us to specify initial and final positions and velocities for a walk cycle, and since the polynomial is easily differentiable, it also allows us to evaluate the velocity at any point as well as the current body position. Thus it allows new motion requests to be "stitched" to the current path rather than executing the entire request for a full cycle. A new target is added in by evaluating the current position and velocity, and using this as the initial state $(p_0, \delta p_0)$ for the new path plotted toward the request of $(p_1, \delta p_1)$. This guarantees that the body motion is C^1 continuous (continuous velocities). It also allows arbitrarily frequent path transitions, which increases the responsiveness and decreases the latency of reacting to the environment.

2.2 Air Path and Target Selection

The air path is one of the most unconstrained parts of a gait, in that the path needs only to allow the foot to clear the ground while moving forward for it to work in our system. The air path and foot placement target are very important however for keeping the robot stable and maintaining the foot position within its reachability space during the entire walk cycle. The air target was chosen so that after the foot is set down it will pass near the neutral position at a specific time in the walk cycle. This can be achieved by evaluating the expected future position of the robot using the current body trajectory $B(t)$. Two points in the future are important for calculating a placement that will pass by the neutral point; The first (t_1) is the body location when the foot is to be set down, and the second (t_2) is the location in the walk cycle when the foot is intended to pass through neutral (usually about halfway through its ground cycle). Since the foot is to be at neutral at t_2, the location relative to the body at that point is known. Using this, the location of that placement can be found relative to the body position at t_1. Thus the robot will reach for the point to put its foot down that will result in that leg position passing through neutral at t_2.

The air target, along with the projected velocity of the ground plane, and the current state of the foot (position, velocity) specify the components of a single spline fully. However, we found that a two segment Hermite spline worked better for the air trajectory, where the middle position is the average (x, y) location of the initial and target positions of the air path, but with z elevated by a constant for foot and toe clearance. The velocity at the center point is given as the average foot velocity required along the entire air path to reach the target point. Finally, the path is continuous in (x, y), but not in z to allow for quick breaking and restoration of contact with the ground. The pickup and set down z velocities are constant parameters for our walk and were chosen empirically. It should be noted that the projection to make the foot pass through neutral is only approximate, since it requires future evaluation of the body path. Due to stitching, this path may be changed by future commands from the behavior system and thus cannot be predicted. Also, once a foot is set down no immediate corrections can be made. Since the path is smooth however, this approach generally works well for keeping the feet passing near the neutral point even during large transitions.

2.3 Other Parameters

In addition to smooth path control, we use pure replacement and linear interpolation on several other walk parameters. Two variables that we varied throughout the walk were the β parameter (fraction of the gait that a leg is on the ground), and the total cycle time of the walk. Depending on the speed requested and the amount of curvature in the path, values for the cycle time varied from 1200 ms at the fastest walk up to 1500 ms at the slowest walk. β varied from 0.75 (one leg was always in the air) to 1 (all feet on the ground while stopped).

Another parameter that proved useful was spread based on angular velocity. We noted that while walking on a tight arc or turning, moving the feet outward from a normal stance allowed more room for motion (leg motion wasn't as constrained by the body), as well as allowing a wider more stable support basis of the feet on the ground. The parameter was varied continuously with angular velocity, although feet would not reflect the change until they were picked up and put down again due to the kinematic constraints for a leg while it was down. The maximum spread was 20%, occurring during fast turning. Very few locomotion systems are capable of varying these parameters during a walk cycle while maintaining kinematic and smoothness constraints, while the spline based motion system presented here supports this quite easily.

2.4 Motion State System

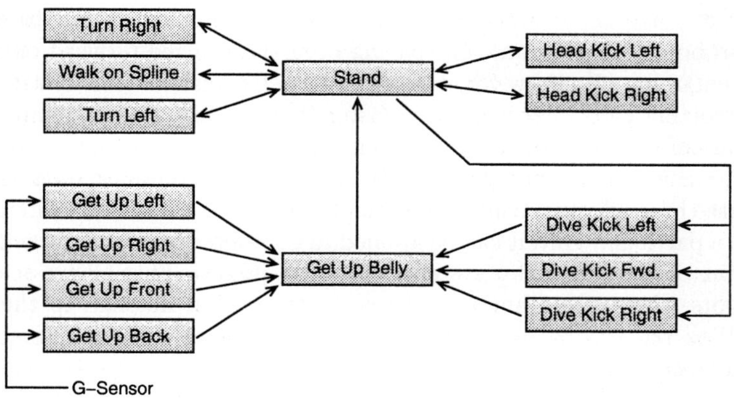

Fig. 2. State diagram of available robot motions.

To put the whole system together, we created a finite state machine model of the family of motions the robot was capable of (see Figure 2). In addition to walking on a path and turning, the robot contained many procedurally defined motions that involved kinematic as well as raw joint angle generating functions. The target state for the motion engine was provided by the action selection system on the robot, and all transitions were constrained by the possible motion on the graph to get to the target states. The one exception was falling, in which case the message from the G-Sensor indicating which side the robot had fallen onto was used to immediately transition to one of the get-up states. The two types of kicks we used to manipulate the ball are head kicks, where the robot dips down and forward and swings its head sideways in the direction it would like to send the ball, and the dive kicks in which the entire robot was controlled to perform a dynamic fall onto the front legs while hitting the ball with its head. Despite the relatively slow motion of the robot, the kick proved highly effective since it used the acceleration of falling as a manipulation tool.

3 Conclusions and Future Work

We presented a complete motion system for an autonomous robot, with special focus on several novel features of the locomotion system. The system was based almost exclusively on kinematics and splines, which allow smooth path generation, and along with a kinematic state machine for the legs enable parametric transitions between motions that do not to pass through the same state. This allows much more general control of the robot, removing restrictive special transition points, all without sacrificing smoothness.

We have a current implementation that can transition four or more times per walk cycle instead of once or twice, and the general approach allows for even more frequent transitions and higher order smoothness. Our current system was tested at the RoboCup-2000 competition in Melbourne, Australia, where we placed 3rd. We demonstrated the fastest quasi-static crawl gait to date on the Sony robots (725 cm/min), out-pacing even the evolved forward only crawl trot (600 cm/min) [3]. However, two of the twelve teams demonstrated non-dynamic trotting gaits that were even faster (UNSW had 1200 cm/min, UPenn had 900 cm/min). A non-dynamic trotting gait lifts two legs at a time but often slides or touches on another part of the robot (front elbows or hind leg toes) so there are three effective contact points [2]. Since our transition system is not specific to a particular gait, it can be applied to guarantee smoothness constraints that neither of the other two systems currently posses. This allows scaling to larger robots where smoothness is necessary to prevent damage to the robot, and it allows transitions between a broad spectrum of walk parameters and gaits for adaptable locomotion.

References

[1] J. Foley, A. van Dam, S. Feiner, and J. Hughes. *Computer Graphics, Principles and Practice.* Addison-Wesley, Reading, Massachusetts, second edition, 1990.

[2] G. Hornby, M. Fujita, S. Takamura, T. Yamamoto, and O. Hanagata. Autonomous evolution of gaits with the sony quadruped robot. In *Proceedings of Genetic and Evolutionary Computation*, 1999.

[3] G. S. Hornby, S. Takamura, J. Yokono, O. Hanagata, T. Yamamoto, and M. Fujita. Evolving robust gaits with aibo. In *Proceedings of ICRA-2000*, 2000.

[4] V. Hugel. *Contribution a la commande de robots hexapode et quadrupede.* PhD thesis, Universite Paris VI, 1999.

[5] H. Kimura and Y. Fukuoka. Adaptive dynamic walking of a quadruped robot on irregular terrain by using neural system model. In *Proceedings of IROS-2000*, 2000.

[6] H. Kitano, M. Asada, Y. Kuniyoshi, I. Noda, and E. Osawa. Robocup: The robot world cup initiative. In *Proceedings of the IJCAI-95 Workshop on Entertainment and AI/ALife*, 1995.

[7] S. Lenser, J. Bruce, and M. Veloso. Cmpack: A complete software system for autonomous legged soccer robots. In *Autonomous Agents*, 2001.

[8] R. B. McGhee. Some finite state aspects of legged locomotion. *Mathematical Biosciences, 2(1/2)*, pages 67–84, 1968.

[9] M. Veloso, W. Uther, M. Fujita, M. Asada, and H. Kitano. Playing soccer with legged robots. In *Proceedings of IROS-98, Intelligent Robots and Systems Conference*, Victoria, Canada, October 1998.

Biter: A Platform for the Teaching and Research of Multiagent Systems' Design Using RoboCup

Paul Buhler[1] and José M. Vidal[2]

[1] College of Charleston, Computer Science, 66 George Street, Charleston, SC 29424
pbuhler@cs.cofc.edu,
[2] University of South Carolina, Computer Science and Engineering,
Columbia, SC 29208
vidal@sc.edu

Abstract. We introduce Biter, a platform for the teaching and research of multiagent systems' design. Biter implements a client for the RoboCup simulator. It provides users with the basic functionality needed to start designing sophisticated RoboCup teams. Some of its features include a world model with absolute coordinates, a graphical debugging tool, a set of utility functions, and a Generic Agent Architecture (GAA) with some basic behaviors such as "dribble ball to goal" and "dash to ball". The GAA incorporates an elegant object-oriented design meant to handle the type of activities typical for an agent in a multiagent system. These activities include reactive responses, long-term behaviors, and conversations with other agents. We also discuss our initial experiences using Biter as a pedagogical tool for teaching multiagent systems' design.

1 Introduction

The field of multiagent systems traces its historical roots to a broad array of specialties and disciplines in the fields of AI, logics, cognitive and social sciences, among others. Within the academic setting, pedagogical approaches are needed that provide opportunities for students to perform meaningful experimentation through which they can learn many of the guiding principles of multiagent systems development. The Biter framework was designed to enable a project-based curricular component that facilitates the use of the RoboCup simulator within the classroom setting.

The RoboCup simulator has many qualities that make it an excellent testbed for multiagent systems' research and for teaching multiagent systems' design. First, the simulator presents a complex, distributed, and noisy environment. Second, in order to win a game, it is necessary to foster cooperation and coordination among the autonomous agents that compose a team. Third, students are engaged by the competitive aspect of RoboCup and many are highly motivated by the prospect of defeating their classmates in a game of simulated soccer. Finally, the RoboCup initiative has generated a wealth of research materials that are easily located and consumed by students.

A. Birk, S. Coradeschi, and S. Tadokoro (Eds.): RoboCup 2001, LNAI 2377, pp. 299–304, 2002.

While all these characteristics make the RoboCup simulator a great platform, there are several aspects which make it difficult for a beginner researcher to use productively.

1. There is a large amount of low-level work that needs to be done before starting to develop coordination strategies. Specifically:
   (a) Any good player will need to parse the sensor input and create its own world map which uses absolute coordinates. That is, the input the agents receive has the objects coordinates as relative polar coordinates from the player's current position. While realistic, these are hard to use in the definition of behaviors. Therefore, a sophisticated player will need to turn them into globally absolute coordinates.
   (b) The players need to implement several sophisticated geometric functions that answer some basic questions like: "Who should I be able to see now?".
   (c) The players also need to implement functions that determine the argument values for their commands. For example: "How hard should I kick this ball so that it will be at coordinates x, y next time?".
2. It is hard to keep synchronized with the soccerserver's update loop. Specifically, the players have to make sure they send one and only one action for each clock "tick". Since the soccerserver is running on a different machine, the player has to make sure it keeps synchronized and does not miss action opportunities, even when messages are lost.
3. The agents must either be built from scratch or by borrowing code from one of the existing RoboCup tournament teams. These teams are implemented for competition, not to be used as pedagogical tools. Their code is often complex, documentation scarce and they can be hard to decipher.

The Biter system addresses each of these issues in an effort to provide a powerful yet malleable framework for the research and study of multiagent systems.

2 The Biter Platform

Biter provides its users with an an absolute-coordinate world model, a set of low-level ball handling skills, a set of higher-level skill based behaviors, and our Generic Agent Architecture (GAA) which forms the framework for agent development. Additionally, many functional utility methods are provided which allow users to focus more directly on planning activities. Biter is written in Java 2. Its source code, Javadoc API, and UML diagrams are available at [1].

2.1 Biter's World Model

In the RoboCup domain it has become clear that agents need to build a world model [3]. This world model should contain slots for both static and dynamic objects. Static objects have a field placement that does not change during the course of a game. Static objects include flags, lines, and the goals. In contrast, dynamic objects move about the field during the game; they represent the players and the ball.

A player receives sensory input, relative to his current position, consisting of vectors that point to the static and dynamic objects in his field of view. Since static objects have fixed positions, they can be used to calculate a player's absolute location on the field of play. The absolute location of a player is used to transform the relative positions of the dynamic objects into absolute locations.

As sensory information about dynamic objects is placed into Biter's world model it is time stamped. World model data is discarded after its age exceeds a user-defined limit. Users can experiment with this limit. Small values lead to a purely reactive agent, while larger values retain a history of environmental change.

Access to world model data should be simple; however, approaching this extraction problem too simplistically leads to undesirable cluttering of code. This code obfuscation occurs with access strategies that litter loop and test logic within every routine that accesses the world model. Biter utilizes a decorator pattern [2] which is used to augment the capabilities of Java's ArrayList iterator. The underlying technique used is that of a filtering iterator. This filtering iterator traverses another iterator, only returning objects that satisfy a given criteria.

Biter utilizes regular expressions for the selection criteria. For example, depending on proximity, the soccer ball's identity is sometimes reported as 'ball' and other times as 'Ball'. If our processing algorithm calls for the retrieval of the soccer ball from the world model, we would initialize the filtering iterator with the criteria [bB]all to reliably locate the object. Accessing the world model elements, with the aid of a filtering iterator, has helped to reduce the overall complexity of student-authored code. Simplification of the interface between the student's code and the world model, allows students to focus more directly on building behavior selection and planning algorithms.

2.2 The Generic Agent Architecture

Practitioner's new to agent-oriented software engineering often stumble when building an agent that needs both reactive and long-term behaviors, often settling for a completely reactive system and ignoring multi-step behaviors. For example, in RoboCup an agent can take an action at every clock tick. This action can simply be a reaction to the current state of the world, or it can be dictated by a long-term plan. Biter implements a GAA [4] which provides the structure needed to guide users in the development of a solid object-oriented agent architecture.

The GAA provides a mechanism for scheduling activities each time the agent receives some form of input. An activity is defined as a set of actions to be performed over time. The action chosen at any particular time might depend on the state of the world and the agent's internal state. The two types of activities we have defined are behaviors and conversations. Behaviors are actions taken over a series of time steps. Conversations are series of messages exchanged between agents. The ActivityManager determines which activity should be called to handle any new input. A general overview of the system can be seen in Figure 1.

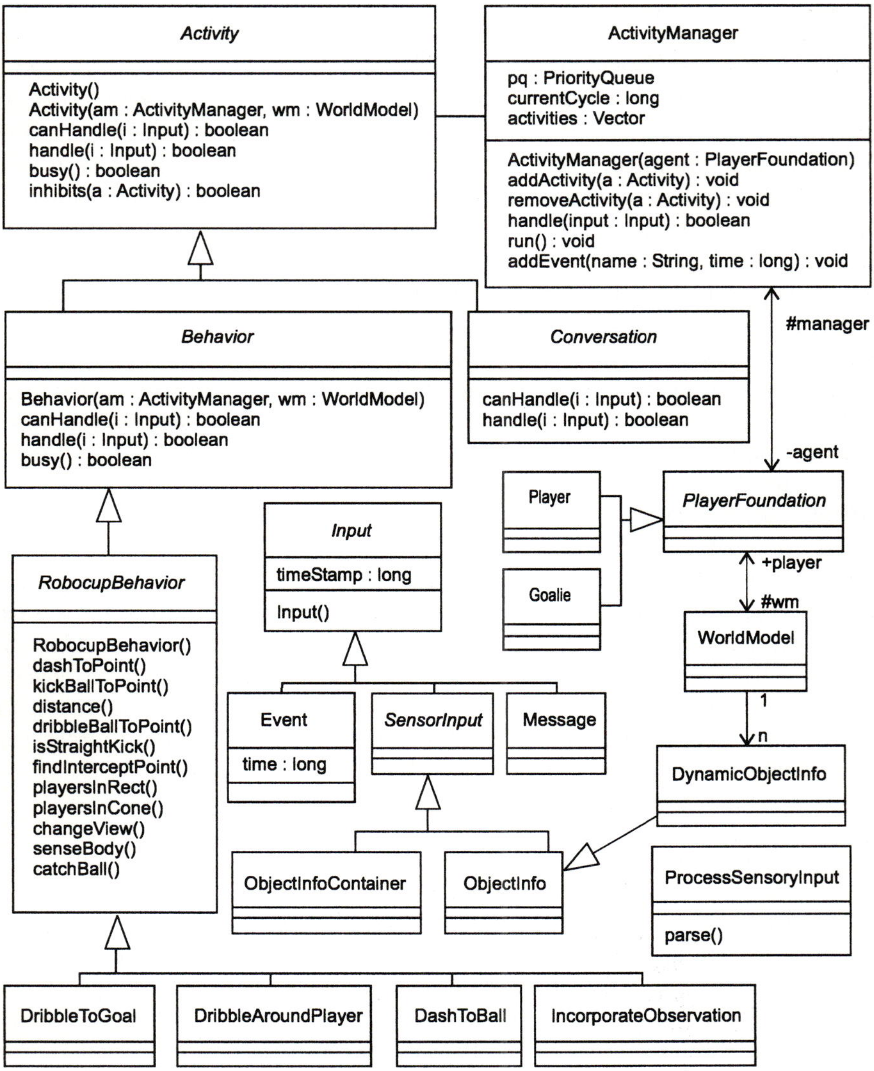

Fig. 1. Biter's UML class diagram. We omit many of the operations and attributes for brevity. Italic class names denote abstract classes.

The `Activity` abstract class represents our basic building block. Biter agents include a collection of activities that the activity manager schedules as needed. A significant advantage of representing each activity by its own class is that we enforce a clear separation between behavior and control knowledge. This separation is a necessary requirement of a modular and easily expandable agent architecture.

The Activity class has three main member functions: `handle`, `canHandle`, and `inhibits`. The `handle` function implements the knowledge about *how* to accomplish certain tasks or goals. The `canHandle` function tells us under which conditions this activity represents a suitable solution. Meanwhile the `inhibits` function incorporates some control knowledge that tells us *when* this activity should be executed.

Biter defines its own behavior hierarchy by extending the `Behavior` class, starting with the abstract class `RoboCupBehavior` which implements many useful functions. The hierarchy continues with standard behaviors such as `DashToBall`, `IncorporateObservation` and `DribbleToGoal`. For example, a basic Biter agent can be created by simply adding these three behaviors to a player's activity manager. The resulting player will always run to the ball and then dribble it towards the goal.

The `Conversation` class is an abstract class that serves as the base class for all the agent's conversations. In general, we define a conversation as a set of messages sent between one agent and other agents for the purpose of achieving some goal, e.g., the purchase of an item, the delegation of a task, etc. A GAA implementation defines its own set of conversations as classes that inherit from the general Conversation class.

The `ActivityManager` picks one of the activities to execute for each input the agent receives. That is, an agent is propelled to act only after receiving a new object of the `Input` class. The Input class has three sub-classes: `SensorInput`, `Message`, and `Event`. A `SensorInput` is a set of inputs that come directly from the agent's sensors. The `Message` class represents a message from another agent. That is, we assume that the agent has an explicit communications channel with the other agents and the messages it receives from them can be distinguished from other sensor input. The `Event` class is a special form of input that represents an event the agent itself created. Events function as alarms set to go off at a certain time.

Biter implements a special instance of `Event` which we call the *act* event. This event fires when the time window for sending an action to the soccer server opens. That is, it tries to fire every 100ms, in accordance with the soccerserver's main loop. Since the messages between Biter and the soccerserver can be delayed their clocks can get skewed over time; therefore, the actual firing time of the `act` event needs to be constantly monitored. Biter uses an algorithm similar to the one used in [3] for keeping these events synchronized with the soccerserver.

3 Experiences with Biter

The University of South Carolina has taught a graduate-level course in multiagent systems for several years. The RoboCup soccer simulation problem domain has been adopted for instructional, project-based use. Without the Biter framework, students spent the majority of their time writing support code that could act as scaffolding from which they could build a team of player agents. Multiagent systems theory and practice took a backseat to this required foundational software construction. At the end of the semester, some teams competed, however the majority of these were reactive agents due in part to the complexity of creating and maintaining a world model.

With Biter available for student use, the focus of development activities has been behavior selection and planning. The GAA allows students to have hands-on experience with both reactive and BDI architectures. Students are no longer focused on the development of low-level skills and behaviors, but rather on applying the breadth and depth of their newly acquired multiagent systems knowledge. Biter provides a platform for flexible experimentation with various agent architectures.

4 Further Work

Biter continues to evolve; new features and behaviors are added continuously. We expect the pace to quicken as more users start to employ it for pedagogical and research purposes. One of our plans is the addition of a GUI for the visual development of agents. We envision a system which will allow users to draw graphs with the basic behaviors as the vertices and "inhibits" links as the directed edges. These edges could be annotated with some code. Our system would then generate the Java code that implements the agent. That is, the behaviors we have defined can be seen as components which the programmer can wire together to form aggregate behaviors. This system will allow inexperienced users to experiment with multiagent systems' design, both at the agent and the multi-agent levels. We also believe the system will prove to be useful to experienced multiagent researchers because it will allow them to quickly prototype and test new coordination algorithms.

References

[1] Biter: A robocup client. `http://source.cse.sc.edu/biter/`.

[2] Erich Gamma, Richard Helm, Ralph Johnson, and John Vlissides. *Design Patterns : Elements of Reusable Object-Oriented Software.* Addison-Wesley, 1995.

[3] Peter Stone. *Layered Learning in Multiagent Systems: A Winning Approach to Robotic Soccer.* The MIT Press, 2000.

[4] José M. Vidal, Paul A. Buhler, and Michael N. Huhns. Inside an agent. *IEEE Internet Computing*, 5(1), January-February 2001.

ViperRoos: Developing a Low Cost Local Vision Team for the Small Size League

Mark M. Chang[1], Brett Browning[2], and Gordon F. Wyeth[1]

[1] Department of Computer Science and Electrical Engineering,
University of Queensland, St Lucia, QLD 4072, Australia.
{chang, wyeth}@csee.uq.edu.au

[2] The Robotics Institute, Carnegie Mellon University,
5000 Forbes Avenue Pittsburgh, PA 15213, USA.
brettb@cs.cmu.edu

Abstract. The development of cheap robot platforms with on-board vision remains one of the key challenges that robot builders have yet to surmount. In this paper we describe the prototype development of a low cost (<US$1000), on-board vision based, robot platform that has been used in robot soccer competitions as a test of its capabilities. The robot platform, named the *VIPER* robot, represents an integration of state of the art embedded technology to produce a robot that has reasonable computational abilities without compromising its size, cost or versatility. We describe the hardware architecture of the system, its motivation through to its implementation. In addition, we reflect upon the performance of a team of *VIPER* robots that participated in the RoboCup2000 competition as a measure of the robustness and versatility of the robot platform. Based on these reflections, we describe what remaining developments are required to evolve the system into a truly capable and versatile research platform.

1 Introduction

Robot soccer is an exciting new research domain aimed at driving forward robotics and multi-agent research through friendly competition. In this paper, we describe the development of the Viper robot and its performance at RoboCup 2000. The Viper robot is a prototype robot platform with on-board vision developed for the annual RoboCup competition. Although primarily intended for robot soccer competition, the Viper robot platform provides a versatile research base for investigating autonomous robot intelligence issues at low cost.

Building a small, versatile mobile robot with on-board vision capabilities for the robot soccer competition is challenging as the system must be both robust to the strenuous RoboCup environment and provide high-performance characteristics while fitting inside a small volume. Although there have been numerous attempts at building cheap robots with on-board vision capabilities, few approaches are suitable for small size RoboCup competition either because of size, cost or some other constraints. We believe that a fully custom design, utilizing

A. Birk, S. Coradeschi, and S. Tadokoro (Eds.): RoboCup 2001, LNAI 2377, pp. 305–311, 2002.

the wide range of embedded processors that are currently available, provides the best approach to satisfy the many constraints required for a successful RoboCup robot platform. In contrast to most other robot hardware (e.g. [1, 5]), the Viper robot divides the problem by using different embedded processors to handle the contrasting needs of vision and motion control processing. As most high end embedded processors are not suited to motion control and vice versa, this approach enables one to choose from a much wider range of processors thus producing a better overall product.

In the next section, we describe the Viper robot platform and the specifics of the prototype Viper system developed for RoboCup2000. Section 3 describes the performance of the system, from a user's perspective, at the RoboCup competition. Section 4 discusses the design approaches of other robots, followed by section 5 which determines what remaining work is required. Finally, section 6 concludes the paper and lists the future direction of the Viper project.

2 The VIPER System

The Viper system consists of the mechanical chassis, motors, the electronics, associated batteries, the software environment on the robot and the development support facilities that run on an off-board PC. The robot is fully autonomous, in the sense that all sensing, computation and actuation are performed on-board. In this section, we describe the physical hardware of the robot, its computational arrangement and capabilities, and the software environment that was developed for the RoboCup competition.

2.1 Mechanical Structure

Mechanically, we desired a robot that would be robust to the RoboCup conditions while offering high-speed maneuverability. Based on our prior RoboCup experiences, the Viper robot was chosen to build upon the same mechanical base of the RoboRoos' robots, a global vision robot soccer team used successfully in previous RoboCup competitions. To conserve space we will not discuss the mechanical details here but instead refer the reader to [2, 7] for further details. Fig. 1 shows the Viper robot with its primary mechanical components labeled.

2.2 Computational Structure

Vision and motor control have very different requirements in terms of electronic hardware. Vision, a processor intensive task, simply requires a fast processor with interface to the CMOS image sensor. Motor control requires a processor with peripheral set that is capable of performing timer-related operations without CPU intervention. Due to these contrasting hardware requirements, the Viper's computational architecture is divided into two parts. A vision board performs all operations related to vision and communication to other robots. Meanwhile, a second processor board performs all motor control operations. The two, distinctly

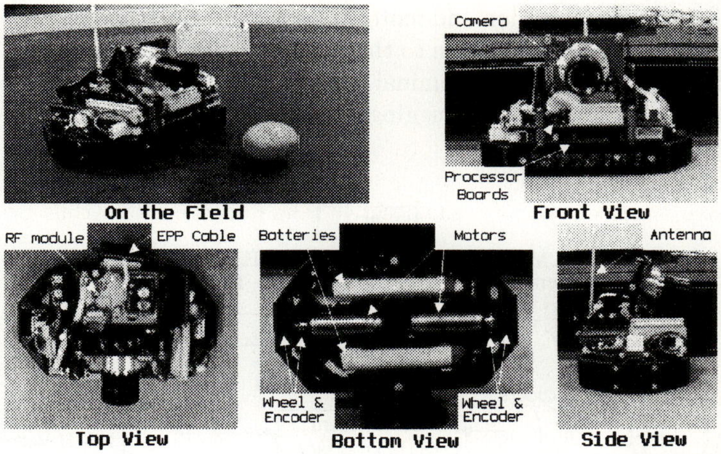

Fig. 1. The major components of the VIPER robot.

different processor systems interact over a 60kbps asynchronous serial link. Fig. 2 shows the conceptual arrangement of the two processor boards. The following sections describe each component of the system.

Camera. The camera chosen for the Viper robot was a Photobit PB-300 CMOS image sensor. A CMOS image sensor was chosen over a CCD sensor for its lower power consumption and fully digital interface. The camera provides color images of 640x480 pixels arranged in a Bayer 2G format through a parallel interface. Due to the SH-3's limited bus bandwidth, the full data rate of 30fps was limited to around 10 fps. The camera supports electronic panning and windowing. Additionally, the chip allows full control over exposure time, color gains, color offsets, horizontal and vertical blanking through a standard synchronous serial (I^2C) interface.

Vision Board. To perform the necessary vision capture and processing requires an embedded processor that has both a capable CPU core and a wide ranging peripheral suite that supports glueless interfacing to high-density memories, Direct Memory Access (DMA) for image capture and other large-scale memory transfers, and serial communications. There are a number of embedded processors currently on the market that fulfil these requirements at low costs. We chose the Hitachi SH-3 processor as a stepping stone to the fully code compatible SH-4/5 processors when they became avaliable.

The SH-3 processor is a 32-bit RISC processor that can perform at up to 80MIPs. The processor has a peripheral suite that facilitates glue-less connection of numerous memory types, three serial connections, and a four-channel DMA controller. Color image data from the image sensor is transferred via DMA. There are three main communication channels on the vision board. An asynchronous serial link connects the vision and motor boards, while another asynchronous serial links provides communication to the PC and other robots though a wireless, half-duplex RF module at the nominal 19.2kbs. Finally, an EPP interface to the PC enables real-time video debugging with a custom written GUI debugging software on a PC.

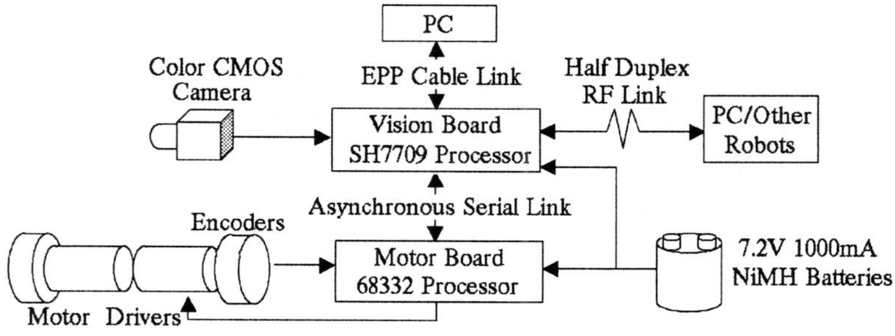

Fig. 2. The processor architecture on the Viper robot

Motor Board. There are a number of embedded micro-controllers designed specifically for motor control primarily for use in the automotive industry. We chose the Motorola MC68332, a 32-bit CISC processor that gives around 5MIPs performance, for the motor controller board given its wide range of peripherals, reliable development system and low cost. The on-board Timer Processor Unit peripheral is a powerful timer utility that requires no CPU intervention to perform a wide variety of timer functions with 16 independent output channels. This powerful timer module drives the two motors through MOSFET H-Bridges with Pulse-Width-Modulation signals and keeps track of the encoders with Quadrature Decode (QDEC) operations. Each QDEC provides 2000 'clicks' per wheel revolution giving a fine grained resolution of 47μm and a velocity resolution of 47mm.s^{-1} when sampled at 1kHz.

2.3 Software Development Environment

The software, both on-board and off-board was written specifically for the Viper robots to achieve optimal performance. For a more detailed description of the Viper software used at RoboCup 2000 please refer to [3]. All software is written in C with some of the boot and time critical code written in assembler. Both processors have a library of device drivers that provide a single layer of hardware abstraction without significantly hindering computational efficiency. In addition to the device driver library, the vision processor has a library of vision related function optimized for the SH-3 processors. As an aid for debugging, we have written a custom GUI software under PC environment to provide real-time video debugging, fast vision calibration via the EPP connection and interaction with the robots via the RF-link.

3 System Performance

In general the robot hardware performs at or beyond our initial expectations. Once built, the hardware proved robust and reliable with no real maintenance required for the operating period of 18 months.

The ViperRoos is capable of processing images at the frame rate of 10Hz, and this includes RGB to YUV color space conversion, subsampling and color segmentation based on a pre-generated YUV lookup table. Blobs are found in the segmented image and categorized into objects based on their dominant color. Beside the usual objects such as the walls, goals, robots, or the ball, the vision board also output an obstacle map for collision avoidance. The motor board performs all motion control routines and the behavior architecture that produces the actual soccer game play of the robot.

As a demostration of the Viper system's capabilities, and as its primary design aim, a team of three Viper robots (two field robots and one goalkeeper) was entered in the RoboCup2000 competition with the team name *ViperRoos*. In any performance metric, the ViperRoos were clearly amongst the best of the local vision teams at RoboCup2000. The robots even proved somewhat competitive against the weaker global vision teams. In terms of raw competition performance, the ViperRoos performed admirably for their RoboCup debut. The team finished the round robin stage with one win and two loses against three global vision teams. In a specialized local vision only competition the robots performed well but no goals were scored either for or against. However, ViperRoos demonstrated their speed and accuracy during the penalty shoot out. The ViperRoos missed one goal in ten attempts and managed to block all shots on goal from the opponents.

4 Comparisons to Other Systems

There are a number of alternative approaches to building small, local vision robots. As with the Viper, most approaches rely on building custom hardware to achieve the desired performance characteristics within the compact size requirements. Probably the best amongst these are the Eyebots [1]. The Eyebot controller board is designed around a Motorola 68332 processor and uses a commercially available CCD camera with a parallel output. As the entire robot system is built around the same processor used for motor control on the Viper robot, the Eyebot is naturally more limited in terms of its processing capabilities.

Outside of the small-size scale there are a number of local vision robots that are either low cost or commercially available. The most notable of these is the Sony Aibo robot dog that is used as the base platform in the RoboCup legged league (e.g. [6]). These robots offer relatively cheap robot vision with the addition of legged motion. The mid-size league consists entirely of on-board vision robots, but these systems are neither cheap nor small. Typically these systems, due to the larger size specifications, use laptops or single board computers to process vision with simpler local processors to perform motor control (e.g. [4]). This is clearly a similar strategy to the Viper robot taken to a larger extreme.

5 Future Work

The heavy use of the Viper system for the RoboCup competition clearly demonstrated the versatile ability of the system. In particular, the choice of the dual processor architecture enabled a wider selection of processors more suited to the differing tasks of vision and motor control. The choice of CMOS imaging technology as opposed to CCDs reduced the complexity of the electrical interface and the power consumption of the camera overall.

At it is a prototype, there are always bugs or deficiencies that need to be improved. Firstly, the vision processor requires an upgrade to achieve the full possible 30fps. At this frame rate, the overall robot performance can be improved significantly. The data throughput rate of the RF link also requires upgrade to achieve efficient inter-robot communication. Using newer RF transceiver technologies such as BlueTooth^{TM} or the single chip transceivers that operate at higher frequencies offer an opportunity to greatly increase the transmission speed and throughput of the RF link.

6 Conclusions

This paper has described the Viper robot system and its performance as a robot platform in robot soccer competition and as a general research vehicle. We believe that the robust performance of the vehicle demonstrates both the virtue of investigating cheap autonomous robots with vision sensing, and our approach to doing so. Specifically, the dual processor architecture presented in this paper not only offers a way to make the most of processor technology that is currently available without breaking the budget but also providing a versatile upgrade path.

There is considerable remaining work to develop the Viper platform into a truly capable and versatile robot base for research and robot soccer purposes. Most of this work focuses on a redesign of the system to incorporate the latest technology that was not available during the initial design phase. Additionally, improvements to the software and development environment are required to make the developer's life easier and to aid in general productivity.

References

[1] Bräunl, T., Reinholdsen, P., Humble, S.: CIIPS Glory - Small Soccer Robots with Local Image Processing. submitted for RoboCup-2000: Robot Soccer World Cup IV, Springer-Verlag, (2001).

[2] Browning, B.: A Biologically Plausible Robot Navigation System. Phd. Dissertation, Dept. of Comp. Sci. & Elec. Eng., University of Queensland, (2000).

[3] Chang, M., Browning, B., Wyeth, G.: ViperRoos 2000. submitted for RoboCup-2000: Robot Soccer World Cup IV, Springer-Verlag, (2001).

[4] Emery, R., Balch, T., Bruce, J., Lenser, S.: CMU Hammerheads Team Description. submitted for RoboCup-2000: Robot Soccer World Cup IV, Springer-Verlag, (2001).

[5] Marchese, F., Sorrenti, D.: Omni-directional Vision with a Multi-part Mirror. Proceedings of The Fourth International Workshop on RoboCup, (2000), pp.289-298.

[6] Veloso, M., Winner, E., Lenser, S., Bruce, J., Balch, T.: Vision-servoed localization and behavior-based planning for an autonomous quadruped legged robot. Proceedings of AIPS-2000, Breckenridge, (2000).

[7] Wyeth, G., Tews, A., Browning, B.: RoboRoos: Kicking into 2000. submitted for RoboCup-2000: Robot Soccer World Cup IV, Springer-Verlag, (2001).

Design and Implementation of Cognitive Soccer Robots

C. Castelpietra, A. Guidotti, L. Iocchi, D. Nardi, and R. Rosati

Dipartimento di Informatica e Sistemistica
Università di Roma "La Sapienza"
Via Salaria 113, 00198, Roma, Italy
{castelp,guidotti,iocchi,nardi,rosati}@dis.uniroma1.it

1 Introduction

One of the major challenges in the design of robots that can act autonomously in unstructured, dynamic and unpredictable environments is the ability to achieve desired goals by executing complex high-level plans, while promptly reacting to unpredicted situations and adjusting the behavior based on new knowledge acquired through the sensors during task execution. Several years of research have focussed on the software architecture by exploiting the sense-plan-act (or deliberative) approach [9], the behavior-based (or reactive) one [1], as well as hybrid architectures [3] which combine advantages of both reactivity and deliberation. However, for an effective application of hybrid approaches a crucial role is played by the organization of information among the layers, which is heavily influenced both by the features of the robotic platform and by the problem at hand.

In this paper we present a hybrid approach that has been used in the implementation of both the players of our University that contributed to the ART team [8] (now competing in the SPQR-Wheeled team) and the players of the SPQR-Legged team [7]. The main features of the approach are: a hybrid architecture that is heterogeneous and asynchronous allowing for an effective integration of reasoning and reactive capabilities; a high level representation of the plans executed by the player; a formal account of the system for generating and verifying the plans. Specifically, we present a novel approach towards the realization of a system that is able to execute the plans that are generated from a formal specification. In this respect the approach is in line with the work on Cognitive Robotics [10,6,2], which aims at the realization of a system based on a specification given in a formal language, such as for example the Situation Calculus. The use of a high-level specification has provided significant advantages also from a practical viewpoint. A significant outcome of the approach is that we were able to implement the players of the SPQR-Legged team (that had a good performance in the Sony Legged League of RoboCup 2000) in a very short time.

A. Birk, S. Coradeschi, and S. Tadokoro (Eds.): RoboCup 2001, LNAI 2377, pp. 312–318, 2002.

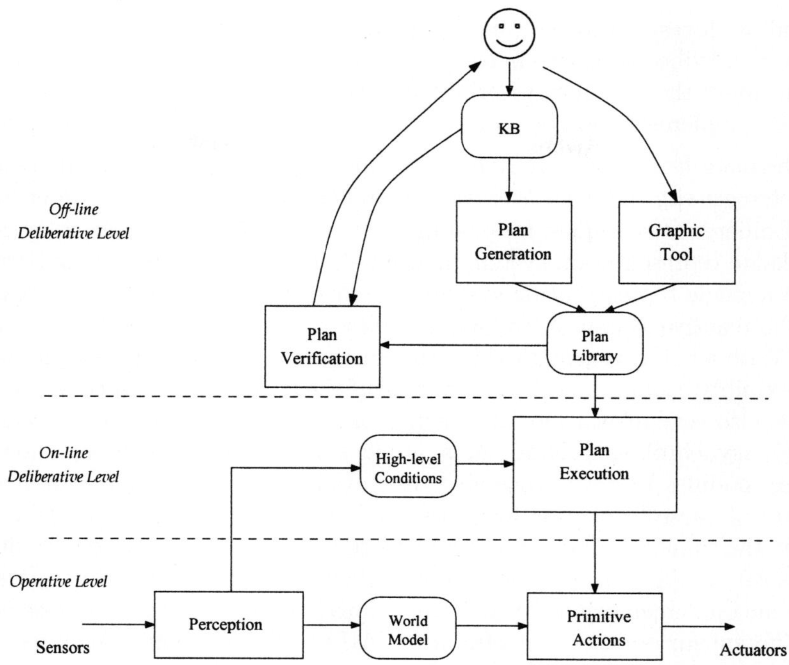

Fig. 1. Layered architecture for our robots

2 Software Architecture

In this section we describe a layered hybrid robot architecture (see also [5]) that has been implemented on different kinds of robotic platforms, namely Sony AIBOs, Pioneer, and home-made wheeled robots. The software architecture for all our robots follows a common schema (Fig. (1)) presenting two layers:

1. An *Operative Level* based on a numeric representation of the information acquired by the robot sensors and of the data concerning the current task;
2. A *Deliberative Level* based on a symbolic representation of the information about the state/situation and of the data concerning the task to be accomplished: the *On-Line Deliberative SubLevel* is in charge of evaluating data during the execution of the task, while the *Off-line Deliberative SubLevel* is executed off-line before the actual task execution.

This layered architecture is based on a different representation of the information. In fact knowing the exact position of the objects on the field is often not necessary in the high-level and therefore a symbolic representation is more appropriate, while the low-level numerical representation is based upon the exact coordinates of the various objects. For instance the high level representation of the position of the ball with respect to the robot could be given by a predicate such as *BallOnTheLeft*, while at the operative level we have the (x, y) position of the ball with respect to the robot. The robot control is also different in the two layers: the deliberative level makes use of action plans representing high-level decisions about the execution of primitive actions; in the operative level instead we have the implementation of primitive actions, that are specific to each platform.

The main features of our architecture for integrating reasoning and reactivity are *heterogeneity* and *asynchronism*. Heterogeneity is important in order to make use of different techniques, for example a fuzzy reactive controller, a logic-based knowledge representation system, and low-level image processing routines. The use of a single representation system, as proposed in [4], for both the reasoning and the reactive system introduces a limitation in the technologies that can be used. With a heterogeneous architecture instead, we provide the designer with all the flexibility required by the complexity of the development process. Asynchronism is also very relevant for an effective integration of reasoning and reactivity. Indeed, asynchronous architectures, unlike synchronous ones [9,10], allow time-critical modules to have processing time available when needed. Moreover, the division of these layers has been designed in such a way that the deliberative level is the same for all our robotic platforms, while the operative level depends on the robotic platform. In fact, all our robots share the same deliberative level, while the implementation of primitive actions and numeric world representations are different for our different platforms (AIBO, Pioneer, home-made base).

The *operative level* is heavily dependent on the robot platform: Saphira controller is used for the Pioneer wheeled robots, and Sony OPEN-R for the AIBO robots. The perceptual module processes sensor data in order to reconstruct a geometrical view of the environment around the robot (e.g. ball, robots, and marker position). The main sensor for all our robots is a color camera, therefore image processing includes color segmentation, feature extraction and world reconstruction, that are very similar for all our robotic platforms. The control module in the operative level implements a set of primitive actions, that are considered atomic at the higher level, but that correspond to a sequence of control commands to send to the robot's actuators. This module is again dependent on the different robots.

The *deliberative level* is mainly concerned with an explicit symbolic representation of the robot's knowledge about the environment. This knowledge is formed by both a general description of the environment provided by the robot's designer and the information acquired during task execution. The world model in this level contains a set of symbolic information corresponding to the data in the geometric world model of the operative level. The deliberative level is formed by two main components: 1) a plan execution module that is executed on-line during the accomplishment of the robot's task and is responsible for coordinating the primitive actions of a single robot; 2) a reasoning module, that is executed off-line before the beginning of the robot's mission, and generates a set of plans to be used to deal with some specific situations. The reasoning system processes a knowledge base containing information on the world for making decisions about actions to be performed for goals achievement. In addition to an automatic plan generation module, the user can also manually design (or modify) a plan by using a graphic tool or validate a plan with respect to the knowledge base by using a plan verification module based on model checking.

3 Representation and Execution of Plans

A *plan* is represented as a *transition graph*, where each node denotes a state, and is labeled with a set of properties, and each arc denotes a state transition and is labeled with the action that causes the transition.

Actions are represented using preconditions and effects. Preconditions are the conditions that are necessary for activating the action and indicate what must be true before and during the action execution. Effects are the conditions that must hold after the execution of the action and characterize how the state changes after the execution of the action: they specify direct effects of an action if executed under given circumstances. A description indicating the overall behavior of the action is associated with each action, and is only concerned with the execution of the action in that particular context. The actions in the graph can be classified into ordinary (i.e. movement) actions and sensing actions. The former cause changes in the environment, while the latter are used for acquisition of information, in order to let the robot take better decisions. In Section 4 we present a more general characterization of actions used for plan generation.

A *Plan Execution Monitor* is in charge of the correct execution of the actions composing the plans. In the monitor's implementation, a plan is stored as a graph data structure. The monitor's task is that of visiting the graph, calling and suspending the actions as necessary. In the formal framework that we have adopted, a number of assumptions are made both to limit the complexity of the language and to enable for automatic reasoning. Consequently, for an effective execution of the plans, additional information concerning the verification of preconditions and effects is needed. In particular, both the duration of actions and the failures of action execution must be taken into account. For example, some preconditions must be constantly verified during the entire execution of the action, while others need to be checked only for the action's activation. In the first case, the action is carried out as long as the condition is true. If the condition becomes false during the execution of the action, the action fails and a recovery action is needed. For instance, in a *PushBall* action (the robot pushes the ball towards the goal) the condition *NearBall* (the robot has the ball next to itself) has to be true during the entire execution of the action. In other cases, the condition is checked only once at the activation and, once the action has been activated, it is carried out, independently of the condition's value during the action. For example, *NearBall* can be considered as an activation condition of a *Kicking* action, that is thus checked only before the action starts.

Following the considerations above, the plan must be marked with additional information, that is external to the formalism, but necessary to the monitor for interpreting and executing the plan. Specifically, the monitor has to know which preconditions have to be verified during the entire execution of the action and which effects determine the state transition.

4 Plan Generation and Verification

We now briefly present the formalism used for the high-level specification of dynamic scenarios (we refer the interested reader to [2] for further details). We have adopted a logic-based formalism, and have exploited both its formal semantics and algorithms for automated reasoning in order to address both plan generation and plan verification in our system. More specifically, the formalism, which is called $\mathcal{ALCK}_{\mathcal{NF}}$, is a *description logic* that has been used for modeling dynamic systems. The representation is based on the use of *concepts*, i.e., unary relations (predicates). Each *state* of the world is labeled by a set of concepts, corresponding to the properties that hold in that state. Moreover, the execution of actions is modeled through *roles*, i.e., binary relations between states.

We represent the robot's *knowledge* about the environment by means of a logical theory ($\mathcal{ALCK}_{\mathcal{NF}}$ knowledge base) $\Sigma = \Gamma_S \cup \Gamma_I \cup \Gamma_P \cup \Gamma_E \cup \Gamma_{DFR}$, where each Γ_x is defined below. In the following, we assume to deal with a set of primitive actions, partitioned into a set of *ordinary* actions, which are assumed to be deterministic actions with (possibly) context-dependent effects, and *sensing* actions, which are assumed to be actions able to sense boolean properties of the environment. We represent primitive actions in $\mathcal{ALCK}_{\mathcal{NF}}$ through a set of atomic role symbols, which we call *action-roles.*

State constraints (Γ_S) State constraints are used for representing background knowledge, which is invariant with respect to the execution of actions. We formalize state constraints as general $\mathcal{ALCK}_{\mathcal{NF}}$ axioms, not involving action-roles.

Initial state (Γ_I) The specification of the initial state in our formalization is given in terms of a formula of the form $C(init)$, where C is a concept, and *init* is the constant denoting the initial state. This axiom can be read as: C holds in the state *init* in every possible interpretation.

Precondition axioms for primitive actions (Γ_P) Precondition axioms specify sufficient conditions for the execution of primitive actions in a state. Precondition axioms are expressed as $C \Rightarrow (\mathit{Exists}\ R)$, where C is a concept representing the precondition and R is a primitive (either an ordinary or a sensing) action. This axiom can be read as: if C holds in the current state s, then there exists a state s' which is the R-successor of s.

Effect axioms (Γ_E) Effect axioms specify the effects of executing an action in a given state. We distinguish between ordinary actions and sensing actions. In our framework, for ordinary actions we specify an axiom $C \Rightarrow \mathit{Effect}\ (R, D)$, meaning that if in the current state the property C holds, then, after the execution of the action R, the property D holds. Moreover, effect axioms for sensing axioms are of the form $\mathit{True} \Rightarrow (\mathit{Effect}\ (R_S, D) \vee (\mathit{Effect}\ (R_S, \neg D))$, and specify the fact that, after the execution of the sensing action R_S, the robot knows whether the property D is true or false.

Frame axioms (Γ_{DFR}) In $\mathcal{ALCK}_{\mathcal{NF}}$ it is possible to enforce different forms of inertia laws, including *default frame axioms* which state that, if in the current state the property C holds, then, after the execution of the action R, the property C holds, if it is consistent with the effects of R. Such a rule can be represented in our framework by the axiom $C \Rightarrow \mathit{Effect}\ (R, \neg(\mathit{Consistent}\ C) \vee C)$, and call the set of default frame axioms in our specification.

Given a dynamic system specification in $\mathcal{ALCK}_{\mathcal{NF}}$ and a goal expressed in terms of a set of concepts, we are able to express the plan generation problem in terms of a reasoning problem in $\mathcal{ALCK}_{\mathcal{NF}}$ [5,2]. and implemented a plan generation/plan verification module based on such an automated reasoner.

5 Discussion

The main goal of the present work is the attempt to provide a new, effective way to drive the behaviour of the system based on a high-level specification of the actions that the agent is able to perform. In fact, a weakness of systems implementing similar approaches [10,6], is that the plans that are derived according to the formal specification are very loosely connected to the underlying execution layer. In this respect we have enriched the plan representation with additional information that can partly be extracted from the plan generation process and partly needs to be given as an additional specification. We have thus been able to implement a plan execution mechanism that relies on this richer representation and can take into account some of the aspects that cannot be adequately treated at the formal level, but are necessary in order to make the implementation effective. One consequence of the approach taken is that, in the design process, it is possible to interleave both the use of automatic plan generation and the direct specification of plans through a graphical interface.

The approach has been successful in two respects. First of all the system has been successfully implemented in our RoboCup teams both in the Legged and in the Middle-size League of RoboCup 1999 and 2000 and it performs well in a task (soccer playing) where the environment is much more dynamic as compared with other experiments described in the literature of Cognitive Robotics Systems (see for example [10,6]). The design of a system based on a formal high level specification has lead us to speed up the software development, when we have been able to port the system developed on a wheeled mobile base to the Sony legged platform.

References

1. Rodney A. Brooks. A robust layered control system for a mobile robot. *IEEE Journal of Robotics and Automation*, RA-2(1), 1986.
2. G. De Giacomo, L. Iocchi, D. Nardi, and R. Rosati. A theory and implementation of cognitive mobile robots. *Journal of Logic and Computation*, 5(9):759–785, 1999.
3. Erann Gat. Integrating planning and reacting in a heterogeneous asynchronous architecture for controlling real-world mobile robots. 1992.

4. S. Hanks and R. J. Firby. Issues and architectures for planning and execution. In *Proc. of Workshop on Innovative Approaches to Planning, Scheduling, and Control*, 1990.
5. Luca Iocchi. *Design and Development of Cognitive Robots.* PhD thesis, Univ. "La Sapienza", Roma, Italy, 1999.
6. Y. Lesperance, K. Tam, M. Jenkin. Reactivity in a logic-based robot programming framework. In *Proc. of AAAI98 Fall Symposium on Cognitive Robotics*, 1998.
7. D. Nardi, C. Castelpietra, A. Guidotti, M. Salerno, and C. Sanitati. S.P.Q.R. In *RoboCup-2000: Robot Soccer World Cup IV.* Springer-Verlag, 2000.
8. D. Nardi et al. ART-99: Azzurra Robot Team. In *RoboCup-99: Robot Soccer World Cup III*, pages 695–698. Springer-Verlag, 1999.
9. N. J. Nilsson. Shakey the robot. Technical Report 323, SRI AI Center, 1984.
10. M. Shanahan. Reinventing Shakey. In *Proc. of AAAI98 Fall Symposium on Cognitive Robotics*, 1998.

Genetic Programming for Robot Soccer

Vic Ciesielski, Dylan Mawhinney, and Peter Wilson

RMIT Department of Computer Science
GPO Box 2476V, Melbourne 3001
Tel: (03) 9925-2926, Fax: (03) 9662-1617
vc@cs.rmit.edu.au

Abstract. RoboCup is a complex simulated environment in which a team of players must cooperate to overcome their opposition in a game of soccer. This paper describes three experiments in the use of genetic programming to develop teams for RoboCup. The experiments used different combinations of low level and high level functions. The teams generated in experiment 2 were clearly better than the teams in experiment 1, and reached the level of 'school boy soccer' where the players follow the ball and try to kick it. The teams generated in experiment 3 were quite good, however they were not as good as the teams evolved in experiment 2. The results suggest that genetic programming could be used to develop viable teams for the competition, however, much more work is needed on the higher level functions, fitness measures and fitness evaluation.

1 Introduction

Genetic programming is a process of generating programs by simulated evolution[3, 4]. This paper considers the question of whether genetic programming would be useful in a complex, changing, uncertain environment such as simulated robot soccer.

Genetic programming has previously been used to develop RoboCup teams [1, 5]. The team evolved in [5] won two games in the 1997 tournament. Both these attempts have been somewhat competitive. Our focus is on finding a good set of high level functions and associated fitness measures prior to working on a team for the competition. Other machine learning techniques have been used previously for RoboCup including agent oriented programming [6], neural networks [7], and decision trees [8].

Strongly typed genetic programming (STGP) is an extension of standard genetic programming. All functions and terminals in a STGP system must return a particular data type, all function arguments must also be restricted to a specific type. The main benefit of STGP is the reduced search space it offers. We have developed a strongly typed genetic programming system which can evolve players for the RoboCup competition, full details can be found in [9].

[1] Mpeg files of some games can usually be found at http://www.cs.rmit.edu.au/~vc/robocup

A. Birk, S. Coradeschi, and S. Tadokoro (Eds.): RoboCup 2001, LNAI 2377, pp. 319–324, 2002.

The overall goal of this work is to determine whether a competitive RoboCup team can be developed by genetic programming. In this investigation we are interested in determining if a team can be generated using just the basic soccer simulator primitives (kick, turn etc), and what fitness measures are useful.

2 Experiment 1 – Basic Robocup Functions

Experiment 1 was the basic starting point for evolving genetic programs to play soccer. Each chromosome represents a control program for a soccer player. A team consists of 11 copies of this player. The only action functions available to programs were those already provided by the soccer server (kick, dash, turn). It was hoped that programs would use these very basic functions and terminals to evolve some useful soccer-playing behaviours.

The program elements (terminals and functions) available to programs in experiment 1 can be found in table 1. Terminals are inputs to a conventional program[4] and will always be leaves of a program tree. Functions form the interior nodes of a program tree.

Table 1. Functions and terminals used in experiment 1. In the type columns, `f` denotes a floating point value and `v` a vector.

Terminal Name	Type	Description
Random Number	f	A random number n where $0 <= n < 360$.
NearTeam	v	Returns the location of the player's nearest teammate.
2ndNearTeam	v	Returns the player's 2nd nearest teammate.
3rdNearTeam	v	The player's 3rd nearest teammate.
4thNearTeam	v	The player's 4th nearest teammate.
NearEnemy	v	The player's nearest opponent.
2ndNearEnemy	v	The player's 2nd nearest opponent.
3rdNearEnemy	v	The player's 3rd nearest opponent.
4thNearEnemy	v	The player's 4th nearest opponent.
BallPos	v	The position of the ball.
MyGoalPos	v	The position of a player's goal (the goal a player must defend).
TheirGoalPos	v	The position of a player's opponent's goal (the goal a player needs to kick the ball through to score).
Function Name	**Returns**	**Description**
$iflte f(a, b, c, d)$	f	If a is less than or equal to b, return c, else return d.
$kick(pow, dir)$	a	Send a kick message to the soccer server with the parameters pow power and dir direction.
$turn(moment)$	a	Turn the player within a range of -180 to 180 degrees.
$dash(pow)$	a	Cause the player to dash in the direction they are facing.
$min(a, b)$ $max(a, b)$	f	Return the minimum or maximum of a and b.
$+, -$	f	Standard arithmetic operators.
$getDist(v)$	f	Return the distance element of a vector.
$getDir(v)$	f	Return the direction element of a vector.
$getCDist(v)$	f	Return the change in distance element of a vector.
$getCDir(v)$	f	Return the change in direction element of a vector.
$execTwo(a, b)$	a	Unconditionally execute both actions and return an action.

2.1 Genetic Operators

The basic genetic operators are crossover and mutation. These operators work on the tree representations of the evolving programs. The crossover operator selects two parents and swaps randomly selected subtrees to get the children, as shown in figure 1. The mutation operator selects a subtree at random and replaces it with a new, randomly generated tree.

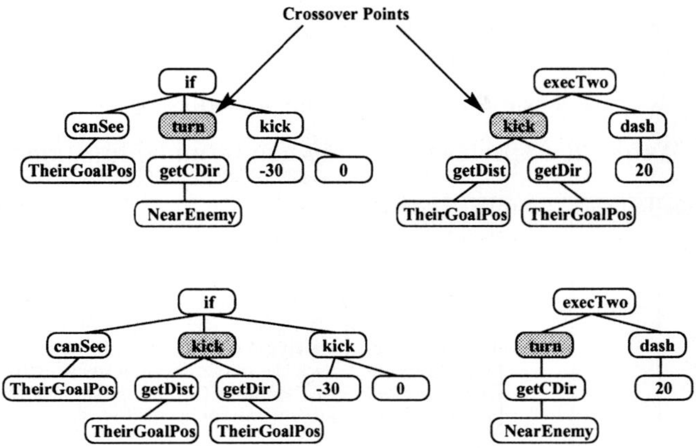

Fig. 1. The genetic crossover operation. The top two trees are the parents, selected at random from the population. The bottom two trees are the children. The chosen subtrees have been swapped.

2.2 Fitness

For this experiment, each generation proceeded in an elimination-round style tournament [2]. Programs were randomly paired up and played off. The winner of the meeting was then advanced to the next round. The fitness of each program was determined by the round in which it was eliminated. The best individuals of each generation were then played off in a best-of tournament.

2.3 Results

The results for this experiment were unexpectedly disappointing. Not only did every program fail to score a proper goal, most failed to even kick the ball. Even less encouraging was the fact that some teams even managed to score goals against themselves. Since very few teams scored goals or executed successful kicks or passes, the winner of a tournament was often determined at random.

One winner was a program with 77 nodes. This program provides its player with one action — kick, it contains no `dash` or `turn` commands and as such cannot cause the player to move at all. The quality of this program is typical of the winners from experiment 1.

Using just low level functions (as was the case in this experiment) did not allow the genetic programming process to evolve players which could display any basic soccer playing behaviours.

3 Experiment 2 - Higher Level Functions

The terminals, genetic operators and fitness evaluation were the same as in experiment 1. The non terminal functions used in experiment 1 (table 1) were replaced by those shown in table 2. These functions correspond to more complex soccer actions than just turning, dashing and kicking. A number of additional, lower level, support functions were also needed. Full details are in [9].

Table 2. Experiment 2, main non terminal functions

Function Name	Returns	Args	Description
$if(a, b, c)$	a	b, a, a	If a is true, execute b, else execute c.
$canSee(v)$	b	v	Return true if the player can see its argument, otherwise return false.
$turnTo(v)$	a	v	If the percept vector is visible, turns the player to face it and runs towards it. If it isn't visible, the player scans for it by turning 90°and waiting for a percept update.
$moveTo(v)$	a	v	This function causes the player to turn towards its argument in the same fashion as *turnTo* (including the scanning if it isn't visible). If the percept vector is visible, the player then runs towards it.
$kickTo(v)$	a	v	If the player can see the ball and is within kicking distance, it will kick it in the direction and distance of its argument, else this function acts like *moveTo* with the ball's position as its argument.
$shorten(v)$	v	v	Shorten the distance v is away from the player by 20%.

3.1 Results

The higher level representations used in this experiment allowed the generated programs to exhibit basic soccer-playing behaviour. Like early programs described in Luke *et al.* [5], the programs evolved exhibited behaviour which might be expected at a children's soccer game (ie. all players believe that they alone can win the game). All programs which made it into the final best-of-the-best elimination-round tournament flocked to the ball, irrespective of the location of teammates and opponents on the field. Most programs could kick the ball, some could not. Those that could not kick the ball surrounded it and typically blocked their opponents' efforts to get near the ball. Few goals were scored in this best-of tournament, mainly due to the ball-smothering behaviours of the players.

While many of the programs evolved in experiment 2 were able to play basic games of soccer, the code which had been evolved was usually not very sophisticated. The majority of the programs had simply made use of the `kickTo` function to run to the ball and kick it in some direction (usually towards the goal). In fact 27 of the 32 best programs played off in the tournament had the `kickTo` function as their root node.

4 Experiment 3 - Augmented Basic Robocup Functions

In experiment 3 the terminals and low level functions in experiment 1 were used along with some additional terminals and functions. The new terminals and functions added for this experiment can be found in table 3. Experiment 3 used the roulette wheel selection method [4], each team played one game against another team in the population. Fitness was calculated as a weighted sum of goals scored and kicks made.

Table 3. Experiment 3, added terminals and functions

Terminal Name	Returns	Description
GameState	f	Returns the current state of the game (play_on, kick_off etc) as a number.
PlayNum	f	Returns the shirt number of the current player.
IsLeft	b	Returns true if the player is on the left hand side team, false otherwise.
IsRight	b	Returns true if the player is on the right hand side team, false otherwise.
ClosestToBall	b	Returns true if the player is the closest to the ball out of the players it can see.
Function Name	Returns	Description
$equal(f, f)$	b	Returns true if the two numbers passed to it are equal.
$or(b, b)$	b	Returns true if either of it's arguments are true.
$and(b, b)$	b	Returns true if both it's arguments are true.
$*, \%(f, f)$	f	Multiplication and protected division.
$if(a, b, c)$	a	If a is true return b, else return c.

4.1 Results

The programs generated during this experiment were able to get to the ball, and then kick it, however the ball did not always go towards the goal. Most of the teams which were evolved did not run directly towards the ball, they instead developed a "swirling" behaviour where the team would run around in small circles as a group, the group of swirling players would move towards the ball eventually kicking in what seemed to be a random direction. Strangely this behaviour was common, it was observed in many independent evolutionary runs which had each started from random populations. Teams which used this strategy were able to kick goals some of the time, however this was only the result of kicking the ball hard in random directions, most teams scored just a many own goals as they did goals. We are currently looking at modifying the fitness function to encourage more goals. Another interesting behaviour which emerged from some of the evolutionary runs, was that of goalie behaviour. Some of the teams had a player which would stand in front of the goals and not move around or follow the ball.

The teams evolved in experiment 3 were not as good as those evolved in experiment 2. However the actual "learning" that took place in experiment 3, was much more advanced than in experiment 2. In experiment 2 the majority of the programs which were evolved simply made use of the `kickTo` function which was available. In experiment 3 however the programs evolved were more complex and contained many different functions and terminals.

5 Conclusions

Experiment 1 showed that very low-level behaviours by themselves are not enough to evolve a set of high-level and complex soccer-playing strategies. Players all but lost the ability to dash by generation 7. No goals were scored, some own-goals were scored, but most players failed to even kick the ball.

Experiment 2 utilised higher-level functions and resulted in players which exhibited 'school boy soccer' abilities, that is, each player followed the ball around as it was kicked back and forth.

Experiment 3 utilised low-level functions, however it had quite a few different functions than in experiment 1. The players exhibited reasonable abilities, however they were not up to the level of the players from experiment 2.

The form of the fitness function and the method fitness evaluation remain unresolved issues. It seems clear that different fitness functions are needed early in the evolutionary process when the teams are not very good and later, when the teams have improved. Further work is under way in this area.

References

[1] David Andre and Astro Teller. Evolving team Darwin United. In Minoru Asada, editor, *Robocup-98: Robot Soccer World Cup II. Lecture Notes in Com puter Science*. Springer-Verlag, 1999.

[2] Peter J. Angeline and Jordan B. Pollack. Competitive environments evolve better solutions for complex tasks. In Stephanie Forrest, editor, *Proceedings of the 5th International Conference on Genetic Algorithms, ICGA-93*, pages 264–270, University of Illinois at Urbana-Champaign, 17-21 July 1993. Morgan Kaufmann.

[3] Wolfgang Banzhaf, Peter Nordin, Robert E. Keller, and Frank D. Francone. *Genetic programming: An introduction on the automatic evolution of computer programs and its applications.* San Francisco, Calif. : Morgan Kaufmann Publishers, 1998. Subject: Genetic programming (Computer science); ISBN: 1-55860-510-X.

[4] John Koza. *Genetic Programming: on the Programming of Computers by means of Natural Selection.* The MIT Press, 1992.

[5] Sean Luke, Charles Hohn, Jonathan Farris, Gary Jackson, and James Hendler. Co-evolving soccer softbot team coordination with genetic programming. In *Proceedings of the First International Workshop on RoboCup, at the International Joint Conference on Artificial Intelligence*, Nagoya, Japan, 1997.

[6] Itsuki Noda. Team gamma: Agent programming on gaea. In H. Kitano, editor, *RoboCup-97: Robot Soccer World Cup I*, 1998.

[7] Itsuki Noda, Hitoshi Matsubara, Kazuo Hiraki, and Ian Frank. Soccer server: a tool for research on multi-agent systems. *Applied Artificial Intelligence*, 12(2-3), 1998.

[8] Peter Stone and Manuela Veloso. A layered approach to learning client behaviors in the robocup soccer server. *Applied Artificial Intelligence (AAI)*, 12, 1998.

[9] Peter Wilson. *Development of a Team of Soccer Playing Robots by Genetic Programming.* Honours Thesis, RMIT, Department of Computer Science, 1998 http://www.cs.rmit.edu.au/~vc/papers/wilson-hons.ps.Z.

“As Time Goes By” - Using Time Series Based Decision Tree Induction to Analyze the Behaviour of Opponent Players

Christian Drücker, Sebastian Hübner, Ubbo Visser, and Hans-Georg Weland

TZI - Center for Computing Technologies
University of Bremen
D-28334 Bremen, Germany
grp-rcintern@tzi.de
http://www.virtualwerder.de/

Abstract. With the more sophisticated abilities of teams within the simulation league, high level online functions become more and more attractive. Last year we proposed an approach to recognize the opponents strategy and developed the online coach accordingly. The coach was able to detect their strategy and then passed this information together with appropriate countermeasures to his team. However, this approach gives only information about the entire team and is not able to detect significant situations (e.g. double pass, standard situations). In this paper we describe a new decision tree induction for continous valued time series, used to analyze the behaviour of opponent players.

1 Introduction

As the RoboCup 2000 in Melbourne showed, the differences between the technical abilities of the major teams aren't as big as in previous world-cups. Due to teams which share their progress with the RoboCup community by releasing their source code shortly after the event - a big thanks to CMU at this point-, every team is able to easily arrange eleven pretty good agents.

Version 7.0 of the soccer server extended the abilities of the online coach further, so that its use becomes even more interesting. It is still the most effective instrument to analyze the opponent, because it possesses all the information about the simulated environment. Therefore it was important for us to continue the development of the online coach which we used at the RoboCup 2000. In [Visser et al., 2001] we describe how the old coach determines the opponent tactical formation with a neural network and how it is able to change our team formation during a match.

This year our main target is to analyze an individual opponent agent to detect which task in their system it fulfills. To determine this, we have to observe how a certain player reacts to certain situations; e.g., under which circumstances does an agent pass the ball to a fellow player? Why does it pass the ball in this situation to just this player?

If we know how an opponent agent reacts in a certain situation, we can use this information to be always one step ahead of the opponent. For example, if we realize that the opponent forwards only try to score if they have enough space, we could prevent them from shooting at the goal by placing our defenders close enough to them.

A. Birk, S. Coradeschi, and S. Tadokoro (Eds.): RoboCup 2001, LNAI 2377, pp. 325–330, 2002.

Similar work has been done over the last couple of years by [Wünstel et al., 2000], [Frank et al., 2000], and [Raines et al., 1999]. Our objective is to find a method that can improve the behaviour of our team during the game and that not only shows us how we could do better next time.

2 Qualitative Abstraction of Time Series

A new method for the qualitative abstraction of multiple time series has been developed at the Center for Computing Technologies [Boronowsky, 2001]. This method seems especially suitable to us in order to deliver patterns which we then use as a basis for learning algortihms. The method has been developed to analyze continuous valued time series. It uses decision tree induction and therefore generates rules about the analyzed time series, in our case the observed players. The special feature of this method is the way the continuous valued attributes are discretized.

Certain rules, generated by the decision tree, can be evaluated automatically by our coach. This information can be used to improve our agents' behaviour during the game. Additionally, all rules can be evaluated manually after the game has ended. The results of the manual evaluation by hand can be used to improve our agents' basic behaviour. A big advantage of the used method is, that the discretization of continuous valued

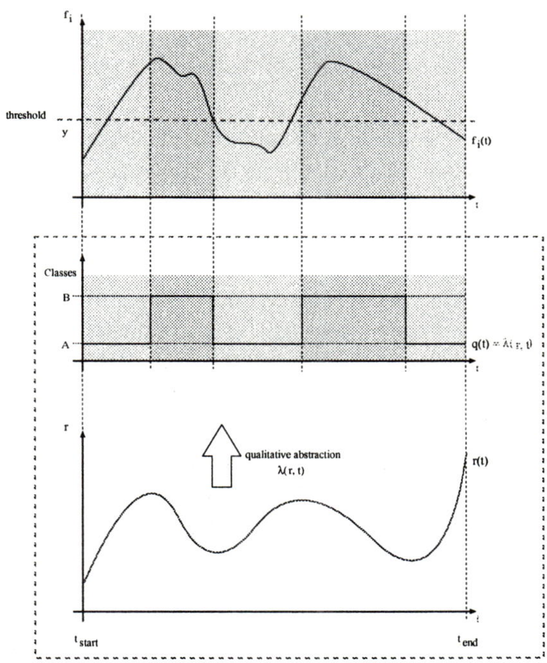

Fig. 1. Qualitative abstraction and horizontal splitting

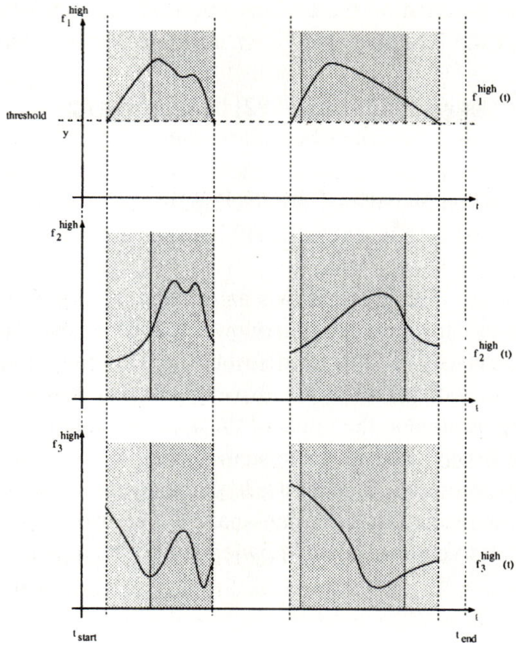

Fig. 2. vertical splitting

attributes is done automatically by our system. Thus, there will be no problem with the decision tree induction not generating a proper result, if an adverse discretization has been chosen by the designer. The discretizations resulting from the method particularly hold valuable information about the opposing team. Discretization of continuous valued time series is the main task of the method. The method uses a very efficient algorithm to solve this problem.

To analyze our opponent, we record several time series F. Supplementary mathematical transformed time series can be added in order to improve the results. One of the time series $r \in F$ must be qualitatively abstracted. This special series gives the classes to be learned, e.g. a time series containing: *goalie leaves goal*, *goalie stays in goal*, and *goalie returns to goal*. By the qualitatively abstracted time series, time slices of all time series are assigned to the classes. One of the time series f_i is split horizontally at a threshold y (fig. 1). To determine this threshold, all possible split points must be evaluated with a special heuristic. The best split point is the one which separates the different classes the best.

All the other time series are split vertically (fig. 2) at all points at which the time series f_i crosses the threshold y. With these separated graphs the process is recursively repeated.

To apply the uniform distribution theorem of the entropy minimalization heuristic, the time series must be approximated by partially linear functions.

The used method is based on the entropy minimalization heuristic as used in ID3 [Quinlan, 1986] and C4.5 [Quinlan, 1993]. An attribute is split at the point at which the heuristic is minimal.

Fayyad and Irani [Fayyad and Irani, 1992] have proved that this can only happen at class boundaries, we call these points boundary points (BP). By using this knowledge, an important increase in efficiency can be reached. But it only works properly for non-overlapping classes. When two classes are overlapping the number of BPs can increase dramatically. In the worst case there can be BP between all examples in the overlapping area.

The method we use in this work allows an efficient splitting of continuous valued attributes in these cases. Boronowsky [Boronowsky, 2001] shows that this problem can be solved by using intervals of uniform distribution. This can be achieved by a linear approximation of the real distribution. He also explains that the entropy minimalization heuristic can only be minimal at the joints of the approximation.

So far the method was used in the system ExtraKT. ExtraKT is a system to aid humans in analyzing time series. It generates hypotheses about coherences in time series which can then be interpreted by a human analyst. The user can also manipulate the decision tree induction to enter his own knowledge into the process.

3 New Coach

In order to improve our abilities to analyze the opposing team, we integrated the system into the online coach. The coach continuously tracks certain aspects of the game, e.g. the positions of the players and their distance to the ball. They are stored in memory forming several time series which are analyzed in regular intervals by the system mentioned above. As a result we receive rules about the opponents behaviour.

In the first step we use the described method to analyze the opponents goal keeper. The goalie is particulary suitable, because his role in the team is fixed. He is the only player who's function is known from the beginning of the game. Additionally, the goalie is a very important player. The knowledge of his strengths and weaknesses can be used to optimize the behaviour and configuration of our forwards, leading to better scoring abilities.

We also want to use the method to analyze our own goal keeper. In this way we can test the quality of our system by cross checking the results with the details of our implementation. Besides we may find some aspects in the behaviour of our goalie which could be improved further.

To prepare the analysis, our coach stores some time series of game aspects which are related to the goal and the goal keeper. This includes

- the distance between the ball and the goal,
- the distance between the ball and the goal keeper,
- the distance between the goal keeper and the goal,
- the number of opponents and team mates within the penalty area,
- the number of team mates which may intercept the ball when kicked towards the goal.

For us, it is the most important task to determine, under which circumstances the goal keeper starts to leave the goal to get the ball and when he starts to return to it. According to this problem we have chosen a suitable qualitative abstraction. This abstraction uses the change in the time series corresponding to the distance of the goal keeper to the goal. This is used to determine whether the goal keeper is leaving the goal, is staying in the goal or is returning to the goal

The situations in which the keeper leaves the goal or returns to it are of special interest for us, because in these moments he can be taken by surprise more easily.

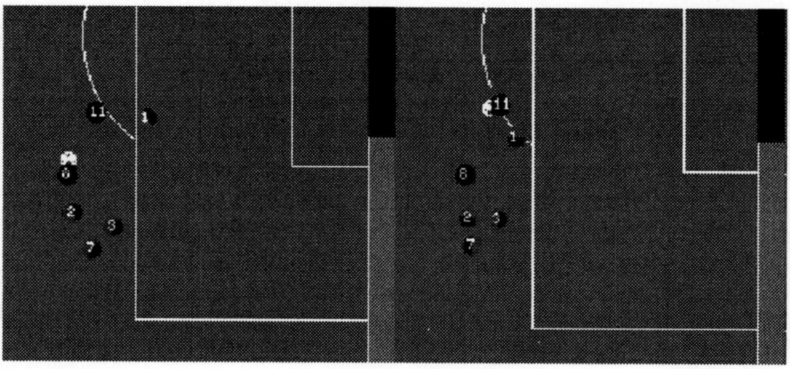

Fig. 3. 2-vs-1 situation

Fig. 3 shows an 2-vs-1 situation (the three defenders beneath, number 2, 3 and 7, don't really affect the play). The goalie (1) decides to attack the ball carrier (8), so as the forward passes the ball to his teammate (11), he gets a big opportunity to score. If we assume that the decision of the goalie is deterministic, he should react in the same way during the next 2-vs-1 situation. The coach should instruct the agents that the supporting player should depart a little more, to get in an even better scoring position.

If the goalie had suspected and intercepted the pass, the coach would have told the agents that the ball carrier should dribble a little longer next time.

The counter actions of the coach have to be taught by a human instructor. They have to be parametrizable, to avoid that every possible situation has to be taught manually; e.g. a goalie, which leaves his goal at a little different distance to the ball, shouldn't cause a complete different behaviour of our agents.

4 Results

To test our method, we applied it to our own team which played in Melbourne. Thus we where able to compare the results with the implemented behaviour of our goalkeeper. The qualitative abstraction used for this tests determines whether the goalie is moving or not. As opponents we used the team FCPortugal. After 500 cycles our coach found rules such as:

```
if  DistGoalGoalie<3.65 and DistBallGoal<19.36
     and DistBallGoalie<13.73
then 1/0.972222
(...)
if  DistGoalGoalie>3.65 and DistBallGoalie<13.73
then 1/0.972222 (*)
(...)
if  DistGoalGoalie>3.65 and DistBallGoalie>13.73
    and TeamMemInPenArea<1 and DistBallGoal>19.44
then 0/0.75 (**)
```

The rules are taken from a test in which our coach produced 10 rules. Rule(*) shows, that our goalie moves at 97% of the time, when he is more than 3.65 away of the goal center an the ball is closer than 13.73 to him. Rule(**) is an example for conditions in which the goalie is not moving. When we compared these rules with our source code, we found out that the values in the rules are similar to those implemented in our goalie.

4.1 Future Work

The next step is to make the discovered rules useable to our team. Therefore we need to find suitable countermeasures and implement a way to automatically transmit them to the players. We also want to use our method on other parts of the team.

Acknowledgement

We would like to thank Michael Boronowsky for the work he has done helping us getting started with his method.

References

[Boronowsky, 2001] Boronowsky, M. (2001). Diskretisierung reellwertiger attribute mit gemischten kontinuierlichen gleichverteilungen und ihre anwendung bei der zeitreihenbasierten entscheidungsbauminduktion. In *DISKI*, volume 246, St. Augustin. infix Verlag.

[Fayyad and Irani, 1992] Fayyad, U. and Irani, K. (1992). On the handling of continuous-valued attributes in decision tree generation. In *Machine Learning*, volume 8, pages 87–102.

[Frank et al., 2000] Frank, I., Tanaka-Ishii, K., Arai, K., and Matsubara, H. (2000). The statistics proxy server. In *The Fourth International Workshop on RoboCup*, pages S.199–204.

[Quinlan, 1986] Quinlan, J. (1986). Induction of decision trees. In *Machine Learning*, volume 1.

[Quinlan, 1993] Quinlan, J. (1993). *C4.5 Programs for Machine Learning.* Morgan Kaufmann.

[Raines et al., 1999] Raines, T., Tambe, M., and Marsella, S. (1999). Automated assistants to aid humans in understanding team behabiors. In *Proceedings of the Third International Workshop on Robocup*, pages S.85–104.

[Visser et al., 2001] Visser, U., Drcker, C., Hbner, S., Schmidt, E., and Weland, H.-G. (2001). Recognizing formations in opponent teams. In *RoboCup-00, Robot Soccer World Cup IV*, Lecture Notes in Computer Science, Melbourne, Australia. Springer-Verlag. to appear.

[Wünstel et al., 2000] Wünstel, M., Polani, D., Uthmann, T., and Perl, J. (2000). Behavior classification with self-organizing maps. In *The Fourth International Workshop on RoboCup*, pages S.179–188.

KENSEI-chan: Design of a Humanoid for Running

Kosei Demura, Nobuhiro Tachi, Tetsuya Maekawa, and Tamaki Ueno

Matto Laboratories for Human Information Systems
Kanazawa Institute of Technology
3-1 Yatsukaho, Matto, Ishikawa 924-0838, JAPAN
demura@his.kanazawa-it.ac.jp

Abstract. This paper presents the design of the humanoid robot KENSEI-chan for RoboCup Humanoid League. Many humanoids have been developed, but no humanoids have a capability of running. Because running needs huge torque, shock tolerance, and weight reduction. KENSEI-chan was designed with these things. KENSEI-chan is planed to have 12 degrees of freedom (DOFs) in two legs, 4 DOFs in two arms, 2 DOFs in the neck, and 1 DOFs in the trunk. Totally KENSEI-chan has 19 DOFs. 2DOFs in the knees and the ankles are Rotary Elastic Joints, which provide shock tolerance and force control.

1 Introduction

The RoboCup Federation will start Humanoid League from RoboCup-2002 Fukuoka Japan. The success of the humanoid league is very important key for the future of RoboCup, and it will have a tremendous impact on robotics research and industries.

Many humanoids have been developed, for example, P3 [1], ASIMO [2] from Honda company, WL-10RV1, WABIAN [3] from Waseda university, H5, H6 [4] from university of Tokyo, Saika-3 [5] from Tohoku university, PINO [6] from JST and so on. However, those humanoids lack the capability of running what soccer needs. Thus, we have begun to develop the humanoid robot named KENSEI-chan that has a capability of running, and following points dominated the design of it.

- **Running Mechanism** Rotary Elastic Joints are used for knee joints. Rotary Elastic Joints are composed of encoders, motors, reduction gears, torsion springs, and leg links. Those joints transform kinetic energy to elastic energy, transform elastic energy to kinetic energy, and provide shock tolerance as well.
- **Infant Proportion** The size of the humanoid is restricted to have the ability of running using motors sold on the market. The height of the humanoid is about 110cm, the approximate size of a 5 year old child.
- **Pretty Exterior Design** The importance of exterior design is addressed by PINO. The design of KENSEI-chan would be pretty like an infant, the audience in the RoboCup games will watch it attentively and affectionately.

A. Birk, S. Coradeschi, and S. Tadokoro (Eds.): RoboCup 2001, LNAI 2377, pp. 331–336, 2002.

This paper describes the design concept of KENSEI-chan and its running mechanism. KENSEI-chan is under developing, and the lower part of the body is almost finished.

2 Running

2.1 Running Mechanism

In this paper, running is defined by periods of ballistic flight with all feet leaving the ground. Running is very similar to the bouncing ball [7]. That is , during the flight phase, a leg is swung forward in preparation for the heel touches ground. After the heel touches ground, heel and toe contact, then toe contact, and flight phase again. During the transitions, kinetic energy is transformed to elastic energy in the elastic elements, such as muscle, ligament and bone tissue in the body and stored. Then elastic energy is reused. The knee acts as a passive spring to store the kinetic energy. The use of elastic energy is up to from 25 % to 50% [8]. Human has a high efficiency running mechanism.

In RoboCup, humanoid robots have to play soccer during half time without replacement batteries. The humanoid which we have developed introduces the mechanism of human running.

2.2 Condition of Running

Let us consider the condition of running to design the humanoids. When a runner runs with an speed of 3.6m/s, vertical ground reaction forces reach 2.1 times body weight and horizontal ground reaction forces reach 0.4 times body weight.

Thus, running humanoids have to have the shock tolerance mechanism and huge torque for the touchdown, moreover angular velocity of each joints would be enough for running.

Table 1 shows the condition of running for KENSEI-chan. Angular velocity data are from a 178cm and 63kg runner runs at 2.8m/s. Necessary Torque T_n are calculated as follows

$$T_n = T_{peak} \times weight\ ratio \times size\ ratio.$$

Where, the weight ration is 0.63, and the size ratio is 0.62.

	Angular Velocity [rad/s]	Peak Torque [Nm]	Necessary Torque for KENSEI-chan [Nm]
Hip (pitch)	5.8	300.0	117.2
Knee (pitch)	11.0	245.0	95.6
Ankle (pitch)	5.4	220.5	86.1

Table 1. The Condition of Running. Torque and angular velocity are given for maximum value in teh sagital plane.

3 Architecture

3.1 Overview

Fig.1 shows the DOFs configuration of the lower part of KENSEI-chan which we had developed. It has 12 DOFs, and Table 2 shows the specification of each joint. We adopted Maxon DC motor RE40 (power:150W, maximum torque 2.3Nm, weight 480g) for the hip pitch, the hip roll, and the knee pitch joints, RE35 (90W, 0.78Nm, 340g) for the hip roll and ankle pitch joints, and RE25 (20W, 0.24Nm, 130g) for the hip yaw and ankle roll joints.

All motors for the pitch are connected to Harmonic Drive (HD) gears. The types of the Harmonic Drive gears are CSF-25-100-2A-GR and CSF-20-160-2A-GR [9]. Harmonic Drive gears are zero-backlash speed reduction system, light and compact in relation to the large torque. The other motors are connected to high ratio hipoid (MHP) gears . The hipoid gears are also high reducation ratio, precise and high performance. We adopted MHP1-0601R, MHP1-1060L from Kohara gear industrial company [10].

Table 2 shows that the maximum torque of each joints are bigger than the torque condition of Table 1. Thus, KENSEI-chan has a capability of running at the speed of 2.5m/s.

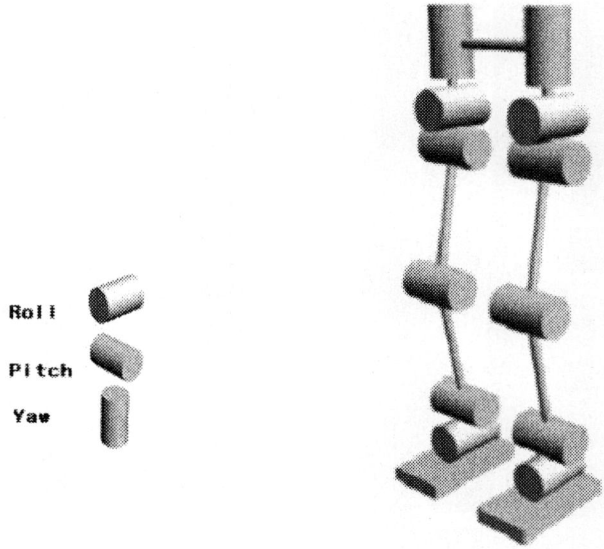

Fig. 1. DOFs configuration

		Motor Power [W]	Gear		Max. Angular Velocity [rad/s]	Max. Torque [Nm]
			Type	Ratio		
Hip	Pitch	150	HD	100	8.4	230.0
	Yow	20	MHP	60	12.0	14.0
	Roll	90	HD	100	8.1	78.0
Knee	Pitch	150	HD	100	11.3	170.0
Ankle	Pitch	90	HD	160	6.0	127.0
	Roll	20	MHP	60	6.0	28.0

Table 2. Robot Joint Specifications

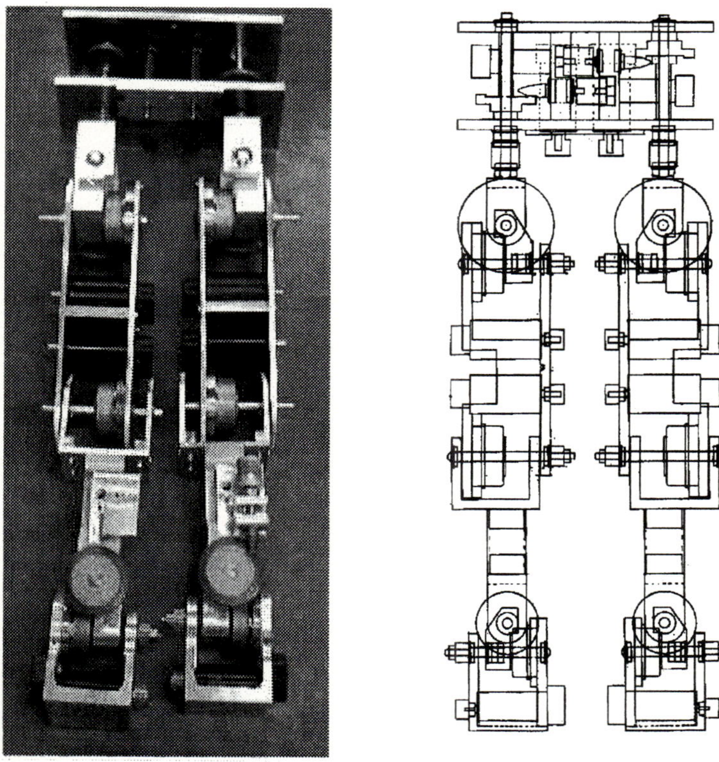

Fig. 2. The mechanical design and a photo of the lower part of KENSEI-chan

3.2 Mechanism for Running

To realize the efficiency running like human, we have developed REJs (Rotary Elastic Joints) for knee joints shown in Fig.3. In REJ, the output of reduction gear is connected to a torsion spring directly, and the torsion spring is connected to the lower leg of KENSEI-chan directly. Thus, the output of gear and the lower leg is connected indirectly via the torsion spring.

The shock of touchdown is stored as the elastic energy of the spring, and the elastic energy is used for the flight phase.

Besides, REJ can measure the torque of the knee joints by a potentio meter which measures the spring deflection.

The basic idea of using the spring for dynamic locomotion is not new. For example, Marc Raibert developed Series Elastic Actuators (SEAs) [11]. They are composed of a spring in series with the output of the actuator, after the gear reduction (ball screw).

However, they are reciprocating motion. About a motor and the combination of the Harmonic drive, the REJs are suitable compared with SEAs. By using the Harmonic drives, the mechanism of the humanoid can be simplified.

4 Conclusions

This paper introduced the mechanism of KENSEI-chan under development for the RoboCup Humanoid League. The lower part of KENSEI-chan has almost completed, the upper part equipped a processing system (Pentium III 1GHz) and various sensors (FSR, CCD camera, gyro, encoders) will be completed by the end of this year.

Fig. 3. Image of Rotary Elastic Joint

References

1. Hirose, M. , Takenaka, T. , Gomi, H., Ozawa, N.: Humanoid Robots. The Journal of the Robotics Society of Japan (JRSJ), vol.15, no.7 (1997) (in Japanese)
2. ASIMO Special Site:http://world.honada.com/ASIMON
3. Yamaguchi, J., Soga, E., Inoue, S. and Takanishi, A.: Development of a Bipedal Humanoid Robot – Control Method of Whole Body Cooperative Dynamic Biped Walking', IEEE International Conference on Robotics and Automation, pp. 368-374 (1999)
4. Nishiwaki, K., Suguhara, T., Kagami, S., Kanehiro, F. and Inaba, M.: Design and Development of Research Platform for Perception-Action Integration in Humanoid Robot : H6, Proc. of IEEE/RSJ International Conference on Intelligent Robots and Systems (IROS'00), vol. 3, pp. 1559—1564 (2000)
5. Konno, A., Kato, N., Shirata, S., Furuta, T., and Uchiyama, M: Development of a Light-Weight Biped Humanoid Robot, Proc. of the 2000 IEEE/RSJ International Conference on Intelligent Robots and Systems, pp. 1565 – 1570 (2000)
6. Yamasaki, F., Matsui, T., Miyashita, T., and Kitano, H.: PINO The Humanoid: A Basic Architecture: Proc. of the Fourth International Workshop on RoboCup, pp.52-61 (2000)
7. Cavagna, G.A. and M. Kaneko: Mechanical work and efficiency in level walking and running. J. Physiol. 268:pp. 467-481 (1977)
8. Kobayashi, K. The Science of Running, pp. 104, Taishukan-Shoten (1990) (in Japanese)
9. Harmonic Drive: http://hdsystemsinc.com/
10. MHP High Ratio Hipoid Gear: http://www.khkgears.co.jp/index.html (in Japanese)
11. Pratt, J.E.: Exploiting Inherent Robustness and Natural Dynamics in the Control of Bipedal Walking Robots, Ph.D thesis, MIT (2000)

Supervision of Robot Control

Albert Figueras[1], Joan Colomer[1], Thor I. Fossen[2], and J. Lluis de la Rosa[1]

[1] Institute of Informatic and Applications. University of Girona, E-17071, SPAIN
figueras@eia.udg.es, http://rogiteam.udg.es

[2] Department of Engineering Cybernetics, Norwegian University of Science and Technology, N-7491 Trondheim, Norway.

Abstract. This work shows how to supervise a set of controllers designed by different specification to be applied at requested constrained trajectory for the autonomous mobile robot. In this paper are explained part of the control structure from local controllers to the agent level, the communication between the high supervisor and the local controller are detailed. Several controllers are designed in order to minimize different cost functions. Non-linear behavior of the robot force to work off-line to model all possible controllers, the on-line process concludes the optimal controller to realize a requested task. This work is applied over the soccer robots in the RogiTeam [1].

1 Introduction

The robot control problem in dynamic scenarios is a specific case of control domain. RoboCup competition is a platform to apply the best controllers in order to achieve the goal. In recent competitions hierarchical structures of control have been applied from embedded control to distributed agent structure. RogiTeam is one of these and uses a distributed architecture of control with local control and distributed agent supervisor [2]. Supervisor requirements are the knowledge capability of the local controllers, this means the possibility to realize a trajectory under some specification and restrictions. M.Samaan [3] B.J.Kuipers[4] have treated this subject and solved the problem by logic. Difference of these approaches is we solve the problem of the reachability of the solution in the continuos signal not discretized in intervals.

Morse[5] Goodwin[6] Narendra[7] have treated switching adaptive control using multiple models, or controllers. Stability analysis and switching problems are treated. Our problem differs of them, in a single controller is applied to execute one action, and the following actions can be interpreted independently without interactions. Stable controllers must be designed in order to assure the reachability of the end conditions.

This work focuses the problem to apply the adequate controller which support the required specification. The possible specifications can be minimum time, energy cost and precision. Environmental conditions, robot behavior and opposite team are dynamic conditions which make changes at behavior and the local controllers to be applied. Agent decides which and when the specification is necessary to apply.

A. Birk, S. Coradeschi, and S. Tadokoro (Eds.): RoboCup 2001, LNAI 2377, pp. 337–342, 2002.

2 Problem Definition

The working robot environment is very dynamic, this mean the trajectories to be executed change continuously due the continuous change of the ball position and the precise goals (figure 1). Different goals can be shot the ball, passing the ball, stop the ball, robot positioning, etc. This implies a necessary control basis adapted at this dynamic environment. That possibility to implement different controllers with different specifications used in specific case. These controllers are registered and reused by introspection when chosen the adequate controller.

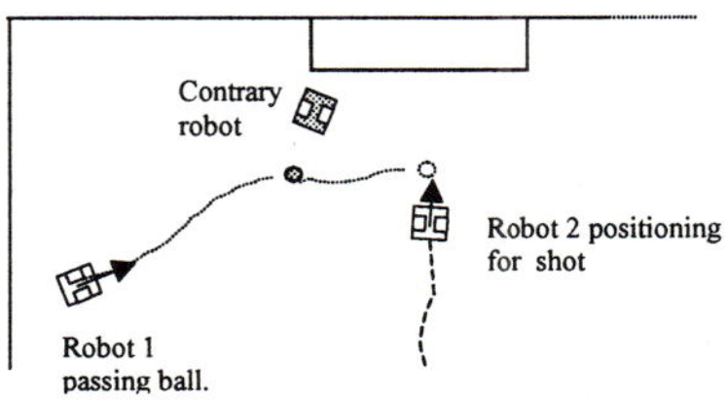

Fig. 1. Different required goals.

The goal is request by the supervisor. Agent supervisor requires a different goal in order the move will be played. Robot position or shot ball can be requested at certain part of the game. To execute the actions a different controller must be applied in each case. The robot answer is, if this request action can be done with one of those controllers that he has. In affirmative case he express the capacity to complete the requested action.

3 Controllers Specifications

The structure of the control is depicted in figure 2 where a high-level supervisor suggests a trajectory under certain specifications. A protocol between high-level supervisor and the local supervisor is necessary to validate the designed trajectory. Is necessary a set of controllers which makes the system at different specifications: Cont=[c1 c2 ... cn]

Where c_1,c_2,... c_n are the group of closed loop controllers designed with different specifications, for example minimum time, energy cost, precision and others as the composition of these specifications.

Design of the controllers subject to the cost functions:

1. Min time of execution the trajectory
   Cost function applied:
   $J(t) = \int_{t0}^{tf} dt$ Where t_f is the time to reach the end point.

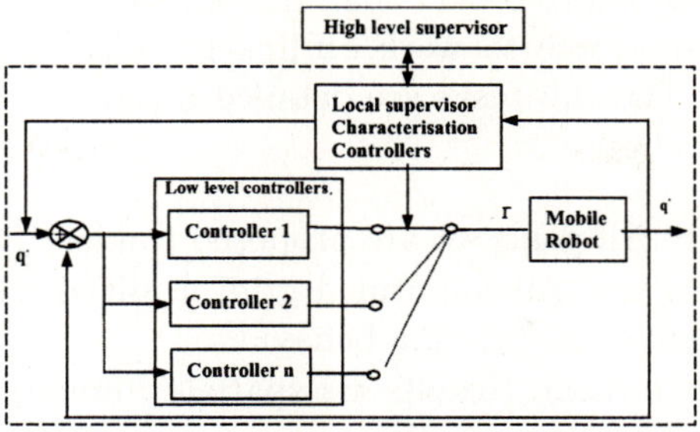

Fig. 2. Hierarchical control structure.

2. Fixed time to execute the trajectory.
   Cost function applied:
   $J(t) = \int (t - t_f)^2 dt$ Exact time is required at specific point in the field, passing ball or shoot condition.
3. Precision in the end point trajectory execution.
   Cost function applied:
   $J = \int_{t0}^{tf} X_f^T Q X_f dt$ Where Q is a semi-definite positive matrix.
4. Accurate precision to executing the trajectory.
   $\|e(\omega)\|_\infty = Sup\,|e(\omega)|\ ,\|e(\omega)\| \leq k$
   Where e(ω) is the error of the reference to the trajectory and Sup is the supremal of the error.

These four cost functions give a minimum of four controllers. Combined weighed two cost functions can give as maximum six new controllers.

4 Solution Proposed

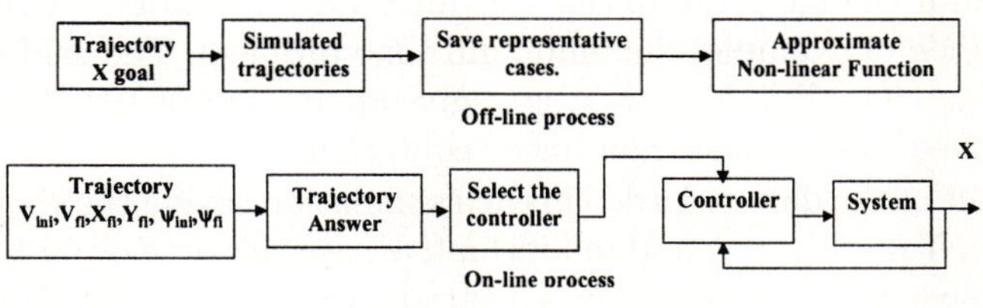

Fig. 3. Working methodology.

The non-linearity of the robot avoid to extrapolate behavior of the system in non analyzed points, this requires analysis of great quantity of representative points. Computational effort is very high due amount of the data therefore the real time requirements force to work off-line process. A methodology used is depicted in figure 3. The sub-tasks are detailed below:

Off-line:

1. *Trajectory X goal:* All goals set are proposed to identify the robot behavior.
2. *Simulated trajectories:* An uniformed grid of initial and final conditions is generated to simulate all possible behaviors.
3. *Save representative cases:* Results are spatial filtered in order to retain only the limits of positive and negative areas to record them in minimum data space.
4. *Approximate Non-linear Function:* All the affirmative cases are approximated by non-linear function.

On-line:

5 *Trajectory* V_{ini}, V_{fi}, X_{fi}, Y_{fi},ψ_{ini}, ψ_{fi}*:* Supervisor decides if can execute a trajectory under X specification. To request this trajectory, V_{ini}, V_{fi}, X_{fi}, Y_{fi}, ψ_{ini}, ψ_{fi} are parameters to pass to the robot.

6 *Trajectory answer:* Search the proposed case in the limit map of the areas, affirmative response says at least one controller can do this action.

7 *Select the controller:* All affirmative cases are evaluated with the non-linear cost function (4). The controller selects the best response, this means the best controller.

8 *Controller:* Apply in the robot the best controller and the required action.

The On-line process is repeated, as many actions are needed.

5 Example

A proposed methodology is applied at mobile robot soccer with the nonholonomic model (see[8]). A complete grid of simulated trajectories are made in Matlab Simulink changing the initial and final state conditions. Only one goal is treated a mode of example, the minimum time problem. The grid into the six parameters V_{ini}, V_{fi}, X_{fi}, Y_{fi}, ψ_{ini}, ψ_{fi}, give an amount of 104000 simulations over the specified model and a non-linear controller.

The amount of the data recorded in each simulation saving the six parameters V_{ini}, V_{fi}, X_{fi}, Y_{fi}, ψ_{ini}, ψ_{fi} and in affirmative case the time or on the contrary negative conclusion forces to apply a spatial over all the 104000 simulations to obtain only a representative values. The result is about 10.000 representative cases, which are the limits of the areas, which determines the positive or negative possibility of realization the trajectory. The certainty to apply this representative cases is about 95% for all situations.

All the affirmative cases before filtered, are approximated by non-linear function. The non-linear function is obtained with non-linear least square method.

Lsqnonlin is a Matlab function, which adjust the function parameters in a minimum square error. The non-linear function to predict the time is:

$$\hat{T} = a1 * log(V_{ini} + 0.2) + a2 * log(V_{fi} + 0.2) + a3 * X_{fi} + a4 * Y_{fi} + \\ +a5 * \sqrt{X_{fi}^2 + Y_{fi}^2} + a6 * \psi_{ini} + a7 * \psi_{fi} + a8 * sin(\psi_{traj}) \quad (1)$$

This function approximates the time needed to execute the required trajectory with absolute mean square error around 0.11% for overall simulated trajectories. When is affirmative the possibility to realize the requested action, this function calculated On-line the approximate time to be executed with the six inputs parameters V_{ini}, V_{fi}, X_{fi}, Y_{fi}, ψ_{ini}, ψ_{fi}(initial and final conditions).

Results:

A sample of eight representative trajectories are tested by this method to analyze their reliance. The predicted possibility give how result seven affirmation cases and one negative case. On the contrary the real execution give six affirmation cases and two negative cases.

The approximate time estimate function has the expression (1), and the parameters are obtained by least square method: a1=-0.22;a2=-1.72;a3=0.55;a4=-0.81;a5=5.42;a6=-0.22;a7=-3.5;a8=-0.9; The first case is the erroneous prediction, it can be identified because the final precision is not attained. The others six realizable cases, the time, the final angle and position are always accomplished. The specification is less than 10% of precision error.

Cases trajectory Position desired (V_{ini},V_{fi},X_{fi},Y_{fi},ψ_{ini},ψ_{fi})	Predicted Possibility	Execution Time	Estimated Time	Final position (V_{fi},X_{fi},Y_{fi},$_{fi}$)
1. (0 0 25 25 0 0.52)	YES	3.3	3.03	(3.5 38 22.06 0.7)
2. (0 24 125 175 1.04 1.04)	YES	4.17	4.42	(23.1 118.07 167.80 0.96)
3. (8 8 250 150 0 1.57)	YES	7.23	6.80	(7.23 240.72 147.21 1.51)
4. (16 0 75 100 1.57 0)	YES	4.25	4.89	(1.25 68.49 92.48 0.04)
5. (16 24 200 75 0.52 0.52)	YES	3.91	3.95	(23.9 190.13 73.66 0.47)
6. (32 0 125 200 0.52 1.57)	YES	7.15	7.11	(0.21 119.1 191.99 1.45)
7. (32 24 250 175 0 0)	YES	5.49	5.60	(22.4 241.02 171.15 0.08)
8. (24 16 25 25 1.57 1.04)	NO			

Table 1. Trajectory times: real times and calculated times of affirmative cases.

The seven executed trajectories are represented in a figure 4 , filtered data and treated with Matlab, due to the problem of the vision acquisition noise. The case number 1 is a erroneous predicted case and impossible to execute under specifications, the robot try to arrive at desired point at new attempt.

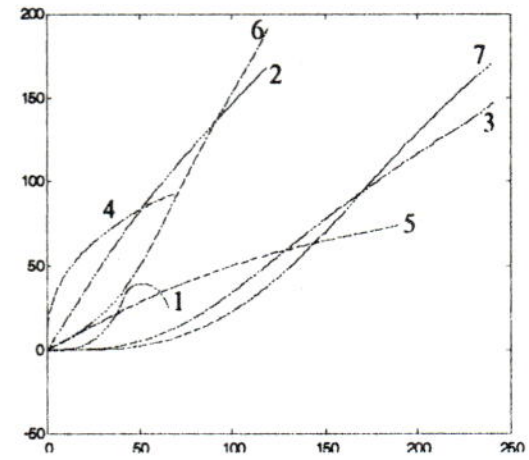

Fig. 4. Realized robot trajectories.

6 Conclusion

In this paper supervision over the family of controllers are treated. Is a part of the pyramidal structure of control, from Intelligent Agent to the local controller. Intelligent Agent decides the action to be taken, at the moment of the game. The controller reads the case base memory to find the approbation to realize the request action. When is possible to realize this action under restrictions the system evolves to the next step. In negative case the intelligent agent proposes another action. So as very important to analyze the proposed action before executed in dynamic scenarios, and more in competition systems, due the non-well executed actions penalize the next situation. Is preferable realize a less optimal action than realize a bad optimal action in competition.

The investigation is running in this direction, choosing optimal controllers, and readjusts them when the play is on. Little deviations of the original behavior, due the battery, friction, etc. make big divergence of the expected results.

References

[1] Oller A., Garcia R., Ramon J. A., Figueras A., Esteva S., De la Rosa J. Ll., Humet J. M., And del Acebo E.: Description Of RogiTeam. ON robocup-97: Robot Soccer World Cup I. 1998 286–294

[2] Oller A., De la Rosa J.Ll., García R, Ramon JA, and Figueras A.:A Micro-Robots Playing Soccer Games: A Real Implementation Based On A Multi-Agent Decision-Making Structure. Int. Journal of Intel. Autom. and Soft Comput.**Vol6 N1** 2000

[3] Samaan M., Duque M., and M'Saad M.: A real Time supervision system for adaptive control. IFAC-Glasgow 1989

[4] B.J. Kuippers and K. Astrom :The composition and validation of heterogenous control laws. Automatica **30(2)**, 1994 233–249

[5] Daniel Liberzon and A. Stephen Morse.: Basic problems in Stability and Design of Switched systems. IEEE Control systems. 1999 59–70.

[6] A.S. Morse, D.Q. Mayne and G.C. Goodwin.: Applications of hysteresis switching in parameter adaptive control. IEEE Trans.Aut.Control. 1992 1343–1354.

[7] K. Narendra and J. Balakrishnan.: Improving transient response of adaptive control systems using multiple models and switching. Proc. of the 32nd IEEE Conf. On Decision and Control, San Antonio, Texas, 1993

[8] J. Vehi, A. Figueras, N.Luo.: Interval PI Velocity Control of a Nonholonomic Mobile Robot. Proc. IFAC Workshop on Digital Control. Terrassa (Spain).2000 433–437

Walkie-Talkie MIKE

Ian Frank[1], Kumiko Tanaka-Ishii[2], Hitoshi Matsubara[1], and Eiichi Osawa[1]

[1] Future University, Hakodate, Japan, {ianf,matsubar,osawa}@fun.ac.jp
[2] Tokyo University, Japan, kumiko@ipl.t.u-tokyo.ac.jp

Abstract. We address the problem of information flow in disaster relief scenarios by presenting an architecture for generating natural language dialogues between large numbers of agents. This architecture is the first step towards real-time support systems for relief workers and their controllers. Our work demonstrates how techniques from the MIKE commentary system for RoboCup soccer can be carried over to the domain of RoboCup Rescue. Thanks to this background, the initial product of our research is a system that explains a RoboCup Rescue simulation not to the agents in the domain themselves but to a watching *audience*. This "commentary" is produced by recreating the actual dialogues most likely be occurring in the domain: walkie-talkie conversations.

1 Introduction

In a disaster relief situation, *information* is crucial. The disaster scene itself will be confusing, complex and quickly-changing. Yet the different groups of agents in the domain (relief workers, ground controllers, and civilians) will each need fast access to different kinds of knowledge.

Our theme in this paper is that thinking only of *information acquisition* in such domains is not enough. Rather, we suggest that it is the *flow* of information between groups of agents that makes the difference between "disaster" and "disaster relief". As a concrete first step towards understanding this information flow, we investigate probably the single most important communication method in the domain: *walkie-talkie conversations.*

Specifically, we demonstrate how walkie-talkie conversations can be generated for the RoboCup-Rescue simulator [1]. That is, we take the current (silent) simulator and, based on the events in the simulated world, produce a "soundtrack" of the exchanges you might expect to hear from *real* relief workers and *real* control centres if the simulation being played out was itself also real.

In effect, the output of our system is a kind of "commentary", and this is no accident, since it builds on the research that produced the MIKE commentator system for the RoboCup simulator league [2] and small-size robot league [3]. We show in this paper that many of the lessons learned in domains such as soccer commentary carry over remarkably well into natural language generation for a disaster relief scenario. Most notably, we demonstrate how the large number of voices in the walkie-talkie conversation is actually an extreme version of decomposing an explanation task among several agents [4]. For such dialogues, and in

A. Birk, S. Coradeschi, and S. Tadokoro (Eds.): RoboCup 2001, LNAI 2377, pp. 343–349, 2002.

the face of the impracticality of constructing prior plans for the discourse, the agents benefit from using different discourse strategies. Most obviously in the disaster relief domain, the controller needs to concentrate on high-level analysis of the overall situation, whereas relief workers on the ground need to pay more attention to the low-level events happening around them.

In its role as a "commentator", we call our system Walkie-Talkie MIKE, and we intend to demonstrate it at RoboCup events, to add atmosphere to RoboCup-Rescue demonstrations in much the same way that MIKE itself adds atmosphere to soccer games. The main motivation for our research, however, is not this "entertainment" but instead the investigation of how best to handle information flow in disaster relief situations. The use of natural language is critical in this respect, as it is "resource sensitive"; it presents information in a way that leaves recipients free to use their hands and eyes for other important tasks. Yet, to our knowledge, this is the first time that research on walkie-talkie dialogues has been carried out. We anticipate that our work will be a first step towards a deeper understanding of natural language dialogues between large numbers of agents in rapidly changing domains.

2 The Basic Walkie-Talkie Architecture

The RoboCup-Rescue simulator is graphically impressive but presents a large amount of information in a small screen. Helping viewers understand this simulation was one of the initial motivations for considering the Walkie-Talkie MIKE project. To a significant extent, we succeeded in processing the information generated by the server with an architecture similar to that of the original MIKE soccer commentator [2]. Space limitations prevent us from giving more than a sketch of our implementation, but we highlight the main similarities and differences below, before giving an example commentary in §3.

Analyser Modules

Walkie-Talkie MIKE uses "analysers" to post text fragments we call *propositions* into a shared memory. For example, fire brigade number 5 heading towards location id number 34 is represented as `(move fire 5 34)`, where `move` is the tag, and `fire`, `5` and `34` are the attributes.

The primary analysers used by Walkie-Talkie MIKE interpret the status of each type of agent in the domain: civilians, the police force, the ambulance and the fire brigade. These analysers have a rather different role to the analysers in soccer. Rather than trying to scrutinise the differences between two teams of agents with very similar abilities, they instead build up a picture of three types of information:

- Interpretation of the agents' actions (summary, analysis and prediction).
- Information that the agents will want to broadcast to other agents in the nearby area or further afield (different from the *actual* messages, if any,

broadcast by the agents themselves, since the protocol for inter-agent communication is not yet stable in the Rescue simulator, and any given set of clients may use semantics that are not known in advance).
- Information that the agents will need to know or be warned of. (All conversations — but especially walkie-talkie dialogues — are two-way processes. As well as simply listening and speaking, it is important to know what kind of information to *request* in order to perform at the best possible level.)

Additional Analysers

In addition to monitoring the actual agents, the RoboCup Rescue domain requires two further analysers. The first of these processes information about buildings and other infrastructure in the simulation. The second analyser is one that we use to represent a "control center".

In the original versions of the Rescue Simulator, there was no "control center". Our ControlCenter analyser therefore represented an imaginary control center of our own creation. However, the most recent version of the server allows for one control center for each type of relief worker (ambulance, police, fire brigade). The actual role of these control centres within the simulation is not yet clear (they are represented as a special type of "building" that can send messages), but we regard the notion of the control center as vital, since it is here where the lower-level inferences produced by the agent and the building analysers are most naturally built up into a higher-level picture of the overall domain. We have designed our ControlCenter analyser with two purposes in mind. Firstly, we require the ControlCenter to be a participant in the dialogue between the domain agents, so that we can experiment with the widest possible range of options for producing walkie-talkie dialogues. Secondly, we want to be able to use the ControlCenter as a real-time supporter and adviser for genuine, human ground controllers at a disaster scene. What we envisage here is something similar to the visualisation provided by the Statistics Proxy Server in RoboCup [5], but also including a natural language component.

Importance Scores

The Control Center also forms the basis for our use of importance scores: each proposition has a score that captures how important the event it describes should to a *ground controller* in the domain. The basic architecture of Walkie-Talkie MIKE then provides two mechanisms for selecting propositions to be uttered. The first, and simplest, is to allow the setting of an independent threshhold for each of the analysers. Any propositions with importance values higher than this value are then flagged and time-stamped. The second mechanism is more sophisticated, and requires the monitoring of *changes* in the values of importance scores over time. This flags and time-stamps any proposition whose change exceeds a configurable threshhold. Together, these mechanisms allow for maximum flexibility when applying the basic Walkie-Talkie MIKE architecture to different tasks (such as advising ground controllers, or generating a commentary).

To support the actual generation of speech, Walkie-Talkie MIKE associates a number of templates with each proposition, and selects randomly amongst these for the propositions that have been flagged for utterance. The actual generation of natural language depends on the application that the system is being used for. In this paper, we give the specific example of generating a "commentary" for the Rescue Simulator, as described below.

3 Walkie-Talkie Dialogues for the Rescue Simulator

Generating a realistic dialogue for the Rescue Simulator with the basic architecture of Walkie-Talkie MIKE requires the specification of a control algorithm for selecting among flagged propositions. We chose to implement this by concentrating on the high-level events identified by the ground controller (for the moment, we assume a single ground controller in our dialogues). The basic operation is thus for the controller to speak whilst there are important events to describe. In any gaps, individual domain events are selected in order of importance and an appropriate agent is chosen to speak a template describing the event.

To increase the realism of the generated dialogue, we also allow the control center to acknowledge statements made by other agents in the domain. Further, before the control center describes an event, we may carry out some "scene-setting" by generating an extra utterance from an agent associated with the event. Finally, we also complement the Natural Language Generator with a small repertoire of sound effects. For instance, when a fire brigade agent is reporting a fire, we augment the audio with the sound of a fire burning.

To give an idea the effect of the resulting dialogues, we present in Figure 1 an example. Each phrase in this dialogue is generated by instantiating one of the NL templates contained in the basic Walkie-Talkie MIKE architecture. For the best overall effect, imagine hearing the exchanges as though they are being carried out over walkie-talkie communications, with bursts of radio static at the beginning and ends of each person's speech.

Already, the commentary of Figure 1 is of a type that rescue workers could simply have playing through an earpiece to pick out important pieces of information (e.g., the comments of the controller, or of other rescue workers they know are nearby). Note, though, that this example dialogue was not generated from an actual simulation, but was constructed by us manually to illustrate a number of key attractions of researching walkie-talkie dialogues in the disaster relief domain:

- **Appropriate Atmosphere**. As a commentary, the style of a walkie-talkie dialogue represents a very natural way of describing the situation to people watching the simulation. The example of Figure 1 could alternatively begin "This simulation is of an earthquake of size 7 on the Japanese scale. Such an earthquake causes structural damage and makes movement very hazardous... If you look here on the display you can see a fire brigade heading towards this fire in the North..." This kind of "explanation" may work in other domains, but it seems out of place in disaster simulation.

```
<Control> All units. The latest estimation of the size of the
earthquake is a seven on the Japanese scale. Expect structural damage
to roadside buildings. Proceed with extreme caution.
<Voice1> Control, Red brigade 3 heading towards fire in North.
<Control> Roger, fire spreading rapidly there. Red squares on our
monitors increasing.
<Voice2, background of flames roaring> Situation very bad
here. <background of victims screaming> people are trapped here.
<Control> Now have seven teams fighting fires in the North.
<Control> Cover in other areas is low. Don't over-commit, people.
<Voice4> We got three people out safely here.
<Voice3> Red 2 drawing back to the West.
<Control> Fire in the North quarter looks to be under control. Blue
squares our monitor there increasing.
```

Fig. 1. Example dialogue based on Walkie-Talkie MIKE templates

This is an issue of *appropriateness*. For instance, televised soccer almost always features announcers, so it is appropriate to simulate them in RoboCup. However, in disaster relief nobody really expects a commentary as such, so it is more appropriate to recreate the voices that would normally be present: the rescue workers themselves and their controllers.

- **High-level vs low-level**. The most recent versions of the MIKE soccer commentator system have demonstrated the benefits of separating the discourse strategies used to present high-level and low-level information. The same distinction also fits naturally into a walkie-talkie dialogue: we have low-level information about what each of the rescue workers or teams are doing, and also have high-level information based on the overall view of the disaster area. Thus, we are able to use the disaster relief domain to further investigate how to effectively combine dialogues generated by different strategies.
- **Continuity**. This is one of the issues that makes soccer commentary surprisingly hard. For instance, if a commentary contains the phrase "Manchester United are playing well today" the subsequent statements can be phrased "*they* are using space well" and "*their* pass success rate is high" rather than "Manchester United are using space well" and "Manchester United's pass success rate is high". Whilst overall continuity is difficult to achieve in RoboCup, a walkie-talkie dialogue in the RoboCup Rescue domain is more forgiving. Listeners hear the burst of static before each phrase and recognise that there may be a context shift coming. It is therefore less critical to maintain continuity.

Note that since each phrase gives a context to itself (emphasised by the burst of static at the beginning and end of each speaker's turn), we are not even limited to having all the voices speaking in the same *language*. The entire audience of the disaster simulation does not have to understand every phrase. It is enough to assume that every utterance is understood by one of the other agents in the simulation (or by the controller).

- **Explaining the Simulation**. In some of the controller's statements in our example commentary, there are explanations of the graphical view of the simulator itself. This is a convenient, indirect way to explain RoboCup-Rescue to an audience. The implicit assumption is that the controller is using the same viewer as we are. Note that even if a human is present to explain the simulator (for example at a RoboCup-Rescue demonstration), Walkie-Talkie MIKE can be used without significant modification as "background colour". Since our goal is not to "explain" but to recreate a part of the domain currently missing (the audio part), the walkie-talkie conversation can be left running while the human talks about various specific features of interest. Walkie-Talkie MIKE is also fairly robust to the possible changes that may be made to the display (for example zooming, highlighting, panning) for the benefit of the watching audience. The walkie -talkie phrase "Control, Red brigade 3 heading towards fire in North" is acceptable even when Red brigade 3 is not visible on the screen, unlike the corresponding "commentary" of "Now you can see Red 3 heading North".

4 Future Work and Conclusions

We have described the basic architecture of Walkie-Talkie MIKE and shown how to make use of this architecture to produce a "commentator" for the Rescue Simulator. In doing so, we demonstrated the feasibility of extending established natural language generation techniques to the disaster relief domain.

We will be demonstrating Walkie-Talkie MIKE as a live "commentator" at the 2002 RoboCup. However, this represents just the first step of our work in this area. In the long term, we expect our work on Walkie-Talkie MIKE to provide new insights into multi-agent natural language generation. In the more immediate future, we plan to use our system to produce natural language systems for summarising and passing high-level information from a control centre to individual (and teams of) relief workers. Building on this, we also plan on producing a more complete system to assist ground controllers in real-time. Another possibility is the broadcasting of information to civilians trapped in a disaster area. For instance, some parts of the world (including areas of Japan) have an infrastructure of public speakers that could be used to broadcast instructions directing citizens to the safest routes away from the most dangerous areas.

When it comes to actually pushing RoboCup-Rescue techniques into practice in the future, it will be necessary for robots to work with humans (and for humans to work with robots). Through our work on Walkie Talkie MIKE, we are starting to appreciate both the problems and the types of dialogues that are important for such collaboration. Although the name "Walkie Talkie" MIKE describes primarily the style of our system's current dialogue, we hope our work will eventually justify a literal interpretation: allowing relief workers and robots to walk and do their jobs whilst talking to them and for them about all the important events that are unfolding in the domain.

References

1. Hiroaki Kitano, Satoshi Tadokoro, Itsuki Noda, Hitoshi Matsubara, Tomoichi Takahashi, Atsushi Shinjoh, and Susumu Shimada. RoboCup-Rescue: Search and rescue for large scale disasters as a domain for multi-agent research. In *Proceedings of IEEE Conference on Man, Systems, and Cybernetics (SMC-99)*, 1999.
2. K. Tanaka-Ishii, I. Noda, I. Frank, H. Nakashima, K. Hasida, and H. Matsubara. MIKE: An automatic commentary system for soccer. In *Proceedings of ICMAS-98*, pages 285–292, 1998. Also available from Electrotechnical Laboratory as Tech Report ETL-97-29.
3. I. Frank, K. Tanaka-Ishii, H. Okuno, Y. Nakagawa, K. Maeda, and H. Nakadai, K. Kitano. And the fans are going wild! SIG plus MIKE. In *Proceedings of the Fourth International Workshop on RoboCup*, Melbourne, Australia, 2000. To be published by Springer in LNCS/AI series.
4. K. Tanaka-Ishii and I. Frank. Multi-agent explanation strategies in real-time domains. In *Proceedings of ACL2000, the 38th Annual Meeting of the Association for Computational Linguistics*, Hong Kong, 2000.
5. I. Frank, K. Tanaka-Ishii, and K. Arai. The statistics proxy server. In *Proceedings of the Fourth International Workshop on RoboCup*, Melbourne, Australia, 2000. To be published by Springer in LNCS/AI series.

Quadruped Robot Navigation Considering the Observational Cost

Takeshi Fukase, Masahiro Yokoi, Yuichi Kobayashi, Ryuichi Ueda, Hideo Yuasa, and Tamio Arai

The Department of Precision Engineering, The University of Tokyo, Hongo, Bunkyo-ku, Tokyo, 113-8656, Japan
{fukase, yokoi, kobayashi, ueda, yuasa, arai}@prince.pe.u-tokyo.ac.jp

Abstract. In this paper, we present a new method to decide the optimal robot's motion with a quadruped robot of Sony Legged Robots League. The method keeps the "observational cost" as small as possible. The task we cope with is navigation. The robot can reach the destination rapidly and robustly with this method in spite of under the unfavorable conditions such as restricted sensor ability, limited CPU power, and inaccurate locomotion. Finally the efficiency of our method is verified by experiments.

1 Introduction

On the soccer field, the robot's decision mainly depends on its location. There are many researches about localization[1, 2, 3]. However, the task which should be finally accomplished is not to localize accurately, but to decide the optimal motion.

There are two researches which provide effective motion decisions with simple observations under the condition of Sony Legged Robots League[4, 5]. In these researches, the robot decides its motion according to the appearance of the landmarks and the ball. However, the relations between the location and the decision are designed arbitrarily. Therefore, the optimality of the decision is not guaranteed.

In this paper, we propose a method to decide the optimal motion with the smallest observational cost. The task we cope with is navigation. This is a task to make a robot reach a target position and orientation which has some allowance called "target region". This task is applicatory to other tasks such as occupying a defensive position. In addition, we can evaluate the optimality by the time to reach the target region.

In this research, we use ERS-1100 that is the unique hardware in Sony Legged Robot League. Following characteristics of ERS-1100 make the task difficult.

- Restricted sensor ability: Insufficient information due to the narrow field of view
- Limited CPU power: A calculation that needs high CPU power is not executable.

A. Birk, S. Coradeschi, and S. Tadokoro (Eds.): RoboCup 2001, LNAI 2377, pp. 350–355, 2002.

– Limited control ability: The discrete locomotion is not adapted for a small move.

The approaches to accomplish the navigation are as follows. First, we use "Motion Planned Map(MPM)" to decide the optimal locomotion commands. Second, as a localization method, we adopt "Sensor Resetting Localization[3] (SRL)" which is developed for Sony Legged Robots League. By using "MPM" and "SRL", we propose a method that provides the motion decision that minimizes the observational cost.

The rest of this paper is constructed as follows; Section 2 discusses the control method we propose in this research. Section 3 provides the results of the experiments and discussions. Section 4 provides the conclusion and future works.

2 Methods

Here we summarize our method to navigate a robot to the target region(Fig.1). At the beginning of every step, the robot's estimated position is updated by SRL. Then, the robot decides its motion by comparing the cost of observation with the risk of moving without observation according to its estimated position and MPM. And it executes the motion. This process is repeated until the robot reaches the target region.

2.1 Motion Planned Map

MPM is a data base which has the information of optimal locomotion commands at any robot's position and orientation. We make this Map by the offline calculation. Therefore, only a low computational power is required to the robot. In this research, the map is represented by grid cells. We made this map with Dynamic Programming. Fig.2 shows the MPM made under the conditions shown in Table.1.

When making MPM, we can also calculate the expectations of the time from each grid to the target region. We express the position-time relation by grid-based form which is the same as MPM. We call this "State Value Map (SVM)", and use with the MPM.

2.2 Motion Decision Method

After updating the self-position, the motion to be executed is decided according to MPM. In SRL, a set of finite discrete points called samples represents the positional and orientational probabilistic distribution of the robot. When all SRL samples are inside the single region as shown in Fig.3(a), the motion which corresponds to the region on the map is selected. On the other hand, we must decide to move or observe when the samples are distributed over more than two regions as shown in Fig.3(b). This decision is made according to the rule shown in Fig.4. That is "If the observational cost is smaller than the loss of time caused by non-optimal decision, then we decide to observe". The expectation of the time to the target region is calculable by the samples' positions and orientation and the value of the SVM there.

3 Experiments

3.1 Conditions of Experiments

The described method was implemented and evaluated under four different conditions shown in Table.2. When the initial position is "fixed and known", all the samples are settled on the actual robot's position on the soccer field. On the other hand, "random and unknown" means that the samples are distributed uniformly.

In addition, we set other conditions as follows:

- **Initial position and Target region:** Shown in Fig.5.
- **Number of samples:** 100(fixed).
- **Observational cost:** 3.0[sec](fixed).
- **Number of trials:** 10 (on each condition)
- **Terminative condition:** Beyond 80% of all the samples are inside the target region.
- **Judgment of success:** The robot actually reaches the target region in 180[sec].

3.2 Result and Discussions

Table.3 shows the result of the experiments. Fig.6 shows the overview of the experiments.

The robot successfully reached the target region at the rate of about 70% on each condition. In Experiment.1 and 3, the numbers of observations were only three or four times. In Experiment 2 and 4, they were larger. This means that the robot selected to observe according to its positional and orientational probabilistic distribution.

In Experiment.1 and 3, the initial position and orientation were the same on each trial. However, there were no two trials on which the robot crossed the same path to the target region. This is due to the errors of robot's locomotion and the localization. The fact that the robot can reach the target region through different paths indicates the robustness of our method.

In failure cases, the robot considered itself inside the target region based on the distribution of samples even though it is located incorrect actual position. This is due to the factors which are unconsidered or unable to model in SRL such as the collision with the walls. In order to improve the success rate, we must clear up the causes of localization errors and model them. This is also an essential problem. One supposable approach is the practical use of information about other objects in addition to the landmarks.

4 Conclusion and Future Works

In this research, we proposed the method to decide the optimal motion with the smallest observational cost. We need not to set the arbitrary rules to decide the robot's motions. The robot can select the optimal motion in spite of the uncertain information of its location. The efficiency of our method was verified by the results of the experiments.

In this method, it is assumed that the probabilistic distribution of the location is modified well enough to select the optimal locomotion by an observation. But the observation causes the loss of time if the robot is located where the recognition of the landmarks is impossible. In this case, it is necessary to predict the efficiency of the observation. In future, the prediction of the observational efficiency should be considered. Furthermore, improvement of localization method and tasks with the ball or dynamic obstacles also will be our future works.

References

[1] P. Buschka, A. Saffiotti and Z. Wasil: "Fuzzy Landmark-Based Localization for a Legged Robot", Proc. International Conference on Intelligent Robots and Systems, pp.1205-1210, (2000)

[2] Masahiro YOKOI, Yuichi KOBAYASHI, Takeshi FUKASE, Hideo YUASA and Tamio ARAI: "Learning Self-Localization with Teaching System", Proc. International Conference on Intelligent Robots and Systems, pp.1211-1216, (2000)

[3] S. Lenser and M. Veloso: "Sensor Resetting Localization for Poorly Modeled Mobile Robot", Proc. International Conference on Robotics and Automation, pp.1225-1232, (2000)

[4] Vincent HUGEL, Patrick BONIN and Pierre BLAZEVIC: "Reactive and Adaptive Control Architecture Designed for the Sony Legged Robots League in RoboCup 1999", Proc. International Conference on Intelligent Robots and Systems, pp.1211-1216, (2000)

[5] Noriaki Mitsunaga and Minoru Asada: "Observation strategy for decision making based on information criterion", Proc. International Conference on Intelligent Robots and Systems, pp.1211-1216, (2000)

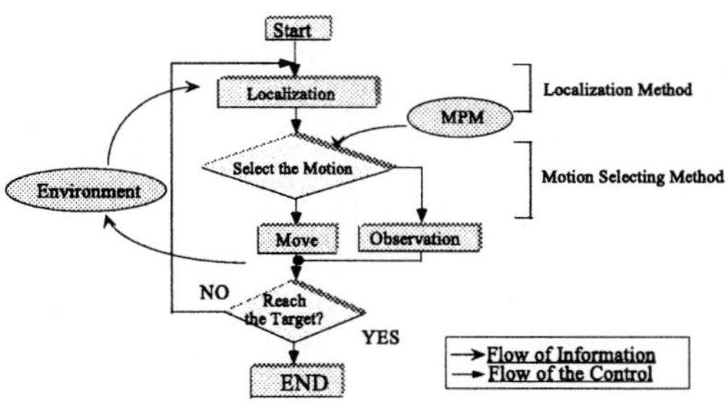

Fig. 1. The flow chart of the control method.

Table 1. The conditions to make Motion Planned Map

Locomotion commands	Forward, Right turn, Left turn (3 motions)
Map size	5000[mm] × 5000[mm] × 360[deg]
Grid size	50[mm] × 50[mm] × 15[deg]
Target region	(2500 ± 300[mm], 2500 ± 300[mm], 0 ± 30[deg]

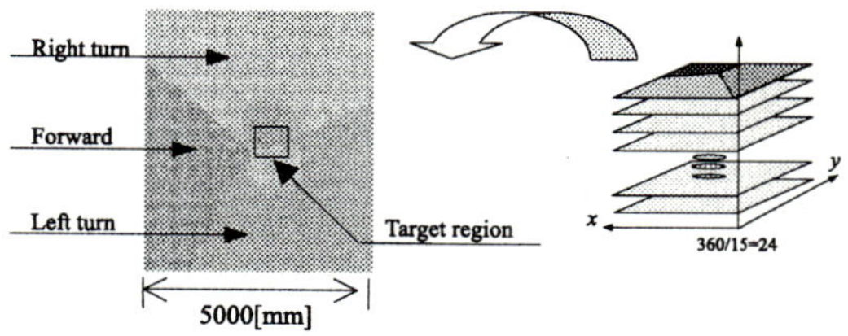

Fig. 2. Motion Planned Map. This map shows the result between 0[deg] and 15[deg]. Actually each orientation creates different maps. That means we get 24 maps.

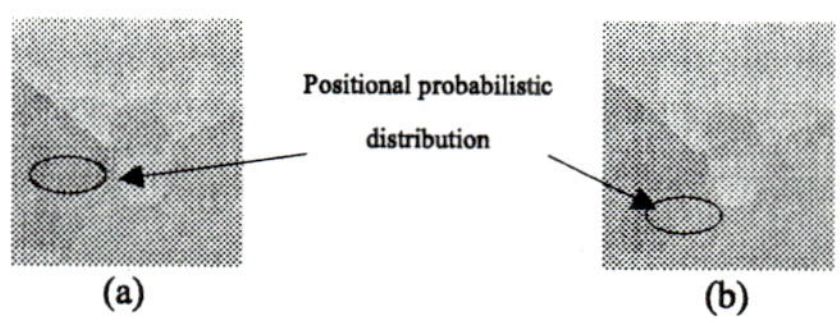

Fig. 3. The motion decision method with the positional probabilistic distribution. In case of (a), the robot move "Forward", but in case of (b), it is needed to decide to move or observe.

Variables

Number of the locomotion command	: n
ID of the locomotion commands	: i
Locomotion command	: a_i
Cost of the locomotion	: t_i
Cost of the observation	: $t_{observation}$
Set of samples of the initial state	: Ω_0
Set of samples after the locomotion a_i	: Ω_i
Number of samples	: N
Sample j at the initial state	: $\omega_{j,0}$
Sample j after the locomotion a_i	: $\omega_{j,i}$
SVM value of $\omega_{j,0}$	: $v_{j,0}$
SVM value of $\omega_{j,i}$	: $v_{j,i}$
Expectation from Ω_0 to the target region	: E_0
Expectation from Ω_i to the target region	: E_i

Algorism

$$E_0 = \sum_{j=0}^{N} p_{j,0} v_{j,0}$$
$$E_i = \sum_{j=0}^{N} p_{j,i} v_{j,i}$$
$$t_{\max} = \max_{1 \leq i \leq n} (E_i - E_0 + t_i)$$
$$t_{\min} = \min_{1 \leq i \leq n} (E_i - E_0 + t_i)$$
$$d = \arg\min_{1 \leq i \leq n} (E_i - E_0 + t_i)$$

if($t_{\max} \leq t_{observation}$) do a_d locomotion

if($t_{\max} \geq t_{observation}$) do observation

Fig. 4. The motion selecting algorism.

Table 2. Conditions of experiments

	Locomotion patterns	Initial position
Experiment.1	3	fixed and known
Experiment.2	3	random and unknown
Experiment.3	5	fixed and known
Experiment.4	5	random and unknown

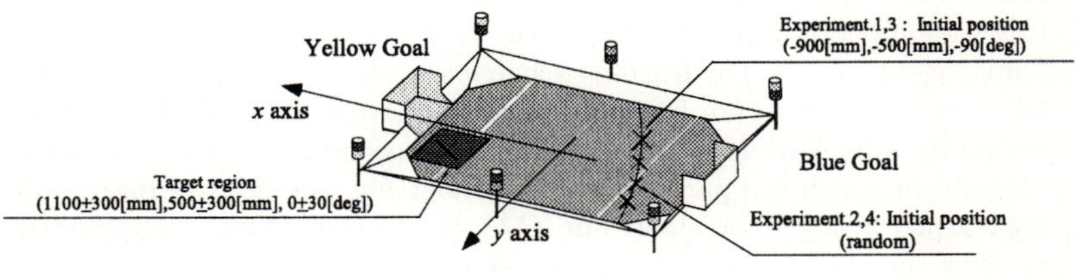

Fig. 5. The initial position and target region of this experiment.

Table 3. The results of the experiments.

	Succeed times	Average of the time[sec]	The number of observation
Experiment.1	7/10	62.3	3.4
Experiment.2	6/10	65.2	5.0
Experiment.3	7/10	53.4	3.5
Experiment.4	7/10	100.4	7.4

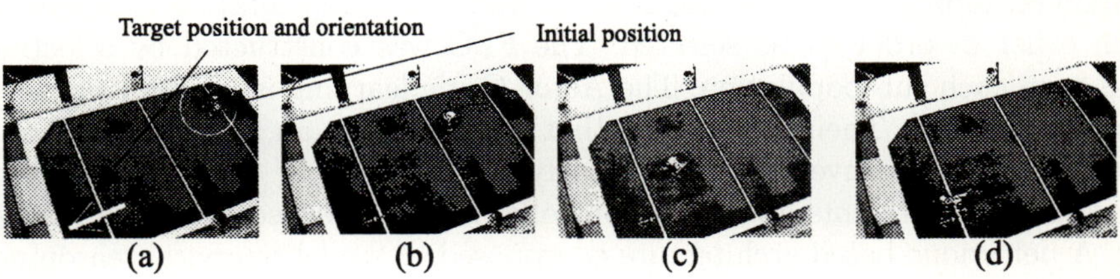

Fig. 6. The overview of this experiment.

Evolving Fuzzy Logic Controllers for Sony Legged Robots

Dongbing Gu and Huosheng Hu

Department of Computer Science, University of Essex
Wivenhoe Park, Colchester CO4 3SQ, UK
dgu@essex.ac.uk, hhu@essex.ac.uk

Abstract. This paper presents an evolutionary approach to learning a fuzzy logic controller(FLC) employed for reactive behaviour control of Sony legged robots. The learning scheme is divided into two stages. The first stage is a structure learning in which the rule base of FLC is generated by a backup updating learning. The second stage is a parameter learning in which the parameters of membership functions of fuzzy sets are learned by a genetic algorithm (GA). Simulation results are provided to show the effectiveness of the proposed learning scheme.

1 Introduction

The behaviour-based control is a strategy that decomposes complex robot tasks into a collection of sensory and action pairs. These pairs tightly couple perception and motor responses and work in a co-operative manner to complete complex tasks. There is an increasing tendency to build up the mapping of sensory and action pairs by using FLC to make robots be reactive and adaptive [8]. Problems in the design of a FLC are parameter setting of membership functions and composition of rules. They are classified into two catalogues: structure identification and parameter identification [7].

Karr first used a GA to evolve the membership functions of a FLC [4], in which a FLC is encoded as an individual. GA-based learning has been used for both structure learning and parameter learning of FLCs in robot reactive control[6]. In [1][3], the authors adopted different evolution strategies where an individual represents a rule, not a FLC. Rules in one generation compete with each other in order to be selected. The FLC was constructed by selecting a rule from each sub-population. The Actor-Critic learning is used for parameter learning in [2][5]. Their difficulty is that it needs both an actor network and a critic network to converge simultaneously.

This paper presents a reactive controller for Sony legged robots to play soccer. A behaviour-based architecture is proposed to model temporal reasoning as a deliberative component and to select a set of reactive behaviours as a reactive component. The main reactive behaviour is to move towards the ball guided by on-board sensors. There is no global visual system to provide relative positions among the robots, the ball and the goal. Only information provided from

A. Birk, S. Coradeschi, and S. Tadokoro (Eds.): RoboCup 2001, LNAI 2377, pp. 356–361, 2002.

the environment is the ball's local sensory information and the reward is not presented until the robot moves to the place close to the ball and faces to the goal. A two-stage learning process is proposed to evolve a FLC for the reactive behaviour. A rule base is first learned through a backup updating provided with initial values of the membership parameters. For Sony legged robots, output actions are discrete commands, each of which can make legged robots move a single step in different directions. The parameter identification for input membership functions is investigated in the parameter learning. A real value GA learning is adopted to optimize membership parameters. In section 2, we will formulate a FLC for Sony legged robots to move towards the ball. The generation of rule base and GA learning mechanisms of the parameter learning are addressed in section 3. The simulation results are given in section 4. Finally section 5 presents brief conclusions and future work.

2 A Fuzzy Logic Controller for the Reactive Behaviour

A reactive control scheme is adopted in the *approaching-the-ball* behaviour for Sony legged robots to play soccer. There are three state variables: the orientation relative to the ball represented by θ, the distance to the ball, and the goal's angle relative to the robot's head direction represented by α, which are important for this behaviour due to lack of the global localization. A tracking algorithm has been implemented for the head to track the ball based on color detection. It leads θ to be expressed by the head's pan angle that can be easily read from the encoder mounted on the pan motor. We use the number of image pixels(h) of the ball in a CCD camera mounted inside the head to estimate the distance. The tracking algorithm will also yield the goal's angle in the process of tracking the ball. Therefore, the input state vector is $S = [s_1, s_2, s_3]^T = [\theta, h, \alpha]^T$. The behaviour is to move the robot to approach the ball by taking action such as $MoveForward$, $LeftForward$, $RightForward$, $LeftTurn$, $RightTurn$, $Backward$, or $Stop$ provided by low-level walking software.

Both sensory information and actions are imprecise, incomplete, and vagueness. For example, the ball size measured in pixels from the CCD camera does not accurately correspond to the distance from the robot to the ball, but just has fuzzy concept such as $Small(S)$, $Middle(M)$, or $Large(L)$. The pan angle measured in degrees from the head motor encoder is accurate to some extent, but it is not the precise direction from the robot to the ball since the ball may not stay in the image center during moving. A set of fuzzy concepts $Left(L)$, $Zero(Z)$, or $Right(R)$ is designed for the pan angle. The goal's angle is also define as $Left(L)$, $Zero(Z)$, or $Right(R)$. The output of this behaviour is one of the seven one-step moving commands that can not be precisely represented in position and orientation due to non-perfect actuators or slippage on the ground.

Three fuzzy sets are designed for each of three input state variables s_n. A quadruple of (a, b, c, d) is used to represent the triangle or trapezoid membership function of a fuzzy set shown as in figure 1. The output action o is the crisp value that can be seen as fuzzy singletons $c^m (m = 1, \ldots, 7)$ in a FLC. There are N

($N = 3 \times 3 \times 3$) rules in total. The membership degree of a fuzzy set F_n^i of the nth input state variable s_n in the ith rule is expressed as $\mu_{F_n^i}(s_n)$. The true value of the ith rule activated by an input state vector S is calculated by Mamdani's minimum fuzzy implication:

$$\alpha_i(S) = \min(\mu_{F_1^i}(s_1), \mu_{F_2^i}(s_2), \mu_{F_3^i}(s_3)) \tag{1}$$

The crisp output o stimulated by the input state S after the fuzzy reasoning is calculated by the center of area method (COA):

$$o(S) = \sum_{i=1}^{N} \alpha_i(S) \times c^{m_i} / \sum_{i=1}^{N} \alpha_i(S) \tag{2}$$

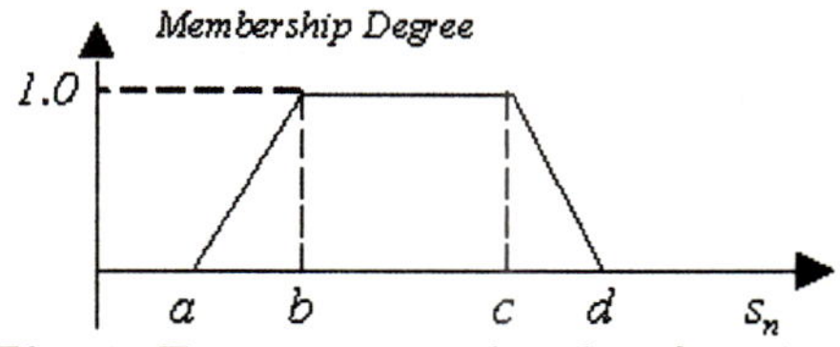

Fig. 1. Fuzzy set membership function

3 Learning FLC

The aim of the *approaching-the-ball* behaviour is to move towards the ball with an appropriate orientation to the goal. The fitness should be designed to guide the evolution of the FLCs to achieve this purpose. The fitness function is defined as:

$$fitness = w_1 \times \sum d_i/l + w_2 \times (A - l) + w_3 \times (B - \theta) + w_4 \times (C - \alpha) + w_5 \times h \tag{3}$$

where w_i ($i = 1, ..., 5$) is the weight, A, B and C are constants, and l is the number of the walking steps used in one trial. The first term in (3) indicates payoffs received during the whole movement. The second term rewards those FLCs that have fewer steps. The last three terms with last step θ, h, and α reflect the final position of the robot.

In the first stage, the rule base is learned incrementally through the backup updating. The fuzzy sets divide the input state space into difference regions and these regions overlap with each other. The points in these regions will be chosen as the learning examples where the corresponding membership degrees are equal to 1. The backing updating means the learning will start from the points where the membership degree of the fuzzy set h = *Large* is equal to 1 to the points where the membership degree of the fuzzy set h = *Small* is equal to 1. The robot will select as its fuzzy rule's outputs those commands with the highest fitness. The current learning examples can make use of the previous learned rules. The robot will become increasingly aware of better control rules and capable of interact with the environment via these rules during the learning process. The areas where the fuzzy sets overlap are not covered in the learning process. But they are implicitly generalized by the rules in the areas where there is no overlap among fuzzy sets. The fine-tuning within the overlapped areas will be proceeded in the next stage through the parameter learning.

In the second stage, GA is used to optimize the parameters in membership functions in order to improve the reactive behavioural performance. One FLC is encoded as one individual. In each generation, a collection of the individuals is maintained to compete with each other to survive. By evolving through genetic operators, the best one will be selected as the optimal FLC for the behaviour control. For a FLC, the quadruple (a, b, c, d) of its input membership functions are real values. The real value encoding of a FLC is shown in figure 2 where the individual p represents the pth FLC in one generation.

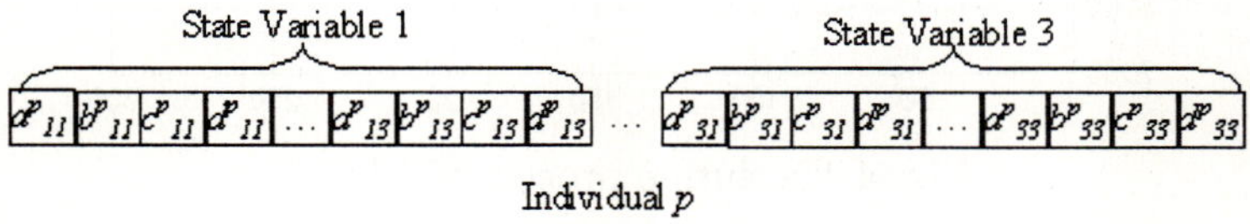

Fig. 2. An individual in a generation

There are three genetic operators employed: reproduction, crossover, and mutation. The real value mutation is proceeded by adding a certain amount of noise to new individuals to produce the offspring with a mutation probability. The learning procedure starts from random formation of initial generation of FLCs. After completing trials of one generation, the FLCs evolve into a new generation and the learning repeats until a GA terminal condition is met.

4 Experimental Results

The experiments are tested in a simulation environment where the models with the exact size of the pitch and ball are built up. The rule base is incrementally built up by the backup updating approach described in section 3. Figure 3 shows the results for the *approaching-the-ball* behaviour controlled by the FLC with the learned rule base. It can be seen that the robot can move to the ball although the robot did not exactly face the goal at final positions.

After the parameter learning by GA, the same situations as in figure 3 are tested for the robot with the evolved FLC. The results are shown in figure 5 where the improvement at final positions can be seen. In the parameter learning, the size of population in one generation is 50. The GA learning process is shown in figure 4 after evolving 300 generations. The upper curve is the maximal fitness values in each generation. The low curve is the average fitness values in each generation. It is shown that the average fitness values converge to the maximum as the generation increases.

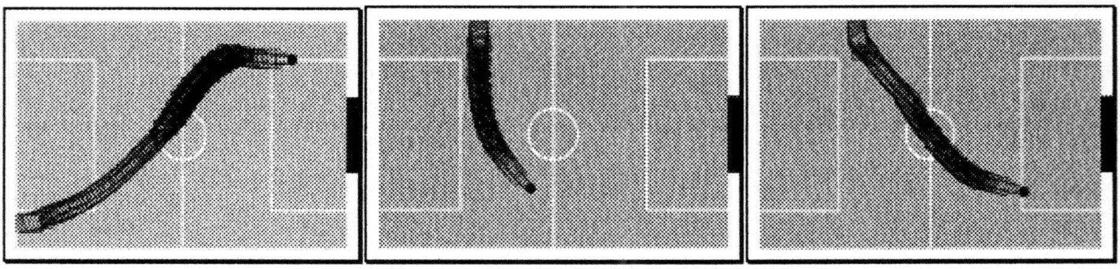

Fig. 3. Simulation results after the structure learning

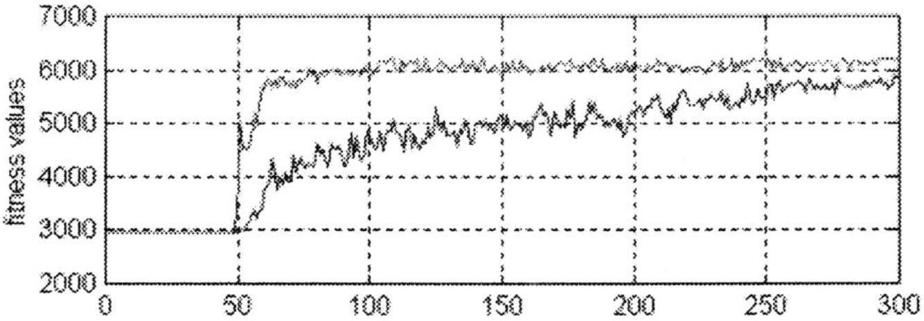

Fig. 4. Evolution process of FLCs

The quadruple (a, b, c, d) for each input fuzzy label should be constrained by their geometric shapes $(a \leq b \leq c \leq d)$ during GA learning. A validation process is employed to check the constraints for each FLC before it is used. Invalid FLCs are given up. Therefore, most of FLCs are not executed due to their invalidation at early stage and the fitness values are unchanged before the $50th$ generation. Figure 6 shows the comparison made before and after the parameter learning. In the structure learning, only the areas where the input fuzzy membership degrees are equal to 1 are learned. The overlapped areas are implicitly generalized by the adjacent areas. The result shown on the left of figure 6 is the case that can not be generalized, and the result on the right shows the compensation made by the parameter learning.

5 Conclusions and Future Work

The learning of FLC for Sony legged robots is addressed through both structure and parameter learning stages in this paper. The strategy is to make use of the heuristic experience and autonomous exploration of robot's environment to yield a good reactive controller.

The structure learning is similar to the human reasoning about rule design. The most of the state space where no overlap exists in the fuzzy sets is covered by the actual crisp conditions. It implies the overlap areas can be generalized by adjacent areas. The further adjustment of the FLC is left for the parameter learning in which the parameters in input fuzzy sets are modified. A real value GA is used to evolve the FLCs. It explores the state space by using crossover and mutation operators, while the offspring with high fitness value is exploited

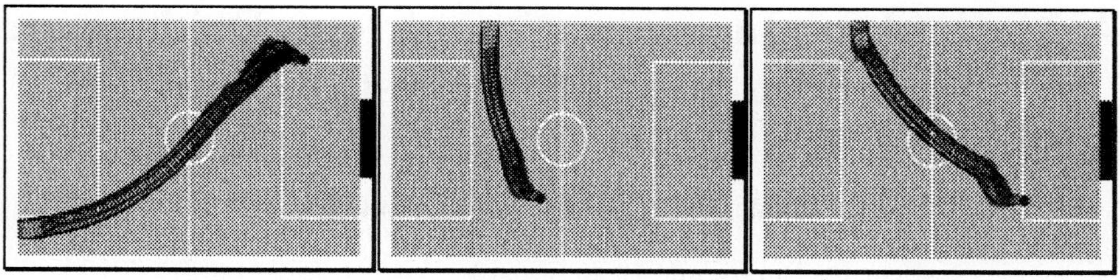

Fig. 5. Simulation results after the parameter learning

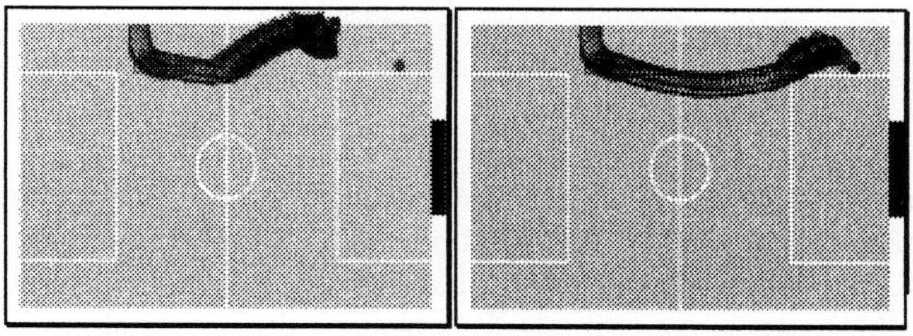

Fig. 6. The comparison between the structure learning and parameter learning

by reproduction operators. The experiment results show that the FLC can be learned by the proposed learning scheme. The proposed method will be tested on real robots in the next stage.

References

1. Bonarini, A., Evolutionary Learning of Fuzzy rules: competition and cooperation, In Pedrycz, W. (Ed.), Fuzzy Modelling: Paradigms and Practice, Kluwer Academic Press, Norwell, MA, 1996, pages 265-284.
2. Jouffe, L., Fuzzy Inference System Learning by Reinforcement Methods, IEEE Transactions On SMC-Part B, Vol. 28, No. 3, August 1998, pages 338-355.
3. Juang, C. F., Lin, J.Y., and Lin, C. T., Genetic Reinforcement learning through Symbiotic Evolution for Fuzzy Controller Design, IEEE Transactions on SMC-Part B, Vol. 30, No. 2, April 2000, pages 290-301.
4. Karr, C. L., Gentry, E. J., Fuzzy Control of pH Using Genetic Algorithms, IEEE Transactions on Fuzzy Systems, Vol. 3, Jan. 1993, pages 129-139.
5. Lin, C. T., Jou, C. P., GA-Based Fuzzy Reinforcement Learning for Control of a Magnetic Bearing System, IEEE Transactions on SMC-Part B, Vol. 30, No. 2, April 2000, pages 276-289.
6. Matellan, V., Fernandez, C., Molina, J. M., Genetic Learning of Fuzzy Reactive Controllers, Robotics and Autonomous Systems, Vol. 25, 1998, pages 33-41.
7. Mitra, S., Hayashi, Y., Neuro-Fuzzy Rule Generation: Survey in Soft Computing Framework, IEEE Transactions on Neural Networks, Vol. 11, No. 3, May 2000, pages 748-768.
8. Saffiotti, A., Ruspini, E. H., and Konolige, K., Using Fuzzy Logic for Mobile Robot Control, in Zimmermann, H. J., editor, Practical Applications of Fuzzy Technologies, Kluwer Academic Publisher, 1999, pages 185-206.

A Method for Incorporation of New Evidence to Improve World State Estimation

Martin Haker, André Meyer, Daniel Polani, and Thomas Martinetz

Institut für Neuro- und Bioinformatik, Universität zu Lübeck
D-23569 Lübeck, Germany
polani@inb.mu-luebeck.de, martinetz@inb.mu-luebeck.de
http://www.inb.mu-luebeck.de/staff/{polani,martinetz}.html

Abstract. We describe an approach to incorporate new evidence into an existing world model. The method, *Evidence-based World state Estimation*, has been used in the RoboCup soccer simulation scenario to obtain a high precision estimate for player and ball position.

1 Introduction

One of the problems which agents in realistic multiagent scenarios are facing is the need for an adequate world model. Set in the context of the RoboCup soccer server, we describe an approach to reconstruct the true world state of an agent and of objects (here: the ball) sensed by it as closely as possible by incrementally incorporating new evidence. In this approach, *EWE* (*E*vidence-based *W*orld state *E*stimation), new evidence, e.g. sensor readings, is incorporated into the current world model striving to maintain a maximum degree of consistency.

EWE in many ways a natural approach for a near-optimal reconstruction of a world state. It is clear-cut, conceptually concise and can be given a well-defined interpretation. We believe it has a strong potential to be generalized to domains beyond the simulated RoboCup world. In fact, many important state reconstruction methods can be regarded as special case of EWE. The EWE concept acts: 1. as a generalization of nonlinear Kalman filtering Lewis (1986) towards model-based filtering; 2. in a geometric context, as a world state description in terms of a generalized interval arithmetic; 3. as a natural extension of Markovian Self-Localization Fox et al. (1999) to spaces beyond 2-dimensional spatial positions; 4. as a continuous implementation of BDI Burkhard et al. (1998); Rao and Georgeff (1995) for world state estimation, since it strives at consistency between hypotheses about the world state and evidence.

2 Evidence-Based World State Estimation

In EWE, a single-valued state variable in the world model is replaced by a (multi-valued) set or a probability distribution of possible values for that variable. This multi-valued representation of the state variable is a model of the set of

A. Birk, S. Coradeschi, and S. Tadokoro (Eds.): RoboCup 2001, LNAI 2377, pp. 362–367, 2002.

hypotheses that the agent carries about the true world state. The arrival of any new evidence imposes restrictions on the set of possible and/or probable hypotheses, discarding some of them altogether and modifying the probability of others. An advantage of the model is that evidence can be used whenever it is available, not being forced to obtain sensor data every step, a difference to many standard filter models.

The EWE concept was implemented specifically for agents participating in the RoboCup simulation league tournaments. Since it is slower than the conventional approaches, efficiency considerations had to be made. For this reason, EWE concept was limited to the estimation of only the most important world state variables, namely player position and ball movement. Originally, EWE was first used to determine ball position and velocity and later extended to also calculate the player position. For presentation purposes we will first describe calculation of the agent position, followed by the mechanism for ball localization. A first version of EWE was already implemented in the team LUCKY LÜBECK, but not yet activated for the Eurobocup Amsterdam 2000 tournament for lack of stability. The first use of EWE in an official tournament was in the Melbourne 2000 event[1].

2.1 Determining the Agent Position

Player position is a two-dimensional vector $p \in \mathbb{R}^2$. In the EWE concept, we use a two-dimensional polygon as corresponding multi-valued representation. It can be interpreted as a probability density distribution for the agent position that attains the value $1/A$ in the inside the polygon and 0 outside (with A the polygon area). Equally possible is the interpretation as the basis for a "polygon arithmetic" as generalization of standard interval arithmetic to the 2-dimensional plane. The probability view leads to slower algorithms, so we restricted ourselves to the polygon arithmetic view.

In EWE, one starts with an a priori model for the position. Data which is sent to the agent by the server, representing the noisy and inaccurate sensor readings, is called *raw data.* Each piece of raw data received represents an evidence which restricts the available possibilities for the true position. When a real-valued quantity s sent to the agent by the server, be it angle or distance, it is converted to a quantized value $\hat{s} = Q(s)$ via a *quantization function* $Q : \mathbb{R} \to \mathbb{R}$ which depends on the type of data Corten et al. (1999). For the soccer server quantization functions the *inverse quantization* $S := Q^{-1}(\hat{s})$ for any given value $\hat{s} \in \mathbb{R}$ is either an interval $[a, b] \subseteq \mathbb{R}$ or empty (if $\hat{s}$ cannot result from quantization). To determine its position, our agent uses its neck angle, its body orientation and flag landmarks. From the raw neck angle $\hat{\phi}_{\text{neck}}$ an interval $\Phi_{\text{neck}} = Q^{-1}(\hat{\phi}_{\text{neck}})$ of possible (true) neck angles is computed. The orientation of the view cone (which is the sum of body orientation and neck angle) is obtained from the observed angle of the field border lines, again yielding an interval Φ_{view}. The interval

[1] We gratefully acknowledge the work of the MAINZ ROLLING BRAINS 1999 team on whose agent code the present work is based Polani and Uthmann (1999).

Fig. 1. Sector defined by an orientation and distance interval reconstructed by inverse quantization of raw server data representing the possible true relative positions of a landmark flag w.r.t. the agent. The solid line represents the form of the area as derived by the raw data. For the polygon algorithms, the area is slightly enlarged to the polygon denoted by the dashed lines.

arithmetic used now mimics the arithmetic procedure to determine agent orientation by single-valued variables. Thus, the angle interval for agent orientation is calculated to give $\Phi_{\mathrm{orient}} = \Phi_{\mathrm{view}} \oplus (-\Phi_{\mathrm{neck}})$. $-$ denotes inversion of an interval at the origin, reversing its orientation (i.e. $-[a,b] = [-b,-a]$, where $a < b$). If the intervals were interpreted as probability distributions, $\oplus$ is the convolution operation, but we implemented $\oplus$ as interval arithmetic addition.

The landmark flag position data sent by the server give quantized values for distance and angle. Reconstruction via the inverse quantization gives an interval for possible true distances and another one for the possible true angles of the flag. With these, the set of possible flag positions relative to the agent can be reconstructed (sector in Fig. 1 marked by solid line). The sector is then extended to the polygon marked by dashed lines to simplify treatment by polygon arithmetic.

The same way the calculation of Φ_{orient} traces the computation of a single-valued variable for orientation, we trace the computation of the agent position from its orientation and the flag data obtained. We appropriately rotate P_{flag} (widening its area to take into account the interval structure of Φ_{view}) and, by shifting the result w.r.t. the absolute flag position, we calculate the polygon area P_{player} representing the possible player positions. This process is done for every flag seen in the current time step, each time yielding a set of agent positions consistent with the observed flag. The "true" position, i.e. the agent position in the soccer server must be consistent with all observations made and thus with the (set) intersection of all these sets. Also here a probabilistic formalism would have been a viable alternative. An estimate for the current position of the player is obtained by averaging the corner positions of the intersection polygon. We found this polygon usually to be quite small and the resulting estimated positions to deviate typically around 5-10 cm from the true positions, which is more than an order of magnitude more accurate than the values obtained by conventional approaches, as averaging the position estimated directly from the raw data.

To estimate the player position, the CYBEROOS team independently implemented a mechanism similar to the one described in this section (private communication, Amsterdam 2000); so did the MAINZ ROLLING BRAINS (private communication, Melbourne 2000). They restricted themselves to the calculation of the agent position. This task can be seen as a special instantiation of the EWE concept. In fact, we used EWE first to estimate the ball state (Sec. 2.2) and only later applied it to the estimation of the player position.

2.2 Ball State Estimation

The generality of the EWE method allows us to determine ball movement in a conceptually similar fashion. For this, the approach of incorporating evidence needs to be applied to the *phase* instead of position space as in Sec. 2.1. The phase space is the set of the ball states which are fully represented by ball position *and* velocity. For simplicity, we limit ourselves to explain a situation where the player is placed at the origin of the coordinate system and looks along the x-axis, the ball moving exclusively and undisturbedly along the x axis, the y-component of its movement vanishing. The equation of movement during a single time step for an undisturbed ball in the soccer server is given by

$$\begin{pmatrix} x(t+1) \\ v_x(t+1) \end{pmatrix} = \begin{pmatrix} 1 & 1 \\ 0 & \lambda \end{pmatrix} \begin{pmatrix} x(t) \\ v_x(t) \end{pmatrix} , \qquad (1)$$

where x is the location of the ball along the x-axis, v_x is its velocity and λ is the decay rate for the ball movement Corten et al. (1999). The ball state is completely specified by the pair x, v_x because of our assumption that $y = 0$.

Denote the time before any evidence has been observed as t. At time t, the ball state is unknown. This means that the ball can be at any x-position inside the field and have any velocity $v_x \in [-v_x^{\max}, v_x^{\max}]$ where $v_x^{\max}$ denotes the maximum ball speed. The set of these possible initial ball states forms a rectangle in phase space (the full rectangle is not shown in Fig. 2 to not clutter the figure). When evidence about the ball is observed, one obtains a distance (and an angle which vanishes in our example). This distance defines a distance interval, restricting the possible x-positions of the ball to an interval. In phase space this defines the strip marked by the dotted lines in Fig. 2 (observation 1). After obtaining this evidence, the velocity, however, is still unknown. Due to the maximum velocity limits, the set of possible ball states in phase space is now delimited by the rectangle with solid lines in Fig. 2 (observation 1).

After one simulation step, by virtue of Eq. (1), for any given state in phase space we can predict the state in the next time step. In particular, the region of phase space that is consistent with observation 1 after a single simulation step is the oblique parallelogram depicted in Fig. 2. With a possible additional observation (no. 2 in Fig. 2), the requirement of consistency of the simulated phase space region with this observation restricts the region of possible ball states to the grey region in the figure. This obviously leads to a strong restriction of the set of possible velocities. During time this leads to a mutual restriction of x and v_x and thus to a quickly shrinking volume of phase space allowing a very precise localization of a ball as far away as 30–40 m after few time steps.

The method relies heavily on the mixing of the x and v_x-components of the feasible phase space region during time. By continued observations, the current

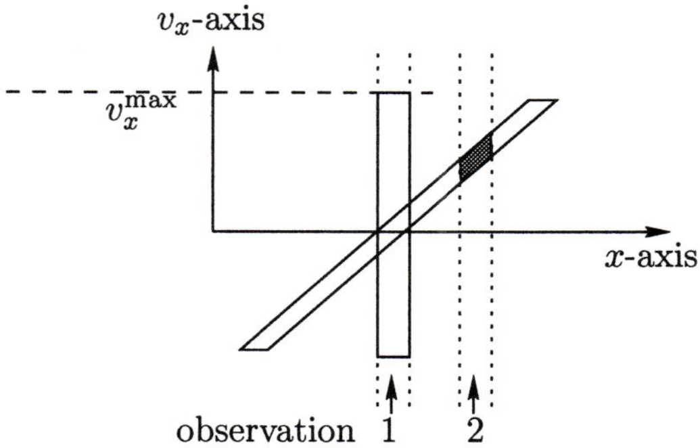

Fig. 2. Identifying position and velocity in phase space via observations

state of the system can be identified with increasing accuracy, similarly to the state identification models in chaotic systems, where under certain circumstances mixing effects also allow the accurate reconstruction of an initial system state.

3 A Simple Scenario

For illustration, we show some results from the method implemented in a regular player, with a full agent world model used in the Melbourne tournament. It therefore has more features than described above; we cannot discuss all particulars of the results presented here. We consider a player that repeatedly runs to the ball placed at the center, kicking it towards the goal.

Figure 3 shows the initial rectangle at the left. The horizontal width of the rectangle shows the initial uncertainty in determining the x-position of the ball. The agent now moves from its initial position at $x = -10$ m towards the ball. In subsequent simulation steps the rectangle is transformed into a parallelogram; new evidence defining vertical strips in the x/v_x-plane then increasingly restricts the set of possible ball states. Note that, since the ball is not moving in the initial phase of the simulation, the slices stay approximately at the same place. Thus the process of incrementally slicing parts from the possible positions is not spread out in x-direction as in the schematic Fig. 2. When the ball is kicked (the kick is modeled in the polygon simulation), the ball velocity jumps up and EWE is used to keep state model and observation synchronized.

Comparison of EWE with previously used ball position estimation algorithms shows that the ball position is indeed estimated to a much higher accuracy. Here, the agent regularly approaches the ball to kick it and therefore regularly obtains a more precise view of the ball. However, further experiments show that even for a ball as far as 30–40 m away from the agent, after a few simulation steps, the error in estimating the ball position quickly drops much below the size of the quantization levels.

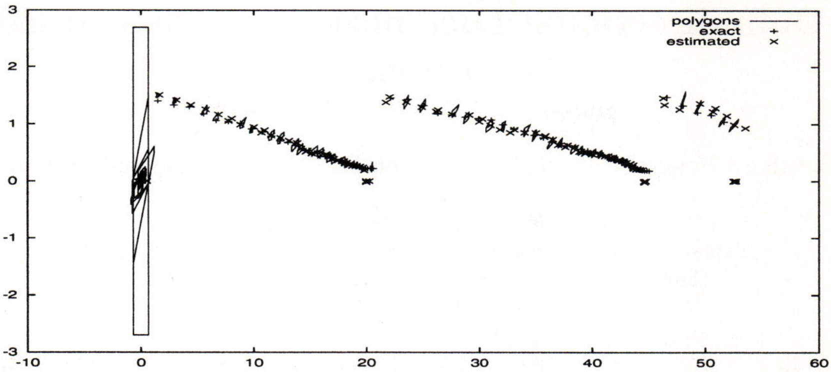

Fig. 3. The x/v_x-plane showing the development of the phase space estimation of the ball position x and velocity v_x.

4 Summary and Future Work

We introduced EWE, a general concept for the incorporation of new evidence into a internal world model and applied it to a scenario from the RoboCup simulation league, using it to estimate the player position and ball movement to a high degree of accuracy. Assuming an accurate model of world dynamics in a one-dimensional setting, the estimation of the ball movement even becomes near-optimal for the soccer server scenario. A number of difficulties with the current implementation, e.g. in presence of unmodeled noise or external influences will be discussed in future. We plan to extend our implementation to allow mixing of more than 2 dimensions and thus to better utilize existing evidence, also in more complex settings.

References

Burkhard, H.-D., Hannebauer, M., and Wendler, J., (1998). Belief-Desire-Intention Deliberation in Artificial Soccer. *The AI Magazine*, 1998(3):87–93.

Corten, E., Heintz, K. D. F., Kostiadis, K., Kummeneje, J., Myritz, H., Noda, I., Riekki, J., Riley, P., Stone, P., and Yeap, T., (1999). *Soccerserver Manual Ver. 5 Rev. 00 beta (for Soccerserver Ver.5.00 and later).* `http://www.dsv.su.se/ johank/RoboCup/manual/`, March 19, 2001

Fox, D., Burgard, W., Dellaert, F., and Thrun, S., (1999). Monte Carlo Localization: Efficient Position Estimation for Mobile Robots. In *Proc. (AAAI-99)*, 343–349.

Lewis, F. L., (1986). *Optimal Estimation - with an introduction to stochastic control theory.* John Wiley & Sons, Inc.

Polani, D., and Uthmann, T., (1999). Between Teaching and Learning: Development of the Team MAINZ ROLLING BRAINS for the Simulation League of RoboCup '99. In *Proc. 3rd RoboCup Workshop, Stockholm.*

Rao, A. S., and Georgeff, M. P., (1995). BDI Agents: from theory to practice. In Lesser, V., editor, *Proc. ICMAS*, 312–319. San Francisco, CA: MIT Press.

Omnidirectional Locomotion for Quadruped Robots

Bernhard Hengst, Darren Ibbotson, Son Bao Pham, and Claude Sammut

School of Computer Science and Engineering
University of New South Wales, UNSW Sydney 2052 AUSTRALIA
{bernhardh,s2207509,sonp,claude}@cse.unsw.edu.au

Abstract. Competing at the RoboCup 2000 Sony legged robot league, the UNSW team won both the challenge competition and all their soccer matches, emerging the outright winners for this league against eleven other international teams. The main advantage that the UNSW team had was speed. A major contributor to the speed was a novel omnidirectional locomotion method developed for the quadruped Sony ERS-110 robot used in the competition. It is believed to be the fastest walk style known for this type of robot. In this paper we describe the parameterised omnidirectional walk in detail. The walk also made a positive contribution to other robot tasks such as ball tracking and localisation while playing soccer. The authors believe that this omnidirectional locomotion could be applied more generally in other legged robots.

1 Introduction

In RoboCup 1999 our own robots, as well as those of our competitors, moved slowly and gingerly, often falling over, even when unprovoked. From our observations in that competition, we concluded that locomotion was a fruitful area for research.

In 1999, the UNSW team, Dalgliesh and Lawther [2], used what they called atomic actions to control the robots movements. The atomic actions were simple macro actions to perform specific effector sub-tasks such as turning, walking forward or walking sideways. These atomic actions needed to be activated sequentially and could not be interrupted once initiated. Consequently robots would often continue to the completion of an atomic action, long after it was appropriate, in the circumstances. The transition between atomic actions was problematic in that robots could lose their balance and fall over because the leg positions were mismatched at the point of transition. While a fall and the struggle to get up are entertaining for the audience, they lose valuable time during a soccer match.

For RoboCup 2000 we decided to address these shortcomings by redesigning the locomotion for the Sony ERS-110 entertainment robot. The rest of this paper will describe the approach and implementation of an omnidirectional walk style that characterised the University of New South Wales entry in the competition. The omnidirectional walk gave the UNSW team a decisive advantage

A. Birk, S. Coradeschi, and S. Tadokoro (Eds.): RoboCup 2001, LNAI 2377, pp. 368–373, 2002.

over the other teams. It is also a good illustration of the advantage a simplifying representation gives in making a difficult task more easily achievable.

2 Gait Selection

Hornby, et al in [4] describe three of the most common gaits used by quadruped robots, the crawl, trot and pace gaits.

The crawl gait operates by moving each leg in turn lifting one at a time. This gait is static in the sense that it keeps the centre of gravity inside the triangle described by the three legs that touch the ground. The trot gait lifts the two diagonally opposite legs alternately. The robot balances along the diagonal joining the two feet that are on the ground. In practice the robot is steadied by either of the other two legs not supporting the robot's weight. That is, the robot falls onto a third leg. The pace gait lifts both legs on the same side of the body simultaneously, alternating between the sides. This is the least stable gait as it is easy for the robot to lose balance and fall over sideways.

We experimented with all three gaits. From the results of early prototype walks, we rejected the pace gait because the robot would not walk forward reliably. It was difficult to synchronise the shifting of weight from side to side with the leg movements and it easily became unstable. The crawl gait was stable but slow in comparison with the trot gait. We therefore decided on the trot gait for the competition.

3 Design Objectives

Our three primary design objectives were to:

1. Drive the robot as if controlled by a joystick with three degrees of freedom. We wanted to be able to move forward or backward, sideways left or right and to turn on the spot clockwise or counterclockwise. This requirement suggested itself because of the shortcomings described above. Ideally we would like to be able to infinitely vary the speed of movement for any of the three degrees of freedom rapidly and simultaneously to allow the robots to react quickly to environmental changes.
2. Move the robot over the ground at a constant speed. This should reduce the strain on the robot motors by not accelerating and decelerating the body. It should also allow a faster walking pace to be achieved after the robot has gained momentum.
3. Keep the camera as steady as possible. The camera is located in the nose of the robot. We observed that, using previous walks, images from the robot's camera showed wildly erratic movements due to the robots head and leg motions. We postulated that a steadier stream of images would assist in object tracking. When approaching a ball, for example, it is desirable not lose sight of it.

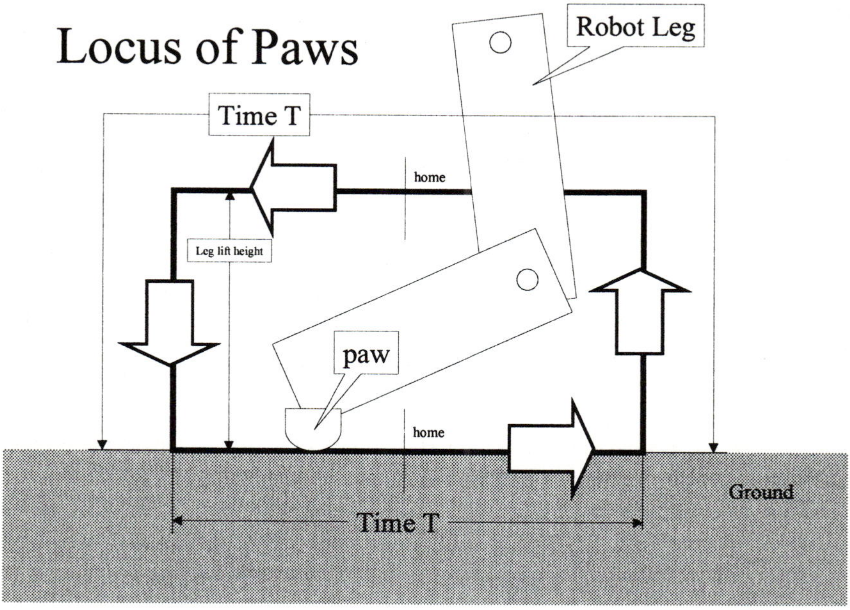

Fig. 1. Rectangular locus of each robot paw

4 The Representation

We chose to describe the leg action of the robots by a rectangular paw motion. The bottom edge of the rectangle describes that part of the path during which the paws make contact with the ground. The sides and top of the locus describe the path used to lift the paw back ready for it to take the next step. The legs that are touching the ground exert the forces that move the robot. In the trot gait diagonally opposite legs touch the ground alternately. If the paws that touch the ground move at a constant velocity, the robot should move at that same constant velocity. This requires that the time taken to move the paw along the bottom edge of the rectangle is equivalent to the total time taken to move the paw along the other three edges as shown in figure 1. Another way to think of the locus action is to see that its effect is similar to a wheel turning.

Design objectives 2 and 3 were achieved in this way. The speed over the ground is constant as long as the size of the rectangular locus does not change and its traversal is at a constant frequency. The robots camera is steadied because the bottom edge of the rectangular locus is a straight line lying in the ground plane. When the robot looses balance in the trot walk there is camera movement until it is arrested by falling on one of the legs that is off the ground. This movement can be minimised by lifting the legs as little as possible during the walk. Unfortunately in practice it was necessary to specify a significant leg lift height to ensure that the spring-loaded claws would clear the carpet. This introduced some unwanted camera movement.

5 Omnidirectional Control

We will now address design objective 1, that is, how to control which way the robot moves. The plane containing the rectangular locus for the paw is always perpendicular to the ground. By changing the inclination of this plane relative to the sides of the robot we can determine whether the robot moves forward/backwards or sideways. For example if the locus plane is parallel to the robot sides, the robot will move forwards or backwards. Whether the robot moves forward or backward is determined by the direction in which the paw moves around the locus. It helps to imagine that the paw moving around the locus acts similarly to a rotating wheel. If we look at the robot side on and it is facing left, then if the paws move in a anti-clockwise direction around the rectangular locus the robot will move forward. If the paws are made to move clockwise the robot will walk backwards. If we incline the locus plane perpendicular to the sides of the robot it will move either left or right in a similar fashion to the way it moves forward/backward. An inclination of the locus plan at another angle will cause the robot to move both forward or backward and left or right calculated by adding the vectors of both components. So far we have assumed that all the locus planes are kept parallel to each other and all the paws move either clockwise or counterclockwise. Before we move on to explain how the robot is made to turn, we should also note that the width of the rectangular locus and the speed at which it is traversed by the paw determines the speed at which the robot moves. There are limitations to these parameters and we will see later how they were tuned to achieve a locally optimal performance.

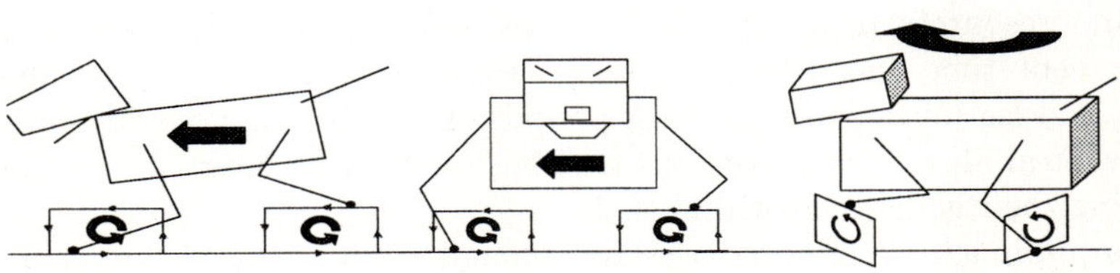

Fig. 2. Locomotion forward, sideways, and turning by moving the paws of the legs in a rectangular locus

How can the robot be made to turn? Using the analogy of each rectangular locus as a wheel we can imagine that if we were able to turn each locus plane around a vertical axis we would have the freedom of movement of a supermarket trolley with each of its wheels motorised. This large degree of freedom may introduce conflicts where one leg can work against another. If we take a plan view of the robot and describe a circle that passes through each of the upper leg joints, then by inclining the locus planes tangentially at each joint we can coordinate them to turn the robot clockwise or anti-clockwise. This constraint gives us the third property of our omnidirectional locomotion, that is, being able to turn on the spot. Figure 2 illustrates the three components of the omnidirectional locomotion.

Again, it is possible to combine components of each of the three movement dimensions (forward/backward, sideways and turning) creating complex waltz like movements in which the robot moves forward, sideways and turns simultaneously.

6 Optimisation

We used 3 control parameters, 1 speed parameter and 8 stance parameters that influence the leg movements for a particular walk style. A machine learning solution to finding the best speed and stance parameter setting suggested itself. We could perform a gradient ascent on the performance measure (for example forward speed) incrementally adjusting the parameters over many test runs. The problem with automating this approach was that it would take considerable time and resources to set up the learning environment. We were on a tight development time-line. Also of concern was the wear and tear on the robot motors and leg joints given the long periods of training that would be required. Hornby, et al [4] report continuous training times of 25 hours per evolutionary run using their genetic algorithm (GA).

The approach we adopted was to manually adjust the parameters after a number of runs of observing and measuring the performance. Unlike gradient ascent or a GA we were able to adjust the parameters using our judgement and knowledge about the robots dynamics. These considerations include:

1. The home position of the legs needed to be adjusted so that the robot will dynamically balance on the two legs that touch the ground.
2. A longer stride required the robot body to be closer to the ground to allow the legs to reach further.
3. The front and back legs do not interfere with each other during the cycle of the gait.
4. The legs needed some sideways spread to allow the robot to move its legs laterally so it could walk sideways.

Besides ground speed we also judged the reliability of the walk and the deviation from the intended path (accuracy) as a part of the performance measure. We found that there was a trade-off between the frequency of the gait and the length of the stride. Maximum ground speed was achieved with a longer stride and a slower gait frequency. This manual approach to tuning parameters used about 10 minutes of robot running time and only about 12 iterations of parameter adjustments to find a good performance setting resulting in the characteristic low forward leaning stance of the UNSW robots.

A further refinement that increased ground speed considerably was the introduction of a canter action. The canter action sinusoidally raises and lowers the robots body by 10mm synchronised with the trot cycle. The parameters were then manually tuned so that the robot was able to reach speeds of 1200cm/min. This compares with 900cm/min achieve in [4] using a GA approach which itself was reported to have improved on the previously fastest hand developed gait of 660cm/min. The camera is not as steady in this type of walk because of the additional canter movement.

7 Conclusions and Future Development

As evidenced from the results of the Sony legged robot league in RoboCup 2000, the parameterised omnidirectional walk performed very well. Often UNSW robots would reach the ball first at kick-off despite the significant distance handicap in the defending position. At no time during any of the competition matches did any of the UNSW robots fall over.

Gradient descent can only discover local optima. It is therefore possible that other combination of parameters may result in better performance. Lack of time prevented a more extensive search. Machine learning using either GA as in [4] or gradient descent could automate this effort given adequate resources. The geometry used to model the robots limbs and the inverse kinematic equations did not take into consideration the interference of the ornamental claws on the Sony ERS-110 entertainment robot. They have the effect of an unwanted lift on the hind legs in particular. Including their geometry may further improve the steadiness of the walk.

Because of the generality of the underlying dynamics of the locomotion, we believe it could be applied to any legged robot that has a similar geometry and capability.

References

1. Beer, R.D., Quinn, R.D., Chiel, H.J. and Ritzmann, R.E. Biologically Inspired Approaches to Robotics, Communications of the ACM, Vol. 40, No. 3, March 1997.
2. Dalgliesh, J. and Lawther, M. (1999). Playing Soccer with Quadruped Robots. Computer Engineering Thesis, University of New South Wales.
3. Hengst, B., Ibbotson, D., Pham, S.B., Dalgliesh, J. Lawther, M., Preston, P., Sammut, C. The UNSW RoboCup 2000 Sony Legged Team, (to be published for RoboCup 2001 by Springer), 2001.
4. Hornby, G. S., Fujita, M., Takamura, S., Yamamoto, T., Hanagata, O. Evolving Robust Gaits with Aibo. IEEE International Conference on Robotics and Automation. pp. 3040-3045. 2000.
5. Stone, P., RoboCup-2000 The Fourth Robotic Soccer World Championships, AI Magazine, Vol. 22, No. 1, pp. 11-38, Spring 2001

An Omnidirectional Vision System That Finds and Tracks Color Edges and Blobs

Felix v. Hundelshausen, Sven Behnke, and Raúl Rojas

Freie Universität Berlin, Institut für Informatik
Takustr. 9, 14195 Berlin,
{hundelsh|behnke|rojas}@inf.fu-berlin.de

Abstract. We describe the omnidirectional local vision system developed for the FU-Fighters, a RoboCup F180 league soccer team. A small video camera mounted vertically on top of the robots looks at a concave parabolic mirror placed above the camera that reflects the field around. The image is sent via a radio link to an external PC for processing. Our computer vision system can find the ball and detect other robots as obstacles. The walls of the field are also recognized and are used to determine the initial position of the robot. In order to be able to process the video stream at full frame rate the movement of all objects is tracked, including the walls of the field. The key idea of our approach is to predict the location of color edges in the next frame and to search for such color transitions along lines that are perpendicular to the edge.

Introduction

We developed a team for the F180 RoboCup league, the FU-Fighters, that has taken part in the competitions held at Stockholm and Melbourne. To make the robots more autonomous, we replaced the global vision system by an omnidirectional local vision system where each robot carries its own camera.

Three tasks have to be accomplished by the computer vision software that analyzes the captured video stream: detecting the ball, localizing the robot, and detecting obstacles. These tasks are non-trivial, since sensor noise and variances, such as inhomogeneous lighting, are present in the images. The image analysis must be done in real time, which is not easy, due to the enormous data rate of video streams. Some teams need to reduce frame rate or resolution to match the available computing power, however, such an approach leads to less precise or less timely estimates of the game status. To be useful for behavior control, the system also needs to be robust. Unexpected situations should not lead to failure, but to graceful degradation of the system's performance.

Local vision is the method used by most teams in the F2000 league as the main sensor. Some of the successful teams adapted the omnidirectional vision approach. The Golem team [4] impressively demonstrated in Melbourne that, using an omnidirectional camera, sufficient information for controlled play can be collected. Another example for the use of omnidirectional cameras is the goalie of the ART team [3,6,7]. In our league, F180, only three teams tried to play in

A. Birk, S. Coradeschi, and S. Tadokoro (Eds.): RoboCup 2001, LNAI 2377, pp. 374–379, 2002.

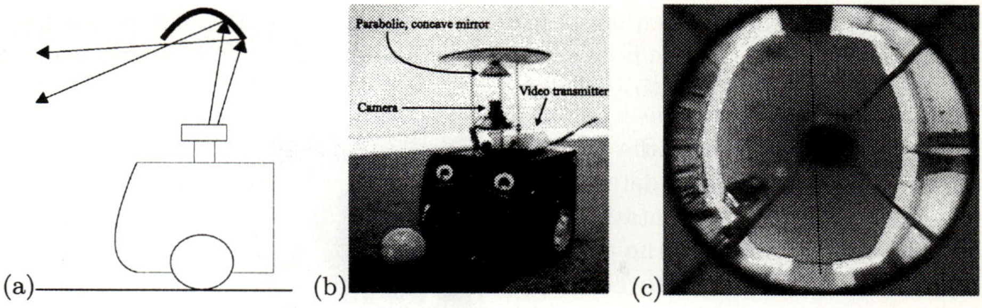

Fig. 1. Omnidirectional camera: (a) principle; (b) physical construction; (c) captured.

Melbourne with local vision. In the smaller form factor the implementation of a local vision system is more challenging than in the F2000 league. Due to space and energy constraints smaller cameras of lower quality must be used and less computing power is available on the robot. Recently, the OMNI team [5] demonstrated controlled play with omnidirectional local vision at the Japan Open competition. This team sends the video stream to an off-the-field computer that contains special purpose hardware for image processing.

We use a similar system design and mount a small video camera and a mirror on top of the robots, as shown in Fig. 1. The camera is directed upwards and looks into a parabolic mirror. The mirror collects light rays from all directions, reflecting them into the camera. The parabolic shape of the mirror produces less distortions, as compared to a spherical mirror. Far-away objects appear larger and are hence easier to detect [1]. The optical mapping preserves the angle of an object to the perception origin. The non-linear, increasing distance function of the camera can be calibrated easily by measuring distances in the image and in the world. To avoid carrying a large computer on the robots, the video stream is transmitted to an external computer via an analog radio link.

The main idea of the paper is to implement a tracking system for the analysis of the video stream produced by an omnidirectional camera that needs to inspect only a small fraction of the incoming data. This allows to run the system at full frame rate and full resolution on a standard PC. For automatic initialization of the tracking, an initial search analyzes the entire video frame.

The paper is organized as follows: The next section describes the initial localization of the robot, the ball and obstacles. The tracking system is described in Section 2. Some experimental results are reported at the end.

1 Initial Search

1.1 Initial Robot Localization

The task of initial robot localization is to determine the robot's position and orientation on the playing field, given a captured omnidirectional image. Two different methods have been developed for localizing the robot. The first is fast, but is only applicable when both goals can be found. The second method is more flexible and robust, but also is slower.

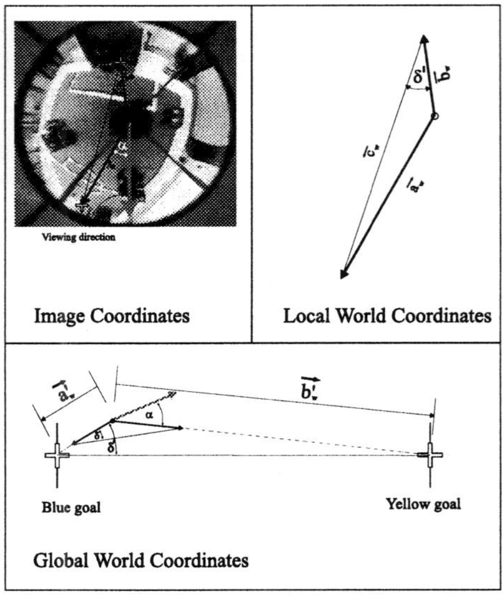

Direct Localization uses the distances to the goals and the angle between the goal vectors to determine the robot's position and orientation. First, the vectors $\boldsymbol{a}$ and $\boldsymbol{b}$ from the perception origin to the goals have to be mapped to local world coordinates. If $\boldsymbol{a_w}$ and $\boldsymbol{b_w}$ denote these mapped vectors, we define $\boldsymbol{c_w} := \boldsymbol{a_w} - \boldsymbol{b_w}$. Now δ' is the angle between $\boldsymbol{c_w}$ and $\boldsymbol{b_w}$. Hence, the position p of the robot lies on a line at an angle of δ' to the line connecting the goals. Knowing the distance to one goal determines p. The orientation is also known, because the angle α at which the other goal appears in the image in respect to the robot's viewing direction is preserved by the optical mapping.

Localization Using Evidence Aggregation consists of two steps. First, plausibility values for several positions are computed using evidence accumulation in a grid. In a second step, these positions are investigated in the order of their plausibility, until the correct position has been found.

If we recognize a goal and can estimate its distance from the robot, a circle segment with radius r_x around goal x is added to the grid. The circles will be drawn more fuzzy for great distances, as the estimation of r_x becomes worse.

Another feature used is the best visible wall that is found by a radial search followed by a Hough Transformation [2]. Lines are sent radially from the robot's perception origin and searched for transitions from the floor to the wall. The

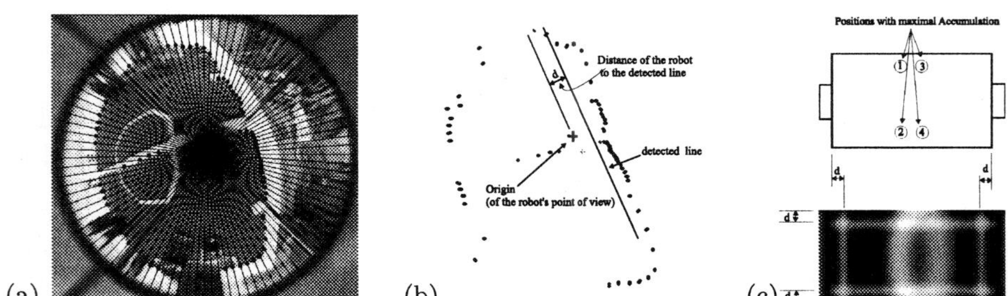

Fig. 2. Using the closest wall for localization: (a) robot next to a wall; (b) detected wall points transformed to world coordinates; (c) grid with goal-circles and wall-lines.

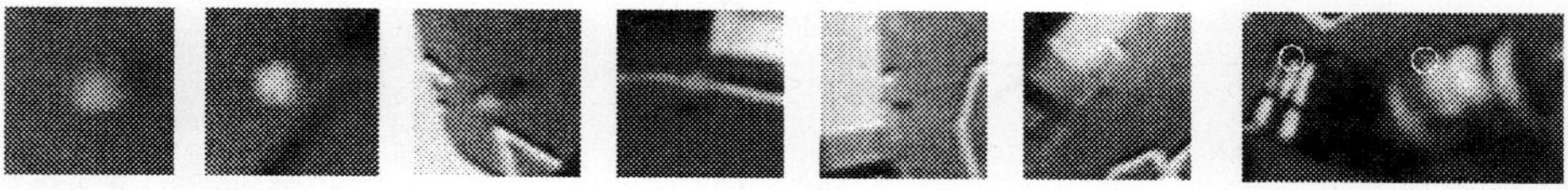

Fig. 3. The ball's appearance varies in color and shape and is similar to robot markers.

transitions found are transformed to local world coordinates, using the cameras' inverse distance function. The corresponding sinusoidal curves are accumulated in parameter space. The most significant local maximum corresponds to the best visible wall. Since it is not known, which wall has been detected, for all walls parallel lines are drawn on the grid at the perceived distance (see Fig. 2).

Local maxima are now candidates for robot locations. They are evaluated by a registration procedure (see next section) that computes a quality measure for a model fit. The best candidate is used to initialize the tracking.

1.2 Initial Localization of Ball and Obstacles

Finding the ball is of essential importance for successful play. Although it seems to be an easy task, detecting the ball clearly demonstrates the difficulties of computer vision. Figure 3 shows some images of the ball. Its appearance varies greatly in size, shape, and color. Furthermore, it can be easily confused with robot markers. The detection of the ball can be summarized in three steps: (a) determine all clusters of the ball's color class and their sizes, (b) discard all clusters for which no transition to the floor can be found, and (c) choose the biggest cluster. Obstacles are detected as large clusters of the dark obstacle color. If clusters are close to each other, they are merged.

2 Tracking Objects

Tracking of color edges and blobs is key to the low computational load of our vision system. The idea is to utilize the fact that the world changes slowly, as compared to the frame rate. This makes it possible to predict the location where certain features will appear in the next frame. If most of these features are found close to their predicted positions, only small parts of the image need to be touched. The differences between the measured locations and the predictions can be used to update estimates of the parameters of a world model.

We use a 2D-model of the field, with the robots and the ball on it, as shown in Fig. 4. The model is matched sequentially to the video frames. For each model line, short orthogonal equidistant tracking lines form a tracking grid (Fig.5(a)).

The ends of each tracking line specify two color classes, according to the expected colors at both sides of the line. Each tracking line is mapped into the image, using the inverse camera function. This is done by mapping the endpoints of each transition line into the image and then reconnecting them with a line. Next, the pixels along the projection line are searched for the appropriate color

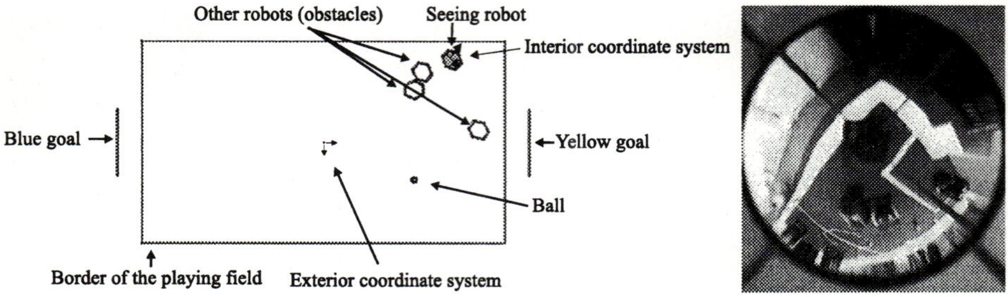

Fig. 4. Model seen from an external point of view.

transition. In Figure 5(b) detected transitions are marked with a black dot. It can be seen that the model does not fit precisely to the playing field in the image, due to a rotation of the robot. Sometimes false transitions may be found, e.g. at field lines. They need to be detected as outliers that must be ignored.

The next step is to calculate a 2-D rigid body transformation that brings the model in correspondence with the found transitions. In the case here, the model should be rotated slightly to the left. To determine the model's transformation, first a rotation and translation is calculated for each track grid independently. Then the results are combined to obtain a transformation for the whole model.

Repeating the above steps while perceiving a sequence of images yields the desired result: the pose of the field seen from the robots point of view is tracked and so the position and orientation of the robot is known by a simple coordinate transformation. Figure 6(a) shows the tracking while the robot rotates.

During initial search candidate positions are evaluated using the tracking mechanism. Given a position of the robot and the angle of a goal, the robot's orientation is calculated. The field can now be projected into the image and the ratio of found transitions can be used as quality measure for model fit. This ratio is also used during tracking to detect situations when the tracking fails and the initial search is needed to localize the robot again.

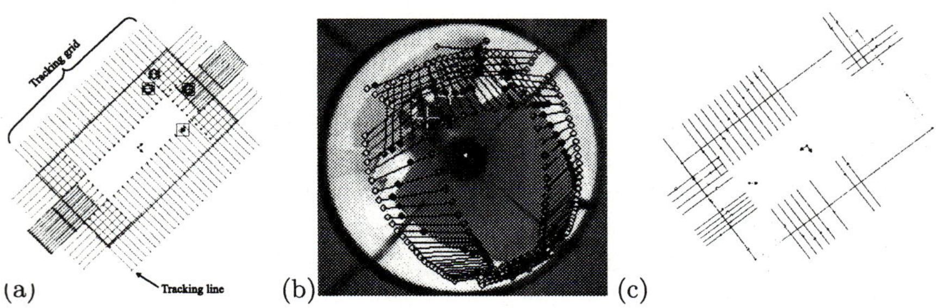

Fig. 5. Tracking of the field: (a) enhanced CAD model; (b) projected into the image with found transitions marked; (c) found transitions in local world coordinates.

(a) (b)
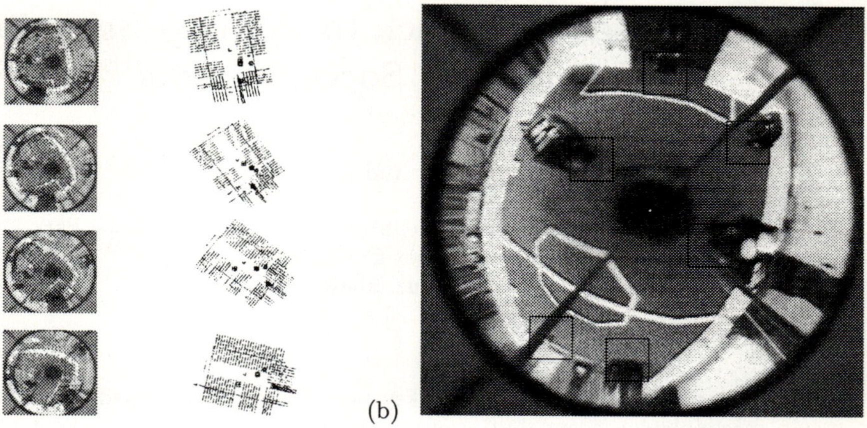

Fig. 6. Tracking of (a) the field while rotating; (b) ball and obstacles.

The system does not only track color edges, but also color blobs, such as the ball or obstacles. The blobs are only searched for in small square windows around their predicted positions, as shown in Fig. 6. If an object cannot be found within its rectangle, initial search is started to localize it again.

Conclusions

We implemented a local vision system for the F180 league that uses an omnidirectional camera. The system fits a world model to the input by finding and tracking color edges and blobs. It is able to process a full resolution, full frame rate video stream on a standard PC. For the production of the Seattle qualification video we used the extracted information about the world as input for behavior control. A single robot was able to drive behind the ball, and to kick it towards the goal, as well as to defend the goal. To control a team of robots, we plan to fuse multiple local views to a single global view. Currently, the image analysis is done on an external PC. As semiconductor technology advances, it will be possible to integrate a small computer on-board the robots.

References

1. Ryad Benosman, Sing Bing Kang (editors), *Panoramic Vision: Sensors, Theory, and Applications*, Springer, 2001.
2. Hough, *Machine Analysis of Bubble Chamber Pictures*, International Conference on High Energy Accelerators and Instrumentation, CERN, 1959.
3. ART, *F2000 team homepage*, http://www.dis.uniroma1.it/~ART/.
4. Golem, *F2000 team homepage*, http://www.golemrobotics.com/gteam/gteam.html.
5. OMNI, *F180 team homepage*, http://robotics.me.es.osaka-u.ac.jp/OMNI.
6. Bonarini, A., (2000) The Body, the Mind or the Eye, first? In M. Veloso, E. Pagello, A. Asada (Eds), Robocup99 - Robot Soccer World Cup III, Springer Verlag, Berlin.
7. Bonarini A., Aliverti, P., Lucioni, M. (2000). An omnidirectional sensor for fast tracking for mobile robots. T. Instrumentation and Measurement. 49(3). 509-512.

A Generalised Approach to Position Selection for Simulated Soccer Agents

Matthew Hunter and Huosheng Hu

Department of Computer Science, University of Essex Wivenhoe Park, Colchester, Essex CO4 3SQ, UK
mchunt@essex.ac.uk, hhu@essex.ac.uk

Abstract. Position selection is a key task that must be carried out by a soccer-playing agent, but is often overlooked in favour of the more active tasks such as ball control. This paper examines the position selection implemented by the Essex Wizards team in the RoboCup Simulator league in recent competitions. The initial approach using task specific behaviours is firstly reviewed. The new approach is then addressed based on modular decomposition for flexibility. Implementation results are given to show the applicability.

1 Introduction

In multi-agent systems, position selection plays a key role for achieving collaborative actions and undertaking complex tasks. The implementation of such systems is frequently focused on a specific domain. In other words, the selection of a good location is domain dependent in a multi-agent system. In the domain of puck gathering by a team of agents [6], for example, position selection involved picking a search area, and sticking to it. If one of the agents fails, a new working area is selected. Within a chosen area the agent is free to move in task specific ways, another layer of positioning. At the other end of the scale position selection may choose very specific locations that are only valid for short periods of time. For example, flocking, following and other team formations require frequent small changes in position [1][2][6].

Position selection is a general problem and therefore any approach should attempt to maintain that generality, while allowing modification for a particular domain. In order to achieve this, it is assumed that the action of position selection is independent of the action of moving. This abstracts away the details of the underlying agent movement and it gives the agent more control over how the position is interpreted. Position selection can present a new target as soon as the environment dictates, there is no need to explicity abort the previous movement.

Currently in the Robotic Soccer domain [5] position selection is done by a player when it has nothing better to do. Considerable effort is put into ball control, but good positioning can maximise chances for getting close to the ball. For example in [8] some mention is made of position selection for specific situations and solutions such as tracking and marking. This could result in the

A. Birk, S. Coradeschi, and S. Tadokoro (Eds.): RoboCup 2001, LNAI 2377, pp. 380–386, 2002.

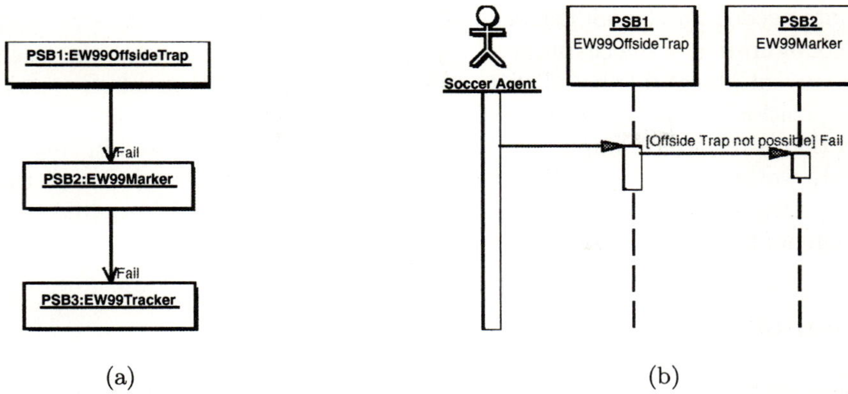

Fig. 1. The failsafe mechanism allows several position selection behaviours to be chained together to obtain more complex behaviour.

position selection being tightly coupled to the rest of the agent making it more difficult to experiment with.

This paper examines the position selection implementation used by the Essex Wizards team in the RoboCup Simulator league in recent competitions. Section 2 reviews the initial approach using task specific behaviours. Section 3 examine the current approach and show the team in action. Conclusions are then drawn and future work discussed.

2 Task-Specific Decomposition

The version used in the RoboCup'99 competition in Stockholm [3] introduced the concept of a Position Selection Behaviour (PSB), i.e. an entity responsible for establishing a target position. The term *behaviour* is used here to reflect the fact that externally each PSB is a sealed unit that simply provides a target position to the user. The soccer agent works at a more abstract level, for example by *defending* rather than specifically *marking*. Seven high-level behaviour classes (BC) were actually implemented for this version, each targeted towards a specific task[3]. A PSB is an instance of a behaviour class. The most important of the BC were the tracker group (Tracker, Ball Tracker and Marker) used defensively, and the support behaviour, based on SPAR [8]. Additional behaviours were used for Offside Traps and moving to fixed locations on the pitch.

All behaviours have an identical external interface so that they can be transparently exchanged for the soccer agent. This in turn lead to the failsafe mechanism which gave the behaviours some control. A chain of behaviours is constructed so that a result can be returned from one of the behaviours in the chain in case some of them fail. In the example shown in Fig. 1(a) a chain of three behaviours is constructed for the defenders, as soon as one PSB finds a result it is returned and the other behaviours are never activated as in Fig. 1(b). Similar chains are needed for the midfielders and forwards. The soccer agent carries out initial configuration of the chains, and in play selects the position using the chain for the current player role based on team formation and player number.

The position selection mechanism used in the Essex Wizards'99 team demonstrated that a relatively simple rule-based system could perform well in robotic soccer. However the task-specific nature of the behaviours themselves and the limited control made available through the failsafe mechanism still required that soccer agents take active part in position selection.

3 Modular Decomposition for Flexibility

To generalise the behaviours and improve the level of control provided, the new version of position selection has been developed to provide more behaviours, more flexibility and reduced commonality between behaviours.

3.1 Behaviour Generalisation

Common components were identified to form the basis of the first two groups of new BC, *Basic* and *Calculated.* In addition the failsafe mechanism was replaced by the *Control* BC group. The relationships are shown in Fig. 2.

The *Basic* BC group consists of, *Ball*, *Fixed*, *Goal*, *Home* and *MyPos*, which represent significant objects in the environment.

The *Calculated* BC group, is responsible for processing the results of one or more behaviours to produce a target location. *Combine*, mixes two positions. *Constrain* limits the result of a behaviour to a given area. *Interpolate* calculates a point that lies a given distance along the line between two positions. *Offset* relocates a position by a given offset. *Protected* restricts a position to be the home area of the player. The *RightAngle* BC was suggested as an advanced marking method [9], the object is to select a target position (T) such that the angle OTB is 90 ° where O is the opponent being marked and B is the ball. *SelectOpponent* uses a series of rules to find the position of an opponent relevant to positioning. This involves finding nearby opponents that are not near teammates. The final behaviour class in the calculated group is *Support*, which attempts to find a good place to receive a pass from the team member with the ball.

The *Control* BC group replaces the previous failsafe mechanism, which was only appropriate where a clear fail condition existed, e.g *SelectOpponent* cannot succeed if no opponents are nearby. On some occasions a behaviour may find a valid result, but an alternative would be better. For this reason control behaviours were introduced. Control behaviours select between one or more positions but do not change the selected position. *Congestion* tests the relative numbers of opponents and team members in a region. *Near* tests two positions to see if they are within a given distance of each other. *OurBall* tests which team has possession of the ball. *TestPosition* tests whether a position is within a particular area of the pitch. The most complex control behaviour class is *Role*, which selects between seven possible roles based on the player's home position.

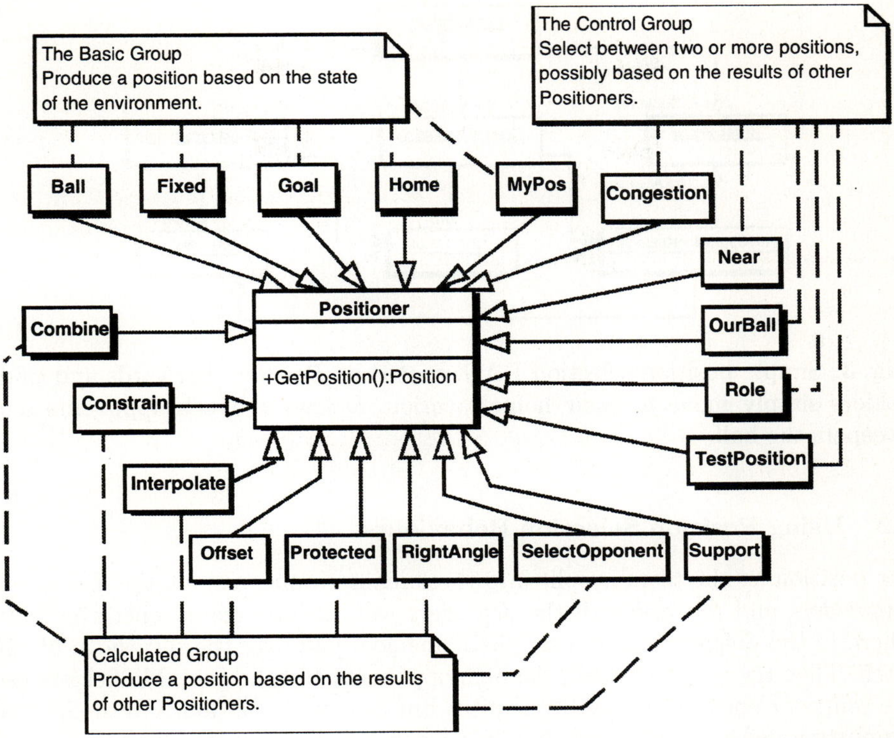

Fig. 2. There are three groups of Behaviour Class, Basic, Calculated and Control.

3.2 Behaviour DAGs

In order to carry out useful tasks it is necessary to combine the individual behaviours described above to form a Behaviour DAG. Figure 3 shows an object diagram of a simple behaviour DAG, rooted at the Position behaviour. Initially the Behaviour Registry contains a prototypical instance of each BC. When a behaviour is requested by name, it will be returned if it is present, otherwise it is generated recursively in terms of other behaviours from the configuration rules read in from a file. As each behaviour is generated, it is added to the registry for future use.

If a behaviour is created from another it inherits the behaviour class and current configuration. If a behaviour is configured to use another behaviour class then it will call on the sub behaviour as needed, for example an Combine behaviour needs two sub behaviours to mix. In this way the Behaviour Graph is built up,possibly consisting of many sub-behaviours.

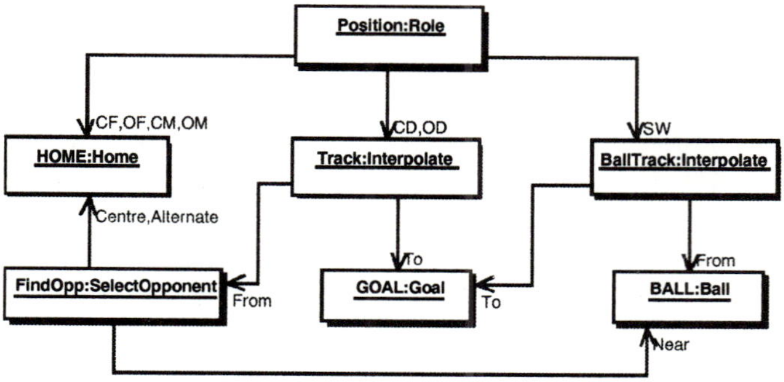

Fig. 3. Simple position selection DAG, rooted at *Position*. Forwards and midfielders simply move to their home location. Defenders track opponents and sweepers the ball.

3.3 Using Position Selection Behaviours

For positioning the team is split into three main groups, namely the *defenders*, *midfielders* and *forwards*. All the *defenders* work as one group, operating anywhere in the defensive zone, near the home goal and over the full width of the pitch. They try to get between the ball, opponents and the goal. Midfielders use the centre or one of the sides of the pitch but can move over much of the length, supporting defenders of forwards as needed. Forwards look for space away from each other and opponents, to create passing and shooting opportunities.

When the fixed plan mechanism was added [4] it became obvious that moving to locations was part of many plans. By having Fixed plans use Positioning they can take advantage of the positioning behaviours to adjust to an opponent.

4 Competition Results

The performance of our team has shown considerable improvement since the changes described, particularly defensively. An example is shown in Fig. 4 where the attacking team is preparing to take a kick in. This situation was initially intended to be a fixed plan, but players were more effective using the standard positioning and so the fixed plan was disabled.

A particular difficulty occurs with the attacking team's players 9 and 10 due to their closeness. Previously this lead to problems when the two attackers separated, only one of them would be marked. As it is even if the attacking players do get the ball none of them will have a good shot on goal.

5 Conclusions

This paper has considered the key role of position selection. A generalised behaviour-based approach to tackling the problem has been described along

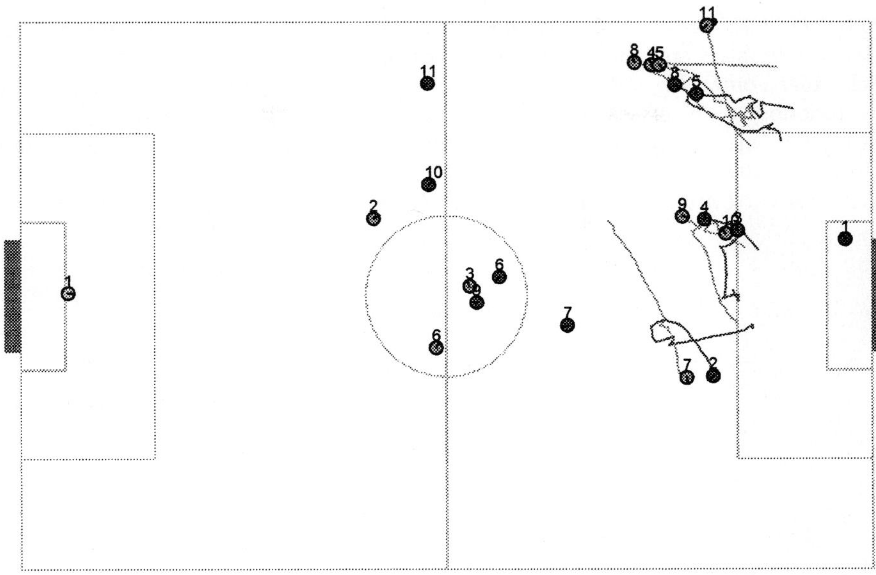

Fig. 4. Essex Wizards 2000 defensive marking in action.

with the practical implementation of these ideas into a working system. This system uses a small number of interchangeable behaviours that are combined to perform rule based position selection in real time. It is anticipated that the mechanism can be directly translated to other leagues with minimal changes.

Acknowledgements We would like to thank Kostas Kostiadis and Nikos Kalyviotis who have made contributions to this project. This project has been supported by RPF grant DDP940.

References

1. Desai J., Kumar V., Ostrowski J.: *Control of changes in formation for a team of mobile robots.* Proc. of 1999 IEEE International Conference on Robotics and Automation, 1556-1561, May 1999
2. Hu H., Kelly I., Keating D. and Vinagre D.: *Coordination of multiple robots via communication.* Proc. of SPIE'98 - Mobile Robots XIII, Boston, 1998, 94-103.
3. Hunter M., Kostiadis K., Hu H.: *A Behaviour-based Approach to Position Selection for Simulated Soccer Agents.* Proceedings 1st European RoboCup 2000.
4. Kalyviotis N. and Hu H.: *A Cooperative Framework for Strategic Planning.* Proc. of the 3rd British Conference on Autonomous Mobile Robotics and Autonomous Systems, Manchester, England, 5th April 2001.
5. Kitano H., Tambe M., Stone P., Veloso M., Coradeschi S., Osawa E., Matsubara H., Noda I., and Asada M.: *The RoboCup Synthetic Agent Challenge.* Proc. of International Joint Conference on Artificial Intelligence (IJCAI97), 1997.

6. Mataric M.J.: *Issues and Approaches in the Design of Collective Autonomous Agents.* Robotics and Autonomous Systems 16:(2-4) 321-331 DEC 1995
7. Schneider-Fontán M. and Mataric M.: *Territorial Multi-Robot Task Division.* IEEE Transactions on Robotics and Automation, 14:(5) 815-822 OCT 1998
8. Stone P.: *Layered Learning in Multi-Agent Systems PhD Thesis.* Carnegie Mellon University 1998
9. Williams D.: *University of Essex Football Club Coach, personal correspondence.*

On the Motion Control of a Nonholonomic Soccer Playing Robot

Giovanni Indiveri

GMD-AiS, German National Research Center on Information Technologies,
Institute for Autonomous Intelligent Systems,
Schloß Birlinghoven, 53754 Sankt Augustin, Germany
giovanni.indiveri@gmd.de
http://ais.gmd.de/~giovanni/

Abstract. A nonlinear control law to steer the unicycle model to a static or dynamic target pose is presented. If the target is static the control signals are smooth in their arguments and the solution guarantees exponential convergence of the distance and orientation errors to zero. The major advantages of the proposed approach are that there is no need for path planning and, in principle, there is no need for global self-localization either.

1 Introduction

The most simple kinematic model of an underactuated wheeled robot is given by the so called unicycle model, namely:

$$\left.\begin{aligned} \dot{x} &= u\cos\phi \\ \dot{y} &= u\sin\phi \\ \omega &= \dot{\phi} \end{aligned}\right\} \tag{1}$$

where x, y are the cartesian coordinates of the center of mass of the robot, ϕ is its orientation with respect to the x axis and u, ω are its linear and angular velocities. The major difficulties in designing motion control laws for underactuated mobile robots are captured by the above (1) simple kinematic model as it is subject to Brocketts Theorem [1]. For a detailed description of nonholonomic car-like system motion control issues refer to [2] [3] [4].

This paper is organized as follows: in section (2) the addressed problem is formally stated, in section (3) a solution for the most simple case (still ball) is described, in section (4) the more general solution is outlined. Section (5) contains some simulation examples and conclusions are drawn in section (6).

2 Problem Statement

Given the unicycle model (1) design a feedback law for the linear and angular velocities u and ω such that asymptotically the $\mathbf{q} = (x, y)^T$ position of the

A. Birk, S. Coradeschi, and S. Tadokoro (Eds.): RoboCup 2001, LNAI 2377, pp. 387–392, 2002.

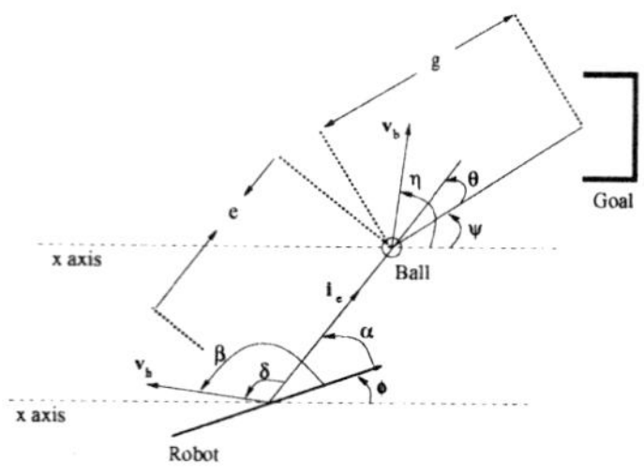

Fig. 1. The model

robot is driven to a given target one $\mathbf{q}_b = (x_b, y_b)^T$ (the ball) along a possibly time-varying reference direction $\psi(t)$. The control law will be designed adopting a polar-like representation of the unicycle kinematic model (1), namely with reference to figure (1) the following variables are introduced: $\mathbf{q} = (x, y)^T$: absolute cartesian robot position; $\mathbf{q}_b = (x_b, y_b)^T$: absolute cartesian ball position; $\mathbf{V}_b$: ball velocity vector; $V_b = \|\mathbf{V}_b\|$: norm of the ball velocity vector (assumed constant); η : absolute direction (assumed constant) of $\mathbf{V}_b$; $e = \|\mathbf{q}_b - \mathbf{q}\|$: robot - ball distance; $\mathbf{i}_e = \frac{\mathbf{q}_b - \mathbf{q}}{\|\mathbf{q}_b - \mathbf{q}\|}$: unit vector pointing from the robot to the ball; $\mathbf{e} = \mathbf{i}_e\, e$: robot - ball "distance" vector; $\mathbf{j}_e = \mathbf{k} \wedge \mathbf{i}_e$: unit vector normal to $\mathbf{i}_e$ and to the z axis unit vector $\mathbf{k}$; g : ball - target (by example, the center of the goal) distance; $\mathbf{i}_u$: unit vector parallel to the robot heading; u : robot linear velocity; $\mathbf{u} = \mathbf{i}_u\, u$: robot linear velocity vector; $\omega = \dot{\phi}$: robot angular velocity; $\psi(t)$: desired asymptotic robot heading. Time varying if $V_b \neq 0$; $\alpha = \widehat{\mathbf{i}_u, \mathbf{i}_e}$: angle between $\mathbf{i}_u$ and $\mathbf{i}_e$; other relevant angular variables defined similarly to α and depicted in figure (1) are ϕ, θ, β and δ. Notice that α and θ are not defined when $e = 0$ and ψ is not defined when $g = 0$. The relevant dynamic equations for the above defined variables may be deduced starting from the cartesian model (1) yielding:

$$\dot{e} = -u\cos\alpha + V_b\cos(\eta - \theta - \psi) \tag{2}$$

$$\dot{\phi} = \omega \tag{3}$$

$$\dot{\alpha} = -\omega + \frac{u}{e}\sin\alpha + \frac{V_b}{e}\sin(\eta - \alpha - \phi) \tag{4}$$

$$\dot{\theta} = \frac{u}{e}\sin\alpha + \frac{V_b}{e}\sin(\eta - \alpha - \phi) + \frac{V_b}{e}\sin(\eta - \psi) \tag{5}$$

$$\dot{\psi} = -\frac{V_b}{g}\sin(\eta - \psi) \tag{6}$$

$$\dot{g} = -V_b\cos(\eta - \psi) \tag{7}$$

$$\beta = \delta + \alpha \quad ; \quad \beta \equiv \widehat{\mathbf{i}_u, \mathbf{v}_h} \tag{8}$$

$$\phi + \alpha = \theta + \psi \tag{9}$$

$$\dot{\eta} = \dot{V}_b = 0 \text{ (assumption)} \tag{10}$$

The above allow to give a more formal statement of the problem, namely find u and ω such that $\lim_{t\to\infty}(e,\alpha,\theta)^T = (0,0,0)^T$. It should be noticed that given the equations (2-6) the above problem statement is well posed if and only if $V_b = 0$. It will be then shown that in the case $V_b \neq 0$ the same design method adopted for the case $V_b = 0$ allows to compute bounded control laws that drive the vehicle in a neighborhood of $(e,\alpha,\theta)^T = (0,0,0)^T$ although not exactly in $(0,0,0)^T$. The singularities occurring when $e = 0$ may be avoided by a proper choice of the control inputs as will be discussed in the following sections.

3 Still Ball Case

The controller designed for this case is based on the one reported in [5]. The idea is quite simple and originates from the following observation: if one wanted to drive a fully actuated, i.e. not subject to any nonholonomic constraint, ideal point to a fixed ($V_b = 0$) target it would be sufficient to impose to this point the velocity:

$$\mathbf{v}_h = \gamma_e \, e \, \mathbf{i}_e + \gamma_\theta \, \theta \, e \, \mathbf{j}_e \quad : \quad \gamma_e, \gamma_\theta > 0 \tag{11}$$

which is given by the superposition of a linear velocity $\gamma_e \, e \, \mathbf{i}_e$ that would drive e exponentially to zero and an angular one $\dot{\theta}\,\mathbf{k} = -\gamma_\theta \, \theta \, \mathbf{k}$ that would drive θ exponentially to zero. As outlined in [5], to cope with the unicycle nonholonomic constraint and guarantee asymptotic convergence of e, α and θ to zero the linear velocity may be chosen to be

$$u = \mathbf{i}_u^T \, \mathbf{v}_h = \sqrt{\gamma_e^2 + \gamma_\theta^2 \, \theta^2} \;\; e \, \cos\beta. \tag{12}$$

and the angular one such that $\dot{\beta} = -\gamma_\beta \, \beta : \beta = \widehat{\mathbf{i}_u, \mathbf{v}_h}$, yielding:

$$\omega = \gamma_\beta \, \beta + \left(\frac{\gamma_e \, \gamma_\theta}{\gamma_e^2 + \gamma_\theta^2 \, \theta^2} + 1 \right) \sqrt{\gamma_e^2 + \gamma_\theta^2 \, \theta^2} \;\; \cos\beta \, \sin\alpha \tag{13}$$

The $\cos\beta$ term in (12) and (13) is responsible for possible backward driving, but may be replaced by any smooth $f(\beta) \geq 0 \; : \; f(0) = 0$ without affecting the convergence properties [5]. Interestingly this design may be extended to $3D$ [6].

4 General Case

In order to extend the above design to the general case $V_b \neq 0$ first of all the field $\mathbf{v}_h$ must be re-designed: in particular one may chose $\mathbf{v}_h$ to be the superposition of the previously given $\mathbf{v}_h$ (11) plus a velocity $\mathbf{v}_r$ that, if implemented alone, would allow the robot to view e and θ as constants. With this idea in mind it follows that

$$\mathbf{v}_h = \gamma_e \, e \mathbf{i}_e + \gamma_\theta \, \theta \, e \, \mathbf{j}_e + \mathbf{v}_r \tag{14}$$

$$\mathbf{v}_r = \mathbf{V}_b + \mathbf{j}_e \, \frac{V_b \, e}{g} \sin(\eta - \psi) = \tag{15}$$

$$= \mathbf{i}_e \, V_b \, \cos(\eta - \theta - \psi) + \mathbf{j}_e \left(V_b \, \sin(\eta - \theta - \psi) + \frac{V_b \, e}{g} \sin(\eta - \psi) \right) \tag{16}$$

By direct calculation it follows that:

$$v_{hx} \equiv \mathbf{v}_h^T \, \mathbf{i}_e \quad ; \quad v_{hy} \equiv \mathbf{v}_h^T \, \mathbf{j}_e \quad \Rightarrow \quad \dot{\delta} = \frac{v_{hx} \, \dot{v}_{hy} - v_{hy} \, \dot{v}_{hx}}{\|\mathbf{v}_h\|^2} \tag{17}$$

$$\dot{v}_{hx} = \gamma_e \, \dot{e} + V_b \, \sin(\eta - \theta - \psi)(\dot{\theta} + \dot{\psi}) \tag{18}$$

$$\begin{aligned} \dot{v}_{hy} &= [\gamma_\theta \, e - V_b \, \cos(\eta - \theta - \psi)] \, \dot{\theta} + \left(\gamma_\theta \, \theta + \frac{V_b}{g} \sin(\eta - \psi)\right) \dot{e} \\ &\quad - V_b \left(\frac{e}{g} \cos(\eta - \psi) + \cos(\eta - \theta - \psi)\right) \dot{\psi} - \frac{V_b \, e}{g^2} \sin(\eta - \psi) \, \dot{g} \end{aligned} \tag{19}$$

where the terms $\dot{e}, \dot{\theta}, \dot{\psi}$ and $\dot{g}$ are given by equations (2), (5), (6) and (7). Notice that $\dot{\delta}$ given above will not depend explicitly from ω. Wanting to impose once again the closed loop dynamics $\dot{\beta} = -\gamma_\beta \, \beta$ on β equation (8) is differentiated with respect to time yielding:

$$\dot{\beta} = \dot{\delta} + \dot{\alpha} = \dot{\delta} - \omega + \frac{u}{e} \sin\alpha + \frac{V_b}{e} \sin(\eta - \alpha - \phi) \quad \Rightarrow \tag{20}$$

$$\omega = \gamma_\beta \, \beta + \dot{\delta} + \frac{u}{e} \sin\alpha + \frac{V_b}{e} \sin(\eta - \alpha - \phi) \tag{21}$$

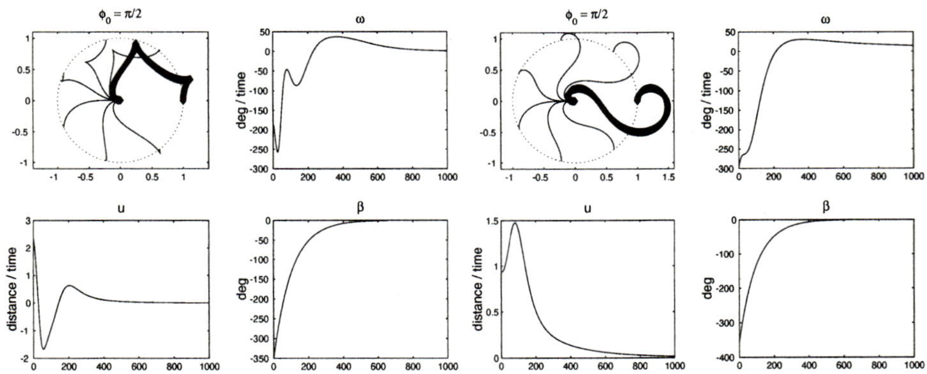

Fig. 2. Static ball ($V_b = 0$) case: paths resulting from different initial positions on a unit circle and initial orientation $\phi_0 = \pi/2$. Gains $\gamma_\theta = 0.3, \gamma_e = 0.1, \gamma_\beta = 1$.

where $\dot{\delta}$ is given by equations (17-19). As expected and anticipated in section (2), the steering input is singular in $e = 0$: indeed while the singularity could be "compensated" in the case $V_b = 0$ by a proper choice of u, i.e. $u \sim e$ in a neighborhood of $e = 0$, when $V_b \neq 0$ there is no way of compensating the terms proportional to V_b/e with a bounded control action. Notice that such terms are also present in the closed loop expression for $\dot{\delta}$. Nevertheless the steering law given by equation (21) may be still adopted in practice and indeed it is

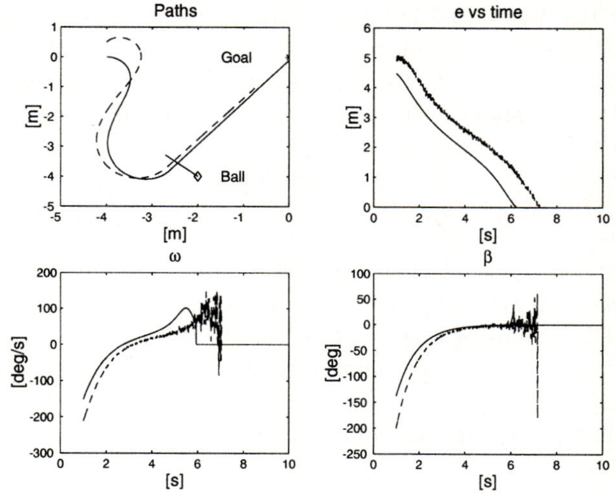

Fig. 3. $V_b \neq 0$ case. Refer to the text for details.

successfully implemented on the GMD RoboCup robots. A possible and simple implementation is to keep the linear velocity u always constant and to clip ω to zero once $e \leq \varepsilon$ for some threshold value $\varepsilon > 0$. As long as $e > \varepsilon > 0$ equation (21) can be implemented without problems and it will steer the vehicle parallel to $\mathbf{v}_h$ given by equations (14) and (16); once in a ε-neighborhood of $e = 0$ the angular velocity is kept null and the linear velocity u is kept constant. The net effect will be to hit the ball along a direction "close" to ψ. The deviation in the kicking direction with respect to the desired value ψ will be a function of the relative speed of the vehicle and the ball, of the angle η and of ε.

5 Simulation Examples

The control laws (12) and (13) relative to the static ball case ($V_b = 0$) have been simulated for several starting poses on a unit circle. Results are shown in figure (2): the first four plots (first two columns from the left) refer to the implementation of the control laws (12) and (13). As expected the linear velocity may be negative (backward drive) resulting in cusps, i.e. points of the plane where $\omega \neq 0$ and $u = 0$. The signals u, ω and β plotted refer to the path drawn with a thicker line. To avoid driving backwards and thus the cusps, the $\cos\beta$ term in (12) and (13) has been replaced with $f = 1$ giving rise to the results shown in the other two columns. Two different simulations of the general case ($V_b \neq 0$) are reported in figure (3). In both cases the robots linear velocity u was kept constant and equal to $u = 1\,m/s$. The ball had velocity $V_b = 0.1\,m/s$ and direction $\eta = 3\pi/4\,rad = 135\,deg$ starting from the initial x, y position $-2\,m, -4\,m$. The goal was positioned in the origin $(0, 0)$ in both cases. The steering law given by equation (21) has been implemented in the region $e > \varepsilon$ with $\varepsilon = 0.3\,m$. As soon as $e \leq \varepsilon$ the robots angular velocity ω was clipped to zero. The control gains where $\gamma_\beta = 1, \gamma_\theta = 0.8, \gamma_e = 1/2$ in both cases whereas the initial robots pose was $(-4\,m, 0, 0)$ for the solid line case and $(-4\,m, 1/2\,m, \pi/3\,rad)$ for the dashed line one. The solid line results refer to the ideal noise free case, while the dashed line results refer to the case in which additive gaussian noise of standard deviation $\sigma_{V_b} = 0.0333\,m/s$ and $\sigma_b = 0.05\,m$ where added respectively on the robots V_b estimate and ball position. This last noise affects in particular g, ψ, α, θ and their derivatives. For the sake of clarity, wanting to compare the two simulations the path driven by the ball is plotted as if there was no impact. The fact that indeed the impact takes place is clearly revealed by the plot of e versus time. The variables e and β are computed and plotted versus time only up to the impact and then set to zero. The high frequency action of ω in the vicinity of the target in the dashed line simulation is to be related to the terms divided by e in equation (21).

6 Conclusions

A nonlinear control law to steer the unicycle model to a desired pose has been presented. If the target is static the control signals are smooth in their arguments and the solution guarantees exponential convergence of the distance and orientation errors to zero. In the case of a moving target a solution has been presented that guarantees exponential convergence of the β angle to a neighborhood of $\beta = 0$.

References

1. Brockett, Millmann, and Sussmann, Eds., *Differential Geometric Control Theory*, chapter Asymptotic Stability and Feedback Stabilization, by Brockett, R. W., pp. 181–191, Birkhauser, Boston, USA, 1983.
2. C. Canudas de Wit, H. Khennouf, C. Samson, and O. J. Sørdalen, "Nonlinear Control Design for Mobile Robots", in "Recent Trends in Mobile Robotics", Yuan F. Zheng editor, vol. 11, World Scientific Series in Robotics and Automated Systems, pp. 121–156, 1993.
3. J.-P. Laumond, Ed., *Robot Motion Planning and Control*, chapter Feedback control of a nonholonomic car-like robot, by A. DeLuca, G. Oriolo, C. Samson, Springer-Verlag, ISBN 3-540-76219-1, available from http://www.laas.fr/~jpl/book.html, 1998.
4. I. Kolmanovsky and N. H. McClamroch, "Developments in nonholonomic control problems," *IEEE Control Systems*, pp. 20–36, December 1995.
5. M. Aicardi, G. Cannata, G. Casalino, and G. Indiveri, "On the stabilization of the unicycle model projecting a holonomic solution," in *8th Int. Symposium on Robotics with Applications, ISORA 2000*, Maui, Hawaii, USA, June 11-16 2000.
6. M. Aicardi, G. Cannata, G. Casalino, and G. Indiveri, "Guidance of 3D underwater non-holonomic vehicle via projection on holonomic solutions," in *Symposium on Underwater Robotic Technology SURT 2000, World Automation Congress WAC 2000*, Maui, Hawaii, USA, June 11-16 2000.

Evaluation of the Performance of CS Freiburg 1999 and CS Freiburg 2000*

Guido Isekenmeier, Bernhard Nebel, and Thilo Weigel

Albert-Ludwigs-Universität Freiburg, Institut für Informatik
Georges-Köhler-Allee, Geb. 52
D-79110 Freiburg, Germany

Abstract. One of the questions one may ask when following research in robotic soccer is whether there is a measurable progress over the years in the robotic leagues. While everybody who has followed the games from 1997 to 2000 would agree that the robotic soccer players in the F2000 league have improved their playing skills, there is no hard evidence to justify this opinion. We tried to identify a number of criteria that measure the ability to play robotic soccer and analyzed all the games CS Freiburg played at RoboCup 1999 and 2000. As it turns out, for almost all criteria, there is a statistically significant increase for CS Freiburg *and* the opponent teams demonstrating that the level of play has indeed increased from 1999 to 2000.

1 Introduction

In robot soccer as in human soccer, the abilities to be developed or improved do not correlate with the overall criterion of success, i.e. goals, in a direct manner. With respect to a representative analysis of the relative performance of two teams playing against each other, the distribution of goals in a single game is subject to too many contingencies and therefore no adequate basis for an evaluation of *team strength*. Even if we consider many games, it is very often impossible to arrive at statistically significant results. For instance, when comparing the average goal rate of RoboCup 1997 with the average goal rate of RoboCup 2000, one notices that this rate has increased from 0.05 goals/minute to 0.1 goals/minute. However, the increase is not significant on the 95% level. Furthermore, for the *CS Freiburg* team the goal rate is approximately the same for 1998–2000, namely, around 0.2 goals/minute [4], and there is no statistically significant difference.

More differentiated data is required in order to assess the progress made by a team on the basis of a statistical analysis. In the case of RoboCup, the task of obtaining relevant data is simplified by a relatively restricted repertoire of possible actions. Asada et al. [1] distinguished three major areas of the RoboCup *physical agent challenge*: *ball moving*, *ball catching* and *cooperative behavior*. Of these, cooperative behavior is currently almost negligible due to the relatively small size of the field and the lack of hardware suitable for receiving the ball. For this reason, we see only very few passes during the games. Similarly, the robots cannot catch the ball. The main focus of this

* This work has been partially supported by *Deutsche Forschungsgemeinschaft* as part of DFG project Ne 623/3-1.

A. Birk, S. Coradeschi, and S. Tadokoro (Eds.): RoboCup 2001, LNAI 2377, pp. 393–398, 2002.

paper will therefore be on the area of *ball moving*, supplemented by *ball possession* and *shooting*.

In order to assess the improvement over the years, we tried to identify a number of different criteria in this skill area and evaluated our team and the opponents for RoboCup 1999 and 2000 according to these criteria. As it turns out, CS Freiburg and our opponent teams showed in most cases a statistically significant improvement, justifying the impression one gets when watching the two sets of games: There is a measurable increase of skills from 1999 to 2000.

2 Approach

The material available for survey consists of video and Web-Player recordings of the matches of CS Freiburg. The restrictions imposed by the manual evaluation of visual recordings make an approach similar to that of Tanaka-Ishii et al. [3] impossible. Based on the "giant set of log data" of the simulator league, they considered "low-level events" to compute values such as *compactness* or *x/y-correlation*, which, in addition, seem less meaningful with respect to the F2000 league's limited complexity (number of players, size of field, etc.). When based on video recordings, human judgment of robot behavior depends on the ascription of intentionality to their actions regardless of their internal state. The lack of cooperative behavior as well as the unclear application of the charging rule suggest concentration on the ball as a focus of such ascriptions.

The following survey thus relies on a distinction of different kinds of ball possession as a global measure of behavior. Ball possession is particularly relevant for the physical agent challenge in contrast to the simulator league because physical components (movement) and vision and sensor fusion (localization) are an integral part of this challenge [2]. The different categories of ball possession are constituted by relating them to the existing repertoire of actions of the CS Freiburg team [4] and assuming that the best possible action which may conform to the observed behavior is executed. In the following, we will distinguish between different kinds of (non-) ball possession:

- **ball free**: no player is at the ball.
- **both at ball**: at least one player of each team is at the ball. The ball is stuck between the players or moves little. It is usually followed by a *ball free* situation.
- **ball possession**: only one team is in possession of the ball, which may be differentiated according to the following criteria:
  - **ball stuck**: at least one player of one team is at the ball. The ball is stuck between two players of the same team or between a player and the wall. Since this is obviously an undesirable state, one can assume that the player(s) is/are prevented from executing an action by the interference of another player or the wall.
  - **active**: a player of one team is at the ball and is executing one of the following actions in a goal-directed manner: *TurnBall*, *DribbleBall*, *BumpShoot*, *ShootGoal*, *MoveShootFeint*, *ShootToPos*, and *TurnAwayBall*.
  - **other**: as *active*, but not goal-directed. This comprises ball possession without visible activity as well as dribbles or shots on the own goal.

Dribbles beginning with a free direct path to the opponent's goal (disregarding the goalkeeper) are distinguished from dribbles beginning with an obstacle, i.e. an opposing player, between the ball and the goal. This distinction allows an evaluation of the defensive positioning of the players of a team, insofar as a team with more mobile players and a better line-up will allow a lower share of dribbles without obstacle. A dribble without obstacle ends with a loss of ball (*ball lost*), with a shot (*with shot*), or with the interference by another player. In the latter case, its description will be continued as a dribble with obstacle. A dribble with obstacle ends with *ball lost*, *with shot*, or the blocking of the ball or player by an opposing player (*blocked*). A player dribbling with the ball in the direction of the opponent's goal and thereby entering the penalty area is assumed to finish this dribble with a shot to the goal, regardless of whether the player does so.

3 Evaluation

Data was gathered from eight hours of video recordings. Situations in a game which were not or only partly recorded were recovered from Web-Player recordings. The net playing time amounts to 247 minutes, of which 104 minutes or approx. 13 minutes per game fall to the eight matches of CS Freiburg in RoboCup 1999 (five in the qualifying round, three in the finals) and 143 minutes or approx. 14.3 minutes per game fall to the ten matches of CS Freiburg in RoboCup 2000 (seven in the qualifying round, three in the finals).

3.1 RoboCup 1999 Vs. RoboCup 2000

In order to get an idea of the development of the game as a whole, the average values of a category for 1999 are compared with those for 2000. Because a comparison of team-specific categories does not yield significant results, only data from categories registered for both teams together is taken into account. The development of the game will thus characterized by the categories *ball free* and *both at ball* (see Figure 1).

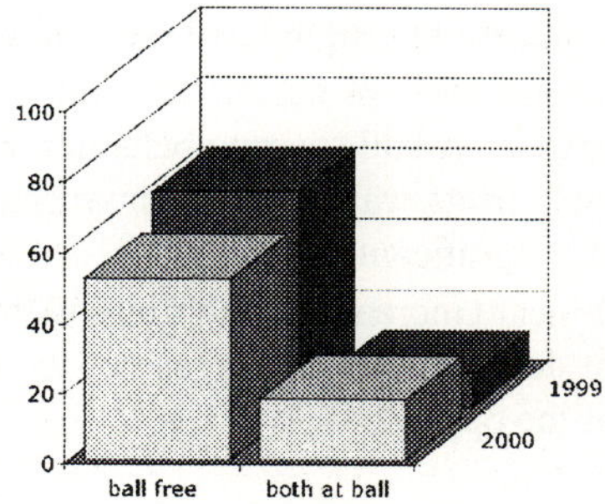

Fig. 1. Ball possession in RoboCup 1999 and 2000

As can be seen from the graph in Figure 1, the amount of time with a *free ball* has decreased. At the same time, the share of *both at ball* situations of the net playing time increased from approx. 10% in 1999 to approx. 18% in 2000.[1] Both changes are probably due to improvements in vision and effectors (e.g. omni-directional movement).

3.2 CS Freiburg 1999 Vs. CS Freiburg 2000

When comparing the amount of ball possession of CS Freiburg between 1999 and 2000, one notices that the amount of time for ball possession in general and for *active possession* in particular is almost the same for 1999 and 2000 (see Figure 2). Interestingly,

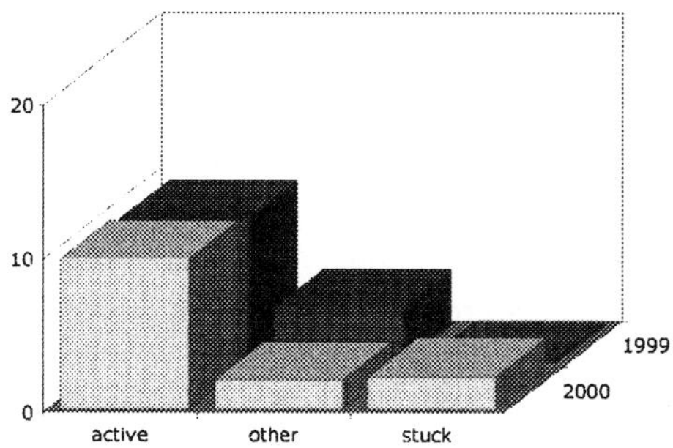

Fig. 2. Ball possession of CS Freiburg in RoboCup 1999 and 2000

however, the CS Freiburg team was able to reduce the amount of time of *other ball possession*, i.e., moving into the wrong direction. In addition, the number of *ball stuck* situations has increased. However, in general, there is no statistically significant change in the area of ball possession.

Because of the new kicking device, a number of statistically significant improvements were visible. For instance, we recorded 14.3 shots per game on average in 2000 instead of 9.8 shots in 1999, and this is statistically significant. Also the average length of a shot increased statistically significant. However, for dribblings only some tendencies were visible, which are not statistically significant (see Figure 3).

The number of situations losing the ball during dribbling without an obstacle in the way has decreased, showing a better ball steering behavior. Furthermore, we could finish a dribbling with an obstacle in its way more often with a shot. However, these two changes were not statistically significant.

There was a statistically significant increase in dribblings with an obstacle in its way. This change, however, does not demonstrate an increase of the skills of CS Freiburg, but a visible better placement of the opponents.

[1] Both of these changes are statistically significant at the 95% level.

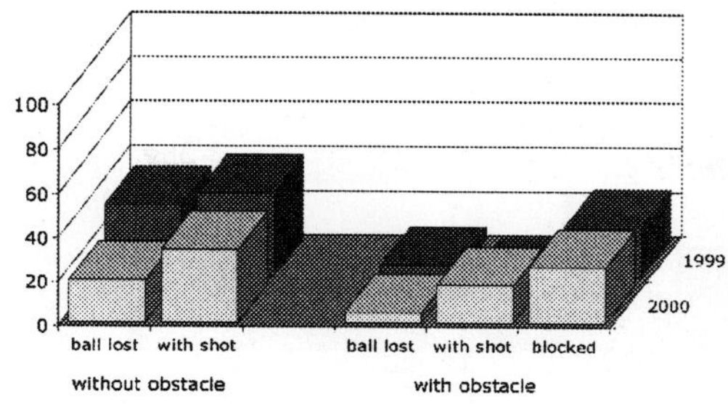

Fig. 3. Dribblings of CS Freiburg in RoboCup 1999 and 2000

3.3 CS Freiburg and Opponents in RoboCup 1999 Vs. CS Freiburg and Opponents in RoboCup 2000

Finally, we will have a look at how much the difference between CS Freiburg and its opponents changed from 1999 to 2000. As mentioned above, we had 14.3 and 9.8 shots in each game on average in 2000 and 1999, respectively. The other way around, our opponents shot 7.2 and 4.6 times on average per game in 2000 and 1999, respectively. While the difference between CS Freiburg and its opponents is statistically significant at the 90% level in both cases, we see a tendency that the opponents are coming closer. A similar statement can be made for shots at the opponents goal. Again we have in both years a statistically significant difference and the opponents come closer.

Reconsidering the ball possession criterion, we get a graph as in Figure 4. In all

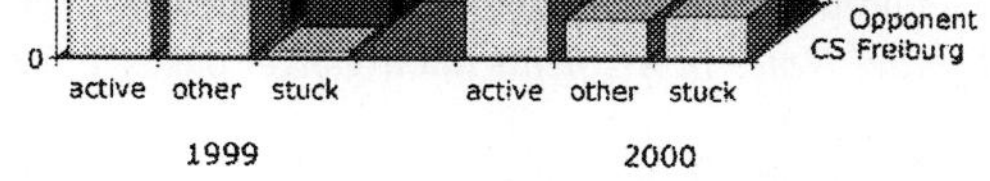

Fig. 4. Ball possession of CS Freiburg and its opponents in RoboCup 1999 and 2000

cases, we have a statistically significant difference between CS Freiburg and its opponents. The most interesting observation is that the *ball stuck* situations increase for both sides and that the *other* situations decrease.

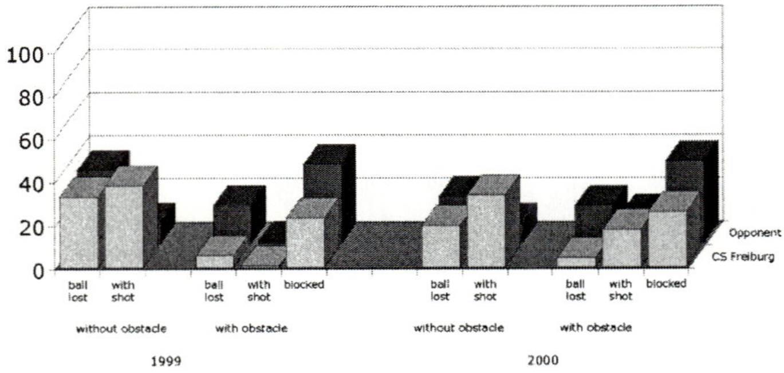

Fig. 5. Dribblings of CS Freiburg and its opponents in RoboCup 1999 and 2000

Finally, when looking at the dribbling capabilites (Figure 5), one notices that from 1999 to 2000 CS Freiburg has not lost its leading edge, e.g., in finishing a dribble with a shot. It is interesting to note, however, that CS Freiburg as well as the opponents much more often finished a dribble with a shot in general.

4 Conclusion

We tried to identify criteria which can be used to measure the progress of robotic soccer teams in order to assess the progress of our own team as well as of the other teams. As it turns out, for most criteria such as ball possession, dribblings and number of shots a statistically significant increase could be identified for CS Freiburg and the opponents. This confirms the impression that the level of play has improved from 1999 to 2000.

References

1. M. Asada, P. Stone, H. Kitano, A. Drogoul, D. Duhaut, M. Veloso, H. Asama, and S. Suzuki. The RoboCup physical agent challenge: Goals and protocols for phase I. In H. Kitano, editor, *RoboCup-97: Robot Soccer World Cup I*, volume 1395 of *Lecture Notes in Artificial Intelligence*, pages 42–61. Springer-Verlag, Berlin, Heidelberg, New York, 1998.
2. H. Kitano, M. Asada, Y. Kuniyoshi, I. Noda, E. Osawa, and H. Matsubara. Robocup: A challenge problem for AI and robotics. In H. Kitano, editor, *RoboCup-97: Robot Soccer World Cup I*, volume 1395 of *Lecture Notes in Artificial Intelligence*, pages 1–19. Springer-Verlag, Berlin, Heidelberg, New York, 1998.
3. K. Tanaka-Ishii, I. Frank, I. Noda, and H. Matsubara. A statistical perspective on the RoboCup simulator league: Progress and prospects. In M. Veloso, E. Pagello, and H. Kitano, editors, *RoboCup-99: Robot Soccer World Cup III*, pages 114–127. Springer-Verlag, Berlin, Heidelberg, New York, 2000.
4. T. Weigel, W. Auerbach, M. Dietl, B. Dümler, J.-S. Gutmann, K. Marko, K. Müller, B. Nebel, B. Szerbakowski, and M. Thiel. CS Freiburg: Doing the right thing in a group. In P. Stone, G. Kraetzschmar, and T. Balch, editors, *RoboCup-2000: Robot Soccer World Cup IV*, Lecture Notes in Artificial Intelligence, pages 52–63. Springer-Verlag, Berlin, Heidelberg, New York, 2001.

Using the Electric Field Approach in the RoboCup Domain

Stefan J. Johansson[1] and Alessandro Saffiotti[2]

[1] Dept. of Software Eng. and Computer Science
Blekinge Institute of Technology
S–372 25 Ronneby, Sweden
sja@ipd.hk-r.se

[2] Center for Applied Autonomous Sensor Systems
Dept. of Technology, Örebro University
S–701 82 Örebro, Sweden
asaffio@aass.oru.se

Abstract. In autonomous robotics, so-called artificial potential fields are often used to plan and control the motion of a physical robot. In this paper, we propose to use an artificial *electric field* to address the problem or real time action selection in embodied, autonomous agents. We attach positive and negative electric charges to the relevant objects in the agent's domain, and use the resulting electric field to estimate the heuristic value of a given configuration. This value is used to select the action that results in the best configuration. This allows us to consider in the same framework both navigation and manipulation actions. We apply the electric field approach in the RoboCup domain, and present results drawn from our experience in the Sony legged robots league.

1 Introduction

There is a concept from physics that has become extremely popular in autonomous robotics: the concept of a potential field. A major initiator of this popularity has been Khatib, who in 1986 proposed an obstacle avoidance method based on the idea that obstacles exert repulsive forces on the robot, while the target exerts an attractive one [4]. At each point of the space, the robot computes a resulting force by summing all the repulsive and attractive forces, and moves as if it were subject to this force. Equivalently, we may say that the robot selects, at each point, the action that brings it to a point of locally minimal potential.

Following Khatib's suggestion, many methods for robot navigation have been proposed based on artificial potential fields (e.g., [1,2,5,6]). All of these methods perform action selection in a very specific case: when the objective is to place the robot at a certain location, and the possible actions consist in robot displacements. It is not obvious if, and how, these methods can be used to solve a more general action selection problem, where the objective may involve the achievement of certain configurations of the objects in environment, and actions may include the manipulation of these objects.

We propose the *Electric Field Approach* (EFA) as a generalization of traditional potential field approaches, which allows us to control both motion and object manipulation. The two main distinctive points of our approach are: the use of the potential

A. Birk, S. Coradeschi, and S. Tadokoro (Eds.): RoboCup 2001, LNAI 2377, pp. 399–404, 2002.

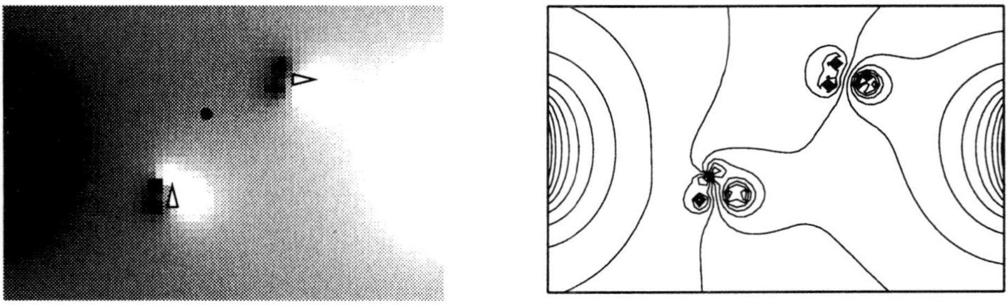

Fig. 1. Electric potential generated by a given configuration of charges (left), which give corresponding equipotential lines (right). Dark areas are negative, white areas are positive. The own net is on the left of the field. The players are indicated by triangles surrounded by charges, the ball (which is uncharged and the probe of this setup) by a circle.

field generated by a set of electric charges as a *heuristic* for action selection in complex domains; and the possibility of the robot to perform actions that *modify* the field. In the rest of this note, we outline the EFA, and show how we have used it to control a team of Sony AIBO legged robots in the RoboCup (RC) domain.

2 The Electric Field Approach

The two main concepts of the Electric Field Approach are *charges* and *probes*. The basic idea is as follows. The overall goals of the system are encoded in terms of positive and negative artificial charges placed on strategically important locations, usually attached to objects (including the robots themselves). We also chose certain probe locations, typically corresponding to the objects which we want to manipulate. *Action selection* is then formulated as the process of trying to increase the electric potential at the probed positions. This can be done by either performing actions that move the probed objects, or actions that move the charged objects. In the RC domain, for instance, the ball is probed, since our main goal it to get it at certain locations (the opponent net) and far from others (the own net). Actions include kicking (moving the probe) and navigating (moving the charges attached on the robot).

In general, charges are attached to areas that are strategically important for the probes. In RC, the opponent net is positively charged, while the own net is negatively charged. These charges are *static* (SC) in the sense that they do not move during the game. But the robots are also surrounded by charges. Positive *Environment-oriented Dynamic Charges* (EDCs) are typically placed on the side of the robot facing the opponent net, while negative EDCs are placed on the side facing the own net. Also, since the robot has better control over the ball when this is in front of it, positive *Agent-oriented Dynamic Charges* (ADCs) are placed in front of the robot. An illustrative situation is shown in Fig. 1.

The electric field encodes our knowledge about the heuristic value of the possible system configurations. Intuitively, we encode knowledge related to the structure of the system by SCs; and knowledge related to the current state by dynamic charges. The principle of superposition integrates this knowledge into an overall electric field. The heuristic value of a system configuration is then evaluated by computing the electric potential at the position of the probe.

This heuristic evaluation can be used for action selection: the goodness of each action a in a given situation s is given by the value of the situation that would result by performing a in s. Intuitively, the placement of the charges will lead the agent to prefer actions that cause the relevant objects to go "at the right places," and to avoid actions that result in a dangerous configurations. In Fig. 1, moving the left robot closer to the ball has a high heuristic value since this would increase the potential at the ball position. In general, action selection can be performed by the simple algorithm in Fig. 2.

```
procedure action_select (situation)
    highest_potential := −∞;
    possible_actions := find_applicable_actions(situation);
    foreach action in possible_actions
            simulate(action, situation);
            potential := calculate_potential(probe);
            if potential > highest_potential then
               highest_potential := potential;
               best_action := action;
    return(best_action);
```

Fig. 2. A skeleton EFA action selection algorithm for single probe environments.

3 Using the EFA in the RoboCup Domain

We have applied the EFA in the "Team Sweden" entry at RoboCup-2000. The robots used were the Sony AIBO legged robots. The architecture implemented in each robot, described in [9], is a layered architecture for autonomy inspired by the Thinking Cap [8]. At the highest layer, a *reactive planner* (RP) selects an appropriate behavior to use based on the information from a *global map* GM. This behavior is executed by the middle layer *Hierarchical Behavior Module* (HBM), that sends back information to the RP when the behavior is either finished, or not possible to perform anymore.

The RP uses the EFA to dynamically decide which behavior, among those implemented in the HBM, should be used at every moment. For instance, the *GoToBall* behavior brings the robot in front of the ball at a close distance. *GoBehindBall* brings the robot close to the ball on the opposite side of the opponent net. *GoBetweenBallNet* brings the robot close to the ball on the side of its own net. *FaceBall* and *AlignWithBallNet* perform local adjustments in order to achieve an adequate posture for kicking. *Kick* and *Steal* respectively kick the ball in front of the robot using the legs, and laterally using the head.

In order to select a behavior, the RP has information about the preconditions, the expected effects, the perceptual needs and the post-conditions of each and one of the behaviors. It then uses an instance of the EFA algorithm given in Fig. 2 to decide which behavior to use. The core of the selection is the *prediction* of the effect of the behavior, and the *heuristic evaluation* of this effect by measuring the potential in the electric field at the selected probe point (the ball).

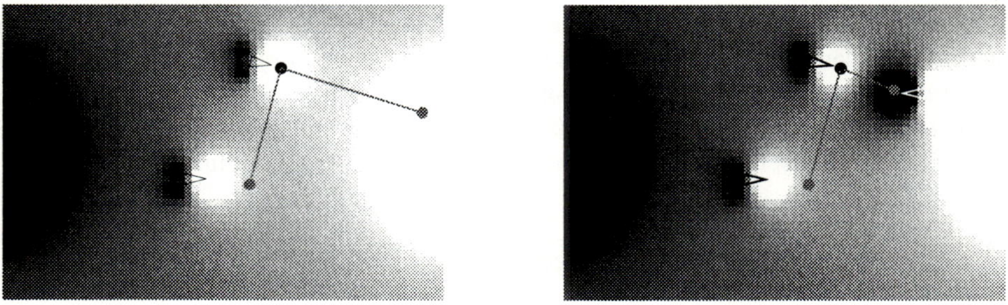

Fig. 3. Choosing between the 'Kick' and 'Steal' behaviors in two different situations. The black dots show the present position of the ball, and the lines and the grey dots, represent the calculated trajectories and predicted final positions (of the ball) respectively when performing the behaviors.

Fig. 3 (left) shows the result of evaluating two possible kicking actions in a given situation. The black circle shows the current position of the ball, the gray circles show the predicted positions of the balls after applying the *Kick* and the *Steal* behavior, respectively. The *Kick* behavior is selected since it would put the ball at a position with a higher potential. Fig. 3 (right) shows how the same example is modified when an opponent player is standing in front of the agent. The predicted position of the ball after executing a *Kick* lays now right in front of the opponent, in an area with a negative potential. Therefore, the *Steal* behavior is selected in this situation, thus passing the ball to the team mate. This example shows how the implemented action selection strategy can achieve some degree of (implicit) coordination.

An important issue is how often the RP should be called. Since the RoboCup domain is highly uncertain and dynamic, the RP cannot simply select a behavior and stick to it until that behavior has completed, since external factors may make this behavior inadequate. On the other hand, continuously re-evaluating the situation may lead to an unstable system, e.g., oscillating between two behaviors that have similar heuristic values. In our domain, we have adopted an extension of the approach proposed in [7]. We select a behavior, and commit to it until either:

– the behavior is close to completion;
– the behavior cannot be safely executed; or
– the situation has significantly changed since the time the behavior was selected.

When any of these conditions is true, the RP is called again and a new behavior is selected. Examples of the last condition are the fact that the ball has moved by a significant amount, or that the quality of the robot's self localization has significantly improved.

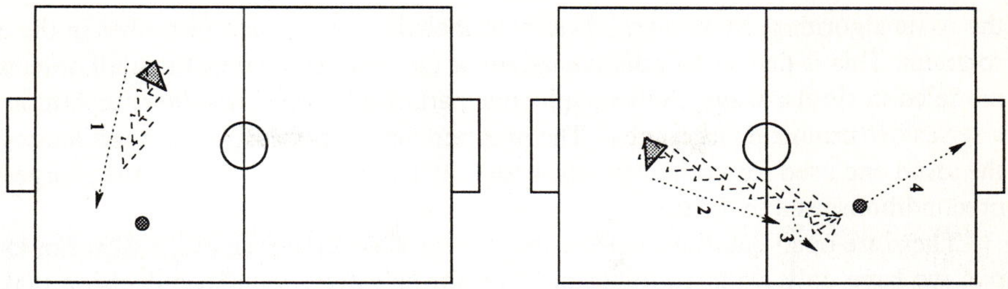

Fig. 4. A simple demonstration of the EFA: (1) *GoBetweenBallNet*, (2) *GoBehindBall*, (3) *AlignWithBallNet*, and (4) *Kick*.

Fig. 4 shows a sequence of decisions produced by the EFA in a simplified (one agent) situation. (Similar sequences occurred in the actual competition, but the complexity of the situations makes them difficult to record and report.) At start, the applicable behaviors are evaluated: since the ball is far from the agent, only the navigation behaviors are applicable. Of these, the *GoBetweenBallNet* results in the highest potential at the ball position (1 in the figure): intuitively, this behavior brings the agent to a defensive position where its positive charges best "shield" the negative potential produced by our own net. While executing the *GoBetweenBallNet,* the ball moves to the other side of the field. This change in the situation triggers a re-evaluation of the behaviors by part of the RP, and *GoBehindBall* is now the preferred one (2). When this behavior is close to completion, the RP is called again. The agent is now close to the ball, so the *FaceBall, AlignWithBallNet, Kick,* and *Steal* behaviors are applicable. Of these, *Kick* and *Steal* would move the ball to places with a lower potential than the current one, while *AlignWithBallNet* would slightly increase the potential on the ball. The robot so performs the *AlignWithBallNet* (3), and then re-evaluates the behaviors. The *Kick* behavior is now the preferred one (4), since it will bring the ball into the opponent net.

The EFA leads itself to efficient implementations, since the field only needs to be evaluated at a few positions and the contribution of all the static charges can be precomputed. The Team Sweden implementation of the EFA, detailed in [3], runs in real time using only a small fraction of the onboard computational resources in the AIBO.

4 Discussion

Our experience has shown that the electric field approach is an efficient tool for higher-level robot control which is general, robust, and maintainable: *General*, in the sense that classical potential field approaches are (at least on a conceptual level) special cases of the EFA. These can be encoded in the EFA by attaching negative charges to the obstacles, a positive charge to the goal, and using the robot itself as a probe. The only actions available in this case are one-step motions of the robot. *Robust*, with respect to errors in the calibration of the charges or in the estimated state of the environment: if these errors are small, the selected action will be close to the optimal one with respect to its effects. *Maintainable*, since we have found that the EFA can be easily adapted to perform new tasks in the same domain by modifying a few elements, and without changing the basic algorithm. Moreover, it is easy to include more objects or probes in the environment. This is due to the additive nature of the field, and the fact that all objects are modeled in similar ways. For example, one part of the RoboCup-200 competition was a series of "technical challenges." The program thaat we used for these challenges was the same one used for normal games, with only a few modifications to the charges and preconditions used in the RP.

There are open questions in the current state of development of the EFA. For example, we have only explored the use of EFA for one-step look-ahead decision making. This may obviously lead to myopic behavior, and it would be worth to study the use of EFA to generate short term plans. Another issue is that we currently account for information gathering actions in a *ad hoc* way. We believe that there are ways to include a uniform treatment of perceptual actions in the EFA framework. A third problem is the current lack of coordination between several robots running an EFA-based action selection mechanism as described here: the robots may decide to head toward the ball simultaneously. We are studying the inclusion of explicit coordination in the EFA through communication. We also intend to test the EFA approach on domains besides the RoboCup.

Acknowledgements

The work reported in this note was partially supported by the Swedish KK Foundation. John Johansson implemented the EFA on the AIBO. The authors are grateful to the members of Team Sweden for worthy insights and collaboration and to the Center for Applied Autonomous Sensor Systems and the Societies of Computations research groups for their support.

References

1. R. C. Arkin. Motor schema based navigation for a mobile robot. In *Proc. of the IEEE Int. Conf. on Robotics and Automation*, pages 264–271, 1987.
2. J. Borenstein and Y. Koren. The vector field histogram: fast obstacle avoidance for mobile robots. *IEEE J. of Robotics and Automation*, 7(3):278–288, 1991.
3. J. Johansson. An electric field approach — a strategy for sony four-legged robot. M.Sc. Thesis MSE-2001-03, Blekinge Institute of Technology, Sweden, April 2001.
4. O. Khatib. Real-time obstacle avoidance for manipulators and mobile robots. *The International Journal of Robotics Research*, 5(1):90–98, 1986.
5. B.H. Krogh and C.E. Thorpe. Integrated path planning and dynamic steering control for autonomous vehicles. In *Proc. of the Intl. Conf. on Robotics and Automation*, 1986.
6. J. C. Latombe. *Robot Motion Planning*. Kluver Academic Publishers, Boston, MA, 1991.
7. S. Parsons, O. Pettersson, A. Saffiotti, and M. Wooldridge. Robots with the best of intentions. In M. Wooldridge and M. Veloso, editors, *Artificial Intelligence Today*, pages 329–338. Springer-Verlag, 1999.
8. A. Saffiotti. The Thinking Cap homepage. http://www.aass.oru.se/~asaffio/Software/TC/.
9. Team Sweden. Official website. http://www.aass.oru.se/Agora/RoboCup/.

A Two-Tiered Approach to Self-Localization

Frank de Jong, Jurjen Caarls, Robert Bartelds, and Pieter P. Jonker

Pattern Recognition Group, Faculty of Applied Science,
Lorentzweg 1, 2628 CJ Delft, The Netherlands
{frankdj,jurjen,robertb,pieter}@ph.tn.tudelft.nl

Abstract. This paper describes a two-tiered approach to the self-localization problem for soccer playing robots using generic off-the-shelf color cameras. The solution consists of two layers; the top layer is a global search assuming zero knowledge, and the bottom layer is a local search, assuming a relatively good estimation of the position and orientation of the robot. The global search generally yields multiple candidate positions and orientations, which can be tracked, and assigned a confidence level using the local search and/or historic information.

1 Introduction

In Robocup, cooperating robots achieve intelligent group behaviour by communicating about scoring opportunities, tactical positioning, opponent movement, etc, which requires a common frame-of-reference. It is a natural choice to take the soccer field as this frame-of-reference, which gives us the task of finding the robot's pose (x, y, ϕ) relative to the field (self-localization).

To recover the pose of the robot we match a model of the world with the observed camera image. The match takes place in image space or in world space. In [1], a survey of both methods of Model to Image Registration is given. In RoboCup, matching is mostly done in world space [2,3,4,5] because the movement of the features in world space is directly determined by the movement of the robot. However, the error in the feature positions in world space is dependent on the distance to the camera.

When matching is done in image space, the robot pose is more difficult to describe using the movement or position of the image features, but verification of a predicted pose can be done cheaply, because we only need to find specific image features close to their predicted positions, instead of doing a general search in the entire image.

Our approach finds candidate poses in world coordinates and verifies these candidates in image space.

2 Self-Localization

Several field features are available for self-localization, such as goals, corner lines and field lines. In our initial version, we relied on the vertical features (goal edges

A. Birk, S. Coradeschi, and S. Tadokoro (Eds.): RoboCup 2001, LNAI 2377, pp. 405–410, 2002.

and corner posts), but they may easily be blocked from view by another robot, or not be visible due to large distance to the camera. We therefore changed to using the field lines.

2.1 Accuracy

When using normal CCD cameras to determine objects' distances from the camera, perspective transformation causes the range measurements to be very sensitive to camera tilt, especially in lines that lie close to the camera horizon. Now although the camera tilt relative to the robot can be fixed and calibrated, the camera tilt relative to the field will be influenced by slight bumps in the field and robot vibrations, leading to big errors in (extrapolated) world coordinates.

To attack this problem, we use three types of techniques. First, we attach a inclinometer to the camera, so that the camera's instantaneous tilt angle can be determined whenever an image is grabbed. Second, we measure the lines with subpixel accuracy. Third, our final line matching happens in image space, where small tilt variations will not have a big influence.

2.2 Two-Tiered Approach

Although a global, zero-knowledge approach could work, it is a waste of effort to start at this zero level each time. However, a local method that assumes only small deviations from the estimated position might go wrong in the case of collisions with other robots, extreme cases of occlusion by other robots, or positions close to the wall (facing the wall).

It is therefore natural to adopt a two-tiered approach, where the local search tracks the robot pose at real-time speeds, and a slow global search runs as a background process, verifying the local search's results, and being put in the foreground whenever the local search fails to find the expected field line pattern. Odometry is used to interpolate between local searches.

The next sections detail the global and local searches used in our approach.

3 Global Search

In short, the global search transforms measured lines in the image to robot coordinates, and, using the field model, generates a number of candidate poses, which are verified in image space.

The global search routine first detects the field-line and field-wall edges. We implemented the edge detector by detecting edges in 3x3 neighborhoods of the labeled image L, in which each pixel is bit-flagged as being a member of one or more classes of objects based on its color information (truncated pie slices in YUV space) [7]. This yields the edge image E.

To find the lines in E, an obvious choice would be the Hough Transform (HT), but since we are dealing with a) relatively high lens radial distortions, b) the presence of the center circle in the soccer field, and c) required subpixel accuracy, we discarded the use of the HT on the entire image.

3.1 Coarse-Fine Approach

Instead, we divided the task in two steps, a) finding line segments, and b) accurately determining the line segments' positions. To find line segments, we divided the image into several small subimages (in our case, 8x6 subimages of 40x40 pixels), small enough to assume that the straight lines are, indeed, mostly straight. Since the first step needs to give only a coarse result, we use the HT of E, taking a small number of bins for the orientation and distance (θ, ρ). Then, using the best two peaks in the HT in each subimage, we run sub-pixel edge detectors perpendicular to the lines found with HT. The edge detector is run on the red component R of the image, since that gives the best contrast between green and white. We then convert the edge positions to calibrated camera coordinates, and fit lines through points from one parent line, with least squares and leave-out. See Figure 1 for an impression of the process.

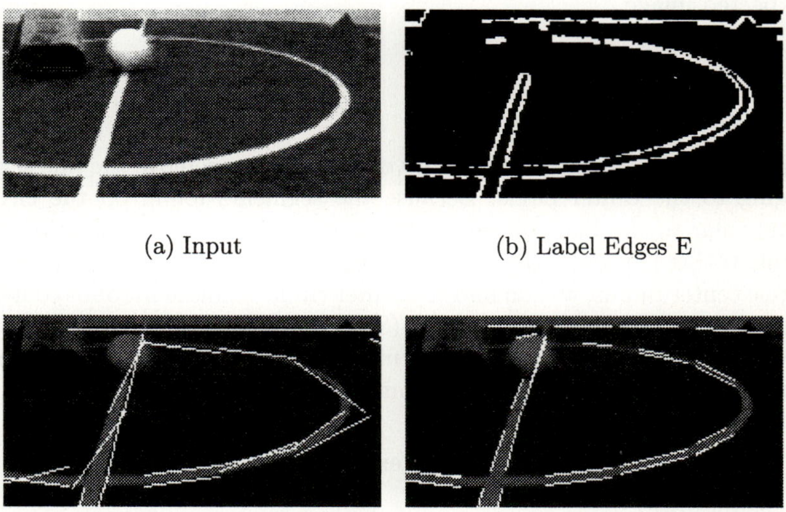

(a) Input (b) Label Edges E

(c) Hough Transform E (d) Accurate Lines

Fig. 1. Several steps of the line detection.

Knowing the field lines in the calibrated image, we then convert the lines to robot coordinates. See Figure 2, left.

We proceed by using that in the field model, all lines (except for the center circle) are in either of two main orientations, so that when projecting the field model on these orientations, we get a very distinct pattern of points, which suggests the possibility of splitting the 3D search problem into three 1D searches.

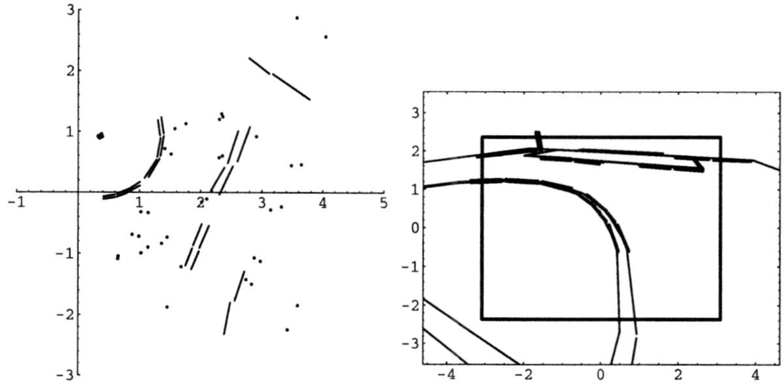

Fig. 2. Left, the measured lines in robot coordinates, and their votes for a center circle. Right, the best match found, with the measured lines overlaid on the expected image.

3.2 Finding Orientation

To take advantage of the orthogonality of the field, we remove the line segments belonging to the center circle. Because the segments found on the circle are perpendicular to the line connecting the circle center to the middle of the line segment, voting left and right of each found line segment, at a distance equal to the circle center radius, we can find the center circle estimate in robot coordinates and remove the lines that belong to it (See again Figure 2, left).

From the remaining segments, the main orientation is determined from the orientations θ_i by finding $\hat{\theta}$ that maximizes

$$P(\hat{\theta}|\theta_i) = \sum_i \exp(\frac{-\Phi(\theta_i, \hat{\theta})^2}{2\sigma_i^2}) \tag{1}$$

where $\Phi()$ is the orientation difference $(mod\pi)$, and $\sigma_i = \lambda d_i$, with λ a normalization constant, and d_i the line segment's distance to the camera (further away objects have higher inaccuracy).

Finding the center circle gives a strong candidate for the robot pose, since we know the main orientations, and the center circle offset.

3.3 Finding Position

We now proceed by matching the projection on the main orientations of the field with the projection of our measured lines on their own main orientation. For the main orientation θ and the orientation perpendicular to it, $\psi = \theta + \pi/2$, we find a number of candidate positions, which follow from minima of a 1D matching cost function.

We combine candidate positions on the two axes exhaustively, and generate the expected line pattern for each candidate pose using standard image formation methods [6,8].

We then calculate the matching cost between the measured lines m_i and the expected line pattern $e_j(\boldsymbol{c_k})$ for each candidate pose $\boldsymbol{c_k}$. For each measured line segment m_i, we determine the distance D from its center point q_i to the closest expected line, and combine these distances to get a matching cost,

$$C(\boldsymbol{c_k}|m_i) = \sum_i atan(\frac{min_{e_j}(D(q_i, e_j(\boldsymbol{c_k})))}{\epsilon}) \tag{2}$$

where ϵ is a normalization constant, and *atan* reduces the influence of erroneous lines. The worst candidates are removed, and the remaining candidates are used as input for the local search step.

4 Local Search

In short, the local search takes an estimate of the robot's pose, calculates the corresponding field line pattern using a model, and tries to match that with lines in the image.

For each expected line pattern, we look for line segments in local neighborhoods in the image. In some locations, the lines will not be found (due to robot occlusion or other errors), so we will find a subset of the expected line segments, displaced slightly due to the error in our prediction.

If not only the estimate of the robot's pose is fed into the local search, but also the estimated pose plus or minus small increments, (e.g. x, $x+\delta x$, $x-\delta x$), we get for x, y, ϕ: 3x3x3 = 27 possible line patterns. The increments are determined by the expected odometry creep error, and the correct parameters are simply those that yield the best match with the measured image.

If none of the candidates generates a sufficiently good match, the global search is activated again.

5 Experiments

We tested the 2-step line detector for the global search on real images, with satisfying results, see Figure 1. Note that the accurate line detector also clips off the line where it no longer satisfies the linearity constraint. The line detector has a bias of 0.02 pixel and σ of 0.06 for synthetically generated images. The subpixel edge positions are found by calculating the second derivative in a 5 pixel neighborhood, and finding the zero-crossing.

The global search usually gives about 20 candidates, which is reduced to about half (or, when the center circle is found, to usually 2 candidates) by matching in image coordinates. If several candidates are found, we keep tracking (locally) the remaining candidates until we have only one candidate left. This candidate is sent to our world knowledge module.

6 Discussion

Both global and local self-localization methods show good results on test images, but have yet to be tested in a real-time environment. We are in the process of implementing the self-localization into the vision system for the upcoming RoboCup event.

In matching the expected and measured line set, we take the lines measured, and try to find support in the expected image, instead of the other way around, because although lines may erroneously be detected as field lines (and therefore introduce errors when matched with the field model), we assume that it is more often the case that field lines will not be detected due to occlusion by the ball or other robots.

The global search routine assumes zero knowledge about the robot's position, and works only if at least 2 perpendicular line segments are visible. When these lines are not available, the global search will yield too many candidates. In these cases, the robot must make for instance rotary movements to increase the chance of observing perpendicular lines.

As an alternative to simply trying the 27 'neighboring' parameters from section 4, we are implementing a direct estimator of the pose error. By writing the line segments' displacements perpendicular to their expected position as a system of linear equations of the small increments (by calculating the image Jacobian [9] on all line-positions), we can find the optimal pose in least squares sense with relatively little effort.

References

1. Claude Duchesne, Jean-Yves Herv, "A Survey of Model-to-Image Registration", Rapport technique GRPR-RT-9913, Groupe de Recherche en Perception et Robotique, Ecole Polytechnique de Montreal, October 1999.
2. Carlos F. Marques, Pedro U. Lima, "A localisation Method for a Soccer Robot Using a Vision-Based Omni-Directional Sensor" , Proc. Fourth International Workshop on Robo-Cup, p159
3. Luca Iocchi and Daniele Nardi, "Self-Localization in the RoboCup Environment", Proc. Third International Workshop on RoboCup, p115
4. Jens-Steffen Gutmann, Thilo Weigel, and Bernhard Nebel, "Fast, Accurate and Robust Self-Localization in the RoboCup Environment", Proc. Third International Workshop on RoboCup, p109
5. Thorsten Bandlow, Michael Klupsch, Robert Hanek, Thorsten Schmitt, "Fast Image Segmentation, Object Recognition and Localization in a RoboCup Scenario", Proc. Third International Workshop on RoboCup, p109
6. Olivier Faugeras, "Three-dimensional Computer Vision", MIT Press, 1996.
7. Pieter P. Jonker, J.Caarls, "Fast and Accurate Vision for Vision-based Motion", Proc. Fourth International Workshop on RoboCup, 1999.
8. Foley, J.D., Dam, A. van, Feiner, S.K., and Hughes, J.F.," Computer graphics, principles and practice, 2nd edition in C", Addison-Wesley, London, 1996.
9. Gregory D. Hager, "Efficient Region Tracking with Parametric Models of Geometry and Illumination", IEEE Trans. on PAMI, Vol. 20, No. 10, October 1998

Miro – Middleware for Cooperative Robotics

Gerhard Kraetzschmar, Hans Utz, Stefan Sablatnög, Stefan Enderle, and Günther Palm

University of Ulm, James-Franck-Ring, 89069 Ulm, Germany

Abstract Developing software for mobile robot applications is a tedious and error-prone task. We suggest the use of object-oriented middleware to remedy the problem. After identifying crucial design goals, we present MIRO, an object-oriented middleware for robots meeting the design goals. We discuss its implementation and demonstrate MIRO's role in the implementation of applications on different kinds of robots.

1 Introduction

Mobile robots are usually composed of heterogeneous hardware components, which are connected using different networking technologies and communication protocols with widely differing bandwidths and interfaces. A large number of different methods for processing sensor information and controlling actuators, for performing computational vision and cognitive tasks like planning, navigation, and user interaction, must be integrated into a well-engineered piece of software. All these issues contribute to the enormous complexity of the mobile robot software development task.

We argue that object-oriented middleware is both necessary and highly beneficial for developing robust and reliable mobile robot software. High-quality robot software and an improved software development process will be key factors in enhancing the state-of-the-art. Also, mobile robot software must be opened up for seemless integration in enterprise information processing frameworks. Object-oriented middleware for robots can take a strong enabling role in this process.

2 Design Goals for Robotics Middleware

Some essential requirements and design goals for object-oriented middleware for robots are:

Thorough Object-Oriented Design: This is self-explaining.

Client/Server Systems Design: We further suggest to adopt a client/server view at least for larger modules of a robot control architecture: the modules, implemented as objects, provide services to other modules (objects) through a set of interfaces.

A. Birk, S. Coradeschi, and S. Tadokoro (Eds.): RoboCup 2001, LNAI 2377, pp. 411–416, 2002.

Hardware and Operating System Abstraction: Object-orientation in combination with the client/server view permits a clean abstraction of sensor and actuator subsystems, as well as low-level system services like communications, and provides a uniform interface design.

Open Architecture Approach: Adopting an open architecture approach, including availability of all source code, is undispensable. Only if applications programmers can access, modify, and fine-tune all components of the software environment, the integration of a heterogeneous set of hardware and software components can be successful.

Multi-Platform Support, Communication Support, Interoperability: The middleware should be available on a wide range of hardware platforms and common operating systems. It should provide a carefully designed set of interfaces for communication between objects and communications transparency, i.e. the programmer should not need to worry much about where objects are actually allocated and which communications protocol she needs to apply. The overall system design must provide sufficient leverage for growth without compromising code readability and maintainability and runtime efficiency in an excessive manner.

Software Design Patterns: The middleware should encourage the use of design patterns by providing a class framework for common, well-understood patterns for common tasks.

Agent Technology Support: Integrating an application into an enterprise-wide information system is often eased by adopting an agent-oriented view. The middleware should allow easy extension to integrate agent-oriented technology.

3 Miro – Middleware for Cooperative Robotics

The MIRO software framework addresses the above design challenges and lays the foundation for implementing robust and reliable robot control architectures. MIRO is rigorously objected-oriented designed and implemented. The MIRO core, containing functionality for primary sensory data processing and actuator control, is entirely implemented in C++ and allows for high runtime efficiency. MIRO builds itself upon widely used, industrial-strength middleware packages, which are open source and available on a wide range of hardware and operating system platforms.

3.1 Miro Architecture

MIRO provides two sets of functionalities. First it contains a set of services to address the sensory and actuatory facilities of the mobile robot platforms. Second it provides a framework of classes that implement design patterns for mobile robot control. Sensors and actuators can be naturally modeled as objects, which can be controlled and queried by their respective methods. Thereby, robots can be viewed as aggregations of sensory, actuatory and cognitive objects, which can trade information and services in an agent-like manner. The general structure of MIRO is illustrated by Figure 1.

One of the essential abstractions designed into MIRO is that objects provide services instead of merely abstract interfaces. This proved to be very important for achieving robot platform independence. For example, it allowed to provide a *motion service* that could be consistently implemented and ported to several mobile robot platforms. Inheritance is used to provide refined interfaces of common sensor/actuator services. This mechanism enables individual service implementations to provide access to the special features available by the specific hardware devices. Figure 1 shows the inheritance graph of the motion service, the primary interface to the robots motory actuator. It inherits the odometry interface, providing methods for the robot's position control. Each individual robot platform's motion service inherits the general motion interface and enriches it with the special features of the platform, like acceleration control, closed-loop motion commands like "*turn 30? left*", etc.

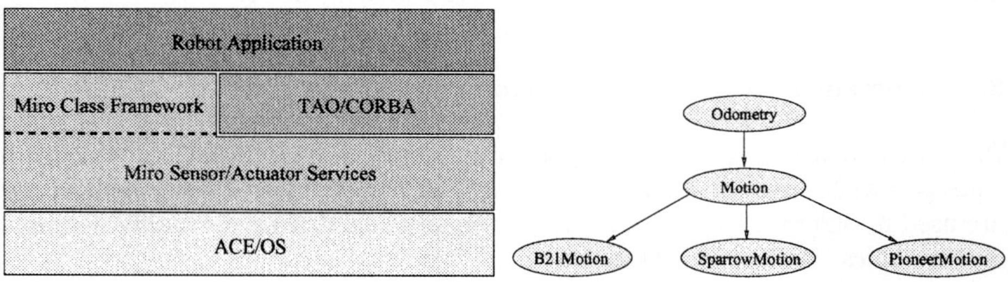

Fig. 1. Left: The Miro Middleware for Robots. Right: *Motion* interface inheritance for different mobile bases.

3.2 Miro Implementation

MIRO makes extensive use of multi-platform libraries for easy portability. The Adaptive Communications Environment ACE provides object-oriented abtraction layers for many operating systems and communications primitives. The TAO package is an implementation of the Common Object Request Broker Architecture CORBA based on ACE [9]. For the construction of GUIs, like monitoring and visualization tools, we use Qt, a graphical user interface application framework, which is available for various Unix derivates including Linux as well as Windows.

To overcome location and programming language dependencies, all sensor and actuator services export their interfaces as network transparent CORBA objects, which can be addressed from any language and platform, for which language bindings and CORBA implementations exist. This enables seamless integration of high-level robot control subsystems, like Lisp-based planners or Java-based user interfaces.

While a traditional method call is the most convenient way of object interaction, the classical synchronous query interface does not scale well for all sensory devices. To give the programmer more flexibility, all sensory services of MIRO also provide an event-triggered communications model for data-driven processing of sensor data. This approach is quite common in real-time environments. The MIRO implementation exploits the CORBA Notification Service [5] for providing this functionality.

Basic support for multirobot control comes naturally with a distributed robot software development environment. Small groups of robots can address each other by exchanging the object references if their respective sensor/actuator configurations are commonly known. This approach is facilitated by the availability of naming service functionality, providing a separate namespace for each robot. Also sharing of sensor data can be achieved via filtered event processing frameworks based upon the notification service. We are currently evaluating such a framework in our Sparrows robot team for RoboCup-2001.

3.3 Current State of Miro Development

Presently, MIRO is implemented for three mobile robot platforms, which are equipped with different sensors, actuators, and computational equipment. They are used in different scenarios, ranging from an office delivery to highly dynamic soccer games [3]. These platforms are:

- A B21 robot, which is equipped with bumpers, IRs, sonars, a laser range finder, and a vision system. It features a synchro drive mobile base and is controlled by two onboard PCs.
- Pioneer-1 robots, which are differential drive robots equipped with sonar sensors only and are controlled either via a laptop mounted on top of a robot or by a host PC via a serial radio link.
- Sparrow-99 robots [7], which are custom-built robots developed in our lab at the University of Ulm. Sensors include sonars and a camera. The mobile base is a differential drive system. The robot also has a pan-tilt unit and — for its special purpose — a kicker. It is controlled via an onboard embedded PC.

Several modules providing higher-level functionality have been implemented using MIRO, including two variants of the Monte-Carlo localization method (one based on distance sensor data, another based on visual features [2]), a hybrid multi-representation world modelling system [6], and a hierarchical system of navigation planners [10] which exploit the world modelling system and provide autonomous navigation.

4 Miro Design Patterns

For the construction of its sensor and actuator services MIRO makes heavy use of the design patterns [1] realized within the ACE framework [8]. For instance, the reactor pattern is used as the central pattern for event demultiplexing when dealing with multiple concurrent event sources, like hardware devices and timers, without causing task switching overhead. The active object pattern is used for long lasting concurrent high-level tasks, e.g. for asynchronous coupling of multiple planners within a hierarchical, reactive planning environment, as described above. The half-sync/half-async pattern is used to shield the application programmers from asynchronous communication provided by most hardware devices and allows for a simpler programming model for less time critical use cases. Exploiting these and various other design patterns helped enormously in factoring out common functionality and, by using them as generalized classes, to gain a stable implementation of the services provided by MIRO.

Apart from low-level hardware and software abstraction, we try to pinpoint working solutions for mobile robot control and provide generalized implementations of them, thereby constructing a class framework of software design patterns for autonomous mobile robot control. For example, classes are available for behavior-based robot control and sample-based Monte-Carlo localization [2, 4]. Current research addresses the identification of further high-level patterns that can be provided uniformly as library or middleware components.

5 Experience with Miro

MIRO is successfully used in several projects striving for autonomous mobile robot control, such as neurosymbolic mapping of indoor environments [6], autonomous self-localization [2], and reactive navigation [10]. Our experiences with MIRO are very positive.

In our application projects, MIRO proved to be robust and reliable. The view of sensors and actuators as services allows flexible integration without major adaptations of interfaces. MIRO is highly portable, as demonstrated by porting MIRO from the B21 platform to the Sparrow-99 platform by a single programmer within one week. The integration of modules or larger functional blocks, which were developed by different programmers, turned out to be very smooth, because the IDL-based interface specifications fostered mutual understanding of concepts and made misconceptions rare. The integration of two mapping modules, a module for self-localization and a hierarchical navigation planner into a working system was established within a few days.

The inherent ease of use of distributed computing concepts was highly appreciated by the developers. This aspect was particularly helpful in defining interfaces for debugging and monitoring. Regarding scalability for multirobot scenarios, we now have an adequate infrastructure to investigate cooperative behavior, fault-tolerant team operation, and realtime cooperative robot control in our team of soccer robots.

6 Conclusions

Because adequate software implementation platforms for mobile robot researchers are not widely available, software development for mobile robots is an interesting and fruitful area for software engineering researchers. We propose the adoption of object-oriented middleware for robots to improve the software development process. MIRO builds upon standard and widely-used middleware packages like ACE, TAO CORBA, and Qt, which significantly eases integration into enterprise information processing frameworks. MIRO is stable and robust and has been applied in several development projects. MIRO is freely available. Researchers interested in using MIRO should contact one of the authors.

References

[1] James O. Coplien and Douglas C. Schmidt, editors. *Pattern Languages of Program Design*. Addison-Wesley, Reading, MA, June 1995.

[2] Stefan Enderle, Marcus Ritter, Dieter Fox, Stefan Sablatnög, Gerhard Kraetzschmar, and Günther Palm. Soccer-robot locatization using sporadic visual features. *Proceedings of the IAS-6 International Conference on Intelligent Autonomous Systems*, 2000.

[3] Stefan Enderle, Hans Utz, Stefan Sablatnög, Steffen Simon, Gerhard Kraetzschmar, and Günther Palm. Miró: Middleware for autonomous mobile robots. In *Telematics Applications in Automation and Robotics*, 2001. submitted.

[4] Dieter Fox, Wolfram Burgard, Frank Dellaert, and Sebastian Thrun. Monte carlo localization: Efficient position estimation for mobile robots. In *In Proc. of the National Conference on Artificial Intelligence*. AAAI, 1999.

[5] T. H. Harrison, D. L. Levine, and D. C. Schmidt. The desing and performance of a real-time corba event service. In *Proceedings of OOPSLA '97*, Atlanta, October 1997. ACM.

[6] Gerhard K. Kraetzschmar, Stefan Sablatnög, Stefan Enderle, and Günther Palm. Application of neurosymbolic integration for environment modelling in mobile robots. In Stefan Wermter and Ron Sun, editors, *Hybrid Neural Systems*, number 1778 in Lecture Notes in Artificial Intelligence, Berlin, Germany, March 2000. Springer. ISBN 3-540-67305-9.

[7] Stefan Sablatnög, Stefan Enderle, Mark Dettinger, Thomas Boß, Mohammed Livani, Michael Dietz, Jan Giebel, Urban Meis, Heiko Folkerts, Alexander Neubeck, Peter Schaeffer, Marcus Ritter, Hans Braxmeier, Dominik Maschke, Gerhard Kraetzschmar, Jörg Kaiser, and Günther Palm. The ulm sparrows 99. In *RoboCup-99: Robot Soccer World Cup III*, Lecture Notes in Computer Science, Berlin, Germany, 1999. Springer.

[8] Douglas C. Schmidt. The adaptive communication environment: Object-oriented network programming components for developing client/server application. In *12th Sun Users Group Conference*, 1994.

[9] Douglas C. Schmidt, Andy Gokhale, Tim Harrison, and Guru Parulkar. A high-performance endsystem architecture for real-time corba. *IEEE Communications Magazine*, 14(2), 1997.

[10] Hans Utz. Quo vadis? Robuste hierarchische Navigation für autonome mobile Roboter. Diplomarbeit, University of Ulm, Ulm, Germany, October 2000. in german.

Building User Models for RoboCup-Rescue Visualization

Yoshitaka Kuwata[1] and Atsushi Shinjoh[2]

[1] NTT DATA CORPORATION, JAPAN.
kuwatay@nttdata.co.jp
[2] International Academy of Media Arts and Sciences, JAPAN.
kaminari@iamas.ac.jp

Abstract. Information visualization is one of the most important fields in RoboCup-Rescue research. As requirements for visualization vary by users and by their purpose of viewer use, we introduce a mechanism called *user model*. User models hold the requirements of typical users. They are used for the selection and customization of information, controls of viewer, and so on. We implemented the mechanism with five user models in RoboCup-Rescue 2D Viewer.

1 Introduction

One of the most significant differences between *RoboCup-Soccer* and *RoboCup-Rescue ([1] and [2])* is their contributions to the real world. RoboCup-Soccer has major impact for AI and robotics research community, as it provides a grand challenge as a benchmark of various AI algorithms. Although RoboCup-Soccer is just a game, the technologies developed for it can be applied a large variety of fields. On the other hand, RoboCup-Rescue can have major impact for the real-world rescue. The technologies developed in RoboCup-Rescue framework can be applied directly to the rescue operation both in planning phases and action phases. In RoboCup-Rescue, the research on the practical use of the technologies is important, as well as theoretical research.

One of the most state-of-the-art technologies in RoboCup-Rescue research field is the information visualization. They are used to show the simulation result to users. We have been developing viewers for RoboCup-Rescue simulation frameworks. There are two goals for our research; A) develop useful techniques of the information visualization, and B) verify and validate the usefulness of the technologies. As the RoboCup-Rescue viewer needs to support wide variety of requirements in real-world rescue operations, we started analyzing the actual requirements. Based on the results, we proposed an architecture of RoboCup-Rescue viewers in [4].

The requirements for visualization vary by users and by their purpose of viewer use. As a matter of fact, it is impossible to design one visualization method suitable for all of needs. For example, designers of agent need to know the details of the status of their agents, such like their damages, HP, internal

A. Birk, S. Coradeschi, and S. Tadokoro (Eds.): RoboCup 2001, LNAI 2377, pp. 417–422, 2002.

status, and so on. On the other hand, urban planners, who design mitigation plan of disasters, don't need to know the details. They would be interested in the statistics of damages of the city.

One obvious solution is to design custom visualizations for each requirement. RoboCup-Rescue simulator is designed to handle more than one viewer simultaneously [3]. Although custom-made viewers can have the best efficiency, the software-engineering process required for the design of customized-viewers can be very complicated and thus the cost can be very high.

Other solution is to design universal mechanisms that make viewers fully customizable for all possible users. Users of such viewers can obtain their own visualization by customizing universal viewer by themselves with their requirements. One drawback of this methodology is the difficulty of customization. For the customization process, the users need to have full knowledge of base models that are built into the viewer.

To simplify the customization, we introduce *User Model Mechanism* for the configuration of viewers.

2 User Model Mechanism

In very complex simulation systems like RoboCup-Rescue, the number of objects and the kinds of objects can easily exceed humans' perceptibility. For example, it is likely that a large number of icons in disaster maps can make the users confused. It will also become physically impossible to display all of simulation objects in one computer display. This problem becomes even more critical in small displays found in portable computers and wearable computers.

One popular solution found in GIS (Geographic Information System) is to rely on *layers*. Figure 1 represents a concept of layer. All of the objects in GIS are stored in one of layers. Users of GIS can select a set of layers to display. Layers are implemented like a pile of transparency sheets. Objects in upper layers hide those in lower layers. Users must choose the appropriate order of layers for the display.

As the number of layer becomes larger and larger, humans tend to have trouble to select suitable layers for their purpose. This is because the adjustment of layers requires deep knowledge and careful discretion in GIS objects. Many commercial geographic data contain more than 100 layers.

The goal of user models is to provide a set of optimal visualization methods for typical users. User model is a kind of database which holds the typical users' requirements for their visualization. Figure 2 represents the mechanism of user model. The user model mechanism controls the followings;

1. Selection of GIS layer or object class for display
   The selection mechanism of GIS layer is similar with GIS. *An user model* have a set of layer selection and the order of layers for display. User model can have the control of layer selection; i.e. in some user models, users don't have ability to change layer configuration.

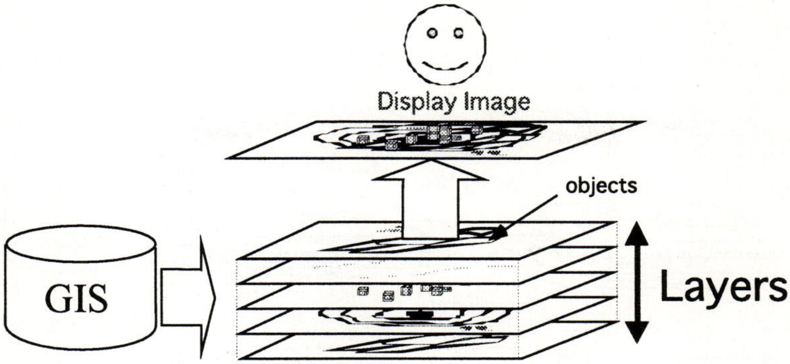

Fig. 1. Layers in GIS: Layers are common idea in GIS. The screen image is generated from a pile of layers. Users can designate layers and orders of layers in the pile.

2. Selection of Display Method
   User models include the selection of display methods, such as display color, selection of display preference. (e.g. use icons or text)
3. Selection of Display Control Methods
   User models have the status of automated viewpoint control (turn on/off). They also have the selection of target-event for auto viewpoint control. (e.g. ignition-event, rescue-event, and so on)

3 Analysis of Users

We had already analyzed the expected users of RoboCup-Rescue framework in [4] in details. The following list represents requirements from each group of users.

1. Development of simulation system
   Agent developers, researchers of various disasters, and researchers of software architectures are possible users of RoboCup-Rescue simulators. They develop algorithms of their agents, disaster simulators and total systems, respectively. In general, they need to know detailed status of the system. For example, agent developers need to find out why one setting of fire-brigade agent works better than the previous settings in certain environments. They will observe the simulation from various aspects of the system, such as the intention of agents, cooperation between agents, and so forth. Thus viewers need to be fully customizable for them, so that it can provide as much information as possible with users' request.
2. Disaster mitigation planning phase
   Urban planners will use RoboCup-Rescue simulator to develop mitigation plans. They need to know the results of their new mitigation plans. In general, mitigation plan is in brief. As the

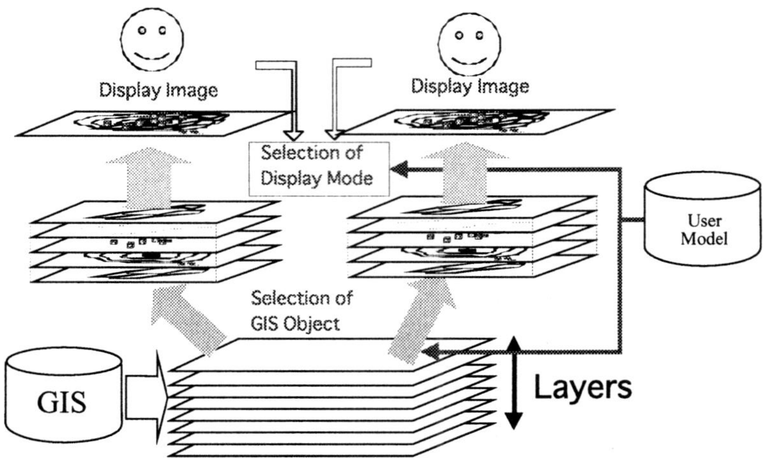

Fig. 2. User Model Mechanism

3. Disaster relief operations
   Rescue supervisors manage rescue resources such as rescue teams, rescue equipments, and so on. In general, resources used for rescue works are insufficient in big disasters. One goal of the resource management is the most efficient use of rescue resources. The other goal is to keep the safety rescue operations. The prediction of disasters by using disaster simulators will make their management easier. The simulation should base the information gathered by rescue team in the field. In operation phases, managers can be confused because of information flood. Thus viewers need to organize information and provide the most important points.
4. RoboCup-Rescue Demonstration
   In competitions, the audience of RoboCup-Rescue wants to observe all of interesting events in simulations such as activities of rescue team agents. Thus the viewpoint control is very important for them. As it is very difficult for humans to control the viewpoint by hands, automated control mechanism is necessary for the demonstration sessions.

4 Implementation

In order to validate the idea of *user model*, we implement the mechanism in RoboCup-Rescue 2D viewer called *logViewer*. *logViewer* is designed for researchers and for demonstrations. Although it is easy to add more user models, the following five models are built; A) Software Architect, B) Simulator Developer, C) Agent Developer, D) Supervisor, and E)Audience.

For the control of visualization, *logViewer* has the following built-in functions.

1. Selection of GIS components
   Users can select the objects on the screen by selecting object class such as buildings, roads, nodes and agents.
2. Selection of display format
   Users can select display format of objects, such as text, icon, color.
3. Automated viewpoint control
   The viewer select the optimal geological location and display scale automatically from the selection of target objects and events. For example, if automated view point control is on and the fire-brigade is selected as target, the viewer tries to follow the movement of fire-brigade.

An example of user model setting is shown as table 1. The columns in the table represent controls described above. Each user model is consisted of the certain combination of these controls. One row represents an user model.

Table 1. An Example of User Model Setting

Controls		User Model				
GIS Attributes		Software Architect	Simulator Developer	Agent Developer	Supervisor	Audience
Building	% Broken	EC	EC	EC	EU	EU
	% Burnt	EC	EC	EC	EU	EU
	Outline	EC	EC	EC	DU	DU
	Details	EC	EC	EC	DU	DU
Road	% Broken	EC	EC	EC	EU	EU
	Outline	EC	EC	EC	EU	EU
	Pathway	EC	EC	EC	EU	EU
	Details	EC	EC	DC	DU	DU
Agent	Rescue Team	EC	DU	EC	EC	EU
	Fire Brigade	EC	DU	EC	EC	EU
	Police Force	EC	DU	EC	EC	EU
	Civilian	EC	DU	DC	EC	EU
View Controls						
Icon Display		EC	DC	EC	EC	EU
Target Object	Rescue Team	DC	DU	EC	EC	EU
	Fire Brigade	DC	DU	DC	DC	EU
	Police Force	DC	DU	DC	DC	DU
	Civilian	DC	DU	DC	DC	DU

EC: Display *E*nabled and user *C*ontrollable
EU: Display *E*nabled and user *U*n-controllable
DC: Display *D*isabled and user *C*ontrollable
DU: Display *D*isabled and user *U*n-controllable

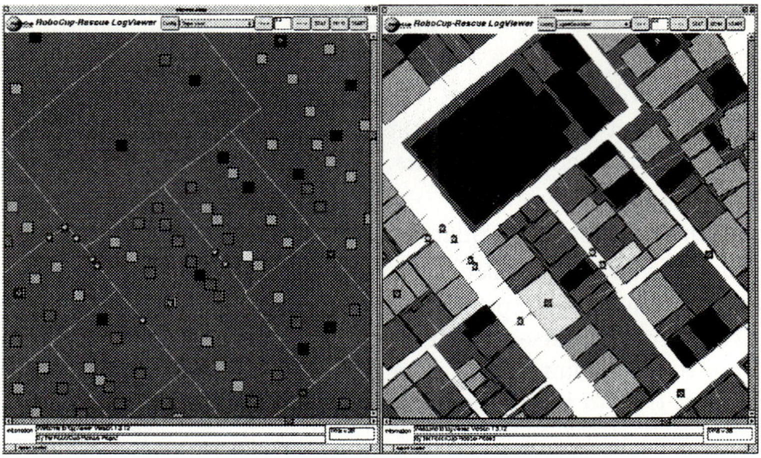

Fig. 3. Screen images of supervisor mode. A) Supervisor mode(left) displays abstract information. B) Agent developer mode(right) displays detailed information of agents.

Figure 3 shows examples of screen images with different user models. In supervisor mode (left), the details of GIS objects aren't displayed. Instead, the activities of agents are highlighted to help making supervisors decisions. On the contrary, the details of buildings, road, agents are displayed in agent developer mode (right).

5 Conclusions and Future Works

We proposed the idea of *user model* for the visualization of complex simulation systems that have many users with various requirements. *User model mechanism* is implemented in RoboCup-Rescue 2D viewer with five user models. In 2D viewer, *user model* is represented a set of visualization controls.

The usability of *user model* is not evaluated. We are planning to carry out some experiments with logViewer.

References

1. Hiroaki Kitano et al., *RoboCup-Rescue: search and rescue in large-scale disasters as a domain for autonomous agents research*, Proc. IEEE SMC, 1999.
2. Hiroaki Kitano et al., *RoboCup-Rescue – Grand challenge for disaster mitigation (in Japanese)*, Kyoritsu-shuppann, May, 2000
3. The RoboCup-Rescue Technical Committee, *RoboCup-Rescue Simulator Manual (Version 0 Revision 2)*, The RoboCup Federation, 2000.
4. Yoshitaka Kuwata and Atsushi Shinjoh, *Design of RoboCup-Rescue Viewers – Toward a Real World Emergency System –*, Proc. RoboCup Workshop, Aug., 2000.

A Modular Hierarchical Behavior-Based Architecture*

Scott Lenser, James Bruce, and Manuela Veloso

Computer Science Department
Carnegie Mellon University
5000 Forbes Ave
Pittsburgh, PA 15213
{slenser,jbruce,mmv}@cs.cmu.edu

Abstract. This paper describes a highly modular hierarchical behavior-based control system for robots. Key features of the architecture include: easy addition/removal of behaviors, easy addition of specialized behaviors, easy to program hierarchical structure, and ability to execute non-conflicting behaviors in parallel. The architecture uses a unique reward based combinator to arbitrate amongst competing behaviors such as to maximize reward. This behavior system was successfully used in our Sony Legged League entry in RoboCup 2000 where we came in third losing only a single game.

1 Introduction and Related Work

Briefly, our behavior system is a modular hierarchical behavior-based system with a unique behavior combinator. By modular, we mean that behaviors can be added, removed, or replaced without affecting other behaviors. By hierarchical, we mean that the system consists of many levels which operate at different levels of detail and/or time scales. Behaviors form the basic modular blocks upon which the architecture is built. Behavior modules can be swapped with similar behavior modules that accomplish similar goals (possibly by very different means). By behavior combinator, we mean a method for selecting which of a set of behaviors should be run together (or combined) to control the robot. We use a unique method of choosing which behaviors to run which allows behaviors to run in parallel while closely approximating the optimal policy (maximal reward) assuming the behavior reward values are calculable.

This behavior system was successfully used in the Sony Legged League of RoboCup 2000 [7] where we came in third losing only a single game. The quadruped robots for this league are generously provided by Sony [4]. The robots are fully autonomous, and have onboard cameras. See our paper in Agents [8] for a description of the overall team.

* This research was sponsored by Grants Nos. DABT63-99-1-0013, F30602-98-2-0135 and F30602-97-2-0250. The information in this publication does not necessarily reflect the position of the funding agencies and no official endorsement should be inferred.

A. Birk, S. Coradeschi, and S. Tadokoro (Eds.): RoboCup 2001, LNAI 2377, pp. 423–428, 2002.

Much work has been done on architectures for behaviors in general and behavior-based architectures in particular. Rodney Brooks has investigated behavior-based architectures in his subsumption architecture [3]. Ron Arkin has produced a thorough examination of behavior-based systems [1]. SRI has created a more deliberative form of behavior-based system in PRS [5]. We build upon the architecture used by the FU-Fighters small size RoboCup team [2]. This architecture is based upon the Dual Dynamics architecture [6].

We make several significant modifications to the architecture that together are the major contribution of this paper:

- We decouple the parts the system as much as possible while maintaining a hierarchical system. Our system can be viewed as a hierarchical analog of the subsumption architecture. We decouple behavior activations from the behavior level above them by introducing intermediate goals which allow the behaviors to calculate their own activation. The intermediate goals (or control sets) act as the only interface between behavior levels. We decouple the selection of behavior level from sensor level by allowing each behavior access to all of the sensors. This avoids unnecessarily restricting the information available to a behavior.
- We introduce a special combinator to select multiple non-conflicting behaviors to be run simultaneously. By non-conflicting, we mean behaviors that do not require the same resource. This means separate parts of the robot can be controlled by separate pieces of code when desired and by a single piece of code when needed.

2 Behavior Architecture

Our behavior architecture is a hierarchical behavior-based system. The architecture is primarily reactive, but some behaviors have internal state to enable sequencing of behaviors and hysteresis. The input to the system is information about the objects seen (from the vision) and an estimate of the robots location (from the localization). Output from the behavior system consists of choosing between walking, kicking, getting up, looking at objects, etc. and selecting appropriate parameters for the motion chosen. Note that looking at objects requires the use of the head and can be run in parallel with walking but kicking (the robot actually heads the ball) and getting up require the use of the whole body. Our system intelligently makes trade-offs between walking/looking and kicking. The behavior architecture consists of three interconnected hierarchies for sensors, behaviors, and control (see Figure 1). The sensor hierarchy represents all that is known about the world. The behavior hierarchy makes all of the robot's choices. The control hierarchy encodes everything the robot can do.

The sensor hierarchy represents the knowledge that the robot has about the world. The lowest level of the sensor hierarchy are real sensors that are filled in from hardware sensors or from the other modules (vision, localization, motion). The other levels are virtual sensors for convenience that are computed based upon the real sensors. All behaviors have access to all of the sensors.

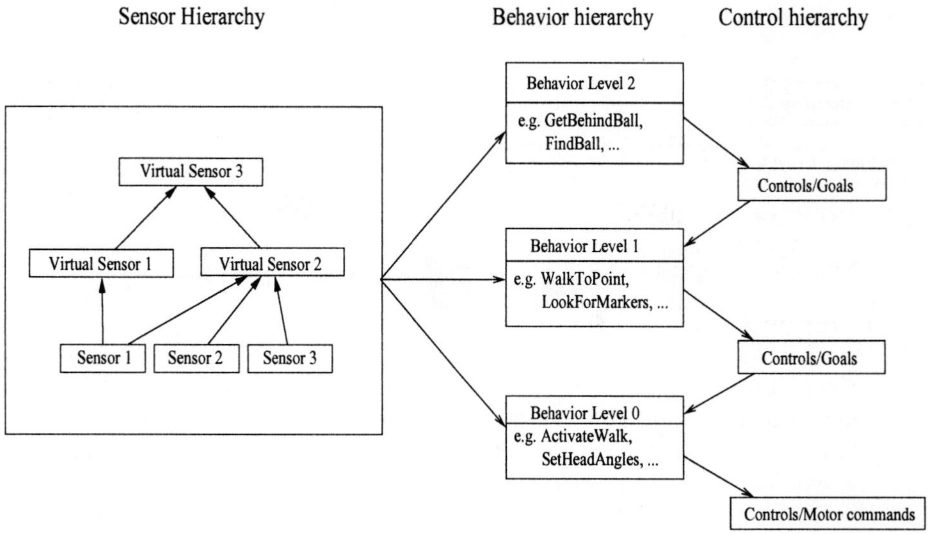

Fig. 1. Overview of the behavior system.

The behavior hierarchy controls what the robot does. Each level k of the behavior hierarchy, which we call a behavior set, is a subsumption architecture. The inputs of the level are the sensors and the control set at level $k+1$. The outputs of the level are fed into behavior level $k-1$ via the control set at level k. Together, the controls sets form the control hierarchy. The control set at level 0 then becomes the action(s) executed by the robot.

Each behavior set takes an abstract description of the task to be performed and creates a more concrete description of actions to be performed. The control hierarchy defines the interface between the various behavior levels. Each control set describes everything the robot can do at some level of detail. Each control set acts as a virtual actuator to the behavior level above it and as a goal set to the behavior level below it.

Behaviors have functions for calculating activation levels and outputs given the sensors and higher level controls. Each behavior looks at the sensors and the goals from higher levels and computes an activation value for the behavior (see middle of Figure 2). These activation values represent predictions of the future reward that will result if this behavior is run. These activations are used by the behavior set (via the combinator) to decide which behavior(s) to run. The processing function for each chosen behavior converts the sensors and control inputs into control outputs.

Each behavior drives a set of control outputs. Some behaviors within the behavior set will drive the same actuator/control, and thus conflict, while some will drive different actuators and can be run in parallel. Both cases occur frequently in our domain, walking/kicking vs. full body kicking motions. Conflicting be-

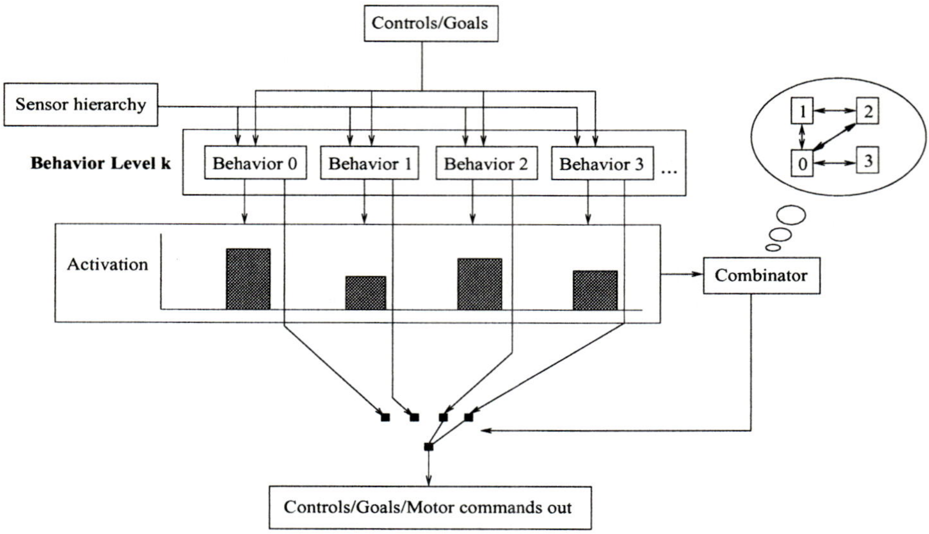

Fig. 2. Detail of a level in the behavior hierarchy.

haviors are mutually exclusive. We represent this constraint as a graph where behaviors are nodes and edges connect behaviors that conflict (see upper right part of Figure 2). We use a special combinator (described below) to choose a set of non-conflicting behaviors with near maximal total expected reward. For the example in the figure, the combinator has chosen behaviors 2 and 3 since this has more reward than executing behavior 0 alone. The chosen behaviors are run and use the sensors, the control set from above, and their memory of what they were doing to choose the controls to write directly into the goals of the next lower level.

3 Selection of Action(s)

The goal of the combinator is to find a set of non-conflicting behaviors that result in the maximal reward. This maximal reward set is the optimal policy under the assumption that the behavior activations are accurate estimates of future reward. Since the reward estimates(activations) and conflict net are given, this is the problem of finding maximal weight cliques in the dual of the conflict graph. Since this problem is NP-complete, we use an approximation algorithm. The basic idea is to find a suitably good approximation iteratively by suppressing weakly activated behaviors with many conflicts and reinforcing strongly activated behaviors with few conflicts.

To do this, we first produce an optimistic estimate for the total reward that can be achieved while running behavior k by assuming that all behaviors that are not in direct conflict with behavior k can be run in parallel. This is calculated by finding the total activation of all behaviors and subtracting the activation of all behaviors in direct conflict. We then treat this estimate as a gradient to change the activation of the kth behavior. Behaviors that might be runnable with more reward than the behaviors they conflict with are reinforced while other behaviors are suppressed. Usually, at least one of the behaviors will have a negative gradient. We follow this gradient over all behavior activations until the activation of one of the behaviors becomes 0. Any behavior whose activation becomes 0 is removed from consideration. This process is repeated until the set of behaviors with non-zero activation contains no conflicts.

Small random perturbations are added to the activations to break any ties. In case all the gradients are positive, we double the amount subtracted for conflicts until one of the gradients becomes negative. In the worst case, it may take $O(n \lg n)$ iterations for the combinator to converge (where there are n behaviors) but for most cases it converges in a few iterations. The set of behaviors with non-zero activation at the end of this process are run completing the execution of the behavior set.

4 Discussion and Future Directions

The behavior system we developed has some interesting features. The system tends to be highly reactive. There is nothing in the system, however, that prevents the use of behaviors with internal state. In fact some of our behaviors use stateful systems internally. The core feature of the system is its modularity. This has many important consequences which are detailed below:

scalable - Since each behavior level is completely separated from the neighboring behavior levels by the control sets, changes to one behavior level do not require any changes to any other part of the system. Note that this is different from the FU-Fighters architecture where higher levels set activations for lower levels. In the FU-Fighters architecture, changes to one level potentially affect all behavior levels above that level and always involve at least two levels.

easy to add/remove behaviors - Behaviors can easily be turned on/off from a configuration file. Because behaviors compete amongst themselves for the right to run, the robot simply uses the best remaining behavior(s) to replace a behavior that has been removed. It is easy to replace a behavior with a different implementation or leave both implementations accessible and selectively disable one from a file.

easy to specialize a behavior - The system makes it easy to specialize behaviors. For example, a quicker but inaccurate shot behavior can be added without any other changes. With an appropriate activation function, the robot will then intelligently decide between quick and accurate shooting. This allows the flexibility to create several specialized behaviors instead of one general purpose behavior that has to handle every special case.

separates concerns when possible - Behaviors only interact through their activation functions and only with conflicting behaviors. So changes to behaviors controlling the head of our robot require no changes to behaviors controlling the legs and vice versa. The ability to separate the control of the head and the legs for most, *but not all*, behaviors allows the system to be conveniently decomposed into mostly independent parts, while preserving the ability for coordinated behavior. This same ability would also allow a multirobot team to work mostly independently except when collaboration, such as passing, requires working together.

Our behavior architecture is best understood as a step in the right direction. The architecture enables new capabilities in behaviors but as always we find that there is room for improvement. The activation functions present an opportunity for learning the future reward that will result from choosing an action now. Oscillations could be avoided by reasoning about the uncertainty present in the activation values and the cost of switching from one behavior to another. The cost of switching behaviors could also be better represented by each behavior providing an expected reward over time so that failure to produce the expected reward could be detected and taken into account.

5 Conclusion

We presented a field tested behavior system aimed at being more modular than existing systems. Behaviors can be removed, replaced, changed, or added extremely easily. The architecture also extends existing behavior-based systems to allow parallel control of multiple parts of a robot in a safe manner without disallowing full robot motions. The system was used on quadruped legged robots in the Sony Legged League of RoboCup-2000 where our team of robots came in third place only losing a single game.

References

1. R. C. Arkin. *Behavior-Based Robotics.* MIT Press, Cambridge, 1998.
2. S. Behnke, B. Frötschl, R. Rojas, et al. Using hierarchical dynamical systems to control reactive behavior. In *Proceedings of IJCAI-99*, pages 28–33, 1999.
3. R. Brooks. Elephants don't play chess. In P. Maes, editor, *Designing Autonomous Agents*, pages 3–15. MIT Press, Cambridge, 1990.
4. M. Fujita, M. Veloso, W. Uther, M. Asada, H. Kitano, V. Hugel, P. Bonnin, J.-C. Bouramoue, and P. Blazevic. Vision, strategy, and localization using the Sony legged robots at RoboCup-98. *AI Magazine*, 1999.
5. M. P. Georgeff and A. L. Lansky. Reactive reasoning and planning. In *AAAI87*, pages 677–682, Seattle, WA, 1987.
6. H. Jaeger and T. Christaller. Dual dynamics: Designing behavior systems for autonomous robots. In S. Fujimura and M. Sugisaka, editors, *Proceedings International Symposium on Artificial Life and Robotics.*, 1997.
7. H. Kitano, M. Asada, Y. Kuniyoshi, I. Noda, and E. Osawa. Robocup: The robot world cup initiative. In *Proceedings of the IJCAI-95 Workshop on Entertainment and AI/ALife*, 1995.
8. S. Lenser, J. Bruce, and M. Veloso. CMPack: A complete software system for autonomous legged soccer robots. In *Autonomous Agents*, 2001.

Localization and Obstacles Detection Using Omni-directional Vertical Stereo Vision

Takeshi Matsuoka[1], Manabu Araoka[2], Tsutomu Hasegawa[3], Akira Mohri[3], Motoji Yamamoto[3], Toshihiro Kiriki[3], Nobuhiro Ushimi[3], Takuya Sugimoto[2], Jyun'ichi Inoue[3], and Yuuki Yamaguchi[3]

[1] Fukuoka University
8–19–1 Nanakuma, Jonan-ku, Fukuoka 814–0180 JAPAN
matsuoka@ctrl.tec.fukuoka-u.ac.jp
[2] Hitachi Information & Control Systems, Inc.
[3] Kyushu University

Abstract. This paper describes an omni-directional vertical stereo vision system for an autonomous mobile robot. The system is composed of two omni-directional cameras and enables both the localization with wide range of view and the obstacles detection by 3D measurement of arbitrary direction. The system architecture and an experimental result are shown.

1 Introduction

Mobile robot needs functions of localization and obstacle detection for completion of a task. Works done on the localization problem bring two approaches. The first approach is based on dead reckoning. The second one consists in determining the robot's position and direction without using the initial configuration of the robot. The robot's configuration is calculated from exteroceptive sensor data[1][2][3]. To assure the precision of localization in this approach, many landmarks need to be observed at the same time. Therefore, omni-directional vision system is proposed[4].

The recognition of high landmarks make localization easily because they are usually visible even in the presence of the obstacles. For avoidance of the obstacles, not only direction of the obstacle but also distance between the robot and the obstacle is necessary. The distance cannot be obtained from single camera. The measurement systems are proposed using both the camera and another sensor, for example, ultrasonic sensor[5], laser range finder[1], etc.. In case that plural robots are moving simultaneously, ultrasonic waves sent by one robot interfere in the function of reception of another robot. Laser range finder is weak in recognition of moving obstacles and cannot sense obstacles in 360 degrees of area in real-time.

In this paper, we propose an omni-directional vertical stereo vision system. This system enables both the localization and the obstacles detection. This system consists of two omni-directional cameras. The system recognizes the object

A. Birk, S. Coradeschi, and S. Tadokoro (Eds.): RoboCup 2001, LNAI 2377, pp. 429–434, 2002.

which is located high. In addition to this, the 3D information of the plural obstacles at the side of the robot is calculated with triangulation in real-time.

2 Omni-directional Vertical Stereo

An omni-directional camera(Fig.1) consists of a hyperboloidal mirror which faces downward along the vertical axis and a CCD camera which faces upward along the vertical axis. These two components are mounted onto the top and the bottom of the acrylic resin cylinder respectively. The mirror is equipped with a needle to suppress diffused reflection.

In the ordinary method of stereo vision, the two cameras are configured horizontally and the baseline of triangulation is in the horizontal plane(Fig.2). This configuration brings following problems.

1. An epipolar line becomes curved line. Therefore the computational cost increases.
2. Accuracy of the 3D measurement depends on the direction of a landmark. In Fig.2, Accuracy of position estimation of point 'b' is worse than the one of point 'a'. The position of point 'c' cannot be calculated because point 'c' is not perceptible by the camera1.

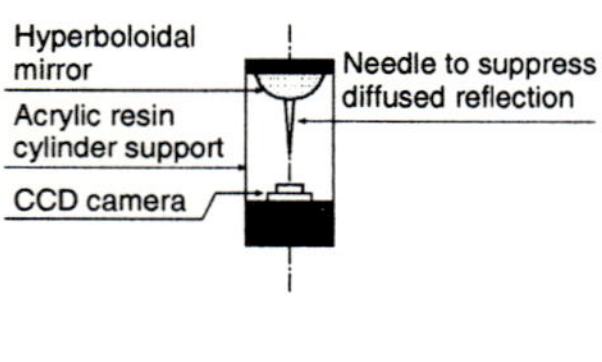

Fig. 1. Omni-Directional Camera

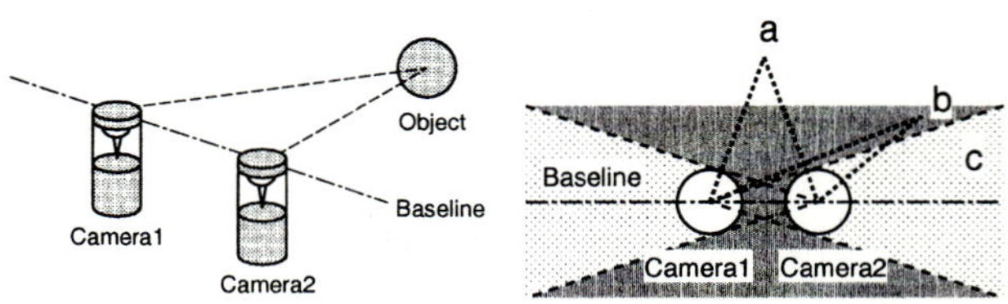

Fig. 2. Horizontal Stereo

To cope with these problems, we developed the omni-directional vertical stereo vision system. Fig.3 shows our system. The system is composed of two vertically arranged omni-directional cameras. The axis of one camera is in line with the axis of the other. The area named 'a' is the area where the object is recognized by both cameras. 3D position of the object in the area 'a' can be calculated by stereo matching. In our omni-directional vertical stereo system, an epipolar line becomes radial straight line. Therefore the computational cost is reduced. Besides, despite of the direction of the landmark, accuracy of the 3D measurement is constant in one horizontal plane.

Fig.4 shows our omni-directional vertical stereo vision. Let OM_i be the hyperboloidal focal point. Let the coordinate value be $(0, 0, bz_i)$. Define $P(X, Y, Z)$ to be an arbitrary point in the 3D space. Let the corresponding points on the image planes of P be $p_i(x_i, y_i)$ Unit vectors $e_i(ex_i, ey_i, ez_i)$ from OM_i to P can be calculated uniquely. Then the following equation is obtained:

$$X = \frac{ex_1 ex_2 (bz_2 - bz_1)}{ex_2 ez_1 - ez_2 ex_1}, \; Y = \frac{ey_1 ex_2 (bz_2 - bz_1)}{ex_2 ez_1 - ez_2 ex_1}, \; Z = \frac{ez_1 ex_2 bz_2 - ez_2 ex_1 bz_1}{ex_2 ez_1 - ez_2 ex_1} \tag{1}$$

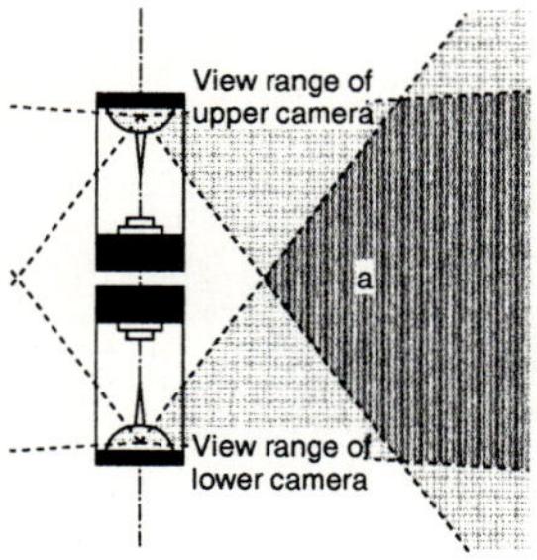

Fig. 3. Omni-Directional Vertical Stereo

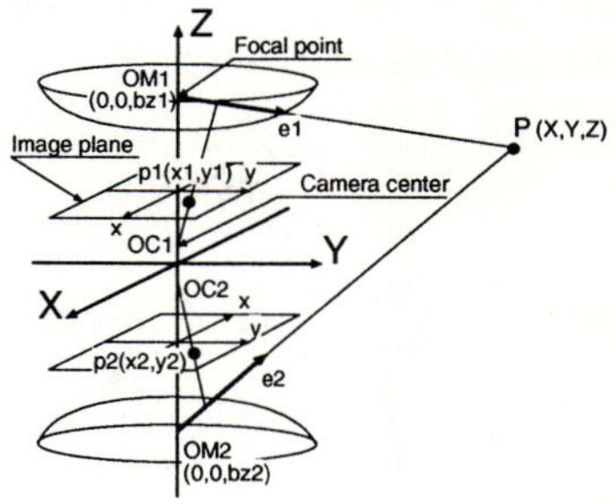

Fig. 4. Camera Coordinate

3 Construction of Omni-directional Vertical Stereo Vision System

We developed the omni-directional vertical stereo vision system and mount it on our autonomous mobile robot. Fig.5 shows our system and the images obtained by the system. Occlusion occurs by the support pillar in the image of lower camera. Refinement of the shape of the pillar would decrease the occlusion.

The vision system is composed of two omni-directional cameras, CPU board (K6–2 300MHz, 64KB memory) and two image capture/processing boards (HITACHI IP5005, 256x240 pixel). The image of each camera is captured by the corresponding image/processing board. Then the boards process the captured images respectively.

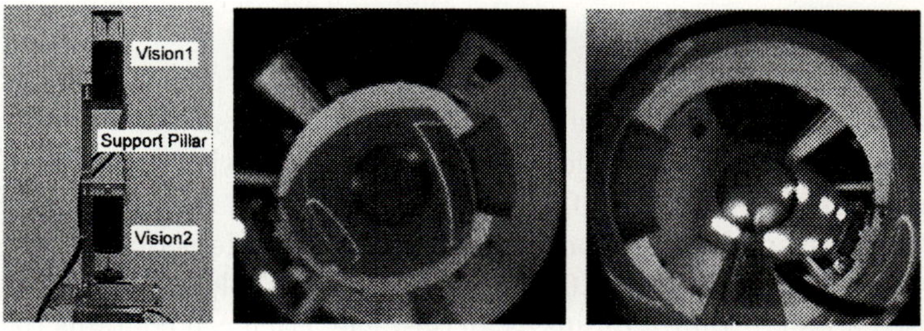

Fig. 5. Omni-Directional Vertical Stereo Vision and Images Taken by the System

3.1 Localization

Our robot moves around in the soccer field shown in Fig.6. Let Σ_B be an internal base frame. Let a, b, c and d be the positions of all posts as landmarks. Let r be the position of the robot. Let r' be the candidate for position of the robot. Localization is made based on the directions of the landmarks. We make two circumscribed circles(Fig.7) where the angle "$a-r'-b$" and the angle "$c-r'-d$" are constant respectively, i.e.,

$$(x-a_1)^2+y^2=r1^2,\ a_1=-2010+\frac{1000}{tan\theta_1},\ r_1=\frac{1000}{sin\theta_1}$$
$$(x-a_3)^2+y^2=r3^2,\ a_3=2010-\frac{1000}{tan\theta_3},\quad r_3=\frac{1000}{sin\theta_3}$$

One of the intersection of two circles is the position of the robot, i.e.,

$$\begin{cases} X=\frac{1}{2}\{(a_3+a_1)-\frac{r_3^2-r_1^2}{a_3-a_1}\} \\ Y=\pm\sqrt{-\frac{1}{4}(a_3-a_1)^2+\frac{1}{2}(r_3^2+r_1^2)-\frac{1}{4}\frac{(r_3^2-r_1^2)^2}{(a_3-a_1)^2}} \\ \quad if\ \ \theta_4\geq\theta_2\ \ then\ \ Y\geq 0\quad else\quad Y<0 \end{cases} \tag{2}$$

A set of singular points exists where the robot cannot localize this point(Fig.8). If the robot is on this circumference, a_1 and a_3 equals to 0 and equation(2) cannot be solved. In this case, other localization method is necessary for example, dead reckoning.

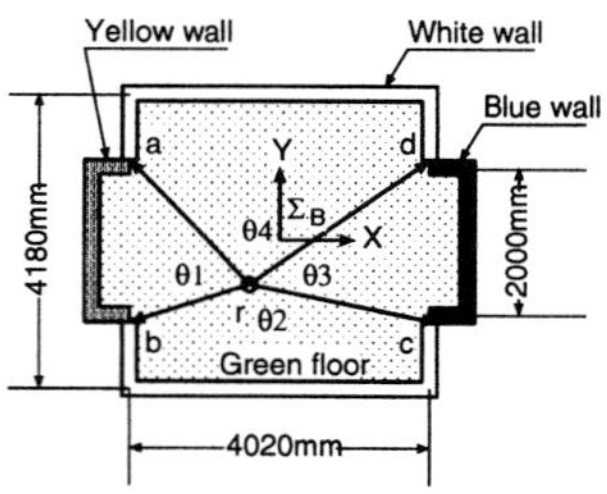

Fig. 6. Soccer Field

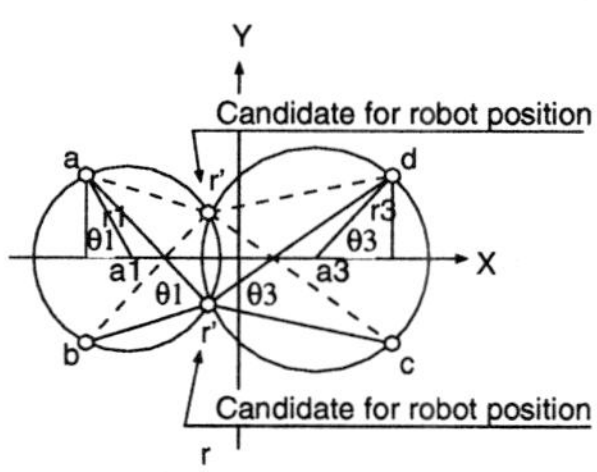

Fig. 7. Localization

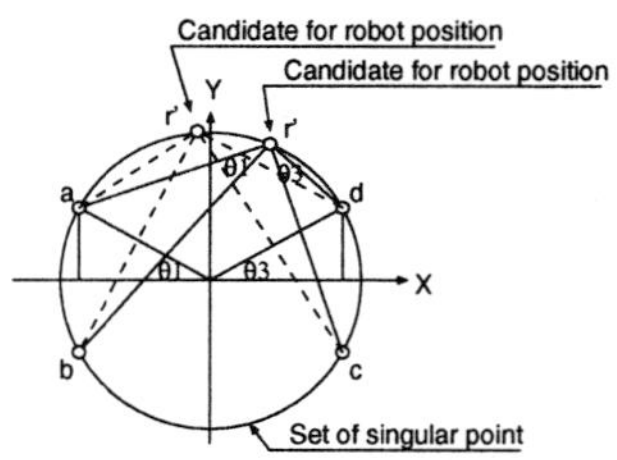

Fig. 8. Singular Points

3.2 Obstacle Detection

We have evaluated the obstacle detection error. Vertical plane which includes the Z axis of our system is divided into regular grids with intervals of $200mm$ along the horizontal axis and $100mm$ along the Z axis. At each point we have measured the obstacle detection error. The obstacle detection error on point $(L,Z)=(200(n+1),100n)n=0,1,2...$ is shown in Table1.

We used a ball of $75mm$ radius as an obstacle. Each cell contains the average of errors along horizontal axis and the average of errors along Z axis. The blank cell means that the obstacle is not recognized by both cameras. In the cells neighboring the blank cells, The error is larger.

4 Localization and Obstacle Detection/Avoidance Experiment

We show an experimental result of localization and obstacle detection/avoidance with our omni-directional vertical stereo vision. We located the obstacle on the

700	–	–	–	–	–	–	42.9	113.6	40.5
	–	–	–	–	–	–	-54.4	-13.9	-37.7
600	–	–	–	–	57.3	136.5	143.3	104.2	25.9
	–	–	–	–	-88.6	1.6	15.0	-2.1	-31.0
500	–	–	–	123.2	170.5	148.4	111.4	69.5	36.4
	–	–	–	-25.6	34.1	28.9	19.3	5.7	-6.4
400	–	–	–	187.7	124.1	125.6	73.2	73.6	43.4
	–	–	–	30.0	11.9	33.5	13.3	8.5	-0.4
300	–	–	146.5	129.4	93.3	114.8	85.3	92.5	53.7
	–	–	31.3	9.1	10.2	35.1	32.2	17.4	4.5
200	–	115.6	101.5	98.2	70.2	87.0	56.5	63.4	32.6
	–	26.8	22.1	15.3	13.3	36.5	22.6	23.1	16.0
100	–	73.6	73.0	82.8	62.0	94.9	50.7	42.1	46.1
	–	16.4	23.2	24.6	26.9	48.6	41.5	36.7	20.5
Z=0	77.0	59.2	48.7	74.8	53.6	63.8	61.9	65.4	49.2
	22.6	17.2	15.1	29.8	20.0	39.1	45.1	37.7	35.1
	L=200	400	600	800	1000	1200	1400	1600	1800

Table 1. Obstacle Detection Error

point $(0, 0)$. The robot moves from the point $(-1500, 0)$ to the point $(1500, 0)$. Calculation cycle of our vision system is $100msec$.

Fig. 9. Robot Movement

Fig.9 shows the snapshot of the robot movement. Fig.10 shows the loci of the robot. The continuous line indicates the estimated position of the robot based on dead reckoning. The broken line indicates the estimated position of the robot based on our vision system. To avoid the obstacle, the robot turned to the left. Then, to reach the goal, the robot turned to the right and went straight. In the neighborhood of the goal, the robot successfully stopped. In this case, robot's movement time is short(7 seconds) and the floor is flat with sufficient coefficient of friction. Therefore, the localization by dead reckoning is more accurate than the one by our system. In the actual usage of the robot, the movement time becomes long and the field is not ideal. Our system can estimate the position and direction of the robot within constant range of error independent of movement time or characteristics of the floor.

Fig.11 shows the distance between the robot and the obstacle obtained by 3D calculation. Fig.12 shows the direction of the obstacle in the coordinate frame attached to the robot. The robot determined the 3D position of the obstacle of any direction with sufficient accuracy.

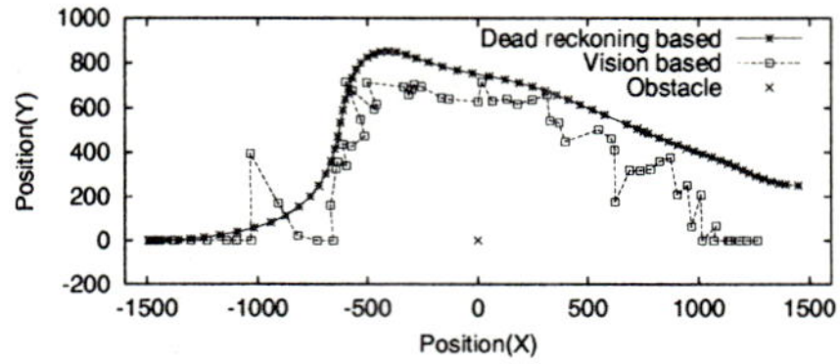

Fig. 10. Loci of the Robot

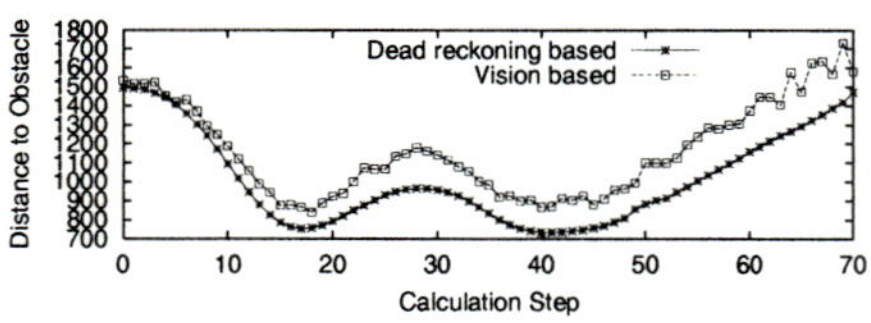

Fig. 11. Distance between the Robot and the Obstacle

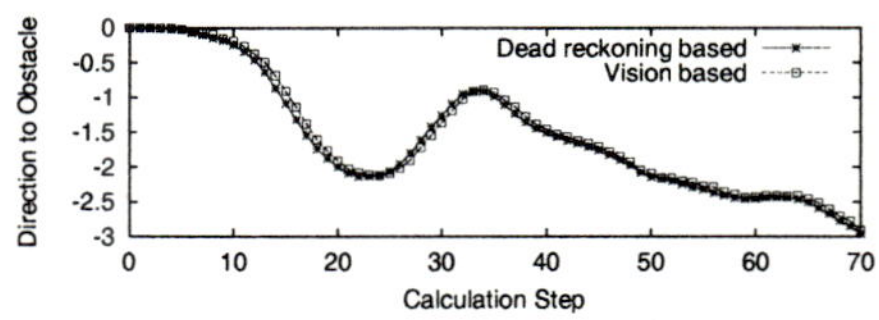

Fig. 12. Direction of the Obstacle in the Coordinate Frame Attached to the Robot

5 Conclusion

Omni-directional vertical stereo vision system has been proposed. Wide range of view enables the usage of many landmarks in one calculation of the localization and increase the robustness of localization. The ability of 3D measurement of arbitrary direction enables efficient motion planning of the robot. This characteristics is useful for cooperation of multi-robots. Each robot recognizes the object and others and plans the optimal motion of the robot.

References

1. A. Clerentin, L. Delahoche, C. PEGARD and E. Brassart, "A localization method based on two omnidirectional perception systems cooperation". *Proc. of 2000 IEEE Int. Conf. on Robotics & Automation*, pp.1219-1224, 2000.
2. T.Yoneda, Y.Shirai, and J. Miura, "Navigation of a Vision-Guided Mobile Robot in an Unknown Environment". *Proc. 1998 JAPAN-U.S.A. SYMPOSIUM ON FLEXIBLE AUTOMATION*, pp.II:569-572.
3. A Yamamoto S. Maeyama A. Ohya and S. Yuta, "An Implementation of Landmark-based Position Estimation Function as an Autonomous and Distributed System for a Mobile Robot". *Proc. of 1999 IEEE/RSJ Int. Conf. on Intelligent Robots and Systems*, pp.1141-1148, 1999.
4. Y. Yagi, H. Nagai, K. Yamazawa and M. Yachida, "Reactive Visual Navigation based on Omnidirectional Sensing –Path Following and Collision Avoidance–". *Proc. of 1999 IEEE/RSJ Int. Conf. on Intelligent Robots and Systems*, pp.58-63, 1999.
5. J. Leonard and H. Durrant-Whyte, "Mobile robot localization by tracking geometric beacons". *IEEE Trans. Robotics & Automation*, Vol.7, pp.89-97, June 1991.

Karlsruhe Brainstormers - A Reinforcement Learning Approach to Robotic Soccer

A. Merke and M. Riedmiller

Institut für Logik, Komplexität und Deduktionssyteme
University of Karlsruhe, 76131 Karlsruhe, Germany

Abstract. Our long-term goal is to build teams of agents where the decision making is based completely on Reinforcement Learning (RL) methods. It requires an appropriate modelling of the learning task and the paper describes how robotic soccer can be seen as a multi-agent Markov Decision Process (MMDP). It discusses how optimality of behaviours of agents can be defined and what difficulties one encounters in developing concrete algorithms which are supposed to reach such optimal agent/team policies. We also give an overview of already incorporated algorithms in our 'Karlsruhe Brainstormers' simulator league team and report some results on learning of offensive team behaviour.

1 Introduction

The robotic soccer domain became very popular during the last few years. Since 1997 there are annual world championships to measure the progress of playing quality between the different approaches pursued around the world. Our group takes the approach of viewing the robotic soccer as MMDP while using techniques from Reinforcement Learning. Section 2 introduces basic definitions of optimality. In section 3 we show what difficulties have to be mastered to use the theoretical optimality criteria defined in section 2. Finally in section 4 we discuss some practical work in the development of our Robocup competition team 'Karlsruhe Brainstormers'.

2 Robotic Soccer as a Reinforcement Learning Problem

A MMDP is defined as a tuple (cf. [3])

$$M_n := [S, A, r, p],$$

where S is the space of all states, A is a cartesian product of action sets $A = A_1 \times \ldots \times A_n$ and p denotes the state transfer probabilities, i.e. $p(\cdot\,|s,a)$ is a probability measure on S depending on the current state s and the joint action $a = (a_1, \ldots, a_n)$. In this paper we will concentrate on two special MMDPs cases: cooperative and zero sum MMDPs. We will combine them to characterise the

A. Birk, S. Coradeschi, and S. Tadokoro (Eds.): RoboCup 2001, LNAI 2377, pp. 435–440, 2002.

robotic soccer environment[1]. In the cooperative MMDP case all agents get the same reinforcement i.e. $r_1 = \ldots = r_n$. In the zero sum MMDP we have two agents with reversed reinforcements, $r_1 = -r_2$.

To define optimality we need the notion of a (total) policy. A total deterministic policy π is a mapping

$$\pi : S \to A = A_1 \times \ldots \times A_n$$

It is important to see that this definition concerns the behaviour of all agents simultaneously and that the policy of a single agent i can be seen as a projection $\pi_i : S \to A_i$ in $\pi = (\pi_1, \ldots, \pi_n)$. Every agent concerns the value $v_i[\pi]$ of a total policy, which is a expectation of the sum of its future reinforcement signals

$$v_i[\pi](s) = v_i[\pi_1, \ldots, \pi_n](s) = E\left[\sum_k r_i(s_k, \pi(s_k)) \mid s_0 = s\right].$$

In the cooperative MMDP case we have $v_1[\pi] = \ldots = v_n[\pi]$ for all π. All we need is an optimal total policy $\pi^\star$ which has the maximal value

$$\pi^\star = \arg\max_\pi v_i(\pi).$$

The existence of $\pi^\star$ is guaranteed ([6]), but this existence doesn't say anything how such a $\pi^\star$ can be computed. We will discuss this peculiarity further in section 3. In the two agent zero sum MMDP we have a total policy $\pi = (\pi_1, \pi_2)$. Borrowing from game theory we can define a minimax policy for each agent. We look for a total policy $(\pi_1^\star, \pi_2^\star)$ which fulfils

$$v_1[\pi_1^\star, \pi_2^\star] = \max_{\pi_1} \min_{\pi_2} v_1[\pi_1, \pi_2] = \min_{\pi_1} \max_{\pi_2} v_1[\pi_1, \pi_2] = -v_2[\pi_1^\star, \pi_2^\star]$$

Existence of such a pair $(\pi_1^\star, \pi_2^\star)$ can be always assured (see [3]). Such policies are in general no longer deterministic, i.e. they map from S to a probability distribution over A.

Let us now turn our attention to robotic soccer. Here we have two teams which perform a zero sum game. Each agent in the same team has the same objective: to score more goals then the agents from the opposite team. This can be expressed as zero sum MMDP of 2 cooperative teams

$$M = (S, A = B_1 \times \ldots \times B_m \times C_1 \times \ldots \times C_n, r, p)$$

where $r = (r_b, \ldots, r_b, r_c, \ldots, r_c)$ and $r_b = -r_c$ express the team respectively zero sum character. Using our previous optimality definitions it is now a straight forward task to define a total optimal policy in the case of two competitive teams (with cooperation within every team). We speak of an optimal total policy $(\pi_1^\star, \ldots, \pi_m^\star, \phi_1^\star, \ldots, \phi_n^\star)$ if

$$v_b[\pi_1^\star, \ldots, \pi_m^\star, \phi_1^\star, \ldots, \phi_n^\star] = \max_{(\pi_1, \ldots, \pi_m)} \min_{(\phi_1, \ldots, \phi_n)} v_b[\pi_1, \ldots, \pi_m, \phi_1, \ldots, \phi_n]$$

[1] A more extensive presentation can be found in [5].

3 Learning in Independent Distributed Systems

In section 2 we discussed how an optimal policy could be defined. What we did not discuss were the difficulties to find such a policy in a distributed way. We will demonstrate this in the cooperative MMDP case considering a possible solution in the zero sum MMDP with two teams as even harder to reach.

To illustrate the problem of action choice coordination imagine a small 1 state (deterministic) system with 2 cooperative agents, each having two actions:

$$M_2 = [\{s\}, \{a_1, a_2\} \times \{b_1, b_2\}, r = (r_1, r_2), p]$$

The reward function is for both agents equal $r_1 = r_2$ and given by $r_1(s, (a_1, b_1)) = r_1(s, (a_2, b_2)) = 2$ and equal 0 in connection with the remaining actions. It is easy to see that we have two optimal policies $\pi^\star(s) = (a_1, b_1)$ and $\hat{\pi}^\star(s) = (a_2, b_2)$. But if agent 1 decides to take the projection $\pi_1^\star(s) = a_1$ with respect to $\pi^\star$ and agent 2 takes the projection $\hat{\pi}_2^\star(s) = b_2$ with respect to $\hat{\pi}^\star$ we get a total policy $\pi = (\pi_1^\star, \hat{\pi}_2^\star)$ which isn't optimal anymore. The problem lies in the lack of coordination. In the rest of this section we will use three differenf agent types to demonstrate problems of multi-agent learning and coordination

- White Box Agents (WBA) also called Joint Action Learners (see [2]) are agents which have knowledge of all the joint actions $a = (a_1, \ldots, a_n)$ performed in every step.
- Black Box Agents (BBA) also called Independent Learners are agents which just know about their own actions. They aren't really aware of the other agents, all the influence through actions of other agents could also be interpreted as environmental noise.
- Gray Box Agents (GBA) are Black Box Agents which can communicate with other agents. It is no further specified how much information can be exchanged. If no communication takes place we remain having BBAs, but if every agent communicates his action we get the special case of WBAs. But we can do even more with communicating agents, we can exchange information about their future intentions.

In figure 1 we can see the connections between the agents models. The WBA case can be identified with the single agent MDP situation. We will demonstrate this using the terminology of Q-learning ([11, 1]). To this end we define

$$q^\star(s,a) := r(s,a) + \sum_{j \in S} p(j \mid s,a) v[\pi^\star](j)$$

in which $\pi^\star$ in an optimal total policy. The value of $q^\star(s,a)$ represents the expected reward[2] for using action a in state s and pursuing an optimal total policy afterwards. The knowledge of $q^\star$ amounts to knowing all optimal policies as we have

$$\pi^\star(s) \in \arg\max_{a \in A} q^\star(s,a)$$

[2] We don't distinguish different r_i here because all agents get the same rewards.

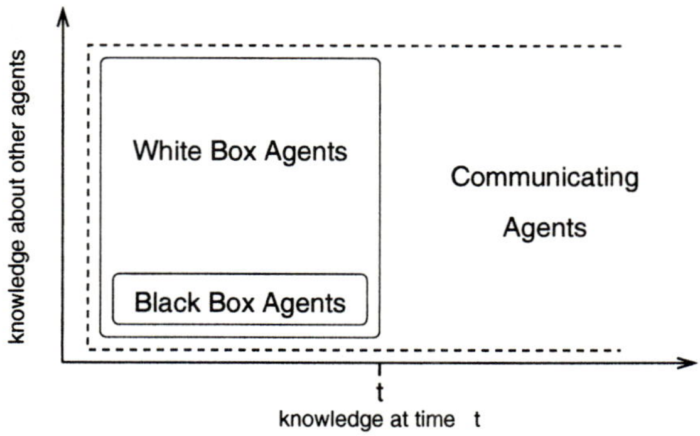

Fig. 1. Illustration of potential knowledge during learning in the different models.

for an arbitrary optimal policy $\pi^\star$. Using Q-learning we are able to compute the $q^\star$ function provided we get access to enough tuples

$$\langle s^t, a^t, r^t, s^{t+1} \rangle$$

where s^{t+1} is the successor state after using the joint action a^t in state s^t while getting reward $r^t = r(s^t, a^t)$. Using Q-learning every WBA can build up his own $q^\star$ function. The problem of choosing an unique optimal total policy $\pi^\star$ remains, so that each agent can use the projection $\pi_i^\star$ of the same optimal total policy. We will not discuss the possibilities of attaining this goal but we just mention that theoretically sorting all optimal total policies and taking the first will do the job.

The major drawback using WBA and joint actions $a = (a_1, \ldots, a_n)$ comes from the fact that the number of such actions grows exponentially with the number of agents. Furthermore such learning is inflexible with respect to changing the number of participating agents. These reasons are the main motivation for using the considerably weaker model of BBAs. But in this model all that we can hope for is to be able to compute

$$q_i^\star(s, b) = \max_{a \in A, a_i = b} q^\star(s, a)$$

for each agent i. This is the maximal possible expected reward if agent i is using action b in state s. Computing the $q_i^\star(s, b)$ constitutes the first problem which has to be solved. The second problem has to do with a coordinated choice of actions from $\arg\max q_i^\star$. If for example $\{b, b'\} = \arg\max q_i^\star$, then we have somehow to decide which of these two actions we'll actually take. Both problems can be solved if the underlying MMDP is deterministic (see [4]). In the case of probabilistic state transitions and the pure BBA scenario, it can be shown, that the computation of the $q_i^\star(s, b)$ values is in general impossible (cf. [5]).

There remains the question of how to deploy communication or some other coordination scheme to solve both problems using BBA agents (i.e. to use some incarnation of GBAs). As far as we know no such algorithm has been published yet. In the next section we will present some empirical work, which uses a sort of implicit coordination while working with Black Box Agents.

4 Our RL Approaches to Robotic Soccer

In [8] we already published some of our work related to single agent reinforcement learning. This includes successful deployment of moves to learn several basic agent behaviors. All these moves were learned using reinforcement learning with neural nets as function approximators.

On the tactical or team behaviour level we didn't use reinforcement learning until very recently. In particular our last year team used planning for players with the ball and a priority list of moves for players without the ball (see [8] for more details). As we already mentioned we work with higher level moves on the tactical level. As the first step of developing a whole team policy with RL we started with the attack behaviour.

We use 7 attackers against 7 defenders and one goalie. At the moment we put all positions of the players as input to an neural network (34 dimensions). Simultanously we also work on a feature extraction scheme, which will enable us to ignore the exact number of defenders and to lower the dimension of the encoding input vector.

Each of our attackers without ball has 10 actions to choose from. He can just go to one of 8 direction, go to his home position or try to intercept the ball. To go on we must say a little bit more about the home positions concept, as this is the main coordinating scheme for our agents. Our team can use different formations (for example 4-3-3) which are stretched over the soccer field with respect to the ball position and the offside lines. As each of our attackers has the choice to go towards his home position we have implicitly a mechanism which makes it easier to avoid two attackers going to the same position. If one attacker gains the ball he uses our planning algorithms which tries to find the shortest pass chain to score a goal.

Our algorithm learns along trajectories which lead to a goal or to the loss of the ball. In the first case we get a positive reward in the second a negative cost. To update the value for every state along a trajectory we use $TD[1]$ (see [10]) to propagate the reward/cost along the whole trajectory. The update of the neural networks are performed with a variant of the backpropagation algorithm called RPROP (see [7]). The results of this learning scheme are very promising. We used our last year team attack and the attack of the FCPortugal team[3] for comparison and our last year defence as a benchmark. The results are presented in the following table

	Brainstormers2000	$TD[1]$ attack
success rate	13%	20%

[3] Our team was runner up and the FCPortugal team was winner of the Simulation League Robocup World Championship in Melbourne 2000.

5 Summary

The soccer domain can be modelled as a multi-agent Markov Decision Process (MMDP). To deal with the multi-agent aspects, we pursue both the investigation of theoretically founded distributed RL algorithms plus the empirical/heuristically motivated way of modified single-agent Q-learning. We report very promising results in using learning of a coordinated offensive behaviour. Still a lot of research questions are open, for example the dealing with partial observability of state information, the definition of theoretically founded and efficient distributed learning algorithms (including opponent modelling) and the search for appropriate features.

5.1 Acknowledgements

We would like to thank our students (D. Eisenhardt, A. Hoffman, M. Nickschas, A. Sinner, O. Thate, D. Withopf) for their active involvement in the development of our team 'Karlsruhe Brainstormers'. Also many thanks go to former members of our development team (S.Buck, S. Dilger, R. Ehrmann, D. Meier). Finally we thank the CMU-Team 99 for providing parts of their world model.

References

[1] Dimitri P. Bertsekas and John N. Tsitsiklis. *Neuro-dynamic programming.* Optimization and neural computation series ; 3. Athena Scientific, 1996.

[2] Caroline Claus and Craig Boutilier. The Dynamics of Reinforcement Learning in Cooperative Multiagent Systems. In *IJCAI*, 1999.

[3] Jerzy Filar and Koos Vrieze. *Competitive Markov decision processes.* Springer, 1997.

[4] M. Lauer and M. Riedmiller. An algorithm for distributed reinforcement learning in cooperative multi-agent systems. In *Proceedings of International Conference on Machine Learning, ICML '00*, pages 535–542, Stanford, CA, 2000.

[5] A. Merke. Reinforcement Lernen in Multiagentensystemen. Master's thesis, Universität Karlsruhe, 1999.

[6] Martin L. Puterman. *Markov decision processes : discrete stochastic dynamic programming.* Wiley series in probability and mathematical statistics : Applied probability and statistics. Wiley, 1994.

[7] M. Riedmiller and H. Braun. RPROP: A fast and robust backpropagation learning strategy. In Marwan Jabri, editor, *Fourth Australian Conference on Neural Networks*, pages 169 – 172, Melbourne, 1993.

[8] M. Riedmiller, A. Merke, D. Meier, A. Hoffmann, A. Sinner, O. Thate, C. Kill, and R. Ehrmann. Karlsruhe brainstormers - a reinforcement learning way to robotic soccer. In A. Jennings and P. Stone, editors, *RoboCup-2000: Robot Soccer World Cup IV, LNCS.* Springer, 2000.

[9] P. Stone, R. S. Sutton, and S. Singh. Reinforcement learning for 3 vs. 2 keepaway. In *RoboCup-2000: Robot Soccer World Cup IV*. Springer Verlag, 2001.

[10] R. S. Sutton and A. G. Barto. *Reinforcement Learning: An Introduction.* MIT Press, 1998.

[11] Christopher J.C.H Watkins and Peter Dayan. Technical Note: Q-Learning. *Machine Leaning*, 8:279–292, 1992.

Interpretation of Spatio-temporal Relations in Real-Time and Dynamic Environments

Andrea Miene and Ubbo Visser

TZI - Center for Computing Technologies
University of Bremen
Universitätsallee 21-23
D-28359 Bremen, Germany
{andrea|visser}@tzi.de

Abstract. With the more sophisticated abilities of teams within the simulation league high level online functions become more and more attractive. Last year we proposed an approach to recognize the opponents strategy and developed the online coach accordingly. However, this approach gives only information about the entire team and is not able to detect significant situations (e.g. double pass, standard situations). In this paper we describe a new method which describes spatio-temporal relations between objects. This approach is able to track the objects and therefore the relations between them online so that we are able to interpret situations over time during the game. This enables us to detect the above mentioned situations. We can implement this in the online coach in order to enrich our team with high level functions. This new method is domain independent.

1 Motivation

The online coach is still the most effective instrument to analyze the opponent, because it can obtain all information about the simulated environment. Therefore, it is important to continue the development of the online coach which we used for the Virtual Werder team in the RoboCup 2000 tournament. In [Visser et al., 2001] we describe how the coach determines the opponent tactical formation with a neural network and how it is able to change a team formation during a match. We showed that it makes sense to recognize strategies and change the own team accordingly.

However, our approach relies on information about the opponent's team in total and is therefore not able to recognize and/or predict 'local' situations. We believe that the detection of the opponents behavior in smaller areas, e.g. a double pass or a standard situation would help to find the appropriate countermeasures. The online coach is the optimal player for the collection of this kind of information and it is obvious that the coach should be able to process the data and find the appropriate tactic for the own team. Also, the analysis should be available online as the developed methods should be able to function in a real-time environment.

In this paper we describe a new method that is able to track moving objects in real time. The idea is to detect spatio-temporal relations between objects (players and ball) in a first step and then learn from this observations whether there is a repeating pattern, e.g. an attack over the wings with a pass onto the penalty point.

A. Birk, S. Coradeschi, and S. Tadokoro (Eds.): RoboCup 2001, LNAI 2377, pp. 441–446, 2002.

Our approach is related to the work from Raines and his colleagues [Raines et al., 2000] who describe a new approach to automate assistants to aid humans in understanding team behaviors for the simulation league. This approach is designed for the analysis of games, off-line after playing, to gain new experiences for the next games. Frank and colleagues [Frank et al., 2000] presented a real time approach which is based on statistical methods. A team will be evaluated statistically but there is no recognition of team strategies.

2 Approach

Object motion takes place in space and time. Therefore, it is useful to describe the behavior of moving objects in terms of spatio-temporal relations. This leads to a domain independent description. Once the spatio-temporal relations between the objects are described one can interpret the behavior of moving objects in the scenario.

2.1 Description of Spatio-temporal Relations

A spatial relation between two objects holds during a time interval and an object motion of continuous direction and speed also has a certain duration. Therefore, the idea is to describe both, the duration of objects motion as well as spatial relations holding between objects via time intervals.

The input data for the approach consists of a series of object coordinates which are updated at moments which occur within a certain time frequency. In a first step the time intervals of continuous object motion (OMI) and of duration of spatial relations between pairs of objects (SRI) have to be generated step by step at runtime.

At each moment and for each object a motion vector describes the objects displacement from the last moment to the actual by length and angle. OMIs are established for each object. As long as the length and angle of the motion vector is similar to the average length and direction of the motion vector already belonging to the actual OMI the interval is extended, otherwise a new OMI is started. The metric values for the average length and angle of the vectors are very precise but not intuitive for a human being. Therefore, length and angle within each OMI are classified to a fixed number of motion directions and speeds. The number of classes to be distinguished depends on the application. For the directions a wind rose is used to specify the classes. Concerning the speeds at least two classes are necessary to distinguish objects in motion from still ones. But in most cases it is useful to distinguish more speeds. The qualitative description of an objects movement consists of a continuous sequence of OMIs covering the whole duration of existence of the object. Obviously, each OMI refers to exactly one object O and has exactly one start moment i_s, end moment i_e, motion direction α and motion speed v: $i_{om} = [O\langle\alpha, v\rangle]_{i_s}^{i_e}$.

Several methods on representing spatial relations between objects have been suggested [Clementini et al., 1997, Guesgen, 1989, Schlieder, 1996] and [Zimmermann and Freksa, 1996]. In this approach a spatial relation between two objects is specified through a direction in which the second object is located and the distance between the objects. On a quantitative level a metric distance and an angle in which direction the related second object is placed is calculated. The procedure of generating time intervals of the duration of spatial relations between two objects (SRI) is the same as for the motion intervals. Due to the fact that there is a maximum distance up to that a spatial relation between two objects is taken into account at all, gaps may occur in the sequence of SRIs referring to a pair of objects. To obtain a qualitative description a fixed number of directions and distances such as *meets*, *near*, *medium* and *far* are distinguished. The wind rose in used again for the direction. Each SRI refers to exactly one pair of objects (O_1, O_2) and has exactly one start moment i_s, end moment i_e, location direction l and displacement d: $i_{sr} = [O_1\langle l, d\rangle O_2]_{i_s}^{i_e}$.

The relationship between the time intervals – OMI and SRI – are described using the seven temporal relations *before, meets, overlaps, starts, during, finishes* and *equals* and their inverse described in [Allen, 1981]. Any scenarios of moving objects can be described by use of OMIs and SRIs which are temporally related.

2.2 Interpretation of Spatio-temporal Relations

The first step to interpret scenarios with moving objects is to identify the concepts a human observer uses to describe the movement of objects. Then the concept is split into OMIs and SRIs and the temporal relations between these time intervals are described. This leads to a definition of the concept in terms of spatio-temporal relations and makes it possible to identify the concept within a scenario.

Elementary concepts describing simple motion events are domain independent and build a basis for the construction of more complex events. Complex events are often domain specific and are described as a combination of simple events.

The entire set of simple events can be divided into groups according to the number of involved SRIs and OMIs and their temporal relations. The following group of simple events includes two objects that are spatial related and may move within duration of the spatial relation, i.e. three overlapping time intervals are involved:

$$i_{sr} = \Big[O_1\langle l_1, d_1\rangle O_2\Big], i_{om_1} = \Big[O_1\langle \alpha_1, v_1\rangle\Big], i_{om_2} = \Big[O_2\langle \alpha_2, v_2\rangle\Big] \qquad (1)$$

To distinguish and recognize the different simple events belonging to one group the attributes motion direction, speed, direction of spatial location and distance have to fulfil certain constraints, e.g.:

- Two objects *meet* each other: $d_1 = meets$.
- O_1 is *approaching* O_2 (*app*(O_1, O_2)): $v_1 > 0 \wedge \alpha_1 = l_1 \wedge v_2 = 0$.
- O_1 is *departing* from O_2 (*dep*(O_1, O_2)): $v_1 > 0 \wedge opposite(\alpha_1, l_1) \wedge v_2 = 0$.

In the same manner it is possible to define constraints for objects moving in parallel or following each other.

In addition to this there are several further groups of simple events that are not mentioned in this paper.

3 Application and Results

In this section we explain the usage of simple events to interpret the behavior of the players in a soccer match to support the coach. Each simple event lasts at least from one simulation cycle to the next. The cycles are snapshots of the match that build the boundaries of the time intervals.

A meaningful simple event is *departing* ($dep(b,p)$), i.e. a player p is passing the ball b. A further evaluation of the balls motion direction leads to a more detailed information whether the ball is played to the front (i.e. in direction to the opposite goal) or backward. This simple event is part of numerous complex situation such as player p_1 passes the ball (b) to player p_2. This situation consists of four simple events, with $p_1 \neq p_2$:

$$se_1 = meets(p_1, b) \wedge se_2 = dep(b, p_1) \wedge se_3 = app(b, p_2) \wedge se_4 = meets(b, p_2) \tag{2}$$

The temporal relations between the simple events are $meets(se_1, se_2)$, $equal(se_2, se_3)$ and $meets(se_3, se_4)$. This example shows how complex situations are constructed by using temporal related simple events. To make this interpretation more specific it is also possible to distinguish different types of players such as defenders, forwards or keeper semantically, i.e. draw conclusions from their behavior rather than taking the information provided by the soccer server.

To explain this in more detail we give an example taken from a match with 4 vs. 4 players. The example situation lasts over 60 cycles. The beginning of the situation is illustrated in fig. 1. For the next cycles we focus on the area high lighted in fig. 1. The ongoing situation is illustrated in fig. 2. The nine detailed images give a brief overview over the sequence of 60 cycles: The forward succeeds to pass the ball. Another forward of team 1 runs for the ball and tries to approach the opposite goal with the ball. This is noticed by the defenders of team 2 which move backwards and then try to stop it. One of them is approaching the forward and unfortunately there is no other forward of team 1 he could pass the ball to. The defender of team 2 takes the ball from the forward and passes it in the opposite direction. The description leads to the following interpretation

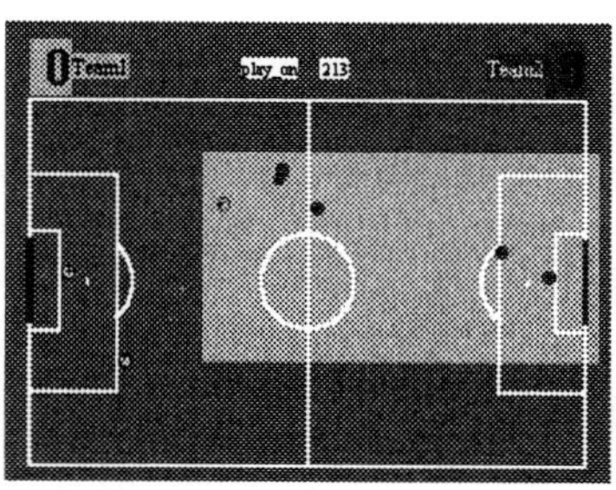

Fig. 1. Two players fighting for the ball

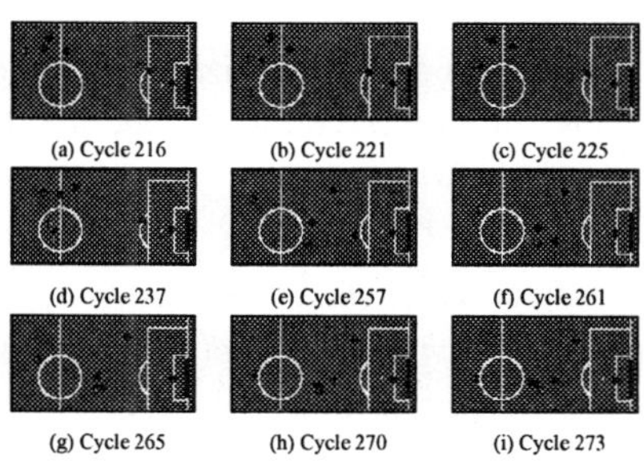

(a) Cycle 216 (b) Cycle 221 (c) Cycle 225
(d) Cycle 237 (e) Cycle 257 (f) Cycle 261
(g) Cycle 265 (h) Cycle 270 (i) Cycle 273

Fig. 2. defense situation

in terms of time intervals of simple events which is close to the interpretation an human observer would probably give:

- Player of team 1 and team 2 are both in meets-relation to the ball. This is interpreted as fighting for the ball (fig. 1).
- In the following time interval the ball is *departing* from both players, i.e. it is passed (fig. 2 a).
- A second player of team 1 approaches the ball (fig. 2 b).
- Then the second player of team 1 reaches the ball (*meets*) (fig. 2 c).
- The player and the ball move in direction to the opposite goal, while the spatial relation between the player and the ball is *meets* or close (fig. 2 d-f and also fig. 3 for a detailed view).
- While the player moves with the ball some players of team 2 are moving backwards. They are building up a defensive position (fig. 2 d-f).
- Then two defenders of team 2 are approaching the player with the ball, until one of them is meeting him (fight for the ball again) (fig. 2 e-g).
- The player of team 2 is still in *meets* relation to the ball whereas the player of team 1 is *close* to the ball but no more meets it, i.e. has lost it (fig. 2 h).
- At last the ball is departing from the player of team 2, i.e. he is passing the ball in the opposite direction to avoid the player of team 1 to score (fig. 2 i).

This defence strategy is typical for team 2 and occurs for several times in the game. To support the coach such situations can be detected by the sequence of time intervals described above. If the situation occurs again, it can be recognized early, so that team 1 could try another way e.g. positioning a second player near to the forward to pass the ball to when the defenders approach. For a more detailed example on how the temporal

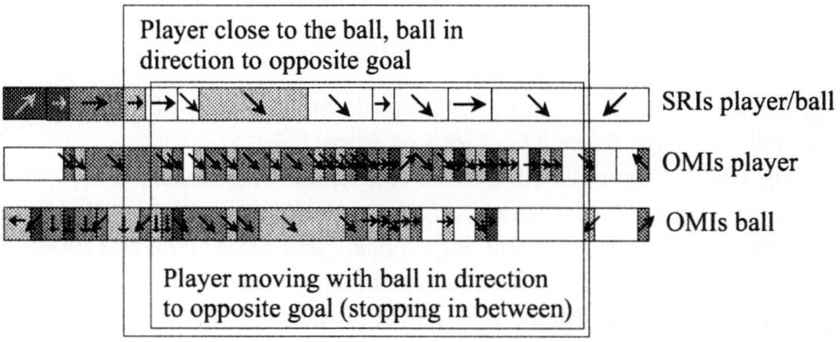

Fig. 3. SRIs and OMIs of the attacking player and the ball.

relations between the SRIs and OMIs are interpreted refer to fig. 3. The diagram focuses on the attacking player. It shows the SRIs between the player and the ball and the OMIs of the player and the ball. Light colors refer to small distances, dark color to large ones, white means still stands resp. meets (spatial). The arrows represent the motion direction resp. the direction in which the spatial related object (in this example the ball from the viewpoint of the player) is placed. Within the entire sequence of SRIs there is an interval in which the distance is either meets or close and the ball is placed in direction to the opposite goal (large rectangle). Within this time interval both objects – player and ball – are moving in direction to the opposite goal except for short interrupts where they stand still. This time interval is *during* the previous described interval (smaller rectangle).

4 Conclusion and Future Directions

We presented an approach of how objects are spatially and temporally related to others and track this over time. We showed that simple events such as *meets*, *departing*, *approaching*, *equals* can be detected and combined to more sophisticated events. We also showed that interpretations of situations can be more specific. We argue that an implementation of this method within the online coach could enhance teams abilities.

Another interesting feature is the ability to analyze games not only online but also off-line. One of the biggest advantages of this approach is the independence from the domain. In the near future, we will also test other domains such as cell tracking in biological systems. Here, the objects are monitored with a camera and the method is able to track the objects over time and describe and store the spatial relations between them as well. However, there are also difficulties with the approach at the moment. For a human being it is relatively easy to follow a RoboCup soccer game on the monitor and see situations fluently and continuously. Because the objects are described on a discrete level it sometimes happens that the approach is not able to detect a continuous flow. For example, while attacking a goal with the ball the player is moving, kicking the ball in the goal direction and so on. However, there will be moments when either the ball or the player have no movement so that the approach terminates the continuous flow and starts a new situation. The next step with the current approach is to detect complex situations over time and learn patterns such as attack over the wing with a pass in the penalty area, double pass etc.

References

[Allen, 1981] Allen, J. F. (1981). An interval-based representation of temporal knowledge. In Hayes, P. J., editor, *IJCAI*, pages 221–226, Los Altos, CA.

[Clementini et al., 1997] Clementini, E., Di Felice, P., and Hernández, D. (1997). Qualitative representation of positional information. *Artificial Intelligence*, 95(2):317–356.

[Frank et al., 2000] Frank, I., Tanaka-Ishi, K., Arai, K., and Matsubara, H. (2000). The statistics proxy server. In Balch, T., Stone, P., and Kraetschmar, G., editors, *4th International Workshop on RoboCup*, pages 199–204, Melbourne, Australia. Carnegie Mellum University Press.

[Guesgen, 1989] Guesgen, H.-W. (1989). Spatial reasoning based on allen's temporal logic. Technical report TR-89-049, International Computer Science Institute (ICSI).

[Raines et al., 2000] Raines, T., Tambe, M., and Marsella, S. (2000). Automated assistants to aid humans in understanding team behaviors. In *Fourth International Conference on Autonomous Agents (Agents 2000)*, Barcelona, Spain.

[Schlieder, 1996] Schlieder, C. (1996). Ordering information and symbolic projection. In et al., S. C., editor, *Intelligent image database systems*, pages 115–140. World Scientific, Singapore.

[Visser et al., 2001] Visser, U., Drücker, C., Hübner, S., Schmidt, E., and Weland, H.-G. (2001). Recognizing formations in opponent teams. In *RoboCup-00, Robot Soccer World Cup IV*, Lecture Notes in Computer Science, Melbourne, Australia. Springer-Verlag. to appear.

[Zimmermann and Freksa, 1996] Zimmermann, K. and Freksa, C. (1996). Qualitative spatial reasoning using orientation, distance, and path knowledge. *Applied Intelligence*, 6(1):49–58.

Stochastic Gradient Descent Localisation in Quadruped Robots

Son Bao Pham, Bernhard Hengst, Darren Ibbotson, and Claude Sammut

School of Computer Science and Engineering
University of New South Wales, UNSW Sydney 2052 AUSTRALIA
{sonp,bernhardh,s2207509,claude}@cse.unsw.edu.au

Abstract. Competing at the RoboCup 2000 Sony legged robot league, the UNSW team convincingly won all their matches. One of the advantages of the team was a new localisation algorithm that is very fast and can tolerate noisy input from the environment as well as unexpected collisions with other objects. This paper describes the algorithm in detail.

1 Introduction

Being able to determine the location of a robot on the field and in relation to other objects is obviously crucial for playing soccer autonomously. In this paper, we present the localisation algorithm used by the UNSW United team in the RoboCup 2000 Sony legged league. The smallest winning margin of the team was 10-0, in the final against the Laboratoire de Robotique de Paris. The decisive factor in the team's superiority was its speed. This was not only due to a novel walk [2], but also to the speed of the vision and localisation systems. Localisation may slow down the robot in several ways. The localisation algorithm itself may waste time, requiring the robot to look around, unnecessarily for landmarks. Furthermore, robots on a soccer field can easily become "disoriented" when their vision is blocked or they are impeded by another robot. If the localisation algorithm is not robust, the robots can spend an inordinate amount of time trying to find themselves again. These considerations led us to develop the localisation algorithm described here. We begin with a description of the environment in which the robot plays. We then give an overview of the UNSW software structure to show how localisation fits into the system, and then we describe the algorithm, itself.

2 The Robots and Their Environment

The robots used in the competition are based on the Sony AIBO entertainment robot. They are slightly modified, by Sony, to include a "debug box" for communicating to a PC, a PCMCIA slot (currently unused), and an NTSC video output port, giving a direct feed from the robot camera.

A. Birk, S. Coradeschi, and S. Tadokoro (Eds.): RoboCup 2001, LNAI 2377, pp. 447–452, 2002.

The Sony robots are remarkably sophisticated, with an on board MIPS R4000 processor, colour camera, gyroscopes and accelerometers, contact sensors, speaker and stereo microphones. Each of the four legs has three degrees of freedom, as does the head. Programs are written in C++ using a proprietary PC-based development environment. The code that controls the robot is loaded onto a memory stick that is inserted into the robot, so that, during a game, they operate autonomously. The field used in the Sony legged league measures two metres wide by three metres long. The field has an angled white border designed to keep the ball within the field.

The game ball is coloured orange and the two goals are coloured yellow and blue. The field also contains six distinct coloured beacons to aid the robots in localising. There is a beacon in each corner, and one at each end of the centre line.

3 System Description

The UNSW United software system consists of four main modules: vision, localisation, strategy and locomotion [2]. Vision is provided by a colour CCD camera with a resolution of 88 x 60 pixels. The robot's field of view is about 40 vertically and 60 horizontally. The vision system's task is to recognize all the objects in the image and calculate the distances, headings and elevations relative to the robot. The strategy module examines the objects that have been found and, taking into account the robot's position and orientation, decides what action to take next. It tells the locomotion module the direction and distance the robot should go.

The localisation module determines where the robot is on the field so that the strategy module is able to decide what to do. Inputs to the localisation module are the set of objects seen in the input image and odometry information. Each object recognized by the vision module has associated with it, the distance, heading and elevation relative to the robot's neck, together with an estimate of confidence in the accuracy of the information about that object. The odometry from the locomotion module has the x, y displacement of the robot and the direction in which the robot has moved.

4 Localisation

Observations of localisation methods in RoboCup 1999 led us to consider several problems:

- noise from the vision system;
- odometry errors, particularly as a result of collisions with other robots and the field border;
- the robots can be relocated by the referees as a result of a rule infraction;
- a slow localisation algorithm can allow the opponent to run away with the ball because the robot is "still thinking" while other action is taking place on the field.

Errors in odometry can occur because of slippage or sticking to the carpet on the field. As there are six robots on the field, robots often collide with each other or with the wall. Currently, we have no method for detecting collisions and consequently the odometry passed to the localisation module during a collision is completely inaccurate. In some situations, a referee may pick up a robot and place it in a location unknown to the robot. This requires the robot to be able to re-localise very quickly in order to make sensible decisions.

4.1 Representation

As mentioned above, the input to the localisation module includes the set of objects recognized in the current camera image, together with their distances, headings and elevations relative to the robot's neck. Localisation tries to determine where the robot is on the field and where the others objects are. It does so by combining the new inputs with a model of the world. Since all beacons and goals are static, the model only needs to store the position of the robot and the ball. At present, the other robots are ignored.

We maintain three variables for a robot: its x and y coordinates and its heading. The left-hand corner of the field, at the team's own end, is designated the origin, with the x-axis going through the goal mouth. The own goal of the robot is set by the higher-level strategy code. Strategies refer to "own goal" and "target goal" rather than "blue goal" and "yellow goal" to capitalise on the symmetry in strategy.

The robot first attempts to localise using only the objects detected in the current image. Being stationary, beacons and goals serve as the landmarks to calculate a robot's position. Because of the camera's narrow field of view, it is almost impossible to see three landmarks at a time, so any algorithm that requires more than two landmarks is not relevant. If two landmarks are visible, the robot's position is estimated using the triangulation algorithm used by the 1999 team [1]. This technique requires the absolute coordinates, distance and heading of two objects relative to the robot.

More information can be gathered by combining information from several images. Thus, the localisation algorithm can be improved by noting that if the robot can see two different landmarks in two consecutive images while the robot is stationary, then triangulation can be applied. Typically, this situation occurs when the robot stops to look around to find the ball. To implement this, we use an array to store landmark information. If there is more than one landmark in the array at any time, triangulation is used. This array is cleared every time the robot moves, that is, when the odometry from the action module is not zero. We could maintain the array and move the objects in the array relative to the robot. However this type of calculation tends to give large errors over time, so this idea was abandoned.

The localisation module also receives feedback from the locomotion module, Pwalk, to adjust the robot's position. The feedback is in the form (dx, dy, dh) where dx and dy are the distances, in centimetres, that the robot is estimated to have moved in the x and y directions and dh is the number of degrees through which the robot is estimated to have turned. We try to get feedback as quickly as possible, which is about every 1/25 second when the robot is moving. Odometry information is clearly not very accurate and small errors in each step accumulate to eventually give very large inaccuracies. Another problem occurs when the robot gets stuck, for example, by being blocked by another robot, Pwalk is not aware of this and keeps feeding wrong information to the Localisation module.

4.2 Stochastic Gradient Descent Algorithm

The methods described above cannot provide reliable localisation because the robot usually sees only one landmark in an image and the robot is moving constantly. We therefore tried to devise a method for updating the robot's position using only one landmark.

The important observation is that if the robot is d cm away from a landmark then it will be on the circle of radius d centred at that landmark. We'll then combine this information with the current position of the robot to estimate the new location of the robot. After determining the robot location, we'll calculate the robot absolute heading. Suppose the landmark has coordinates (xb, yb) and its distance and heading relative to the robot are db and hb respectively, as shown in Figure 1. We draw a line between the (xb, yb) and the estimated current position of the robot in the world model, (x_cur, y_cur). The robot's perceived position, relative to the landmark, is the point on that line d cm away from (xb, yb) and on the same side as the current robot position, in the world model. The localisation algorithm works by "nudging" the estimated position in the world model towards the perceived position relative to the landmark.

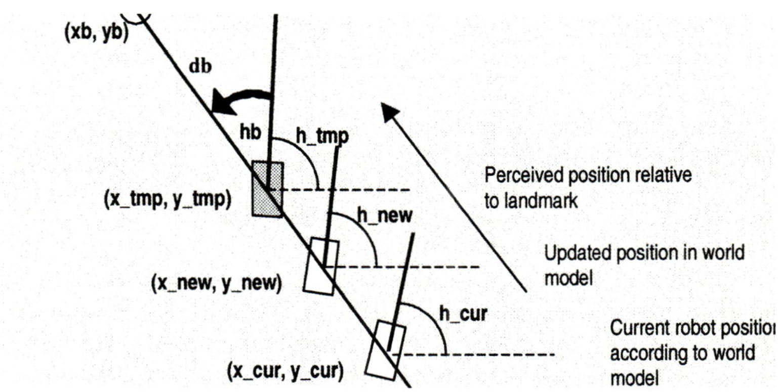

Fig. 1. Update of position based on single beacon

If the error measure in location is defined to be the square of the difference in the robot's current position and that indicated by its sensors, we effectively use gradient descent to update the position and orientation of the robot. The learning rate parameter is varied based on confidence measures, as described later. There is a constant stream of sensory data providing new location information. In order to make immediate use of this data, the localisation algorithm does not wait for updates over a sample of the data, but uses stochastic gradient descent, updating the robot position incrementally after each observation. With a camera frame rate of about 26 frames/second, this algorithm converges quite quickly and accurately. One landmark update overcomes many of the problems caused by odometry error. Even when the robot's movement is blocked and the odometry information is incorrect, if the robot can see one landmark it will readjust its position based on that landmark. Because we use a low trot gait, the robot can see goals and beacons most of the time, if other robots do not obscure them.

A confidence factor is associated with each dynamic object in the world model. This is used when incorporating new data into the existing model. The confidence factors were devised by the 1999 team [1] and were used extensively before the one beacon update approach was developed. Previously, only triangulation was used and therefore it was necessary to store information about an object over a long period of time. To avoid the problem of error accumulation, the confidence factors of the objects were decayed after every camera frame. With the new approach and the high camera frame rate, localisation converges to an accurate world model very quickly. Therefore, the confidence factors are of less importance than they used to be.

4.3 Updating the Robot's Position

In this section, we give details of the method by which the robot's estimated position in the world model is updated. As described in the previous section, the position is updated by "nudging" x, y, and heading variables in the direction of the values as perceived by the robot relative to a landmark, such as a beacon. The update may be conservative, giving more weight to the current position, or it may be more aggressive, weighting the perceived position more. The degree of conservatism is controlled by a rate parameter. At present, this is computed as follows:

$$\begin{aligned} if(CurrentCF > 600)\ rate &= 0.4 \\ else\ rate &= 1.0 \end{aligned} \tag{1}$$

where $CurrentCF$ is the current confidence factor for the position of the robot. As mentioned above, (x_cur, y_cur) is the current robot location in the world model. (x_tmp, y_tmp), the perceive position of the robot relative to the beacon, is the intersection between the line connecting $(x\_cur, y\_cur) - (xb, yb)$ and the circle radius d centered at the beacon. We compute the robot's new x position as:

$$x\_new = x\_cur * CurrentRate + x\_temp * BeaconRate \tag{2}$$

Where

$$BeaconRate = \frac{BeaconCF * rate}{BeaconCF * CurrentCF} \tag{3}$$

BeaconCF is the confidence factor of the beacon seen and

$$CurrentRate = 1 - BeaconRate \tag{4}$$

The new y position is updated similarly. Updating the heading is slightly more complicated since the heading is a circular measure that is always in the range ± 180. h_tmp is calculated so that if the robot is at (x_tmp, y_tmp) with absolute heading h_tmp then the relative heading of the seen beacon and the robot is hb.

$$\begin{aligned}
&if(h_cur - h_tmp >= 180)\ h_tmp+ = 360\\
&else\ if(h_tmp - h_cur >= 180)\\
&\quad h_cur+ = 360\\
&\quad h_new = h_cur * CurrentRate + h_tmp * BeaconRate\\
&if(h_new > 180)\ h_new- = 360\\
&if(h_new <= -180)\ h_new+ = 360
\end{aligned} \tag{5}$$

The confidence for the new robot position (x_new, y_new, h_new) is calculated as follows:

$$NewCF = CurrentCF * CurrentRate + BeaconCF * BeaconRate \tag{6}$$

5 Conclusions and Future Work

Stochastic Gradient Descent Localisation provides a very efficient localisation method for real time, autonomous robots with a limited computational capacity. Its main advantage is the ability to localise based on information from one beacon. This feature overcomes the noisy input from the camera as well as the noise in the output from vision module. Odometry errors are corrected automatically whenever the robot sees a landmark. The algorithm is very fast, with a constant complexity, allowing the robot to react quickly during a game. Thus, the UNSW robots rarely stop to look around to localise, as was the case with many other teams.

The current localisation algorithm is passive in the sense that the robot never looks for beacons intentionally. The robot simply tries to knock the ball to the target goal, and at the same time, localise based on whatever landmarks it happens to see. The algorithm might be improved if the robot were to actively look for beacons, for example closest to the currently estimated position of the robot, to confirm its current position.

References

1. Dalgliesh, J. and Lawther, M. (1999). Playing Soccer with Quadruped Robots. Computer Engineering Thesis, University of New South Wales.
2. Hengst, B., Ibbotson, D., Pham, S.B., Dalgliesh, J. Lawther, M., Preston, P., Sammut, C. The UNSW RoboCup 2000 Sony Legged Team, (to be published for RoboCup 2001 by Springer), 2001.
3. Mitchell, T.(1997): Machine Learning, McGraw-Hill.

Recognizing Probabilistic Opponent Movement Models

Patrick Riley and Manuela Veloso

Computer Science Department, Carnegie Mellon University, Pittsburgh, PA 15213-3891

Abstract. In multiagent adversarial domains, team agents should adapt to the environment and opponent. We introduce a model representation as part of a planning process for a simulated soccer domain. The planning is centralized, but the plans are executed in a multi-agent environment, with teammate and opponent agents. Further, we present a recognition algorithm where the model which most closely matches the behavior of the opponents can be selected from few observations of the opponent. Empirical results are presented to verify that important information is maintained through the abstraction the models provide.

1 Introduction

Multiagent domains can include team and adversarial agents. Our work is driven by the goal of significantly improving the performance of teams of agents through their adaptation and effective response to different adversarial teams.

We use the concrete simulated robotic soccer platform, which is a rich multiagent environment including fully distributed team agents in two different teams of up to eleven agents. Extensive work has been done on adapting the teammates' behavior to their opponents, mainly at the individual low-level of positioning and interactions between a small number of agents [5]. The procedure for these positioning adaptations does not change throughout a game, limiting the degree to which the team adapts to the opponent team's behaviors.

This paper reports our work on going one additional major step towards the adaptive response of teammates to the opponent, by gathering and responding to the opponents' behaviors throughout the game. We specifically focus on responding effectively after the game is stopped, in what are known as setplay situations. Several preset setplay plans [5] have been introduced which indeed provide great opportunities to position the teammates strategically and have been shown to impact the performance of a team. In this work, we contribute *adaptive* setplays which change and improve throughout a game as a function of and in response to the opponent team's behavior.

We use a "coach" agent with a centralized view of the world, but whose only action is to communicate with its teammates, but only at occasional times. In the robotic soccer domain, these times are when the game is stopped for some reason. This type of scenario is one instance of the *Periodic Synchronization Domains* [6] where team agents can periodically synchronize their team strategies.

Our coach agent compiles the necessary overall view of how the opponent team behaves. The coach communicates to its teammates a team plan which is executed and monitored in a fully distributed manner. The coach agent is equipped with a number

A. Birk, S. Coradeschi, and S. Tadokoro (Eds.): RoboCup 2001, LNAI 2377, pp. 453–458, 2002.

of pre-defined opponent models. These models are probabilistic representations of predicted opponents' locations. The models can then be matched to observed movements of the opponent agents. The coach continues to gather observations and when the game is stopped, e.g., due to an out-of-bound call, the coach rapidly takes advantage of the short available time to create a team setplay plan that is a function of the matched modeled opponents' behavior. The plan is generated through a hill-climbing search in plan space using an evaluation function that embeds the predictions of the opponent model perceived to be the most likely during the game. The plan generation, in addition to the recognition of the opponents' model, notably uses the model *to predict* the opponent agents' behaviors. The coach also observes the execution of the plan generated in order to update the selection of an appropriate model for the current opponent. This paper focuses on the structure of an opponent model and the algorithm for recognizing the best-matching model. A more complete overview of the system can be found in [3] and more details on the planning can be found in [4].

2 Opponent Models and Model Selection

Conceptually, we want an opponent model to represent how an opponent plays defense during setplays. We expect that a wide range of decision making systems can be roughly captured by a small set of models. In order to handle the uncertainty in the environment and to allow models to represent more information than just a simple predicted location, we use probabilistic models. Given the recent history of the ball's movement (from the start of the set play for example) and the player's initial location, the model should give, for each player, a probability distribution over locations on the field.

Let p be the number of players on a team. Let F be the set of positions on the field, discretized to 1m. We represent the ball movement as a sequence of locations on the field (an element of F^*). A location for each player is a sequence of positions on the field (an element of F^p).

An opponent model defines a probability distribution for each player over the end locations on the field for that player, given the ball movement $B \in F^*$ and a set of starting positions $S \in F^p$ for all the players. If R is a probability distribution over locations on the field, an opponent model M is a function:

$$M: \underbrace{F^*}_{\text{ball movement}} \times \underbrace{F^p}_{\text{initial pos.}} \rightarrow \underbrace{R^p}_{\text{probability distribution for each player}} \tag{1}$$

In particular, for an opponent model M, the probability for player i being at end location ℓ_i ($P_i[\ell_i|B,S]$) is calculated by choosing the ith probability distribution output by the model and calculating the probability of the location ℓ_i according to that distribution.

Notice that each player's distribution may depend on the starting positions of *all* the opponent players. In order to avoid recursive modelling, we do not allow the opponents distributions to depend on our own player locations. The ball movement B is the planned ball movement. The starting positions for the players are their current locations. We have a probability distribution for each player, which is the probabilistic obstacle for planning.

Given a set of opponent models, it is still a challenge to decide which model best describes the opponent. We assume that the opponent has chosen one of our models at the beginning of the game and is then independently generating observations from that model. We can then use a naive Bayes classifier.

We maintain a probability distribution over the models. The original distribution (the prior) is set by hand. Then, whenever a planning stage is entered, the model with the highest probability is used. Upon observing a plan execution, we use observations of that execution to update our probability distribution over the models.

What this means is that we start with a probability distribution over the set of models $\{M_1, \ldots, M_m\}$ (known as the *prior*) and then we get an observation. An observation is a triple of starting locations for all the players $S \in F^p$, a ball movement $B \in F^*$, and ending locations for all of the players $E \in F^p$. We want to use that observation to calculate a new probability distribution, the *posterior*. That distribution then becomes the prior for the next observation update.

We make the following assumptions, in order to simplify our equations:

1. The players movements are independent. That is, the model may generate a probability distribution for player x based on everyone's starting locations. However, what the actual observation is for player x (assumed to be sampled from this probability distribution) is independent from the actual observations of the other players.
2. The probability of a particular starting position and ball movement are independent of the opponent model. This assumption is questionable since the ball movement (i.e. plan) generated depends on the opponent model.

Consider one updating cycle of getting an observation $\omega = (S,B,E)$. We want $P[M_i|\omega]$.

$$
\begin{aligned}
P[M_i|\omega] &= \frac{P[\omega|M_i]P[M_i]}{P[\omega]} = \frac{P[S,B,E|M_i]}{P[\omega]}P[M_i] = \frac{P[E|S,B,M_i]P[S,B|M_i]}{P[\omega]}P[M_i] \\
&= P[E|S,B,M_i]\frac{P[S,B]}{P[\omega]}P[M_i] \quad \text{(assump. 2)} \\
&= \underbrace{P[\ell_1|S,B,M_i]P[\ell_2|S,B,M_i]\ldots P[\ell_p|S,B,M_i]}_{\text{what opponent model calculates}} \underbrace{\frac{P[S,B]}{P[\omega]}}_{\text{norm. constant}} \underbrace{P[M_i]}_{\text{prior}} \quad \text{(assump. 1)}
\end{aligned}
$$

The term labeled "norm. constant" is a normalization constant, i.e. it does not depend on which model is being updated, so it does not need to be explicitly calculated. We calculate the remaining terms and normalize the result to a probability distribution.

3 Empirical Evaluation

In order to verify that the abstracted form of models given here are still sufficient to reliably capture variation, we implemented several different movement recipes for the defense. We then implemented models to capture each of these effects. This is a nontrivial task, both because of the great deal of uncertainty in the environment, and the abstraction which the models provide over the basic actions. An empirical evaluation was necessary to verify that our model implementations match the effects of the players and that the naive Bayes update can quickly recognize models of this form.

Besides the choice of the models, a decision must be made about how to generate observations from the stream of data being received about the opponent positions. Clearly, if an observation triple is generated every cycle we will be violating the independence assumption of the naive Bayes update, as well as giving the models little information (in terms of the ball movement) with which to work. For this testing we decided to create an observation for processing every time the agent who is controlling the ball changes. An agent is considered to be controlling the ball if (i) the agent is the closest player to the ball and (ii) the player can kick the ball. A given observation can cover anywhere from approximately 5 to 50 cycles.

In all of the models used, the distribution of each players final position is represented by a 2-dimensional Gaussian with equal variance in all directions. The standard deviation is an affine function of time (since the beginning of the setplay); we tested with several different slopes. The mean is computed as discussed below. This is likely not the optimal distributional representation, but it suffices for recognition task here.

We used five models for the empirical evaluation. Naturally, the mean of each player's final distribution is computed relative to the initial position as follows:

No Movement At the initial position of the player
All to Ball Moved towards the ball at a constant speed
All Defensive Moved towards the defensive side of the field at a constant speed
All Offensive Moved towards the offensive end of the field at a constant speed
One to Ball This model is slightly different from the others. The ball's movement is broken down into cycles. At each cycle, whichever player is closest to the ball is moved slightly closer. Note that since the ball can move faster than the players, which player is closest to the ball can change several times during the ball's movement. The final positions of the players are the means of the distributions.

Before looking at how well each model can be recognized out of this set of models, it is important to understand how "separable" the models are. That is, if two models make similar predictions in most cases, it will be difficult to recognize one model over the other. Clearly, getting more observations should improve the recognition accuracy.

We will develop the concept of separability in three stages. First we will consider the separability of two distributions, then the separability of multiple *sets* of distributions, and finally define empirically the separability of a set of models.

Start with two distributions A and B over two dimensions. The question we are interested in is if we are seeing an observation probabilistically generated from A, what is the probability that the naive Bayes update (starting with a uniform prior) will return with A being the most likely distribution? This is equivalent to the probability mass in A of the area where the probability distribution function (i.e. p.d.f.) of A if greater than the p.d.f. of B. Of course, we are also interested in the case where B is the distribution generating observations, and in general these probabilities can be different.

Now consider seeing multiple observations instead of a single one. Once again, we are interested in the probability that A will have the highest posterior after the naive Bayes update on all of the observations. This is challenging to solve analytically, but Monte Carlo simulation can estimate this probability.

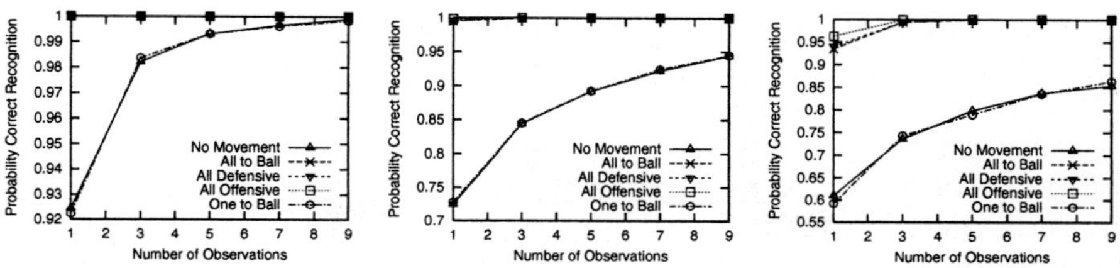

(a) St. Dev. Rate 0.1 (b) St. Dev. Rate 0.3 (c) St. Dev. Rate 0.5

Fig. 1. Separability of the models (Error bars are one standard deviation)

This concept of separability extends naturally to the update over the distributions of all the players and over all of the models. Given a distribution for each player from each model, we can use Monte Carlo simulation to compute the probability that the correct set of distributions will be recognized for any given number of samples.

However, the models are functions from starting positions of the opponents and a movement of the ball to distribution of the opponents, not simply distributions themselves. We use an empirical test to estimate separability of *models*. For each observation $O_1, \ldots O_k$ from real game play, we repeatedly generate a series of artificial observations $(S,B,E_1)\ldots(S,B,E_n)$ (where n is the number of observations for which we want to estimate the probability of correctness). The sequence of ending positions $E_1 \ldots E_n$ are sampled from the distributions output by the correct model (the model for which we want to estimate the probability of the naive Bayes being correct). For each sequence of artificial observations, the update is performed. We can then estimate the probability that the correct model will be the most likely. Averaging over all observations, we can estimate the probability of a model being correctly recognized, given n observations.

Figure 1 shows, for each model, the probability of that model being recognized as a function of the number of observations. The three graphs are for three different choices of the slope of the line controlling the increase in the standard deviation of the distributions over time. One can see that if the models perfectly capture the opponents, after only a few updates, the recognition accuracy should be quite high (more than 80%). Also note that since a lower rate of standard deviation increase means less overlap in the distribution, lower rates give better separability.

The testing of recognition accuracy is much simpler compared to estimating the separability of the models. For each number of observations n, we examine contiguous sets of n observations from real game play (where the opponents are using a recipe which should reflect predictions of the models). For each set, we perform the naive Bayes update and check whether the correct model has the highest posterior. From this sequence of binary answers, we estimate the probability of the model being recognized.

Figure 2 summarizes the recognition accuracy of the models for each of the choices of the standard deviation increase rate. With standard deviation rate of 0.3 or 0.5, an accuracy of 94% is obtained with just four observations. The fact that the recognition is much lower for a standard deviation increase rate of 0.1 suggests that the true standard deviation of the players' positions is higher than predicted by the model.

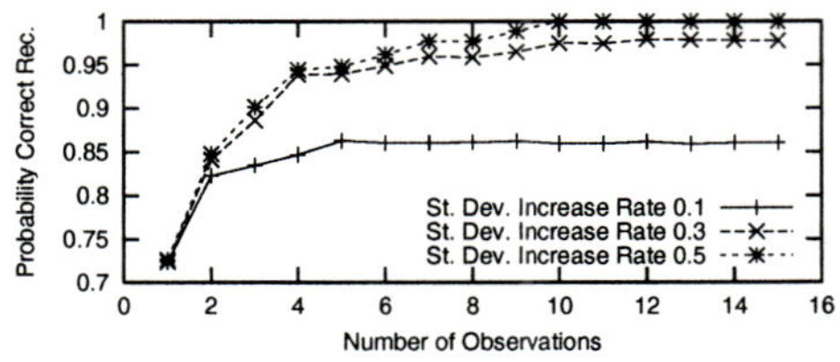

Fig. 2. Summary of Recognition Accuracy

4 Discussion

Similar work is by Intille and Bobick[2], who give a framework for recognizing plays in American football. The plays are given in terms for goals for the agents and limited temporal constraints between actions. Further, similar to Charniak and Goldman [1], they use Bayesian style networks to incorporate observed information.

One important difference between ATAC and other plan and behavior recognition work is that we use the recognized information for purposes for predicting and adapting to *future* behavior. We assume that the opponents will behave similarly to how they performed in the past, and use that information to develop a plan.

We fully implemented this system for RoboCup2000, and it was evident that our team benefited from adaptive setplays. We hope to explore automated learning of the models and providing more flexibility to take advantage of unpredicted opportunities. We plan to continue to explore the area of generating and using opponent models further exhibit adaptive, intelligent responses in our agents.

Acknowledgments: The authors thank Michael Cleary and Deb Dasgupta at Draper Laboratory. This research was sponsored in part by Grants Nos F30602-00-2-0549 and F30602-00-2-0549 and by an NSF Fellowship. The content of this publication reflects only the position of the authors.

References

1. E. Charniak and R. Goldman. A Bayesian model of plan recognition. *Artificial Intelligence*, 64(1):53–79, 1993.
2. S. Intille and A. Bobick. A framework for recognizing multi-agent action from visual evidence. In *AAAI-99*, pages 518–525. AAAI Press, 1999.
3. P. Riley and M. Veloso. Coaching a simulated soccer team by opponent model recognition. In *Agents-2001*, 2001.
4. P. Riley and M. Veloso. Planning for distributed execution through use of probabilistic opponent models. In *IJCAI-2001 Workshop PRO-2: Planning under Uncertainty and Incomplete Information*, 2001.
5. P. Stone, P. Riley, and M. Veloso. The CMUnited-99 champion simulator team. In Veloso, Pagello, and Kitano, eds, *RoboCup-99: Robot Soccer World Cup III*, pages 35–48. Springer, Berlin, 2000.
6. P. Stone and M. Veloso. Task decomposition, dynamic role assignment, and low-bandwidth communication for real-time strategic teamwork. *Artificial Intelligence*, 110:241–273, 1999.

Design and Implementation of a Soccer Robot with Modularized Control Circuits

Kuo-Yang Tu

Department of Electronic Engineering
Lunghwa University of Science and Technology
300 Wan-Shou Rd., Sec. 1, Kueishan, Taoyuan Hsien,
TAIWAN, R. O. C.
tuky@cc.hwh.edu.tw

Abstract. RoboCup is designed to evaluate various theories, algorithms, and architectures for new challenges of Artificial Intelligent (AI) research. New challenges will import new functions into a soccer robot. Thus, a soccer robot with modularized control circuits is proposed to conveniently add new functions for further challenges. At present, the designed soccer robot, based on circuit function, has five modules: processor module, radio communication module, input/output module, supersonic module and extensible module. Five modules are together connected via Industry Architecture Standard bus (ISA bus) in extensible module. In addition, the motion planning of a soccer robot for kicking the ball is formularized to shoot goal. Experiment is also included.

1 Introduction

RoboCup is proposed as a standard problem to evaluate various theories, algorithms, and architectures for Artificial Intelligent (AI) research [1]-[2]. Its beginning is the ending of the Deep Blue, a computer chess team, beating Kasparov, human grand master chess. The Deep Blue team's win implies that computer chess lost its objective, the challenges of AI research. As a result, RoboCup replaces computer chess as AI challenges.

RoboCup has currently chosen soccer as its standard task. In RoboCup2001, there are five leagues, Small size, middle size, simulation, four-leg robot (AIBO) and RoboCup Rescue. The RoboCup Rescue league is first beginning in this year. In addition, many exhibition topics, such as humanoid robot, RoboCup Junior and Industrial robot, have progressed for the ultimate goal of RoboCup: a team of autonomous humanoid robots plays with the human World Cup champion team under the official regulation of FIFA. From low-level control to high-level planning, RoboCup offers an integrated research, including mechanical, electronic, and information engineering, and covering broad topics such as non-holonomic control system, motion planning, robot localization, visual servoing control, multi-agent systems, etc. Not only in the present soccer but also for the ultimate goal, RoboCup faces many unknown challenges. New challenges will compel soccer robots to have new functions. Soccer robots always need to add

A. Birk, S. Coradeschi, and S. Tadokoro (Eds.): RoboCup 2001, LNAI 2377, pp. 459–464, 2002.

or replace a function in itself. Therefore, this research is devoted to proposing a soccer robot with modularized control circuits.

Modularizing control circuits will be convenient to adjusting its function for the RoboCup competition. The modularizing is based on control circuits' functions for adding or replacing easy. The control circuits of the designed soccer robot are divided into five modules, including processor module, radio communication module, input/output module, supersonic module and extensible module. Five modules are connected by Industrial Standard Architecture (ISA) bus, which is a general bus used in IBM PC. The ISA bus will make the soccer robot extend its functions using PC's hardware. In addition, for demonstration, experiments show the designed soccer robot can be controlled to kick the ball toward the opponent goal.

The rest of this paper is organized as follows. The next section introduces how to modularize the control circuits of a soccer robot. Section 3 describes the formularizing of the motion planning of a soccer robot for kicking the ball toward the opponent goal. Experimental results are presented in section 4. Section 5 gives some conclusion and discussion.

2 A Soccer Robot with Modularized Control Circuit

It is necessary to adjust functions of a soccer robot for facing a new challenge considered in RoboCup. Easily adding or replacing a function stimulates this research to Modularize control circuits in a soccer robot. However, how to modularize control circuits has many considerations, such as circuit connections or circuit functions. The circuit connection consideration results in less connection line between the different modules, but the circuit function consideration has the advantage of easily combining together every module. Easy combination is the objective of modularizing control circuits of a soccer robot. Thus, modularizing is based on the circuit functions of the designed soccer robot.

Fig. 1 is the control circuits' block of the designed soccer robot. Based on its functions, the control circuits are divided into five modules: processor module, radio communication module, input/output module, supersonic module and extensible module. The processor module contains one microprocessor 80196, two ROMs for saving program code, and one flash RAM for reserving information after the soccer robot is powered off. The radio communication module is based one single chip processor 8951. It is used to implement radio protocol, to treat ultrasonic sensors, and then to communicate with microprocessor 80196 via one programming parallel port controller 8255. The input/output module contains one LCD module, used as the output for showing soccer robot's situations, and one 4*4 keyboard, used as the input for sending commands to the soccer robot. In addition, this module also includes one timer/counter controller to receive the encoder signal from two DC motors and then to record the move position of the soccer robot. Fig. 2 is the photograph of the soccer robot, which includes five modules.

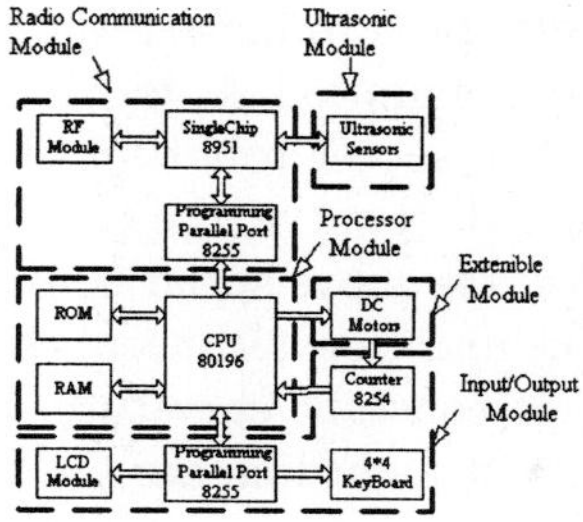

Fig. 1. The control circuit of the designed soccer robot.

Fig. 2. The photograph of the designed soccer robot.

3 Motion Planning of a Soccer Robot

The motion planning is based on Dubins' [3] and Reeds&Shepp's [4] cars to develop the algorithm, which is applicable to the robots for soccer play. Dubins' car is the optimal path under only considering forward move, but Reeds&Shepp's car is the consideration under both forward and backward move. Here, the motion planning of a soccer robot does not intend to find the optimal path, i. e. the short path, because of the following two reasons:
1). Finding the optimal path will exhaust computer time and then result in insufficient time for the other tasks in the soccer robot system, such as image processing and play strategy planning. 2). During competition, the environment of the soccer play is dynamic, and has many obstacles.
For the crowded and dynamic environment, the optimal path may be always varying, and the soccer robot system then sinks into solving this path. The main consideration of the soccer robot's motion planning is the strategy whether is fit for RoboCup competition.

The competition strategy is defined to first rotate the soccer robots at appropriate orientation before they want to execute their tasks. This competition strategy has the advantage of quickly changing the soccer robots' tasks to deal with opponent behaviors and then to avoid losing score or to have win. Thus, assume that the soccer robot's motion satisfies A1 as follows. A1: Before starting to move, a soccer robot has been at appropriate orientation. In addition, a soccer robot usually needs the fastest speed to move toward its goal position. This idea also stimulates us to propose assumption A_2 for the motion planning of a soccer robot in the following. A_2: The soccer robot moves at a constant speed. Note that as a soccer robot to move on a curve, the constant speed will result in it at a constant curvature. Besides, the move speed is usually the maximal speed.

Under the ideas of Dubins and Reeds&Sheep, the optimal path is composed of straight lines and curves at constant curvature. It is well know that the velocity commands of path on a straight line are

$$\begin{aligned} u_\ell &= V \\ u_\gamma &= V \end{aligned} \tag{1}$$

where u_ℓ and u_γ are the velocity command of left and right wheels, respectively, and V is the velocity of the robot following the straight line. For a curve at a constant curvature, the velocity commands are

$$\begin{aligned} u_\ell &= V + WV/R \\ u_\gamma &= V - WV/R \end{aligned} \tag{2}$$

where R is the rotational radius. The motion planning of the soccer robot is based on the path formed by straight lines and curves. Fig. 3 shows the idea of path planning for finding the straight line L_1 and the curve C_1 at the curvature R. The motion planning will solve u_ℓ and u_γ from Eqs (1) and (2).

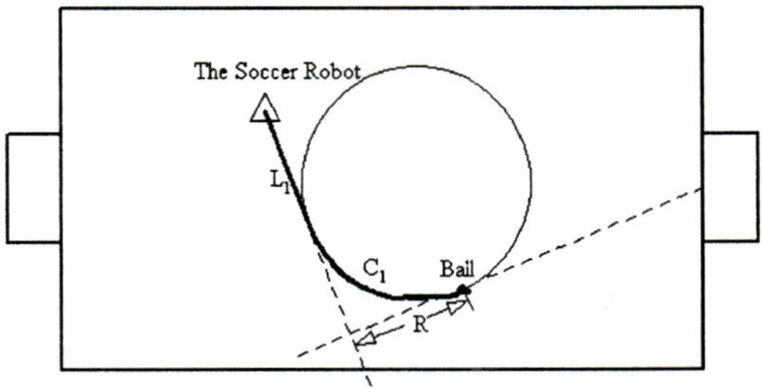

Fig. 3. The motion planning of the soccer robot.

4 Experiment of the Motion Planning

Assume that the center of the play field is the coordinate position (0.0, 0.0). As the soccer robot starts at (0.0, -0.6), Fig. 4 shows the soccer robot is controlled to follow the planned path and then to kick the ball forward toward the opponent goal. Fig. 5 shows the left and right motors' velocity, including desired and actual velocity. The time between two velocity commands is 30ms. In Fig. 4, the car and ball positions are shown one time after passing eight commands.

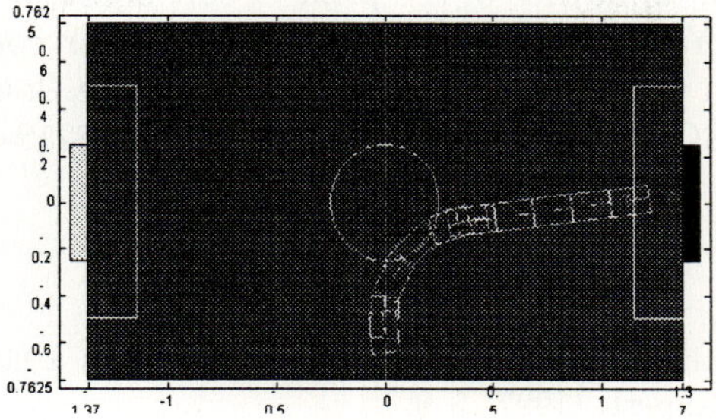

Fig. 4. As starting at (0.0, -0.6), the soccer robot is controlled to kick the ball forward toward the opponent goal.

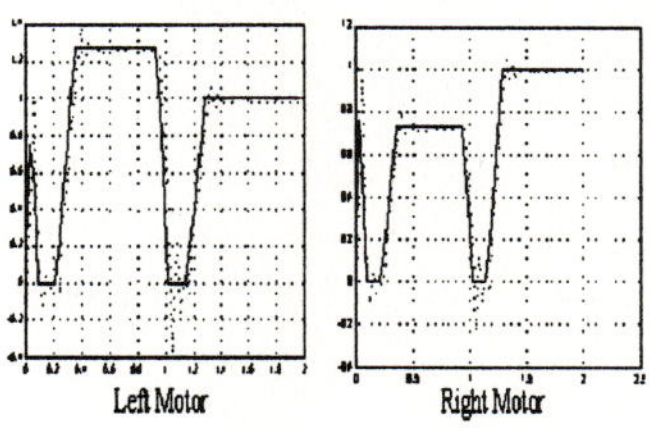

Fig. 5. The motor velocity as the soccer robot starts at (0.0, -0.6) (Solid line: desired velocity; Dotted line: actually velocity).

5 Conclusion

In this paper, modularizing control circuits is proposed to build a soccer robot. The control circuits of the soccer robot are divided into five modules based on their functions. Five modules are connected via ISA bus, a standard bus of IBM PC. This feature will make the soccer robot easy use IBM PC hardware and then helpfully extend it to develop commercial products. In addition, the motion planning of the soccer robot first divides its passing path into the segments of straight lines and curves. And u_ℓ and u_γ at every segment are then solved by eqs (1) and (2). The solved u_ℓ and u_γ are used to control the soccer robot to pass the desired path. Experiment shows the proposed motion planning is useful for soccer robots.

6 Acknowledgments

Many thanks to all current and former members of our team, especially Deray Xuu who finishes all the layout of the print circuit board of the designed soccer robot, and Ritzen Yang who develops the program for controlling the soccer robot. This research was supported by National Science Council, Taiwan, R. O. C. under Grants NSC 89-2218-E-146-003 and NSC 89-2218-E-146-005.

References

1. Hedberg, Sara: Robots playing soccer ? RoboCup poses a new set of AI research challenges. IEEE Expert, (1997) 5–9.
2. Kitano, Hiroaki, M. Asada, I. Noda, and H. Matsubara: RoboCup: Robot world cup. IEEE Robotics & Automation Magazine, (1998) 30–36
3. Dubins, L. E.: On curves of minimal length with a constraint on average curvature and with prescribed initial and terminal positions and tangents. Amer. J. Math., vol. 79, (1957) 497–516.
4. Reeds, J. A. and R. A. Shepp: Optimal paths for a car that goes both forward and backward. Pac. J. Math., vol. 145, (1990) 367-392.

Cooperation-Based Behavior Design

Hui Wang, Han Wang, Chunmiao Wang, and William Y.C. Soh

Division of Control & Instrumentation, School of EEE
Nanyang Technological University
Nanyang Avenue, Singapore 639798
{p147513815, hw, p149510969, eycsoh}@ntu.edu.sg

Abstract. Robot soccer explores such a research topic that multiple agents work together in a Real-time, Cooperative and Adversarial (RCA) environment to achieve specific objectives. It requires that each agent can not only deal with situations individually, but also present cooperation with its teammates. In this paper, we describe a robot architecture, which addresses "scaling cooperation" among robots, and meanwhile allows each robot to make decisions independently in real-time case. The architecture is based on "ideal cooperation" principle and implemented for Small Robot League in RoboCup. Experimental results prove its effectiveness and reveal several primary characteristics of behaviors in robot soccer.

1 Introduction

Our work is based on Small Robot League in RoboCup [1], which includes two five-robot physical teams that play a twenty-minute game on a field with the size of a table tennis board. It presents many challenges for physical agent [2]:

(1) For an individual agent, it should deal with the infinite motion states related to the moving objects in a real-time case.

(2) For a collection of agents, they are expected to play like a team, which means extra abilities beyond simple reactive behaviors, such as situation recognition, communication and cooperative behaviors.

In this paper, we aim to explore the cooperation mechanism in a RCA environment. Our approach is to present an agent architecture, in which cooperation is embedded. Section2 describes the agent architecture. Section3 presents an implementation of this architecture for Small Robot League. Experiment results in Section4 prove its effectiveness and reveal some main features of behaviors in robotic soccer. In the end, we will summarize important conclusions and discuss many crucial problems for further study.

2 Architecture Overview

Before we define an architecture for agents surviving in a RCA environment, we must understand what determines our design. Arkin indicates , in [3], that "If the roboticist intends to build a system that is autonomous and can successfully

A. Birk, S. Coradeschi, and S. Tadokoro (Eds.): RoboCup 2001, LNAI 2377, pp. 465–470, 2002.

compete with other environmental inhabitants, system must find a stable niche or it (as an application) will be unsuccessful.", which shows a logical way for us to decide what an agent should be, that is, environment and task require agents to have many essential functions for their survival; the agent architecture is therefore requested to fully support those functions' strong points and suppress their weak ones.

2.1 Ideal Cooperation

Although cooperation has various types, there should be some criteria to evaluate their performance since cooperation is observable for us. Four standards are therefore presented, and if certain cooperation can meet all of them, it can be called *ideal cooperation.*

We believe that for a team of agents living in a RCA environment, if they execute ideal cooperation, the cooperation should be *reliable*, *scaling*, *voluntary* and *evolutionary*. "Reliable" has double-fold meanings: first, when the same situations appear, a certain cooperation behavior will be executed repeatedly; second, if one teammate stops working due to failures, the team performance will degrade elegantly and if one more teammate is added, the team performance will improve. "Scaling" means cooperation can be local (*micro-cooperation*) or global (*macro-cooperation*). Micro-cooperation is often achieved by a few agents within a relatively small area, where they can affect the same situation directly. Macro-cooperation is always concerned with all the team members scattered around the space. "Voluntary" means agents can independently reason the opportunities of cooperation, that is, an agent can be influenced by its teammates, but never dominated by any of them. It prevents serious damages to the whole team caused by partial failures of individual agents. "Evolutionary" means the performance of certain cooperation can be improved by training.

2.2 Agent Architecture

This section presents an architecture for agents surviving in RCA environment with which an agent can make its own decision, meanwhile display cooperation with others. A high level block diagram of the architecture is shown in Fig. 1 (Shadowed area represents one agent, which is "connected" with its teammates and the environment by dashed lines).

The "Receiver" is divided into two parts: "Communication Unit" and "Perception Unit", which monitor the outside environment and extract useful information for decision-making. The difference between them is that perception unit is responsible for sensing the external states, but the communication unit receives the message from other teammates. A point is the output of receiver is the information interpreted rather than raw data. The function of *opponent and teammate modeling* is often realized in interpreter.

Decision machine is the kernel of an agent responsible for deciding what should be done in the next step(s). It has two functional units: "Information

Fusion Unit" and "Behavior Selection Unit". The former collects all the information related to the agent and performs situation recognition. The latter selects suitable behaviors according to the output of Information Emergence Unit and sends out the desired action(s). "Behavior Library Unit" and "Internal State

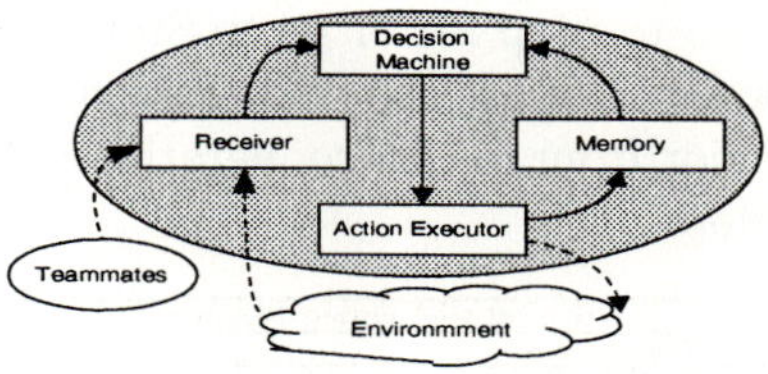

Fig. 1. Agent general architecture

Unit" consist of Memory of a robot. Internal State Unit saves the agent's personal data temporarily (it is always updated in each step). Behavior Library is a long-term memory whose contents are preset in design. **Cooperation in a team is just expressed as specific behaviors for each agent.**

Action Executor translates the selected behaviors into concrete control commands and performs them.

3 Implementation in Robot Soccer

Robotic soccer is a good example of RCA environment. We implement the architecture presented in Section 3 for RoboCup Small Robot League. The key in implementation is to construct a behavior library based on cooperation and design a decision process that can estimate the cooperation opportunities independently.

3.1 Behaviors Based on Cooperation

The behaviors of a robot are organized in a layered form, and each layer may have *behavior group* with different priorities. The higher level behaviors are supported by those in the lower levels. Table 1 is a summary.

3.2 Decision-Making

Although a robot will consider other teammates when it takes actions, the robot will never be dominated by others, that is, every robot may influenced by its teammates, but such influence is achieved by message or prediction rather than orders. As for our robot team, we adopt such a belief to design the process of decision that the behaviors of dealing with the ball have higher priorities than the others. The entire decision process is broken into two phases as follows:

Phase 1 - For all of the team members

Step 1.1: Determine which robot is to deal with the ball immediately (it is called Robot A).

Step 1.2: Robot A selects the suitable behavior.

Step 1.3: Robot A broadcast its intention to its teammates.

Phase 2 - For each of the left robots,

Step 2.1: Check the message from Robot A.

Step 2.2: Determine whether to help Robot A by modeling the opponents and teammates. If YES, go to Step 2.3 or go to Step 2.4.

Step 2.3: Select behaviors from micro-cooperation behavior group.

Step 2.4: Select behavior from macro-cooperation behavior group.

Step 2.5 Return to Step 1.1.

Level 0 – Basic Motion			
Name	**Function**	**Related behaviors**	**Feature**
Direct Move	Move to a position without consideration of orientation	(None)	Pure reactive without intentions
Quick Stop	Stop as soon as possible at any speed	(None)	
Turn	Tune direction in an exact angle	(None)	
Spin	Turn around as fast as possible	(None)	
Level 1 - Advanced Motion			
Name	**Function**	**Related behaviors**	**Feature**
Approach	Include the ball within the controllable area as soon as possible	Direct Move; Quick Stop; Turn	Pure reactive without intentions
Smooth Move	Move with desired speed and direction	Direct Move; Turn	
Level 2 –Individual Behavior			
Name	**Function**	**Related behaviors**	**Feature**
Kick	Push the ball to in a desired direction	Direct Move; Spin	Intentio nal but no consideration about others
Block	Hinder the motion of opponents	Smooth Move	
Intercept	Get the ball when it's moving	Smooth Move	
Dribble	Control the ball in a controllable area	Smooth Move	
Unstuck	Leave corners, borders of the play ing ground or opponents when necessary	Smooth Move	
Level 3.1 – Micro-cooperation Behavior Group			
Name	**Function**	**Related behaviors**	**Feature**
Pass	Push the ball to an open area where another teammate stays.	Kick; Smooth move	Intentional and with considera tion about other teammates in a local area
Block-to-Kick	Help others finish kicking by blocking opponents	Kick; Smooth move; Block	
Block-to-Dribble	Help others finish dribbling by block opponents	Dribble; Smooth move; Block	
Maximize ball control	Look for the chance to get the ball by wait at a desired place	Smooth move;	
Level 3.2 – Macro-cooperation Behavior group			
Name	**Function**	**Related behaviors**	**Feature**
Scatter	Return to the default area	Smooth move	Intentional and with consideration about other teammates in a local area
Formation change	Rearrange the robots' default positions	(None)	

Table 1. Layered behaviors based on cooperation

4 Experiments

4.1 Experiment Design

In order to check the effectiveness of our cooperation-based architecture, we present six teams with different abilities to compete each other. They are defined as in Table 2. During the development, we find out that there are several inherent weak points in a pure physical platform [4]. Therefore, a simulation platform is implemented to test the influence of certain behavior on overall team performance. The same Basic Motion listed in Table 1 is used by each team, however the performance is different. A match has two halves, each of which lasts 10 minutes. "Long" response time means 20% more time added in decision; "Common" motion accuracy means 20 % random noises added in robot's motion (for both position and orientation). Micro- and Macro-cooperation are the same for the teams that have them. Each team has the same default initial position for each team member. However, for each game, the initial positions of robots are slightly changed by adding 5% white noises. It intends to make various beginning situations.

Name	Response time	Motion Accuracy	Strategy
Team1	Long	Common	1)Chase the ball and push it to the opponent goal; 2)Only one robot deals with the ball at any time; 3)Always the robot that is nearest the ball is selected to deal with it.
Team2	Short	Good	(Same as Team1)
Team3	Long	Common	Micro-cooperation only
Team4	Long	Common	Micro-cooperation and macro-cooperation
Team5	Short	Good	Micro-cooperation only
Team6	Short	Good	Micro-cooperation and macro-cooperation

Table 2. Design of the six teams in experiment

4.2 Results and Discussion

The results of totally 50 games are recorded (refer to Table 3). We use the Ratio of Win (RW), Average Goal (AVG) and Average Loss (AVL) to evaluate the team performance.

Opponent	RW (%)	AVG	AVL
Team1 V.S. Team2	0	0.2	8.3
Team1 V.S. Team3	14	1	8.1
Team1 V.S. Team4	12	0.7	8.0
Team1 V.S. Team5	0	0.1	10
Team1 V.S. Team6	0	0	9.5
Team2 V.S. Team3	92	5.5	1.5
Team2 V.S. Team4	86	4	1.3
Team2 V.S. Team5	32	1.4	4.7
Team2 V.S. Team6	28	1.1	4.2
Team3 V.S. Team4	48	1.2	1.3
Team3 V.S. Team5	2	0.3	9.3
Team3 V.S. Team6	0	0.25	9
Team4 V.S. Team5	0	0.4	11
Team4 V.S. Team6	0	0.1	8.7
Team5 V.S. Team6	46	3.1	3.3

Table 3. Competition Scores

Summarily, we order the teams according to their performance as follows: Team6 $\geq$ Team5 > Team2 >Team4 $\geq$ Team3>Team1. Here, "$\geq$" means "slightly better than"; ">" means "obviously better than".

From the experiment results, it can be summarized as: (1) All the teams with better performance in Basic Motion defeat the teams with common performance. (2) All teams with cooperation behaviors defeat the teams without cooperation behaviors. (3) The teams with both micro-cooperation and macro-cooperation behaviors are just slighter better than the teams with micro-cooperation behaviors only.

(1) means **a behavior that is frequently used has great influence on the team's overall performance**. Basic Motion represents the personal ability of robot, which is the foundation of other behaviors and therefore used at any time. So, we must try our best to enhance the capacities of individual robots when developing a soccer robot team.

(2) means **cooperation is an essential mechanism for robots in RCA environment if they intend to complete goals more efficiently and robustly**.

(3) may lead to a misunderstanding that macro-cooperation is not so effective as macro-cooperation. We cannot get such a general conclusion since the scale of team is not considered in experiments, which may be a potential factor. A more reasonable explanation to (3) is **the behaviors closely related to dealing with the ball directly impose heavy affects on team performance**.

5 Conclusions and Future Work

Beginning with a cooperation-embedded architecture, we construct a team of robots able to make decisions independently to look for cooperation opportunities in both local areas and the global area. Experiment results preliminarily prove the effectiveness of our design.

We have stated that ideal cooperation should be evolutionary. How to improve the cooperation performance is our next objective. A common method is to use machine learning. Previous researches put focus on simulation [5] and middle-size physical robot with local vision [6]. There is still few report about success in small-size robot application. The possible challenges include: "How to evaluate cooperation performance quantitatively?", "How to applied learning in situation recognition?" and "How to obtain learning samples in physical environment effectively?"

References

1. Kitano, H., Asada, M., Kuniyoshi, Y., Noda, I., Osawa, E., and Matsubara, H. RoboCup: A Challenge Problem for AI. AI Magazine , 18(1): pp73-85, 1997.
2. Minoru Asada, Peter Stone, Hiroaki Kitano, Barry Werger, Yasuo Kuniyoshi, Alexis Drogoul, Dominique Duhaut, Manuela Veloso, Hajime Asama and Sho'ji Suzuki. The RoboCup Physical Agent Challenge: Phase I. Applied Artificial Intelligence (AAI), Volume 12, 1998
3. Ronald C. Arkin. Behavior-based robotics. pp51, MIT Press, 1998.
4. Hui Wang, Han Wang, Yew Tuck Chin, William Y. C. Soh, A Multi-Platform for Robot Soccer Team Development. Tech. Report of Robot Soccer of NTU, 2000
5. Peter Stone. Layered Learning in Multi-Agent Systems: A Winning Approach to Robotic Soccer. MIT Press, 2000.
6. M. Asada, S. Noda, S. Tawaratsumida, and K. Hosoda. Vision-based behavior acquisition for a shooting robot by using a reinforcement learning . In Proc. of IAPR / IEEE Workshop on Visual Behaviors-1994, pp 112-118, 1994.

Multi-platform Soccer Robot Development System

Hui Wang, Han Wang, Chunmiao Wang, and William Y.C. Soh

Division of Control & Instrumentation, School of EEE
Nanyang Technological University
Nanyang Avenue, Singapore 639798
{p147513815, hw, p149510969, eycsoh}@ntu.edu.sg

Abstract. Robot soccer is a challenging research domain, which involves multiple agents (physical robots or "softbots") to work together in a dynamic, noisy, cooperative and adversarial environment to achieve specific objectives. The paper focuses on the design and implementation of an effective system for developing a small-size physical soccer team for the competition of RoboCup. There are three goals preset in our research, that is, robust individual behaviors, controllable cooperation and behavioral learning. The system is therefore required to collect data in real-time situations and repeat experiments accurately, which are found difficult to be achieved in pure physical or simulated case. A distinct feature of our system is that it is a close combination of three functional parts: physical test platform, simulation platform and measurement platform, which support each other to overcome many defects inherent in traditional physical system or simulation system.

1 Introduction

In recent years, robot soccer has become a popular research topic for robotics and artificial intelligence. In order for a robot team to perform a soccer game, various technologies should be incorporated, such as, robotics, multi-agent collaboration, real-time image processing, sensor fusion, and machine learning [1]. Our research focus is to develop an autonomous physical team to explore the cooperative behaviors in robot soccer.

The small-size robot soccer game contains two teams, each of which has five players. Our team is composed of three functional parts: vision module, intelligent control module and real robots. It is designed according to a well-defined processing cycle : Vision module receives the raw image sequence from a camera hung over the playing field and processes it, giving out the motion states of the ball and all robots. The result is sent to a control module, which evaluates the world state and decides what to be done in the next step based on its specific strategy. Robots' actions are motion commands (encoded in a packet) broadcasted by the control module through wireless communication. Each robot decodes its own part from the command packet based on a unique ID code and then executes corresponding actions.

A. Birk, S. Coradeschi, and S. Tadokoro (Eds.): RoboCup 2001, LNAI 2377, pp. 471–476, 2002.

In this paper, we first explain why a multi-platform system should be adopted in developing a small-size robot team and then discuss how to solve a critical issue in such a system. Details of implementation of the multi-platform system are also presented to embody some principles. Finally, experiences and results in our work are summarized and future work is presented.

2 Obstacles in Soccer Robot Team Development

Since there is still not a formal way to design and implement a cooperative small-size soccer robot team, our work heavily depends on experiments. There are three objectives in our development. Firstly, single robot is able to perform individual behaviors in a robust way, that is, it can repeat certain actions in specific situations; the second is to investigate what mechanisms can produce a controllable rather than "emergent" cooperation in a robot team, that is, in an ideal case, each robot can work together intentionally to help its teammates; finally, to study how to apply machine learning to enhancing the performance of robots including their individual behaviors and team behaviors. In order to obtain these goals, our system is at least required to have the following abilities:

- Collect data and/or perform analysis in real-time situation.
- Repeat experiments in controlled mode, such as in the same system specifications.

The first requirement is because soccer is a time-critical game. A playing strategy should deal with dynamic rather than static situations. The second requirement is because we intend to develop a collection of robots that are able to act like a team, which means the robots can present cooperative behaviors in typical situations. Such cooperation not only includes those so-called "emergent" cooperation [2][3], but also "controllable" cooperation. The key difference between them is that emergent cooperation can be achieved by adding simple rules in robot's individual behaviors, which are useful to avoid some explicit conflicts, such as a robot vies with teammates to deal with the ball, but not sufficient to guarantee that a player can help its teammates to complete a task intentionally. However, such intentional cooperative behaviors are fairly helpful in soccer games [4].

In our development, we have found that if we rely on a pure physical platform, it is hard to meet the above two requirements. Firstly, our system has to run in a real-time case. Any extra on-line data collection and analysis task may affect the system performance, which will lead to incorrect experimental results. Secondly, many unexpected factors limit the accuracy of system specifications. It is sometimes a tough task to suppress such random influences.

Therefore, a concept of "multi-platform system" is introduced, which includes several functional platforms that undertake well-defined sub-tasks in development. In our case, a simulation platform and an assistant measurement platform are added to overcome the above obstacles in a pure physical system. Different from traditional ways, our simulation platform doesn't work only for testing some ideas. It will serve to develop effective algorithms applied in physical environment. It enables us to account for specific statistics in real-time situation, which are extremely useful in comparing the feasibility and performance of each strategy. Another advantage of simulation is that it is easy for us to control the experiment conditions, that is, in each trial, we can only focus on one or several factors.

3 Challenge in a Multi-platform Development System

Since there are more than one experiment platform in our system to develop physical robots, a subsequent critical problem is "How to keep coherence between them?", that is, a question must be answered: "If an algorithm is developed in simulation platform, can it be used in physical environment without any radical change? If not, what should be noticed?". In fact, some scholars once developed simulation platform to study the intelligent robots and try to put the results into practice. However, there was few report about the successful application based on those simulation results ([6] and [7] may be two exceptions.). We believe that there are two main reasons: 1) Overly specific assumptions are applied in simulation, so that the results are hardly extended to general cases and 2) There is no clear work decomposition and rigorous development process for those platforms. In our system, the simulation platform is not an independent part. It is designed to facilitate the physical test platform to develop robots' high-level strategy. The point is each platform only undertakes the work, at which they are good. In addition, their interactions are rigidly defined (refer to Fig. 1):

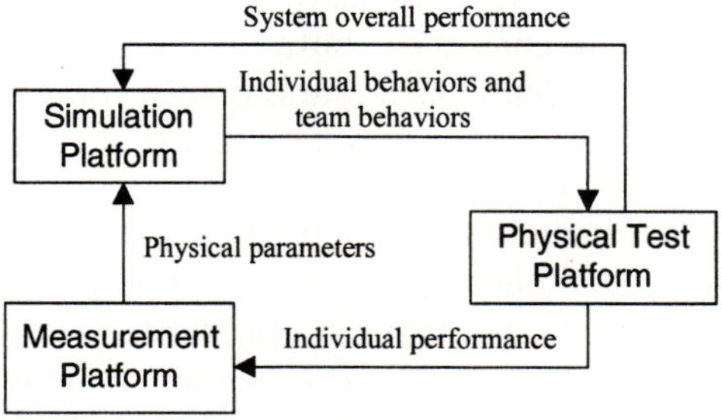

Fig. 1. Interactions among three functional modules in our multi-platform development system.

In addition, there are three methods taken to keep the coherence between simulated and physical experiments: First, the kinetic model of robot in the simulation platform is the same as that in the physical platform. All related motion parameters, such as max/min speed and max acceleration, are kept the same. They are measured on measurement platform. Second, the same system architecture (refer to Fig. 2) is adopted in both platforms. It provides convenience for simulation program to test the performance of each component in the physical system by altering some parameters, so that we can determine which factor of what component how to affect the system's overall performance. Finally, algorithms and software frame in high-level strategy are the same in both platforms. It makes sure that the computation complexity is identical and thus the structure of decision process developed in the simulation platform can be easily transferred to the physical environment.

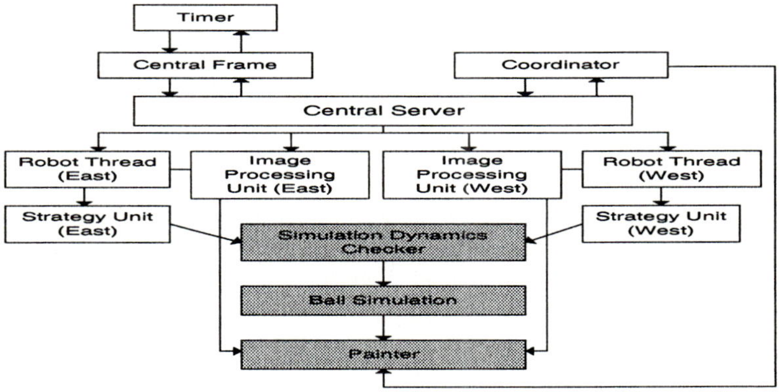

Fig. 2. Structure of simulation platform.

4 Implementation of a Multi-platform Development System

Fig. 3(a) and 3(b) are the photos of physical test platform and simulation platform. In this section, we will present related details of implementation in the view of function rather than entity. For both physical and simulated cases, our system adopts a uniform modeling as follows to keep coherence among multiple development platforms:

- A single robot is represented as a set of physical parameters: size, max. speed, max. acceleration and a table mapping wheel speeds to motion commands, all of which are determined by measurement platform and applied in simulation.
- Moving objects (robots and the ball) in soccer game are represented as a motion state set and behavior set, where motion state is a vector [x, y, dx, dy, theta, ID], that is, position, velocity, orientation and identification; behavior is a situation-action pair (a situation is a designer-determined motion state subset; action is only for robot, which has four basic types, that is, forward/backward linear move with quick start, immediate stop, turning in a fixed angle and spin).

System software frame are summarized in Fig 2. It is implemented based on a typical multi-thread client-server model: (1) Central Server waits for the inputs from user through its interface - Central Frame and will inform each of five clients, that is, two Robot Thread, two Image Processing Unit and Painter. (2) Coordinator is basically responsible for monitoring the game state, such as goal.

Its additional special function is to operate methods in Central Server so as not to hang on when calling objects for long time. Timer thread keeps track of time and informs Central Frame when time is up or after a second has passed. (3) Image Processing Unit processes data captured from the soccer field (i.e. the Painter). It calculates motion state of each robot (including the opponents) and ball. The Robot Thread from the same team will then pick up this processed data. (4) Robot Thread contains the number of robots requested by the user. Whenever the user activates the game, it will first try to obtain information from the Image Unit. If new data are available, it will then proceed to prompt each Strategy Unit to use these data for strategy planning. The data returned by the robots will then be kept in a vault waiting to be picked up by the Painter. (5) Simulation Dynamics Checker checks the boundary, collision and calculates the next robot step according to the speed and desired direction the strategy unit has set. Ball Simulation calculates the ball position in the next step. (6) Painter thread provides the animation and it holds on to the Simulation Dynamics Checker and Ball Simulation. Therefore, as soon as it receives the next step data from Robot Thread, the Simulation Dynamics Checker and Ball Simulation will digest these data before showing the animation.

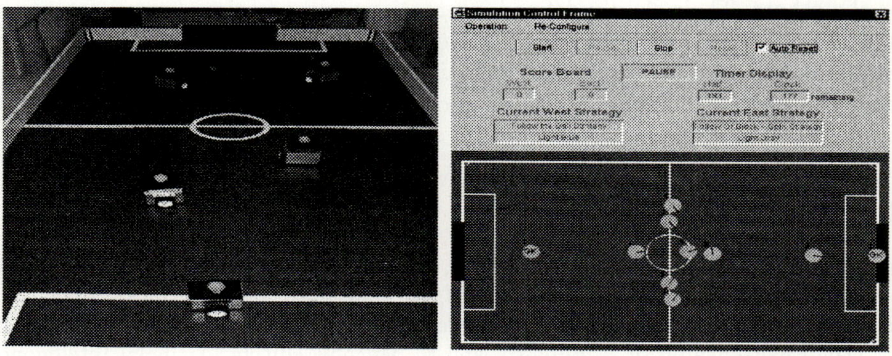

(a) Physical test platform (b) Simulation platform

Fig. 3. Development platform for soccer robot team

The shadow parts in Fig.2 are absent in physical platform, that is, Painter, Simulation Dynamics Checker and Ball Simulation only serve for simulation. The animation in simulation is embodied as robots' motion in physical environment. To develop a simulation platform is for behavior design and selection, we pay special attention to many special features:

- Convenient to add noise and evaluation of the performance of various filters used in image processing unit of physical system.
- Many choices in extra actions, robot's speed and response time.
- Easy to develop strategy.
- Collect and show data of robots in real-time case (refer to Fig. 4).

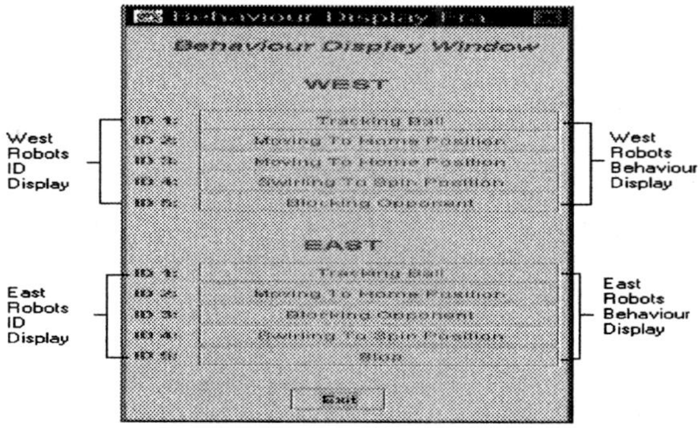

Fig. 4. Robot behavior display frame

5 Conclusions

This paper describes a design and implementation of a multi-platform system for building a robot soccer team. The development task is decomposed into three interrelated platforms, each of which complete an objective and provide help for the others. Their rigidly defined interactions and additional requirements in internal structure guarantee the effectiveness of this system. At present, partial results are obtained in individual behavior design, and perception (image processing in physical system). However, more efforts are still needed to investigate the mechanism in team cooperation.

References

1. Kitano, H., Asada, M., Kuniyoshi, Y., Noda, I., Osawa, E., and Matsubara, H, RoboCup: A Challenge Problem for AI, AI Magazine , 18(1): pp73-85, 1997.
2. Ronald C. Arkin and Tucker Balch, Cooperative Multiagent Robotic System, In David Kortenkamp, R.P. Bonasso, and R. Murphy, editors, Artificial Intelligence and Mobil Robots, Cambridge, MA, 1998. MIT/AAAI Press.
3. B. Werger, The spirit of bolivia: Complex behavior through minimal control, In Proceedings of RoboCup 97, IJCAI 97, 1997.
4. M. Veloso, P. Stone, and M. Bowling, Anticipation as a key for collaboration in a team of agents: A case study in robotic soccer, In Proceedings of SPIE Sensor Fusion and Decentralized Control in Robotic Systems II, volume 3839, Boston, September 1999.
5. P. Stone and M. Veloso, Using decision tree confidence factors for multiagent control, In H. Kitano, ed, RoboCup-97: The First Robot World Cup Soccer Games and Conferences. Springer Verlag, Berlin, 1998.
6. P. Stone and M. Veloso, Task decomposition, dynamic role assignment, and low-bandwidth communication for real-time strategic teamwork, Artificial Intelligence 110(2), pp241-273, June 1999.
7. Raffaello D'Andrea, Jin-Woo Lee, Andrew Ho man, Aris Samad-Yahaja, Lars Creman, and Thomas Karpati, Big red: The cornell small league robot soccer team, In M. Veloso, E. Pagello, and H. Kitano, editors, RoboCup-99: Robot Soccer World Cup III. Springer Verlag, Berlin, pp49-60 2000.

On-line Navigation of Mobile Robot Among Moving Obstacles Using Ultrasonic Sensors

Nobuhiro Ushimi[1], Motoji Yamamoto[1], Jyun'ichi Inoue[1], Takuya Sugimoto[2], Manabu Araoka[2], Takeshi Matsuoka[3], Toshihiro Kiriki[1], Yuuki Yamaguchi[1], Tsutomu Hasegawa[1], and Akira Mohri[1]

[1] Kyushu University
6-10-1 Hakozaki, Higashi-ku, Fukuoka 812-8581, Japan
[2] Hitachi Information & Control Systems, Inc.
[3] Fukuoka University

Abstract. This paper proposes a realistic on-line navigation method of the mobile robot in dynamical environment where multiple obstacles are always changing their velocities. Considering characteristics of actual sensor system, a method to estimate the velocity of moving obstacles is presented. The estimated velocity and measured distance from the nearest obstacle are used to plan a velocity of mobile robot based on a new idea of Collision Possibility Cone(CPC). Then an on-line navigation method is proposed by using CPC and feasible velocity space of mobile robot. Simulational examples show an effectiveness of the new navigation.

1 Introduction

Navigation of mobile robot to travel for given destination autonomously in various environment is considered as one of the most important capability, thus many researchers have studied about the navigation problem. Particularly, the study on on-line sensor based motion planning or control is recently active, and some algorithms are proposed[1][2]. Most of these studies treat the case of multiple static obstacles, and mainly focus theoretical global convergence of the proposed algorithms. For more applications to real world, the mobile robots are expected to surely reach desired point even in the dynamical environment where many obstacles are moving. This problem is considered as an on-line sensor-based navigation problem among multiple moving obstacles.

Considering such dynamical environment, some on-line motion planners are presented. Tsoularis[3] gives a path from start point to goal point neglecting moving obstacle's paths, then plans velocity pattern of the robot along the given path to avoid moving obstacle by changing the velocity. Another important approach for moving obstacles is the idea of time-state space which adds time axis to normal configuration space[4].

These methods, however, take much time for planning, thus are not appropriate for on-line use. Fiorini et al. propose an on-line method based on the idea of Velocity Obstacle[5]. Moving obstacles are mapped into a two-dimensional

A. Birk, S. Coradeschi, and S. Tadokoro (Eds.): RoboCup 2001, LNAI 2377, pp. 477–483, 2002.

"velocity space". Then velocity of mobile robot is directly planned using Velocity Obstacle in velocity space. Most of studies of on-line motion planning for multiple moving obstacles assume that the velocity and position of each moving obstacle can be measured using robot's sensor systems. In the situation of multiple obstacles moving crowdedly, it is, however, very difficult to distinguish each obstacle using sonar sensor or laser range finder, because the sensor systems have only limited space resolution. Furthermore, most on-line navigation methods assume that velocity of moving obstacle is constant during sensor cycle. It is not appropriate for the dynamical environment such as multiple moving obstacles where the obstacles may change their velocities at any time.

In this paper, a realistic on-line navigation problem is discussed, where multiple obstacles are moving around and their paths, velocities and sizes are not given in advance. The paper also deals with the problem of velocity changes during the sensor cycle. The available information is assumed to be only distance information to obstacles at every sensor cycle considering the use of sonar sensor. Our navigation method basically selects the best velocity of mobile robot considering the worst case in the view point of danger for collision with moving obstacles at every sensor cycle.

2 On-line Collision Avoidance for Moving Obstacles

This section describes assumptions on mobile robot and moving obstacles. Then, an estimation method of relative approaching velocity of obstacle is presented.

2.1 Assumptions

- Mobile robot is an omni-directional vehicle.
- The mobile robot and moving obstacles move on a flat floor.
- The mobile robot does not communicate with moving obstacles.
- The mobile robot and moving obstacles are assumed to be approximated by cylinders.
- The mobile robot has n_r ultrasonic sensors. Each sensor detects distance from the robot to moving obstacle every sensor cycle if the obstacles are in the detectable area. The maximum range of the area is denoted by L_s.
- Moving obstacles possibly change their velocities v_o and accelerations a_o at anytime within maximum values of $\pm v_{o\max}$ and $\pm a_{o\max}$.
- Maximum values $v_{o\max}$ and $a_{o\max}$ of moving obstacles are smaller than mobile robot's maximum velocity $v_{r\max}$ and acceleration $a_{r\max}$.

2.2 Estimation of Approaching Velocity for Obstacle

The ultrasonic sensor basically detects only distance d_k from mobile robot to a nearest moving obstacle inside the sensor detectable area (sector-k) every sensor cycle(See Fig.1). When the distance data changes from $d_k(i)$ to $d_k(i+1)$ during

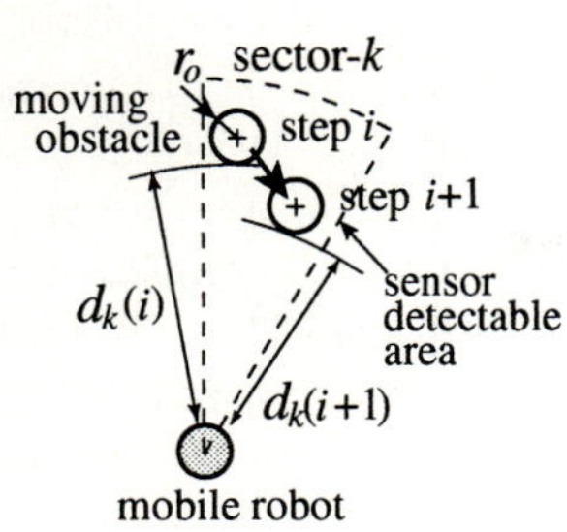

Fig. 1. Sensor detectable area (sector-k)

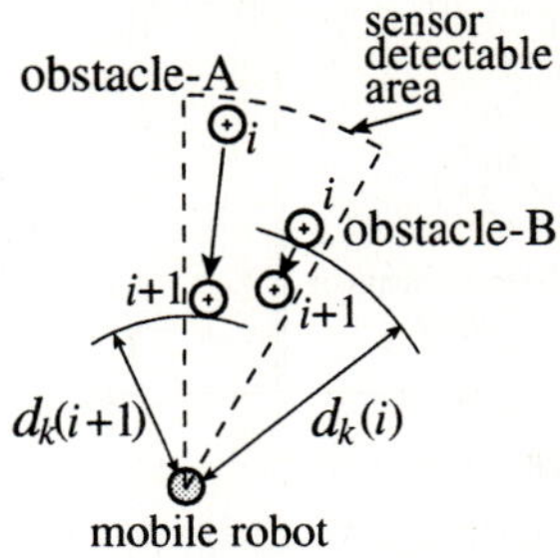

Fig. 2. Multiple moving obstacles in the same area

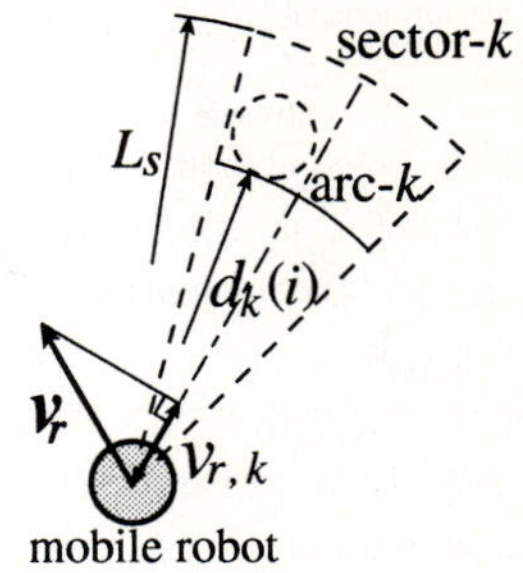

Fig. 3. Velocity $v_{r,k}$ to the direction of sector-k

sensor cycle T_s, the relative approaching velocity of moving obstacle in the direction of sector-k may be written $v_{ap,k}(i+1) = (d_k(i+1) - d_k(i))/T_s$ (where i is the step number of sensor cycle). This is, however, incorrect for the case that multiple moving obstacles are in a same sensor area, because the sonar sensor can not distinguish the individual obstacles as shown in Fig.2. In the figure, two distances $d_k(i)$ and $d_k(i+1)$ are the result of two different obstacles.

To cope with the problem, we modified the estimation of approaching velocity of obstacles by

$$v_{ap,k}(i) = v_{r,k} + v_{o\ \max} \quad (\text{if} \quad d_k(i) < L_s) \tag{1}$$

where $v_{r,k}$ denotes projection of velocity $\boldsymbol{v}_r$ to the direction of sector-k as shown in Fig.3. This estimation basically assumes most dangerous case of velocity and moving direction of moving obstacles for the mobile robot.

3 Collision Possibility Cone

An idea of Collision Possibility Cone(CPC) is introduced to guarantee of collision free with moving obstacles using distance information and estimated approaching velocity from sonar sensor. In this section, the robot (r) and the obstacles (o) are described by circle (see STEP 0 in Fig.4). The mobile robot's radius and velocity are denoted with r_r, v_r. On the other hand, a smallest radius of the moving obstacles is denoted by $r_{o\ \min}$(if radius is unknown, it is defined with zero). The procedure to construct CPC is described as follows.

STEP 0 Detect distance $d_k(i)$ from mobile robot to moving obstacle if a moving obstacle enters into sensor detectable area.

STEP 1 Calculate the possible existing area of the moving obstacle using the detected distance as in STEP 1 of Fig.4.

STEP 2 Enlarge the area with r_r, then denote the extended area $\bar{O}$. The mobile robot is then represented by point robot $\bar{R}$.

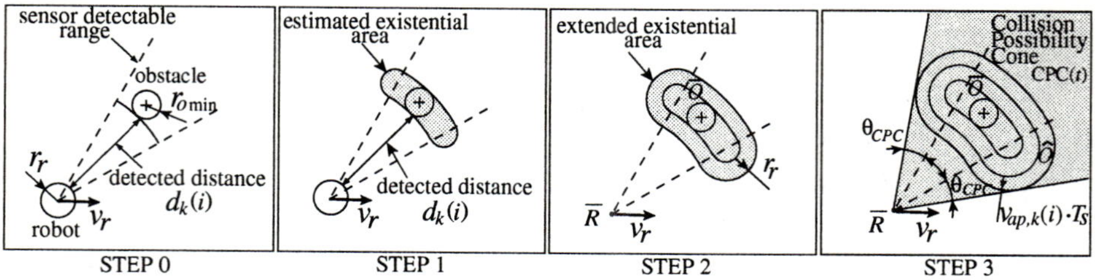

Fig. 4. Collision Possibility Cone (CPC)

STEP 3 Enlarge $\bar{O}$ with $v_{ap,k}(i) \cdot T_s$, then denote the extended area $\hat{O}$. Where T_s is sensor cycle of robot. Make two tangential lines to the extended area $\hat{O}$ from the point robot $\bar{R}$. The resultant cone region surrounded by the two lines is CPC. Where θ_{CPC} is an extended angle from the original sensor detectable area.

CPC indicates admissible collision free velocity of mobile robot. If the end point of the velocity vector located out the region of CPC, the mobile robot will never collide with moving obstacle. Note that collision free is guaranteed at least until next sensor step.

4 On-line Motion Planning Problem

This section formulates an on-line motion planning problem. Then, a navigation strategy based on the idea of CPC is proposed.

4.1 Formulation of On-line Motion Planning

The dynamics of the mobile robot is generally described by the following equation

$$\ddot{\boldsymbol{x}}(t) = \boldsymbol{f}(\boldsymbol{x}, \dot{\boldsymbol{x}}, \boldsymbol{u}) \tag{2}$$

$$\boldsymbol{u} \in U, \quad |\dot{\boldsymbol{x}}| < v_{r\,\max} \tag{3}$$

where $\boldsymbol{x}$ is position vector, $\boldsymbol{u}$ is actuator input vector, U is admissible input set, and $v_{r\,\max}$ is maximum velocity of mobile robot. The on-line motion planning problem is to generate velocity for minimizing traveling time from an initial point to goal point avoiding moving obstacles. Where the moving obstacles change their velocities. The dynamics of moving obstacle is assumed to be unknown considering general application. However, we assume the velocity and acceleration of moving obstacles are limited as described in section 2.1.

4.2 On-line Velocity Planning

When the mobile robot moves with velocity $\boldsymbol{v}_r(t)$, the possible velocity at $t + \Delta t$ is described with

$$\boldsymbol{v}_r(t + \Delta t) = \boldsymbol{v}_r(t) + \ddot{\boldsymbol{x}}(t)\Delta t \tag{4}$$

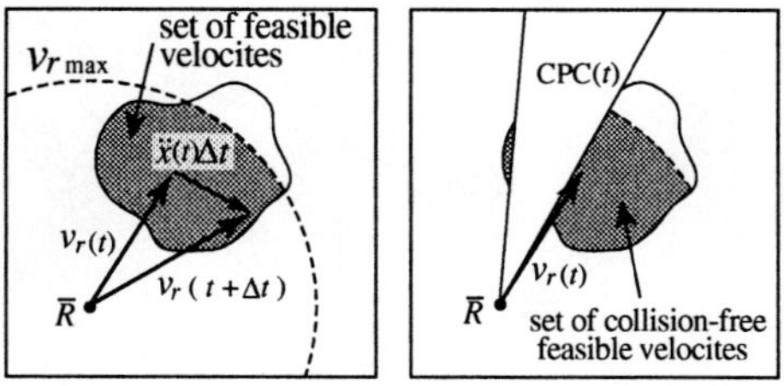

Fig. 5. Sets of feasible velocities and collision-free feasible velocities

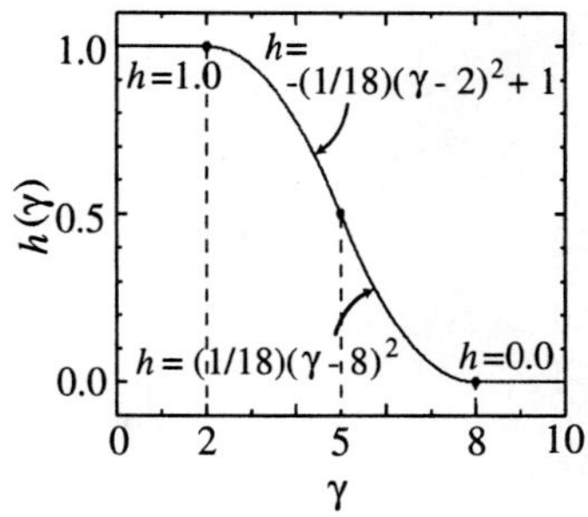

Fig. 6. Weighting function $h(\gamma)$

by dynamics of mobile robot and input constraints of mobile robot. Where $\ddot{\boldsymbol{x}}(t)$ is an element in the set of feasible velocity (gray region in Fig.5(left)) which is satisfied with equation (2) and admissible input (3). By removing the region of CPC(t) from the feasible velocity set, we get collision-free feasible velocity set as shown in Fig.5(right). We plan the mobile robot's motion by selecting velocity which minimizes the traveling time in the collision-free feasible velocity set.

4.3 Collision Distance Index

If the mobile robot current position is far from moving obstacle, the selected velocity by CPC is no danger of collision with obstacle by considering maximum velocities of moving obstacles. To prevent the conservative property, we introduce the following "collision distance index". It is defined by the detected distance $d_k(i)$ and the relative approaching velocity of obstacle $v_{ap,k}(i)$ by (1). The index is used for changing the extended angle θ_{CPC} of CPC which means danger of collision in the meaning of velocity. We define the collision distance index by

$$\gamma = \frac{d_k(i)}{v_{ap,k}(i) T_s} \tag{5}$$

Then using the index, the extended angle θ_{CPC} of CPC is modified by

$$\tilde{\theta}_{CPC} = h(\gamma)\theta_{CPC} \tag{6}$$

where a function $h(\gamma)$ is shown in Fig.6 as an example. The function $h(\gamma)$ is selected such that the value of the function is 1 when γ is small, and becomes asymptotically 0 in accordance with increasing γ.

5 Simulation of On-line Navigation

This section shows some simulational examples to confirm the effectiveness of the idea of CPC and collision distance index. In the simulations, the constraints

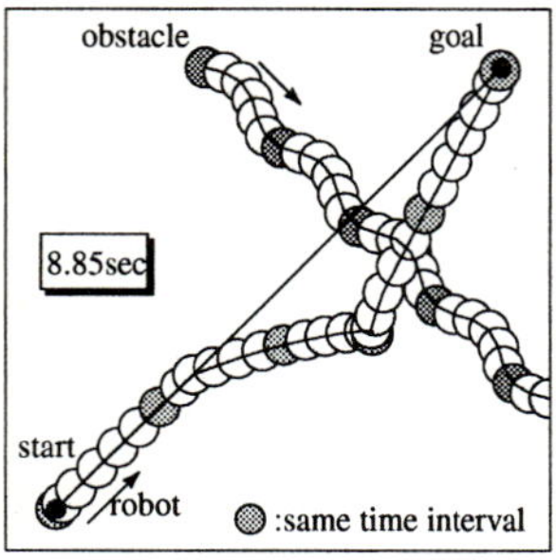

Fig. 7. Path 1 for changing velocity of moving obstacle without considering collision distance index γ

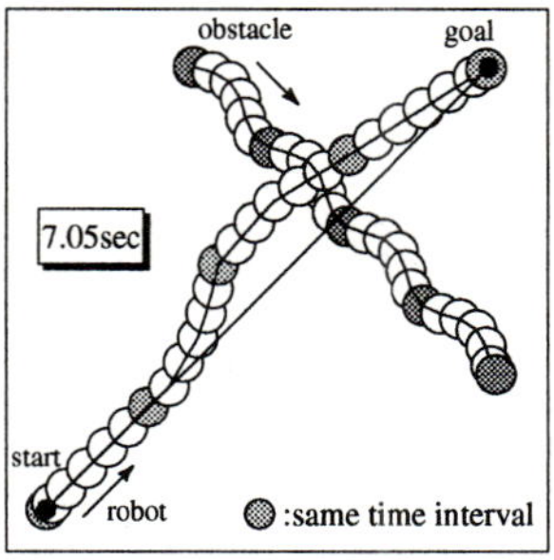

Fig. 8. Path 2 for changing velocity of moving obstacle considering collision distance index γ

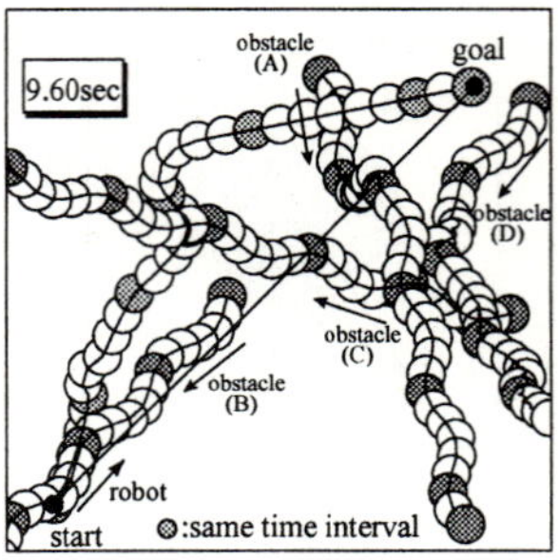

Fig. 9. Path 3 for the effectiveness of proposed navigation method in the dynamical environment

on mobile robot and moving obstacles are $v_{r\,\max} = 2.0$m/s, $v_{o\,\max} = 1.5$m/s, $L_s = 6.0$m, and $T_s = 0.3$sec. Then, the size of test field is 12m $\times$ 12m, and the radius of robot and obstacles are $r_r = 0.4$m and $r_o = 0.4$m. The mobile robot has 12 ultrasonic sensors with ring shape. In the following example, the robot does not know the velocities and paths of obstacles in advance, but only knows the distance from the nearest obstacle by its sensor system every sensor cycle.

Simulation results with one moving obstacle are shown in Fig.7 and Fig.8. Circles in the figures represent positions of robot and obstacle at same time interval. Each traveling time to goal point is shown in shaded box. In Path 1 (Fig.7), the velocity of moving obstacle is always changing, and the path of obstacle is a meandering one. By estimating the maximum velocity of obstacle in the idea of CPC, the mobile robot could arrive at goal point without collision, even for the case that the velocity of obstacle changes between sensor cycle. In Path 2 (Fig.8), the collision distance index γ is considered, whereas it is not considered in the simulation of Path 1. We find that the consideration of the collision distance index γ results in more efficient path by this example.

To confirm the effectiveness of the proposed navigation method for more complicated case, we simulate an example of four moving obstacles in Fig.9. All moving obstacles always change their velocities. Also for this example, the mobile robot successfully reaches its destination without collision.

6 Conclusions

In this paper, a realistic on-line navigation problem in dynamical environment with multiple moving obstacles is discussed. By considering the characteristics of actual ultrasonic sensors, an estimation method of velocity for moving obstacles is presented. For the on-line collision avoidance of moving obstacles, the idea of CPC is presented. New on-line motion planner is proposed by using the idea of

CPC. The method is based on a selection of admissible velocities in the restricted two dimensional velocity space. For an over conservative property of the proposed navigation method, the collision distance index is introduced. An effectiveness of the proposed method is shown by simulational examples.

References

1. V. Lumelsky and T. Skewis: Incorporating Range Sensing in the Robot Navigation Function. IEEE Transactions on Systems, Man, and Cybernetics, Vol.20, No.5 (1990) 1058–1069
2. I. Kamon, E. Rivlin and E. Rimon: A New Range-Sensor Based Globally Convergent Navigation Algorithm for Mobile Robots. Proceedings of the IEEE International Conference on Robotics and Automation (1996) 429–435
3. A. Tsoularis and C. Kambhampati: Avoiding Moving Obstacle by Deviation from a Mobile Robot's Nominal Path. The International Journal of Robotics Research, Vol.18, No.5 (1999) 454–465
4. T. Tsubouchi, T. Naniwa and S. Arimoto: Planning and Navigation by a mobile Robot in the Presence of Multiple Moving Obstacles and Their Velocities. Journal of Robotics and Mechatronics, Vol.8, No.1 (1996) 58–66
5. P. Fiorini and Z.Shiller: Motion Planning in Dynamic Environments Using Velocity Obstacles. The International Journal of Robotics Research, Vol.17, No.7 (1998) 760–772

11monkeys3 Team Description

Keisuke Suzuki, Naotaka Tanaka, and Mio Yamamoto

3-14-1, Hiyoshi Kohoku-ku, Yokohama, 223-8522 Japan
Institute of Science and Technology, Keio University
robocup@yy.ics.keio.ac.jp

Abstract. For these three years, we have been trying to build an accurate world model for the agent to select a best/better behavior. This required a lot of resources of the system, though it is still impossible to get the exact world model. So we stood back to the point: Why are we trying to get an accurate world model. It is only because we believed that an accurate world model would help the agent determine the best/better behavior to win the soccer game. If the agent could select best/better behavior with less information, our requirement is fulfilled. So, this year we have worked to build a world model that has minimum but enough for the agent to determine its behavior.

1 Introduction

This is the 3rd year since 11Monkeys first joined the RoboCup. On the first year, Shuhei Kinoshita started developing the team based on the three- layered structure. And on the second year, we tried to build a better world model, and used the immunenetwork algorithm to determine the agent 's behavior. Using the immune-network didn't really work well, though it was an interesting theme to work on. This year, immune-network is taken away, and our theme is to let the agent determine its behavior by the minimum information. For the past two years, we were trying to build a world model that has a considerable precision. Because we thought that accurate world model will allow the agent to select the best/better behavior. This logic may be true, but it required a lot of the system's resource. And it is impossible to get the exact world model, when we think of the real world. So, this year, we decided to build a system with less calculation to get an accurate world model, but still can select a best/better behavior.

2 World Model

So, how much information does the agent need to determine its behavior? In the past, agents in 11Monkeys had information of the whole soccer field. But do they really need to know what's going on in the whole field? It may be better than not having it, though it is not that necessary enough to waste the system's resource. And because the world model was not always accurate, it often caused conflict between the agents. This lead to a problem, the agent selected an unsuitable

A. Birk, S. Coradeschi, and S. Tadokoro (Eds.): RoboCup 2001, LNAI 2377, pp. 484–486, 2002.

behavior due to the conflict information. In the new world model, we narrowed the agent's sight, little by little, to find out the suitable area an agent 's world model has to cover.

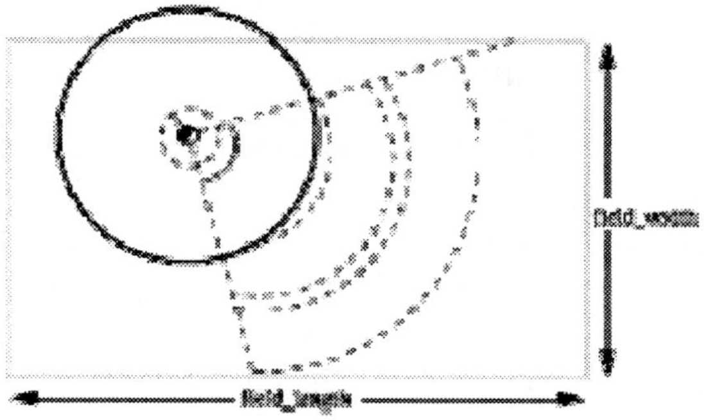

Fig. 1. Difference between the new world model and the old model: In the old model, the agent was trying to keep information of the hold field, but in the new model, agent keep the precise information of the circled part, and doesn't really care about the whole field.

We changed the diameter of the circle, i.e. the world model/the agent's view area, and did the simulation. What we found out was: When diameter less then 5m, agent needs accurate information. This area contains enough information for the agent to determine its behavior. When diameter is over 40m, there is less effect to the game whether the agent has an accurate information or not.

3 Other Changes from the Former Version

In the former version, we had some problems with the goalies. When the goalie couldn't decide which direction to kick out the ball, especially when the agents gathers near the goal, it used to determine the direction randomly. Because of this randomness, it sometimes kicked the ball into its own goal. To get rid of this problem, we allowed the goalie to move around the goal, to escape from getting into the crowded part. And we also made a limit to the direction to kick out the ball. Now we don't have the own-goal problem.

4 About the Coach Client

Our team is not using the coach client for now. Using the coach client may be effective to win the competition. Though we think that in the real world, existence of something like a coach client is rare. And when the coach client is used, each agent is no more an autonomous agent. We want to make research on autonomous agents, so that is why we aren't using the coach client for now. We might use it in the future.

5 Result

As a result of this year's development, it became a little bit better then the former version. When we compete the new team and the old one, the probability that the new team win is about 62.3%. This number may prove that it has got better than the last year.

3T Architecture for the SBCe Simulator Team

Eslam Nazemi, Mahmood Rahmani, and Bahman Radjabalipour

Electrical and Computer Engineering Faculty
Shahid Beheshti University
nazemi@cc.sbu.ac.ir
{m-rahmani, b-radjabalipour}@ce.sbu.ac.ir

Abstract. In this paper we briefly describe the process of action control used by the agents of the SBCe Simulator team. It is based on 3T Architecture, which helps to coordinate planful activities with real-time behavior for dealing with dynamic environments.
Our main goal in this project was to achieve some features of teamwork in the multi-agent system of simulated RoboSoccer such as Combination Play and Dynamic Positioning. To achieve this, SBCe used some real soccer rules and knowledge. SBCe's low level structure is based on the CMUnited-99.

1 Introduction

The design of intelligent agents is an important research direction within Multi-Agent System (MAS), where the behavior of a society of agents is described by modeling the individuals and their interactions from a local, agent-based perspective. Thus, finding appropriate architectures for these individuals is one of the fundamental research issues within agent design. Our main goal in this project is to achieve an architecture that supports some teamwork features of the multi-agent environment, such as Dynamic Positioning [2] and Combination Behavior. To achieve this, the SBCe uses the 3T Intelligent Control Architecture [5]. Also it uses some real soccer based on PCB (Pressure-Cover-Balance) System and MACP (Multi-Agent Combination Pattern) [6]. We argue that these techniques are very efficient in the coordination of agents in a team.
Section 2 discusses about the team development. Section 3 describes the agent architecture. Section 4 deals with the main concepts of our team, namely PCB and MACP. Finally we present future work and conclusions.

2 Team Development

Team Leader:

Eslam Nazemi
- Iran
- Faculty member

A. Birk, S. Coradeschi, and S. Tadokoro (Eds.): RoboCup 2001, LNAI 2377, pp. 487–490, 2002.

Team Members:

Mahmood Rahmani
- Iran
- Undergraduate student
- In charge of the design and implementation of the Virtual Soccer Team "SBCe"; the body of his B.Sc. project at the university.

Bahman Radjabalipour
- Iran
- Undergraduate student

3 Agent Architecture

The 3T architecture separates the general agent intelligence problem into three interacting layers or tiers (and is thus known as 3T). Skill layer that is a dynamically reprogrammable set of reactive skills coordinated by a skill manager. Sequencing layer that is a sequencing capability that can activate and deactivate sets of skills to create networks that change the state of the world and accomplish specific tasks. Planning layer that is a deliberative planning capability that reasons in depth about goals, resources and timing constraints.

We have modified the 3T that acts as a simulated robot. This is being accomplished by augmenting the skill layer of the architecture with a set of primitives or retrieving commands that are accepted for SoccerServer [1].

We have put our Tactic layer in the Sequencing layer. The Tactic layer activates a specific set of skills in the skill level (reactive layer). The Tactic layer will terminate actions, or replace them with new actions when the monitoring events are triggered, when a time out occurs, or when a new message is received from the deliberative layer indicating a change of plan.

We have developed our Strategy layer in the Planning layer that operates at the highest possible level of abstraction so as to make its problem space as small as possible. Figure 1 shows the relationship of the three layers and their interactions.

We have developed Skill Manager which acts as an interface between a set of situated skills and the rest of the architecture. Situated skills represent the architecture connection with the SoccerServer. For example, one might develop a situated skill for kicking the ball. Such a skill will only be useful if the ball is currently kickable. In other situations the skill will fail. The Skill Manager also handles the interface with the sequencing system by providing all of the communications and asynchronous events that the Sequencer (Tactic layer) needs in order to stay coordinated with the skills.

The SBCe's low-level skills are almost entirely based on the source code of CMUnited-99 team [3] with a substantial modification in order to be able to include the Skill Manager.

To accomplish tasks that the agent routinely performs, the architecture has a sequencing system. In our case, the Sequencer is a tactic interpreter. In it's simplest form a tactic is a description of how to accomplish tasks in the game.

The Sequencer Interpreter runs a continuous loop installing new goals on its agenda, deinstalling old goals that have been achieved or that have failed, activating or deactivating skills and watching for responses from event skills. On each cycle of interpreter, for each skill that needs activation or deactivation, an activate or deactivate message is sent to the Skill Manager which acknowledges the messages and activates or deactivates the skills. Usually the Sequencer will have requested activation of event skills; and in those cases on each cycle it listens for a response from the Skill Manager whether an event has occured or not.

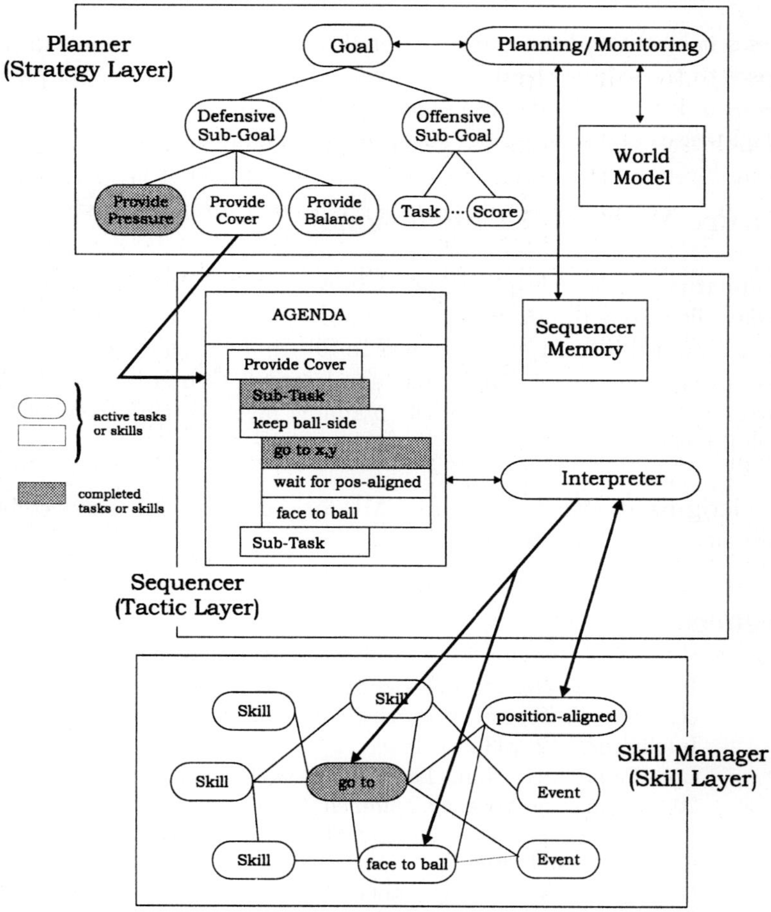

Figure 1. 3T Intelligent control Architecture

4 Defensive and Offensive Systems

SBCe uses the defensive positioning system called "PCB" and the offensive system called "MACP".

In the PCB system the fastest agent to the ball possessor provides **Pressure** on the ball. The second closest agent must provide **Cover**. Other agents will provide **Balance** to the defense. In this system, in general, a defender stands **goal-side and ball-side** of its mark and a midfielder stands ball-side.
There are a number of different ways in which an off-ball player can provide support for an on-ball attacker. These are called the Multi-Agent Combination Pattern or Combination Play. For example, in the attacking positions, the basic two-agent combination pattern can be divided into four basic categories, including passes to the side of the defender:

- Passes behind the defense (slotted or through pass — diagonal run).
- Passes to the side of the defense (give-and-go — overlap — wall pass).
- Passes in front of the defense (square balls — drop passes).
- Faked passes and other tricks (take over — dummy run).

5 Future Work and Conclusions

The modularity of the architecture provides several benefits. The bottom-up development allows one to add functionality incrementally to the agent.
Having achieved this framework, we will in future investigate the integration of other AI disciplines. For example machine learning techniques can be investigated from case-based reasoning in the planning layer to reinforcement learning in the skill layer [4].

Acknowledgment. We like to thank CMU for their low level source code which was used as a base of our team.

References

1. Ituki Noda, Hitoshi Matsobara, Kazuo Hinaki and Ian Frank, SoccerServer: A Tool for research on multiagent system- Applied Artificial Intelligence, 12: 233-250, 1998.
2. Luis Paulo Reis, Nuno Lau, FCPortugal Team Description, RoboCup 2000 Simulation League Champion, 2000.
3. Peter Stone, Patrick Riley and Mannuela Veloso, Cmunited-99 Source Code, Accessible from http://www.cs.cmu.edu/ pstone/RoboCup/CMUnited99.sim, 1999.
4. R. Peter Bonasso and David Kortenkamp, An intelligent agent architecture in which to pursue robot learning. In proceedings of the MLC_COLT'94 Robot Learning Workshop, 1994.
5. R. Peter Bonnaso, David Kortenkamp, David P.Miller and Marc Slack, Experiences with an Architecture for Intelligent, Reactive Agents. Intelligent Agents II, Springer: 187-202,1995.
6. Shel Fung, Basic Coaching Manual, Accessible from http://www.usc.mun.ca dgraham/manual, 1999.

Architecture of TsinghuAeolus

Jinyi Yao, Jiang Chen, Yunpeng Cai, and Shi Li

State Key Lab of Intelligent Technology and System,
Deparment of Computer Science and Technology,
Tsinghua University , Beijing, 100084, P.R.China
{Yjy01, Chenjiang97, Caiyp01}@mails.tsinghua.edu.cn
lishi@s1000e.cs.tsinghua.edu.cn

Abstract. RoboCup Simulation Server provides a wonderful challenge for all the participants. This paper explains key technology implemented by Tsinghuaeolus RoboCup team played in RoboCup environment, including basic adversarial skills, which is developed by using Dynamic Programming in combination with heuristic search algorithm, and reactive strategy architecture. Tsinghuaeolus was the winner of RoboCup2001.

1 Introduction

RoboCup has gone through five years, with annual matches becoming more and more attractive. Tsinghuaeolus was established in early 2000. Our goal is to do research on muti-agents environment, especially in RoboCup. How to make agents react as what we intend? How to coordinate them? How to teach and direct them online? All these challenges have been pushing us forward. We have attempted to apply some learning methods in individual players' basic skills. But in high-level strategy decision-making, we put more effort on analytical approaches that seems comparatively easier and more effective than many learning methods.

2 Basic Skills

Basic skills include interception, dribble and kick, etc. We experience that these skills are very important to win a match. In Seattle it seems many teams have done it to near perfection. To make an agent competent in these skills alone is easy for most teams. Our main effort is put on adversary management.

2.1 Adversarial Dribbling

This skill is totally hand-coded. There are two rules which are strictly followed. First, the ball is kept in the kickable area every cycle. Second, the ball is kept out of the kickable areas of all opponents. The first rule is to ensure that the ball can be adjusted in time when an opponent is closing in. Only after applying these rules, will the agent consider how to move forward.

A. Birk, S. Coradeschi, and S. Tadokoro (Eds.): RoboCup 2001, LNAI 2377, pp. 491–494, 2002.

2.2 Adversarial Kicking

When meeting the ball, an agent may hope to kick it out in some desired angle with some desired speed. To complete this task, several steps to adjust the ball might be necessary, such as pulling the ball back first. Meantime, given an opponent or the sideline is close to the left of the agent, then he had better to do these adjust-kicks with the ball to the right. Our approach can be divided into two steps. First, putting the later limitation aside, the task is simplified to adjust the ball by minimum times. Only ball's relative position and velocity are considered here. The position is discretized, and velocity is dealt as from one discretized position to another one. Alternatively, the inputs are translated to kicks that move ball from one position to another. Then Reinforcement Learning is used to evaluate all possible kicks between two discretized positions. Learning process is completed in only several seconds. Second, the former evaluation is taken as heuristic knowledge. With the help of that knowledge, a searching algorithm, similar to A*, is used to find a possible routine for the ball to both avoid the nearby opponents or sidelines and move effectively as intended (See Figure 1). This approach proved to be successful in the games.

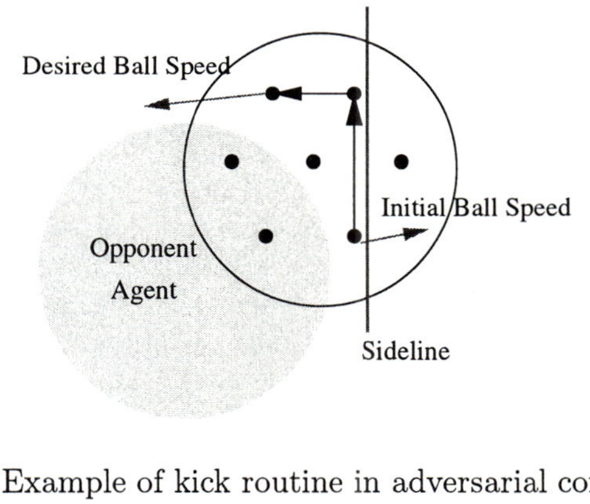

Fig. 1. Example of kick routine in adversarial conditions

3 Strategy Architecture

Real-time decision-making in a match is a complex task as so many factors have to be considered. We structured the whole task into several modules, including communication, visual control, handle-ball, offense positioning, defense positioning, etc. More attention was paid to the later four. The whole strategy architecture is shown in Figure 2. The following explains the typical components in the architecture.

3.1 Action Generation

The task of the action generator is to provide possible actions. The less actions provided, the easier the following decision would be. But the best action

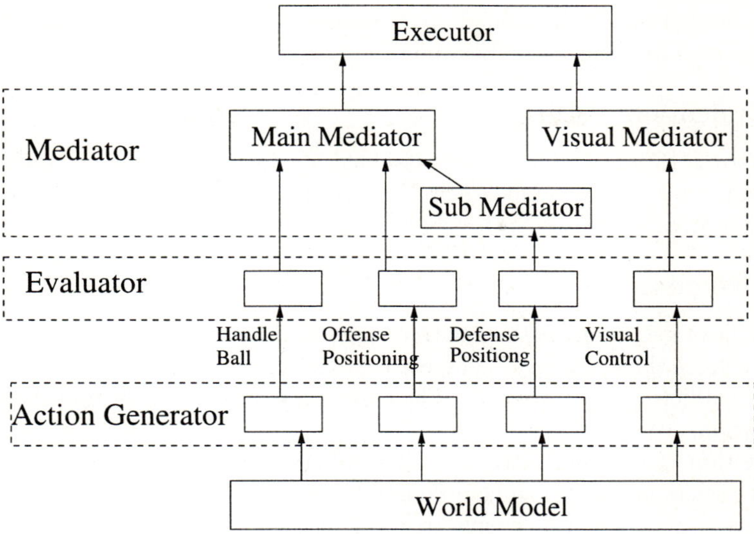

Fig. 2. Strategy Architecture

shouldn't be left out. Here we explain how the action generator for passing works. Given that an agent want to pass the ball in the game, he may have many choices. Because it's a tremendous state space consisted of 22 players. In other words, there are so many pass routes, each of which can be represented by a pass angle and kick power. We developed an analytical approach to slashes the useless routes. The angle of pass routes is first discretized. Then along a given angle, the ray that starts from the position of the ball can be divided into many segments, each of which is "controlled" by an agent in the field. "Control" means the agent is nearest to the arbitrary point on this segment of the line (See Figure 3). So when passing along a specific angle to an agent, there exits a velocity section reciprocal to his controlling segment. Alternatively, only if the ball is kicked out in a speed belonging to this velocity section, then it can be reached first by this agent. There is a presupposition that all the agents are homogeneous. Now in a given state that the ball is ready to be passed, we can

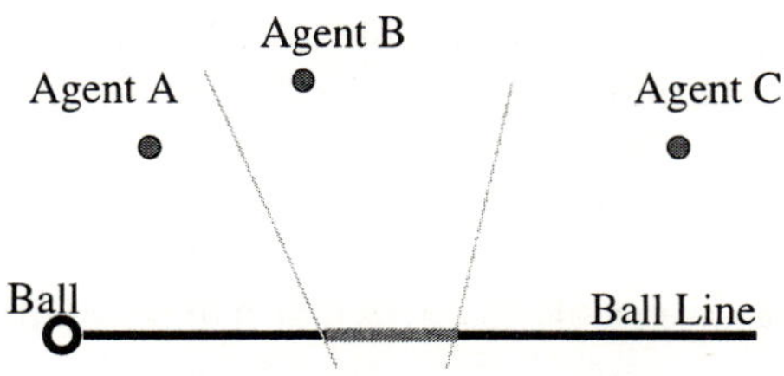

Fig. 3. Example of "control" (The green and bold segment is controlled by Agent B).

get a collection of all the possible pass routes. The following strategy is just to evaluate them and select the best one.

3.2 Evaluation

Each input action is evaluated and assigned a priority that means how valuable it is . Neural networks do much help here. Human prior knowledge is transferred into codes directly.

3.3 Mediation

The mediator gets many actions with reciprocal priority assigned by the previous evaluator from the lower layer. There are two kinds of relationships between these actions, conflicting or compatible. The task of the mediator is to find the optimal combination.

When actions to be mediated are all conflicting with each other, the task is simply to select the best one. Handle-ball belongs to this kind. But more difficult situations may occur in defensive strategy decision-making. For example, three defenders versus three forwards. For each defender, there are four choices: defend against one of the three forwards or keep one's own formation position. So there are totally $3 * 4 = 12$ actions are submitted. We think it's wasteful for two defenders to defend the same forward, and it's impossible for one defender to defend against two forwards. In other words, these actions are conflicting. It's obvious that a combination of compatible actions can includes three actions at most. A fast algorithm is developed for the mediator to find the optimal combination. With the help of this mediator, a reasonable defensive system can be constructed when facing an opponent team's attack.

4 Conclusions

Coordination and Counterplot are two basic themes in mutiagent system as RoboCup. With limited communication , it's a convenient and effective way to assume that other agents would reason as oneself. Based on this, analytical approaches are widely applied in Tsinghuaeolus. The architecture of our agents is wholly a reactive and layered one, which seems to be competent for adversarial planning and coordination in mutiagent system. But with the work going on, some obstacles and limitations appeared with this architecture. For example, the information exchange between neighboring layers is commonly only from the lower layer to the upper one. It would obviously be much better if there exits mutual exchange. Furthermore, it seems rather difficult in high-level strategy. We haven't still done any work about online coach and heterogeneous agents, which we'll consider in the future.

Acknowledgements. We would in particular like to thank the RoboCup Committee for providing such a wonderful platform. It was a great event in Seattle. We would like to thank CMU for their published source code which was used for reference at our initial stage of development.

ATTUnited-2001: Using Heterogeneous Players

Peter Stone

AT&T Labs — Research
180 Park Ave., Florham Park, NJ 07932, USA
pstone@research.att.com
http://www.research.att.com/~pstone

1 Introduction

ATTUnited-2001 is a successor of the CMUnited teams: CMUnited-97, CMUnited-98, CMUnited-99, and ATT-CMUnited-2000. It is built mainly upon CMUnited-99 [4]. It also incorporates the team action architecture from ATT-CMUnited-2000 [3], but not the dynamic planning of set-plays capability [2]. The main research focus of ATTUnited-2001 is the affect of heterogeneous players in the RoboCup simulator [1].

Heterogeneous players were introduced to the RoboCup simulator for the first time this year, in version 7.0 of the simulator [1]. In particular, in any given game, each team is able to select from identical pools of players, including the default player type from years past and six randomly generated players. At startup, teams are configured with all default players. However, the autonomous on-line coach may substitute in the randomly generated players for any player other than the goalie. The only restriction is that each random player type can be assigned to at most 3 teammates.

The random players are generated by the simulator at startup time by adjusting five parameters, each representing a trade-off in player abilities:

1. Maximum speed vs. stamina recovery
2. Speed vs. turning ability
3. Acceleration vs. size
4. Leg length vs. kick accuracy
5. Stamina vs. maximum acceleration

For the remainder of this paper, the parameters will be referred to by their associated numbers above.

At the outset, it was not known whether using heterogeneous players would be advantageous. The research reported here was designed to generate a prediction of the advantage of including a player with a specific player type on a team, and as a secondary goal, characterizing the utility of playing a given player in a given position (location on the field).

2 Setting the Stage

To experiment with heterogeneous players, the ATTUnited-2001 players had first to be made aware of their player type and made to alter their play according

A. Birk, S. Coradeschi, and S. Tadokoro (Eds.): RoboCup 2001, LNAI 2377, pp. 495–498, 2002.

to their type. The players' decisions that are affected are include kicking skills, stamina management, and prediction of player locations, among others [4]. As all of these decisions were already parameterized based on the players' physical ability, this change was a straightforward modification to the existing code.

3 Establishing a Potential Effect

3.1 Isolating the Parameters

The initial question to be examined was whether there was any potential whatsoever for heterogeneous players to improve team performance. In order to examine this question, we endeavored first to determine the parameters' effects in isolation. That is, we matched a team with non-default players (the "tested" team) against a team comprised of all default players. The players on the tested team were able to adjust their skills to their player type, but were otherwise identical to the players on the opposing team.

All the players on the team being tested were created with one of the 5 player parameters set to its maximum value. The rest of the parameters, as well as those for the opponent players, were kept at the default values. We assume that the effect of each parameter is linear within the range of possible values, thus requiring only that we test the endpoints of each range. Since keeping the tested parameter at its minimum value corresponds to having all default players, there is no expected advantage (or disadvantage) in that case.

The following results were observed. In all cases, the parameter number being varied is listed first, followed by the tested team's average score and then that of the default team. The number of games run is listed in parentheses.

1. 0 – 4.3 (19 games)
2. 2.1 – .5 (10)
3. 1.7 – .8 (12)
4. 2.3 – .8 (16)
5. .2 – 6.8 (5)

These results indicate that, considered independently and assuming monotonic functions of parameters to advantage, players are better off with parameters 1 and 5 set to their default (low) values and parameters 2–4 set to their maximum (high) values.

3.2 Bounding the Advantage

The experiments reported in Section 3.1 suggest that there is a potential for heterogeneous players, when selected optimally from the within the parameter space, to improve a team's performance. To bound the potential advantage, we tested a team comprised of players with their most beneficial parameter settings (as indicated in Section 3.1) against a default team.

We ran the test under two conditions. First, the tested team's players were fully aware of their player types and reacted accordingly (knowledge). Second, while the players were still given the advantageous capabilities, they acted as if they were default players (no knowledge). As before, the tested team's score is shown first and the number of games run is indicated in parentheses.

knowledge: 2.9 – .35 (17)
no knowledge: .9 – 1 (11)

These results suggest that heterogeneous players can improve a team's performance by *at most* 3 goals per game. However, the advantage is only observed if they are aware of their types and adjust their skills accordingly.

4 Distributing the Players

The bound discovered in Section 3.2 was obtained assuming that *all* players on the team are given the *optimal* player parameters. But in practice, some player types are better than others, and only 3 players are allowed to assume any given type. In this section we address the question of whether it would be more advantageous to have the strongest players on defense or on offense.

Since the team's default formation uses 4 defenders and 6 midfielder/attackers ("attackers"), we began by testing a team composed of 4 "optimal" defenders (i.e. players with the best parameters from Section 3.1) and 6 default attackers against a team composed of 4 default defenders and 6 optimal attackers[1]. In this case, the first team out-scored the second by an average of 1.4 – .9 over 19 games, suggesting that it it is preferable to put the strongest players on defense.

To verify this result, we tested a team with optimal defenders and default attackers, and then, conversely, a team with default defenders and optimal midfielder/attackers, both against an all-default team. In these experiments, we observed the following results.

Optimal defenders: 2.2 – .6 (32 games)
Good attackers: 3.3 – 1.6 (12)

In this case, although the goal difference per game was roughly equivalent in the two cases, the ratio of goals scored to goals suffered is much more favorable for the team with strong defenders. Therefore, we conclude that, all else being equal, the strongest players should be placed in defensive positions.

5 Establishing a Real Effect

In a real game, the 6 player types from which to select are generated randomly. To this point, it was still not clear whether selecting from among these player types could give a team any expected advantage over using all default players.

In order to establish a real effect, we assumed that the 5 player type parameters were linearly independent, and weighted them based on the goal differences in the experiments reported in Section 3.1. Doing so led to the weights for parameters (1,2,3,4,5) of (-4.5,1.5,1.0,1.5,-5.0).

[1] For the purpose of the experiments reported in this paper, the players are not allowed to exchange positions with teammates.

Multiplying the weight vector by the vector of percentage values of the parameters (i.e. $\frac{val - val_{min}}{(val_{max} - val_{min})}$), we then assigned the best player types to the players 3 at a time, starting with the defenders. The resulting heterogeneous team out-scored the default team by an average of 2.2 – .8 over 10 games, and in 9 of the 10 games, the heterogeneous team won.

We therefore conclude that using heterogeneous players can improve a team's performance by *at least* 1.4 goals per game.

6 Weighting the Parameters

Finally, to more precisely weight the parameters, we conducted a series of experiments in which all players on the test team were set to the same random type and played against the default team. Still assuming a linear model and independence among the parameters, we then found a least-squares fit from the parameters to the average score difference. Doing so led to a parameter weight vector of (-2.8, 4.2, 0.8, 1.4, -1.7) with a bias term of -0.6.

7 The Competition

In RoboCup-2001, ATTUnited-2001 used the weight vector determined in Section 6 to rank the player types, and assigned them 3 players at a time starting with the best player types as defenders (provided the predicted score difference for the player type is positive.). Overall, the team won 6 games, tied 2, and lost 1, failing to qualify for the quarterfinals based only on an unfavorable goal difference. In the coach competition, the coach was able to improve the performance of another team by 2 goals (against a constant opponent) simply by assigning the player types as indicated above.

Acknowledgements

ATTUnited-2001 was based closely on CMUnited-99, which was co-developed with Patrick Riley and Manuela Veloso, as well as parts of ATT-CMUnited-2000 co-developed with David McAllester. Thanks to Sanjoy Dasgupta for consultation regarding the best fit of the parameters to the data in Section 6.

References

1. M. Chen, E. Foroughi, F. Heintz, S. Kapetanakis, K. Kostiadis, J. Kummeneje, I. Noda, O. Obst, P. Riley, T. Steffens, Y. Wang, and X. Yin. Users manual: Robocup soccer server manual for soccer server version 7.07 and later. Available at `http://sourceforge.net/projects/sserver/`.
2. P. Riley and M. Veloso. Adaptive team coaching using opponent model selection. In *Fifth International Conference on Autonomous Agents*, 2001.
3. P. Stone and D. McAllester. An architecture for action selection in robotic soccer. In *Proceedings of the Fifth International Conference on Autonomous Agents*, 2001.
4. P. Stone, P. Riley, and M. Veloso. The CMUnited-99 champion simulator team. In M. Veloso, E. Pagello, and H. Kitano, editors, *RoboCup-99: Robot Soccer World Cup III*, pages 35–48. Springer Verlag, Berlin, 2000.

Behavior Combination and Swarm Programming
University of Virginia Team Description

Keen Browne, Jon McCune, Adam Trost, David Evans, and David Brogan

University of Virginia Computer Science, Charlottesville VA 22903, USA
robocup@cs.virginia.edu — http://wwfc.cs.virginia.edu

Abstract. The RoboCup Simulator League provides an excellent platform for research on swarm computing. Our research focuses on group behaviors emerge from collections of actors making decisions based on local information. Our RoboCup simulator team is designed around an architecture for experimenting with behavioral primitives defined over groups and mechanisms for combining those behaviors.

1 Introduction

A fundamental challenge for computer scientists over the next decade is to develop methods for creating, understanding and validating properties of programs executing on swarms of communicating computational units. The key difference between swarm programming and traditional programming is that the behavior of a swarm program emerges from the total behavior of its individual computing units. Unlike traditional distributed programming, swarm programs must deal with uncertain environments and mobile nodes. The scientific challenges of swarm computing are to understand how to create programs for individual units based on the desired behavior, and conversely, how to predict high-level behavior based on the programs running on the individual computing units.

We built the Wahoo Wunderkind team during the summer of 2001 for the RoboCup 2001 competition. We concentrated on creating a modular architecture using the principles of swarm programming. We prioritized the construction of a platform for conducting research over producing a competitive team. We began with source code from the Mainz Rolling Brainz Team from the 2000 competition because it was well documented, freely available, and modular in design [1]. Our team is based on a clear modular structure, uses a `MasterControl` which has a `WorldModel`, a `Body` to execute `Actions`, and a facility to evaluate recommended actions from many behavior modules.

2 Swarm Programming

Our work focuses on creating and reasoning about swarm programs in principled ways. We believe this is best done by constructing swarm programs by combining primitives defined over groups of agents. We are developing a library of primitive

A. Birk, S. Coradeschi, and S. Tadokoro (Eds.): RoboCup 2001, LNAI 2377, pp. 499–502, 2002.

operations for swarm programs. A primitive defines a functional behavior of a swarm of devices, such as "disperse" or "flock". These primitives are described formally in terms of constraints on the state of the devices in the swarm. For example, disperse is described by the constraint that no two devices are within a certain distance. In addition to the functional behavior, the non-functional behavior of an implementation of a primitive is formally described. With respect to robot soccer, this technique captures uncertainty in the agents' environments, resilience to malfunction of other agents, and the unpredictability of opponents' actions.

We are investigating different mechanisms for composing swarm primitives to create new behaviors. The goal is to be able to reliably predict the properties of the resulting program. In addition, we hope to synthesize programs by combining primitives with known functional and nonfunctional properties. In this manner, we will take a functional description of a swarm program along with formal models of its execution environment and devices, and select a combination of primitive implementations from a library based on the desired properties.

Our agent consists of a `WorldModel`, `Body`, and behaviors, where each behavior is implemented as a module. `MasterControl` has a three-part architecture, divided into *listen*, *think*, and *act*. Listen takes sensory data received by the agent through the `Body` and inserts the data into the `WorldModel`. Think loops over each Behavior, determines the behavior's usefulness and then returns a recommended Action. Act commands the body to execute a recommended action. `WorldModel` stores all the information that comes from the server. The objects in the world have attributes such as position, velocity, and age. The world model can be logged in an XML format for debugging with a replayer. The flexibility of XML allows the future development of more complex debugging without deprecating our existing replayer that was based on the MRB replayer.

The `Body` represents a layer of separation between the higher order control and the simulated robot. The `Body` has the ability to execute a complex action and return sensor information from the server. All commands sent to the server pass through the `Body`. This structure proved useful for debugging, since it allowed us to isolate unusual behavior by distinguishing between commands that were generated improperly and commands that were never sent. Additionally, this structure will allow our design to support a future implementation on physical robots.

3 Behavior Modules and Actions

A behavior module consists of a behavior, perception, and a recommendation. Each behavior represents a primitive desire or action such as dispersing, centering, or pushing up on the ball. Some behaviors are designed to establish and maintain global properties of the team. For example, disperse strives to maintain the property that no two players are within a certain distance of each other. Other behaviors deal with special situations, often associated with the closest player to the ball, such as passing or shooting. Each behavior has two main methods, one to determine its usefulness and the other to generate a recommendation. A behavior uses perceptions purely as an optimization. The check for usefulness is done to eliminate unnecessary calculations. For example, if a player does not have the ball, it need not calculate the best direction to kick. If a behavior is useful, additional perceptions are used to calculate information about the world state in order to calculate parameters for the suggested action.

A perception is a collection of functions that operate on the world model to interpret the state of the game. The complexities of different perceptions vary considerably, from merely checking the play mode to performing intercept calculations. The perceptions are written as utility libraries, so that multiple behaviors can share perceptions. If efficiency becomes a problem, we intend to implement a caching facility in the perceptions.

The recommendations from each module are compiled into a list and returned to `MasterControl`. `MasterControl` then uses combination logic, based on the principles of swarm programming, to select the best recommendation, and thus the best behavior for a particular situation. The `ActionList` from the winning recommendation is sent to the `Body` to be transmitted to the server.

The recommendation produced by each behavior module contains a grade and an action list. An action list is a list of actions to be executed in one server cycle. A grade is representing using a floating point number ranging between 0 and 100 where 0 represents impossibility and 100 is the highest possible grade. All of our actions have an associated duration. For instance, the kick command takes a complete server cycle, so the duration of the kick action is one. On the other hand, turn neck actions can be sent more than once per server cycle, hence, they have a duration of zero. We have implemented a combination method that strips the zero-duration actions out of losing recommendations and appends them to the winning recommendation. One result of this combination is that a player can dribble the ball up the field and turn his neck to be sure the ball stays in his line-of-sight. This makes the dribble and scan behavior strictly independent.

4 Strategy

We put most of our efforts into developing a defense. The principles of swarm programming proved to be very useful in balancing the various requirements of a defensive player. We have independent modules that seek to establish different, often conflicting, properties. For example, don't be too close to teammates, center on the ball, try to push the offside line up the field. Vector summing is an effective tool for combining defensive positioning behaviors.

Programming the offensive objectives of a soccer agent with swarm programming practices presents some new challenges. Swarm programming is best suited to large numbers of relatively simple actions. For the commands accepted by the soccer server, this is an excellent application. However, an attack on goal by a single player is best implemented as a complex set of conditionals, usually including a long list of actions to be executed sequentially. Behaviors of this type have a unique structure, and present a particular challenge in the development of combinational methods. We are investigating how behavior based control can be combined with long range planning.

5 Level of Detail

Our future plans involve enhancing the numerical grading system with a more analytical method. As we developed our team, the number of modules continued to grow. We found that too much time was being spent balancing grades, so that modules could cooperate effectively.

One technique aimed at reducing the complexity of the numerous sources of uncertainty in simulations like soccer is called simulation level of detail (SLOD). SLOD would allow us to effectively control groups of soccer players as they coordinate and compete in a complex environment. SLOD would allow us to effectively control groups of soccer players as they coordinate and compete in a complex environment. The technique builds abstractions to represent the limitations of a player's maneuverability and passing ability [2]. Low-level, individual, abstractions are built using such data-centric methods as principle components analysis and temporal differencing. A hierarchy of SLODs, developed using optimization methods and heuristics, allows us to describe individual players and even groups of players in progressively simplified ways. These simplified versions of players expedite the evaluation of strategy search algorithms and provide a common framework for describing heterogeneous agents. The shared player description framework allows higher levels of control to abstract the implementation details of lower levels, thus supporting reusability and interchangeability. For example, a defensive strategy that entails a lot of movement from the defenders need only evaluate the energy level of the defensive players (a simple SLOD) without considering the overall state of each defender. SLODs fit well with swarm programming as we can consider swarm primitives that operate at different levels.

6 Summary

Our team demonstrates that it is possible to produce complex behavior in a modular and structured way. Our architecture is designed to make it easy to develop new primitive behaviors and adapt how behaviors are combined. Our team was not competitive in the 2001 tournament, but the framework upon which our team is constructed should facilitate research in swarm computing.

References

1. Uthmann, T., Meyer C., Schappel B., Schulz F.: Description of the Team Mainz Rolling Brains for the Simulation League of RoboCup 2000. RoboCup-2000: Robot Soccer World Cup IV. Springer Verlag, 2000.
2. Faloutsos, P., van de Panne, M., and Terzopoulos, D.: Composable Controllers for Physics-based Character Animation. ACM SIGGRAPH 2001, pp 251 – 260.

ChaMeleons-01 Team Description

Paul Carpenter, Patrick Riley, Gal Kaminka, Manuela Veloso, Ignacio Thayer, and Robert Wang

Computer Science Department
Carnegie Mellon University
Pittsburgh, PA 15213
{carpep+, pfr+, galk+, mmv+}@cs.cmu.edu
{iet, rywang}@andrew.cmu.edu

Abstract. The following paper describes the ChaMeleons-01 Robocup 2001 simulation league team. Development was concentrated in two main areas: the design and implementation of an action-selection architecture and the development of the online coach. The architecture was designed in such a way to support the integration of advice from an external agent. Currently, the only external agent the ChaMeleons-01 support is an online coach offering advice using the standard coach language. The online coach offers advice in the form of passing rules, marking assignments, and formation information.

1 Introduction

The ChaMeleons-01 team is a new team developed for the Robocup 2001 simulated robotic soccer competition. Although two members remain from past Carnegie Mellon simulation teams, the addition of new team members and the desire to pursue new research directions prompted the decision to start a team from scratch. The ChaMeleons-01 coding effort started with the world model from the CMUnited-99 publicly released source code. Although some time was spent on the skills development, most of the effort was focused on building a team that is highly coachable and able to easily adapt to different opponents. Implementing a coachable team requires not only the development of a coach but also considerations in the design of the team with regards to how advice will be integrated into the action selection process. In section two, the architecture designed and implemented to support the coach will be described and in section three, the implementation of the coach will be described.

2 Action-Selection Architecture

The ChaMeleons-01 use a hierarchical, behavior-based architecture. The major feature of the design is that it facilitates the addition of advice from outside sources. This is accomplished through the use of a Behavior Manager (BM) and different Behavior Arbitrators. Presently, the only advice the players use is from

A. Birk, S. Coradeschi, and S. Tadokoro (Eds.): RoboCup 2001, LNAI 2377, pp. 503–506, 2002.

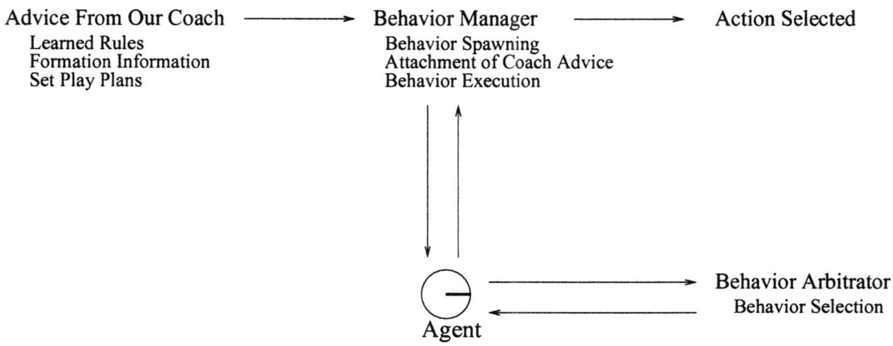

Fig. 1. Action Selection Architecture

the online coach but other players or human operators could potentially provide advice. Figure 1 shows how each of these components are related.

A behavior generates all of its children when it begins execution. For example, every time *passBall* executes, it generates a *passToPlayer* behavior for every teammate whose position is known. If a behavior from the list of coach advice matches one of the children types of the executing behavior, the BM attaches it to the list of children.

All behaviors inherit from the same abstract behavior class. Each behavior provides several methods to aid in the action selection process. Most of the agents decision making rely on a behavior's *isApplicable* method. For example, the *handleBall* behavior is applicable only if the ball is kickable. Other methods include *isChildOf* which returns whether or not a given behavior has the same type as one of its children and *probabilityOfSuccess* which estimates the probability that a behavior will execute successfully. Behaviors range from the lowest level behaviors like *dash*, *kick*, and *turn* to higher level behaviors such as *dribbleBall*, *passBall*, and *shootBall*.

The BM creates and executes behaviors while also attaching the matching coach advice when appropriate. Because players will tend to consider a lot of the same actions from one cycle to the next, the BM maintains a pool of the most recently used behaviors. When a behavior attempts to execute another behavior, the call goes through the BM. The BM first checks if the behavior currently exists in memory before it allocates a new behavior. The BM logs each request to execute a behavior and whether or not the behavior executed successfully. Logging the entire call chain aids in locating bugs while debugging.

The Behavior Manager also integrates all advice from the coach. For an executing behavior, the BM searches through the coach advice and adds to the list of children any behaviors that match one of the children types. For example, if a coach recommends that player seven pass to player ten, when player seven begins to execute the *passBall* behavior, the action recommended by the online coach, *passToPlayer* corresponding to pass to player ten, will be added to the list of children it considers. The BM also flags the behavior as originating from the pool of advice behaviors. It is important to note that simply because a coach offers advice to a player, there exists no obligation on the player's part to follow the advice. A behavior arbitrator ultimately decides among the considered behaviors.

A behavior arbitrator selects which behavior to execute when more than one behavior is applicable. The arbitration method can be any heuristic and it does not have to be the same for all behaviors. Simple heuristics include executing the first behavior it considers that is applicable or always executing coach advice before considering anything else.

The *handleBall* behavior uses a much more complex arbitration scheme. In this case, behaviors are classified by descriptors such as *passForward*, *passBack*, *passToMostOpen* with each descriptor having its own priority. Behaviors that could potentially result in a goal have the highest priority, e.g., shooting the ball, passing to someone that has a great shot, while other behaviors have lower priorities. The prioritization is currently done by hand but it is feasible to learn the priorities using machine learning techniques.

Whether or not to actually execute a behavior or to consider the behavior at the next highest priority level depends on the behavior's probability of success. Each priority level contains a threshold for the probability of success; if the threshold is not met, the next highest level is considered.

The *handleBall* arbitration method can also determine whether or not to enable a particular behavior that has not met the required threshold. For example, if a player has the ball near the opponents goal and the probability of success for a shot is just below the threshold, the best thing to do may be to hold the ball for a few cycles to see if a better shot develops. Although the capability of enabling behaviors exists, it was not used in Robocup-2001.

3 Online Coach

The ChaMeleons online coach (also known as OWL) competed in the online coach competition as well as providing coaching for the ChaMeleons team during the team competition. Besides substitution of heterogeneous players, the coach generates advice using the standard coach language as described in the Soccer Server Manual[5].

OWL uses a simple heuristic approach to choose heterogeneous types. Based on empirical data gathered in a fixed training environment as well as the role of the agent, the coach evaluates the player types on their kicking ability, interception ability, and affect on stamina. The player types with the highest heuristic value are substituted for the default players. All substitutions are made at the beginning of the game.

As far as the opponent's types, the coach carefully tracks error bounds in order to evaluate whether each opponent agent's actions (dashes and kicks) are consistent with a particular type. If a type is not consistent, that information is transmitted to the player through 'info' messages in the standard coach language. Similar analysis is done based on the player size, since (in normal cases), two object cannot overlap.

Based on the setplays planning developed for ATTCMUnited2000[2], the coach creates a plan for ball and player movement for each setplay situation. The primary difference is that the plan is expressed as a set of condition-actions rules using the standard coach language. Further information about setplays can be found in [3, 4].

OWL also has the ability to learn formations based on observations of a team playing. The learning is done in two stages. First, for each player, a small rectangle is found which encloses most of the points in which that player was throughout the game. Then the rectangles are shifted in a hillclimbing search on the field. The evaluation function favors positions where the average distance between the rectangles for players are close to the same as the average distance observed throughout the game.

OWL can then send this formation to the team using the 'home' action of the standard coach language. In addition, if OWL is given the expected formation for the opposing team, OWL can assign marks to the defenders.

The final important module of the coach deals with learning to imitate another team's passing behavior. First, a team to imitate is chosen by hand. In the online coach competition, we chose to imitate Brainstormers as they played against Gemini. OWL extracts all of the successful passes made by both teams when playing each other. Then, using C4.5, rules are extracted to predict where a pass will go to. The input conditions are the starting location of the pass and the distance and angle of all players. Note that the starting and ending locations of the passes are discretized using Autoclass C[1].

These rules are then translated into coach language advice in the following fashion. For the team to be imitated (Brainstormers for the coach competition), the rules are translated into 'pass to region' actions with the appropriate conditions. For the opponent (Gemini for the coach competition), the rules are translated into 'mark line region' directives instructing the team to try and block the predicted pass.

References

[1] P. Cheeseman, J. Kelly, M. Self, J. Stutz, W. Taylor, and D. Freeman. Autoclass: A bayesian classification system. In *ICML-88*, pages 54–64, San Francisco, June 1988.

[2] Patrick Riley, Peter Stone, David McAllester, and Manuela Veloso. ATT-CMUnited-2000: Third place finisher in the RoboCup-2000 simulator league. In Stone, Balch, and Kreatzschmarr, editors, *RoboCup-2000: Robot Soccer World Cup IV*. 2001.

[3] Patrick Riley and Manuela Veloso. Planning for distributed execution through use of probabilistic opponent models. In *IJCAI-2001 Workshop PRO-2: Planning under Uncertainty and Incomplete Information*, 2001.

[4] Patrick Riley and Manuela Veloso. Recognizing probabilistic opponent movement models. In *Proceedings of the RoboCup International Symposium*, 2001.

[5] RoboCup Federation, `http://sserver.sourceforge.net/`. *Soccer Server Manual*, 2001.

Cyberoos'2001: "Deep Behaviour Projection" Agent Architecture

Mikhail Prokopenko[1], Peter Wang[1], and Thomas Howard[2]

[1] Intelligent Interactive Technology
CSIRO Mathematical and Information Sciences
Locked Bag 17, North Ryde, NSW 1670, Australia
{mikhail.prokopenko, peter.wang}@cmis.csiro.au
[2] London, U.K.,
thoward@tibco.com

The RoboCup Simulation League presents developers with a variety of challenges. Arguably, the following list includes the most critical topics:

- distributed client/server system running on a network;
- concurrent communication with a medium-sized (25) number of agents;
- fragmented, localised and imprecise (noisy and latent) information about the environment (field);
- heterogeneous sensory data (visual, auditory, kinetic);
- asynchronous perception-action activity;
- limited range of basic commands/effectors (turn, kick, dash, ...);
- limited window of opportunity to perform an action;
- autonomous decision-making under constraints enforced by teamwork (collaboration) and opponent (competition);
- conflicts between reactivity and deliberation;
- no centralised controllers (no global vision, etc.);
- evolving standards, rules and parameters of the Simulation.

Clearly, these challenges are not uniform and may require a complex agent architecture, covering not only simple situated behaviour but also more involved strategic skills. Pure behavioural approaches to artificial intelligence often feature situatedness of agents reacting to changes in environment and exhibiting emergent behaviour, instead of reliance on abstract representation and inferential reasoning [2,3]. Tactical and strategic reasoning, however, would seem to require domain knowledge and a degree of multi-agent cooperation beyond the reach of situated behaviour-based agents. Over the last few years, it has become apparent that a unifying architecture, combining cognitive (top-down) and reactive (bottom-up) approaches, cannot be achieved by simply connecting higher and lower layers. It has been suggested in recent literature that a "middle-out" architecture [1] is required. The approach adopted in [1] follows the *behavioural programming* principle (and the situated automata framework [3]) in "taking symbolic descriptions of tasks and predictably translating them into dynamic descriptions that can be composed out of lower-level controllers".

Our principal target is a systematic description of increasing levels of agent reasoning abilities, where a given behaviour can "grow" into an expanded level,

A. Birk, S. Coradeschi, and S. Tadokoro (Eds.): RoboCup 2001, LNAI 2377, pp. 507–510, 2002.

while every new level can be projected onto a deeper (more basic) behaviour. More precisely, we develop a scalable *Deep Behaviour Projection* (DBP) agent architecture, and enhance Cyberoos agents incrementally. The DBP framework introduced in our previous work [9] achieves:

- realisation of the behavioural programming principle (translating symbolic descriptions into the low-level specifications or reactive behaviours);
- encapsulation of an emergent behaviour (a feedback between the emergent behaviour and symbolic descriptions);
- behaviour depth or duplication (both embedded and emergent levels), unlike "shallow" behaviour subsumption;
- behaviour projection and resultant functional interchangeability, mediated internally within the agent.

The feedback connection from an emergent behaviour to a (meta-)action theory, and then to a derived behaviour on a higher level, extends the behavioural programming principle. Thus, a behaviour can be present in the architecture in two forms: implicit (emergent) and explicit (embedded). This duplication (or depth) allows the agent to "mediate" among related behaviours. The results reported in [6,7,8,9] formalise the DBP approach at the situated and tactical levels and introduce new systematic models and formal correctness results.

The developed hierarchy can be briefly summarised as follows:

$$
\begin{array}{lll}
\langle C, S, E, & sense : C \to S, & response : E \to C, \\
 & timer : C \to S, & command : S \to E, \\
 & tropistic\_behaviour : S \to E, & \\
I, & hysteretic\_behaviour : I \times S \to E, & update : I \times S \to I, \\
D, & domain\_update : I \times S \times D \to D, & \\
 & domain\_revision : I \times S \times D \to D, & \\
 & domain\_projection : I \times S \times D \to I, & \\
T, & decision : I \times S \times T \to T, & combination : T \to 2^H, \\
P, & engage : I \times S \times T \times P \to P, & tactics : P \to 2^T \ \rangle
\end{array}
$$

where C is a communication channel type, S is a set of agent sensory states, E is a set of agent effectors, I is a set of internal agent states, D is a domain model, T is a set of agent task states, P is a set of agent process states, and H denotes the set of instantiations $\{(i, s, e) : e = hysteretic\_behaviour(i, s)\}$.

The following list illustrates the hierarchy with informal examples from soccer, while highlighting some obvious biological parallels:

- tropistic behaviour: Sensors → Effectors (obstacle avoidance, chase, goalkeeper catch; *plants following sunlight, spiders catching prey*);
- hysteretic behaviour: Sensors & Memory → Effectors (intercept, resultant pass, dribble; *movement of a school of fish, ants tracing pheromones*);
- task-oriented behaviour: Sensors & Memory & Task → Effectors (defensive blocks, backing-up, maintaining triangles; *lions hunting, patrolling territory*);
- domain-oriented memory projection: Sensors & Memory & Domain → Memory, followed by behaviour-based actions ("blind" pass, off-side check; *whales communication during a hunt*).

The Deep Behaviour Projection framework underlies this hierarchy and ensures that more advanced levels capture relevant behaviour more concisely than their deeper projections. Our primary application domain is the RoboCup Simulation — an artificial multi-agent world [4], where the DBP framework provided a systematic support for design and full implementation of Cyberoos [6,8]. Previous generations of Cyberoos developed under the DBP approach, captured certain types of situated behaviour (Cyberoos'98) and some basic classes of tactical behaviour (Cyberoos'99). Cyberoos'2000 exhibited emergent tactical teamwork. Cyberoos'2001 is the fourth "generation" designed in line with this framework.

Cyberoos'2001 agents incorporate a domain model into the architecture. The idea of having a "world model" directly represented in the architecture is intuitively very appealing. However, we believe that "world model" should grow incrementally instead of being inserted and glued to other elements. In other words, our preference is to observe and encapsulate instances of the emergent behaviour which potentially make use of the domain model. Most definitely, the RoboCup Simulation domain is rich enough to produce such examples — one considerable motivation among others is sensory stalls, when an agent does not receive a visual information during a simulation cycle. In particular, Cyberoos'2001 agents extrapolate their domain model each simulation cycle with the *domain_update* function, and revise it with the *domain_revise* function whenever new information becomes available (new visual inputs or team-mate communications). The partition between update and revision operators corresponds to the well-known distinction between belief update and belief revision [5].

Previous generations of Cyberoos did not use world models and inter-agent communication — they relied entirely on deep reactive behaviour and emergent tactics. For the first time in our four years long experiment, we can now directly observe and rigorously compare behaviours produced by agents using domain models and purely reactive agents. *This comparative analysis is the primary topic of our research.*

The described hierarchical framework has enabled a systematic incremental implementation and testing of software agents and their modules. In particular, the framework allowed us to correlate enhancements in agent architecture with tangible improvements in team performance. The Cyberoos'98 team took 3rd place in the 1998 Pacific Rim RoboCup competition. Cyberoos'99 finished in the top half of the RoboCup-1999, while Cyberoos'2000 were 4th in the Open European RoboCup-2000 ($\sim$ 16 teams) and shared 9th place at the RoboCup-2000 ($\sim$ 40 teams). Cyberoos'2001 shared 9th place at the RoboCup-2001 as well — this time among 44 teams. In addition, the DBP framework was used for development of an intelligent multi-agent system for *networked multimedia appliances*. The system is aimed at providing customisable user support in operating and controlling a flexible and changing network of multimedia appliances, while satisfying several (possibly conflicting) users' preferences [10].

It has been illustrated elsewhere [6,7,8,9] that an agent's complexity (be it *Tropistic* or *Hysteretic* or *Domain-Oriented* agent) is relative: for any elaborate agent type, it is possible to define more *concisely* another agent type with the same external behaviour. Therefore, an agent based on the DBP architecture will always have a choice as to which one of related hierarchical levels should assume control to better suit external environment. This selection process serves, in fact, as the "middle-out" layer between any two levels of the DBP agent architecture — replacing the need for a generic hub connecting "reactive behaviour" and "cognitive skills". Although, in general, it does take time to implement and experiment with the DBP-based agents, the resulting depth and flexibility in the agents behaviour appears to be worthwhile. The main lesson of the Cyberoos chronicle is, hence, the identification and study of the dialectic relation and necessary feedbacks between emergent behaviour and the agent architecture. More precisely, an emergence of essentially new behavioural patterns always indicates a need for new elements in the agent architecture.

Acknowledgements
The authors are grateful to Marc Butler and Ryszard Kowalczyk for their comments on various aspects of the RoboCup Simulation and multi-agent modelling, and for their contributions to previous generations of Cyberoos.

References

1. Michael S. Branicky. Behavioural Programming. In AAAI Tech. Report SS-99-05, the AAAI-99 Spring Symp. on Hybrid Systems and AI, 29–34, Stanford, 1999.
2. Rodney A. Brooks. Intelligence Without Reason. In Proceedings of the 12th Int'l Joint Conference on Artificial Intelligence, 569–595 Morgan Kaufmann, 1991.
3. Leslie P. Kaelbling and Stanley J. Rosenschein. Action and planning in embedded agents. In Maes, P. (ed) Designing Autonomous Agents: Theory and Practice from Biology to Engineering and Back, 35–48, 1990.
4. Hiroaki Kitano, Milind Tambe, Peter Stone, Manuela M. Veloso, Silvia Coradeschi, Eiichi Osawa, Hitoshi Matsubara, Itsuki Noda and Minoru Asada. The RoboCup Synthetic Agent Challenge. In Proceedings of the 15th International Joint Conference on Artificial Intelligence, 1997.
5. Pavlos Peppas, Abhaya Nayak, Maurice Pagnucco, Norman Foo, Rex Kwok and Mikhail Prokopenko. Revision vs. Update: Taking a Closer Look. In Proceedings of the 12th European Conference on Artificial Intelligence, 1996.
6. Mikhail Prokopenko, Ryszard Kowalczyk, Maria Lee and Wai-Yat Wong. Designing and Modelling Situated Agents Systematically: Cyberoos'98. In Proceedings of the PRICAI-98 RoboCup Workshop, 75–89, Singapore, 1998.
7. Mikhail Prokopenko. Situated Reasoning in Multi-Agent Systems. In AAAI Technichal Report SS-99-05, the AAAI-99 Spring Symposium on Hybrid Systems and AI, 158–163, Stanford, 1999.
8. Mikhail Prokopenko and Marc Butler. Tactical Reasoning in Synthetic Multi-Agent Systems: a Case Study. In Proceedings of the IJCAI-99 Workshop on Non-monotonic Reasoning, Action and Change, 57–64, Stockholm, 1999.
9. Mikhail Prokopenko, Marc Butler and Thomas Howard, On Emergence of Scalable Tactical and Strategic Behaviour, In Proceedings of the RoboCup-2000 Workshop, Melbourne 2000.
10. `http://www.cmis.csiro.au/iit/Projects/RoboCup/rc_network.htm`, 2001.

Essex Wizards 2001 Team Description

Huosheng Hu, Kostas Kostiadis, Matthew Hunter, and Nikolaos Kalyviotis

Department of Computer Science, University of Essex
Wivenhoe Park, Colchester CO4 3SQ, UK
{hhu,kkosti,mchunt,nkalyv}@essex.ac.uk

Abstract. This article presents an overview of the Essex Wizards 2001 team participated in the RoboCup 2001 simulator league. Four major issues have been addressed, namely a generalized approach to position selection, strategic planning and encoded communication, reinforcement learning (RL) and Kanerva-based generalization, as well as the agent architecture and agent behaviours.

1 Introduction

The simulation league continues to be the most popular event in RoboCup, in terms of the number of teams participated annually and the team strategies being adopted. In general, all teams competed in this league are faced with several major research challenges: multi-agent coordination, agent modelling, real-time performance and learning. In order to satisfy all the necessary timing constraints for simulated soccer agents, a multi-threaded implementation has been adopted in the Essex Wizards team so that the agents can perform various computations concurrently and avoid waiting for the slow I/O operations [4]. Moreover the behaviour-based approach plays a key role in building the Essex Wizards team. A decision-making mechanism based on reinforcement learning enables co-operation among multiple agents by distributing the responsibilities within the team. The focus of our Essex Wizards 2001 team is on adaptive position selection, flexible strategic planning, multi-agent learning and real-time agent architectures. We briefly outline our research focus here in terms of these four aspects.

2 Generalized Approach to Position Selection

In the Robotic Soccer domain, position selection is often seen as a baseline case since it is often the last resort. If a player is positioning then that is usually because it has nothing better to do. Some teams use position selection in specific situations such as ball tracking and marking. This results in the position selection being tightly coupled to the rest of the agent, making it more difficult to experiment with. Our approach has been to provide a general interface for selecting a position whatever the situation while still allowing specific cases to be exploited where appropriate.

A. Birk, S. Coradeschi, and S. Tadokoro (Eds.): RoboCup 2001, LNAI 2377, pp. 511–514, 2002.

Three issues have been addressed in order to improve the position selection mechanism for the Essex Wizards 2001 team. The first issue was to improve individual behaviour classes and reduce them to more general components. The second was to provide more flexible control within position selection. Finally the interface between the soccer agent and the position selection mechanism has been modified. In fact, a generalized behaviour-based approach has been deployed to tackling the problem and the practical implementation of these ideas has been integrated into a working system [1]. This system uses a small number of interchangeable behaviours that are combined to perform rule-based position selection in real time.

The current positioning is purely rule-based, which can be tricked by opponent strategies that are different from those the rules were designed to deal with. If both teams use similar strategies they can interfere with each other in unexpected ways. More rules can be added to deal with different strategies, although detecting which strategy an opponent is using may be difficult. Although individual behaviours could be augmented with learning modules, a better alternative may be to learn the behaviour tree, requiring that the tree be dynamically changeable. We have considered the position selection mechanism as a specialized programming language, and if it can be modified to allow the program to be altered while it is running it could find uses in other domains such as planning.

3 Strategic Plans and Encoded Communication

In the RoboCup domain, co-operation is the key to success. However, in order to achieve co-operation between the agents, planning has a key role to play. We have focused on the framework and design issues regarding strategic planning. The framework for our strategic plans consists of three major components, namely *Triggers, Actions* and *Abort* conditions [2]. The *Triggers* are used as signals to allow or forbid the actions or plans that are predefined. The *Actions* are a combination of low-level and high-level behaviours that are executed sequentially. The *Abort* condition is a safeguard to ensure the conditions of the environment are suitable for the strategic plan (SP) being executed.

Strategic planning is a relatively simple but very useful method. Although it has been used successfully only in the RoboCup environment so far, there is no reason why it can not be used in the other domains. Given that adequate knowledge of the domain is acquired and careful design and implementation of the strategic plans has been done, then the performance of a MAS (Multi-agent System) that uses those strategic plans can be increased significantly. Having a SP for a situation that happens frequently is a good idea. Having more than one SP for the same situation is even better. The only problem is that the agent should choose which one to execute. For this reason, how to implement more flexible and adaptive strategic planning has been investigated [3].

In RoboCup communication plays a very important role, since it can enhance dramatically the world model of each player. Therefore the more information is communicated between the players the better the team performance. The use of an effective communication model is imperative, however the effectiveness of any communication model in RoboCup is limited by the size of the message that can be sent (512 bytes). We have demonstrated that the use of encoding can maximize the amount of information and increase the capacity of each message at least 2 times. Moreover encoding also results in hiding the actual information contained in each message from opponent agents.

4 RL and Kanerva-Based Generalization

RL (reinforcement learning) has been adopted in our behaviour-based decision making process since it provides a way to program agents by reward and punishment without needing to specify how a task is to be achieved [4]. Each time the agent receives sensory inputs, it determines the state of the environment and then chooses an action to execute. The action itself changes the state of the environment and also provides the agent with either a reward if it does well or a punishment if it does badly. The agent should choose actions that maximize the long-term sum of rewards. It should be noticed that the agents in our implementation not only have different roles and responsibilities, but also have different sub-goals in the team. Hence, every individual agent tries to reach its own goal state, and cooperation emerges when the goals of all agents are linked together. The ultimate goal, scoring against the opposition, becomes a joint effort that is distributed among team members.

The complexity of most multi-agent systems prohibits a hand-coded approach to decision making. The problem of learning in large spaces is tackled through generalization techniques, which allow compact representation of learned information and transfer of knowledge between similar states and actions. In large smooth state spaces, it is expected that similar states will have similar values and similar optimal actions. Therefore it should be possible to use experience gathered through interaction with a limited subset of the state space and produce a good approximation over a much larger subset. The combination of Kanerva coding and reinforcement learning has been investigated in order to build the K-RL multi-stage decision-making module [5]. The purpose of K-RL is twofold. Firstly, Kanerva coding is used as a generalization method to produce a feature vector from the raw sensory input. Secondly, the reinforcement learning component receives this feature vector and learns to choose a desired action.

5 Agent Architecture

To achieve real-time performance, we have adopted a modular approach in the overall agent implementation [4]. In such a design, there are five function modules, namely *Sensors*, *Behaviours*, *Actuators*, *World Model* and *Parameters*. Based on information from the *Sensors*, *Parameters* and *World Model* modules, the *Behaviours* module in each agent decides on the best course of action. This involves both low-level behaviours such as *moving* and *kicking*, and high-level ones such as selecting where to move to and which teammate to pass to.

At the lowest level any decisions made by the agent must be reduced to the core primitive actions provided by the server, i.e. *Kick*, *Turn* and *Dash*. In order to provide the options for high-level behaviours, extended primitives have been implemented such as *Advanced Kick* that moves the ball to a position where the desired kick can be made; *Move* that mixes turns and dashes to reach the desired location. The high-level tactical behaviours are built on top of low-level primitive behaviours, and are currently implemented as a hybrid of Q-learning and rule-based decision-making, including *Intercept* that involves predicting the location of the ball for interception and moving to that location; *Clear Ball* that involves kicking the ball, using the *Advanced Kick* behaviour; *Send Ball* that occurs when the agent attempts to get the ball to a position from which a teammate can score; *Pass Ball* that generates a good pass based on the locations of teammates and opponents on the field; and *Position Selection* that examines the current view of the pitch and suggests a good place to move to, which is a non-trivial task, requiring information about the current role of the agent and the state of the pitch.

6 Conclusions and Future Work

This article presented the main features of the Essex Wizards 2001 multi-agent system for the simulated RoboCup competition. The four major research issues of our team are addressed, namely adaptive position selection, flexible strategic planning, multi-agent learning and real-time architectures. In the next stage of our research, we will investigate how to improve these features in order to maximise the team's performance.

Acknowledgements This research is financially supported by the Essex University Research Promotion Fund DDP940.

References

1. M. Hunter and H. Hu, A Generalised Approach to Position Selection for Simulated Soccer Agents, Proc. of the RoboCup 2001 International Symposium, Seattle, Washington, 4-10 August 2001
2. N. Kalyviotis and H. Hu, A Co-operative Framework for Strategic Planning, Proceedings Towards Intelligent Mobile Robots (TIMR) '01, Manchester, 2001.
3. N. Kalyviotis and H. Hu, ASPF: an Adaptive Strategic Planning Framework in RoboCup, Proc. of IASTED International Conference on Robotics and Applications, Clearwater, Florida, USA, 19-22 November 2001.
4. K. Kostiadis, M. Hunter and H. Hu, The Use of Design Patterns for Building Multi-Agent Systems, IEEE Int. Conf. on Systems, Man and Cybernetics, Nashville, Tennessee, USA ,8-11 October 2000
5. K. Kostiadis and H. Hu, K-RL: Kanerva-based Generalisation and Reinforcement Learning for Possession Football, Proceedings of IEEE/RSJ International Conference on Intelligent Robots and Systems, Hawaii, USA, November 2001

FC Portugal 2001 Team Description: Flexible Teamwork and Configurable Strategy*

Nuno Lau[1] and Luis Paulo Reis[2]

[1] IEETA - Electronics and Telecomm. Eng. Institute - Aveiro University, Portugal
lau@det.ua.pt, http://www.ieeta.pt/

[2] LIACC - Artificial Intelligence and Comp. Sci. Lab. - Porto University, Portugal
lpreis@fe.up.pt, http://www.ncc.up.pt/liacc/

Abstract. FC Portugal is a cooperation project between the Universities of Aveiro and Porto in Portugal. FC Portugal 2001 is our second step towards the creation of a flexible RoboSoccer team, with tactical changing abilities, that may be coached at any level, before and during the games, by human or automatic coaches. Although having the best goal average in the competition (scoring 150 goals in 13 games), the team was not able to score against the good defenses of Tsinghuaeolus and Brainstomers and finished third in RoboCup 2001.

1 Introduction

FC Portugal 2001 is the result of the evolution of our previous team, FC Portugal 2000, which was RoboCup 2000 champion. For a description of FC Portugal 2000 please read [1,2]. The team achieved third place in RoboCup 2000, scoring a total of 150 goals and conceding only 5 goals. FC Portugal was stopped in RoboCup 2001 by the amazingly powerful defenses of Tsinghuaeolus and Brainstormers and although dominating spatially the games against these two teams, was not able to score against neither of them.

This paper summarizes FC Portugal 2001 main innovations and describes the team' results in RoboCup 2001. Section 2 describes our flexible team strategy and the team ability to change its tactic with and without the automatic coach intervention. Section 3 is concerned with our individual decision mechanisms and to some innovations in our low-level skills. The paper ends with the results achieved in RoboCup 2001 and its discussion.

2 Team Strategy

One of the main developments of FC Portugal 2001 team is the improvement of its ability to be coached both before and during the game. Before the game,

* We would like to thank the financial support from IEETA/University of Aveiro, LIACC and FEUP/University of Porto and ICEP.

A. Birk, S. Coradeschi, and S. Tadokoro (Eds.): RoboCup 2001, LNAI 2377, pp. 515–518, 2002.

the strategy is defined through configuration files and during the game the tactics used may be changed by the automatic coach or even without the coach intervention by a set of rules known by all players.

The definition of strategic behaviors has been enhanced with new parameters that effectively control player movements. Active behaviors have also been extended to support different playing styles. Some of the concepts of Coach Unilang [3]; like game pace, team aggressiveness, team mentality, etc have been included in order to turn the team adaptable to the opponents strategy.

For each game a human coach specifies the strategy to be taken by the team. This strategy consists of several tactics that should be activated according to the opponent team' behavior and statistical information gathered during the game. Each tactic consists of several global parameters that control the team behavior and of several formations that are active in different situations (like defense, attack, etc). For each formation the human coach can specify the strategic and active behavior of each player. During the game players follow the coach advice to change tactics, formations and player types. Players may also decide tactical changes by themselves, based on statistical information sent by the coach and teammate and opponent modeling techniques.

FC Portugal 2001 extended FC Portugal 2000 flexible tactical approach and now, the team configuration files are far more flexible, enabling to change completely the team behavior for a given game. The information contained in the team main configuration file (strategy.conf) is summarized in Fig. 1. The first section is concerned with Player Type definition at three levels: strategic, ball possession and ball recovery. The second section enables us to give players information about the opponent team (at several levels) and about the type of game and tournament the team is involved. Other important information concerns the type of coach and its powers and the team default strategy for the game (used unless the coach states the opposite). This default strategy includes the tactic to be used depending on the game result, time and other game parameters. Please see [3] for a better explanation of FC Portugal tactic definition.

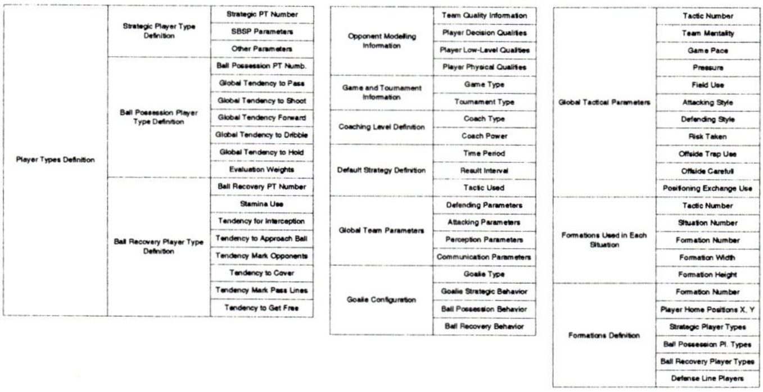

Fig. 1. FC Portugal 2001 Strategy Definition

3 Individual Decision and Basic Skills

Our team considers 3 different modules in the player's behavior decision mechanism. In strategic situations the player moves to a position that maximizes his utility for the team using the SBSP module [1,2], ball possession module is activated when the player can kick the ball, ball recovery module is responsible for behavior selection in active defending situations.

The ball possession module considers the following actions: shoot, pass, forward, dribble and hold. Each of these actions has many options in a given situation: where to shoot?, whom to pass? etc. Our decision mechanism is based on the evaluation of each option by the composition of several high level metrics. For example, passes are evaluated through 10 metrics: positional value, receiver catching probability, out of bounds probability, opponent interception possibilities, initial congestion, final congestion, pass distance, expected pass reception, confidence in receiver position and receiver shooting position. These metrics are evaluated for each of the passing options, and the best pass is selected. Similar reasoning is applied to the other ball possession behaviors and comparison of best utilities determines which is the final player decision.

The ball recovery module selects one of the following behaviors: ball interception; ball passive interception; approach ball position; cover goal; cover center or mark pass line. The selection is based on the current situation. Each player calculates a positioning matrix that includes ball position and velocity, offside lines, and all players positions, velocities, distances to ball, number of cycles to intercept the ball. The decision is based on the analysis of the effect of each of the items of the positioning matrix. Coordination is achieved through prediction of teammates' decisions based on mutual knowledge of their decision rules.

FC Portugal 2001 also improved some of the basic skills of our previous team. The optimization kick was enhanced to optimize the difficulty of opponent interception while kicking and in the first cycles after the ball is released. Dribbling ability was changed in order to turn it faster and safer from opponent steals. A new algorithm for ball interception has been developed. In this algorithm the interception point is chosen taking into account how risky is the interception, i.e., players choose risky interception points in disputed balls and safer interceptions when they are clearly the balls owners.

4 Results and Discussion

FC Portugal 2001 dominated round robin qualification groups of RoboCup 2001, scoring 138 goals without conceding a single goal (Tab. 1). The team won the first game of double elimination by 8-0 against YowAI 2001 (RoboCup Japan 2001 Champion). The next game against Uva TriLearn showed FC Portugal ability to perform dramatic tactical changes. On the second half of the game, around 4000 cycles, FC Portugal was losing 1-0 and even though the team seemed to have the power to reverse the game, it was with lots of difficulties to pressure the opponent and score a goal. At this time the team changed to an aggressive tactic

Table 1. Scores of FCPortugal 2001 in RoboCup 2001

Group E - Round Robin 1	Score
TUT-Groove (Japan)	9 - 0
RoboLog2k1 (Germany)	8 - 0
RMIT-Goannas (Australia)	32 - 0
11Monkeys3 (Japan)	29 - 0
Group C - Round Robin 2	
ATTUnited01 (USA)	22 - 0
FC Tripletta (Japan)	4 - 0
AT Humboldt. (Germany)	13 - 0
UTUtd (Iran)	16 - 0
Helli-Respina 2001 (Iran)	5 - 0
Double elimination	
YowAI 2001 (Japan)	8 - 0
Uva Trilearn 2001 (Netherlands)	4 - 1
Tsinghuaeolus (China)	0 - 3
Brainstormers 2001 (Germany)	0 - 1
Total Score	150 -5

Team	Games	Total
1 Tsinghuaeolus (China)	12	90-1
2 Brainstormers 2001 (Germany)	16	79-5
3 FC Portugal 2001	13	150-5
4 Uva Trilearn 2001 (Netherlands)	13	46-14
5 FC Portugal 2000	12	121-13
6 Wright Eagle (Japan)	12	109-6
7 YowAI 2001 (Japan)	11	64-15
8 FC Tripletta (Japan)	12	30-16

using a 443 formation (without a goalie) for attacking. Afterwards FC Portugal dominated the game clearly and the final result (4-1), showed the importance of having tactical changing abilities. RoboBase [4] statistics show that the ball stayed mostly in the FC Portugal's attack (42%) and midfield (42%) with little time in defense (16%). In the next game, against Tsinghuaeolus, FC Portugal 2001 was, again, loosing by 1-0 near the end of the game. The team had already tried several different tactics, with clear spatial dominion (final results are of 36% in attack, 55% in midfield and 9% in defense), the risky tactic the team chose at the end allowed Tsinghuaeolus to score 2 more goals. Our last game was against BrainStormers, perhaps the best defense in this year championship. Again, our team dominated spatially (46% attack 33% midfield 21% defense) but the German team scored a goal that gave them the victory. It should be noted that the 4 top teams of RoboCup 2001 showed different approaches to the game and all of them proved to be really effective. This shows that research in the simulation league is far from coming to a dead end and a lot of different techniques are still to be evaluated.

References

1. Luis Paulo Reis and Nuno Lau, FC Portugal Team Description: RoboCup 2000 Simulation League Champion, in P.Stone et all editors, RoboCup-2000: Robot Soccer World Cup IV, Springer LNAI 2019, pp.29-40, 2001
2. Luis Paulo Reis, Nuno Lau, Eugenio Oliveira. Situation Based Strategic Positioning for Coordinating a Simulated RoboSoccer Team, in M.Hannebauer et al eds, Balancing React. and Social Delib. in MAS, Springer LNAI 2103, pp.175-197, 2001
3. Luis Paulo Reis and Nuno Lau. COACH UNILANG - A Standard Language for Coaching a (Robo)Soccer Team, in A.Birk et all editors, RoboCup-2001: Robot Soccer World Cup V, Springer, LNAI, 2002 (to appear)
4. John Sear. RoboBase - Extensively Repository and LogPlayer Supporting Immediate Remote Access, in A.Birk et all editors, RoboCup-2001: Robot Soccer World Cup V, Springer, LNAI, 2002 (to appear)

Helli-Respina 2001 Team Description Paper

Bahador Nooraei B., Siavash Rahbar N., and Omid Aladini

Allameh Helli High School
National Organization for Development of Exceptional Talents
Kamali St., South Karegar Ave., Tehran, Iran
helli@schoolnet.or.ir,
www.schoolnet.or.ir/~helli/robocup

1 Introduction

One of the most important problems for development of intelligent agents is adaptation to the environment. In this paper we briefly describe Helli-Respina soccer simulator team that uses a new self-adaptive method named Dynamic Multi-Behavior Assessment (DMBA). By using built-in behavior manager named dynamic behavior transformer method lets the agent can choose the best algorithms to decide during the game. This system always tries to choose a set of available algorithms to get the best result against each opponent. The main objective in this research is how to choose a set of algorithms dynamically to get the best result against an opponent.

2 Dynamic Multi-behavior Assessment

Helli-Respina agent architecture is based on DMBA[1]. Several objects called assessors are connected to the agent's kernel named Core. Each assessor processes the game with a particular algorithm and tells its decision about current state of the game and the probability of successfuly taking that action. Core has a built-in assessor manager that every server cycle asks each assessor about its decision and chooses the one with higher success rate. If Core finds out that one assessor can recommend good decisions against an opponent, it may increase the probability of choosing that assessor's decision in the following states of the game. Two types of assessors can be connected to the Core: with ball assessors and without ball ones. For instance, when an agent has the ball, Core asks about decision of current state of the game from with-ball assessors.

Core has built-in behavior evaluators [1] that evaluate the decency of an algorithm against the opponents. If an assessor could not take good actions with its decision-making methods, Core decreases the probability of accepting that assessor for the next states of the game. This method lets the agent throw away algorithms that cannot decide good actions against the opponents. For example, a problem occurred to CMUnited99 [2] simulator team in Robocup 99 against of Zeng99 [3] when CMUnited players in breakaway modes mistook a defender for the goalkeeper and missed some goal opportunities. To avoid such problems,

A. Birk, S. Coradeschi, and S. Tadokoro (Eds.): RoboCup 2001, LNAI 2377, pp. 519–521, 2002.

after losing some breakaway occasions, Respina players decrease the probability of choosing breakaway assessor's decision, and make it a useless one.

All of our high-level behaviors except the formation selecting module are assessors so we can add every new high-level method to our agents and if it works well, Core will pay more attention to it due to the results it achieves against the opponent team.

The complete details about Dynamic Multi-Behavior Assessment are available in our paper submitted to the 5th RoboCup symposium [1].

2.1 Sharing Assessors Decency Knowledge

When Core of a player transforms assessors' decencies, it tries to share its information about the opponents with other players, to help the teammates to use assessors as well as possible. The decency of assessors may differ between two player with different tasks or different types in a heterogeneous environment, so each player may share the information about its assessors and players with the same types and tasks. Another usage of sharing decencies with the other teammates is that due the dynamic positioning of players, a player may sometimes be in a position in the field that is not for its task. For instance, a midfield player may be placed in position of a striker; so the player, temporarily, changes its assessors' decencies with received information from a teammate striker in the past. Also this feature is for times that a role exchange will happen; Because of decreasing stamina and also heterogeneous environment, it's necessary.

2.2 Pre-training Assessors' Decencies

Fast adaptation to the environment is very important for soccer agents. So for better performance in real-time mode, agents must use pre-trained values for their assessors. After training agents against different teams, the overall achievement of an assessor will be the default decency value for that assessor. For instance, if an assessor that analyses the perceptions with decision trees fails in many tests and trainings, the default decency value will be lower and vice versa.

Our team uses a classification of assessors' decency values based on previous competitors (especially previous year teams) that lets faster regulating decency values.

2.3 Our Implementation of DMBA

Our architecture for agents is a fully multi-threaded system and each assessor is a thread. Each cycle, after telling their decision to the Core, assessors suspend themselves until the next cycle. The multi-threaded architecture lets complete implementation of DMBA system that needs a parallel processing of assessors each cycle. Because of optimum implementation of base assessors, if some assessors with high load of game processing can not finish their decision making through one cycle, Core can choose an assessor from available ones.

3 Conclusion and Future Works

Our opinion is that DMBA is sufficient agent architecture and learning method for both adaptation and learning areas. Our future works concern designing more assessors, better pre-training of assessors' decencies and creating new opponent modeling assessors. A high load of our future plans concern the coach standardized language and players' teamwork and coordination. As other types of learning methods like Q-Learning and Neural-Networks can process the game like assessors, we plan to create more agent learning methods as assessors under the design of DMBA. Some ideas like a positioning method named Liquid Positioning (LQP) and another method based on LBG algorithm are not described in this paper and used for RoboCup 2001 competitions.

Respina 2001 team with DMBA architecture got the first place in Scientific Evaluation Challange of Soccer Simulation competitions in RoboCup 2001.

References

1. Bahador Nooraei B., Omid Aladini and Siavash Rahbar N. : Dynamic multi-behavior assessment:an approach to unsupervised machine learning, (2001)
2. Peter Stone, Manuela Veloso and Patrick Riley: The CMUnited99 soccer simulator team, (1999)
3. Hironori Aoyagi, Hiroki Shimora, Takuya Morishita, Tomomi Kawarabayashi, Takenori Kubo, Kyouiti Hiroshima, Junji Nishino: The Zeng 99 soccer simulator team, (1999)

Lazarus Team Description

Anthony Yuen

Dalhousie University, Department of Computer Science, Halifax, Canada
yuen@cs.dal.ca

Abstract. This paper describes the Lazarus team that participated in the RoboCup Simulation League in 2001. It is characterized by tight formation of players during play. The purpose is to provide data for training the coach in recognizing team formations using neural networks.

1 Introduction

Being able to recognize patterns and structures in highly dynamic environments is an essential requirement in such areas as speech recognition and computer vision. In general, a software system is repeatedly *trained* against a series of labelled data samples, a process called *supervised learning*. The software system typically uses an *artificial neural network* [3] as the backend to do the learning. Once the system is trained, it can then be used on unlabelled test cases.

In the domain of face recognition, a typical task involves taking a series of pictures of a group of people with different facial expressions and orientations, such as front, back, left, and right. The system is trained against some of these pictures. The goal is then to classify the remaining pictures, that is to say, which picture belongs to which person.

The RoboCup Soccer Simulation [6] presents a challenging and interesting problem for the *online-coach*: to recognize the opponent team's **formation**, if any. The term formation used here is loosely based on real soccer formations. For example, **4-3-3** denotes the formation of 4 *defenders*, 3 *mid-fielders*, and 3 *forwards*.

The problem is challenging due to the fact that the players are constantly moving and changing positions on the soccer field. In fact, during the game, the players closest to the ball tend to deviate from their positions. Moreover, some teams, such as FCPortugal [5], uses dynamic positioning during play. On the other extreme, some teams do not pay attention to team formation at all. To keep the problem more managable, we will restrict ourselves to teams that do keep formations and change formations only occasionally (e.g. half-time).

The Virtual Werder team [2] participated in the RoboCup 2000 competition and made use of an online coach agent for the purpose of changing team formations dynamically. Their coach used neural network to recognize the opposing team formation and then chose an appropriate counter-formation from a fixed pool of known formations. Our approach follows a similar path.

A. Birk, S. Coradeschi, and S. Tadokoro (Eds.): RoboCup 2001, LNAI 2377, pp. 522–525, 2002.

2 Team Development

Team Leader/Member: **Anthony Yuen**

- Masters Thesis Graduate Student
- Institution: Dalhousie University
- Department: Computer Science
- Country: Canada

3 Player Description

Lazarus is a soccer simulation team based on the publicly available FCPAgent [4], which in turn is based on the CMUnited99 [7] code. As a result, a lot of development time is saved for the basic skills such as ball dribbling and shooting. This team has been deliberatly written in such a way that the players keep a strict team formation throughout the game. Only the currently **active** players deviate from their predefined formation position. An active player is a player closest to the ball. If it currently has the ball, then it decides to either *dribble* or *pass* to a teammate. Otherwise, the player engages in *interception* to retrieve the ball.

In addition to a formation position, each player (excluding the goalie) is assigned a **role**. A role is one of: *Sweeper*, *Defender*, *Mid-Fielder*, *Captain*, *Forward*, and *Striker*. The meaning of each role is loosely based on real soccer roles. The role of a player determines several player attributes such as aggressiveness and is dependent on the current team formation and player position.

4 Coach Description

A **coach** agent is a special client program that can connect to the soccer server and oversee and ongoing soccer match. The coach receives noiseless information (e.g. coordinates, velocities, staminas... etc) about every object in the soccer simulation. Therefore, at any given moment, the coach has complete knowledge of the ball, its team players, as well as the opponent team players. However, the coach-to-player communication is restricted to when the ball is out of play using free-style message formats. If the coach uses the new standard coach language [1], communication is restricted to one *info*, *advice*, and *define* message every 30 seconds (300 simulation cycles). The Lazarus coach uses the free-style message format.

The coach is assigned the task of recognizing team formations. To tackle this problem, the coach uses neural networks and the backpropagation learning algorithm [3]. For the initial training, the prototypical **4-3-3** formation is used to train the network. Due to the computational intensive nature of the backpropagation algorithm, the training is done offline. During each simulation cycle, the coordinates of each player in the soccer field are recorded in a format suitable for the neural network. The logged data is then used to train the network to classify the formation as 4-3-3 consistently. This process is repeated for several other common formations.

5 Research

By observing the top-level teams, it is clear that they employ some kind of team formation system, usually conventional and sometimes proprietary. Moreover, formations do not change from cycle to cycle – it would be rather chaotic otherwise. Instead, it is common for teams to switch from an offensive formation to a more defensive one at half time when it is leading the match.

The Virtual Werder team [2] demonstrated that by using an online coach to switch team formation, their team performed better than without a coach. Their coach uses a *single* neural network to classify approximately fifteen different formation systems, including the Catenaccio. In the same vein, the Lazarus coach also advises players to switch to more appropriate *counter-formations* during play. However, a **layered** approach is used in the learning and classification systems.

In the layered approach, one neural network is used to recognize one particular formation system. For example, the 4-3-3 neural network outputs a *normalized* confidence value to indicate how closely the inputs correspond to the 4-3-3 formation. Similarly, the 4-4-2 network recognizes the 4-4-2 formation. Separate networks are trained in similar ways. The end result is a collection networks for different formations.

To utilize these networks, a larger, layered neural network is formed. The inputs are as before. The second layer consists of the trained networks, now used as *nodes*. A third layer of *hidden* nodes is added. The final layer is the output layer that indicates which formation system is recognized based in the inputs.

After this phase is accomplished, the online coach can then perform team formation classification in *real-time*, during a soccer match. This is possible because the execution of a learned neural network is very fast. How this information is used is implementation-dependent.

6 Conclusion

Traditional pattern recognition domains deal with static data samples such as facial images. To make the task more challenging, we can consider the problem of recognizing faces *in motion*. Similarly, we can compare the task of analyzing team formation from a particular **snapshot** of a RoboCup simulation versus that of recognizing team formation with the players *in motion*.

A human can normally recognize a team formation *visually*. A computer program, on the other hand, cannot **see** what is happening in the soccer field. It needs to rely on some advanced machine learning techniques in order to perform this task. It is hoped that the layered neural network approach is a step in the right direction.

7 Future Work

It is unclear whether the layered network approach to recognizing team formations is better compared to the single network approach. More experiments are needed to provide a conclusive result. In addition, the usefulness of an online coach in RoboCup Soccer Simulation is yet to be determined. The new coach competition in RoboCup 2001 should provide some insights to this question. This research will become part of a Masters Thesis by the author of this paper.

References

1. M. Chen, E. Foroughi et al, *RoboCup Soccer Server Users Manual*, http://www.cs.cmu.edu/~galk/rc01/manual.ps.gz, 2001
2. C. Drucker et al, *"Recognizing Formations in Opponent Teams"*, http://www.virtualwerder.de, 2000
3. J. Hertz, A. Krogh, and R.G. Palmer, *"Introduction to the Theory of Neural Computation, Volume 1"*, Addison-Wesley Publishing Company, Redwood City, California, USA, 1991
4. R. Luis, N. Lau, and L. Lopes, *FCPAgent Source Code*, http://www.ieeta.pt/robocup/documents/FCPAgent.tgz, 2000
5. R. Luis, N. Lau, *"FCPortugal Team Description: RoboCup2000 Simulation League Champion"*, in *RoboCup-2000: Robot Soccer World Cup IV*, Springer Verlag, Berlin, 2001
6. I. Noda et al, *"Soccer Server: A Tool for Research on Multiagent Systems"*, in *Applied Artificial Intelligence*, vol 12, pp. 233-250, 1998
7. P. Stone, M. Veloso, and P. Riley. *"The CMUnited-99 Champion Simulator Team"*, in *RoboCup-99: Robot Soccer World Cup III*, Springer Verlag, Berlin, 2000

RoboLog Koblenz 2001*

Jan Murray, Oliver Obst, and Frieder Stolzenburg

Universität Koblenz-Landau, AI research group
Rheinau 1, D–56075 Koblenz, GERMANY
{murray,fruit,stolzen}@uni-koblenz.de

1 Introduction

Outline. A formalism for the specification of multiagent systems should be expressive enough to model not only the behavior of one single agent, but also the collaboration among several agents and the influences caused by external events. For this, *state machines* [4] seem to provide an adequate means. Therefore, the approach of the team *RoboLog Koblenz 2001* employs techniques from software engineering and artificial intelligence research by using UML statecharts and implementing them systematically with *logic and deduction* in Prolog [3].

The current work concentrates on formal agent design. The decision process of soccer agents can be made more flexible by introducing *utility functions* for rational behavior as proposed in [5]. Furthermore, it is desirable to be able to check whether the multiagent system satisfies certain interesting properties. Therefore, the formalism should also allow the verification or the formal analysis of multiagent systems, e.g. by *model checking*, as described in [7].

The RoboLog team. The RoboLog team participated in the simulator competitions in 1999 (Stockholm) and 2000 (Melbourne). 3 people, Jan Murray, Oliver Obst and Frieder Stolzenburg (team leader), form the core of the team. There are currently 6 additional members, namely Joschka Bödecker, Björn Bremer, Marco Dettori, Marion Levelink, Jana Lind, and Karsten Sturm. However, most of them joined the group very recently. As in previous years [8], the team is implemented in two parts. The kernel, hosting the soccer server interface and low-level functions, is implemented in *C++*, while the control program for the team behavior is written in *Prolog*.

In Section 2, we describe our approach with an explicit state machine and its concrete implementation [2]. This is part of the current team, later on this shall lead to verification [7] and more sophisticated decision making [5] of agent systems. Building agents for a scenario such as the RoboCup also requires the careful and efficient programming of low-level facilities (see Section 3). Ball interception and marking are important features (already efficiently incorporated into our team), but also the ability of more abstract, qualitative reasoning.

* This research is partially supported by the grants *Fu 263/6-1* and *Fu 263/8-1* from the German research foundation *DFG*.

A. Birk, S. Coradeschi, and S. Tadokoro (Eds.): RoboCup 2001, LNAI 2377, pp. 526–530, 2002.

2 Declarative Agent Design

Structured State Machines. Statecharts are a part of UML [4] and a well accepted means to specify dynamic behavior of software systems. They can be described in a rigorously formal manner, allowing for flexible specification, implementation and analysis of multiagent systems [7]. In statecharts, states are connected via transitions with conditions and actions annotated. Since states may be simple, composite or concurrent, the behavior of agents or their state machines, respectively, cannot be described by sequences of simple states (as for finite automata), but of configurations.

A *configuration* c is a rooted tree of states, where the root node is the topmost initial state of the overall state machine. A configuration must be completed by the following procedure: if there is a leaf node in c labeled with a composite state s, then the initial state of s is introduced as immediate successor of s; if there is a leaf node in c labeled with a concurrent state s, then the tree branches at this point. In the current implementation of our team, an explicit state machine is built in. It processes the transitions, performing micro-steps in this case. Several transitions can be executed in parallel if they stem from concurrent regions, forming a macro-step [6] then.

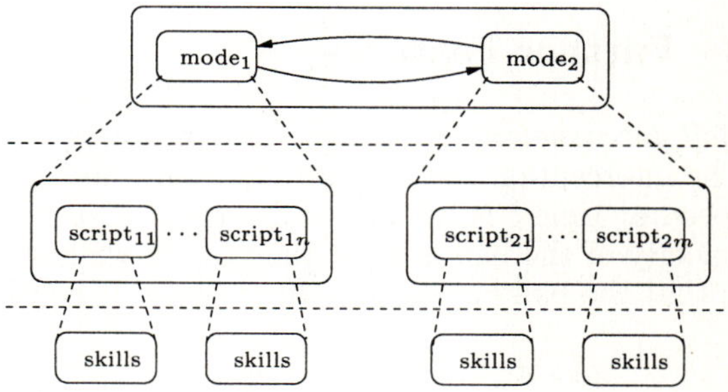

Fig. 1. Three-level approach.

Implementation with Logic. The agent specification can effectively be transformed into running Prolog code. The resulting program consists of several Prolog modules, which reflect the layered specification of the agent. The actual implementation, which also contains an explicit state machine module, is described in [2]. The soccer agents are designed with a three-level approach (Figure 1): the *mode level* contains the most abstract desires an agent has (e.g. setup, attack, defend); the *script level* provides plan skeletons that are used as long as the mode is not changed (e.g. marking, passing, role exchange); the *skill level* hosts basic actions (e.g. kick, dribble).

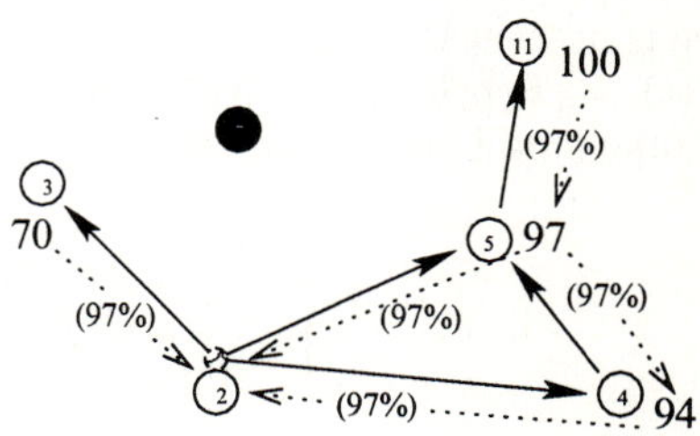

Fig. 2. Searching for pass partners.

Extension by Utility Functions. The design of adaptive agents with the method presented so far is possible, but it is a good idea to extend the approach by a more adaptive action selection mechanism and to facilitate a more explicit representation of goals of an agent. Until now, the measure for the expected success are Boolean conditions annotated at transitions. Actions with firing conditions are expected to be successful. If an agent has more than one option to execute actions, the first applicable one is selected. A more adaptive action selection mechanism should evaluate the usefulness of applicable options and execute the most useful one. Therefore, [5] proposes the use of *utility functions* in order to provide a mechanism to evaluate the utility of a script in the current situation taking the commitment to selected options into account.

In our team, utility functions are used for finding the best pass partner. The utility function for pass partners prefers teammates closer to the opponent goal and further away from the own goal. The search for a pass partner is executed recursively for a fixed number of steps, so that a player prefers pass partners who can pass to teammates in a better position. Only in the last step of the search or if no teammate in a better position can be found, the value for the player's position is taken. In other cases, the preference value of the best pass partner is used, so that players with good passing opportunities are preferred. Since we want the players to prefer direct passes to a teammate over a chain of passes to the same teammate, we discount each pass in the preference value of the pass partner (see also Figure 2, where player 2 is about to pass to player 5).

3 Further Features

Ball Interception. An important feature of a soccer agent is *ball interception.* The interception time can be computed effectively. In the soccer server, the ball speed at time t is calculated as $v(t) = \mu \cdot v(t-1)$ with $\mu < 1$. Therefore, if the velocity of the ball is v_0 at time $t = 0$, we have $v(t) = v_0 \cdot \mu^t$. For the distance s, that the ball has moved after t steps, it holds:

$$s(t) = \sum_{i=1}^{t} v_0 \cdot \mu^{i-1} = v_0 \cdot \frac{1-\mu^t}{1-\mu}$$

In the sequel, we assume that an agent can move in any direction with a fixed velocity v_1. At time $t = 0$, the ball is at position $\vec{a}$ in the Cartesian coordinate system with the agent at its origin. Let $\vec{b}$ (with $\|\vec{b}\| = v_0$) be the velocity vector of the ball in this coordinate system. Then, after t steps the position of the ball is $P(t) = \vec{a} + s(t) \cdot \vec{b}$.

Clearly, the agent can reach the ball at any time t with $\|P(t)\| \leq v_1 \cdot t$. Since there is no closed form for t, we apply *Newton's method* in order to find the zeros of $f(t) = \|P(t)\| - v_1 \cdot t$. We compute the following sequence t_n, until $|f(t_n)| < \epsilon$ for some small threshold $\epsilon > 0$:

$$t_n = \begin{cases} 0, & \text{if } n = 0 \\ t_{n-1} - \frac{f(t_{n-1})}{f'(t_{n-1})}, & \text{if } n > 0 \text{ and } f'(t_{n-1}) < 0 \\ 999, & \text{otherwise} \end{cases}$$

This procedure eventually yields the first of at most three zeros $t > 0$. There exists at least one zero; it is found at latest after t_n has been set to 999, which avoids oscillation. If there are three zeros, then Newton's method will find the smallest one. This follows from the fact that the acceleration a of the ball (the derivative of v) is negatively proportional to v. A similar (but different) method for computing the interception time has been described in [9].

Stable Marriage. Close marking can be very helpful in a variety of situations. By blocking the path between the ball and an opponent, the agent prevents direct passes. But for close marking to work it is important that the agents select the opponents to mark in an intelligent way. Otherwise situations arise in which opponents are marked by more than one agent while others remain completely free. In addition each player should mark an opponent that is as close as possible to avoid stamina consuming dashes across the field.

This mapping of teammates to opponents can be seen as an instance of the *stable marriage problem* [1]. The ranking of the opponents by each player is based on the distance from the player to the individual opponents. It is then possible to generate a solution which assigns the closest possible opponent to each player. If all players have complete knowledge of the positions of teammates and opponents on the field, each opponent will be marked by exactly one player.

Qualitative Reasoning. For the classification of situations occurring during a match, mapping of quantitative numbers to qualitative data is helpful to describe situations in a shorter and more natural way for the programmer. Usage of qualitative predicates for distances and directions is already supported by our C++ interface.

With the *soccer server* version 7, a new application for qualitative reasoning becomes obvious: the *online coach* can define regions, and inform or advise players about the current situation or actions they should perform. These regions can be viewed as a qualitative abstraction from the quantitative data the online coach gets. To effectively use the defined regions, the coach agent has to reason about the implications of its advice to other players. Besides the development and implementation of a coach using qualitative spatial data, qualitative reasoning about velocities and object movement is one of our next planned activities.

References

[1] D. Gale and L. Shapely. College admissions and the stability of marriage. *American Mathematical Monthly*, 1962.

[2] J. Murray. Soccer agents think in UML. Diplomarbeit D 610, Fachbereich Informatik, Universität Koblenz-Landau, 2001.

[3] J. Murray, O. Obst, and F. Stolzenburg. Towards a logical approach for soccer agents engineering. In P. Stone, T. Balch, and G. Kraetzschmar, editors, *RoboCup 2000: Robot Soccer World Cup IV*, LNAI 2019, pages 199–208. Springer, Berlin, Heidelberg, New York, 2001.

[4] Object Management Group, Inc. *OMG Unified Modeling Language Specification*, 1999. Version 1.3, June 1999.

[5] O. Obst. Specifying rational agents with statecharts and utility functions. In *Accepted paper at RoboCup International Symposium (RoboCup 2001)*, 2001. To appear.

[6] A. Pnueli and M. Shalev. What is in a step: On the semantics of statecharts. In T. Ito and A. R. Meyer, editors, *International Conference on Theoretical Aspects of Computer Software*, LNCS 526, pages 244–264, Sendai, Japan, 1991. Springer, Berlin, Heidelberg, New York.

[7] F. Stolzenburg. Reasoning about cognitive robotics systems. In R. Moratz and B. Nebel, editors, *Themenkolloquium Kognitive Robotik und Raumrepräsentation des DFG-Schwerpunktprogramms Raumkognition*, Hamburg, 2001.

[8] F. Stolzenburg, O. Obst, J. Murray, and B. Bremer. Spatial agents implemented in a logical expressible language. In M. Veloso, E. Pagello, and H. Kitano, editors, *RoboCup-99: Robot Soccer WorldCup III*, LNAI 1856, pages 481–494. Springer, Berlin, Heidelberg, New York, 2000.

[9] P. Stone and D. McAllester. An architecture for action selection in robotic soccer. In *Fifth International Conference on Autonomous Agents*, 2001.

Team Description Mainz Rolling Brains 2001

A. Arnold, F. Flentge, Ch. Schneider, G. Schwandtner, Th. Uthmann, and M. Wache

Department of Computer Science
Johannes Gutenberg-University Mainz
D-55099 Mainz, Germany

1 Introduction

The Mainz Rolling Brains 2001 team is based on our last year's team. Our modular design as described in [1] has proved to be efficient and flexible. Thus the team could easily be adopted to the soccerserver's new features and some of the weak spots of our team could be eliminated.

We use a three-layers concept for our agents: a technical layer (purely technical matters like server communication), a transformation layer (the tools and skills a player might use) and the decision layer. The decision layer consists of various modules for different tasks (goal shot, pass, ballhandling, positioning and standard situation). These modules rate the adequateness of their respective actions and compete for the control over the player. More detailed information about our team architecture can be found in [1] and [2].

One of this year's main focuses was positioning which has been completely revised. We will describe our new position module below. We made extensive use of our new tool "FUNSSEL" (a debugging tool including an extended soccer monitor) which made it possible to further improve our technical skills and some of our modules as well. Due to the lack of space we will concentrate in the following on the new positioning module, FUNSSEL and our method of self-localisation.

2 Self-Localisation

As the agents in the simulation league do not know their absolute position on the field, they have to calculate it from the visual data they get. This data contains information about distance and angle (relative to the agent's body orientation) of the objects in the agent's field of vision. There are several flags in and around the field (of which the exact positions are known) we can use to calculate the position of the agent. Visual data is not exact, i.e. the soccerserver provides distances and angles only with an intentionally added error. Therefore, our position is in an area with its borders given by these errors. To calculate the position of our agents we create polygons with five corners approximating this

A. Birk, S. Coradeschi, and S. Tadokoro (Eds.): RoboCup 2001, LNAI 2377, pp. 531–534, 2002.

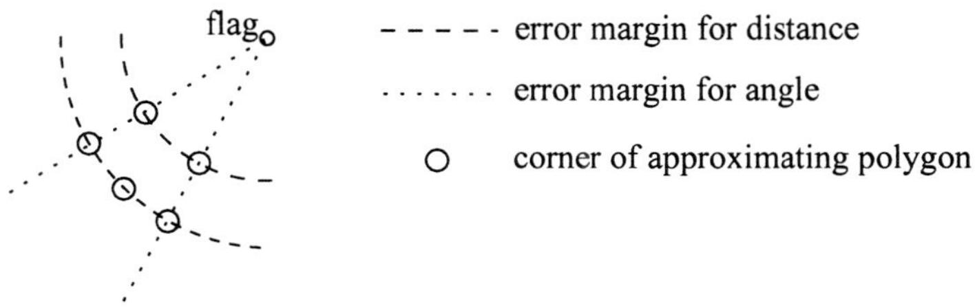

Fig. 1. A polygon approximating the area of the player's position

area as shown in Figure 1 for each visible flag. Since the error depends on the distance these polygons get the smaller the closer the flag is.

We intersect each two polygons and intersect the resulting polygon with the next. Polygons are intersected by iterating over the edges of the second polygon and keeping that part of the first polygon being inside the half-plain which is generated by the current edge and contains the second polygon. Finally, we calculate the agent's position inside the resulting polygon as the average of the polygon's corners. If there is no intersection of all polygons we use the weighted sum of the positions derived from all visible flags.

3 Positioning

We introduced a new positioning module this year giving each agent a very well defined role during a game. This should ensure that at least one agent feels responsible for an area of the field and that an attacking opponent is always attacked by at least one agent. The different roles of the agents are divided into four major groups: the defense, the defensive midfield, the offensive midfield and the attack. Currently we play with 3 defenders, 2 agents belong to each defensive and offensive midfield and 3 attackers.

Each agent has two standard positions, one in a more defensive, the other in a more offensive context. Depending on the ball position he places himself somewhere between these two positions. Positions are described as coordinates in a virtual 100 x 100 area. To get the actual position the player should take on the field these position values are projected onto a certain area of the soccer field. This area is determined by ball position, the opponent's offside line and a given minimum and maximum value for the area. This positioning system allows to generate certain tactical behaviours. For example, in defensive situations defensive midfield players are always placed in one line with the ball and therefore these players will almost always feel responsible for attacking the opponent with the ball while the whole defense stays covering the area between ball and own goal. In specific situations additional rules are triggered, e.g. if the defensive midfield agents fail to get the ball, the positioning behaviour is switched, i.e. one of the defenders attacks the opponent, and a defensive midfield tries to get behind the attacking defender to back him up. These rules have to be chosen very

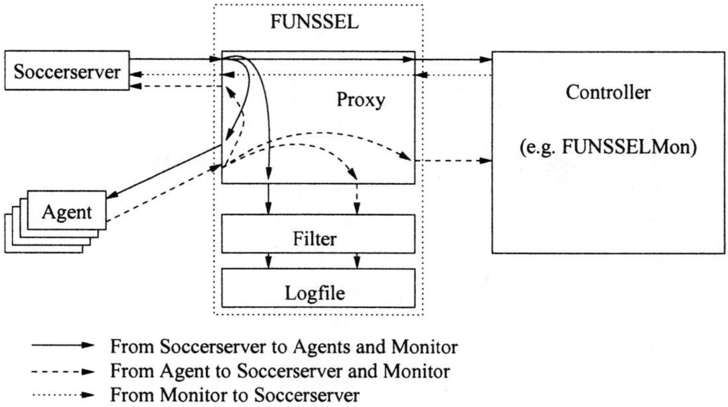

Fig. 2. Scheme of FUNSSEL functionality

carefully, so that they altogether lead to a complex behavior with well organized interactions between the players.

4 FUNSSEL - A Visualisation and Debugging Tool

FUNSSEL ("Flexible Utility for Network based Soccer Simulation using Extended Language") developed by the Mainz Rolling Brains is a powerful utility to debug simulation league agents. It consists of two single programs, *FUNSSEL* and *FUNSSELMon*.

FUNSSEL is a kind of proxy server that is placed between the soccerserver and both the agents and the monitor. FUNSSEL passes standard soccerserver commands in the usual way, while special commands allow redirection of commands directly to the monitor (e.g. FUNSSELMon) without sending anything to the soccerserver. All this is implemented using an extended version of the standard soccerserver protocol to make it easy to use the special monitor commands. It enables the players to report information about their current worldmodel and tactics or even to control the game. The whole communication is logged through a filter for a later replay. The logging of the players additional output may increase the logfile size dramatically. That means filtering out unneeded information from the soccermonitor protocol and compressing the data is necessary to reduces the logfile to a normal size. For a structural overview of FUNSSEL and the related processes, have a look at Figure 2 showing the data flow of different command types.

To use all the features described above, a special monitor is needed - the FUNSSELMon. Acting like a normal Soccermonitor it also offers extended functionality to control FUNSSEL and to gain more information about the game. Using special player commands FUNSSELMon can display some agent's worldmodel and special tactical information as graphics on the playing field as well as simple logfile entries. While watching a game or logfile different colors can be assigned to players to display data of several agents simultanously. Other commands allow players to pause the game, do a drop ball or a freekick or to use the Soccermonitor set player command, all being useful for training or debugging certain game situations. To have a closer look at some technical issues during the game, a zoom function is available which allows to open separate windows for different regions, each at its own scale. FUNSSELMon can display technical information like the heterogenous player types, stamina values for all players, the kickrange and a viewcone showing the current vision of one or more players.

The main advantage of this concept is that it is nearly independent of the soccer server version used. FUNSSEL is not dependent on FUNSSELMon, therefore other monitor programs could be designed to provide for example a batch mode to train the agents and run games automatically.

5 Conclusion and Outlook

The newly developed debugging tool FUNSSEL offers a lot of assistance in monitoring and analysing soccer simulation league agent's behaviour. It clearly has improved our understanding of what the agents "see", on how they decide what to do and last but not least why they do not act like they are supposed to. Thus it helps to increase the strength of our agents' soccer team. For the next future we are planning to make use of machine learning techniques (e.g. Kohonen Feature Maps) for analysing games and changing tactical behaviour.

Acknowledgements: We wish to mention all current members of the Mainz Rolling Brains: Axel Arnold, Jochen Ditsche, Felix Flentge, Manuel Gauer, Marc Hoeber, Christian Knittel, Claudia Lautensack, Christian Meyer, Birgit Schappel, Christoph Schneider, Goetz Schwandtner, Thomas Uthmann, Martin Wache. We also would like to thank our former team leader Daniel Polani and our sponsors SerCon and F^3G.

References

1. Schappel, B., Schulz, F.: Mainz Rolling Brains 2000. In: Stone, P., Balch, T., Kraetzschmar, G. (ed.): RoboCup 2000: Robot Soccer. World Cup IV. Lecture Notes in Computer Science, Vol. 2019. Springer-Verlag, Berlin Heidelberg New York (2001)
2. Uthmann, Th., Meyer, C., Schappel, B., Schulz, F.: Description of the Team Mainz Rolling Brains for the Simulation League of RoboCup 2000. `http://www.rollingbrains.de/mrb2000/MainzRollingBrains2000.ps`

Team Description of NITStones2001

Tetsuya Esaki, Taku Sakushima, Yoshiki Asai, and Nobuhiro Ito

Department of Intelligence and Computer Science,
Nagoya Institute of Technology
Gokiso-cho,Showa-ku,Nagoya 466-8555,JAPAN
esaki@egg.ics.elcom.nitech.ac.jp

Abstract. In this paper, we describe the feature of the NITStones2001 team participated in the RoboCup 2001 simulation league. In a Multi-Agent System, it is important that an agent cooperate with the others appropriately. We consider the cooperative activity of an agent as action by a group of the agent who aims at achievement of the common goal, and construct the cooperative agent and the dynamic grouping agent model.

1 Introduction

One of the important problems in a multi-agent system is how an agent cooperate with the others in order to achieve the goal of the system[1]. In this paper, we describe the design of cooperative activity and the construction of the agent in RoboCup simulation league team "NITStones2001"[2].

In the case of an autonomous distribution robot like RoboCup's agent, the environment information which each agent recognizes for the restriction of observation capability is different. So, agents can not act by the same circumstances recognition. Moreover, because of restriction of communication capability, it is difficult for agents to understand mutual intention by a lot of information exchange. Furthermore, the circumstances and the role of the agent often change. Therefore, the realization of cooperative activity is a difficult problem.

In order to cooperate under such environment, it is important that agents understand the intention at each other. Therefore, Many researches which realize cooperative activity have been made, such as a cooperative protocol, organization structure, and cooperation by learning. However, many approach realized the cooperative activity as a result, and there was nothing that clarified the cooperative activity. Then, we define the cooperative activity clearly and design the cooperative activity by construct the agent which realizes it.

We consider the cooperative activity as action of the agent group which aims at achievement of the common goal(subgoal) and define these agents as "group". In case our team agents cooperate, they act holding clearly the intention of co-operating with each other by forming this group dynamically. By formation of group, we design cooperative activity clearly and realize it. Therefore, we define the agent model which can determine the goal according to change of a circumstance, and define the agents which shares the goal as a group. Furthermore, we

A. Birk, S. Coradeschi, and S. Tadokoro (Eds.): RoboCup 2001, LNAI 2377, pp. 535–538, 2002.

construct the agent based on this definition, and design the cooperative activity by forming the group.

2 Agent Model

In the beginning, we consider an agent as follows: An agent knows the goal which the whole system aims at, and has several subgoals in order to achieve it. He recognizes the circumstances from the environment information which self gained, determines the subgoal from recognized circumstances, and acts to achieve it.

In case the agents cooperate, we consider that they act aiming at achievement of the common subgoal. That is, if the agents can share the subgoal, we consider that they can cooperate. Therefore, we construct the agent model which can determine a goal according to change of a circumstance. We define an agent as follows:

$$ag_i =< id, S_i, A_i, P_i, I_i, f_i, g_i, h_i > \tag{1}$$

Here id is ID of $agent_i(ag_i)$, S_i is a set of the internal state which $agent_i$ recognizes, A_i is a set of the action which $agent_i$ can take, P_i is a set of the subgoal which $agent_i$ has, I_i is a set of the input(the observation information), f_i is the state determination function, g_i is the goal(subgoal) determination function, h_i is the action determination function. An agent determines the self internal state s_t with the function $f(s_{t-1}, i_t)$ from the input $i_t(\in I_i)$ from environment and the past internal state $s_{t-1}(\in S_i)$. Next he determines the subgoal p_t with the function $g(s_t, p_{t-1})$ from s_t and the past subgoal $p_{t-1}(\in P_i)$. Afterward he determines the action $a_t(\in A_i)$ with the function $h(s_t, p_t)$ from s_t and p_t.

This agent model clarifies an agent's intention by making the subgoal hold clearly. Moreover, by taking the past state and subgoal into consideration to the state determination and the subgoal determination, this model has the robustness over lack and noise of the observation information.

3 Cooperation and Group

In this paper, we define the agents which share the goal as a group, and regard the cooperative activity as action of a group. That is, by forming the group of the agent who aims at achievement of the common subgoal, we design the cooperative activity. Therefore, an agent can act on the common intention of "cooperating" clearly.

However, in a environment where the circumstances changes frequently like RoboCup, each agent's subgoal may change from each agent's condition, or the difference in environmental recognition. Moreover, although a subgoal sharing is done by communication, an agent can not check whether a subgoal is shared certainly in the environment which has restriction in communication capability. When there is such the possibility, it is dangerous to act expecting other cooperation. Then, we define a group as follows:

$$gp_i =< id, \mathcal{AG}(p_i), L_i, n_i > \quad (2)$$

Here *id* is ID of $group_i(gp_i)$, $\mathcal{AG}(p_i)$ is a set of the agent who aims at achievement of the common subgoal, n_i is the number of the maximum agents of $group_i$, and L_i is a set of the link between agents. A link shows whether two agents can communicate, and the agents in a group have link each other. Therefore, an agent can form the agent group which can communicate with the agent in a group mutually, and has the common subgoal. Moreover, the scale of a group is restricted by defining the maximum number n_i.

In this definition, the agent belonging to a group has the common subgoal, and can communicate with each other. Each agent in a group holds the group information, such as the subgoal and a member of the group. And when the self subgoal or a group member has changed, the group information is updated by broadcasting the new group information in the group. By holding this group information with each other, an agent checks the intention of cooperation with each other and acts on it.

4 Agent Architecture

In our team, we constructed following agent architecture in order to realize the agent model of Section 2.

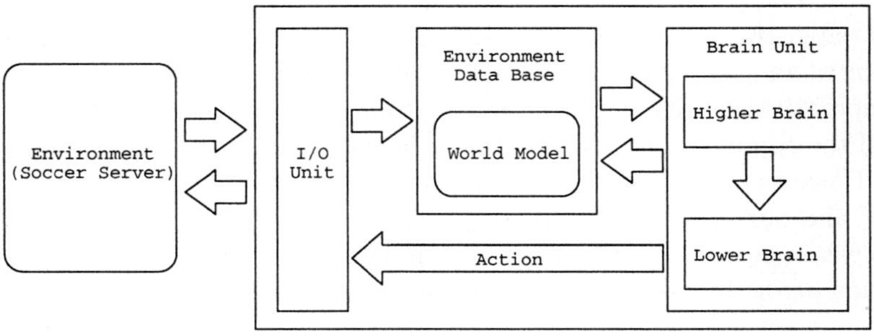

Fig. 1. Agent Architecture

An agent receives the observation information (i_t) with *IO Unit*, processes the information which he can use, and constructs a world model ($s_t \leftarrow f(s_{t-1}, i_t)$) to *EnvironmentDataBase*. The state determination and the action determination are performed by *Brain Unit*. *Brain Unit* has the subsumption structure constituted by *HigherBrain* and *LowerBrain*. *HigherBrain* determines the subgoal ($p_t \leftarrow g(s_t, p_{t-1})$), and *LowerBrain* determines the action ($a_t \leftarrow h(s_t, p_t)$). Each *Brain* is implemented by the thread, and it can determine the subgoal and the action according to change of a circumstance.

5 Dynamic Group Formation

Our team designs the cooperative activity by forming the group of Section 3 dynamically. The agent tries to form the group, when he has the subgoal which needs a cooperation. At first, the group is formed of one agent, the agent who forms the group broadcasts around the group information, such as the subgoal and a member of the group. The agent who receives the group information and has the same subgoal broadcasts new group information including himself in order to join the group. Thus, change of the group and maintenance are also performed by updating the group information fundamentally. The image of the formed group is shown in Figure 1.

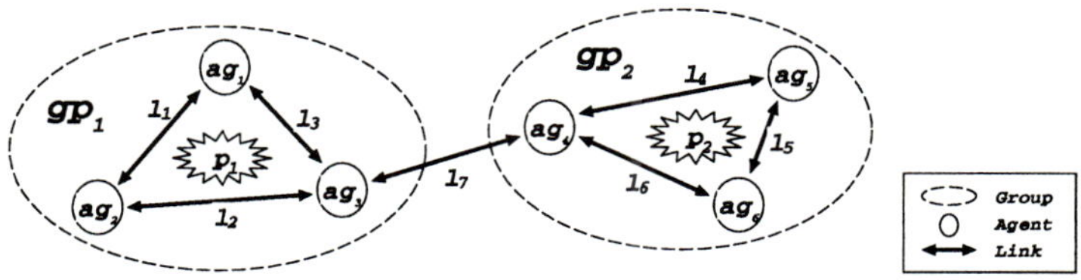

Fig. 2. The image of the formed group

6 Conclusion

In this paper, we described the design of the cooperative activity and the construction of an agent in soccer simulation team "NITStones2001". We considered the cooperative activity as action by the group of the agent who aims at achievement of the common goal, and designed the cooperative activity by defining the group. However, since the group formation algorithm in our soccer agent was not perfect, and our program was not adjusted enough for RoboCup2001, sufficient results were not able to be achieved. Therefore, we want to complete the group formation algorithm and evaluate the cooperative activity from now on.

Acknowledgment

This work was partly funded by the research grants from the Hori Information Science Promotion Foundation and the Tatematsu Foundation.

References

1. Nobuhiro Ito. A Description-Processing System for SoccerAgents, RoboCup98 : Robot Soccer World Cup II.
2. NITStones2001 Web Site : http://adam.elcom.nitech.ac.jp/NITStones01/

Team YowAI-2001 Description

Koji Nakayama and Ikuo Takeuchi

Department of Computer Science
The University of Electro-Communications
Chofu, Tokyo 182-8585, Japan
robocup@takopen.cs.uec.ac.jp

1 Introduction

The team YowAI-2001 inherits the low-level individual skill and the world modeling technique of YowAI-2000 developed by Takashi Suzuki and improved by Sinnosuke Asahara. Its objective is "cooperation by short shouts" instead of cooperation by communicating large amount of non human-like messages among agents.

Our former team YowAI-2000 puts emphasis only on the quality of low-level individual skill and world modeling, but it does not involve explicit cooperation among agents. For example, YowAI-2000 uses neither say command nor hear command at all. It uses a fixed positioning strategy. However, YowAI-2000 won the victory in RoboCup Japan Open 2000, and obtained the 6th place in RoboCup 2000 held in Melbourne.

From these result, it can be said that each agent's individual low-level skill and world-modeling accuracy have prior importance to cooperation. However, it is also true that it is difficult to achieve further improvement in the performance of a multi-agent system only by the improvement in agent's individual low-level skill and world-modeling accuracy. Therefore, YowAI-2001 tries to adopt action decisions based upon tactics and strategy of skillful human soccer players. Here, the most important criterion of YowAI-2000 is that it uses only communication by short shouts as seen (or heard) in human soccer play.

The main purpose of YowAI-2001 as a research topic in distributed AI is that how far man-like communication can realize cooperation, without using detailed numerical or symbolic information such as absolute coordinate and elaborate plan sequence. We believe that powerful cooperation in RoboCup should be realized only by minimum utterance.

2 Team Play by Shouts

The communication between agents should be transmitted by the soccer server. If an agent who wants to communicate publishes say command to the server, the server transmits hear command to agents within 50 meter distance from the utterer. Detailed coordinates and numerical information can be told because of the comparatively loose current regulation that a say command is a sequence of less than 512 (almost any kind of) ASCII characters.

A. Birk, S. Coradeschi, and S. Tadokoro (Eds.): RoboCup 2001, LNAI 2377, pp. 539–542, 2002.

However, say commands used in YowAI-2001 are much simpler as shown below. We are now extending the vocabulary based upon common soccer tacitics and strategy.

```
(say SwordFish shotteru 4)
```

This message means that an opponent player is pressing behind the player No. 4. Here, "SwordFish" is the nickname of YowAI-2001 in the Takeuchi Laboratory of the University of Electro-Communications.

In actual soccer, players use minimun communication methods such as a shout and eye contact. This shows that the efficiency of communication between players is important in soccer. For example, the shout "AGERO" is used in the meaning "push up a defense line." So, players do not say concretely, "push up a defense line." Becase the length of the word which man can understand immediately is one or two. If players say concretely, the player who is said cannot understand immediately and cannot take action. Note the important thing that players can communicate tactically enough without adding any concrete word.

The essence of shout is that gives a tactical self/partner's situation immediately and gives a hint which is selection of judgment or action for partner. Selection of judgment or action is not forced. The shout must be short, the given tactical situation have to be highly abstracted concept with an understanding common to teammates.

Since the shout is highly abstracted, both of player who uses the shout have to refer to own world model or the strategy plan which the whole team should have in common. Therefore, in order to perform cooperation only by the shout, it is necessary to make each agent's intelligence into high level from the case where extensive information communication is possible. In addition, also have to consider quality of a shout. For example, as for the shout "pass" any teams are used abundantly. However, the tactical judgment of the player who said, "pass" may be worng. As it becomes a strong team, the reliability of a thing which "pass" means becomes higher.

2.1 shotteru

"SHOTTERU" is shouted when one or more opponent players approach behind our player who is going to receive the ball. The player who hears this voice uses its ball keeping skill, and prevents taking a ball if he is in the opponent's field and showing his back to the opponent's goal. Other players occasionally move to supporting position and say "HATAKE" which will be described immediately below. This simple message may raise various action decisions depending on the players condition.

2.2 hatake

"HATAKE" is shouted from a player who is in the position who can receive a short pass from a "SHOTTERU" player rather safely. Then, the "SHOTTERU" player who hears this voice will pass to the direction of the voice if he decides it is reasonable.

2.3 gotzan

"GOTZAN" is shouted by a player who is in a better position for shooting than who is now controlling the ball.

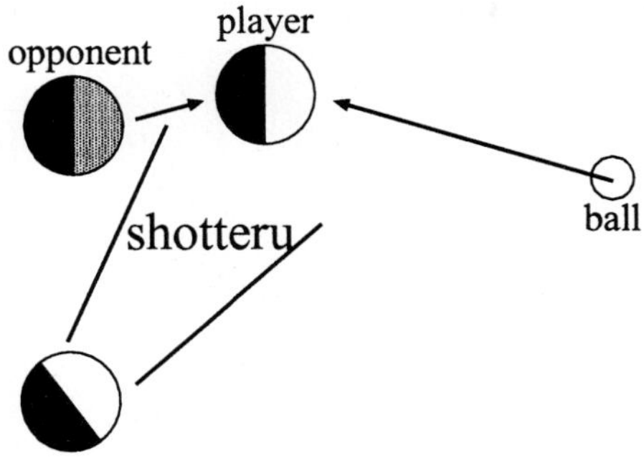

Fig. 1. shotteru

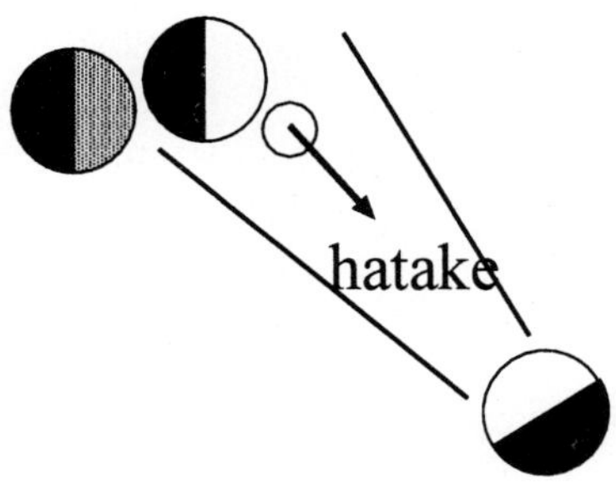

Fig. 2. hatake

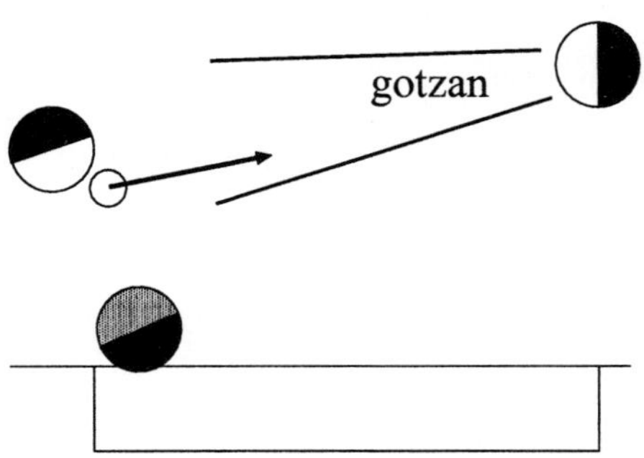

Fig. 3. gotzan

3 Future Work

We have to extend the vocabulary of shouts. The vocabulary shown above is obviously too small to make agent's communication more effective and "organic". We shall limit the usage of numerical value to minimum, that is, only small positive integers.

It should be noted here that improvement of individual tactical and strategic ability is important to make such short shouts work well. This improvement will be our main future research.

The Dirty Dozen Team and Coach Description

Sean Buttinger, Marco Diedrich, Leo Hennig, Angelika Hoenemann, Philipp Huegelmeyer, Andreas Nie, Andres Pegam, Collin Rogowski, Claus Rollinger, Timo Steffens, and Wilfried Teiken

Institute of Cognitive Science, University of Osnabrueck, Germany

Abstract. The Dirty Dozen team consists of the player agents and an online coach. The players' low-level-skills and world-model are based on the publicly available code of the CMU-99 team [1]. The team behavior is specified using a new strategy formalization language (SFL), an extension of the standard coach language. This allows to modify the behavior easily and makes the whole team coachable, even for online coaches developed by different teams. We introduce the main concepts of SFL and one possible system with modules that interprete and run the SFL specifications. We also give an outline of the online coach.

1 Introduction

Specifying the behavior of agents in multi-agent-systems is often possible only for domain experts [4]. This paper proposes a new method for specifying a team's strategy in a way that makes adjustments by a human designer or an online coach easy. The Strategy Formalization Language (SFL) is an extension of the standard coach language (Clang) [2] which enables coaches and players of different teams to communicate with each other. SFL extends Clang in terms of expressibility, low-level- and high-level-concepts. The Dirty Dozen team is implemented as an SFL-System which is described below.

2 Strategy Formalization Language System (SFLS)

In order to easily handle coach-messages, SFL completely subsumes Clang and thus retains its syntactical structure. There are five main message types of which we will focus on two in this paper, namely info- and advice- messages. These are syntactically the same, just the semantics differ in that info-messages contain knowledge about the behavior of one of the teams, and advice-messages contain directives about how to behave in certain situations. The syntax of these messages is:

$(\{info \| advice\}\ TOKEN_1\ TOKEN_2 \ldots TOKEN_n)$

where TOKENs are

$(T\ CONDITION\ DIRECTIVE_1\ DIRECTIVE_2 \ldots DIRECTIVE_n)$

T denotes the time-to-live or timespan in which the token will be valid. CONDITION denotes a situation and the DIRECTIVEs denote directives for a set of players. Basically, these tokens are production rules. In the following some of the SFL-extensions to CONDITION- and DIRECTIVE-tokens are described.

A. Birk, S. Coradeschi, and S. Tadokoro (Eds.): RoboCup 2001, LNAI 2377, pp. 543–546, 2002.

2.1 Language Concepts

Clang is not sufficient to implement the behavior of a team. Its concepts are too general, so SFL contains several concepts that are important for implementing teams. It also makes many Clang-advices that may be uttered by an online-coach more elegant and concise.

Due to the extendibility of Clang, SFL was created by adding new concept primitives with clear semantics. First of all, it introduces abstractions of the constant uniform-numbers used in Clang to denote players: variables and symbols for situation-specific player denotation (e.g. ClosestPlayerToBall). These extend the practical expressibility of Clang, because it was not possible to concisely express "mark the ballowner". One would have to list all possible ballowners. In SFL this becomes "If opponent X has the ball, mark X."

A lot more complicated in Clang are advices that refer to different players in certain situations, e.g. the closest player to the ball. Since there is no concept for distances in Clang, the advice would have to simulate distance by several increasing circles around the ball. In SFL it's a short statement using the primitive symbol "ClosestPlayerToBall" instead of constant uniform numbers.

Other SFL-concepts extend the set of situation specification primitives, e.g. by making conditions depend on the stamina of an agent, the ball's velocity or the fact that the ball is interceptable. The set of actions was also extended with new features like "interceptball" and parameters for existing actions, e.g. velocities for positioning-actions and modes for passing.

Since Clang was continually extended, some concepts of SFL, like relative positions, were already included before publishing this paper. For the complete syntax of SFL see the final documentation [3].

2.2 Implementation

The actual implementation of our team is just one possible implementation about how to handle the SFL-specification. The Strategy Formalization Language System works as follows.

The behavior of our team is specified by a set of advice-tokens, i.e. rules with a condition and several effects. In each cycle a module, called matcher, evaluates the conditions of all rules with the world-model of the agent. Rules whose conditions are true in the actual cycle and that contain directives for the agent are labeled as "active".

Most of the time several rules are active. Thus, another module, called selector, has to decide which rule should be executed. The basic idea is that rules are heuristically evaluated on each of the three Clang levels (actions, directives and conditions), being assigned three fitness values that are summed up. So, certain actions seem more promising, e.g. interceptball has a higher fitness than marking on the action-level. Directives refer to different sets of players and the more specific a player set, the higher the directive's fitness on the directives level. E.g. the fitness of a directive that refers to the whole team has less fitness than a subset, which again has less fitness than a situation-specific player-symbol.

Since the rules specified in the team-implementation are fixed, the fitness-assignment is done manually, but automated assignment will be straight-forward.

The third level is based on the conditions and is basically a way to save world knowledge from the rules and incorporate them into the selector. So it is not necessary to specify in a rule that the agent should only mark an opponent if no teammate is already there. Certain common sense heuristics can be used on this level to assign a fitness value to each rule. This has not yet been implemented, so the SFL-rules in our team contain certain amounts of this common sense knowledge explicitly.

The selector will then execute the rule with the highest fitness, unless one of the rules contains the "force"-flag which denotes that this rule should always be executed if its condition is true.

The action is then handed over to the effector-module which encapsulates the strategic actions and executes the appropriate low-level-skills. In the case of a pass-action for example, the effector has to take care that the agent gets close enough to the ball or is turned in the proper direction.

Coach advices are added to the rulebase as they arrive via the communication channel and will be evaluated in the following cycles like the other rules. As of now, the assigned fitness is high, yet fixed. So it is possible for the team designer to specify rules that will not be overwritten by coach advices by assigning a higher fitness than the coach advice fitness.

3 The Online Coach

One of the main concerns of designing the Dirty Dozen online coach was to make it compatible with foreign teams. This was made possible by Clang at RoboCup 2001 for the first time. We use the coach to contribute in two ways: a) opponent/team modelling and subsequentally advising the agents to change their behaviour appropriately and b) facilitating coordination thanks to the coach's central role in communication and the global and noiseless information that it receives.

The coach looks for repeting patterns in the setplay[1] of the opponent, e.g. player positions during goalie-kickoffs. This information is sent to the agents so that they can position themselves strategically in the next setplay. Unfortunately, this requires a lot of inference on the agents' side and one can not rely on the team's designers having incorporated this mechanism into their players.

So a more promising way is to send not only information, but explicit advice about how to behave in certain situations. This encapsulates the inference about tactics and formations on the team level within the coach. E.g. our coach assigns tasks as to which agent should mark which opponent in defense situations. Since the coach does not know a priori which players are defenders, it has to model both teams in order to identify opponent forwards and its own team's defenders. The assignment of defenders to forwards is based on their spatial distribution on the field, so that each player is assigned an opponent that tends to be closeby in most defensive situations. Experiments showed that it is not wise to change only parts of the tactics, because most teams use well-tuned tactics that the coach might not be able to detect in the first place. In the marking assignment the coach therefore also advices new formation positions for the defenders based on the most likely position of the assigned opponent players. So the defensive formation is tuned to work with the marking assignments. Note that this would be hard to achieve in a distributed way if each agent had to decide for itself which player to mark, because the agents have only incomplete and noisy visual information and would have to communicate in order to avoid marking opponents twice or not at all.

4 Future Work

Considering the rather small set of primitive features in both Clang and SFL it seems that playing soccer successfully is not about having the best concepts, but about the domain knowledge regarding how to handle the existing concepts in given situations. So we will elaborate on the selector module to choose between rules more dynamically. This will also lead to a better integration of coach advices which are handled with fixed fitness values now. Because the rule-conditions typically formulated in SFL are more detailed and specific than Clang can express, the rule-selection needs more work.

Until now the players can only handle explicit advices from the coach and ignore opponent modelling information. An inference mechanism that also uses this information, i.e. can infer how to behave given certain information about the opponent's tactics, could also be useful for implementing the coach itself.

Additionally, as the coach contest showed, the coach needs to observe how its own messages influence the team's behavior and react appropriately. The opponent modelling has to be supported by modelling its own team. This year, all coaches kept on sending messages throughout the whole game, although it was obvious that the information only confused the foreign coached team. The online coach and the standard coach language are rather new concepts in the RoboCup domain and need more experiments and evaluation.

References

1. Peter Stone, Patrick Riley, and Manuela Veloso: The CMUnited-99 champion simulator team, In Veloso, Pagello, and Kitano, editors, RoboCup-99: Robot Soccer World Cup III , pages 35–48. Springer, Berlin, 2000.
2. Mao Chen, Ehsan Foroughi, Fredrik Heintz, ZhanXiang Huang, Spiros Kapetanakis, Kostas Kostiadis, Johan Kummeneje, Itsuki Noda, Oliver Obst, Patrick Riley, Timo Steffens, Yi Wang and Xiang Yin: Soccerserver Manual v7. 2001.
3. Sean Buttinger, Marco Diedrich, Leo Hennig, Angelika Hoenemann, Philipp Huegelmeyer, Andreas Nie, Andres Pegam, Collin Rogowski, Claus Rollinger, Timo Steffens, Wilfried Teiken: ORCA project report, 2001, (to appear)
4. Scerri P. and Ydren J.: End User Specification of RoboCup Teams, in RoboCup-99: Robot Soccer World Cup III, Springer. 2000

UQ CrocaRoos: An Initial Entry to the Simulation League

Gordon Wyeth, Mark Venz, Helen Mayfield, Jun Akiyama and Rex Heathwood

School of Computer Science and Electrical Engineering
University of Queensland, Australia
`wyeth@csee.uq.edu.au`

Abstract. The UQ CrocaRoos entered the RoboCup Simulation League for the first time in 2001. The team demonstrated many of the basic functions required to play the simulated game: vision, action and multi-agent planning. The design of the team extends many of the concepts developed for the real robot leagues. This paper describes the fundamental architecture of the team, and the principles for developing the team in the future.

Introduction

The University of Queensland has had and continues to have a presence at RoboCup, first with the RoboRoos [5] and with the ViperRoos [1]. This year the University has also entered a Simulation League team, the CrocaRoos.

The CrocaRoos is furthering the work on the applicability of MAPS (Multi Agent Planning System) and introducing a coach agent. By using MAPS in the CrocaRoos it is hoped to further research into its use on agents with incomplete world models [1], to test its scalability moving from the current five agents to the eleven of the simulator league and explore its ability to work with autonomous agents with limited communication skills.

Multi Agent Planning System (MAPS)

MAPS technology was developed for use in the University of Queensland's small-size league team, the RoboRoos. MAPS has demonstrated very promising results as a general coordination system in both competition and testing environments [2, 3, 4]. It improves coordination among agents by choosing individual goals for each agent that will improve the probability of achieving the team's goal. MAPS coordinates agents through the superposition of potential fields. Each field reflects the probability of positive or negative influence of an environment attribute on the team's goal in the near future. The summation of all the fields maps the best choice of an agent to perform an action and where it should perform that action.

A. Birk, S. Coradeschi, and S. Tadokoro (Eds.): RoboCup 2001, LNAI 2377, pp. 547-550, 2002.

In the small-sized league, MAPS utilizes a world model obtained from an overhead camera, and as such has a complete world model. Even with this world view MAPS considers the world from each player's perspective when making decisions that accomplish coordinated multi-agent plans [4]. Each robot receives a complete world model from the vision system as well as the MAPS command, which when combined with its own reactive navigation routines, enable it to navigate or kick to the desired location.

In the simulation league, MAPS is run on each agent independently relying on the partial world model created from the sensory data from the Soccer Server. MAPS functions in the same as it does in the small-size league, but instead of sending plans to all robots it only sends one, its host's. The problems faced here are similar to those of the ViperRoos [1].

CrocaRoos Agent Base

As MAPS already exists for the small-size league, the CrocaRoos agents are designed to integrate it. As MAPS uses grids of potential fields, the simulated agents had to be able to generate a world model that would create these grids. Figure 1 shows the components of the CrocaRoos agent.

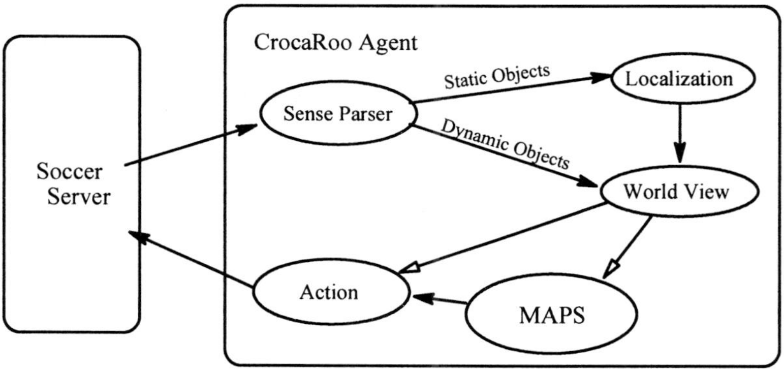

Figure 1: The CrocaRoos Agent Composition.

Developing a World Model

Ultimately the action of the CrocaRoos agent depends on the developed world model. Data from the **see** message received from the server is split into two data structures containing information about two types of objects, static and dynamic. The agent uses the static objects, lines and flags, to locate itself within the MAPS grid. The player determines its direction from lines it sees, and its position on the field from the flags it sees. It averages its position gained from flags seen within ***unum_too_far_length*** to create a grid location.

Dynamic objects, the ball and other players, are located with respect to the sensing agent. All dynamic objects seen remain in an agents memory, those objects sensed in a **see** message are updated, while the memory of unseen objects decays over a series of cycles.

MAPS uses the generated world view and returns an action command, KICKTO, GOTO or DEFEND and the location of where this action is to take place. The Action component then determines the best way to carry out this action.

Communication

The CrocaRoos currently communicate to enhance each agent's world model. Every two **sensebody** cycles each player, in turn, broadcasts its world model. Players hearing this message adjust, where appropriate, their own world model.

Coach Agent

The CrocaRoos coach agent has been developed to comply with the standard coach language. While functionality has been implemented to allow the coach to give reasonable advice on marking players, the primary functionality used in the 2001 simulation coach competition was the allocation of player types. In order to effectively select player types the CrocaRoos coach first allocates players to positions based on their distance from the oppositions goal. Four player positions are allocated: one striker, three offence, three midfielders and three defense. The allocated striker is the player who has the smallest average distance from the opposition's goal. The remaining players are also allocated based on their average position from the opposition's goal.

Once the player positions have been assigned, the coach will select which of the seven possible player types should be allocated to each position. The striker type is chosen with the aim of maximizing speed while still maintaining an accurate kick. The offence type is chosen to maximize speed, while maintaining an acceptable stamina. The midfielders are expected to be required to run for larger portions of the game and are therefore allocated the player type with the maximum stamina. The defense type is selected with the aim of maximizing the distance the player is required to be from the ball before gaining possession of it.

Future Work

At present, MAPS in the CrocaRoos agent only uses dynamic objects it has seen in the last few cycles and then only the static information. To get better performance from MAPS future work will be on determining predicted locations of seen objects, and probable locations of objects outside of the sensory fields.

Current communication between players is also limited to aging static information of the caller. Future work in this area will be to include action plans along with world models, as well as developing algorithm as to which agent is best suited to providing that information.

References

1. Chang, M.M., Browning, B., Wyeth, G.F.: ViperRoos 2000. In: Stone, P., Balch, T., Kraetzschmar, G. (eds.): RoboCup 2000: Robot Soccer World Cup VI Lecture Notes in Artificial Intelligence, Vol. 2019. Springer-Verlag, Berlin Heidelberg New York (2000) 527--530
2. Tews, A. and Wyeth, G.F. MAPS: A System for Multi-Agent Coordination. In: Advanced Robotics, Vol 14 (1). VSP / Robotics Society of Japan (2000) 37--50
3. Tews, A. and Wyeth, G.F. Multi-Robot Coordination in the Robot Soccer Environment. Proceedings of the Australian Conference on Robotics and Automation (ACRA '99), March 30 -- April 1, Brisbane. (1999) 90--95
4. Tews, A. and Wyeth, G.F. Thinking as One: Coordination of Multiple Mobile Robots by Shared Representations. International Conference on Robotics and Systems (IROS 2000) 1391—1396
5. Wyeth, G.F., Tews, A., Browning, B. UQ RoboRoos: Kicking on to 2000. In: Stone, P., Balch, T., Kraetzschmar, G. (eds.): RoboCup 2000: Robot Soccer World Cup VI Lecture Notes in Artificial Intelligence, Vol. 2019. Springer-Verlag, Berlin Heidelberg New York (2000) 527--530

UvA Trilearn 2001 Team Description

Remco de Boer, Jelle Kok, and Frans Groen

Faculty WINS, Department of Computer Science
Intelligent Autonomous Systems Group
Kruislaan 403, 1098 SJ Amsterdam, The Netherlands
{remdboer,jellekok,groen}@science.uva.nl

Abstract. This paper describes the main features of the *UvA Trilearn 2001* soccer simulation team. *UvA Trilearn 2001* is a new team that participated for the first time in 2001. It has been built from scratch and does not contain any code copied from other RoboCup teams. Topics that will be discussed include our architecture, world model and synchronisation method. On a higher level we will talk about an optimal scoring policy and about our fast-play strategy which makes use of heterogeneous players. *UvA Trilearn 2001* finished 5th in the German Open 2001 and reached 4th place at the RoboCup 2001 World Cup.

1 Introduction

The *UvA Trilearn 2001* soccer simulation team was built by two masters students from the Intelligent Autonomous Systems Group at the University of Amsterdam for their graduation project. For two reasons we decided not to copy any code from other RoboCup teams. Firstly, we found that the released source codes did not conform to regular software standards, i.e. they were not well structured and scarcely documented. We therefore thought that it would be more time-consuming to reuse the code than it would be to write it all ourselves. Secondly, we also felt that the lower levels used by the top teams from recent years could be improved in several ways. We thus decided to build a new team from scratch. Much of the initial effort has therefore gone into getting the lower levels to work. We have spent a lot of time on this particular task since we felt that the lower levels in the architecture would be the most crucial for the success of the team. Furthermore low-level imperfections can not be compensated for by high-level behaviour added later on. We had several ideas for improvements in the low-level methods used by top teams from the past, which we have successfully implemented. This has among other things led to an advanced synchronisation method and accurate estimation techniques for positions and velocities. The higher levels were added later on. This resulted in a fast-play strategy and an optimal scoring policy. During the project we have also focused much attention on software engineering issues [2,6] to facilitate future use. This has led to highly modular object oriented code and to a multi-level log system for quick debugging (similar to [8]). Much effort has also gone into documenting our code for future release using the documentation system Doxygen [4]. The main features of *UvA Trilearn 2001* will be discussed in the remainder of this paper.

A. Birk, S. Coradeschi, and S. Tadokoro (Eds.): RoboCup 2001, LNAI 2377, pp. 551–554, 2002.

2 Architecture

Agents are capable of perception, reasoning and acting. We use a multi-threaded architecture which allows our agents to use a separate thread for each of these three tasks. In this way we minimize the delay caused by I/O to and from the server so that agents can spend the majority of their time thinking about their next action [5]. We use a 3-layer architecture for each agent. The threads used for perception and acting can be seen as the bottom layer which hides the soccer server details from the other layers. On top of this we have the *Skills Layer* which uses the functionality offered by the layer below to implement the different skills of each player (e.g. marking). The *Control Layer* then selects the best possible action from the *Skills Layer* depending on the current world state.

3 World Model

The world model of each agent is a probabilistic representation of the world state based on past perceptions. For each object an estimation of its position and velocity is stored (among other things) together with a confidence value which indicates the accuracy of the estimation. We have experimentally compared different methods for estimating positions and velocities of the dynamic objects in the simulation and used the best of these methods based on statistical results. The world model is updated when new information about the world is received by the agent and thus always contains the last known information. Objects which are not seen are also updated based on their last observed velocity. The confidence in the estimate decreases however for each cycle in which the object is not seen. Our agents also use communication to improve the accuracy of their world model. Depending on the position of the ball, the agent that has the best view of the field communicates his world model to nearby teammates. The teammates that hear this message then update their world model using the communicated information about the parts of the field that are currently not visible to them. Experiments have shown that on average the world model of an *UvA Trilearn 2001* agent contains up-to-date information about 17 players on the field. In addition, the world model also contains various methods which use the low-level world state information to derive higher-level conclusions.

4 Synchronisation

Synchronisation is an important issue in the soccer server, since actions that must be executed in a given simulator cycle must arrive at the server during the right interval [3]. We use an advanced synchronisation method which in each cycle determines the optimal moment to send an action to the server. This moment depends on the arrival times of the different server messages and thus changes from each cycle to the next. Furthermore our method guarantees that the action sent is always based on the latest information from the server. After sending an action it is checked whether the server has actually performed it. When this is not the case (i.e. the action did not arrive in time) no action is sent in the next cycle to avoid a *clash*[1]. Experiments have shown that our synchronisation method clearly outperforms several others used in the past. During a full-length match consisting of 6000 cycles the number of *holes*[2] for each agent is less than 0.2% of the total number of cycles and the number of *clashes* is less than 0.01%.

5 Optimal Scoring

Our agents use an optimal scoring policy [1] which determines the optimal shooting point in the goal together with an associated probability of scoring when the ball is shot to this point. This probability depends on the position from which the ball is shot and on the position of the goalkeeper. The optimal scoring problem was solved by decomposing the problem into two subproblems. The first subproblem is to determine the probability that the ball will enter the goal when shot to a specific point in the goal from a given position. This probability was estimated by performing an experiment in which a player was placed straight in front of the goal and the ball was shot to the middle of the goal from different distances. From the resulting data set it was possible to determine the average deviation of the ball given the distance that the ball had travelled. Using this function we could calculate the probability that the ball would enter the goal when shot at a certain point from a given position. The second subproblem was to determine the probability of passing the goalkeeper in a given situation. This was estimated by performing an experiment in which a player shot the ball in a straight line while a goalkeeper would be placed in different positions relative to this player. A data set was formed by recording several values for each situation together with a boolean indicating whether the goalkeeper had intercepted the ball or not. From these data it was possible to extract a function which determined the probability of passing the goalkeeper. Combining the solutions to both subproblems led to the optimal scoring policy used by the agents.

6 Team Strategy

The main philosophy of *UvA Trilearn 2001* is to keep the ball moving quickly from each player to the next and preferably in a forward direction. In addition, our agents often try to pass the ball into the depth in front of the wing attackers at the side of the field thereby cutting through the opponents' defense and disorganizing their team. The effectiveness of this strategy is greatly enhanced by the use of heterogeneous players on the wings. Although these players become tired more quickly, it is their faster speed which makes the difference. We have developed a formula for determining the quality of a heterogeneous player. For

[1] A *clash* occurs when two actions are sent during one cycle. This situation is undesirable since the server then randomly chooses one for execution.

[2] A *hole* occurs when no action is sent during a cycle. This is also undesirable since missing an action opportunity may lead to the opponents gaining an advantage.

each player type offered by the server, this formula returns a utility value based on the player parameters. In this way we can determine which player types are best suited for which positions on the field. The standard formation used by the team is a 4-3-3 formation. Inside this formation each player chooses a strategic position determined by a weighted sum of his home position in the formation and the current position of the ball which serves as an attraction factor [7,9]. In general, the behaviour of each agent depends on where the ball is located on the field. We have divided the field into different areas and the action that an agent chooses depends on the area in which the ball is located.

7 Results

UvA Trilearn 2001 has participated in two international robotic soccer competitions. The first one was the German Open held in Paderborn in June of this year. In this competition we used a very simple high-level strategy due to the fact that up to that point we had spent most of our time on the lower levels. Despite this, we still managed to reach 5th place in the tournament. In the next two months we worked on our high-level strategy and on the optimal scoring policy. In the beginning of August we then participated in the RoboCup World Championship in Seattle. During this competition it was clear that the team had improved a lot since Paderborn. *UvA Trilearn 2001* finished in 4th place.

References

1. Remco de Boer, Jelle Kok, Nikos Vlassis: An Optimal Scoring Policy for Simulated Soccer Agents in the RoboCup Simulation League. To appear. Currently at http://gene.wins.uva.nl/~jellekok/robocup (2001).
2. Frederick P. Brooks Jr.: The Mythical Man-Month: Essays on Software Engineering. 20th anniversary edition, Addison-Wesley (1995).
3. Marc Butler, Mikhail Prokopenko, Thomas Howard: Flexible Synchronisation within the RoboCup Environment: a Comparative Analysis. In Proceedings of the RoboCup-2000 Workshop (2000).
4. Dimitri van Heesch: Documentation System Doxygen. At http://www.doxygen.org (1997).
5. Kostas Kostiadis, Huosheng Hu: A Multi-Threaded Approach to Simulated Soccer Agents for the RoboCup Competition. In Proceedings of the 16th IJCAI RoboCup Workshop (1999), 137-142.
6. Roger S. Pressman: Software Engineering: a Practitioner's Approach. 4th edition, McGraw-Hill (1997).
7. Luís Paulo Reis, Jośe Nuno Lau, Luís Seabra Lopes: FC Portugal Team Description Paper. At http://www.ieeta.pt/robocup/archive.htm (2000).
8. Patrick Riley, Peter Stone, Manuela Veloso: Layered Disclosure: Revealing Agents' Internals. In Proceedings of the Seventh International Workshop on Agent Theories, Architectures and Languages (ATAL-2000).
9. Peter Stone: Layered Learning in Multi-Agent Systems. PhD thesis, Computer Science Department, Carnegie Mellon University, Pittsburgh, PA (1998). Available as technical report CMU-CS-98-187.

Zeng01 Team Description: Formation Decision Method Using Game Theory

Takuya Morishita[1], Hiroki Shimora[1], Kouichirou Hiratsuka[1], Takenori Kubo[1], Kyoichi Hiroshima[1], Raiko Funakami[1], Junji Nishino[2], Tomohiro Odaka[3], and Hisakazu Ogura[3]

[1] Graduate School of Engineering, Fukui University, 3-9-1 Bunkyo, Fukui City, Fukui, Japan, zeng@rook.fuis.fukui-u.ac.jp

[2] Faculty of Electro-Communications, The University of Electro-Communications, Chofugaoka 1-5-1, Chofu City, Tokyo, Japan

[3] Faculty of Engineering, Fukui University, 3-9-1 Bunkyo, Fukui City, Fukui, Japan

Abstract. The combined team, Zeng01, have players developed by several subject of investigation. We have analyzed better player combination based on game theory. We look up the formation property through changing a opposite team strategy on every games. Comparing formation property on every games, we decide own team formation.

1 Introduction

Zeng01 team consist of many players who are accomplishments of each researchers[1, 2, 3, 4]. Decision making based on game tree search, each study is parameter tuning by GA, realtime learning by GA, circumstantial judgment routine by GA, optimization of recognitive routine by GA, flexible agent making by fuzzy decision making system. Each program (player agent) was developed individually. Player agents are developed with each role such as FW, MF, DF, and GK, to be constituted team easily. Accordingly, our team have many independent player agents that was created by different programmers respectively. These players are four kinds of goalkeepers, eighteen kinds of Defenders, twenty one kinds of Mid-Fielders and twelve kinds of Forwards.

It is necessary to decide a formation and combination of players. It is difficult to decide player combination suitability.We prepared the test team in order to decide formation. Probing Zeng01 team property, we decide our team formation and player combination. In this paper, we proposed a method for decising formation using game theory[5].

2 Basic Skills

Most of programs in our team was based on the published source code (CMUnited99[6]). In this year, client programs was changed to use a library which was developed newly. This library was based on the old library that Zeng's

A. Birk, S. Coradeschi, and S. Tadokoro (Eds.): RoboCup 2001, LNAI 2377, pp. 555–558, 2002.

goalkeeper had used. An action (i.e. KEEPBALL, DRIBBLE, SHOOT, PASS, etc) was pushed into queue. Accordingly, it is easy to execute complex continuous action.

3 Implementation and Experimentation

We prepared a test team which has two strategies; center-breakthrough strategy and side-breakthrough strategy. Playing game with the mixed strategy, we look up the formation and/or team characteristic.

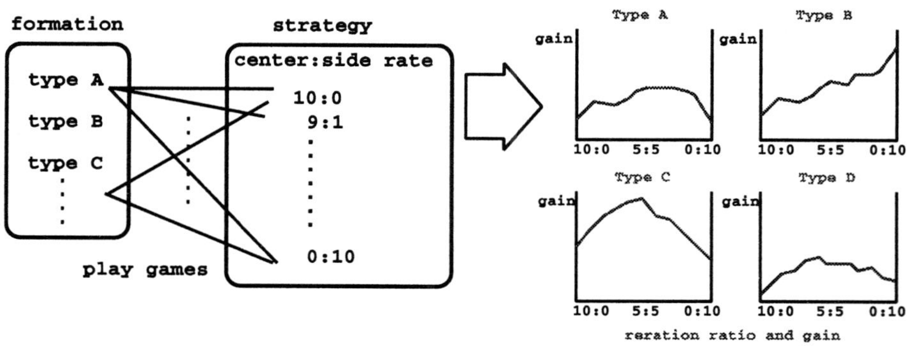

Fig. 1. Relation of mixed ratio and estimate value

Figure1 shows relation of ratio and gain obtained from games with changing strategy ratio. Doing games with changing strategy ratio, against team formation type, we obtain the relation with each formation type. Existence of saddle point and higher gain imply better formation.

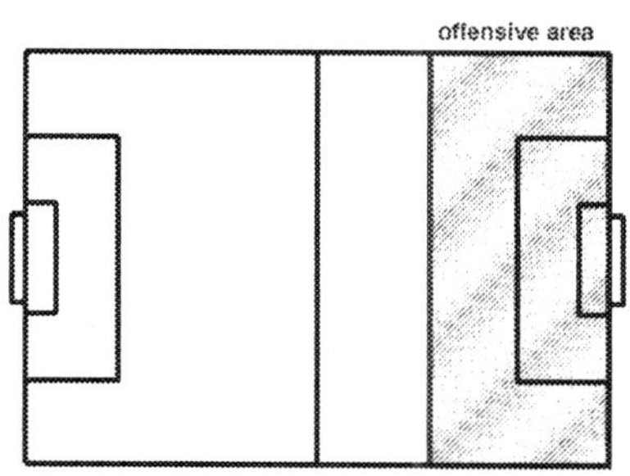

Fig. 2. Offensive area (over 20m–line)

If there was ball in a offensive area, the estimation of the match increased. The offensive area was shown in Figure2.

Table 1. Results of competition in Seattle (Zeng01 is left side)

Opponent	Score	Offense Ratio(%)	Control Ratio(%)
ATTUnited01	0-0	28.0	61.1
MRB	0-0	19.1	55.3
DrWeb	4-0	62.4	56.1
AnderLecht	7-0	63.8	61.2
RoboLog2k1	0-0	17.6	51.4
living_systems	2-0	25.4	57.2
WrightEagle2001	0-0	20.6	55.8
Brainstormers01	0-4	20.7	39.5
Gemini	4-0	50.7	69.0

Table1 shows results in Seattle. Two ratios are calculated by formulas following:

$$OffenseRatio = \frac{T_0}{T_{all}}$$

$$ControlRatio = \frac{T_1}{T_{all}}$$

T_0:The time of a ball existed in the offensive area(Fig.2).
T_1:The time of keeping control of a ball.
T_{all}:The time of whole game.

Let us begin our analysis by comparing score and shown ratios. It is likely that there is a close relation between these ratios and results of matches. In this result, although this statistics are not completely reliable, it seems reasonable to suppose that those ratios used to estimate match. We use these validation in order to decide team formation. In Seattle, Zeng01 formation was decided by human hand coding. The reason for this is that this work was not completed yet and an examination was not performed.

4 Summary and Future Works

In this paper, we proposed formation decision method using game theory. In the future, we will be working on the improvement of the functionality of the team, and we will finish implementation completely. Moreover, we will apply this evaluation method to decide formation with GA.

References

[1] Nishino, J., Kawarabayashi, T., Morishita, T., Kubo, T., Shimora, H., Aoyagi, H., Hiroshima, K., Ogura, H.: Zeng99 : Robocup simulation team with hierarchical fuzzy intelligent control and cooperative development. In Veloso, M., Pagello, E., Kitano, H., eds.: RoboCup-99:Robot Soccer World Cup III, Berlin, Springer Verlag (2000)

[2] Morishita, T., Kawarabayashi, T., Nishino, J., Aoyagi, H., Kubo, T., Shimora, H., Hiroshima, K., Ogura, H.: Zeng00 : The realization of mixed strategy in simulation soccer game. In Stone, P., Balch, T., Kraetzschmar, G., eds.: RoboCup-2000:Robot Soccer World Cup IV, Berlin, Springer Verlag (2001)

[3] KUBO, T., NISHINO, J., ODAKA, T., OGURA, H.: Evolutional acquistion of condition recognizer in the robocup soccor agent. In: Proceedings of Fuzzy, Artificial Intelligence, Neural Networks and Soft Computing (FAN'99). (1999) 155–158

[4] Kubo, T.: Gnez: Adapting knowledge to the environment with ga. Technical report, RoboCup (2000)

[5] Smith, J.M.: Evolution and the Theory of Games. Cambridge University Press (1982)

[6] Stone, P., Riley, P., Velose, M.: The CMUnited-99 champion simulator team. In Veloso, M., Pagello, E., Kitano, H., eds.: RoboCup-99:Robot Soccer World Cup III, Berlin, Springer Verlag (2000)

4 Stooges

Jacky Baltes

Centre for Imaging Technology and Robotics
University of Auckland
Auckland, New Zealand
j.baltes@auckland.ac.nz

1 Introduction

The 4 STOOGES are a small sized ROBOCUP team, which grew out of the ALL BOTZ, a team which competed previously at ROBOCUP-99 and ROBOCUP-00.

One of the main differences is that the 4 STOOGES are a team of fully autonomous robots. All sensors, actuators, and power supply (CMOS camera, and battery) as well as all processing is housed on the robot itself.

The localization, path planning, and task planning problems for local vision teams are more difficult than those for global vision robots. However, an autonomous robot is more realistic for applications outside of the ROBOCUP domain. Therefore, the development of robust solutions to the above mentioned problems is very important to drive the development of practical mobile robotics.

This was the second year that the 4 STOOGES entered the ROBOCUP competition.

In 2001, the main improvement has been the use of demosaicing to quadruple the resolution of the image capture to 160x120 without increasing the processing overhead.

This demosaicing performed extremely. The robots constantly were able to recognize the ball at distances up to 50cm.

2 Team Development

Team Leader: Jacky Baltes

Team Members:

Jacky Baltes
- Uni. of Auckland
- New Zealand
- Team Leader
- attended

Matthew Painter
- Uni. of Auckland
- New Zealand
- Student
- attended

Andrew Thomson
- Uni. of Auckland
- New Zealand
- Student
- attended

Peter Yu-shan Cheng
- Uni. of Auckland
- New Zealand
- Student
- attended

Web page http://www.citr.auckland.ac.nz/~jacky

A. Birk, S. Coradeschi, and S. Tadokoro (Eds.): RoboCup 2001, LNAI 2377, pp. 559–563, 2002.

3 Hardware Platform

In 2001, we redesigned the hardware platform. Instead of toy cars with Ackerman steering, we used a differential drive design. However, true to the spirit of the 4 Stooges and All Botz, we used cheap components. The driving mechanism is based on a twin gear box available from Tamiya, which can be bought for less than $20.00.

The cars are controlled by an Eyebot controller. The Eyebot controller is a MC68332 processor based controller, developed by Prof. Thomas Bräunl from the University of Western Australia.

The Eyebot controller includes a parallel interfaces, a serial interface, A/D converters and 12 servo control outputs and two motor control circuits. It also provides an interface for the Radiometrix serial transceiver. The biggest advantage for small robotics projects is that a CMOS camera is available for the controller. The Eyebot controller also comes with some libraries for image capture and processing.

Fig. 1. An image of one of the robots of the 4 Stooges. The Eyebot controller is mounted on the top.

Using the standard RoBios developed by Thomas Braunl, the CMOS camera provides a 80x60 image with 24 bit colour information at a maximum of three frames per second. The colour information is not very good, since the camera uses a cheap Bayer RGB pattern [2].

In 2000, we developed a new image capture routine which was able to capture images at up to 30 frames per second [1].

In 2001, we used demosaicing to quadruple the resolution of the camera. Our demosaicing technique is described elsewhere in this book in more detail. The main idea is to infer missing colour information from the Bayer pattern.

4 World Model

Localization of the robots is a difficult problem for the local vision teams in the RoboCup small sized league. Other leagues using local vision added special corner or side markers to make localization easier. Since the small sized league is intended mainly for global vision teams, it does not make any such concessions.

The only two distinguishing features in the domain are the two goals on either side of the playing field. The problem is that the 4 Stooges are not able to see the goals in most positions.

Since the 4 Stooges do not have shaft encoders, they can not use dead reckoning for the periods where the robots are not able to observe the goals. Instead, the 4 Stooges use the walls to recailbrate their relative short term motion. Note that by observing the wall, only the orientation angle can be accurately determined.

The 4 Stooges recailbrate their orientation whenever they are able to observe a goal. The orientation is updated using the relative change of orientation from walls and lines on the field. If no feature can be detected by the robot, a "best guess" of the motion of the robot is used.

5 Skills

The code for the field players is almost identical. The only difference is their home position. The team strategy employed by the 4 Stooges uses one defense player, a left striker, and a right striker.

The goal keeper starts out on the goal line and attempts to stay on this line as much as possible. It uses a simple feed forward control to keep the goalie between the ball and the goal. If it can not see the ball, it moves back towards the home position in the center of the goal, since this provides it with the best view of the playing field.

The strikers always approach the ball if they can see it. If a striker can see the goal and the ball, it will attempt to shoot at the goal (if the direct line through the ball points at the goal) or will try and maneuver itself behind the ball. If the robot can not see the goal, but has recently recalibrated its orientation, it will use this information to either kick the ball into the opponent's direction (if it is facing the opponent's goal) or to dribble the ball (if it is facing its own goal). In all cases, it attempts to keep the ball away from the opposition.

6 Special Team Features

The original Eyebot controller was limited to five frames per second, because of the large interrupt latency for each byte transferred.

The 4 STOOGES changed the camera interface routine to use a carefully timed polling and were able to increase the frame rate to 30 frames per second. Since the source code for the Eyebot BIOS is not freely available, we had to disassemble the BIOS and "hack" the BIOS to support the higher frame rate.

Another features of the 4 STOOGES was that a lot of the image processing code was based on the video server of the ALL BOTZ, a global vision team also from the University of Auckland.

7 Conclusion

This year was the second year that the 4 STOOGES entered the competition. Clearly, a local vision team is solving a more challenging problem than a global vision team. It comes therefore as no surprise that the 4 STOOGES were not competitive against their global vision opponents.

Of more interest to the 4 STOOGES were the results of the local vision derby which was held for the second time this year. The games between the local vision teams have significantly improved from last year. Our robots were able to see the ball consistently if it was less than 50cm away.

This improvement has been mainly due to the improved resolution and speed of our image capture routines. Instead of looking for four or less pixels (the size of the ball in x 80x60 image), we are now looking for 16 pixels. This means that we are much less susceptible to noise in the image.

Current development for the 4 STOOGES centers around improved team play. We plan to add wireless communication to our robots so that they can communicate during the game.

References

1. Jacky Baltes. 4 stooges. In Peter Stone, Tucker Balch, and Gerhard Kraetszchmar, editors, *RoboCup-2000: Robot Soccer World Cup IV*, pages 519–522. Springer Verlag, Berlin, 2001.
2. Bryce E. Bayer. Color imaging array. U.S. Patent 3,971,065.

5dpo Team Description

Paulo Costa*, Armando Sousa**, Paulo Marques***, Pedro Costa***, Susana Gaio***, and António Moreira†

Faculty of Engineering, University of Porto (FEUP)
R. Dr. Roberto Frias, S/N / 4200-465 Porto / Portugal
{paco, asousa, pamarques, pedrogc, sgaio, amoreira}@fe.up.pt
http://www.fe.up.pt/robosoc

Abstract. The 5dpo team presented a solid set of innovative solutions. The overall workings of the team are presented. Mechanical and electronic solutions are explained and closed loop working is discussed. Main innovative features include I-R communications link and circular bar code for robot tracking. Low level control now presents a dynamics prediction layer for enhanced motion control. Team strategy is also new and a multi-layered high level reasoning system based on state charts allows for cooperative game play.

1 Robot Design and Communications

The mechanical design of our robots for 2001 is essentially the same as the one used last year in the Euro-RoboCup Amsterdam 2000 and Padderborn Euro-RoboCup 2001. Electronic design has undergone a serious streamlining featuring a single hardware electronics board.

As in previous implementations, our robots are fitted with two differential wheels. Separated stepper motors drive the wheels independently. There is no third wheel and a pair of pods sustains the robot. A castor would result in a more complex mechanical design as well as an increased uncertainty in its dynamical model.

Embedded Ni-Cd batteries presently power the robots for more than 1 hour. The motors are driven by two H-bridges that are directly powered from the batteries. The on board controller for the motors is an 8-bit RISC microcontroller (Atmel AVR90S8515). All digital circuitry gets its power from a switching regulator.

2 Communications

There is another microcontroller, an Atmel AVR90S2313 that handles all real time communications. There are two small single frequency Radio Frequency

* Team Leader and Professor at FEUP
** Lecturer at FEUP
*** Researcher at FEUP
† Professor at FEUP

A. Birk, S. Coradeschi, and S. Tadokoro (Eds.): RoboCup 2001, LNAI 2377, pp. 563–566, 2002.

(RF) modules (433 MHz and 418MHz). The software can automatically switch the active RF module. To minimize the latency of the communications channel, we transmit a lot of small packets. An additional communication channel is provided through an InfraRed (IR) link. We successfully used the IRDA HSDL-1001 chips to communicate with the robots in spite of the heavy lighting conditions. The emitting LED's are at the camera level, pointing down. On the receiving side an automatic gain controller and adequate filtering are necessary to get reliability. The HSDL-7001 chip is used to perform Encoding/Decoding. Alignment problems and also power concerns led to having 3 emitting nests of a total of 8 LED's HSDL-4220. These LED's were submitted to bursts of 1.6 μsecond under a current of 0.5 Amps. The information transmitted via IR and RF is the same but packing is different so two different serial ports are used to transmit the information form the external global controller. All transmission methods send a packet that specifies the target speed for each wheel for all robots. Each packet also has the On / Off state of the kicking device and a one-byte position of the ball handler actuator but only for one robot (to keep packet size small). This ball handler is a guiding bar based on a servo that either guides the ball or allows the previously spinning kicking device to hit the ball thus creating a very powerful kick.

3 Robot Software

Robot embedded software for both microcontrollers has undergone a redesign process during the porting of the code from assembly to C. The C code is now compiled with the freeware AVR-GCC program. This was done in order to allow for code understandability and maintainability.

Communication channels are now RF and IR. Special packets allow for remote robot configuration such as robot number, future receiving frequency, etc.

Embedded software accepts orders for both motors in several categories such as speed and number of steps to be ordered and if the orders are to be carried immediately or after the current command has ended.

4 Vision and Control

The vision system is based on two general consumer PAL camcorders, one for each mid-field. A different PC processes each image with a vision board based on a BT878 vision acquisition system. This allows for parallel processing of the images whilst using cheap hardware and also maintaining standard Operating System device drivers. The image processing software has been ported to Delphi under Windows. Socket communication has been introduced in order to provide communication of the 2 processed images to another independent program, the decision module that may be in another PC.

The global state of the system is updated based on the vision. A fuzzy system classifies the color of each pixel and aggregates them in contiguous groups. The observed groups are matched and that information is incorporated in the state resorting to a set of Kalman filters tuned to the dynamics of each kind of object. By observing the present system state as well as a global mid-term strategy, short-term orders are generated and sent to the players. Robot intelligence is based on such a three layered state diagram engine. This means that each robot has a Role that may require a number of Tasks that, in turn, may require a mesh of different Actions.

The loop is closed through the cameras, as seen in Figure 1. It must be stressed that while the "sampling" frequency of this loop is 25 Hz, there is some intrinsic lag that degrades its optimal performance. The PAL signal takes 20 ms to deliver each frame, then some time is lost processing it, the decision subsystem must decide the new course of action and there is also the time spent sending the orders over the RF channel to the robots. System overall performance benefits from a prediction math layer where low level control is calculated taking into consideration known system lag.

The vision system uses a round-shaped bar code to keep track of which robot is which (as in Figures 2 and 3). Around the circular color of the team (yellow or blue) there is room for 6 bits of black or white sectors of 60 degrees. As only 5 robots are needed on the field at the same time, from all the possible 6 bit combinations, the chosen combinations were those that had the maximum transitions and those that are not a rotation of previously chosen codes. Using the team color of the center of the bar code and going through the pixels at a known radius from the center, we are able to retrieve a binary code of a robot as well as its orientation. Orientation noise is less then 4 degrees for our system.

Team tactics are quite stable. In any case, the defense robots stand in alignment in such a way that the robot that holds the ball can't easily shoot for goal. Team behavior is based on robots changing roles on certain conditions. Certain robot roles and actions have priority over others and thus co-operation is based on well-defined priorities. This prevents robots from the same team from harassing each other.

Speeds for the stepper motors are risen above the maximum stable speed using current calculated speed and known motor acceleration performance. This means that under certain conditions the stable speed of 0.5 m/s will be surpassed to over 1.2 m/s.

5 Final Remarks

The new IR link proved it self very useful to allow for extra practice time during the competition and also as a redundant link during the games.

Driving the stepper motors on the non-stable region is also very important to get near the ball quickly and allows for impressive speed displays with our inexpensive motors.

The prediction layer for the low-level control allows for very fast and exact dynamic control.

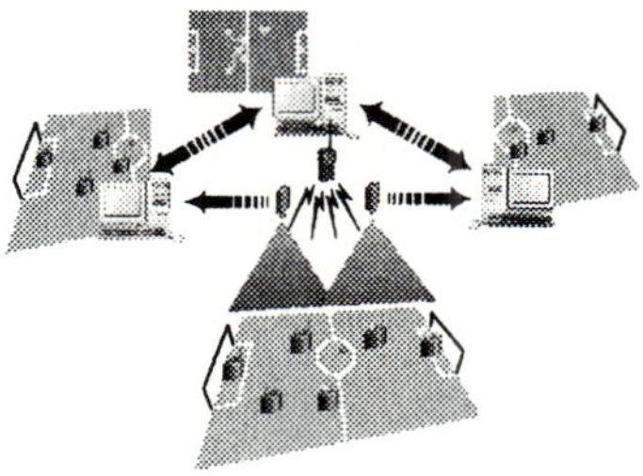

Fig. 1. Diagram of the closed loop workings of the team

Fig. 2. Human and robot team also showing robot's bar codes

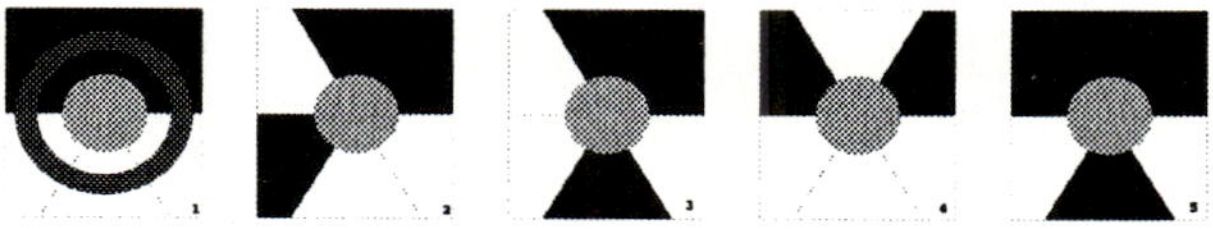

Fig. 3. Actual chosen bar codes and a sample scan circle on number 1

The use of two cameras allowed for much less ball occlusion and the global state obtained through Kalman filtering proved very reliable and easily merges all available information.

The circular bar code also worked fine simultaneously identifying without ambiguity the number of the robot as well as its orientation.

The priority-driven three layered control system for the high level reasoning of the team proved powerful yet very easy to maintain and debug.

The overall performance of the team is very good and several small improvements will be carried out but we are planning a whole new platform with omni-direccional drive for the future.

CM-Dragons'01 - Vision-Based Motion Tracking and Heteregenous Robots

Brett Browning, Michael Bowling, James Bruce, Ravi Balasubramanian, and Manuela Veloso

School of Computer Science, Carnegie Mellon University, Pittsburgh PA 15213, USA
{brettb, mhb, jbruce, bravi, mmv}@cs.cmu.edu,
http://www.cs.cmu.edu/~robosoccer/

1 Introduction

At Carnegie Mellon, we have developed several small-size robot teams that have helped us to investigate a variety of aspects of the small-size RoboCup competition. The CM-Dragons'01 is our new team complete with new hardware and sensing and behavior-processing algorithms. Although still in the development phase, a number of modules have been developed that we feel can contribute to the RoboCup community. In this paper, we briefly describe our vision and tracking modules, the new robot hardware and our new communications modules. Our primary interest is on presenting our advances in modeling and prediction using an Extended Kalman-Bucy Filter (EKBF) that tracks the ten robots and the ball through vision. We identify that Kalman-Bucy filters are susceptible to white noise caused by misidentifications. Within CM-Dragons'01, we developed a new approach, Improbability Filtering, that addresses this problem in a computationally efficient yet principled manner.

Throughout this paper, we assume that the reader is familiar with the general structure of small-size robot soccer applications (for further details, please see other articles in this book or for example [6]).

2 Vision

The image processing module consists of the low-level CMVision color segmentation library and high-level vision algorithms that analyze the colored regions reported by CMVision to identify robots, their orientation, and the ball. CMVision is public software available under GPL [2]. The vision algorithms work as follows. CMVision segments pixels using a 3D lookup table, which is a subsampled version of the color space with 8 levels for Y, and 32 levels each for U and V. The YUV table is generated off-line using a GUI editor. Identically segmented pixels are collected into horizontal runs, which in turn are combined into regions. The run unification stage calculates statistics for use by the high-level vision algorithm.

Identifying robots is a two-stage process, namely: (i) suitable colored regions, as determined by the bounding box, are mapped into world coordinates using a

A. Birk, S. Coradeschi, and S. Tadokoro (Eds.): RoboCup 2001, LNAI 2377, pp. 567–570, 2002.

precalculated camera model, and given confidence ranking based on their area and field position; (ii) robot identification and orientation is accomplished by scanning for the closest orientation markers (pink patches) and identity markers (white squares) within a suitable distance. The orientation markers are offset forward on the robot so that the orientation has only one solution.

Running at 60Hz with interlaced 640x240 YUV 4:2:2 images, the whole vision module takes about 40% of the processor on an Athalon 1.3GHz PC. With the loop closed, considering vision to action, the system has a latency of about 100ms corresponding to 6 to 7 fields. Additionally, vision suffers the usual problems of intermittency, noise due to pixelization, and occasional misidentifications. Tracking such data over time to produce robust position and velocity estimates is the job of the tracking software discussed next.

It is our experience, that global vision processing, as permitted in this league, offers difficult challenges. Robot identification and orientation need to be particularyly robust to false positives. We describe next an algorithm to achieve this robustness. The behaviors can only produce successful results if the state of the world can be processed and modeled with reliability and accuracy.

3 Tracking

To track the ten robots and ball, and to make predictions regarding the future state of these objects, we use multiple independent EKBF's combined with Improbability Filters (ImpF). EKBF's are widely used state estimators that provide optimal state tracking for non-linear systems with Gaussian noise components [4], [7]. We used independent EKBF's, based on the assumption that tracking and dynamics noise between objects was negligible, to track the robots and the ball. Due to space considerations we will not show the full equations that describe the tracking module. Instead we refer the reader to [1] and our earlier work [5]. In short, the EKBF's are a straight implementation of the algorithm with a dynamical model of the robots that reflects their kinematic motions, acceleration limits and velocity commands. Similarly the ball EKBF uses a dynamics model that reflects the physics of its motion including the friction of the field and the motion along the inclined walls.

EKBF's provide an optimal estimator of the system state provided the tracked system has only Gaussian noise components. With the appropriate choice of parameters, EKBF's are robust to moderate violations of the Gaussian noise assumption. However, they are not robust in the short term to white noise in the form of misidentifications. In such cases the filter jumps across the field causing havoc to the robot behaviors and ball prediction mechanisms. We devised and implemented a new approach, called Improbability Filtering (ImpF) [1], to reject false-positives thereby overcoming the white noise problem.

The ImpF rejects false-positive, cases where the vision module reports confidently but incorrectly, by determining the likelihood of observing the reported observation given the current model as specified by $\widetilde{x}_k$ and P_k the estimated state and covariance, respectively. The observation likelihood is just the conditional probability density function (pdf) $P\left[z_k|\widetilde{x}_k, P_k\right]$ evaluated for the reported observation. Given the Gaussian model stored by the EKBF we have :

$$C_k = H_k^T P_k H_k + R_k \tag{1}$$

$$P\left[z_k|\widetilde{x}_k, P_k\right] = \frac{1}{\left(2\pi|C_k|\right)^{\frac{n}{2}}} e^{\frac{-1}{2}\left(z_k - H_k\widetilde{x}_k\right)^T C_k\left(z_k - H_k\widetilde{x}_k\right)} \tag{2}$$

All terms are identical to their use in [7] and are described more fully in [1]. We have H_k as the Jacobian of the observation function $z'_k = h\left(\widetilde{x}_k\right)$ with respect to the state variables, z_k is the observation, R_k is the observation noise covariance, $\widetilde{x}_k$ and P_k are the state estimate and covariance, k is the time step, and n is the number of state variables (4 for the ball and 6 for the robots).

Observations that are sufficiently likely, as determined by a threshold set to reject a majority of false-positives, are accepted and incorporated into the EKBF estimates while low likelihood observations are ignored. This approach operates efficiently to remove white noise from the vision output and produces good, robust tracking behavior.

We use the EKBF's for both filtering and prediction. Clearly, the EKBF will filter small Gaussian noise from the vision output and with the ImpF it will also remove false-positives. In addition to filtering, we use the EKBF dynamics equations to create future state estimates and their associated covariances. Essentially, the dynamics update is applied repeatedly corresponding to the number of time steps into the future the estimate is required. The prediction capability allows us to overcome latency issues. In short, we predict ahead the estimated latency and use the predicted values for the robot behaviors and motion control.

4 Hardware

The CM Dragons are a heterogeneous team consisting of two different types of hardware platforms; a fast differential drive robot and an omni-directional robot. Both robots use identical electronics consisting of a processor board with associated power circuitry and a communications board. The differential drive robot, hereafter the diffbot, is capable of high acceleration, enough to break stiction while in motion, with a top speed of 2.5m.$^{-1}$. Given the dimensions of the field and the limited acceleration enforced by wheel slip, the robot can barely approach its top speed in normal play. The omni-directional robot, hereafter omnibot, uses three roller wheels from North American Roller Products in a Y arrangement. The omnibot has lower acceleration capabilities but a higher top speed with maximum figures of 3m.s^{-1} and 3m.s^{-2}, respectively.

The robots use identical electronics; a processor board and communications board. The processor board, the brain of the robot, is based on a TMS320LF2407 16-bit, 30 MIPS DSP from Texas Instruments. The DSP offers both computational power, low power consumption, and lots of functionality via its on-chip peripherals. In particular, the on-chip timer module facilitates generation of dead-band PWM signals to drive the motors with minimal shoot-through current losses. The communications board is based around the 50kbps commercially available lynx^{TM} HP-II modules. These dedicated transmitters and receivers operate in the International Science and Medicine (ISM) band from 900MHz-1GHz with eight selectable operating channels. These devices provide two key advantages over the common Radiometrix module; a) the ISM band is away from the heavily used 418/433MHz band and b) the modules offer multiple channels on a single chip. The reliability of the modules is excellent with our testing reporting packet losses of 1% or less at 38.4kbps where a packet consisted of a start byte, id byte, frame byte three data bytes and a stop byte. Software wise, the robots currently act as dumb slaves and operate velocity PID loops where the set point for the PID loops is obtained from the received RF packets.

5 Conclusions and Future Directions

CM-Dragons'01 represents a new research effort on teams of heterogeneous robots. We briefly presented our EKBF-based tracking system and the novel ImpF algorithm used to overcome white noise caused by misidentifications.

At RoboCup-2001, the different efforts within CM-Dragons'01 were in a preliminary phase of development and their integration was not fully operational. Much work remains to develop the CM-Dragons platform into a truly integrated and competitive multi-robot system. In particular, we are finishing the robot platforms and high-level vision primitives. Finally, our future aims are to step towards developing automated strategies where the team changes playing mode as a function of the overall state of the game.

References

1. Browning, B., Bowling, M., Veloso, M. "Rejecting false-positives using Improbability Filtering and the Kalman Filter", International Conference on Robotics and Automation 2002 (ICRA 2002), submitted.
2. Bruce, J., Balch, T., and Veloso, M. "Fast and Inexpensive Color Image Segmentation for Interactive Robots", Proceedings of IROS-2000, October, Japan, 2000.
3. D'Andrea, R, Kalmar-Nagy, T, Ganguly, P, Babish, M. "The cornell robocup team", Stone, P, Balch, T., Kraetzschmar (Eds), RoboCup-2000: Robot soccer World Cup IV, Springer Verlag, Berlin, 2001, 41–51.
4. Kalman, R. E., Bucy, R. S. "New results in linear filter and prediction theory", Journal of Basic Engineering, March, 1961, 95–108.
5. Kwun, H, Veloso, M "Reactive visual control of multiple non-holonomic robotic agents", Proceedings of the International Conference on Robotics and Automation ICRA-98, Belgium, 1998.
6. Veloso, M., Bowling, M., Achim, S., and Stone, P. "The CMUnited-98 champion small robot team", In Asada, M, and Kitano, H. (eds), RoboCup-98: Robot Soccer World Cup II, Springer Verlag, Berlin, 1999, 77–92.
7. Welch, G, Bishop, G, "An introduction to the Kalman filter", TR 95-041, Technical Report, Department of Computer Science, University of North Carolina, NC, USA, 2001. Available online at *http://www.cs.unc.edu/~welch*

FU-Fighters 2001 (Global Vision)

Raúl Rojas, Sven Behnke, Achim Liers, and Lars Knipping

Free University of Berlin, Institute of Computer Science
Takustr. 9, 14195 Berlin, Germany
{rojas|behnke|liers|knipping}@inf.fu-berlin.de
http://www.fu-fighters.de

1 Introduction

Our F180 team, the FU-Fighters, participated for the third time at the RoboCup competition. This year we used a heterogeneous team, consisting of improved differential drive robots and new omnidirectional robots. We designed new electronics and added prediction and path planning to the behavior control. Our team won fourth place in the SmallSize league competition.

2 Team Development

Team Leader: Raúl Rojas (college professor)
Team Members:
Sven Behnke (scientific staff): general design
Achim Liers (scientific staff): electronics, mechanics
Lars Knipping (scientific staff): behavior control
Bernhard Frötschl (scientific staff): web site, organization
Mark Simon (student): global vision, user interface
Kirill Koulechov (student): behavior control, microcontroller programming
Lars Wolter (student): behavior control, user interface
Oliver Tenchio (student): mechanics

3 Mechanical and Electrical Design

For RoboCup 2001 we built a new generation of omnidirectional robots as shown in Figure 1. The robots are equipped with three DC-motors that have an integrated 19:1 gear and an impulse generator with 16 ticks per revolution. They drive special wheels which allow for omnidirectional movement [4,5]. Further, the robots use a rotating kicking device. We also used specialized robots as defenders or for offense when appropriate. They are also shown in the figure.

For local control we developed a new microcontroller board that is based on the Motorola HC12, as shown in Figure 2. This controller features 8KB RAM, 2KB EEPROM, 128KB flash, several timers, four PWM-units, digital I/Os, eight analog inputs, two RS-232 serial lines, and a CAN interface. The board can drive with pulse-width modulation four DC-motors and captures the impulses

A. Birk, S. Coradeschi, and S. Tadokoro (Eds.): RoboCup 2001, LNAI 2377, pp. 571–574, 2002.

Fig. 1. Different robots designs. From left to right: classical two wheeled design, two wheeled defender, and robot with omnidirectional drive.

generated by them. Further, four servos can be steered and eight on/off switches can be used. A mezzanine board contains a radio transceiver SE200 working in the 433MHz band that can be tuned to 15 channels in 100kHz steps.

The robots receive commands via a wireless serial link with a speed of 19,200 baud. The host sends 10-byte packets that include address, control bits, motor speeds, and checksum. The microcontroller decodes the packets, checks their integrity, and sets the target values for the control of the motor speeds.

Our robots are powered by 8 Ni-MH rechargeable mignon batteries. We implemented locally a PID-control of the motor speeds.

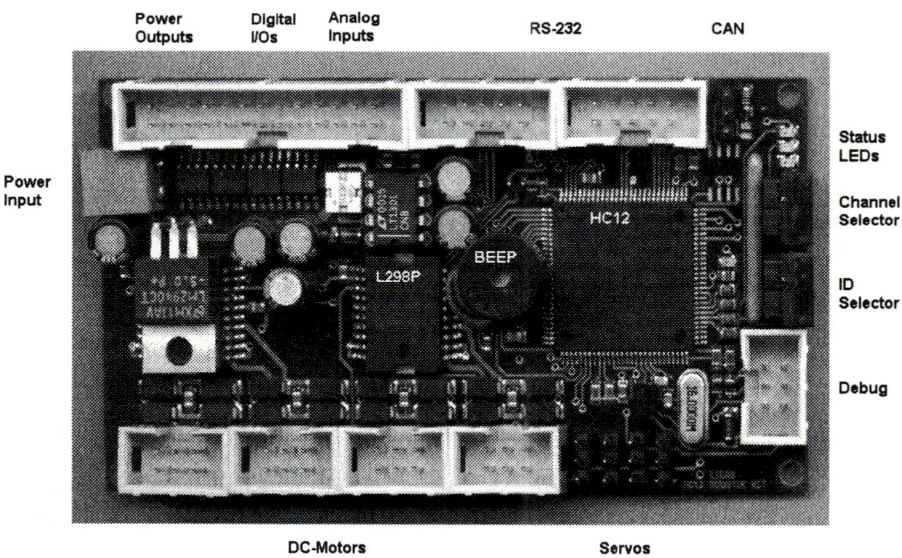

Fig. 2. HC12 microcontroller board.

4 Computer Vision and Prediction

The only physical sensor for our behavior control software is a S-VHS camera that looks at the field from above and outputs a video stream in NTSC format that is captured by a PC. The computer vision system running on the PC finds and tracks the ball and the robots. We used an improved version of the system described in [2]. Unfortunately, the information about the world is extracted with a delay caused by communication, mechanical constraints, and the computer vision system. The feedback (the result) of an action decision is perceived typically between 100ms and 150ms after the decision has been made. This causes problems when robots move fast, producing overshooting or oscillations in movement control. One way to deal with the delay would be to move more slowly, but this is often not desirable.

Our approach to solve that problem is to predict the movement of the robots for the next few frames. We feed the last robot positions and orientations (relative to the current robot position and orientation) and action commands to a feed forward neural network that is trained to predict the robot position and orientation for a point in time 130ms away. We train the network with recorded data before the game. The predicted positions and orientations are used for behavior control. This approximately cancels the effects of the delay and allows for fast and exact movement control.

5 Hierarchical Reactive Behavior and Path Planning

We use a hierarchy of reactive behaviors to control the robots. Simple behaviors are arranged in layers that work on different time scales. Fast primitive behaviors, such as taxis are implemented in the lowest layer. More complex, but slower behaviors are produced by the higher layers of the system. A more detailed description of our control architecture is given in [1,3].

Since the field is very crowded and significant contact with other robots must be avoided, we implemented a path planner on the second layer of the behavior control system. The path planner finds a path from a start point (the current robot position) to a goal (the desired robot position). Using dynamic programming and best-first search it computes the cheapest path on a grid that avoids obstacles such as other robots or the defense area. Figure 3 illustrates the behavior of the path planner. After the path has been found, the first point A where the path significantly turns is determined and communicated as target position to the lowest control level. As the robot moves towards A, the path is constantly re-planned and the turn point A moves towards the goal.

6 Future Work

In order to reduce the control delay, we plan to implement a larger portion of our behavior hierarchy directly on the robots. To allow for faster and more exact movement, fast local sensors for robot motion and obstacles are needed. We also

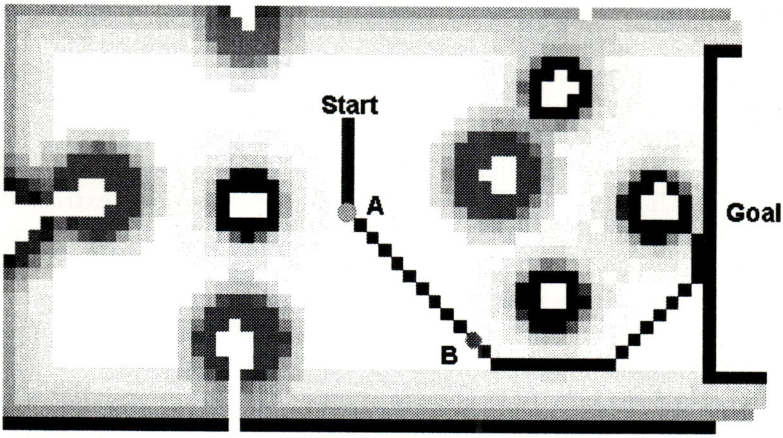

Fig. 3. Pathplanner. The shading of cells corresponds to cost. Expensive dark cells are caused by obstacles, such as other robots and walls, and by the defense area that should not be visited by field players. The dark blue cells show the path with the smallest aggregated costs that is computed by a best first search that visits only parts of the grid. White cells need not be considered. Point A is the first significant turn on the path. This point is communicated to the lowest layer in behavior control together with the direction towards point B. The robot drives from start towards A such that it can next turn to B. As the robot approaches the goal, the path is updated and points A and B move closer to the goal.

plan to develop more sophisticated ball handling mechanisms and controlled kicks that are needed for complex behaviors such as passing.

References

1. Behnke, S., Frötschl, B., Rojas, R., Ackers, P., Lindstrot, W., de Melo, M., Schebesch, A., Simon, M., Sprengel, M, Tenchio, M.: Using hierarchical dynamical systems to control reactive behavior. *In: Veloso, M., Pagello, E., Kitano, H. (eds): RoboCup-99: Robot Soccer World Cup III*, pp. 186-195, Springer, 2000.
2. Simon, M., Behnke, S., Rojas, R.: Robust Real Time Color Tracking. *In: Stone, P., Balch, T., Kraetszchmar (eds): RoboCup-2000: Robot Soccer World Cup IV*, pp. 239-248, Springer, 2001.
3. Behnke, S., Rojas, R.: A Hierarchy of Reactive Behaviors Handles Complexity. *In: Hannebauer, M., Wendler, J., Pagello, E. (eds.): Balancing Reactivity and Social Deliberation in Multi-Agent Systems*, pp. 125-136, Springer, 2001.
4. Reshko, G., Mason, M., Nourbakhsh, I.: Rapid Prototyping of Small Robots, *Technical Report*, CMU, 2000.
5. Palm Pilot Robot Kit. The Robotics Institute, CMU, http://www.cmu.edu/~pprk.

FU-Fighters Omni 2001 (Local Vision)

Raúl Rojas, Felix von Hundelshausen, Sven Behnke, and Bernhard Frötschl

Free University of Berlin, Institute of Computer Science
Takustr. 9, 14195 Berlin, Germany
{rojas|hundelsh|behnke|froetsch}@inf.fu-berlin.de
http://www.fu-fighters.de

1 Introduction

Currently, the Small Size League is the only RoboCup competition that permits the use of external sensing systems. Most teams use a color camera that is mounted above the field to determine the positions of the robots and the ball. Since this simplified setup is not compatible with the idea of autonomous robots, we decided to build a second F180 team for the World Championships 2001 in Seattle, consisting of (only) three robots using local omnidirectional vision, the "FU-Fighters Omni".

Despite of the small number of robots the team was quite successful, since we scored several goals playing in the regular SmallSize league (having played together with the ViperRoos team). We even had a tie with a global vision team (4:4). We won the local vision contest, an extra competition between four local vision teams.

This paper describes the overall system and the principles that yielded the success.

The remainder of this paper is organized as the following: Section 2 gives an overview of the system. Section 3 describes how visual perception is performed in our implementation. This covers color segmentation, robot self-localization, ball, goal and obstacle detection and tracking. Finally, Section 4 summarizes the results, and describes our future goals.

2 System Overview

We used the same omnidirectional robots as in the global vision FU-Fighters team [4], but with a vision system mounted on top the robots. It consists of a small camera directed upwards, looking into a concave mirror that produces an omnidirectional view of the environment. Unlike most teams that use convex mirrors we took a concave mirror from a cheap flashlight.

Each robot is connected to an external computer via a wireless analog video link. The computer grabs the images and analyzes the video stream to extract information about the status of the game. The extracted local views are transmitted through a LAN to a fusion module that merges them to a global view. The reactive behavior control system now can either use a global view or a local

A. Birk, S. Coradeschi, and S. Tadokoro (Eds.): RoboCup 2001, LNAI 2377, pp. 575–578, 2002.

view, as appropriate. For example, if a robot cannot detect the ball, but other robots can, the first can use the calculated global ball position. We adapted the global vision behavior to the special needs of local vision. For instance, if a robot cannot find the ball, then it searches the playing field looking for the ball. Here localization is an important issue. The scanning is coordinated among the robots, ensuring that the search is well distributed over the playing field. If a robot can detect the ball and the opponent goal, then the robot does not care about localization but focuses on relative movements. It tries to move behind the ball, and to direct the kicking mechanism towards the opponent goal.

3 Visual Perception Techniques

3.1 Core Techniques

Color Segmentation Colors are a helpful simplification in RoboCup that ease the detection of objects. To use colors for classification it is necessary to specify a method to obtain the degree of membership of a particular instance of color to a color class. Since the classification must be performed very often, we decided to use the fastest classification possible with a software solution, a lookup table (LUT). Each color (we use 15 bit RGB values) can belong to different color classes. Each entry of the LUT consists of an 32-bit value, where each bit indicates class membership.
The LUT is defined by software tools that allow the user to comfortably select regions in images, whose colors then can be added to a specific color class.

Finding Color Regions When searching the ball, obstacles or goals, we search for all pixel-connected regions whose pixels satisfy a predefined color check-mask. Once we have found an object, we only search within a small rectangular region centered at the object in consecutive camera images. To segement regions, we use a fast region growing algorithm. It starts with a seed pixel and consecutively adds pixels (in the directions up, left, down, right) to the region until the color specification is not satisfied any more. To avoid the algorithm to stop growing because of the presence of noise, that may separate regions which belong together, the algorithm only stops growing when more than $k = 3$ consecutive pixels have been found that violate the color specification.

Finding Color Transitions along a Line For robot self-localization we use the transition of the green playing field to the white walls as features. To detect the borders we have developed a very fast (300000 lines of length 100 pixel) algorithm that finds a predefined color transition along a line. We also use the algorithm when searching for objects (i.e ball, opponents, goals) to discard candidate regions found by the above region growing algorithm. We check for each region whether it has a transition to the green floor (all objects with contact to the floor have this transition, with few exceptions, when an object is located next to a marking line for example). For a more detailed description of the algorithm see [1][3].

3.2 Robot Self-Localization

We have experimented with three different localization methods. The fist two are described in [1] and were not used in Seattle. The third, which was applied in Seattle, is based on generating range scans by using omnidirectional vision. We radially stretch out lines from the center of the omnidirectional image and search for transitions from the green floor to the white walls, as shown in Fig.1a). Then, we use the rotation search/least-squares method described in [2] to match the range scan to a model of the environment. However, our case is easier than in [2] since we have a known environment and the model of the environment is precise and polygonal (Fig.1c)).
Figure 1 shows the obtained range scan that was produced by transforming the transition pixels to a local coordinate system. This transformation uses a calibrated monotonic function that maps distances between an image point and the image center to the distance between the corresponding world point and the robot.
A special adaption of the localization has been made for the goal keeper. Here we create a range scan by searching for transitions from yellow to white (if our goal is the yellow one).

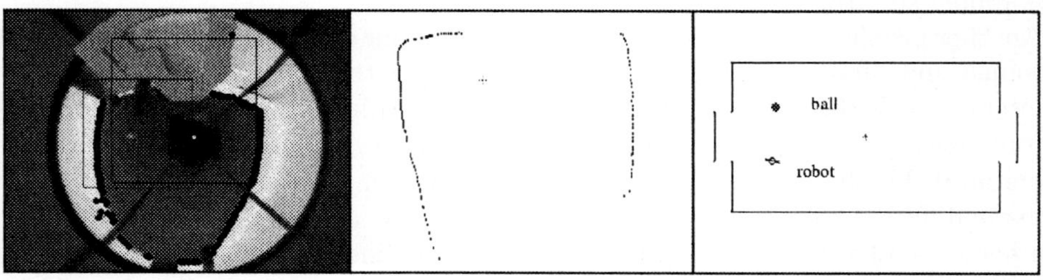

a) b) c)

Fig. 1. Detection of ball, goal and localization of the robot: a) Transition search, ball and goal detection; Black dots correspond to found transitions between the wall and the field. Rectangles (their centers) mark the ball and the goal. b)Obtained range scan after transforming the points; c)Localization of the robot in global coordinates;

3.3 Active Vision and Visual Attention

Consider the following two situations:
Situation S1: The robot has lost orientation and does not detect the ball.
Situation S2: The robot can see the ball and the opponent goal.
In situation S1 it is important that the robot finds the ball. This can be accomplished by two means. The first is that it searches the ball. The second is that

another robot sees the ball and informs it. When deciding on the first possibility (searching) the robot has to know, where it is located on the playing field, since otherwise it would crash with the border when searching. Therefore, in situation S1 localization is an important issue.
In contrast, in situation S2 the robot has not to care about localization but can concentrate all the computing power to exactly determine and track the position of the ball and the goal (in particular a free gap in the opponent goal) *relative* to its own position. Therefore visual attention and active vision are very important topics.

4 Results and Future Work

We were able to localize the robots and track the ball and goals with a rate of 25 fps. The localization was correct in about 95 % of all cases. However, the precision of localization must be improved further. Here, the primary problem is the distance mapping of a point in the real world to a pixel in the image. This mapping is imprecise due to misalignments of the optical system that are enforced by robot movements. To cope with the above problem, we plan to develop an automatic calibration method, which autonomously adapts to changing mapping situations.
Another problem is color classification. Although our color classification was very robust and successful compared with other teams, the fact that the user has to interact with the program to specify color classes infringes the idea of an autonomous system. Thus, we will focus our research on finding auto-configurative methods. Furthermore, we want to develop a vision system that is able to fuse many different techniques for localization, detection and tracking of objects. On a higher level such a vision system will allow to define different visual behaviors, enabling the user to specify how active vision and visual attention is used in different situations. We want the system to be flexible, such that RoboCup is just one application the system can cope with.

References

1. v. Hundelshausen, F.: An Omnidirectional Vision System for Soccer Robots, diploma thesis, Department of Computer Science, Free University of Berlin, http://www.inf.fu-berlin.de/~hundelsh, 2001.
2. Lu, F. and Milios, E.: Robot Pose Estimation in Unknown Environments by Matching 2D Range Scans, IN: CVPR94, pp.935-938, 1994.
3. v. Hundelshausen, F., Behnke, S., Rojas, R.: An Omnidirectional Vision System that finds and tracks color edges and blobs, In: Birk, A., Coradeschi, S., Tadokoro, S. (eds): RoboCup-01: Robot Soccer World Cup V (this volume), Springer, 2001.
4. Rojas, R., Behnke, S., Liers, A., Knipping, L.: FU-Fighters 2001 (Global Vision), In: Birk, A., Coradeschi, S., Tadokoro, S. (eds): RoboCup-01: Robot Soccer World Cup V (this volume), Springer, 2001.

Owaribito – A Team Description

Shinya Hibino[1], Yukiharu Kodama[1], Yasunori Nagasaka[2],
Tomoichi Takahashi[2], Kazuhito Murakami[1], and Tadashi Naruse[1]

[1] Aichi Prefectural University, Nagakute-cho, Aichi, 480-1198 JAPAN
[2] Chubu University, Kasugai, Aichi, 487-8501 JAPAN

Abstract. In this paper, we describe image processing method for high-speed robot ID recognition with black and white markers, path generation algorithm which controls left and right wheel velocity independently, strategy and communication software with error correcting code (ECC).

1 Introduction

An autonomous robot is a product of the system engineering that is built on the accumulation of the wide range of technologies such as mechanical engineering, electronic engineering, information engineering and so on. One of the main issues of the robot system design is the balanced design of the robot components. We explore the balanced design as well as the study of component technology.

We have been developing the Owaribito system since 1998. The hardware of the robot [1] is not changed since then, while the software is newly developed. In this paper, therefore, we describe the main features of the Owaribito's software, i.e. image processing, path generation, strategy and communication software.

2 Image Processing

The size of the image given by the global camera is 640×480. From the image, we extract object regions such as a ball and robots by using color information. We used the method developed by CMU [2] for the extraction .

Top surface of each robot of us has a black and white plate shown in Figure 1. By discriminating the shape of top surface, the robot ID and the forward direction can be simultaneously detected. That is, we prepare, as a template, the table of positions and directions of n equally divided points on the circle with radius r. The r and n are determined to satisfy that each point corresponds to a pixel. In case of the Owaribito, n is 44. Applying the template to each robot region on the image, we find a point whose pixel value changes from black to white. Consulting with the table, the forward direction of the robot can be found. Moreover, finding the point changing from white to black gives an angle of the arc of white region. It determines the robot ID. In case of $n = 44$, the resolution of angle is about 8 degree. This method is so simple that it can detect IDs and directions in high-speed. In the real system, the image noise cannot be avoidable so that we also use color tags to stably detect IDs and directions.

A. Birk, S. Coradeschi, and S. Tadokoro (Eds.): RoboCup 2001, LNAI 2377, pp. 579–582, 2002.

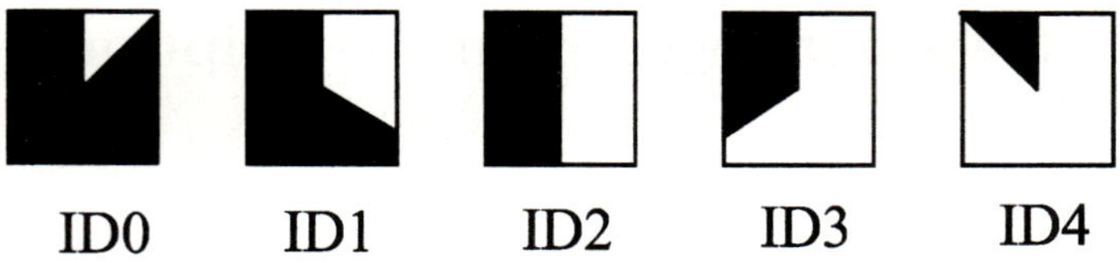

Fig. 1. Top surface of robot

3 Path Generation

Since our robot has two wheels, the degree of control freedom is two and the control of direction is limited. Considering this and the fact that the goal position where each robot should go changes from hour to hour, it is realistic the path generation which the robot approaches the goal position by asymptotically changing his direction. We call this a **direction converging path generation**. Figure 2 shows the paths generated by this method. In the figure, paths are superimposed on the field image. Each double circle with a number is a current position of our robot and the end point of each path is a goal position. The goal position with triple circle is a ball. The dotted circles are opponent robots. If opponent robots stand on the line connecting the current position of our robot and his goal position, subgoals are put near the opponents to avoid collision. In the case, the robot goes to the subgoals at first and then goes to the goal. Robots with number 2 and 3 have a subgoal and robots with number 0 and 1 do not. The goal keeper (number 4) does not move in this figure. Note that the path is generated only to determine the velocity for next Δt time step[1]. The path is newly generated every Δt time step.

Our path generation algorithm is given as follows.

step1 Let $p(=(x,y)), (v_l, v_r), \mathbf{u}$ be position, velocity (of left and right wheel) and forward direction vector of a robot, respectively. Let t be the current time. Calculate the curvature κ and robot velocity $v = \frac{v_l+v_r}{2}$. See literature [3] for detail.

step2 Get a goal position $p' = (ox, oy)$. This is given by the strategy algorithm.

step3 Let $\overrightarrow{pp'}$ be a vector directed from the current position to the goal position and let θ be an angle between vectors $\overrightarrow{pp'}$ and $\mathbf{u}$.

step4 Give a curvature at the time $t + \Delta t$ by $\kappa_{new} = \theta \times n_a / R_A$, where R_A is a constant and n_a is a variable depending on subgoal(s). (This equation generates a path which has a large curvature at first and a small as approaching to the goal. See Fig. 2.)

step5 Putting $dr = 1/\kappa - 1/\kappa_{new}$, calculate a velocity variation of robot $|dv| = |S/dr|$, where S is a constant. Let $v_m(\kappa)$ be a maximum robot speed when a curvature κ is given. Give new velocity by $v_{new} = v + dv$ if $v_{new} < v_m(\kappa_{new})$, otherwise by $v_{new} = v - dv$. Then, calculate a new position at the time $t + \Delta t$.

[1] Image processing speed determines the Δt. In our system, $\Delta t = 33msec$.

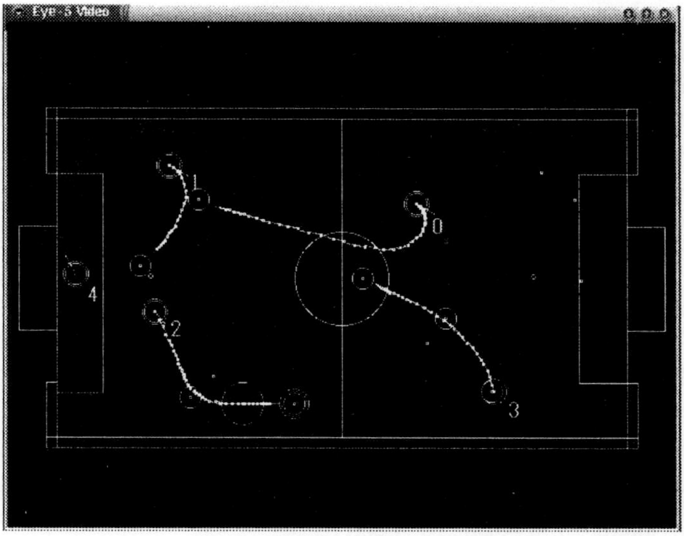

Fig. 2. An exapmle of path generation

step6 Calculate repeatedly the steps from *step1* to *step5* and check whether the path reaches the given goal or not. (Fig. 2 shows the result of this calculation.) If the path reaches the goal, it is OK. If not (if over M times repeated), recalculate these steps by changing the constant R_A until the path reaches the goal. This computation gives the robot velocity of next Δt time period.

4 Strategy

Current Owaribito system has only a simple rule checker. The checker checks a multiple attacker, a multiple defender, a pushing, a contacting with the goalie and a 15 second stopping rule. Pushing wall or a mate is also checked. If a robot will fall in violation, an avoidance operation is applied. The aviodance operation is simple, it is a move to the direction that does not violate the rule.

5 Communication

We use a radio modem system of Futaba Corp. for communication between robots and a host computer. It is a 2.4 GHz band spread spectrum system. It has a direct communication mode, which realizes a low communication latency but a low communication quality. User is responsible for the communication quality.

The host processor dispatches a robot command, which gives wheel velocities of all robots, every 50 msec. The command consists of 17 bytes packet, i.e.

2 bytes for header, 10 bytes for velocities, 5 bytes for ECC, which supports communication quality. The communication latency is only 9 msec.

To achieve high communication quality, we realized a communication monitoring system. With an another receiver on the soccer field, the host processor can monitor, with high accuracy, the same command as the one which the robots received. Comparing the sent command with the monitored command, send-error can be detected. Moreover, since an accurate time that the command arrived at the robots can be calculated, more accurate path control can be possible.

In the RoboCup-01 competition, we monitored the communication throughout our matches. The result is shown in Table 1. It is clarified that the packet with ECC is sufficient for the communication in RoboCup if the competitions are managed with the system which a committee of RoboCup-01 did.

Table 1. Statistics of communication data

owaribito *v.s.*	total bytes	normal bytes	corrected bytes	errored bytes
fu_fighter	138960	138411	548	1
fuomni/viper	271950	270805	1145	0
field ranger	265020	263894	1125	1
robosix	181200	180504	696	0

6 Concluding Remarks

The Owaribito realized the high-speed detection of robot ID and direction, direction converging path generation, communication monitoring and so on. Implementing these functions gave a stable operation of the robot system in the RoboCup-01.

Owaribito has not a kicking device now, however, it is absolutely neccesary. We will participate in RoboCup-02 at Fukuoka with the robots having the kicking device and a omni directional drive mechanism and implementing the high level technology such as pass play operation.

References

1. T. Naruse, T. Takahashi, K. Murakami, Y. Nagasaka, et al, " OWARI-BITO" in M. Veloso et al eds., Lecture Notes in Artificial Intelligence 1856, RoboCup-99: Robot Soccer World Cup III, pp. 675-678, Springer, 2000
2. J. Bruce, T. Balch, M. Veloso "Fast and Inexpensive Color Image Segmentation for Interactive Robots", Proc. 2000 IEEE/RSJ Int. Conf. on Intelligent Robots and Systems, pp. 2061 - 2066, 2000
3. R. D'Andrea and J. Lee "Cornell Big Red Small-Size-League Winner", AI magazine, Vol. 21, No. 3, pp. 41 - 44, 2000

RoboCup 2001 (F180) Team Description: RoboSix UPMC-CFA (France)

Francis Bras, Ryad Benosman, André Anglade, Seko Latidine, Olivier Martino, Aude Lagardere, Alain Boun, Alain Testa, and Gilles Cordurié

Université Pierre et Marie Curie
4 place jussieu
75252 Paris cedex 05, boite 164
France
rbo@lis.jussieu.fr
http://www.robosix .com

Abstract. This paper describes the team RoboSix UPMC-CFA that participated to the German open 2001 and RoboCup 2001 in the F180-league. Wel will present the mechanical and electrical design of the robots. We will then introduce the vision choices explaining the colometric and geometric calibrations stages. Finally we will explain the implemented behaviour and the path planning associated to it.

1 Introduction

RoboSix UPMC-CFA is a team composed of students from different sections of University Pierre and Marie Curie. The team has one electrical engineering student belonging to the engineering school IFITEP and four computer vision students belonging to the DESS IE. The team has three adivisors that have been involved in all Robocup's championships since Nagoya in 1997. The RoboCup UPMC-CFA 2001 team is a team that was created in september 1999. Our main motivation is to use Robocup as a test platform for new approaches in computer vision.

2 Mechanical Design

Our robots are designed to fit the new F180 size regulation. They are built out of aluminum frames that protect their inner parts. The robots have a differential drive with two active wheels that are not in the middle to allow a more complicated and unpredictable trajectories and a better twisting to shoot the ball. The maximum speed has been pushed to more than 2 m/s, by far more than needed by the vision. The robots carry a directional kicker, but due to a lack of time they were not used. The team has also an omnidirectional goalie, to avoid the problem of losing time while repositionning the player.

A. Birk, S. Coradeschi, and S. Tadokoro (Eds.): RoboCup 2001, LNAI 2377, pp. 583–586, 2002.

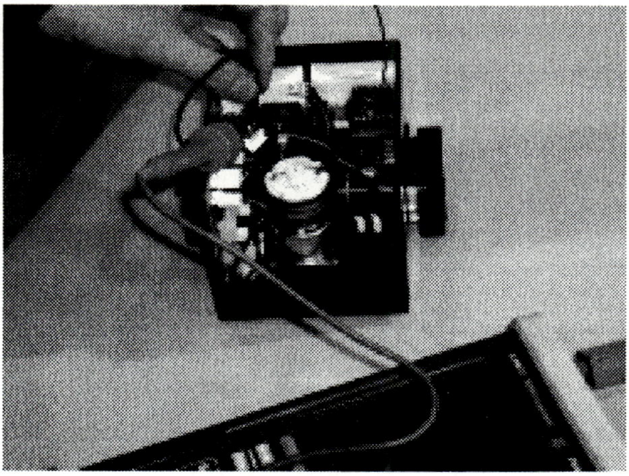

Fig. 1. Inside UPMC-CFA robot

3 Electrical Design

The microcontroller used is an Atmel 89S53 that has the advantage of beinging programmable in situ with 12kb flash memory running at 24 Mhz. For radio communication we used the classical Radiometrix working in the band 433Mhz and 418Mhz. The robots are powered by two Maxon motors with an integrated 19:1 gear, we used a dual H bridge motor driver L6202. The batteries were specially assembled to fit in the robots and are from GP Batteries our sponsor. Finally a Altera 7032 generates the pulse with modulation for the motors. All the described devices were mounted on a single custom card (see Fig.1) including the stabilized power supply.

4 Computer Vision

4.1 Hardware Description

We used a PCI-framegrabber and a progressive CCD camera. The chosen camera has the advantage of needing only one cable for data and power directly connected to the Pc. We capture RGB images of size 640*480 at a rate of 30 fps. The system is running under Windows NT on a pentium 4 1.5 GHz.

4.2 Color Calibration and Regions Labelling

The output signal of the camera is RGB. Compared to the last year approach, we developped a new color calibration concept that was used by our both middle and small size teams. Classically a color is defined as a three component vector containing the values for the red, green and blue color planes. Generally the first calibration stage is manually made where colors are selected, for each color we compute a mean and a standard deviation. The image is then filtered, an image containing only the desired colors is generated. Last year a second detection based on HSV color space was applied only on the detected zones to reduce computation and increase robustness. This year the color calibration relies only on RGB. Instead of using a 3D cube to determine a color, we used 3D quadrics that are projected on three planes in the best oriented 3D coordinates system giving the best separation between close colors, this method will be published soon. The system is robust to change of lighting and has been succesfully tested.

4.3 Recognizing Robots

Fig. 2. Caraterizing the robots and their orientations

The coding of the robots is kept the same as last year and relies on the same bar code. Each robot has a number of thick and thin bars (see Fig.2). The bars are chosen so that we can have the information on the orientation and at the same time allowing to retreive the robot's position with a very high accuracy. The geometric calibration is widely explained in [1] and allows us to locate our robot on the table. The aim is to be able to retreive from the images' coordinates the position of the robots according to a coordinates system on the table taking into account the height of the observed object. It is based on projective geometry and more specifically colineations [2].

4.4 Behaviour

The behaviour of the robots is motivated by different flags corresponding to specific situations and robots' positions. The behaviour relies on the path finding procedure that uses a delaunay triangulation to find the best path taking into account the length of the best paths. The behaviour of the team relies on the same flags as last year. It analyzes the situation and assigns new position to be fullfilled. The path planning determines for each target which robot is the closest and generates the best path. Using such a system at the contrary of what has been used last year, no robot has an assigned position, even the goalie if wanted could switch position with a player if needed for an action. The behaviour of the robot varies according to situations for example if the team is winning or not.

5 Conclusion

We find robocup a very good tool for pedagogy and research. Unless we increase the size of the field, it is going to be harder and harder to have a cooperative game and score against defensive teams. This year, the team's level has highly improved, we see more and more passes. We are considering new techniques to allow more dynamic passes for next year..

References

1. Ryad Benosman and Jerome Douret: A simple and accurate camera calibration for the F180 RoboCup league RoboCup Symposium 2001, Seattle, USA
2. J.G Semple and G.T Kneebone Algebraic Projective Geometry, Oxford University Press 1952

Rogi Team Description

J.Ll. de la Rosa, B. Innocenti, M. Montaner, A. Figueras, I. Muñoz, and J.A. Ramon

Institute of Informatic and Applications. University of Girona, E-17071, SPAIN
pepl1uis,bianca,mmontane,figueras,imunoz,jar@eia.udg.es
http://rogiteam.udg.es

Abstract. This paper resumes the main features introduced to RoGi team for RoboCup 2001.

1 Introduction

RoGi team started up in 1996 at the first robot-soccer competition as the result of a Ph.D. course in multi-agent systems. Since 1997 it has participated at the international workshops held in Japan, Paris, Stockholm and Melbourne. The main goal has always been the implementation and the experimentation on dynamical physical agents and autonomous systems.

This year the team has dealt with new mechanics, an addition of roles and minor changes on the vision system.

2 Team Description

The system that implements micro-robot soccer is made up of three parts: robots, vision system and control system. The vision and control systems are implemented in two different computers and they are connected by means of a LAN, using TCP/IP protocol. This allows a faster communication between them. The control system analyzes data coming from the vision system and takes decisions based on it. The decision is split up into individual tasks for each robot and is sent to them via radio link.

3 Robot Description

The robots have on-board 8 bit Philips microprocessors 80C552 and RAM/EPROM memories of 32kBytes. The robots receive data from the host computer by means of an FM receiver. The FM receiver allows working with two different frequencies 418/433 MHz in half-duplex communication. The information sent by the host computer is converted to RS-232C protocol.

12 batteries of 1.2 V supply the energy to motors and electronics. There are two power sources, one DC-DC switching regulator, which provides stable 12 V to motors, and another one that guarantees 5 V to the IC.

A. Birk, S. Coradeschi, and S. Tadokoro (Eds.): RoboCup 2001, LNAI 2377, pp. 587–590, 2002.

The microprocessor do position and speed control, having as input the desired final position and the actual position, and as output the PWM of each motor. A voltage sensor is used to know the actual speed of the robot.

In addition, a single metal piece has been designed to sustain motors, batteries and electronics, making the robot more robust to collisions.

Fig. 1. New robot

4 Vision System

A specific hardware has been designed to perform the vision tasks, merging specific components for image processing (video converters, analogue filters, etc.) with multiple purpose programmable devices (FPGAs). A real time image-processing tool is obtained, which can be reconfigured to implement different algorithms. According to RoboCup Small Size League Rules, each robot has use a yellow or blue marker mounted at the center of their top surface. In order to provide angle orientation, additional color markings are allocated on top of the robots.

This year the identification algorithm has been modified and improved. Adding this changes to last year modification, the vision system has became more stable and reliable.

5 Control System

The control system is a multi-agent environment in which the agents decide the best action to do. Each agent takes first, a reactive decision based on data given by the vision system, and second, a deliberative one considering the teammates decisions. Once all the agents get an agreement, each one transforms the actions in orders understandable by the robot, and sends them via radio link.

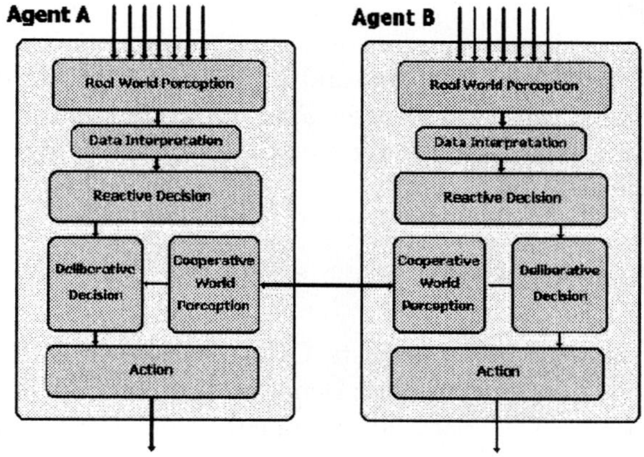

Fig. 2. Schematic structure of agents.

Reactive Decisions Step: Using a global real world perception (as seen in Figure 2) and a set of capabilities, agents take a private/local decision. Capabilities are action that agents can perform and depend on roles, that means that agents are different from each other by the role they play at the field. The world perception is the result of high level abstraction using fuzzy sets of data given by the vision system. With this knowledge of the environment and its capabilities, all the players take the reactive decisions using fuzzy logic. The agent converts this reactive action into its actual belief.

Deliberative (Cooperative) Decisions Step: Deliberative reasoning is implemented by communicating the former reactive belief. Every agent knows the beliefs of the other teammates. Beliefs contain the reactive decision, the certainty of this decision and the identification of the player-agent (reactive-belief, certainty, ID-player). When two or more agents realize their beliefs are in conflict, they use this certainty coefficient in a consensus algorithm based on a positional method to resolve it. In this positional method, players are specialized since they have assigned roles. One possible effect of their specialization is that they prefer to stay in certain position on the playground. Agents take advantage of this feature and modify their vision of the co-operative world according to the positions of team players. These modifications lead the agents to reinforce or to change the former reactive decision. As a result, collisions among playmates are significantly reduced comparing to non-adaptive perception of the cooperative world.

This year, in order to improve tactics new roles have been added, so that the team play can be adapted to opponents tactics.

6 Conclusion

This year, team most important feature has been the modification of robots. With the new body, motors and associated electronics, robots go faster and resist collisions in a better way. Adding new roles to new body the whole team behavior has been improved.

7 Acknowledgements

This work is partially funded by projects: OCYT AE00-0122-DIF "Demostración de Fútbol Robótico / Football Robotic Demonstration", PROFIT "Aplicación de agentes inteligentes para el desarrollo y automatización del comercio electrónico / Intelligent Agents Application on the development and automation of electronic commerce", CICYT DPI2000-0658 "Diseño de Agentes Físicos Dinámicos. Aplicaciones Futuras / Dynamical Physical Agent Design: Future Applications", "Laboratori de l'Equip Internacional de Robots Mòbils de Girona (RoGi), as special action 2000ACES00018 of the Catalan Government and "Plataforma de Experimentación de Robots Móviles de Girona" as special action DPI2000-2100-E.

References

1. Innocenti B., de la Rosa J.Ll., Vehí J., Muñoz I., Montaner M., Fàbregas M., Figueras A., Ramon J.A.: *"Examples of Dynamical Physical Agents for Multi Robot Control"*. 2 Worshop Hispano-Luso de Agentes Físicos Madrid 2001. Madrid - España.
2. De la Rosa J.Ll., Muñoz I., Innocenti B., Figueras A., Montaner M., Ramon J.A.:*"Preliminary Studies of Dynamics of Physical Agents Ecosystems"*.4th Robocup Workshop. Melbourne - Australia.
3. De la Rosa J.Ll., Innocenti B., Oller A., Figueras A., Ramon J.A., Muñoz I., Montaner M.:*"An Example of Dynamical Physical Agents"*. Eupropean Robocup Workshop. Amsterdam - Holland.

Roobots

Jason Thomas, Kenichi Yoshimura, and Andrew Peel

University of Melbourne, Melbourne Vic 3010, Australia,
{jrth,kyosh,apeel}@cs.mu.oz.au,
http://www.cs.mu.oz.au/robocup/index.html

Abstract. This paper describes the Roobots small league RoboCup team from the University of Melbourne which competed in Robocup 2001, Seattle. The highlights from our team included an omnidirectional drive mechanism, an integrated dribbling and kicking mechanism, a robust vision system and communications system and a BDI agent oriented stategy module.

1 Overview

The Roobots are the second entry in the small-league RoboCup competition by the University of Melbourne, and incorporate a number of substantial changes to the team from RoboCup2000: the MuCows. This paper focuses on these innovations, while refering the reader to the MuCows team description [2] for details of those elements of the team which were re-used.

The Roobots played four matches in Robocup 2001, with scorelines: won 4:2 against Sharif; drew 0:0 against RoGI; won 10:0 against Canuck; lost 0:8 against Cornell Big Red. The first match was notable for the substantial time when the Roobots entire team was rendered stationary by a series of software bugs. These bugs weren't detected before the tournament as the Roobots hadn't played a full game, which was an embarrassing failure of systematic testing and project management. We were lucky not to lose the match. The RoGI match saw the Roobots survive without any software lockups, but revealed other shortcomings in the system: a field zoning bug, an overly cautious obstacle avoidance system, and robot shutdowns due to activation of the over-current protection circuit in the LiIon batteries. The overly timid play resulted in very few scoring opportunities, and the game finished without a goal being scored. The Canuck game saw the Roobots performing near their potential, while the Canuck team experienced intermittent problems with their radio communications. The final score of 10:0 was reached shortly after half-time. The final match against Cornell was a highlight of the tournament, and was significant as the first match between omni-directional teams. The game was very fast moving, and generally considered by the spectators to be one of the more exciting games of the competition. The singular ball-control skills of the Cornell team contributed to a number of their 8 goals, while a further defect in our software led to several penalties for 'two defenders in the defence zone'. Also, near the beginning of the match, we received a red card for pushing, and so played the remainder of the match

A. Birk, S. Coradeschi, and S. Tadokoro (Eds.): RoboCup 2001, LNAI 2377, pp. 591–594, 2002.

with only four robots. Nonetheless, the Roobots were faster, and our AI system proved to be quite robust; we are satisfied with the performance of the team. The Roobots had a smaller goal difference then RoGi, who consequently advanced to the quarter finals.

The particular strengths of our team were well built robots, which were both reliable and fast; a reliable radio system and vision system, which are re-used from the MuCows; a flexible AI system, which allowed us to code up a new strategy within a 5 minute timeout during the Cornell game; a solid goalkeeper behaviour; and a sophisticated graphical user interface. The weaknesses of the team were in poor tuning of the control, poor debugging of the code, and not having many of the match-specific skills, such as getting the ball off the walls and out of the corners. These shortcomings are largely attributable to our timetable: we built our robots just in time for the qualification video on June 1, which only left 2 months for tuning the control and implementing the game playing behaviours.

The Roobots team is partly run as a number of projects for final-year undergraduate mechatronics and computer engineering students. The other contributors to the team are the technical staff of our department and masters students. The timetable of RoboCup proved to be a major difficulty as the undergraduate year in Australia commences in March, which leaves only 5 months for the students to get up to speed on the project, and even less for projects involving hardware developments.

1.1 Team Composition

The full team composition for the roobots was:

A. Peel (TL)	N. Barnes	A. Senyard	J. Chow
J. Thomas (PM)	A. Blair	K. Nguyen	T. Pham
K. Yoshimura	E. Kazmierczak	M. Ashmore	L. Truong
D. Hornsby	D. Walter	D. Hruszecky	
J. Horvath	A. Howard	J. Quach	
T. Weichert	H. Meffin	A. Walsh	
A. Corman	D. Shaw	A. Meehan	

2 Robot Design

The robots are omni-directional, with the wheels independently driven by 11Watt Maxon motors with optical encoders. There are two LiIon 14V 1400mAh battery packs: one for the drive system which provides about 90 minutes of use, and the other for the electronics. The control electronics is on a surface mount circuit board with a Hitachi H8 16-bit processor, H-bridge circuits, flash ROM, 2 accelerometers, serial ports, and LCD interface. The radio link is an Innomedia 2.4GHz spread-spectrum module with an adaptor board which converts its interface to that of the Radiometrix modules used by many of the other teams. The radio link proved to be very reliable.

The robots have a unique mechanism for rolling and kicking the ball, which allows both single-shot kicks of a ball already on the roller, and 'continuous' kicking by spinning the kicking bar. In practice, the roller mechanism wasn't used in the competition as we didn't have time to develop competent behaviours to make use of it.

The target wheel velocities are transmitted to the robots via the radio link. The robot implements a PID loop for each wheel in software. The robots do not use any dead-reckoning, nor do they servo to position.

For detailed photographs and descriptions of the hardware and system architecture, the reader is referred to the Roobots website [1].

3 Software Design

Figure 1 illustrates the architecture of our system. The modular-based architecture with clear interface definitions proved highly successful facilitating modification of the existing system, easy customization for different hardware configurations, reusing existing components (eg. vision system), and development of new modules (eg. strategy).

To take full advantage of the architecture, we developed our Artificial Intelligence (strategy) module using an Agent-Oriented programming language, JACK [3]. In this environment, a system is modelled in terms of agents incorporating the Belief/Desire/Intention (BDI) model of artificial intelligence. The JACK intelligent agents are autonomous software components that have explicit goals to achieve or events to handle (desires). Each agent holds a library of plans which explain step-by-step instructions (intentions) for achieving a goal or handling a given event. The agent pursues its given goals (desires) by adopting the appropriate plans (intentions) according to its current set of data (beliefs) about the state of the world. We also used an extension of JACK called SimpleTeam [4] [5], which provides team-modelling support.

We developed four different formations (eg. 1-2-2 formation. 2 defenders and 2 attackers) and incorporated motivational calculations to temporarily allow robots to break the current formation, reflecting an opponent teams strategy (eg. an offensive robot temporary participating in defensive activities after realizing an opponent team is employing offensive strategy).

To integrate the high-level reasoning component, which runs at a lower frequency, we developed a short-term goal selection module (Jack Interface Module) and a behaviour module, both of which run at the same the frequency as the rest of the system. The behaviour module provides primitive soccer skills (eg. kick ball, dribble ball, and mark object) and interacts with our path-planning module.

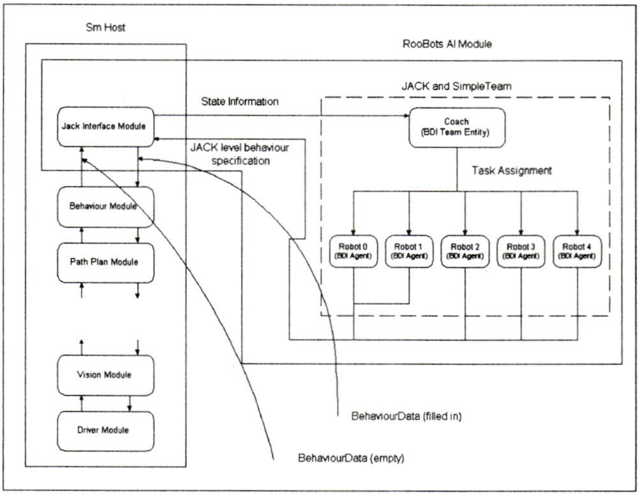

Fig. 1. AI Module

4 Conclusions

Based on our performace in 2001 we believe we have a good base for 2002. We will be focusing our efforts on improvements to: ball control hardware, motion control, tracking, and our strategic and tactical abilities. We will be following more stringent software engineering procedures in order to maintain high quality code thoughout development as lack of quality control was our greatest flaw.

5 Acknowledgements

The authors would like to thanks the Advanced Engineering Centre for Manufacturing (AECM) for generously sponsoring the Roobots in 2001, enabling us to construct the new robots and send a large team of 11 people to Seattle.

References

1. Roobots website http://www.cs.mu.oz.au/robocup/F180/index.html
2. Howard, A.: MuCows Team Description In: Robocup-2000: Robot Soccer World Cup IV Lecture Notes in Computer Science, Springer, pages to appear (2001)
3. P. Busetta and R. Rönnquist and A. Hodgson and A. Lucas: JACK Intelligent Agents - Components for Intelligent Agents in Java: Agent Oriented Software Pty. Ltd. http://www.agent-software.com
4. R. Rönnquist and C. A. Lucas and A. S. Hodgson: Specification of coordinated team behaviour: Proceedings of the IJCAI Team Behaviour and Plan Recognition Workshop 1999
5. K .Yoshimura and R. Rönnquist and E. Sonenberg: An approach to specifying coordinated agent behaviour: Proceedings of the Third Pacific Rim International Workshop on Multi-Agents 2000 Melbourne, Australia

Sharif CESR Small Size Robocup Team

Mohammad Taghi Manzuri[1], Hamid Reza Chitsaz[1], Reza Ghorbani[2], Pooya Karimian[1], Alireza Mirazi[2], Mehran Motamed[1], Roozbeh Mottaghi[1], and Payam Sabzmeydani[1]

[1] Department of Computer Engineering, Sharif University of Technology, Tehran, Iran
manzuri@sharif.edu
[2] Department of Aerospace Engineering, Sharif University of Technology, Tehran, Iran,
http://ce.sharif.edu/robocup/small

1 Introduction

Robotic soccer is a challenging research area, which involves multiple agents that need to collaborate in an adversarial environment to achieve specific objectives. Here we describe the Sharif CESR small robot team, which was participated in Robocup 2001 small size league in Seattle, USA. This paper explains the overall architecture of our robotic soccer system. Figure 1 shows a picture of our soccer robots.

Fig. 1. Two robots of Sharif CESR small size robocup team.

2 Mechanics

The Sharif CESR team consists of four identical field players and a goalkeeper. Each robot uses two DC Faulhaber DC motors with a 3.71:1 reduction gear box and two incremental encoders with resolution of 512 pulses per revolution of motor axis. The algorithm to estimate the velocity from the encoder output is

A. Birk, S. Coradeschi, and S. Tadokoro (Eds.): RoboCup 2001, LNAI 2377, pp. 595–598, 2002.

implemented on the robot hardware. For each robot we have designed and built a kicker with a small pneumatic actuator and a special mechanism for transferring actuator's force to the ball. There are some situations that many robots come around a ball. To handle such situations our kicker is able to shoot the ball over the robots toward the opponent goal or for the offensive robots.

3 Software

3.1 Vision

As specified in RoboCup F180 rules [Robocup Rules, 2001], teams can decide between local or global vision. We considered the global vision and used the output signal of a mounted PAL camera, as the input signal of a Matrox Corona frame grabber. We developed a visual C++ program to input the images, frame by frame into a system running under Microsoft Windows. The frames are captured in 768x576 24bit RGB color pixels at 25 frame per second rate. As all the objects (i.e. ball and robots) are color coded we find them by finding specific color regions. For fast conversion of RGB color values to the corresponding color number, the program uses a lookup table. This lookup table is created when program initialized. The table can be filled using different algorithms.

We preferred to use other color spaces, specially HSV to fill the lookup table. For finding objects in each frame we start from the top left pixel in a given window and make an object list by combining each pixel to the same color pixels in the top of left.

First, a dynamic window resizing method has been implemented to speed up the processing [Simon et al., 2000], but after developing a faster method for processing the whole of a frame, we decided not to use it.

As an important part of the program, we used a Blob filter to filter the unwanted blobs which don't meet the desired specification' such as number of pixels and to merge too near blobs.

The next step was to fill the World Model. World Model is a virtual structure, which contains all of the information about the real world. World Model is the only shared object between all parts of our program and, this model performs the communication between these parts.

3.2 Decision Making

In this part, the World Model which is filled with the information from vision part, is transferred over the network using UDP/IP protocol to another computer which runs the decision making algorithms.

Due to the latencies in vision, network transferring, decision making, wireless modules and on robot hardware, we observed an overall system latency of about 120ms (3 frames). For solving this problem we designed and added a prediction part to our program, which fill the world model by updated ball and robot positions. The location of ball is estimated by calculating its speed and direction.For this aim, a simple linear filter has been implemented. The prediction of the team mate robots is done by running the previous commands sent to robot in the latency time on robot position and direction.

A Coach-Agent design is implemented to both use the power of global vision and ease the complexity of control algorithms for 5 robots. In this design we have a global coach that determines the strategy of the team. After this, each robot agent is provided with some specific inputs, like the information of the World Model, the coach decisions and etc. Based on these inputs each agent makes some decisions, and the decisions will be passed to low level control part.

For decision making, we use position estimation functions which estimate and predict the events going to be happened. For example how much time needs each robot (team mate or opponent) to reach the ball, and etc. We use physical formulas for these predictions and estimations.

3.3 Low Level Control

For controlling the robots we have some low level skills which will send the commands needed to do that skill to the robot. We have used rotational and translational speed instead of left and right wheel speed. The robot will always try to reach the desired direction by setting the needed rotational speed. A heuristic trajectory generation algorithm is used to move the robot to arrive the target with a specific final direction. For obstacle and wall avoidance a function is used to correct the angles and to avoid obstacles.

4 Hardware and Control Section

In the small size league, speed together with accuracy plays a major role in well implementing of the decisions made by the off-board intelligent planner software. The overall speed and accuracy of a robot is based on its fast and accurate accelerating, stopping and turning which yields to an accurate movement on a pre-calculated path. Therefore, a powerful motion controller for controlling mechanical parts is essential. This problem becomes even more critical where the control board should be small in size, low in power consumption and meet some other factors as well. according to these factors, we investigated more than 10 different hardware design schemes and finally implemented the following design.

We will first describe the board's hardware in which we explain the computation core, electrical design and the telecommunication mean. Subsequently, the on-board software and related hardware description codes will be described. This section will be closed by the explanation of what should be done in near future.

4.1 Hardware

Computation Core As the computation core, we made use of a TMS320C25-50 Digital Signal Processor (DSP) on which the digital PID, or adaptive and Neural Network control algorithms can run. Currently the PID is implemented, and we plan to use the computational power of DSP to apply Neural Network and adaptive control algorithms. The DSP works under 40MHz clock frequency and consequently is able to work at 10 MIPS. The DSP boots from a 64Kb EEPROM and uses two 32Kb SRAMs as the external memory. An FPGA is also applied to take care of the subsidiary tasks in order for the DSP to be able to use its total processing power for control algorithms.

Electrical Design In the electrical and motor drive section we use a 5v regulator for providing the digital circuitry with power and we have an L298, Dual H-Bridge motor driver, for the two motors. We have 3 LEDs on the board. As they show the status of the robot, they are used for fast debugging. In order to close the control loop for the robot drive motors, an encoder for each of them is used as the feedback.

Telecommunication For the communication between the decision making module and the robots, we have used RadioMetrix' RX2/TX2 modules. This brought us the advantage of simplicity, together with the appropriate size. Beside these advantages we have some problems with these modules because they are not sufficiently fast and reliable.

4.2 On-board Software

The robot's main program, fed into DSP, was implemented in C which has three major components: A component for control algorithm to take care of the digital PID, a component to parse the incoming commands, and a communication component to deal with the UART protocol.

On FPGA side, the source was written in Verilog to do subsidiary tasks such as generating the PWM, counting the motor encoders, etc.

References

[Simon et al., 2000] Simon, M., Behnke, S., Rojas, R.: Robust Real Time Color Tracking. Robocup 2000: Robot Soccer World Cup IV, Lecture Notes in Computer Science, Springer, (2001) 239–248

[Robocup Rules, 2001] The Robocup Federation. Robocup Regulations and Rules, http://www.robocup.org

The Team Description of the Team OMNI

Daisuke Sekimori[1*], Nobuhito Mori[2], Junichi Ieda[2], Wataru Matsui[2], Osamu Miyake[2*], Tomoya Usui[2*], Yukihisa Tanaka[2*], Dong Pyo Kim[2*], Tetsuhiro Maeda[2], Hirokazu Sugimoto[1], Ryouhei Fujimoto[1], Masaya Enomoto[1*], Yasuhiro Masutani[2*], and Fumio Miyazaki[2]

[1] Akashi College of Technology, Akashi, Hyougo 674-8501, Japan
[2] Osaka University, Toyonaka, Osaka 560-8531, Japan
soccer@robotics.me.es.osaka-u.ac.jp
http://robotics.me.es.osaka-u.ac.jp/OMNI/

Abstract. The Team OMNI has developed a robot system with omni-directional vision and omni-directional mobility according to the rule of the RoboCup Small Size League. This paper describes the hardware configuration, software configuration, and simulator of this robot system.

1 Introduction

The Team OMNI is a research project aiming at the fundamental technology of decentralized autonomous robot system and participating in the RoboCup Small Size League. This team was formed in 1998 and participated in RoboCup Japan Open 2000, 2001 and RoboCup 2001.

2 Hardware Configuration

We are aiming at complete local vision using on-board camera considering importance of study of autonomous decentralized intelligence. A method of sharing processes with an off-field computer is adopted because it is difficult to execute image processing and complex computation with the on-board computer at the current stage.

The team is organized by a goalkeeper and four field players. Each player consists of the robot unit and the main computer unit, and both units are connected by two radio links (**Fig.1**). The main computer unit executes image processing and decides actions for the robot unit, and sends commands to it. The robot unit controls own motors based on commands from the main computer. The communication among the players is done by the Ethernet LAN connecting the main computer units.

Omni-directional vision and omni-directional mobility are good solutions of meeting rapidly changing situation in soccer games. We have successfully crammed all of these devices into the regulation size. The team name is derived

* The RoboCup 2001 competition participants

A. Birk, S. Coradeschi, and S. Tadokoro (Eds.): RoboCup 2001, LNAI 2377, pp. 599–602, 2002.

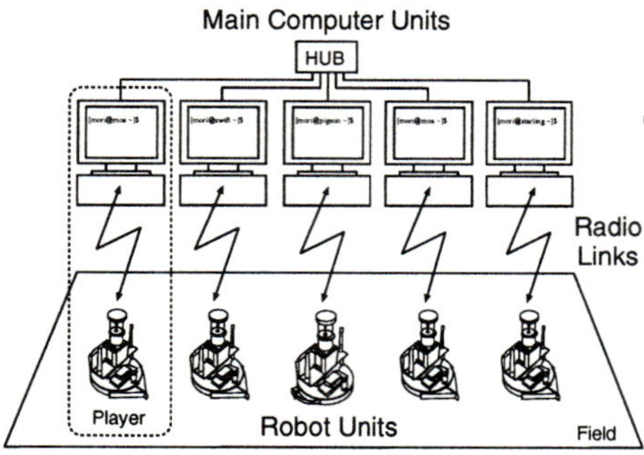

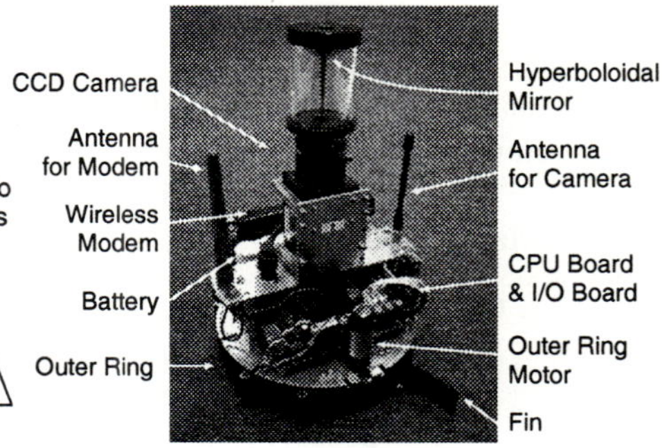

Fig. 1. Concept of the robot system **Fig. 2.** Overall view of the robot unit

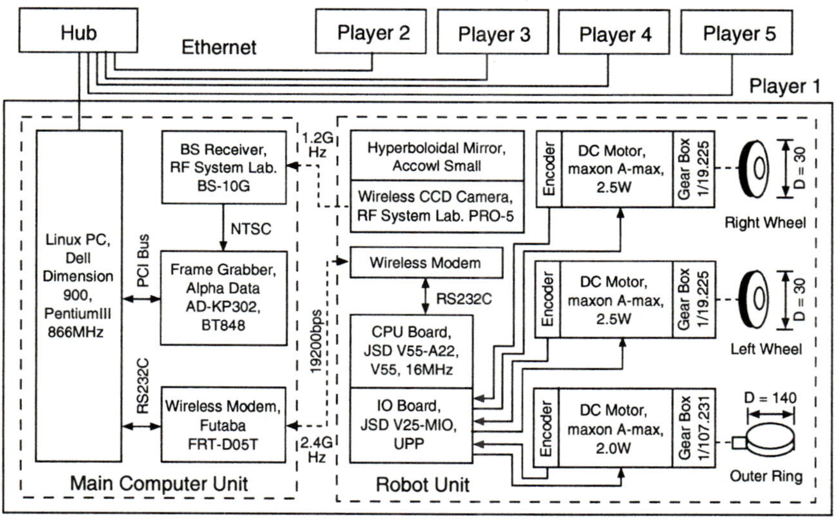

Fig. 3. Configuration of the robot system

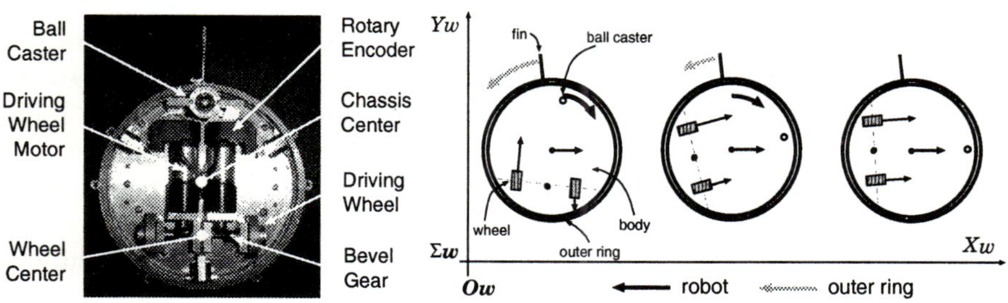

Fig. 4. Concept of the omni-directional mobile mechanism

from these abilities. An overall view of the robot unit and the configuration of the system are shown in **Fig.2** and **Fig.3** respectively. Two driving wheels of the robot unit are positioned 34mm away from the chassis's center(the distance between the driving wheels is 68mm.), the outer ring are equipped circling around the chassis. The outer ring can generate arbitrary translational and rotational velocities simultaneously by controlling motors for driving wheels and outer ring appropriately with the on-board computer(**Fig.4**). The outer ring also manipulates a ball with a fin equipped on it. The robot unit has a diameter of 140mm and a height of 225mm. The maximum translational velocity of the robot unit is 400mm/s. The maximum rotational velocity of the outer ring is 71.5rpm and able to give a ball the speed of 800mm/s maximum.

3 Software Configuration

Programs in the robot unit are developed with C language (Turbo C++) on MS-DOS. The main processing 9ms cycle is interrupted by the software servo routine with 3ms cycle time.

In the main computer unit, standard Linux is used as OS, but the time slice is changed from 10ms to 1ms. In addition, for the asynchronous processing, multi-thread programs are written in C language with the Linux Threads library[1]. Ovals and arrows in **Fig.5**(a) represent the threads and the data transferred among the threads through global variables respectively. The '`vision`' extracts color region. The '`ball`' estimates the current position of the ball. The '`location`' estimates a self-position. The '`fspace`' detects a free space around the robot. The '`kernel`' decides actions of the robot. The '`communicate`' sends commands to the robot unit. The '`playerNet`' updates information from the other robots. The '`gui`' displays information of the robot on X server.

The video signal from the camera on the robot unit is processed with a simple frame grabber and the CPU in the main computer unit. To extract colors and label regions, the modified CMVision[2] is employed. Performance of color classification is improved by representing color region in pillar shape with arbitrary base instead of cuboid shape in the YUV color space. The resolution of a captured image is 320×240 pixel^2. Obstacle avoidance and self-localization based on omni-directional imaging of floor region are developed, which are utilized in the actual competitions[3].

During game, the players exchange information such as distance and direction of a ball from itself, and self-position in the world coordinates. Then, their actions are decided based on role assigned beforehand or distances between the ball and all players, and position of a ball in the field without negotiations.

The main processing cycle in the main computer unit is 40$\sim$60ms.

4 Simulator

To reduce the difficulty in the experiments using the actual robot system, a simulator on which simulates actions of the multiple robots has been developed. It

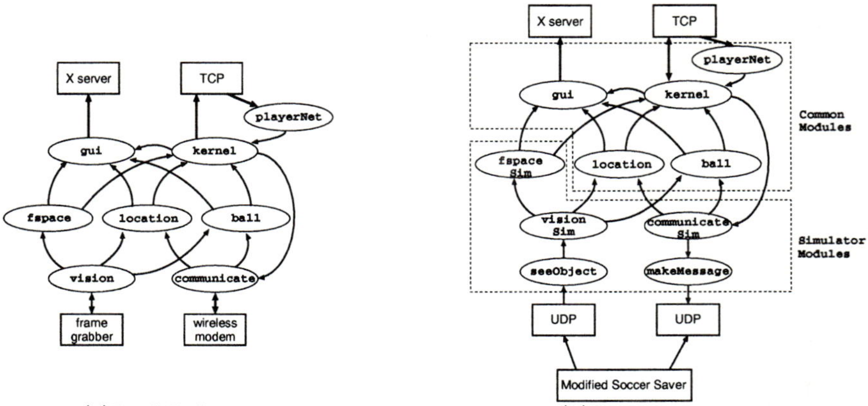

(a)Real Robot (b)Simulator

Fig. 5. Configuration of multi-threads

is based on the soccer server of the RoboCup Simulation League. The programs for the real robots can be executed on the simulator by substituting simulator modules for hardware modules in the main computer programs(**Fig.5**(b)). However, modification is considerably necessary in order to use it for the actual robots because physical models built in the original soccer server do not fully express the actual environment. For each player, omni-directional vision and omni-directional mobility are considered. Moreover occlusion phenomenon, which is an important problem for the local vision, is modeled. Currently, two computers with Pentium III 866MHz CPU and 64MB memory are necessary for simulating 5 versus 5 game.

5 Conclusion

The Team OMNI participated in the RoboCup2001 competition with the above-mentioned system. Unfortunately, the system did not work properly because of multi-path interference of wireless video link. We are planning to construct a pseudo local vision system using global vision in order to avoid the inconvenience. Moreover while we will improve each basic tasks (such as dribbling, passing, trapping and shooting) and advance study of complicated cooperation, we will aim at establishing the general decentralized autonomous robot system utilizing the features of omni-directional vision and omni-directional mobility.

References

[1] Linux Threads library web site `http://pauillac.inria.fr/~xleroy/linuxthreads/`
[2] CMVision web site `http://www-2.cs.cmu.edu/~jbruce/cmvision/`
[3] D.Sekimori, T.Usui, Y.Masutani, F.Miyazaki: "High-Speed Obstacle Avoidance and Self-Localization for Mobile Robots Based on Omni-Directional Imaging of Floor Region", The 5th RoboCup International Symposium, 2001

UQ RoboRoos: Achieving Power and Agility in a Small Size Robot

Gordon Wyeth, David Ball, David Cusack, Adrian Ratnapala

School of Computer Science and Electrical Engineering
University of Queensland, Australia
wyeth@csee.uq.edu.au

Abstract. The UQ RoboRoos have been developed to participate in the RoboCup robot soccer small size league over several years. RoboCup 2001 saw a focus on the mechanical design of the RoboRoos, with the introduction of an omni-directional drive system and a high power kicker. The change in mechanical design had implications for the rest of the system particularly navigation and multi-robot planning. In addition, the overhead vision system was upgraded to improve reliability and robustness.

1 Overview

The RoboRoos are one of the longest serving teams in the RoboCup Small Size league. The robots competed at RoboCup '98 where they were runner-up and in RoboCup '99 where they were eliminated in the quarter finals, despite a 56-1 goal difference during the round robin stage. In RoboCup 2000, the team was eliminated in the semi-finals after a 36 - 2 goal difference in the round robin stage. In RoboCup 2001, the team was eliminated by a "golden goal" in extra time during the quarter finals. Again, the team performed well in the round robin stage of the contest, achieving a 31 - 2 goal difference.

RoboCup 2001 saw a new look mechanical design for the RoboRoos. New robot bases were built featuring an omni-directional drive system. This base supported a new style of kicking device for the Small Size league: a cross-bow kicker capable of propelling the ball at 5m/s. These new features and their impact on the RoboRoos system is the focus of this team description. The drive system and kicker design are described after an overview of the system architecture. Following the description of the mechanical improvements, a brief description of improvements to the vision system is given. The conclusion focuses on the future of the RoboRoos.

1.1 System Architecture

Figure 1 shows the architecture of the RoboRoos system, which has remained constant since its inception. The vision system was extensively revised in 1999 with

A. Birk, S. Coradeschi, and S. Tadokoro (Eds.): RoboCup 2001, LNAI 2377, pp. 603-606, 2002.

excellent results [5], and was revised again in 2001. The new system uses the YUV color space, adapted from many of the principles and techniques used in the RGB color space described in [5]. This change and a new marking scheme on the robots has lead to improvements in accuracy, robustness and reliability that will be described in a later section.

The team planning system, MAPS, has been the focus of ongoing development [2, 3, 4], and represents one of the most significant research results of the RoboRoos development. MAPS forms cooperative strategies by observing team agents at the current point in time and choosing appropriate actions to increase the likelihood of cooperation in the near future. These actions are transmitted to the robots, along with the current state of the field, over the RF communications link. The RF communications link has received significant attention to ensure reliability under adverse conditions. The robots use the information from the communications link to set the current goal for navigation, and to build representations of obstacle maps for path planning. The navigation module for the robots has been another area of significant research effort [1], and provides the smooth controlled motion that is the signature feature of the RoboRoos.

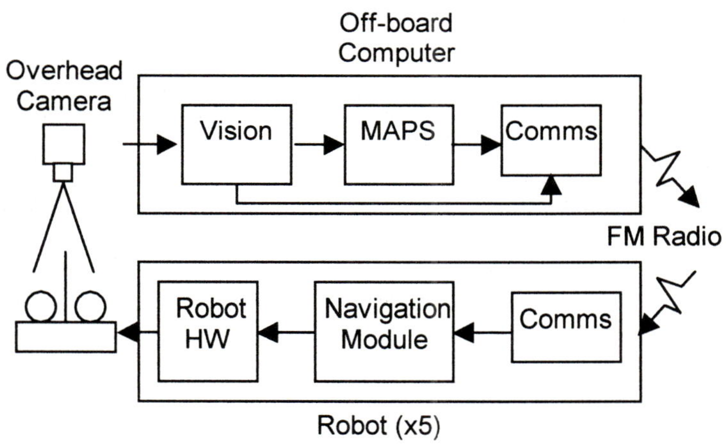

Figure 1. The RoboRoos system in overview.

1.2 Omni-directional Drive System

During RoboCup 2000, the team from Cornell illustrated an omni-directional drive system that allowed their robots to traverse in any direction while rotating at the same time. The maneuverability offered by their drive configuration showed clear advantages for soccer playing robots. During 2001, the RoboRoos team used the same principles for the design of new drive systems. The result is shown in Figure 2. The system is capable of accelerations up to 2.5 ms^{-2} while traversing at 2.2 ms^{-1}, which is comparable to the speeds used in practice by the original RoboRoos field robots. The overall height of drive system and the height of the centre of gravity are also comparable to the original RoboRoos, keeping the load of the robot evenly distributed over the wheels during periods of acceleration. This ensures that the wheels have limited slip, and path integration information from the built-in encoders can maintain an accurate estimate of robot position and orientation between camera updates.

Figure 2. The omni-directional drive system of the RoboRoos 2001 robots. The custom made spherical wheels spin freely about one axis and are driven in the other. Each wheel set is made as a pair so that as one wheel breaks contact with the ground the other makes contact.

1.3 Crossbow Kicker

From analysis of games in previous tournaments, it was clear that existing kicker technology played a pivotal role in the success of the best teams. Our analysis indicated that a kicker capable of kicking the ball at 5 ms^{-1} was required to score goals against the best defenses in the competition. Designs using existing technologies (such as rotating blades, solenoids and servo drives) are not capable of transferring sufficient energy without a significant cost in weight and volume. The solution adopted by the RoboRoos is to have a crossbow mechanism, where the crossbow string is a cotton coated elastic cord. The cord pulls against the kicking blade which is the element that transfers the cord's stored energy to the ball. Energy is stored in the cord by driving the kicking blade using a rack and pinion mechanism from a DC motor. The blade can be locked in a number of positions by the trigger mechanism, so that kick strength can be controlled. The current design enables kicks between 2 and 5 ms^{-1}. During the course of the competition, the kicker developed some mechanical failures, which meant that it could not be used at its full strength in the quarter-final in which the RoboRoos were defeated.

1.4 Vision

The RoboRoos vision system underwent a number of changes in 2001. The prime change was the use of YUV data rather than the RGB data. This change improved the robustness of the vision system further, particularly in relation to the classification of color markers used by some teams for robot identification. The RoboRoos black and white marking scheme was changed to help determine robot orientation. Whereas the previous robot shape allowed orientation to be determined from the overhead image of the robot, the symmetry of the new RoboRoos meant that a long white marker had to be placed on the black top of the robot. This scheme improved determination of orientation from ±7 degrees to ±1 degree, leading to improved efficiency in navigation and accuracy in shooting.

2 RoboRoos 2002

The RoboRoos introduced a number of changes in the team in 2001, most of which had not been fully tested or explored before the competition. In 2002, the RoboRoos will continue to test and develop the new mechanisms and improve the software that controls them. In addition, the RoboRoos will also experiment with ball control devices such as dribbling bars and explore the possibilities of a possession oriented game.

References

1. Browning B., Wyeth G.F. and Tews A. (1999) A Navigation System for Robot Soccer. Proceedings of the Australian Conference on Robotics and Automation (ACRA '99), March 30 - April 1, Brisbane, pp. 96-101.
2. Tews A. and Wyeth G.F. (2001) MAPS: A System for Multi-Agent Coordination. Advanced Robotics, VSP / Robotics Society of Japan, accepted for publication.
3. Tews A. and Wyeth G.F. (2000) Thinking as One: Coordination of Multiple Mobile Robots by Shared Representations, Proceedings IROS 2000.
4. Tews A. and Wyeth G.F. (1999) Multi-Robot Coordination in the Robot Soccer Environment. Proceedings of the Australian Conference on Robotics and Automation (ACRA '99), March 30 - April 1, Brisbane, pp. 90-95.
5. Wyeth G.F. and Brown B. (2000) Robust Adaptive Vision for Robot Soccer, Mechatronics and Machine Vision, Research Studies Press, pp. 41 - 48.

ViperRoos 2001

Mark M. Chang and Gordon F. Wyeth

Department of Computer Science and Electrical Engineering,
University of Queensland, St Lucia, QLD 4072, Australia.
{chang, wyeth}@csee.uq.edu.au

Abstract. The ViperRoos are a team of three autonomous local vision robots that were developed to play robot soccer. The team participated for the second time at the RoboCup 2001. This paper describes the on-board intelligent and the simulator implemented to aid the team development. The paper concludes with future direction of the ViperRoos.

1 Introduction

The ViperRoos are a team of soccer playing robots entering the F180 league at RoboCup. Unlike most other F180 teams, the Viper robot relies on on-board vision, rather than an overhead camera, as its primary means of perceiving the world. Each Viper robot is completely autonomous, and this makes ViperRoos different from some other local vision teams that use off-board PCs to perform the necessary vision processing and team planning. Therefore, much of the ViperRoos development devotes to hardware design to achieve faster on-board vision processing and reliable communication.

Since the basic design of both the hardware and the software has already been published in [1]. This paper concentrates on the progress of the continuing development of the team. The next section describes the DP-MAPS, a planning system derived from the MAPS which has been successfully developed by RoboRoos[3]. The third section explains the SimViper simulator, with its purposes and advantages towards team development. Section 4 concludes the paper with some thoughts for future development.

2 Intelligence

The on-board intelligence of the ViperRoos has been improved from the previous system. The behavior-based reactive system has been modified to perform the low level navigation only. While the high-level decisions for the Viper robots are now generated by DP-MAPS (Distributed Perception/Processing Multi-Agent Planning System). The DP-MAPS is derived from the MAPS system used by RoboRoos[3]. On the RoboRoos, the MAPS received the world view generated by the vision system using the images captured by the overhead camera. The cooperative action for each team member is obtained by constructing potential field using predefined parameters. The MAPS system has been proven to be

A. Birk, S. Coradeschi, and S. Tadokoro (Eds.): RoboCup 2001, LNAI 2377, pp. 607–610, 2002.

effective at several RoboCup competitions, the system is described in detailed in [2]. Clearly, the RoboRoos system is capable of feeding the MAPS system a global world model, and this allows MAPS to generate suitable action for each robots from the perspective of the whole team. Since ViperRoos utilize local vision as its only perception, the information that can be extracted from the on-board camera is limited. For the ViperRoos to achieve effective team cooperation, the perceiving information from all robots must be shared in order for DP-MAPS to produce suitable actions for each team member to achieve the team goal. At the current stage, perception information is not shared among the team, however, this is an intermediate stage for the development of DP-MAPS.

The vision system on the Viper robots gives information of the objects seen by the camera in a local sense. These coordinates are then transformed into global positions using the on-board localization information. For the DP-MAPS system, these global positions are further converted into large grids. Depending on the processing bandwidth available and the accuracy required, the grid size can be adjusted. DP-MAPS functions as a very high level planning system for the robot, it outputs only several commands such as search, defend, kick, and kick-to. Each command is accompanied with grid position to indicate where the robot should carry out the action. The navigation system handles the lower level robot behavior, it modifies the received DP-MAPS commands to accommodate obstacles avoidance and collisions. In addition, the position information is recalculated from the information provided by the proximity map, which is also generated by the vision system. Fig. 1 shows the software system of the Viper robot. The advantage of using DP-MAPS is that the robots can perform task cooperatively as long as their knowledge are shared. Ideally, the system will give similar performance as a global vision with accurate vision and reliable communication system.

3 Simulation

A kinematically accurate simulator for ViperRoos has been developed to ease the development phase. The SimViper software is implemented to simulate a soccer field which allows virtual Viper robots to play on it. Fig. 2 shows a typical 3 on 3 game with proximity map option turned on for debugging. The planning and the navigation codes on the robots can be taken from the robot and executed as a virtual robot without modification. Just like the real robot, the simulated robot requires vision input to see the world. SimViper generates a set of vision data according to the current state of the field, and this set of information is in the same format as the vision data produced by the real vision modules. The accuracy of the vision data can be altered to simulated the uncertainties that are usually presented in the real environment to some extent. This is a very useful feature, because it allows the developer to focus on the game strategy by running the robot with "perfect vision". For navigation, the simulator interacts with the virtual robot by interpreting the motor output and updating the robot position in the virtual field, the simulator also provides necessary encoder feedback.

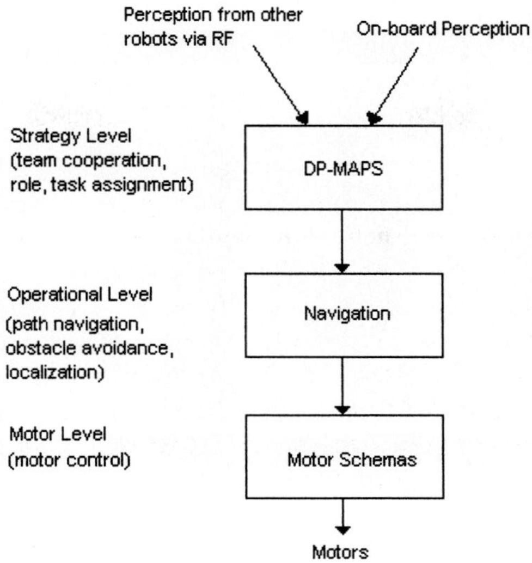

Fig. 1. The ViperRoos software system.

SimViper provides a GUI debugging environment for the ViperRoos. System parameters that are usually difficult to obtain can be observed easily. This is important for DP-MAPS, because the performance of DP-MAPS can only be observed while the game in progress. Therefore, SimViper is a convenient tool for calibrating the parameters for DP-MAPS with the option of diplaying potential field on the screen. The most important purpose of implementing SimViper is that the simulator reduces the time and effort that are usually associated with real robots, this includes programming the robots (because all codes reside on-board the robots), battery charging and time-consuming vision calibration. Although the system has to be tested on the real robots eventually to make final adjustments, we believe that the simulator is a convenient tool for the development of the ViperRoos.

4 Conclusions and Future Work

This paper has described the new on-board intelligence for the ViperRoos system and the SimViper simulator. The development on local vision teams in F180 league can depend largely on the hardware technology. There is considerable remaining work to develop the ViperRoos into a truly ccompetitiverobot soccer team. This work requires both the redesign of the system to incorporate the latest technology and the improvement on the on-board intelligence.

At the time of submitting this article, the final testing phase of the new vision module is being carried out. The new vision module consists of a new color

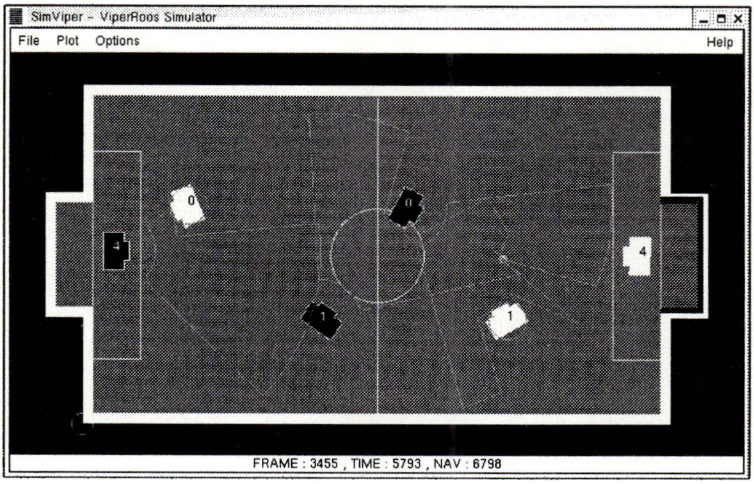

Fig. 2. SimViper - ViperRoos Simulator.

camera that provides additional features not found on our previous camera, this includes subsampling, interlaced and yuv output. In addition, the new vision processing board has the capability of performing image processing and DP-MAPS at 360MIPS. The increased processing power can be allocated to handle higher resolution images, faster frame rate or complex vision processing routines. Once the new vision module is completed, we wish to focus our development on issues such as inter-robot communication and special kicking and/or dribbling hardware.

References

1. M. Chang, B. Browning, G. Wyeth: ViperRoos 2000. In RoboCup-2000: Robot Soccer World Cup IV, LNAI 2019, editor, P. Stone, T. Balch, G. Kraetzschmar, Springer-Verlag, 2001, pp. 527-530.
2. A. Tews, G. Wyeth: MAPS: A System for Multi-Agent Coordination. Advanced Robotics, VSP/Robotics Society of japan, Volume 14 (1), pp. 37-50.
3. G. Wyeth, A. Tews, B. Browning: UQ RoboRoos: Kicking on to 2000. In RoboCup-2000: Robot Soccer World Cup IV, LNAI 2019, editor, P. Stone, T. Balch, G. Kraetzschmar, Springer-Verlag, 2001, pp. 555-558.

AGILO RoboCuppers 2001: Utility- and Plan-Based Action Selection Based on Probabilistically Estimated Game Situations

Thorsten Schmitt, Sebastian Buck, and Michael Beetz

Forschungsgruppe Bildverstehen (FG BV) – Informatik IX
Technische Universität München, Germany
{schmittt,buck,beetzm}@in.tum.de
http://www9.in.tum.de/robocup/

Abstract. This paper describes the *AGILO RoboCuppers* [1] the RoboCup team of the image understanding group (FG BV) at the Technische Universität München. With a team of four Pioneer I robots, all equipped with CCD camera and a single board computer, we've participated in all international middle size league tournaments from 1998 until 2001. We use a modular approach of concurrent subprograms for image processing, self localization, object tracking, action selection, path planning and basic robot control. A fast feature extraction process provides the data necessary for the on-board scene interpretation. All robot observations are fused into a single environmental model, which forms the basis for action selection, path planning and low-level robot control.

1 Introduction

The purpose of our RoboCup activities is to provide a software and hardware platform for (1) pursuing research in robot vision, probabilistic state estimation, multi-robot cooperation, experience based learning, and plan-based control; (2) supporting undergraduate and graduate education in computer vision, artificial intelligence, and robotics; and (3) performing software engineering of large dynamic software systems.

2 Hardware Architecture

We aim at realizing the AGILO RoboCup team on the basis of inexpensive, off the shelf, easily extendible hardware components and a standard software environment.

The team consists of four Pioneer I robots [1] each equipped with a single onboard linux computer (Pentium 200 MHz CPU, 64 MB RAM, 2.5" hard disk, on-board ethernet, and VGA controller, and an inexpensive BT848-based PCI video capture card). They are supported by an additional PC which resides outside the playing field. It is used to fuse the observations made by the individual team members and to monitor the

[1] The name is derived from the Agilolfinger, which were the first Bavarian ruling dynasty in the 8th century, with Tassilo as its most famous representative.

A. Birk, S. Coradeschi, and S. Tadokoro (Eds.): RoboCup 2001, LNAI 2377, pp. 611–615, 2002.

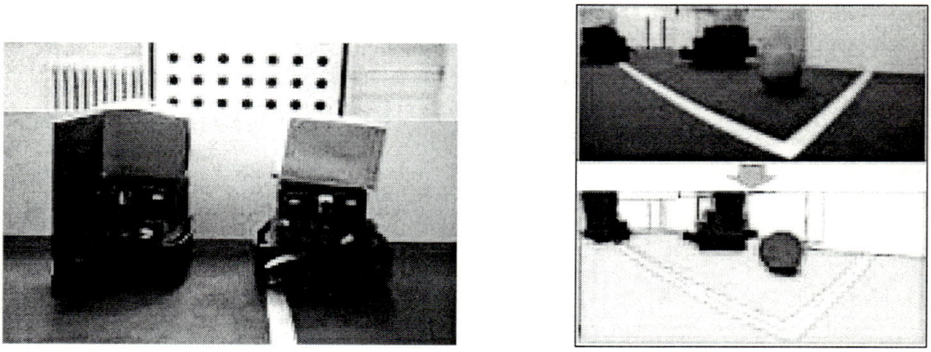

Fig. 1. (a) Theodo (goal keeper), Hugibert (attacker) and (b) what they perceive of the world around them.

robots' data and states. All robot computers and the additional PC are linked via a 10 Mbps wireless ethernet (5.8 GHz) [3, 2].

Fig. 1 (a) shows two[2] of our Pioneer 1 robots. Inside the robot a Motorola microprocessor controls the motors, reads the wheel encoders and the seven ultrasonic sonars, and communicates with the single board computer EM-500, which is mounted within a box on the topside of the robot. PC and robot are connected via a standard RS232 serial port. A PAL color CCD camera is mounted on top of the robot console and linked to the S-VHS input of the video capture card. Gain, shutter time, and white balance of the camera are adjusted manually. Currently RGB-16 images are captured with a resolution of 384 $\times$ 172 (bottom 60% of half PAL resolution). For better ball guidance we've mounted a simple concave-shaped bar in front of each robot. A custom made kicking device enables the robot to kick the ball in direction of the robot's current orientation.

3 Fundamental Software Concepts

The software architecture [9] of our system is based on several independent concurrent modules. The modules are organized hierarchically into main, intermediate, and basic modules. The main modules are image (sensor) analysis, information fusion, action selection, path planning, and robot control. Beside the main modules the system employs auxiliary modules for monitoring the robots' states. The key software methods employed by the AGILO RoboCuppers software are:

1. vision-based cooperative state estimation for dynamic environments,
2. synergetic coupling of programming and experience-based learning for movement control and situated action selection, and
3. plan-based control of robot teams using structured reactive controllers (SRCs).

[2] All attackers and defenders have identical shape and equipment.

3.1 Self Localization, Object Tracking, and Data Fusion

The vision module is a key part of our system. Given a raw video stream, the module has to estimate the pose of the robot and the positions of the ball and opponent robots. Low/medium-level image processing operations are performed with the help of the image processing library HALCON 5.2.3 (formerly known as HORUS [7]).

We have developed a probabilistic, vision-based state estimation method for individual, autonomous robots. This method enables a team of mobile robots to estimate their joint positions in a known environment and track the positions of autonomously moving other objects. All poses and positions of the state estimation module contain a description of uncertainty, i.e. a covariance matrix. The state estimators of different robots cooperate to increase the accuracy and reliability of the estimation process. This cooperation between the robots enables them to track temporarily occluded objects and to faster recover their positions after they have lost track of them. A detailed description of the self localization algorithm can be found in [8] and the algorithms used for cooperative multi-object tracking are explained in [13, 12].

Our vision algorithms can process up to 25 frames per second (fps) on a 200 MHz Pentium PC. The average number of images processed during a match is between 12 and 17 fps. This is due to computational resources being shared with the path planning and action selection modules.

3.2 Experience Based Learning for Situated Action Selection, Path Planning, and Movement Control

Another major field of our research activities is *automatic robot learning based on experiences gained from exploration.* Experience based learning provides a powerful tool for the automatic construction of high-performance action selection and low-level robot control. In this respect experience based learning can effectively complement other methods for developing such controllers, in particular the hand coding of controllers. We use learning from experience in several parts of our system such as low level robot control, path planning and action selection.

In low level robot control we represent the state of a Pioneer I robot as a quintuple $\zeta = \langle x, y, \varphi, v, w \rangle$, where x and y are coordinates in a global system, φ is the orientation of the robot and v and w are the translational and rotational velocities, respectively. The low-level robot controller accepts commands of the form $\xi = \langle v, w \rangle$. A neural network maps the desired state changes to low level robot commands:
$Net : \langle \zeta_0, \zeta_{target} \rangle \mapsto \xi_0$. To train this network we measure a huge number of state changes according to different executed low level commands [6]. Doing so our neural controller is based on nothing but experience *not* making any assumptions.
In order to find the optimal path planning algorithm for our RoboCup robots we statistically evaluated different methods and found out that there is *no optimal* algorithm but a number of navigation problem classes each performed best with a certain algorithm/parameterization [6]. These classes are defined with the help of a feature language. In order to select the best method for the given situation we've learned a decision tree [11]. The training data is obtained from accurate robot simulations where a huge number of path planning problems were performed with different algorithms each.

The selection of an appropriate action is performed on the basis of a fused environmental model. A set of possible actions A such as go2ball, shoot2goal, dribble, block... is defined. For all robots and each of those actions a_i success rates $P(a_i)$ and gains $G(a_i)$ are estimated [5]. From all promising actions $P(a_i)$, which exceed a predefined threshold Θ_{a_i} the one a_{ex} with the highest gain is chosen to be carried out.

3.3 Plan-Based Action Control

While our situated action selection aims at choosing actions that have the highest expected utility in the respective situation it does not take into account a *strategic* assessment of the alternative actions and the respective *intentions* of the team mates. This is the task of the plan-based action control.

In order to realize an action assessment based on strategic consideration and on a considerations of the intentions of the teammates, we develop a robot soccer playbook, a library of plan schemata that specify how to perform individual team plays. The plans, or better plays, are triggered by opportunities, for example, the opponent team leaving one side open. The plays themselves specify highly reactive, conditional, and properly synchronized behavior for the individual players of the team.

The high-level controller is realized as a structured reactive controller (SRC) [4] and implemented in an extended RPL plan language [10]. The high-level controller works as follows. It executes as a default strategy the situated action selection that we have described in Sec. 3.2. At the same time, the controller continually monitors the estimated game situation in order to detect opportunities for making plays. If an opportunity is detected, the controller decides based on circumstances including the score and the estimated success probability of the intended play whether or not to perform the play.

Our research goals in the development of the high-level controller include the development of a computational model for plan-based control in very dynamic multi-robot applications and for the integration of learning processes into the plan-based control.

References

[1] ActivMedia Robotics. http://www.activmedia.com/robots/.

[2] Delta Network Software GmbH. http://www.delta-network-systems.com/, http://www.delta-net.de.

[3] RadioLAN Inc. http://www.radiolan.com.

[4] M. Beetz. Structured reactive controllers. *Journal of Autonomous Agents and Multi-Agent Systems*, 4:25–55, 2001.

[5] S. Buck and M. Riedmiller. Learning Situation Dependant Success Rates of Actions in a RoboCup Scenario. In R. Mizoguchi and J. Slaney, editors, *Pacific Rim International Conference on Artificial Intelligence (PRICAI)*, number 1886 in Lecture Notes in Artificial Intelligence, page 809. Springer-Verlag, 2000.

[6] S. Buck, U. Weber, M. Beetz, and Th. Schmitt. Multi robot path planning for dynamic evironments: A case study. In *International Conference on Intelligent Robots and Systems (IROS)*, 2001.

[7] Wolfgang Eckstein and Carsten Steger. Architecture for Computer Vision Application Development within the HORUS System. *Journal of Electronic Imaging*, 6(2):244–261, April 1997.

[8] R. Hanek and T. Schmitt. Vision-based localization and data fusion in a system of cooperating mobile robots. In *International Conference on Intelligent Robots and Systems (IROS)*, 2000.

[9] M. Klupsch. Object-Oriented Representation of Time-Varying Data Sequences in Multiagent Systems. In N.C. Callaos, editor, *International Conference on Information Systems Analysis and Synthesis (ISAS'98)*, pages 833–839, Orlando, FL, USA, 1998. International Institute of Informatics and Systemics (IIIS).

[10] D. McDermott. A reactive plan language. Research Report YALEU/DCS/RR-864, Yale University, 1991.

[11] R. Quinlan. Induction of decision trees. In *Machine Learning 1 (1)*, 1986.

[12] D. Reid. An algorithm for tracking multiple targets. IEEE Transactions on Automatic Control, 24(6):843–854, 1979.

[13] T. Schmitt, R. Hanek, S. Buck, and M. Beetz. Cooperative probabilistic state estimation for vision-based autonomous soccer robots. In *5th International Workshop on RoboCup (Robot World Cup Soccer Games and Conferences)*, volume 5 of *Lecture Notes in Computer Science*. Springer-Verlag, 2001 submitted.

Artisti Veneti: An Heterogeneous Robot Team for the 2001 Middle-Size League

E. Pagello[1,2], M. Bert[1], M. Barbon[1], E. Menegatti[1], C. Moroni[1], C. Pellizzari[1], D. Spagnoli[1], and S. Zaffalon[1]

[1] Intelligent Autonomous Systems Laboratory
Department of Informatics and Electronics
University of Padua, Italy
[2] Institute LADSEB of CNR
Padua, Italy
epv@dei.unipd.it

Abstract. We illustrate our new team *Artisti Veneti* a new entry in the Middle-size league from The University of Padua (Italy). The team is composed of heterogeneous robots that use only vision as perception system. The vision systems have been designed separately for each robot. Our players are coordinated in the frame of ADE (Artisti Veneti's Development Environment), a multi-thread distributed real-time Environment working under Linux OS. Cooperative abilities, like exchanging a ball, can be achieved through the use of efficient collision avoidance algorithms using roles swapping triggered using an enhanced reactivity approach.

1 Introduction - A Short History of Our Team

Artisti Veneti Team has been founded in 2001, to participate to RoboCup01 in Seattle, by Enrico Pagello, who have coordinated locally the Padua Branch of ART Team, in 1998, 1999, and 2000. The "robotic" members of our team include five players, *Bart*, *Homer*, *Lisa*, *Nelson* and Barney. *Bart and Homer (*designed at our Lab. on 1999), have played with ART'99, by scoring a total of 5 goals in 9 games. *Homer* won a Technical Challenge against Friburgh University. *Bart*, *Homer*, and *Lisa* (designed at our Lab. on 1999/2000), have played successfully with ART-2000, which won the Second Place Cup of RoboCup-Euro2000, and in Melbourne. This year two new players have been added to our robotic team: *Nelson* (based on a Pioneer2 basis) and *Barney*, a Golem robotic platform of Golem Robotics.

A. Birk, S. Coradeschi, and S. Tadokoro (Eds.): RoboCup 2001, LNAI 2377, pp. 616–620, 2002.

Fig. 1. Our heterogeneous robot team. (From left to right) Nelson, Lisa, Barney and Bart (its clone Homer is not displayed).

2 Our Five Players: Bart, Homer, Lisa, Nelson, and Barney

Artisti Veneti has been conceived as an heterogeneous robot team [1]. *Lisa, Bart, Homer* and *Nelson*, even if built upon a similar base (a Pioneer 1 or a Pioneer 2) have several differences, even in the mechanical structure. In *Bart*, and its clone *Homer*, the lower part of the chassis is the original one of the Pioneer, while the chassis of *Nelson* and *Lisa* have being totally reshaped. Eventually, *Barney* is an omnidirectional robot, with a chassis with a circular symmetry..

Homer & Bart use a Pioneer1 base, Intel P233 MMX processor, 64 MB RAM, BT848 FrameGrabber. The kicker used to catch the ball is pneumatic: two pistons independently commanded allow to turn the shovel to the right or to the left. They mount standard perspective cameras.

Lisa uses a Pioneer1 base, AMD k6-3 500 MHz processor, 64 MB RAM, BT848 FrameGrabber. Lisa mounts a special kicker inspired by the Kicker used by Galavron, ART's Goalkeeper. It allows both frontal and lateral kicks. Lisa uses an omnidirectional vision sensor designed by ourselves [5].

Nelson is a new player based on a Pioneer 2 base, Intel Pentium 800MHz, 64 MB RAM, BT848 FrameGrabber. The chassis has been totally rebuilt. Nelson mounts an omnidirectional vision system designed by ourseves [5]. The kicker is fitted with a third piston to raise the ball to perform lobs.

Barney, an holonomic platform with an omnidirectional vision sensor, has been borrowed from Golem Robotics. We have developed a new set of behaviours for the robot and integrated it in the heterogeneous team.

[1] For a more detailed description of our team please refer to the web site *www.dei.unipd.it/~robocup*, where you can download a longer version of this document.

3 Our Vision System

The goalkeeper, Nelson and Barney mount omnidirectional vision systems, Bart and Homer mount standard cameras. The cameras are the only sensors exploited by our robots to locate themselves, the ball and the opponents.

The omnidirectional vision system is composed by a camera pointed upward, looking at a multi-part mirror. In [5] we give details on the design of the mirrors and how a custom designed mirror permits to identify new effective behaviours.

The low-level processing of the image is the same for all sensors[2]. In the Robocup domain the colours of the objects in the field of play are precisely coded. Usually, to associate a colours to a pixel, the pixel value is confronted with a set of thresholds in the colour space. This approach has two shortcomings. First, the thresholds are set by hand, resulting in a skillful and time consuming process. Second, defining a lower and an upper thresholds for a colour, delimits a box in the colour space, that usually poorly describes the disposition of the object's pixels in the colour space. To overcome these problems, we propose a new way to associate a colour to a pixel. At the set-up stage, the operator selects the region of the image containing the object and associates it a colour. The pixels of the object form a cluster in the colour space. The convex hull containing these pixels is created. A colour is then assigned to a pixel only if this pixel is inside one of the convex hulls. This process is performed in real-time using a look-up table. Our colour segmentation algorithm presents two main advantages. First, the convex hulls fits more closely the actual distribution of the image pixels in the colour space, with respect to the previously mentioned boxes. Second, the set-up process is much faster and reliable because the trial and error procedure in setting the thresholds is avoided.

4 Our Planning System

In our previous works, limited global planning capabilities are obtained by balancing between deliberative and reactive capability [1], whereas in our team we have used an enhanced reactivity approach. At the reactive level, our robots are programmed using a *behavior-based* approach. Three different robot roles [6] has been introduced by specifying a set of behaviors [2]. The three roles are: **attacker**, **midfielder** and **defender**. Each basic behavior is realized as a thread in ADE. ADE (Artisti Veneti's Development Environment), is a multi-thread distributed real-time Environment working under Linux OS. ADE has been inspired by the coordination environment used by ART [6].

ADE allows to create a set of processes structured as threads. Each thread can communicate, through message passing, with other threads of the same process and also with other processes running on other processors. Each thread can allow or deny itself to other threads of lower priority. When a segmentation fault happens, it is possible to kill the thread that has caused the error, or to

[2] Except for Barney that uses the original vision software of Golem Robotics.

restart it with or without a new initialization. Each thread can be suspended for a while, and kept ready to be resumed later.

A *measure of quality Q*, able to triggers the proper role, was introduced time ago, at IAS Lab. of Padua Univ., for evaluating how much work must be done by a robot to get the ball in the best position to score [2]. Our robots are able to show a cooperative action, like a *ball exchange*, by coordinating their basic behaviors through the dynamic assignment of the above three roles realized by a set of behaviors that exploit some smart collision-free motion strategies, based on the computation of field vectors [9]. During the 2001 Seattle competition, we had several situations in which the robots swap their roles, and succeed to *exchange the ball*, as an emergent behavior like it was in Stockholm'99 and Amsterdam'00.

5 Conclusions and Acknowledgements

We illustrated our new team *Artisti Veneti*, based on five players. *Bart*, *Homer*, *Lisa*, *Nelson* and *Barney*. *Bart* and *Homer* played together successfully with ART team at both RoboCup'99 and Euro-2000 competitions. *Lisa* played at Euro-2000 for the first time as a reserve goalkeeper. All the three have also played with ART at RoboCup'2000. Two new players, *Nelson* and *Barney* have been added to the team this year. Our *vision systems* rely on omnidirectional vision, [4], [5], for both the goalkeeper, and the new players. Our team plays according to an Emergent Behavior Engineering approach [7], [8], implemented in the frame of ADE, a new multi-thread distributed real-time Environment.

Artisti Veneti is supported by a Special Research Project on "Cooperative Multi-robot Systems" granted by The University of Padua. Additional founding are given by MURST, by the Institute LADSEB of CNR, and by ENEA (Parallel Computing Project). We acknowledge all the students at the Eng. School of Padua Univ., who have contributed in the past to the researches illustrated in this paper by participating to the PaSo-Team (in the Simulation League) and to the Padua Branch of ART Team (in the Middle-size League). We thank F. Garelli, who has contributed to develop ADE Environment. We thank Padova Ricerche s.p.a. We thank the Golem Robotics for the Golem platform they lent us. We like also to acknowledge all the members of ART'98/'99/00, in cooperation with whom we developed several research approaches that are now used in our new team *Artisti Veneti.*

References

[1] Carpin, S., Ferrari, C., Pagello, E., Patuelli, P. *Scalable Deliberative Procedures for Efficient Multi-robot Cordination.* in M. Hannebauer, E. Pagello, and J. Wendler (Eds.) *Balancing Reactivity and Social Deliberation in Multi-Agent Systems,* Springer 2001,

[2] Ferraresso, M., Ferrari, F., Pagello, E., Polesel, R., Rosati, R., Speranzon, A., *Collaborative Emergent Actions between Real Soccer Robots,* T. Balch, G. Kraetzschmar, and P. Stone (Eds.) Eds., *RoboCup-2000: Robot Soccer World Cup IV.* L. N. on A. I., Springer 2001

[4] Menegatti, E., E. Pagello, Wright, M., *A new omnidirectional vision sensor for the Spatial Semantic Hierarchy.* Proc. of The Int. Conf. on Advanced Intelligent Mechatronics (AIM'01), Como (Italy)- July 2001

[5] Menegatti, E., Nori, F., E. Pagello, E., Pellizzari, C., Spagnoli, D., *Designing an omnidirectional vision system for a goalkeeper robot.* A. Birk, S. Coradeschi, and P. Lima (Eds.) Eds., *RoboCup-2001: Robot Soccer World Cup V.* L. N. on A. I., Springer 2001

[6] Nardi, D., Adorni, G., Bonarini, A., Chella, A., Clemente, G., Pagello, E., and Piaggio, M. *ART99 Azzurra Robot Team.* in M. Veloso, E Pagello, H. Kitano Eds., *RoboCup-99: Robot Soccer World Cup III.* L. N. on A. I., Springer 2000, pp. 695-698.

[7] Pagello, E., Ferrari, C., D'Angelo, A., Montesello, F. *Intelligent Multirobot Systems Performing Cooperative Tasks.* Invited paper of Special Session on "Emergent Systems - Challenge for New System Paradigm". IEEE/SMC, Tokyo, 1999, pp. IV-754/IV-760

[8] Pagello, E., D'Angelo, A., Montesello, F., Garelli, F., Ferrari, C. *Cooperative behaviors in multi-robot systems through implicit communication.* Robotics and Autonomous Systems, Vol. 29, No. 1, pp. 65-77, 1999

[9] Polesel, R., Rosati, R., Speranzon, A., Ferrari, C., Pagello, E. *Using Collision Avoidance Algorithms for Designing Multi-robot Emergent Behaviors.* Proc. of IEEE/JRS Int. Conf. on Intelligent Robots and Systems (IROS'2000), Takamatsu, Japan, 2000, pp. 1403-1409

Basic Requirements for a Teamwork in Middle Size RoboCup

M. Jamzad[1], H. Chitsaz[1], A. Foroughnassirai[2], R. Ghorbani[2], M. Kazemi[1], V.S. Mirrokni[1], and B.S. Sadjad[1]

[1] Computer Engineering Department
Sharif University of Technology, Tehran, Iran
{chitsaz,nassirae,ghorbani,kazemi,mirrokni,sadjad}@linux.ce.sharif.ac.ir
jamzad@sina.sharif.ac.ir/∼ceinfo
[2] Mechanical Engineering Department
Sharif University of Technology, Tehran,Iran.
http://www.sharif.ac.ir/∼mechinfo

Abstract. In this paper we describe some mechanical, hardware and software aspects of our robots specially used in teamwork. The pneumatic design of the gripers and kicker enables the robot to make a good control of ball when dribbling and also passing the ball in short and long distances. The teamwork software enables our robots to perform a cooperative behavior by dividing the field to three major regions and assigning each robot a role such as defender, forward and middle in these three regions. The server can order the robots to change their role when needed and make them return to their original state after a short time.

1 Introduction

In a successful soccer robot play, all robots should behave in a cooperative manner [4]. This can be done if each robot has some minimum performance capabilities. Which is, they must have the mechanical capability to pass the ball in short and long distance, good dribbling ability and some sensing devices to inform the robot of the special situation going on around it. In addition, the software should be able to implement a cooperative behavior among robots with a certain strategy. In this paper we describe some aspects about mechanical design of our robots, sensing devices such as touch sensors and infrared sensors installed around robot body, and finally we give a brief description of our software for teamwork.

2 Robot Pneumatic System

The mechanical design, and controlling software of our robot [1], enables it to dribble the ball in various situations. However, there are some cases that fixed position griper will be a limit to fast dribbling, specially when we want to rotate about robot center while having the ball in gripers. For example, according to

A. Birk, S. Coradeschi, and S. Tadokoro (Eds.): RoboCup 2001, LNAI 2377, pp. 621–626, 2002.

the RoboCup rule practiced in 99, the gripers should be positioned in such a way that in all times if a line is connected between the head of two gripers, at most $\frac{1}{3}$ of ball will be located within the griper area. To maintain this rule, we have designed variable length gripers such that each griper can possess 3 lengths: $0, 6.6$ or $9.5cm$. Therefore we will have four cases presented in figure 1 as follows: (a) Both gripers have length zero. (b) Right griper is $9.5cm$ and left griper has zero length. (c) Left griper is $9.5cm$ and right griper has zero length. (d) Both gripers have $6.6cm$ length.

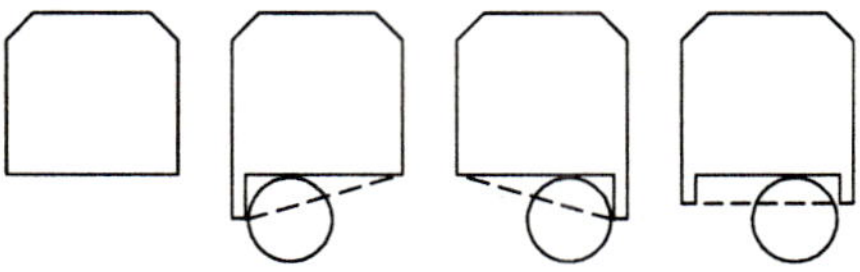

(a) (b) (c) (d)

Fig. 1. Ball position in robots' gripers in four situations

During the game, when the robot does not have the ball, the gripers are in position (a). The reason for this is because the robot size will become smaller and it can maneuver around safer. Gripers are set in position (d), when the robot is moving with the ball in its gripers.

One major advantage of our robot is its capability to rotate left or right round any point on the field or on the robot body, while holding the ball within its gripers [1]. However, if both gripers are fixed in $6.6cm$ in fast rotations the robot loses the ball. In these situations we make the gripers to stay in position b or c.

The displacement of each griper is done by a pneumatic actuator with a maximum course of $10cm$.

On the other hand, to enable the robot to perform a reliable teamwork by means of short or long and quick passes, the kicking power needs to be controlled. The pneumatic actuator we used has a course of $3cm$, where, with a special mechanical design, the actuator linear displacement was transformed to an arm movement (the kicker). The maximum speed of kicking is 3 to 4 $\frac{m}{sec.}$. By this mechanism, we can make the robot pass the ball with desired speed. This is done in software, by changing the settings of time interval in turning on and off the solenoid valve of pneumatic actuator.

3 Robots' Sensing Devices

Infra-red (IR) Sensors, Touch Sensors, and a Video Camera are used in our robots. There are 12 Infra-red sensors around the robot two of which are on the two gripers. Each IR sensor has a transmitter and a receiver. Each sensor has a limited and straight view sight and outputs 1 if there is anything within its 10cm distance in its sight of view and outputs 0 otherwise. By this information, the robot can detect nearby barriers and obstacle and avoid bumping into them. But as there are many situations in which there are more than one obstacle near the robot, and the robot needs to know the direction of actual stuck, there is a need to a touching mechanism. Thus, we use 24 touch sensors around the robot two of which are in front of the two gripers. Each touch sensor outputs 1 if something touches it and 0 otherwise. Finally, the main sensor in our robots is the Sony Handycam color camera. We use a low-price frame-grabber based on BT878 chip in our robots to capture the image.

4 Robots' Algorithms

4.1 Algorithms for Going to a Point with Dribbling

One of the aspects, that is important for robot movement, is its ability to go to a certain destination while doing obstacle avoidance. In the following, we describe how our robots can perform this task.

- *Finding wheels setting for different velocities:*
  In our software we defined an instruction named *go_angle_distance*(α, d) for setting the velocity of wheels. By changing the parameters of this instruction, we can force the robot to make a curved shape move toward a point which is in distance d from the current position of the robot. This move is done in such a way that when the robot is reached the desired point, the robot front side would have an angle α with respect to the robot front in its initial stage, as it is shown in figure 2. In other words, if d is zero, it means that robot has rotated α degrees around point O.
  Since our robot does not have a PID control, we had to find the PWM settings for the left and right wheel motors to make the robot move certain distances with fixed angles. The PWM setting for all other distances and angles were calculated by linear interpolation of the above settings.
- *Tracing an object and go toward it:*
  The following loop shows how the robot can go toward an object o:

```
while (distance(o) > near_distance){
  if (distance(o) < near_distance + 10(cm))
    go_angle_distance(angle(o), distance(o));
  else
    go_angle_distance(angle(o), near_distance + 10(cm));
}
```

  Where distance(o) and angle(o) are the distance and angle from an object o and near_distance is a constant used for close distances.
- *Dribbling other robots*
  Dribbling is a mean of collision avoidance for mobile robots. A robot can not plan to dribble other robots which are too far from it. However, a robot

should be able to dribble other robots which are within a close distance to it. In our dribbling procedure, we assume a bounding box around the opponent robot that should be dribbled [2].
As it is seen in figure 2, α and β are the left and right angles of our robot with respect to the left and right side of the opponent robot. The dribbling procedure is described as follows:

```
while (distance(o) > near_distance){
  // near_robot_distance means the close distance within which
     we dribble.
  if ((distance(robot) < near_robot_distance) &&
     (distance(robot)< distance(o))) {
    if (fabs(left_angle(robot)) < fabs(right_angle(robot)))
      go_angle_distance(left_angle(robot) - 15,
      distance(robot));
      // It is easier to dribble from left side angle
    else go_angle_distance(right_angle(robot) + 15,
    distance(robot))
  }
  else{
    // The same code given in "Tracing an object and go toward
       it" section.
  }
}
```

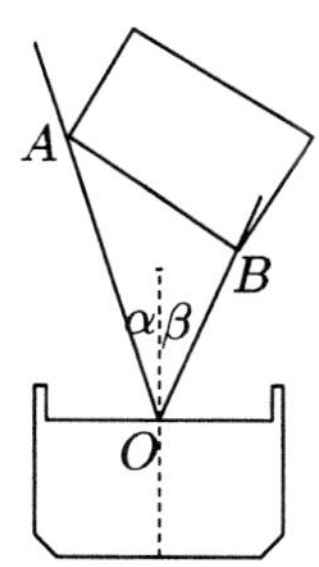

Fig. 2. Left and right angles of our robot with respect to an opponent robot.

However the above code is a general one without much details. In practice, there are special cases when IR sensors, touch sensors and other software measures indicate cases that we have to modify the above algorithm.

4.2 Teamwork

Teamwork can be seen in different aspects. In all cases, robots should be positioned in the field with appropriate distribution. We can virtually divide the field into regions where each region can be the territory of one or more robots. Since we are allowed to have only 3 players in the field, we decided to have a forward, a middle and a backward player with defined territories.

Fig. 3. A picture of our player robots.

The server implements dynamic task assignment to all robots to make sure each robot remains in its own territory. In practice depending on the game situation, robots can change their role, which is, a defender can become a forward if needed and then return to become defender. The most efficient method for task assignment is dynamic task assignment. In this method, server assigns the tasks to the robots so that each robot will be closer to its own territory. In our teamwork strategy we consider the following main situations: (a) None of our robots can see the ball. (b) At least one of our robots can see the ball, and the ball is not in their control. (c) One of our robots is carrying the ball.

In situation (a), all robots will go to their original region after some time and start searching for the ball in those regions.

In case (b), the server selects the robot which is closer to the ball and order it to go and get the ball. Other robots which are seeing the ball, will go toward the ball, but stay away from it within a certain distance. This behavior will prevent our robots to fight for the ball and make enough room for only one robot to control the ball. Another reason for this behavior is that, if the robot which was going to get the ball could not get it for any reason, then other robots will have a good chance of getting the ball.

In situation (c), when one robot is carrying the ball, at first, the other two robots will trace the ball (rotate round and keep seeing the ball in their image center), and then after some time, they will go to the backward and middle regions. For example, if the backward player owns the ball, the forward player goes to the middle of the field and middle player goes to the backward region. We use a LANEscape WL2420 wireless LAN card working in 2.4GHz frequency as the communication mean and an Ad-Hoc architecture.

In addition, we keep a history of ball positions and remember the last position the ball was seen. In situations when no robot can see the ball, the server can inform the last position of the ball to the robot which is closer to it and order that robot to go for the ball in that region. However, we think a good team work requires a perfect localization method, which can be achieved by a CCD camera in front and an omnidirectionl view on top of robot [3]. Figure 3 shows a picture of our player robots.

References

1. M. Jamzad, A.Foroughnassirai, et al, *Middle sized Robots: ARVAND*, RoboCup-99: Robot Soccer world Cup II, Springer (2000), pp 74-84 .
2. M. Jamzad, et al, *a fast vision system for middle size robots in RoboCup*, submitted to RobCup Symposium, Seattle-2001.
3. A. Bonarini, P. Aliiverti, M. Lucioni, *An Omnidiretional vision sensor for fast tracking for mobile robot.*, IEEE Instrumentation and Measurement technology Conference, IMTC 99, Piscataway, NJ, 1999, IEEE Computer press.
4. S. Coradeschi, et al, *Overview of RoboCup-99*, AI magazine, Vol. 21, No. 3 (2000), pp11-18.

Clockwork Orange: The Dutch RoboSoccer Team

Matthijs Spaan[1], Marco Wiering[2], Robert Bartelds[3], Raymond Donkervoort[1], Pieter Jonker[3], and Frans Groen[1]

[1] Intelligent Autonomous Systems Group, University of Amsterdam, the Netherlands
[2] Intelligent Systems Group, Utrecht University, the Netherlands
[3] Pattern Recognition Group, Delft University of Technology, the Netherlands

1 Introduction

The Dutch RoboSoccer Team, called "Clockwork Orange"[1] is a Dutch research project in which 3 universities participate: the University of Amsterdam, the Delft University of Technology, and University Utrecht. In this research project we study multi agent systems in general and multi robot systems in particular, as we participate in the RoboCup middle-size league. Due to severe hardware problems Utrecht University could sadly not contribute to the team this year.

2 The Robots

Our team consists of two types of robots: six Nomad Scout robots (used by Delft and Amsterdam) and one Pioneer 2 robot (used by Utrecht). The Nomad Scout (see figure 1(Left)) has odometry sensors and one wide angle colour camera for sensing and a pneumatic driven kick device as effector. Its 16 ultrasonic sensors and its tactile bumper ring are not used. The Pioneer 2 (see figure 1(Right)) uses a laser range finder, 16 ultrasonic sonar sensors, odometry sensors, and a camera for sensing. It uses the same pneumatic kick device as effector. For communication between robots we use a Breezecom wireless Ethernet system.

3 Vision

Our vision research is mostly directed at the problems of recognition and self-localisation. We can detect and track all objects typically found in a RoboCup environment[5]. This is done via color detection by circular arcs in an adapted HSI space.

We have implemented 2 different ways for self-localisation, a local and a global method (see [3] for details). Both are based on detecting the lines of the field and the goals. The global method first splits the screen into multiple regions of interest and does a rough Hough transform on each of these. Then the method

[1] Nickname of the human Dutch national soccer team of the seventies.

A. Birk, S. Coradeschi, and S. Tadokoro (Eds.): RoboCup 2001, LNAI 2377, pp. 627–630, 2002.

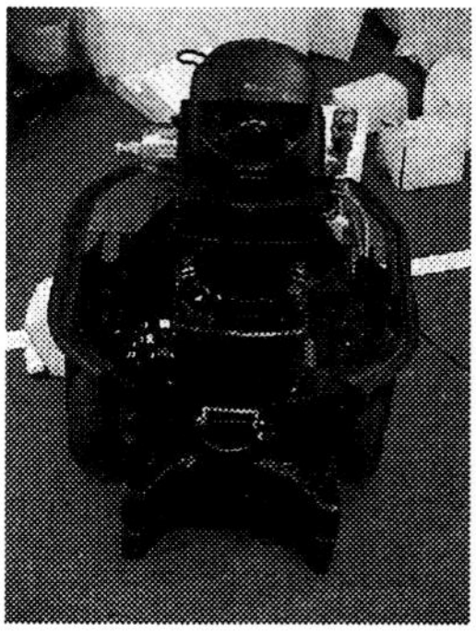

Fig. 1. Our robots. Left: a Nomad Scout robot. Right: the Pioneer 2 robot.

searches for the precise lines using linear edge detectors in the vicinity of the lines found by the Hough transform. These are matched to the world model giving an estimate of the position. Because of the symmetry of the field multiple candidates are found. We then use Multiple Hypothesis Tracking to follow all the candidates over time, updating them for our own movement. Since there are multiple candidates now we need a quick method to verify these and to correct for changes.

The local method uses a first order Taylor approximation of the displacement of the line, independent in x, y and ϕ, to calculate the displacement in robot coordinates. It uses the least square error to give an overall result on the match. We can repeat the method to see if it converges. Next we check all the candidates, verifying their heading and distance to the goals and the overall result given to us by the local method. The loop is repeated until one candidate remains, who is used to update out position. If no candidates remain we switch back to global self-localisation.

4 World Model

To enable a distributed form of control, the world model is also distributed. Each of the robots in our team locally maintains an world model consisting of the robot's own position and the positions of the other objects in the game (i.e. other robots and the ball). To communicate the world model between the robots we need an absolute world model for which we need to know our own position. This position is based on odometry and vision-based self-localisation. Both have their advantages and disadvantages. Odometry is very accurate for small distances but is not suitable for longer periods of driving and completely loses its value when the robot collides. Vision self-localisation can give us quite a good position estimation at any time (assuming it sees enough lines) but is very expensive in time and resources and its positions arrive with considerable delay. The world model combines these two sensors to get an estimation of our position, taking into account the lag of the self-localisation [4].

The other objects are detected by the vision system as shapes, which the world model matches to the known objects which have the same shape and are closest to the observed shape. From this information we can form a relative model of the world, which we can easily transform knowing our own position, into an absolute one. This information is shared with the other robots using the wireless Ethernet, so each robot can not only keep track of objects it perceives itself but also of objects perceived only by its teammates.

5 Team Skills

The advantage of a absolute shared world model is that team coordination is greatly simplified. Sharing the same world model enables the team members to reason not only about their own actions, but they can also predict the next action a team member will take. Communication on the team coordination level is used to improve robustness and is almost a necessity since the team is heterogeneous both in hardware and software. The Utrecht Pioneer 2 robot uses a subsumption architecture while the Amsterdam/Delft Nomad Scouts operate on a hybrid architecture. Our team uses a global team behavior (a strategy) from which each robot derives its individual role.

Our team skills module determines the current team behavior using a simple finite state machine. Each player has its role defined by the team behavior. There are about half a dozen different roles available, all of which have an attacking, defending or intercepting purpose. Utility functions based on potential fields [7, 6] are used to distribute the roles among the team members. This technique is similar to the ones used by other participants in the middle-size league [2, 9]. Each robot calculates its own scores for the roles useful at the moment, communicates them and then calculates the scores of its team mates. When no scores from other team mates are received the locally calculated scores can be used. This introduces a level of redundancy which can improve the team performance. When everybody knows the scores for the roles of each team member the roles can be easily distributed.

The exact specification of an role is not (and can not be) the same on both architectures. The following describes the Amsterdam/Delft approach (see [4] for details) which is similar to the one Tambe [8] described for use in the simulation league. A Markov decision process is used to search through state space of a robot's future states. First a number of possible actions and their parameters with respect to the current situation and individual behavior are generated. For each of these actions the probability of success is calculated. The impact on the world of an action and the expected concurrent actions of team mates and enemy players is evaluated. Modeling team mate actions is relatively easily compared to modeling your opponent with reasonable accuracy. The evaluation criteria both include a role evaluation (using potential fields) as well as concepts like ball possession. Now the action with the maximum expected utility can be chosen and the process can be repeated to look further ahead.

6 Behavior Based Control

The Pioneer 2 robot of Utrecht University uses an extended version of the subsumption architecture [1] in which particular behaviors such as *Get_Ball, Dribble*, and *Score* compete for controlling the robot. All behaviors can react in correspondence to the world model or to direct sensor data provided by the camera and laser range finder. In future work, we extend the system by integrating learning as part of the behavior-based control architecture of the robot. This is done by providing the robot with different implementations of the same behavior. These behaviors are tested during online play against an opponent. If a particular behavior is successful against an opponent team, the behavior gets strengthened, and is therefore used more often. This allows us to learn a sequence of behaviors which has the highest expected probability of success against an opponent.

7 Conclusion

We were able to maintain a distributed world model using vision-based self-localisation. This information was used to create high level team coordination. Clockwork Orange reached the quarter finals of RoboCup 2001 to be defeated by the world champion. In total seven games were played, resulting in three victories, one draw and three losses.

References

[1] R.A. Brooks. Achieving artificial intelligence through building robots. *MIT AI Lab Memo 899*, 1986.

[2] C. Castelpietra, L. Iocchi, M. Piaggio, A. Scalzo, and A. Sgorbissa. Communication and coordination among heterogeneous mid-size players: ART99. In *Proceedings of the 4th International Workshop on RoboCup*, pages 149–158, 2000.

[3] F. de Jong, J. Caarls, R. Bartelds, and P. P. Jonker. A two-tiered approach to self-localization. In *RoboCup 2001*. Springer-Verlag, 2002.

[4] F.C.A. Groen, J. Roodhart, M. Spaan, R. Donkervoort, and N. Vlassis. A distributed world model for robot soccer that supports the development of team skills. In *Proceedings of the 13th Belgian-Dutch Conference on Artificial Intelligence (BNAIC'01)*, 2001.

[5] P. P. Jonker, J. Caarls, and W. Bokhove. Fast and accurate robot vision for vision-based motion. In *RoboCup 2000: Robot Soccer World Cup IV*, pages 72–81, Melbourne, Australia, 2000.

[6] O. Khatib. Real-time obstacle avoidance for manipulators and mobile robots. *International Journal of Robotics Research*, 5(1):90–98, 1986.

[7] J. Latombe. *Robot Motion Planning*. Kluwer Academic Publishers, 1991.

[8] M. Tambe and W. Zhang. Towards flexible teamwork in persistent teams: Extended report. *Journal of Autonomous Agents and Multi-Agent Systems*, 3(2):159–183, 2000.

[9] T. Weigel, W. Auerbach, M. Dietl, B. Dümler, J. Gutmann, K. Marko, K. Müller, B. Nebel, B. Szerbakowski, and M. Thiel. CS Freiburg: Doing the right thing in a group. In *RoboCup 2000: Robot Soccer World Cup IV*. Springer-Verlag, 2001.

CMU Hammerheads 2001 Team Description

Steve Stancliff, Ravi Balasubramanian, Tucker Balch, Rosemary Emery,
Kevin Sikorski, and Ashley Stroupe

Robotics Institute, Carnegie Mellon University
Pittsburgh, PA, 15213
{stancliff, ravib, tucker.balch, rosemary.emery, k.sikorski,
ashley.stroupe}@cmu.edu
http://www.cs.cmu.edu/~coral/minnow

1 Introduction

The CMU MultiRobot Lab focuses on the study of team behavior in dynamic and uncertain environments. In order to further this research, we have developed an inexpensive autonomous robot platform, the Minnow. The CMU Hammerhead robot soccer team competes in the middle-size RoboCup competition using four of these Minnow robots.

The challenges of robot soccer require the robot players both to be individually capable and also to work together with their teammates. Some of the research areas which have been furthered by our work in robot soccer include color vision, multiagent sensor fusion, cooperative behaviors, and communication in noisy low-bandwidth environments.

The 2001 CMU Hammerheads represent the evolution of the previous year's team [1], using largely the same hardware and software architecture. The major changes in hardware are increases in processor speed and memory, and a change to digital cameras. On the software side, most of the low-level behaviors from the 2000 team are retained, and the emphasis has been on refining high-level behaviors and improving coordination and teamwork.

This paper describes the hardware and software architecture of the Minnow platform used in the CMU Hammerhead robot soccer team. Section 2 describes the hardware components of the robot platform. The software and control architectures are presented in Section 3. Finally, our strategies for teamwork and coordination are presented in Section 4.

2 Robot Hardware

The CMU Hammerheads are based on the Minnow robot platform (Figure 1). The Minnow is an inexpensive, fully autonomous robot with wireless communication and color vision. The underlying mechanical platform is a commercially-available Cye robot from Probotics. This platform was chosen for its low cost and highly accurate odometry. The platform consists of a two-wheeled drive unit and a passive trailer. The trailer provides ample area for mounting of hardware. The

A. Birk, S. Coradeschi, and S. Tadokoro (Eds.): RoboCup 2001, LNAI 2377, pp. 631–634, 2002.

Fig. 1. Minnow robot platform.

drive unit includes a microcontroller which, in addition to controlling the motors, provides odometry and bump sensing. This microcontroller communicates with the host computer through an RS-232 connection.

The primary computing power of the Minnow is provided by a standard laptop computer mounted on the Cye trailer. A laptop was chosen over an embedded system in order to reduce platform development time and also for the convenience of having an onboard keyboard and video screen for calibration and debugging.

Vision is provided by a digital camera which is attached to the laptop computer through a USB port. The USB connection provides both signal and power; thus, the power for all computation and vision is provided by the laptop's batteries. The camera employs a wide-angle lens to allow the robot to see most of the playing field at once. The camera is mounted on a shaft above the drive unit, with a slipring for the electrical connections to the trailer, allowing the drive unit to rotate freely under the trailer.

Inter-robot communication is provided by wireless ethernet conforming to the IEEE 802.11 standard, consisting of PCMCIA cards on the robots and a wireless access point. In a noisy environment such as the RoboCup competition, the wireless access point provides a greater robot-to-robot communication range in comparison to a point-to-point network.

3 Software

3.1 Vision Software

The Minnows employ a color vision system as their primary means of information gathering. The camera images are processed by the CMVision software package [2] to detect colored regions (Figure 2). Because a wide-angle lens is used on the camera, the images must be dewarped to remove the spatial distortion caused by the lens [3]. CMVision is able to achieve high framerates by dewarping only the regions of interest rather than the entire image.

Fig. 2. An image before and after processing with CMVision.

3.2 Control Software

The robots are controlled using the Clay [4] architecture of TeamBots [5]. TeamBots is a Java-based package for the simulation and control of multiagent systems. The strength of TeamBots is that it allows for prototyping in simulation of the same control software that is used on the robots. The simulation environment (Figure 3) allows for the interaction of multiple robots, each running a different control system. This flexibility allows for the rapid prototyping of control schemes and allows for the early development of software in parallel with hardware.

Clay is a group of Java classes that can be easily combined into motor-schema based control systems, generally represented by finite-state machines. In this approach, each state corresponds to a suite of activated behaviors for accomplishing a task [6]. Transitions between states are initiated by real-time perception. The activated behaviors create vectors to represent the desired trajectory of the robot. Clay is therefore similar to potential field navigation. It differs, however, in that only the vector at the robot's current position is calculated, rather than the entire vector field.

One advantage of the behavior-based approach for the Minnow robot is that it allows the control system to quickly locate and track the ball without having to explicitly deal with the position of the trailer. A planning approach would require the control system to plan with non-holonomic constraints, which is computationally difficult and time-consuming and thus inappropriate for a highly dynamic environment such as soccer.

4 Teamwork and Coordination

The CMU Hammerheads accomplish teamwork primarily through the sharing of information. The robots communicate to each other the location of the ball and their current operating state.

Communication of ball position between the robots improves the performance of the team by increasing the effective sensor range of individual robots and also by increasing the accuracy of ball-position estimates. The global ball-position

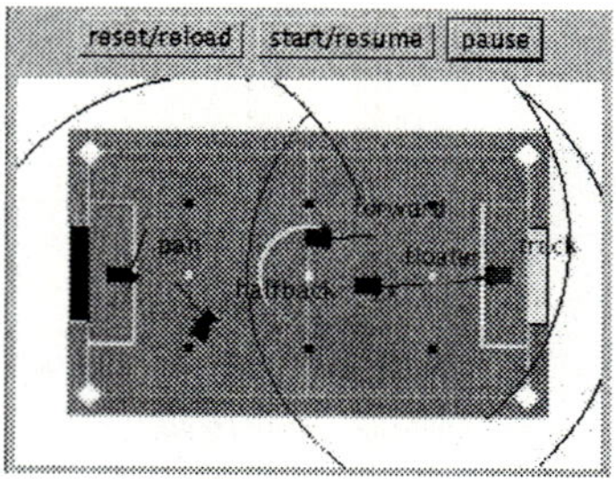

Fig. 3. TeamBots simulation.

estimate is used to guide a robot which does not see the ball so that it does not have to perform a blind search for the ball. The location of the ball is represented probabilistically, with each robot's ball-position estimate represented as a Gaussian. The Gaussians produced by the different robots are multiplied to provide the global estimate of the ball position [7].

Communication of player-state information is used to enable the robots to better coordinate their actions. For instance, when the goalie leaves the goal box, the fullback will position itself in front of the goal. Also, the two offensive players will switch roles dynamically depending on the location of the ball and which team has possession of it.

References

[1] Rosemary Emery, Tucker Balch, Rande Shern, Kevin Sikorski, and Ashley Stroupe. Cmu hammerheads team description. In Tucker Balch, Peter Stone, and Gerhard Kraetzschmar, editors, *Proceedings of the 4th International Workshop on RoboCup*, Melbourne, Australia, 2000.

[2] Jim Bruce, Tucker Balch, and Manuela Veloso. Fast and cheap color vision on commodity hardware. In *Workshop on Interactive Robotics and Entertainment*, pages 11–15, Pittsburgh, PA, April 2000.

[3] Hans Moravec. Robot spatial perception by stereoscopic vision and 3D evidence grids. Technical Report CMU-RI-TR-96-34, Robotics Institute, Carnegie Mellon University, September 1996.

[4] Tucker Balch. Clay: Integrating motor schemas and reinforcement learning. Technical Report GIT-CC-97-11, College of Computing, Georgia Institute of Technology, March 1997.

[5] Tucker Balch. *Behavioral Diversity in Learning Robot Teams.* PhD thesis, College of Computing, Georgia Institute of Technology, December 1998.

[6] Ronald Arkin and Douglas MacKenzie. Temporal coordination of perceptual algorithms for mobile robot navigation. *IEEE Transactions on Robotics and Automation*, 10(33):276–286, June 1994.

[7] A. Stroupe, M.C. Martin, and T. Balch. Merging probabilistic observations for mobile distributed sensing. Technical Report CMU-RI-TR-00-30, Robotics Institute, Carnegie Mellon University, December 2000.

CoPS-Team Description

R. Lafrenz, M. Becht, T. Buchheim, P. Burger, G. Hetzel, G. Kindermann, M. Schanz, M. Schulé, and P. Levi

Institute for Parallel and Distributed High Performance Systems (IPVR)
Applied Computer Science - Image Understanding
University of Stuttgart, Breitwiesenstr. 20-22, 70565 Stuttgart, Germany
robocup@informatik.uni-stuttgart.de

Abstract. The control software of the robot soccer team CoPS is designed as a multi-agent-system. The basis for a cooperation between the robots is a suitable environment model based on uncertain sensory data and communication.

1 Introduction

In the last two years, the CoPS-team (Cooperative Soccer Playing Robots) competed successfully in the RoboCup tournaments. The robots of the team are shown in fig. 1.

Fig. 1. The CoPS team of soccer robots (March 2001)

A. Birk, S. Coradeschi, and S. Tadokoro (Eds.): RoboCup 2001, LNAI 2377, pp. 635–638, 2002.

The control software of our robot team is designed as a Multi-Agent-System ([1], [2]). Although applied to many domains, the full potential of this paradigm developes especially in situations where decisions have to be made upon uncertain data or partial information.

2 System Architecture

In our architecture, not only a whole robot, but also major software components of a robot are modelled as agents. Each single robot has the internal structure shown in fig. 2.

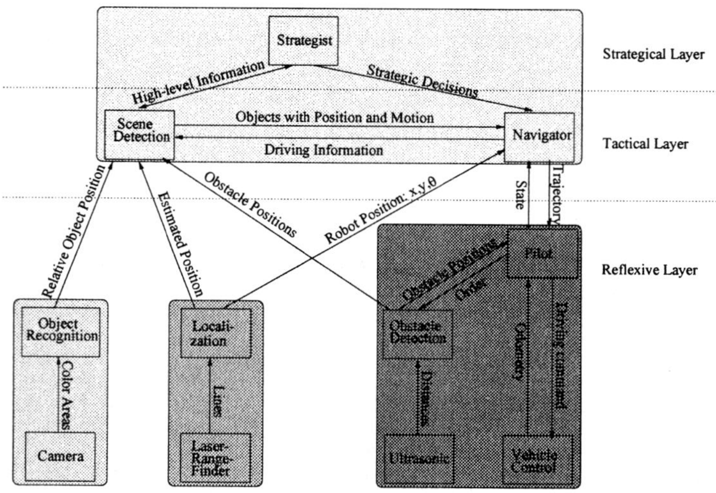

Fig. 2. System architecture

Logically, the software is structured in three layers: reflexive, tactical, and strategical. These layers consist of elementary agents which are described in section 3. The assignment of an agent to a layer helps to design and understand the system.

The Reflexive Layer The elementary agents in the reflexive layer control the hardware of the robot and perform sensing and acting, e.g. *Camera* or *Vehicle Control*. Other kinds of *elementary agents* in the reflexive layer fuse the data of different type or transform sensory data into information, e.g. the *Object Recognition*.

The *Pilot* has to carry out the orders given by the Navigator. As an agent, the Pilot is responsible for leading the robot to a given destination. It has to avoid collisions, and needs access to the results of the Object Recognition, which are provided by the later described Scene Detection. It is able to perform minor corrections to the path planned by the Navigator.

The Tactical Layer In the tactical layer we find two agents, the *Scene Detection* and the *Navigator*. The Scene Detection is responsible for storing and updating the information about the current situation. As an agent, the Scene Detection collects all relevant information form the sensory agents. This local environment model represents the scene, as seen from a single robot. In order to obtain robustness and a better understanding of the current scene, communication within the team is necessary. The communicated information is fused with the local information and also stored in the Scene Detection.

One of the basic requirements for that is the ability to distinguish between team-mates and oppenents. By assuming same shape and colors for robots of both teams, this distiction is impossible without cooperation, in other words, indirect information. But, by exchanging position information for all robots within the team, the recoginized objects can be identified as team-mates or opponents. In addition, the position estimation of a team-mate can be verified.

The Scene Detection answers requests of the Navigator. The navigator generates abstract driving commands (like search ball, dribble ball, etc.) which are executed by the pilot.

The Strategical Layer In our concept, there is only one agent placed in the strategical layer: the *Strategist*. It reacts to infrequent external events (e.g. start of game, goals) and decides on the way of further playing. Thus it decides on the long-term strategies of the game, for example offensive or defensive playing mode. Therefore, it communicates with two other agents, which are placed in the tactical layer: the Scene Detection and the Navigator. In addition, the Strategist is able to communicate with the strategists of other players in the team. Sometimes a single player is not able to recognize the current situation. Therefore we provide communication beyond the boundaries of a single robot. In principle, any elementary agent is able to communicate with any other agent. Hence, very complex and powerful negotiations are possible.

3 Structure of Elementary Agents

The elementary agents are the autonomous function moduls of the system. They get their functionality from their plans. These plans control the interplay of the sensing and acting. The *body* of the elementry agent is supervised by the *decision unit.*

Fig. 3 shows the structure of an elementary agent. The *decision unit* is responsible for negotiation between the agents and for internal configuration. Data is transposed through seperate data channels. The configuration of this channels is done by the *decision unit. State* contains the current state of the elementary agent. *Model* administrates the statical or lernable knowledge. The internal cycles range from simple and fast react behavior (controller) to complex planning of each step using the knowledge of the *model.*

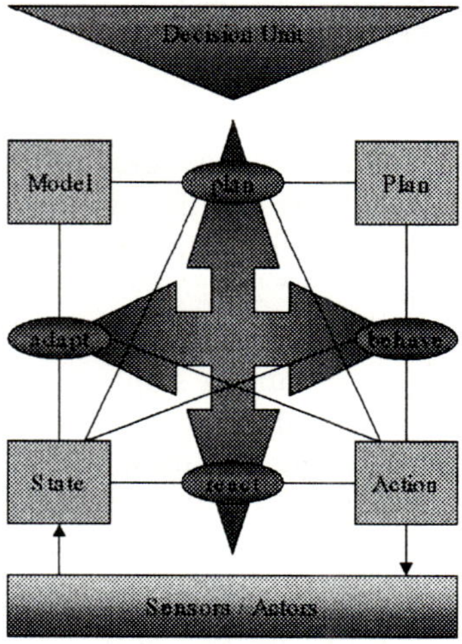

Fig. 3. Structure of an *elementary agent*

4 Conclusion and Outlook

In this paper, we have described the system architecture of our soccer robots. Currently, our research interests concentrate on cooperative environment modelling, cooperative behavior, and we plan to include reinforcement learning in our system.

References

[1] M. Becht, M. Muscholl, and P. Levi. Transformable multi-agent systems: A specification language for cooperation processes. In *Proceedings of the World Automation Congress (WAC), Sixth International Symposium on Manufacturing with Applications (ISOMA)*, 1998.

[2] R. Lafrenz, N.Oswald, M.Schulé, and P. Levi. A cooperative architecture to control multi-agent based robots. In *Proceedings of PRICAI*, 2000.

Fun2Mas: The Milan Robocup Team

Andrea Bonarini[1], Giovanni Invernizzi[1], Fabio Marchese[2], Matteo Matteucci[1], Marcello Restelli[1], and Domenico Sorrenti[2]

[1] DEI – Politecnico di Milano, Milan, Italy
{bonarini,inverniz,matteucc,restelli}@elet.polimi.it
[2] DISCO – Università degli Studi di Milano - Bicocca, Milan, Italy
{marchese,sorrenti}@disco.unimib.it

Abstract. We present Fun2maS, the Milan Robocup Team. In its implementation we have faced many aspects: hardware (electronics, mechanics), sensors (omnidirectional vision), behaviors (fuzzy behaviors management and composition), multi-agent coordination (strategy and tactics), and adaptation of the team behavior. We could fully exploit the characteristics of all the components thanks to the modular design approach we have adopted. All the modules have also been designed to be used in generic applications, and we have already adopted most of them for surveillance, mapping, guidance, and document delivery tasks.

1 Introduction

Fun2maS (*"[Let's] Fun to Multi–Agent Systems !!"*), the Milan Robocup Team, is born from the experience done with the Italian National Team (*ART* [1]) since 1998. In this paper, we summarize our results concerning: development of robot bases, omnidirectional vision [4][2][9][10], fuzzy behavior management systems [6], architecture for multi-agent coordination [8], and techniques to adapt the behavior of the team to that of the opponents [7].

2 Hardware

We have implemented both the mechanical and the electronic aspects of our robots. In 1998, we developed the robot base Mo^2Ro (*Modular Mobile Robot*), with two independent traction wheels on the front and a ball pivot on the rear. These robots (*Rullit* and *Rakataa*, first and third from the left in figure 1) weight about 30 Kg each, and can run up to 1 m/s. Due to inertia it is almost impossible to control faster robots with the same kinematics and mass distribution, so we have developed *IANUS*, a robot base with two independent, central traction wheels and most of the mass placed on them. The name comes from the Roman god with two faces, since this robot is designed to work indifferently in one of the two main directions. In the present implementation, *Roban* (the second from left in figure 1) can run up to 1.5 m/s. We have implemented on the same basic structure also *Ruffon*, our new goal keeper, unmounted in the picture. The last

A. Birk, S. Coradeschi, and S. Tadokoro (Eds.): RoboCup 2001, LNAI 2377, pp. 639–642, 2002.

Fig. 1. The Fun2maS team

robot on the right is *Roby(Baggio)*: it has two independent wheels each mounted on a rotating turret, so to obtain holonomic movement, which fully integrates with our omnidirectional vision system. All the robots are equipped with control and power cards developed in our lab, and a regular PC board, hosting low-cost frame grabber and wireless ethernet. One of the main concerns in the development of these robots has been the cost, which is as low as about 1,100 euros each. We are currently working to improve our mechanical and pneumatic kickers, mainly to obtain special kicking effects, differentiated also on the same robot, in the case of *Roban* and *Roby(Baggio)* which can kick from different sides. We are also working on a new low-level control architecture, based on a network of distributed microprocessors, that should replace the present ISA-bus cards.

3 Omnidirectional Vision

Sensors play a crucial role for the performance of the robot and strongly influence its design [2]. We have focused on omnidirectional vision, where a camera pointed upward to a revolution mirror can provide a 360 deg image of the robot surroundings, whose characteristic depend on the mirror design. Our first mirror was conical with a spherical center, thus including in the image both distant and close objects. Our second mirror [10][9] is multi–part; we have: in the center the ground image without distortions (square angles appear as square angles), from the second part markers on the robots distant up to 11 meters, from the outer part the position of the ball and other robots when close by. This design exploits at best all the available resolution, and it also makes self–localization fast to compute. The color image is sampled in "receptors", small sets of 4 pixels (in cross configuration) whose color features are averaged, to obtain robust interpretation [4]. The number of receptors is much smaller than the number of pixels, so that computation is much faster than that needed by complete images, computing angular position and distance of all the elements on the field at 20 Hz. Moreover, each color is defined by collecting patches from the actual images, instead of defining a single interval for each color feature, thus providing a better characterization. We have adopted this approach also in other systems, operating in indoor environments, that can adapt the color interpretation according to the role the corresponding blobs have in the image.

4 BRIAN & SCARE

BRIAN (*Brian Reasonably Implements an Agent Noddle*) [6] is the fuzzy behavior management system that we have implemented. To each behavior are associated two sets of fuzzy predicates representing its activating conditions (*CANDO conditions*), and motivations(*WANT conditions*). These last may come from evaluation of the current situation, or from our distributed planner, and are used to weight the actions proposed by behaviors. We have developed reactive behaviors (such as `GoToGoalWithBall`) reaching a good performance in a context where agents operate most of the time selfishly. In order to achieve coordinated behaviors we are providing our robots with a set of general basic behaviors which can be instantiated by *SCARE* into higher level behaviors for the whole team, by considering *strategies* and *roles*.
SCARE (*Scare Coordinates Agents in Robotic Environments*) [8] implements all the relevant aspects of the multi–agent cooperation. Agents are assigned to *jobs* defined in a multi–agent task structure (*schema*). This makes it possible to perform coordinated actions such as ball-passing and opponent blocking (*doubling*). The job assignment is done by meta–agents which evaluate *opportunity* and *necessity* of undertaking actions, and consider the respective *predisposition* of agents. Meta–agents are dynamically distributed in the robot network, by taking into account quality of communication and available computational resources. If communication works fine, and one computer has enough resources, the control can be centralized on that computer, whereas, if the network is partitioned, each sub–net (at worst, each robot) has its own specialized, coordination meta–agent. This architecture enables the dynamical reconfiguration of the distributed planner, thus increasing robustness of the whole system.
To support communication among *SCARE*, *BRIAN* and the other components of the multi–agent system, we are substituting *ETHNOS*, the real-time communication system used in ART [1], with *DCDT* (*Device Communities Development Toolkit*) an event–driven middleware that gives to any module (including the low level controllers) the capability to exchange messages with any other, in a peer-to-peer setting.

5 Adaptation

We have worked on adaptation in Robocup since last year, when we obtained our first results in a simulated environment [5][3]. The team having even a single robot adapted to a new environment (e.g., opponents with different strategies) was able to score more goals, conceding less. The reinforcement learning algorithm we have implemented evolves populations of behaviors associated with WANT conditions. It tests different motivations for each behavior in different situations, and evaluates the contribution of each behavior to the global action. We are now working on another adaptation mechanism to be implemented at the SCARE level, so that the team behavior can change dynamically, by changing the mentioned weighting parameters in SCARE, according to alterations in environmental conditions, opposing strategy, faults, etc. Also in this case, we have defined a reinforcement learning algorithm, currently under extensive test.

6 Conclusion

We have briefly summarized our activities in the development of a Robocup team, which span from hardware to adaptation. We adopted an the integrated and modular approach to the an effective team. We also made an effective effort to develop general tools usable in most multi–robot domains. We have adopted modules implemented for Robocup in other applications. In some cases the solutions found for Robocup have been extended to face different aspects of the problem (e.g., person detection). In other cases, results obtained in different domains have been applied also to Robocup; for instance, adaptation to different floor types, implemented for navigation tasks in DEI-Polimi building, can help to automatically tune the vision system also in Robocup.

References

[1] G. Adorni, A. Bonarini, G. Clemente, D. Nardi, E. Pagello, and M. Piaggio. Art'00 - azzurra robot team for the year 2000. In M.Veloso P.Stone, editor, *RoboCup 2000: Robot Soccer World Cup IV*, Berlin, D, In Press. Springer Verlag.

[2] A. Bonarini. The body, the mind or the eye, first? In M. Asada M. Veloso, E. Pagello, editor, *RoboCup 98–Robot Soccer World Cup III*, pages 695–698, Berlin, D, 2000. Springer Verlag.

[3] A. Bonarini. Evolutionary computing and fuzzy rules for knowledge acquisition in agent-based systems. *Proceedings of IEEE*, In press, 2001.

[4] A. Bonarini, P. Aliverti, and M. Lucioni. An omnidirectional vision sensor for fast tracking for mobile robots. *IEEE Trans. on Instr. and Meas.*, 49(3):509–512, 2000.

[5] A. Bonarini, C. Bonacina, and M. Matteucci. Fuzzy and crisp representation of real-valued input for learning classifier systems. In P. L. Lanzi, W Stolzmann, and S. W. Wilson, editors, *Learning Classifier System: new directions and concepts*, pages 107–124. Springer-Verlag, Berlin, D, 2000.

[6] A. Bonarini, G. Invernizzi, T. H. Labella, and M. Matteucci. A fuzzy architecture to coordinate robot behaviors. *Fuzzy Sets and Systems*, Submitted.

[7] A. Bonarini and M. Matteucci. Learning context motivation in coordinated behaviors. In *Proceedings of IAS-6*, pages 519–526, Amsterdam, NL, 2000. IOS Press.

[8] A. Bonarini and M. Restelli. An adaptive architecture for the implementation of cooperative strategies for heterogeneous agents. *IEEE Transactions on Robotics and Automation*, Submitted.

[9] P. Lima, A. Bonarini, C. Machado, F.M. Marchese, C. Marques, F. Ribeiro, and D.G. Sorrenti. Omni-directional catadioptric vision for soccer robots. *Journal of Robotics and Autonomous Systems*, Submitted.

[10] F. M. Marchese and D. G. Sorrenti. Omnidirectional vision with a multi-part mirror. In G. Adorni and W. Hoeke, editors, *Proceedings of EuRoboCup 2000"*, Amsterdam, NL, 2000. On CD-ROM.

Fusion

Takeshi Matsuoka[1], Motoji Yamamoto[2], Nobuhiro Ushimi[2], Jyun'ichi Inoue[2], Takuya Sugimoto[3], Manabu Araoka[3], Toshihiro Kiriki[2], Yuuki Yamaguchi[2], Tsutomu Hasegawa[2], and Akira Mohri[2]

[1] Fukuoka University
matsuoka@ctrl.tec.fukuoka-u.ac.jp
[2] Kysushu University
[3] Hitachi Information & Control Systems, Inc.
http://mari.is.kyushu-u.ac.jp/~fusion

Abstract. This paper introduces team "Fusion" and its soccer robots for RoboCup-2001. The team is organized by Kyushu University, Fukuoka University, Hitachi Information & Control Systems Inc. considering technical requirements of RoboCup F2000 league. The soccer robots have been originally designed and made for the league. It is first attempt to join RoboCup for the robots and the team.

1 Introduction

Team "Fusion" is organized by Kyushu University, Fukuoka University Hitachi Information & Control Systems, Inc. [1], and it is named intending unification of multiple technical fields and people of each organizations. The RoboCup is a standard problem for robotics and artificial intelligence and it needs overall technical abilities for various fields. That's also a challenging research theme. Therefore we decided to participate in RoboCup, especially F2000 league [2], because the league involves most extensive technical problems such as mechanism of mobile robot, image processing, cooperative motion control, etc.

By participating RoboCup, we mainly study a problem of "Optimal Planning in Dynamic Environment". This research topic contains important problems for intelligent mobile robots such as motion planning and control, collision avoidance of dynamical or static obstacles, and self-localization to apply the robots for every day environment. This paper describes a hardware and software architecture of developed intelligent mobile robots to study the research topic and to join RoboCup F2000 league.

2 Hardware Architecture

We have newly developed soccer robots for RoboCup F2000 league. The robots are shown in Fig.1, one is goal keeper robot (second robot from the left one in the figure) and the others are field player robots. For quick motion, the robots are made with powered wheel type (two independently driven wheels type), and

A. Birk, S. Coradeschi, and S. Tadokoro (Eds.): RoboCup 2001, LNAI 2377, pp. 643–647, 2002.

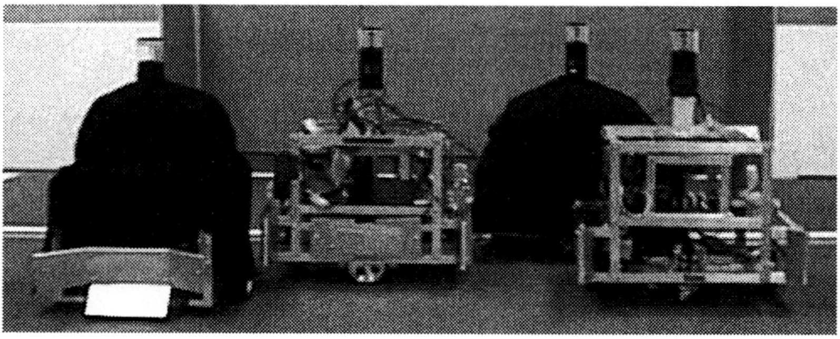

Fig. 1. Our Robots

are made with light weight as much as possible. (See bellow specification.) Each robot has an omni-directional camera system, a kicking device and a communication system. The overall hardware architecture of the robot is shown in Fig.2. To reduce the cost of the robot, an I/O card including pulse generator for motor amplifier, counter for encoder, parallel interface (8255 PPI) is originally developed. The I/O card is connected with CPU board by printer port.

- Dimension: 30cm×40cm×50cm
- Weight: 8 kg
- Maximum Speed: 2 m/s

2.1 Processing System

Considering extension of function such as sensors and actuators, we select a standard industrial CPU board with K6-2 300MHz or celeron 700MHz processor and 128Mbyte RAM as a robot controller having 5 PCI slots. The standard short-size PCI back plane unit containing the cpu board and other interface boards makes the control module compact and light.

2.2 Sensor System

Sensor system, especially a vision system is a significant element for intelligent mobile robots. For wide scope sensing, we use an omni-directional camera system. The omni-directional camera is composed of a CCD camera and a hyperboloidal omni-directional mirror, in which the camera optical axis is in line with the vertical axis of the mirror. This camera system is used for self-localization and recognition of ball, goal and other robots. For the recognition, we use IP5005 (Hitachi) as an image processing board. This board has features of fast transaction speed and versatile image processing ability. All of our robots have this vision system.

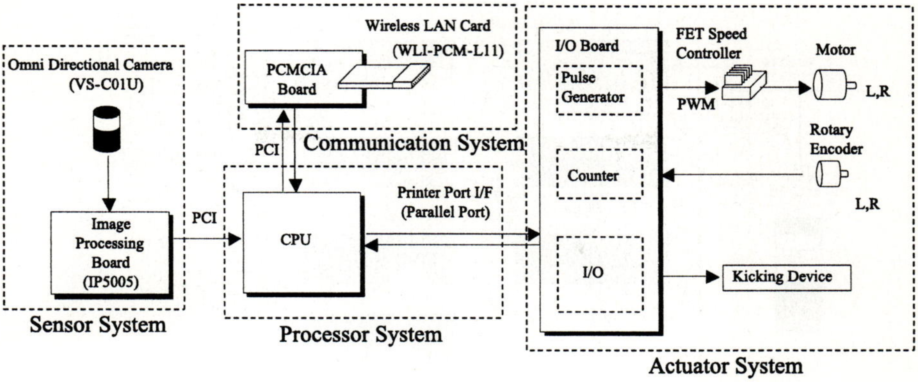

Fig. 2. Hardware architecture

2.3 Actuator System

In our robot, two wheels are independently driven by two DC motors with reduction gears. The DC motors are controlled by an originally designed I/O board and FET speed controller which is for R/C car (Keyence A-07RZ). The robot has a kicking device which is consists of an air cylinder, an electro-magnetic valve, and an air tank. This enables powerful action of shooting and passing.

2.4 Communication System

Communication among field-player robots is important for cooperation movement such as passing, post-playing and positioning. To communicate with other robots, all robots of our team are equipped with wireless LAN cards.

3 Software Architecture

Software architecture of our robots is illustrated in Fig.3. It is composed of six basic tasks running in parallel. We use ART-Linux which is an extension of Linux as a real-time OS, developed by Youichi Ishiwata at ETL[3]. Our robots recognize situations (ball and other robots) by vision system and local sensors installed on the robot, and then communicate with other robots about the information. Each robot determines its action according to the information of field, and negotiate with other robots. During the game, the robot memorizes the data of self-position, the position of the ball, the positions of other robots, the output to I/O board and communication. After the game, we analyze these data for optimal motion planning of robot.

3.1 Image Processing

Each robot is equipped with an omni-directional camera. The vision task extracts colors of the ball, other robots, and the goals in the image from the camera by

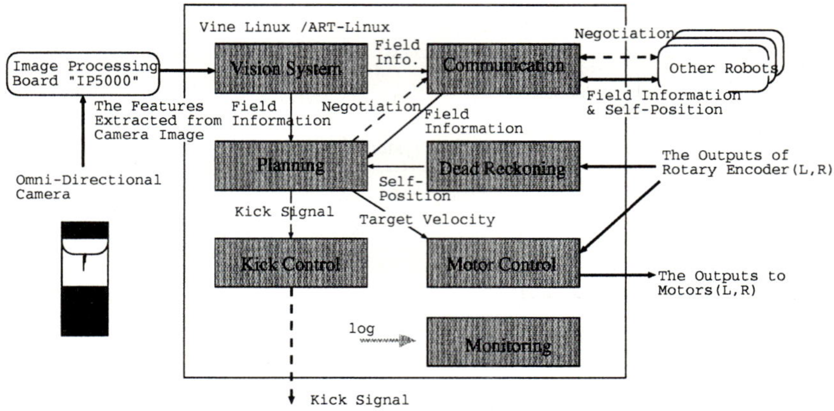

Fig. 3. Software architecture

a simple image processing in real time. Then the task calculates the robot's self-position, the location of the ball, and the location of other robots from the location of the extracted colors in the image. If it is difficult to estimate the robot's self-position with the vision system by the influence of the error of image processing, the robot estimates its self-position with dead reckoning.

3.2 Communication

Our robots share the information of the field by using communication: i.e., each location of the robots on their side, the location of the ball, and the location of the opponent robots. Each robot can use the information which may be difficult to get by local sensors of the robot. In addition, our robots may negotiate each other to achieve more appropriate positioning or passing.

3.3 Planning

Using the most reliable information from all information on the field, the robot decide one action based on if-then rules and negotiation with other robots. The actions are "move to position", "move to ball", "dribble", "pass", and "shoot". Each robot has different if-then rules corresponding to its role: i.e., goal keeper, center forward, left wing, and right wing. For example, if a robot is in the neighborhood of ball, it will move to the ball in order to keep or clear the ball. Simultaneously, the other robots will move to the appropriate position to support the robot.

4 Conclusions

In this paper, the team "Fusion" and the architecture of our robots was introduced. We have developed original robots and studied a motion planning problem based on reliable information of the field and negotiations with other robots. Key feature of our robot is lightweight which enables quick motion and low damage when collision. By participating in this RoboCup, we want to evolve our research on intelligent mobile robots.

References

1. Team Fusion web page
   http://mari.is.kyushu-u.ac.jp/~fusion
2. RoboCup Official HomePage
   http://www.robocup.org/
3. ART-Linux home page
   http://www.etl.go.jp/etl/robotics/Projects/ART-Linux/

GMD-Robots

Ansgar Bredenfeld, Vlatko Becanovic, Thomas Christaller, Horst Günther, Giovanni Indiveri, Hans-Ulrich Kobialka, Paul-Gerhard Plöger, and Peter Schöll

Fraunhofer Institute for Autonomous Intelligent Systems (AIS)
Schloss Birlinghoven, 53754 Sankt Augustin, Germany
bredenfeld@ais.fraunhofer.de
http://www.ais.fhg.de/BE

1 Introduction

The overall research goal of our RoboCup middle-size league team is to increase both, the controlled speed of mobile robots acting as a team in dynamic environments and the speed of the development process for robot control systems. Therefore, we started in 1998 to develop a proprietary fast robot platform and in parallel the development of an integrated design environment.

This team description gives some details on the current state of our hardware platform (September 2001), the control software architecture we use and the design environment we constructed to specify, simulate, run and test our robots. In addition, we point our some specific skills of our robots.

2 Hardware Platform

Our robot hardware is a custom-built 3 degree of freedom platform (2 actuated wheels and a 360 panning camera). We use two 20 Watt, high-quality Maxon motors that are mounted on a very solid, mill-cut aluminium frame. A piezo-gyroscope senses relative changes of heading. Obstacle avoidance is supported by four infrared range detectors and a surrounding ring of switches integrated into protective bumpers. Our robots kick the ball with a pneumatic device which is integrated in a two finger ball guidance. Fig. 1 shows two of our robots.

The goalie has a slightly different sensor configuration with respect to the field players; namely the infrared range detectors are all mounted pointing towards the inside of its own goal rather than in the forward direction. In addition, the kicking device of the goalie has no ball guidance but a simple plate to kick the ball.

The computer system of each robot consists of a Pentium PC notebook connected to two C167 micro controller subsystems for sensor interfaces and actuator drivers. The communication between the PC and the micro-controllers is via CAN bus. The PC communicates with other robots via wireless LAN.

Our vision system relies on the well-known Newton Lab's Cognachrome system for ball and goal detection. Since it is mounted on a 360 degree panning

A. Birk, S. Coradeschi, and S. Tadokoro (Eds.): RoboCup 2001, LNAI 2377, pp. 648–652, 2002.

unit, we are able to perform "radar-like" scans of the robots surrounding. The angle encoder of the panning unit delivers a precise relative angle of each camera picture.

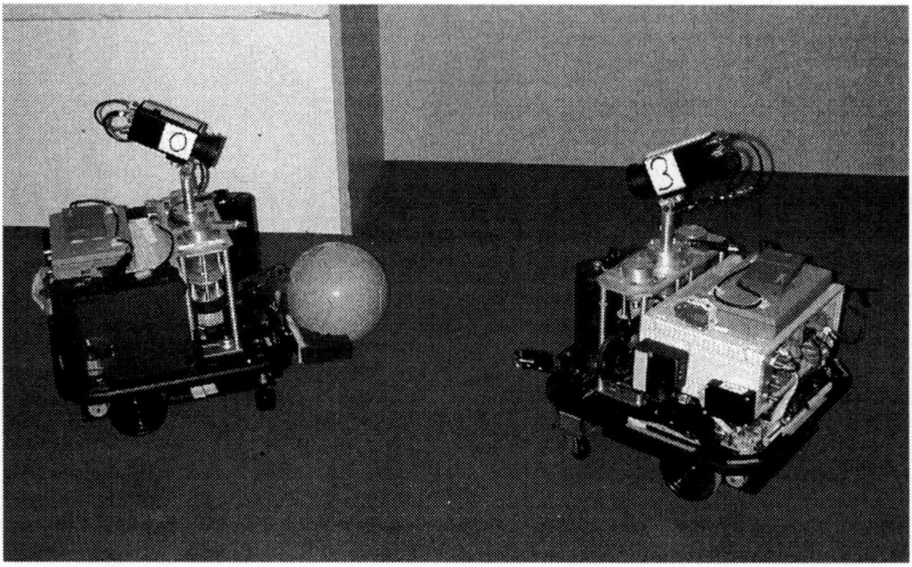

Fig. 1. Two robots of the GMD-Robots team.

3 Software Architecture

Our approach to robot programming is based on Dual Dynamics (DD) [1], a mathematical model for robot behaviors which we developed. It integrates central aspects of a behavior-based approach with a dynamical systems representation of actions and goals. Robot behaviors are specified through differential equations, forming a global dynamical system made of behavior subsystems which interact through specific coupling and bifurcation-induction mechanisms. Behaviors are organized in levels where higher levels have a larger time scale than lower levels. Since the activation of behaviors (activation dynamics) is separated from their actuator control laws (target dynamics), we named our approach "Dual Dynamics". An important feature of DD is that it allows for robust and smooth changes between different behavior modes, which results in very reactive, fast and natural motions of the robots. Through the distribution of the overall control task on several simple controllers, we obtain a very robust system behavior.

We couple the behavior systems of the robots by a team communication mechanism. This allows to establish real-time point-to-point connections between all robots of a team. Since the behavior systems of the robots are specified as an interrelated set of DD-models, we are able to identify in advance the data communication flow occurring during run-time. This allows to synthesize a dedicated communication layer which efficiently distributes variables shared by several behavior systems between the robots without any unnecessary protocol overhead. The variables shared between different behavior systems are simply selected in the graphical specification of a DD-model (see below).

4 Design Environment

The successful design of robot software requires means to specify, implement and simulate as well as to run and debug the robot software in real-time on a team of physical robots. Our integrated design environment [2][3] allows to specify behavior systems as DD-models on a high-level of abstraction and to synthesize all code artifacts required to make these models operative in practice: a simulation model, a control program for the real robot including the team communication layer and the configuration files necessary to the real-time monitoring and tracing tools.

Specify: The DD-Designer specification and code generation tool comprises a graphical editor to enter the specification of a DD-model in terms of sensors, actors, sensor filters and behaviors. Sensor filters and behaviors of a model are further detailed using the equation editor of DD-Designer. Since the robots of our team have different behavior systems, DD-Designer supports concurrent development of a set of behavior systems. This includes the specification of team communication between these different models. We use a multi-target code generator to refine DD-models to a hyperlinked, indexed HTML documentation and all implementation code required by the simulator DDSim, the robots and the real-time monitoring tool beTee. DD-Designer is based on a framework generated from a high-level object-oriented meta-model specification [4].

Simulate: The Java simulator DDSim is specifically tailored to simulate a team of robots with different behavior systems on a virtual RoboCup field. The sensor equipment of each robot may be different and is flexibly specified by an XML configuration file. The simulation models of the robots are Java classes generated by DD-Designer.

Run: The code for the real robot implements the DD-model in C/C++ language. This code is again directly derived from the high-level specification edited in the DD-Designer. Since both artifacts, simulation model and robot control program, are derived from the same specification, we avoid all problems that occur if a migration from a mathematical simulation model to a robot control program has to be performed manually.

Test/Analyze: The trace and monitoring tool beTee allows to capture and analyze internal states of a simulated or real robot in real-time [5].

5 Behavior Skills

Dual dynamics behavior systems use symbolic sensors to represent the percepted environment of the robot. We do not maintain a global world model. Self localization is performed based on odometry and gyroscope data. Since this data is noisy and subject to be disturbed by bumping robots or slipping wheels, we compensate odometry errors by improving the self-localization of our robots using *weighted* Monte Carlo sampling [2]. This approach re-adjusts accumulated odometry data using the positions of the two goals relative to the robot as estimated by the vision sub-system.

In one of our behavior models, a neural network is used to anticipate whether the ball will be lost in the near future. Kicking is then triggered by the behavior system dependent on the pose of the robot and the activation of its behaviors.

A second behavior system is based on a nonlinear control law for the unicycle kinematic model [6]. Such law is designed to steer the vehicle on a static or dynamic target (the ball) along a specified direction (the opponents goal). If the target is static (still ball) the control signals are smooth in their arguments and the solution guarantees exponential convergence of the distance and orientation errors to zero. The major advantages of this approach are that there is no need for path planning and, in principle, there is no need for global self-localization either.

The behavior system of the goalie consists essentially of a two-dimensional controller, which tries to maintain a fixed distance to the back of the goal and a certain angle to the visible ball [7]. If the robot should be hit by opponent robots thus losing its position in front of the goal, a homing behavior is triggered in order to recover the correct position.

All these behavior systems are developed, simulated and tested using our integrated design environment. Although the environment was originally targeted to design control systems using the Dual Dynamics architecture, it proved to be flexible enough to specify and simulate "classical" controllers and to integrate them seamlessly into a behavior-based architecture [7].

References

1. Jaeger H. and Christaller Th.: Dual dynamics: Designing behavior systems for autonomous robots. Artificial Life and Robotics (1998) 2:108-112
2. Bredenfeld, A., Christaller, Th., Jaeger, H., Kobialka, H.-U., Schöll, P.: Robot Behavior Design Using Dual Dynamics. GMD-Report Nr. 117 (2000)
3. Bredenfeld, A., Christaller, Th., Göhring, W., Günther, H., Jaeger, H., Kobialka, H.-U., Plöger, P., Schöll, P., Siegberg, A., Streit, A., Verbeek, C., Wilberg, J.: Behavior engineering with "dual dynamics" models and design tools. In RoboCup-99: Robot Soccer World Cup III / M. Veloso, E. Pagello, H. Kitano (Editors). LNCS 1856. Springer-Verlag (2000)
4. Bredenfeld, A.: Integration and Evolution of Model-Based Prototypes. In Proc. of the 11th IEEE International Workshop on Rapid System Prototyping (RSP 2000) (2000) 142–147

5. Kobialka, H.-U., Schöll P.: Quality Management for Mobile Robot Development. In Proc. of 6th International Conference on Intelligent Autonomous Systems (IAS-6) (2000) 698–703
6. Indiveri, G.: On the motion control of a nonholonomic soccer playing robot. In this book.
7. Bredenfeld, A., Indiveri G.: Robot Behavior Engineering using DD-Designer. In Proc. of IEEE International Conference on Robotics and Automation (ICRA 2001) (2001) 205–210

ISocRob 2001 Team Description

Pedro Lima, Luis Custódio, Bruno Damas, Manuel Lopes, Carlos Marques, Luis Toscano, and Rodrigo Ventura

Instituto de Sistemas e Robótica, Instituto Superior Técnico
Av. Rovisco Pais, 1 — 1049-001 Lisboa, PORTUGAL
{pal,lmmc}@isr.ist.utl.pt
http://socrob.isr.ist.utl.pt/

Abstract. This paper describes the ISocRob team current status, new features demonstrated in RoboCup 2001, and the project long term scientific goals. An evolution of the team prior software architecture, full demonstration of multi-sensor navigation during games, a new goalkeeper and field robots endowed with dribble capability are new features that will be briefly described.

Fig. 1. ISocRob team robots. The goalkeeper is the second from the left.

1 Introduction

The ISocRob team, shown in Fig. 1, has been participating in RoboCup events since 1998. This participation is part of the team scientific project **SocRob** (*Soccer Robots* or *Society of Robots*), under which several graduation and post-graduation students have been developing their Final Year Projects, M.Sc. and Ph.D. theses. The project involves the Intelligent Control and Artificial Intelligence groups of the Lisbon location of the Institute for Systems and Robotics

A. Birk, S. Coradeschi, and S. Tadokoro (Eds.): RoboCup 2001, LNAI 2377, pp. 653–656, 2002.

(ISR/IST), and aims at developing a formal framework for teams of cooperating robots, demonstrable in real world scenarios. The current research topics within the project include Multi-Agent Systems, Multi-Sensor Navigation and Object Location, Reinforcement Learning and Stochastic Games. Recently, most of the researchers involved in the project joined other researchers at ISR/IST in a long-term project on Robotic Rescue (http://www.isr.ist.utl.pt/labs/rescue/).

Current research work covers the development of appropriate software and functional architectures for multi-robot teams, an approach to relations between teammates (e.g., passes, one robot advances and another covers its position left free) based on reinforcement learning applied to stochastic games and the application of Monte Carlo and sensor fusion methods to self-localization, cooperative ball and cooperative localization of opponent robots from omnidirectional images. The implementation work for Seattle 2001 Middle-Size League concerns a considerable update of the current software architecture, as well as behavior fine tuning and introduction of features not demonstrated before in games by the team, such as dribbling and multi-sensor navigation, including the vision-based self-localization method that granted us the *Engineering Challenge Award* in RoboCup 2000 Workshop [2].

2 ISocRob Current Status

The ISocRob robots are based on Nomadic SuperScout II robots, each of them a two-wheel differential drive robot equipped with sixteen sonar sensors radially distributed around the robot, one Pentium 233MHz based motherboard (PCM-5862), 64MB of RAM, 8GB of hard drive, one PCI and one PC104 bus connectors, one m68k based daughterboard with three-axis motor controller, sonar and bumper interface, and battery level meters, as well as two 12V batteries of 18Ah capacity. After RoboCup 99 (the team has only participated in Eu-RoboCup, during the year 2000), one Ultrak KC7500CP color 1/3" CCD PAL camera with an Ultrak KL0412DS lens (4mm, F1.2), in the robot front, an omnidirectional vision assembly consisting of one MicroVideo MVC26C color CCD camera under an 11cm diameter mirror, designed to obtain an image representing the orthographic projection of the points at the floor level (apart from an affine transform), a pneumatic kicking device, based on Festo components, plus two 0.85l bottles for pressurized air storage (70mm diameter by 220mm length), one Lucent WaveLAN/IEEE Turbo 11Mbps (Silver) wireless Ethernet modem connected through a PC104/PCMCIA bridge and one Bt848 based (Zoltrix TV-max) frame grabber board, with S-VHS and Composite video inputs were added to the robot platforms.

The total weight of each robot is about 30Kg for the goal-keeper and 35Kg for the remaining ones, which complies with the RoboCup middle size league rules limit of 80Kg, and its height is about 64cm which is below the rule limit of 80cm. According to the rules, there are two geometric restrictions imposed to the convex hull of the robot projection onto the floor plane, with all actuators extended to their maximum reach: the area must not exceed 2025cm^2, and any single cut must be no longer than 63cm. The field robot dimensions, shown in Fig. 2-left, are $R \simeq 207$mm, $W \simeq 240$mm, and $L \simeq 190$mm. Therefore, after some straightforward calculations [3], the area and maximum cut of the field robot convex hulls are found to be $\simeq 1992\text{cm}^2 < 2025\text{cm}^2$ and $\simeq 62.2\text{cm} < 63\text{cm}$, respectively. The goal-keeper robot dimensions (Fig. 2-right) are $R \simeq 207$mm, $a \simeq 220$mm, $b \simeq 510$mm, $c \simeq 604$mm, and $\delta \simeq 47$mm. Therefore, similar calculations lead an area of $\simeq 2018\text{cm}^2 < 2025\text{cm}^2$ and a convex hull of $\simeq 60.4\text{cm} < 63\text{cm}$.

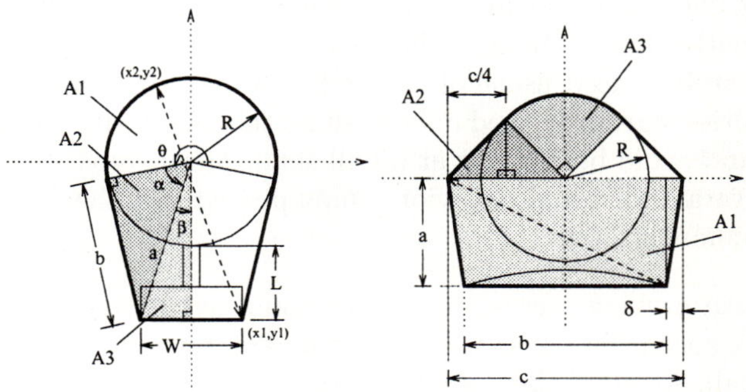

Fig. 2. Diagram showing dimensions used for the convex hull footprint calculation: the central circle denotes the SuperScout platform body while the structure attached below denotes the kicking device fully extended (field robot). Left: field robot; right: goal-keeper robot.

The current software architecture consists of two main layers: the **H**ardware **A**bstraction **L**ayer HAL) and the micro-agents. The HAL encapsulates the interface to the operating system multi-thread mechanisms and to the robot hardware, such as motor control, odometry, sonars, bumpers, and battery status, BTTV frame grabber driver (including an API that allows full-rate vision processing through a dual-buffering technique), multicast network communication, pneumatic kicker hardware, the X Window System and the Blackboard — the communication medium among the architecture components (intra-robot) as well as between physically distinct robots (inter-robot). The micro-agents layer includes concurrent processes that interface the Blackboard with the sensors, the actuators, the communications and the state machine representing a *role*.

3 New Features in RoboCup 2001

The most significant problem we experienced in past competitions, that hindered our efforts to demonstrate some of our scientific advances, was the lack of a robust, fast calibrating and reliable color segmentation. Therefore, in parallel with the development of other new features, our top priority was to attain a color segmentation procedure which is friendly (i.e., through a graphical interface where the result of the adjustments is immediately visible), automatic (e.g., using a *k-means* clustering [1] or thresholds adjustable through virtual sliders) and as robust and reliable as possible (we are considering replacing the current MicroVideo MVC26C cameras, and/or migrating to the YUV or RGB color spaces, instead of the HSV space previously used). A new version of the software architecture was also used in RoboCup 2001, significantly upgrading the current one, mainly by direct connections between perception and actuation micro-agents, instead of the current solution, under which connections are established through the state machine micro-agent. Furthermore, software primitives are now efficiently shared among different robot roles (e.g., Attacker, Defender). We also plan an increased usage of the architecture data sharing among teammates capabilities, currently used only to share the ball position estimates of the team robots and avoid both going at a ball simultaneously [3].

Besides several individual behavior tuning procedures, the following new behaviors were demonstrated:

- **dribble** - in previous years, our robots could avoid other robots at a relatively high speed (0.6 m/s) using their sonars, but only when they did not have the ball. An algorithm based on geometric and kinematic considerations to dribble field robots while keeping the ball and head towards the opponent goal is under test, with good initial results.
- **new goalkeeper** - ISocRob new goalkeeper, partially demonstrated in EuRoboCup 2000, moves parallel to the goal line, as most current robots in the middle-size league. One of its relevant characteristics is that its state machine switches between two operation modes: following the ball if it is moving away or parallel to the goal, intercepting it when it is moving towards the goal, both based on the up camera/catadioptric system. Currently, we are working on providing the goalkeeper robot with local navigation capabilities that include self-localizing near the goal using Markov localization.

References

1. R. Duda and P. Hart, *Pattern Classification and Scene Analysis*, John Wiley and Sons, Inc, 1973
2. C. Marques and P. Lima, “A Localization Method for a soccer Robot using a Vision-Based Omni-directional sensor”, *Proceedings of RoboCup 2000 Workshop*, Melbourne, Australia, 2000, *to be published in RoboCup-2000: Robot Soccer World Cup IV*, P. Stone, T. Balch, G. Kraetzschmar (Eds).
3. R. Ventura, L. Toscano, C. Marques, L. Custódio, P. Lima, *ISocRob-2000: Technical Report*, Internal Report of ISR/IST, RT-701-00, RT-402-00, October 2000

MINHO Robot Football Team for 2001

Fernando Ribeiro, Carlos Machado, Sérgio Sampaio, and Bruno Martins

Grupo de Automação e Robótica, Departamento de Electrónica Industrial
Universidade do Minho, Campus de Azurém
4800 Guimarães – Portugal
{fernando.ribeiro, carlos.machado}@dei.uminho.pt,
sergio.sampaio@mail.telepac.pt, brunomartins@bitmega.com

Abstract. This paper describes an autonomous robot football team. The work is being carried out since 1998. It describes the hardware used by the robots, the sensory system and interfaces, as well as the game strategy. Data acquisition for the perception level is carried out by the vision system, and the image processing system is described. Two cameras are used requiring sensorial fusion. With this architecture, an attempted is made to make the autonomous robots more real world intelligent. These robots have a kicker with controlled power, which allows passing the ball to a teammate with controlled distance and direction.

1 Introduction

This team already participated on previous RoboCup editions and has as main goal to develop a robotic team capable of playing football as comparable as possible to humans. Previous participation's allowed this team to learn and improve the robots and these are becoming more and more reliable. The most important improvements are: (a) new mechanical platform more reliable and robust; (b) kicker with controlled strength; (c) faster motor control with the use of dynamic algorithms; (d) self localization with one camera only; (e) use of two cameras per robot, requiring sensorial fusion but allowing an improved perception level; (f) improved color segmentation and acquisition. The next step is to improve the cooperation between the robots. For all these reasons, this team of robots is playing better than ever.

2 Hardware Architecture

A new robot was built from scratch, following the know-how acquired on previous competitions, and it was tested in a Portuguese championship confirming all the changes were successful. Now, the other three robots are being built.

The hardware is split in two separate parts: head and tail. The head contains all the power electronics, the electronic control boards, the computer and the vision system. The tail contains the chassis, the actuators (two motors and kicker), the motor gear-boxes with encoders, the batteries and a device for ball control, hereby known as fork.

A. Birk, S. Coradeschi, and S. Tadokoro (Eds.): RoboCup 2001, LNAI 2377, pp. 657–660, 2002.

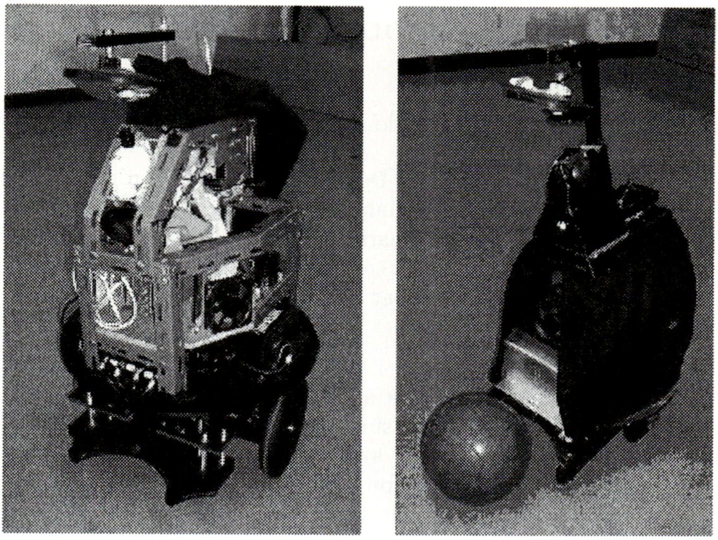

Fig. 1. Robots from MINHO team (new robot on the left, old robot on the right)

The old robots are still in use as testing platforms and are able to play football. Their description can be read in [1].

Table 1. Specifications of the new robot

Diameter	50 cm
Height	75 cm
Weight	38 kg. (incl.12Kg batteries)
Battery Power	4 batteries of 12V / 7Ah
Drive	2 wheel differential
Speed	1.4 m/sec
Motors	6V motors, 8500 rpm
Gearbox	ratio of 1/64
Encoder Resolution	64 counts/ver (9,8 mm per count)

2.1 The Head

The computer consists of a typical motherboard, with a Pentium II microprocessor running at 233 MHz, with 32 Mbytes of Memory, and a 4 Gbytes hard disk. The two serial and one parallel ports are used to communicate with the motors, kicker and encoders. Three motherboard slots are used; the VGA graphics card, the wireless network board from Lucent Technologies (using a 2.4 GHz frequency) and a booktree bt848 chip based frame grabber where the two cameras are connected to. The top camera points to a convex parabolic shape mirror mounted over the camera and points downwards, and the front camera points forward. While the top camera gives an image from all around the robot, the front camera shows an image where game entities are larger and not distorted.

2.2 The Tail: Actuators

The only actuators used by these robots are the motors and the kicker. This robot uses differential drive with two extra wheels to support the structure. The motors are capable of 8500 rpm when fed continuously giving an output (after a gearbox reduction of 1/64) of 134 rpm on the 200mm diameter wheels, allowing speeds of up to 5 Km/hour. It is important to point out that the collision avoidance routines are extremely reliable and efficient.

The kicker consists of an electric wire coil, through which a current is passed, forming a magnetic field strong enough to move forward an iron core, which kicks the ball. The software controls the amount of time the current passes through, allowing to kick the ball to a desired distance. The formula to calculate the time of kick is:

$$time\_of\_kick = \frac{desired\_distance\_in\_metres + 5.5}{0.225}$$

The shorter the distance, the slower the ball moves, the more unreliable the ball direction gets. But for distances above 1,5 meters the direction is perfectly acceptable.

These robots have a device to minimally control the ball hereby called fork, consisting of a round reentrance of only 7cm (allowed by the rules). For safety and efficiency reasons, an infrared sensor is used between these two fork fingers which detect the presence of the ball when it obstructs the signal. The software contains a rule, which permits the kicker to be actuated ONLY if the RED ball is detected in the fork.

2.3 The Tail: Sensors

These robots use only two different physical sensors: an encoder coupled on each motor (to measure the number of turns of each motor) and an infrared transmitter and receptor, to detect the presence of the ball on the fork. In case the robot loses any of these sensors it is still able to play the game. All the information on the field needed to play a football game is percept by the vision system, rather than by any other physical sensors. The vision system apparatus was described above (Hardware: head), and therefore only the image processing software will now be described.

3 Image Processing and Game Strategy

Around about 20 frames per second are achieved on each camera, and the image is delivered on the YUV format, with a resolution of 384 × 288 pixels. Every pixel is converted into one of the 8 known colors with 32 levels, according to a Look up Table, built when calibrating the colors in a real football field. Then, the maximums of red and blue or yellow (depending on side attacking) are calculated and a threshold is applied, to avoid noise. All gray pixels are considered obstacles. On the image, virtual bumpers are used, which consist of 9 squares located

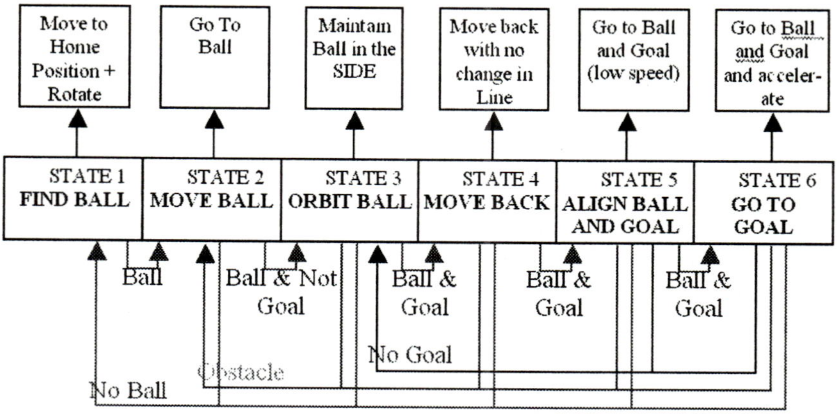

Fig. 2. Game Strategy

around the robot and above a certain amount of gray inside, they consider an obstacle in that location.

The game strategy is described in the figure above and consists of searching the ball, move towards the ball, once near the ball orbiting until the goal is in the field of view, move backwards to orient the robot, align the ball towards the goal, go towards the ball. Automatically, when the kicker "feels" the ball (with the infrared sensor) and the image "sees" the goal in front, it kicks the ball (known as instinct routine).

Self-localization is also achieved by analyzing the image of the front camera. A few transitions from green to white are seek on the image and according to the inclination of the line which passes by the transition coordinates, the location is calculated.

4 Conclusion

This team main areas of research are image processing and color segmentation, co-operative autonomous behaviors in multi-robot systems, self localization and distributed sensor fusion. Several improvements have been made in these new robots what concerns electronics, control, color segmentation and software strategy. These robots have been tested in some real games and proved to be mechanically reliable, with consistent play and scored many goals.

References

1. Sérgio Sampaio, Carlos Machado, Ilídio Costa and António Ribeiro: RoboCup_99 Robot World Cup Soccer Games and Conference IJCAI'99, Teams Description Small and Middle Leagues, pages 175–180, ISSN: 1401–9841, July 1999, Sweden.

Osaka University "Trackies 2001"

Yasutake Takahashi, Shoichi Ikenoue, Shujiro Inui, Kouichi Hikita, Yutaka Katoh, and Minoru Asada

Dept. of Adaptive Machine Systems,
Graduate School of Engineering Osaka University,
Suita, Osaka 565-0871, Japan
{yasutake,ikenoue,inui,khikita,yutaka,asada}@er.ams.eng.osaka-u.ac.jp

Abstract. This is the team description of Osaka University "Trackies" for RoboCup2001. The hardware and software architecture are presented.

1 Introduction

The Robot World Cup Soccer Games and Conferences (RoboCup) are a series of competitions and events designed to promote the full integration of AI and robotics research. The robotic soccer provides a good test-bed for evaluation of various research, e.g. artificial intelligence, robotics, image processing, system engineerings, multi agent systems.

Osaka University "Trackies" has participated the RoboCup since the first one 1997. We have changed our robot system from a remote controlled vehicle to a self-contained robot because the remote brain system often suffered from much noise of camera image and motor command transmissions. On the other hand, we adopt an idea that the cooperative behaviors without any planning to emerge cooperation in the dynamic environment caused by multi robots in a hostile environment.

This paper presents the employed hardware, namely robots and computers, and a fundamental approach to control robots.

2 Hardware

Fig.1 show pictures of our mobile robots we designed and built. Fig.2 shows an overview of the robot system. The system consists of a motor driving unit, a vision system, and a control unit.

2.1 Control Unit

We use an on-board computer as a controller for the robot. It is a standard PC parts, and commercially available. Each PC has a 233 MHz Intel Pentium MMX CPU, 128 MB RAM and 160MB Silicon Disk Drive (as a hard disk drive). The operating system is Debian GNU/Linux, and we install the minimum system in the hard disk drive in order to communicate with external file server. The

A. Birk, S. Coradeschi, and S. Tadokoro (Eds.): RoboCup 2001, LNAI 2377, pp. 661–664, 2002.

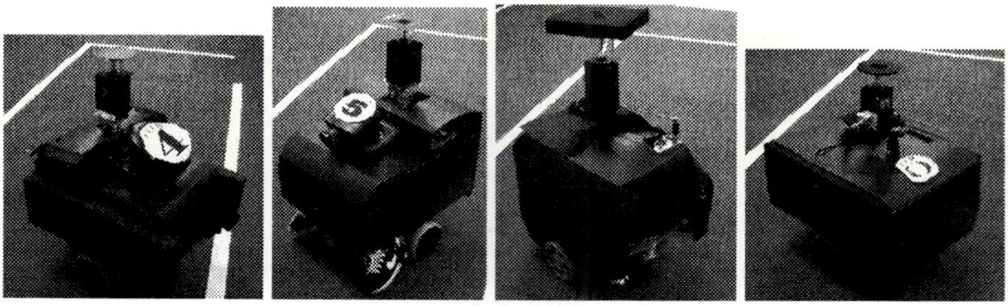

Fig. 1. Our mobile robots

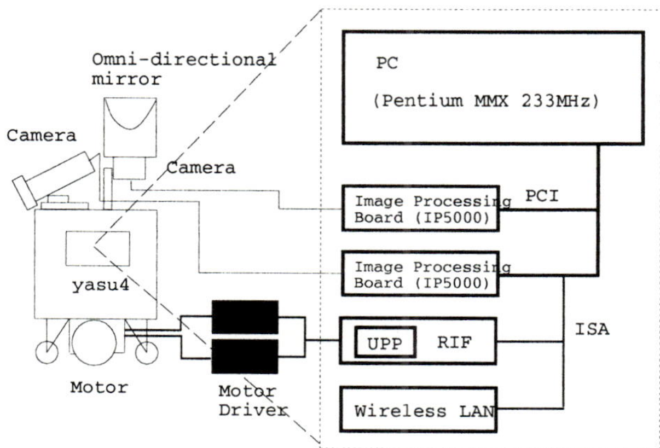

Fig. 2. An overview of the robot system

software development of our mobile robot is done on the external file server. For the communication between a robot and an external file server, wireless PCMCIA Ethernet cards (AT&A WaveLAN/IEEE) in PCMCIA to ISA adaptors are used. Each computer has one or two image processing board (Hitachi IP5005) or frame grabber for vision system, and one Robot Interface Board which has an Universal Pulse Processor for motor control.

2.2 Vision System

The robot has an omni-directional camera system with or without a normal camera system. Two robots have SONY EVI-D30 which has a motorized pan-tilt unit. An omni-directional camera system consists of a color CCD camera and a omni-directional mirror and is mounted on the robot as the camera optical axis is aligned with the vertical axis of the mirror. A simple color image processing is applied to detect the ball, goals, field, and obstacle areas in the image in real-time (every 33ms).

2.3 Motor Driving Unit

The driving mechanism, which is originally designed at Tsukuba University and now commercially available as a vehicle unit with motor, wheels, and control drivers, is a PWS (Power Wheeled System); the vehicle is fitted with two differential wheels. The wheels are driven independently by separated DC motors, and two extra free wheels ensure the static stability. Robot Interface Board generates PWM pulses, which controls the speed of the wheel. The pulses are sent to the motor drivers which drive the wheels.

3 Software

Our robot team consists of four robots. They all share almost the same basic hardware, but they differ in their behavior programming. The basic idea for cooperative behaviors among the teammates is that cooperative behaviors without any planning or intention of cooperation emerges in the highly dynamic, hostile environment provided by RoboCup. We design each robots' behaviors such as "use the environment directly," "replace computation with rapid feedback," and "tolerate uncertainty before trying to reduce it."[1]

3.1 Strategy Learning for a Team

Team strategy acquisition is one of the most important issues of multiagent systems, especially in an adversary environment. RoboCup has been providing such an environment for AI and robotics researchers. A deliberative approach to the team strategy acquisition seems useless in such a dynamic and hostile environment. We have presented a learning method to acquire team strategy from a viewpoint of coach who can change a combination of players each of which has a fixed policy at RoboCup Symposium 2001[2]. Assuming that the opponent has the same choice for the team strategy but keeps the fixed strategy during one match, the coach estimates the opponent team strategy (player's combination) based on game progress (obtained and lost goals) and notification of the opponent strategy just after each match. The trade-off between exploration and exploitation is handled by considering how correct the expectation in each mode is.

3.2 Evolutionary Behavior Selection

We have presented a behavior selection mechanism with activation/termination constraints at RoboCup Symposium 2001[3]. In this method, each behavior has three components: policy, activation constraints, and termination constrains. A policy is a function mapping the sensor information to motor commands. Activation constraints reduce the number of situations where corresponding policy is executable, and termination constrains contribute to extract meaningful behavior sequences, each of which can be regarded as one action regardless of its duration. We apply the genetic algorithm to obtain the switching function to select the appropriate behavior according to the situation. As an example, a simplified soccer game is given to show the validity of the proposed method. Simulation results and real robots experiments are shown, and discussion is given.

References

[1] Barry Brian Werger. Principles of minimal control for comprehensive team behavior. In *Proceedings of the 1998 IEEE International Conference on Robotics and Automation*, pages 3504–3509, 1998.

[2] Yasutake Takahashi, Takashi Tamura, and Minoru Asada. Strategy learning for a team in adversary environments. In *RoboCup 2001 book*, page (to appear), 2001.

[3] Eiji Uchibe, Masakazu Yanase, and Minoru Asada. Evolutionary behavior selection with activation/termination constraints. In *RoboCup 2001 book*, page (to appear), 2001.

ROBOSIX UPMC-CFA: RoboCup Team Description

Ryad Benosman, Francis Bras, Frederic Bach, Simon Boulay, Emmanuelle Cahn, Sylvain Come, Gilles Cordurié, Jarlegan Marie Annick, Lapied Loic, Cyrille Potereau, Franck Richard, Samedi Sath, Xavier Vasseur, and Pascal Vincent

Laboratoire des Instruments et Systèmes d'Ile de France,
Université Pierre et Marie Curie,
T 22, coul 22-12, 2eme,
4 place Jussieu,
75252 Paris cedex 05,
Boite 164, France
rbo@lis.jussieu.fr

Abstract. This paper describes the Robosix team of the University Pierre and Marie Curie. The team is composed of five robots built inside the institute based on an omnidirectional motion system associated to a catadioptric sensor developped by the Laboratory of Instruments and System. The team started developping the robots during the year 1999 in prevision to participate to the RoboCup 2001, the team is partially composed by member of the French middle size team that participated to RoboCup98 in Paris. The robots computational system is based on a pc card with a celeron 633Mhz running under a light version of Linux. The localization system relies on an assocation of multiagent panoramic vision system developped inside the LIS lab.

1 Introduction

Our research group started working on the RoboCup contest since the first edition of RoboCup in 1997 where we participated in the small size league. We consider Robocup to be a challenging competition as it is a good test plat-form to test new algorithm in distributed panoramic vision which is one of the main research subject at the Laboratoire des Instruments et Systemes of the University Pierre and Marie Curie. A RobCup team needs several knowledge in different computer science domains, like networking, mechanical and eletrical engineering, computer vision and artificial intelligence. The team is the a consortium of different sections of the university and is composed of fifteen students from these differents fields and three advisors. Robosix is an association created to participate to the Robocup in different leagues, up to now Robosix handles both the small size and middle size activities, putting together the knowledge of both leagues. Most of the vision control and path planning algorithms are the same

A. Birk, S. Coradeschi, and S. Tadokoro (Eds.): RoboCup 2001, LNAI 2377, pp. 665–668, 2002.

between both leagues. Following our particpation in the middle size competition in 1998 and our numerous participations in the the small size league our goal has always been participating to this league as it presents problems more linked to distributed vision and intelligence that does not exist in the small size league, also because it is one of the main research subject of our research team. Our credo is to developp robots using unexpensive easy and very modulable components.

2 Robots Architecture

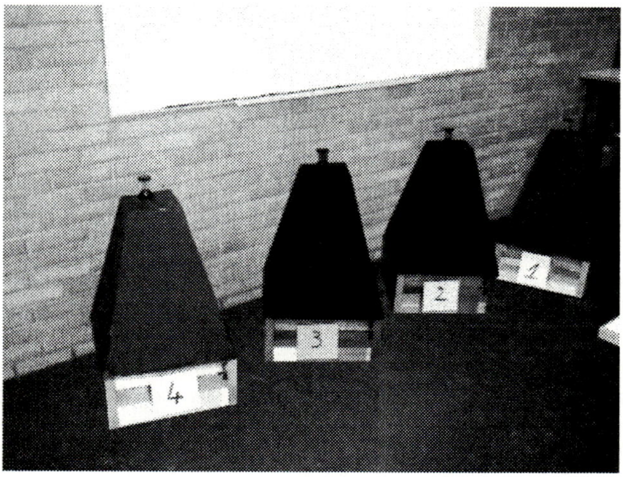

Fig. 1. Robosix middle size team

The robots architecture is based on an omnidirectional plat-form. All the robots have the same mechanical design. The robots computational system is based on a pc card based on a celeron 633Mhz with 128Mb RAM. The operating system is based on a linux Kernel taken from the Suze 7.1 version of linux that was lightened, the linux kernel is the 2.4 using the GlibC 2.2. The robots are size is 50 × 50 × 80. The robots have a pneumatic shooting system. A ST10 micro-controller handles the speed and control direction of the electrical motor. The electrical motor were given to us by MDP and Maxon are alimented under 48V.

3 Vision System

The vision system of our robots is based on a catadioptric device developped in the Laboratoire des Instruments et Systèmes and a colour camera card of 380

TV lines. We initially used the sensor we developped in 98 which desciption can be found in [2] and we finally mounted the one that we developped this year. The catadioptric system developped is under the process of being patented and then due to confidentiality we will not be able to give a scheme of it. The acquisition card is a pc-tv card which is the cheapest available running under linux. The output signal of the camera is RGB. A color is defined as a three component vector containing the values for the red, green and blue color planes. A first stage is manually made where colors are selected, for each color we compute a mean and a standard deviation. We used a bettered version of RGB calibration to allow fast processing. The image is then filered, an image containing only the desired colors is generated. In a second time a second detection based on HSV color space is applied only on the detected zones to reduce computation. A scalar is assigned to each color corresponding to the objects seen, we obtain then a 8 bits images where for example the field is coded zero. A fast labelling algorithm is then applied giving for each detected region its gravity center and the number of pixels it contains. The same low-level image processing has been used succefully used by our small size team.

4 Path Planning and Localisation

Our panoramic device is sharply calibrated so that each robot can estimate the distance of each detected colour zone present in the image. From these 2D information a voronoi diagram is computed giving the best paths, then an A^* algorithm determines the best path to choose. The path planning is the same one of our small size team.

5 Behaviour

The behaviour of our robots is again partially based on the one we used last year in the small size league which proved to be quite efficient against agressive but also defensive teams, but this time we added to each robot a local behaviour giving a priority between what we call "reflex" behaviour and group actions, like passes, etc ... The behaviour of the robots is motivated by different flags corresponding to specific situations and robots positions after analysis of the situation. The field is cut into different zones each having a priority. The knowledge and the analysis of these flags gives an analysis of the situation of the game. Each robot according to its position, the position of other robots and the ball will have a specific behaviour. The behaviour is based on micro and macro orders. Each robot has simple behaviours each one corresponding to an action. The robots use a prediction algorithm that gives the position of the ball and the opponents in the next two images. The behaviour of the robot varies according to situations for example if the team is winning or not.

6 Conclusion

This paper presented the RoboSix team, its composition giving the headlines of what has been developped. We are preparing this event since last year in parallel with our small size team that started from scratch last year. We will make soon available the source codes of what have been developped on www.robosix.com

References

1. T Sogo, H Ishiguro, M Trivedi N-Occular stereo for real time human tracking Panoramic Vision : Theory System and Applications editors R Benosman, S B Kang Springer Verlag, 2001 NY
2. R.Benosman, E.Deforas, J.Devars , A new catadioptric sensor for the vision of mobile plat-form, IEEE Workshop on Omnidirectional Vision, CVPR, Hilton Head Island, 12 June 2000, pp 112-166

S.P.Q.R. Wheeled Team

Luca Iocchi, Daniele Baldassari, Flavio Cappelli, Alessandro Farinelli, Giorgio Grisetti, Floris Maathuis, and Daniele Nardi

Dipartimento di Informatica e Sistemistica
Università di Roma "La Sapienza"
Via Salaria 113, 00198, Roma, Italy
`<lastname>@dis.uniroma1.it`
`http://www.dis.uniroma1.it/~wheeled`

Abstract. This paper presents the design and implementation issues that have been addressed for the realization of the S.P.Q.R. Wheeled team of soccer robots. The team is formed by "senior" soccer robots that have participated to previous RoboCup competitions as well as by new robots developed for completing the team. By exploiting our previous experience in designing and developing soccer robots, we have developed innovative techniques for vision, localization, planning and coordination.

1 Introduction

S.P.Q.R. is the group of the Faculty of Engineering at University of Rome "La Sapienza" in Italy, that is currently involved in developing two RoboCup teams: the **S.P.Q.R. Legged** team [6] in the *Sony Legged League* and the **S.P.Q.R. Wheeled** team in the *Middle Size League*.

The S.P.Q.R. Wheeled team is formed by three "senior" soccer robots that have participated at previous RoboCup competitions within the ART team [7], by a Golem robot, and by a new robot that has been designed and built within our group for the role of goalkeeper. The main feature of our team is the design approach we have adopted for developing cognitive soccer robots, that can be programmed by using a high-level specification of their capabilities, the environment and the task they can execute. The layered architecture we have realized allows for sharing the same high-level modules within both the Legged and the Wheeled teams.

2 Team Development

Team Leader: Luca Iocchi.
Team Members: Daniele Baldassari, Flavio Cappelli, Alessandro Castaldo, Alessandro Farinelli, Giorgio Grisetti, Floris Maathuis, Daniele Nardi, Giampaolo Pucci, Davide Troiani.
Country: Italy.
Sponsors: Facoltà di Ingegneria, University "La Sapienza" and NETikos Web Mobility.

A. Birk, S. Coradeschi, and S. Tadokoro (Eds.): RoboCup 2001, LNAI 2377, pp. 669–672, 2002.

Fig. 1. S.P.Q.R.-Wheeled team

3 Robotic Bases

The robots of our team are based on different mobile platforms. Three of the players are modified Pioneer1 or Pioneer2 [1] robotic bases, equipped with a fixed CCD color camera and with a kicking system formed by two independent air-pressure driven kickers that allow for directional kicks. One of these robots has a kicking system that can raise the ball from the ground. The fourth player is a Golem robot[2] with omnidirectional motion system and omnidirectional vision system obtained by a mirror especially designed for the RoboCup environment. The goalkeeper is a new robot that has been designed and built within this project. Motor control is based on a self-made board connected to the on-board PC on the ISA bus. In this way, the control of the robot is performed by the same CPU (currently a Pentium III 800 MHz) used for high level processing. This has been possible by using the real-time extension of the Linux operating system and by the implementation of a motor control driver that run in real-time mode and thus it is guaranteed to be scheduled every 5 ms. The vision system of our new robot is formed by a moving stereo vision camera and a digital camera with an emisperical lens having field of view greater than 180 degrees. By putting the digital camera on the top of the robot looking downwards, the goalkeeper is able to see 360 degrees for about 5 meters (to the half of the field form the goal line). Image quality is very good both for the use of a digital camera and for the lack of a mirror that always introduces some kind of distortion.

[1] `http://www.activrobots.com/`

[2] `http://www.golemrobotics.com/`

Therefore we have different robots with different capabilities to use according to the situation at hand (e.g. the characteristics of the opponent team). All these robots use: 1) a conventional PC for on-board computing; 2) a wireless high bandwidth connection for communication and coordination among the robots during the game, and as a development tool that allows the programmer operating on a standard platform to obtain accurate information about the situation on-board; 3) high quality CCD color cameras with a resolution 380 TV lines and a low-cost PCI frame grabber based on the BT848 and one digital camera with IEEE1394 interface.

4 Software Architecture

In order to realize a robotic team with heterogeneous platforms three main components have been developed: a heterogeneous and asynchronous layered architecture allowing for an effective integration of reasoning and reactive capabilities of the robot; a high level representation of the plans executed by the robots; a formal account of the system for generating and verifying the plans. The use of a high-level specification has provided significant advantages also from a practical viewpoint. First of all we are able to share the high-level of our architecture among different platform (including the Sony legged robots that compete in the Legged League). Second, we are able to effectively program our players in a very flexible and effective way. All our robots share the same software architecture in which the main components, that are perception, localization, navigation, planning and task execution, and coordination, are described in the following paragraphs. We have also realized a new development tool (based on global vision) that is used for monitoring the behavior of the robots during a match or an experiment, and thus for evaluating the implemented techniques for localization, navigation, and coordination.

Perception. The perception of our robots is mainly based on color cameras used as vision based range sensors, that are able to acquire range information for the objects that are in the environment. The use of a vision system is very important since images are the richest font of information that a robot can acquire from the environment, however, with respect to other range sensors (such as sonars and laser range finders), it requires an additional effort for calibration and the implementation of routines for color segmentation, feature extraction and world reconstruction. In addition to the image processing procedures, that were already developed (see [5]), we have designed a new easy-to-use tool for calibrating the vision system: internal and external parameters and color filter.

Localization. Localization in RoboCup is an important aspect for effective task accomplishment and Hough Localization [5] has proved to be very effective in this environment. We have extended our previous method with the implementation of a probabilistic extension of the Hough Localization [4] that, by using an Extended Kalman Filter, has provided good improvements in the integration of the map matching process with odometry data.

Navigation. A new navigation system has been implemented for our robots. We have defined a new global navigation function that takes into account moving objects and it is used for generating the trajectories for the robot. By taking into account direction and speed of moving objects our path planner is able to generate more effective trajectories for avoiding them. For instance, in situations in which another robot is crossing the robot's trajectory the path planner usually chooses to pass behind it or even to stand still until the other robot is passed.

Planning and Task Execution. A central issue for our research in robotics is the design of cognitive mobile robots, where a deliberative layer including an explicit representation of the robot's knowledge on the environment is added, with the aim of increasing the robot's performance in accomplishing complex tasks. We have developed both a set of primitive actions to provide the robots with individual basic skill and a framework [2,3], for high-level description of the environment, automatic plan generation and verification, and monitoring during task execution.

Coordination. Coordinating a team of heterogeneous robots is an important research issue, and the highly dynamic, uncertain and noisy RoboCup environment provides for an interesting testbed for experimentation of distributed approaches to multi-robot coordination. Coordination among the robots in our team has been designed on a distributed coordination protocol [1], that is implemented on every robot of the team and allows for the exchange of a few data containing information about the status of every robot. The protocol deals both with roles (*GoToBall*, *Defend*, etc.) and with strategies (defensive, offensive). While roles are dynamically assigned to the various players during the game, depending on the configuration present on the field, the strategic level was demanded to an external selection (the human coach).

References

1. C. Castelpietra, L. Iocchi, D. Nardi, M. Piaggio, A. Scalzo, and A. Sgorbissa. Coordination among heterogenous robotic soccer players. In *Proc. of International Conference on Intelligent Robots and Systems (IROS'2000)*, 2000.
2. Giuseppe De Giacomo, Luca Iocchi, Daniele Nardi, and Riccardo Rosati. A theory and implementation of cognitive mobile robots. *Journal of Logic and Computation*, 5(9):759–785, 1999.
3. L. Iocchi, D. Nardi, and R. Rosati. Planning with sensing, concurrency, and exogenous events: logical framework and implementation. In *Proc. of KR'2000*, 2000.
4. Luca Iocchi, Domenico Mastrantuono, and Daniele Nardi. A probabilistic approach to Hough Localization. In *Proceedings of the IEEE International Conference on Robotics and Automation (ICRA2001)*, pages 4250–4255, 2001.
5. Luca Iocchi and Daniele Nardi. Self-localization in the RoboCup environment. In *RoboCup-99: Robot Soccer World Cup III*, pages 318–330, 1999.
6. D. Nardi, C. Castelpietra, A. Guidotti, M. Salerno, and C. Sanitati. S.P.Q.R. In *RoboCup-2000: Robot Soccer World Cup IV*. Springer-Verlag, 2000.
7. D. et al. Nardi. ART-99: Azzurra Robot Team. In *RoboCup-99: Robot Soccer World Cup III*, pages 695–698. Springer-Verlag, 1999.

Team Description Eigen

Kazuo Yoshida, Ryoichi Tsuzaki, Junichi Kougo, Takaaki Okabe, Nobuyuki Kurihara, Daiki Sakai, Ryotaku Hayashi, and Hikari Fujii

Keio University,3-14-1, Hiyoshi, Kohoku-ku, Yokohama, Kanagawa 223-8522, Japan

Abstract. EIGEN participates in RoboCup at World Meeting for the first time. Our team provides three robots of the same type and one goalkeeper robot. Our aim is to realize a cooperative team play among robots as well as high performance of robot relating to image processing, decision making of behaviors, and motion control. As a first step, we realized a motion control for a field-player robot by the Fuzzy control method and cooperative behaviors by a wireless LAN.

1 Hardware Architecture

1.1 System Composition

We made three robots of the same type and one goalkeeper robot as shown in Fig.1. The system composition of robot is shown in Fig.2. These robots consist of sensing system, communication system, actuator system and processing system. This architecture is designed based upon the concept of System Life. Each system is described as follows.

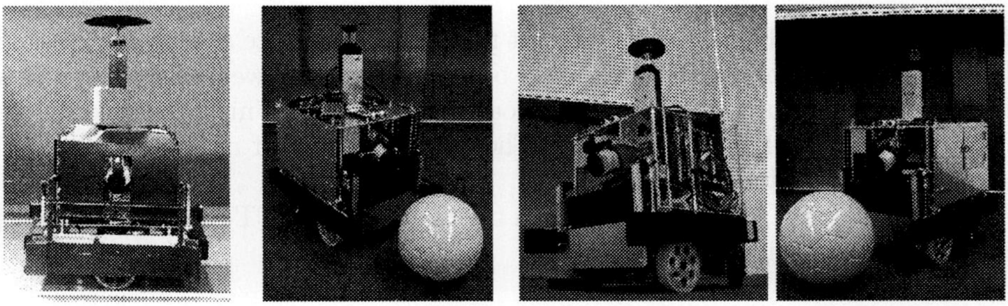

Fig. 1. EIGEN robots

1.2 Sensing System

The robots have a conventional camera system and an omni directional camera system for visual sensing system. For the conventional camera system, we chose a color CCD camera (Panasonic WV-CP464). And the omni directional

A. Birk, S. Coradeschi, and S. Tadokoro (Eds.): RoboCup 2001, LNAI 2377, pp. 673–676, 2002.

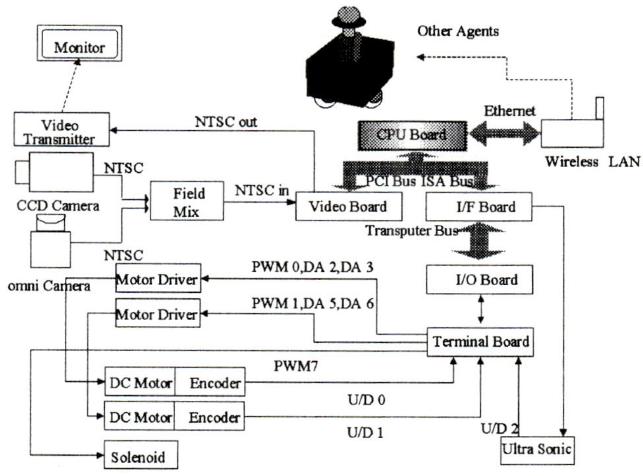

Fig. 2. Schematic diagram of the hardware architecture

camera system consists of a hyperbolic mirror and a color CCD camera (Sony FCB-IX10). All robots have two vision systems. Field-player robots process simultaneously two images from the conventional and the omni direction cameras with Video Mixing Device. In order to control each camera system, we use IP-5005BD cards (HITACHI).

Besides, the robots have an ultra sonic sensor for obstacle avoidance and two encoders for the distance measurement based on internal information.

1.3 Actuator System

Motor Control System: It is difficult to make an interface among the PC and the other devices by using commercial boards. Therefore, we developed an ISA interface board. The robots have I/O board and H-bridge Unit developed by the other laboratory of Keio University. The rotation speed of motor is controlled with PWM from these boards.

Kick System: Kick device consists of DC solenoid (ACT SD18AA-06). The robot uses the device on DC 24V.

1.4 Communication System

We utilize a wireless LAN device for communication among our robots. The wireless LAN device, NetHawk RF-100E is directly connectable to the exiting Ethernet LAN system.

1.5 Processing System

For a processing system-board, each robot possesses a half size CPU board with Pentium 2 450MHz processor and 64Mbytes RAM. Instead of HDD, Silicon Disk

Drive or Disk On Chip 144MB(M-Systems) is adopted. They are stronger for vibrations than HDD.

2 Software Architecture

2.1 Image Processing

EIGEN robots have a conventional camera and a omni directional camera which can watch both goals without turning by the omni directional camera. Hence, the robot calculates self-position by sensing position of two goals. Our procedure for Image Processing is described as follows.

1. Extract colors from camera image (orange: ball, blue: goal, yellow: goal, etc...)
2. Binarize the extracted images.
3. Labeling and Filtering by Area.
4. Calculate distance and angle from position of two goals.

Each robot calculates the distance and angle of each object by the area of object, width, height and gravity.

2.2 Path Planning

We presents a method of decision making on the basis of gSystem-lifeh concept. System Life is the concept to realize a flexible and adaptive intelligent system enabling a symbiosis with environment. The concept is proposed as a new system design concept to realize the adaptation and flexibility which are acquired by the seamless integration of the elements relating to information on the system object. This method is applied to various methods of the human-system, and is as follows concretely.

The soccer robot recognizes environmental information by real time vision system using images from two CCD cameras, and shares some information with other robots by communication system using wireless LAN. Each robot evaluates the self-condition by the information of processing module, and realizes various behaviors in real world by an activating module. These modules being integrated by integrator, the robot can perform intelligent soccer play. The information from each module is fed to the integrator, and the optimal behavior at present condition is selected as an output information.

In this method, it is expected that robots flexibly cope with dynamically changeable environment by means of designing each module separately and evaluating the situations under the information of purpose. The usefulness of the system architecture is verified by numerical calculations and an experiment using real robots.

3 Conclusion

In this paper we presented EIGEN team robots for RoboCup 2001 contest. This description of team EIGEN paid attention to the hardware and software of the robots.

From the viewpoint of hardware, it is characteristic that all EIGEN robots have both conventional and omni directional cameras. They can recognize the global map by two cameras.

For software, it is characteristic that each robot evaluates the self-condition by the information of processing module, and realizes the optimal behavior by the integrator on the base of gSystem Lifeh concept.

References

1. H.Kitano, M.Tambe, Peter Stone and et.al. "The RoboCup Synthetic Agent Challenge 97" In Proc. of The First International Workshop on RoboCup, pages 45-50, 1997.

The Ulm Sparrows 2001

Hans Utz, Gerd Mayer, Dominik Maschke, Alexander Neubeck, Peter Schaeffer, Philipp Baer, Ingmar Baetge, Jan Fischer, Roland Holzer, Markus Lauer, Alexander Reisser, Florian Sterk, Günther Palm, and Gerhard Kraetzschmar

University of Ulm, Neural Information Procssing,
James-Franck-Ring, 89069 Ulm, Germany

1 Introduction

The Ulm Sparrows team is a student-oriented, interdisciplinary research effort at the University of Ulm. The team is active in both the simulation and the middle-size league, but simulation efforts are targeted towards supporting the middle-size league robot team. We competed in the RoboCup World Championships in Paris (1998) [5] and Stockholm (1999, quarterfinals) [7], the European RoboCup Championships in Amsterdam (2000, semifinals) [4], plus a number of national events in Germany. For time and budget constraints, we could not participate in Melbourne in 2000.

Due to its strong foundation in student activity, the team was facing a major overhaul of its personnel in the past two years. Except for two undergraduates, all former team members graduated or finished their Ph.Ds. and left the team. However, we could attract a number of young, dedicated, and promising students, and with Hans Utz and Gerd Mayer we found new first-year Ph.D. students who will coach the robot team. In addition, we made progress in ensuring continuing team support by successfully applying to research funding in the Priority Programme of the German Research Foundation (DFG). This paper describes some early results of the newly formed team and some promising proofs of concept of The Ulm Sparrows Robot Team 2001.

2 Sensory and Actuatory Skills

In 1999, we introduced a new robot hardware platform, the *Sparrow-99* soccer robot, which was developed in our department. The *Sparrow-99* is a differential drive robot, equipped with sonar, a camera mounted on a pan-tilt unit and a pneumatic kicker. The computational power consists of a PC/104 computer, and, attached to it via CAN-Bus, a C167 controller for low-level control tasks (motors, odometry etc.). The hardware platform basically remained the same for the 2001 RoboCup team. However, we carefully extended our robot design to overcome some deficiencies that we discovered in past competitions.

Since the speed of the robots became a more important aspect in tournaments, we improved our low-level control software, in order to achieve higher speeds, better traction control, and more reliable odometry. Our robots also received new, more powerful CPUs (PIII 700), which allow for more sophisticated

A. Birk, S. Coradeschi, and S. Tadokoro (Eds.): RoboCup 2001, LNAI 2377, pp. 677–680, 2002.

vision processing, in order to match the requirements imposed by increased robot speed.

We also installed lightweight cameras with a wider view angle, in order to simplify the pan-tilt unit. In fact, due to the wider view angle, it is not any more necessary to use the tilt capability, which simplifies various steps in our image processing routines. The new setup still allows us to study active vision problems introduced by the base-orientation independent viewing angle.

New hardware components include infrared distance sensors and radio ethernet cards. The infrared sensors have been added in order to study sensor fusion (of vision and IR) for ball handling and to allow a more accurate kicking. The wireless LAN allows the explicit communication between the individual robots and enables us to build powerful logging and monitoring facilities.

3 Research Interest

The research interests of The Ulm Sparrows include robot control architectures, sensor fusion, world modeling, vision, and learning and adaptivity. The latter three are the focus of our project funded by the DFG Priority Programme. Core issues are learning of behaviors, situation classification, the association of situation and action, and the integration of learning functionality on various levels of abstraction in the robot control software. However, this project is just about to start, and work so far concentrated on laying the foundations.

3.1 Implementation Framework

We ported our distributed robot control software Miro (Middleware for Robots) [3] to the *Sparrow-99* robot. Miro utilizes the CORBA [6] standard and multiplatform libraries such as ACE and Qt to meet the needs of developing distributed software for (multi)robot scenarios. Miro provides interoperability and independence of specific robot platforms, communication protocols, operating systems, and programming languages. With Miro, we established a common implementation platform for all robots used within our lab, enabling researchers to easily evaluate their work on multiple platforms and within multiple scenarios. Miro is currently available under GPL for the B21 (pre-rFLEX) robot by RWI, the Pioneer-1 robot by ActivMedia, and the custom-built *Sparrow-99* robot.

For RoboCup, we investigate a distributed publisher/subscriber communications model based upon the CORBA Notification Service for sharing of preprocessed sensor information and features in a realtime scenario.

3.2 World Modeling

In 2000, we integrated a visual feature-based Monte-Carlo Localization (MCL) module. Compared to more common implementations of MCL, which are based on the evaluation of streams of laser scans, typically providing several hundred distance estimates up to ten times per second, our localization procedure used in RoboCup is based on sporadic visual features [2]. The RoboCup environment provides only few visual features that are easily detectable in the visual channel, like goal posts, corners, and lines, and due to camera geometry and specifications, there exist many postures where none of them are visible. To make localization more robust, we implemented visual distance estimators. Currently, we integrate feature vectors based on Hough transforms as well as further sensor modalities such as sonar.

We are redesigning our vision system in order to extract more visual features, such as shape. This additional information will allow for more reliable object recognition, leading to a more accurate dynamic world model.

The sensory information sharing framework enables us to incorporate sensory information gained by the other team members into the robot's world model without introducing dependencies between the players that jeopardize the autonomy of the individual soccer robot. It remains our main goal to build a team of robots as cooperating individuals, as it was described in [5]. Sensory features are distributed asynchronously via an event channel. If a player becomes unavailable, the lack of its sensory information might have an impact on the accuracy of the other robots world models, but not on their ability to make decisions.

3.3 Learning and Adaptive Behaviour

One of the most challenging tasks of autonomous mobile robots research is to enable robots to adapt to changes in the environment or the behavior of opponents. We consider adaptivity and learning to be prerequisites for true robot autonomy. Furthermore, we believe that both software platform and robot control architecture can hinder as well as ease the implementation of learning methods.

As a first step, we increased the runtime configurability of our software. We can tune every parameter of the robot soccer software dynamically. It is possible to specify (and alter on the fly) the entire behaviour engine. Secondly, the logging facilities introduced by the distributed control architecture enable us to acquire substantial amounts of data which can be used in offline training.

Looking at the RoboCup scenario, there are two different problem domains for learning and adaptivity. First, the RoboCup environment, like field dimensions, the colors and the robots individual configuration have very static properties. On the other hand, the game itself is highly dynamic and an even more hostile testbed for learning algorithms. Actually, adopting the robot's sensors and actuator to a specific field and the according lighting conditions of an individual tournament is an intermediate goal of our team. An interesting approach we successfully applied elsewhere [1] is to use the redundancy amongst different sensor modalities to adapt the model of a particular sensor. The long term goal is to master the learning challenges of the dynamic parts of the scenario. This is a far more sophisticated task, due to time constraints and limitations on the size of the sample set. But by solving the static part of the problem, we are confident to master the architectural challenges to investigate learning and adaptive behaviour within the very heart of the RoboCup challenge.

4 Summary

From a scientific perspective the RoboCup challenge is a perfect scenario to study various difficult problems in mobile robotics. We consider the research aspects together with the competitive aspects of the RoboCup challenge as highly synergetic.

References

[1] Stefan Enderle, Gerhard K. Kraetzschmar, Stefan Sablatnög, and Günther Palm. Sensor Interpretation Learned by Laser Data. In *Proceedings of EUROBOT-99*, 1999.

[2] Stefan Enderle, Marcus Ritter, Dieter Fox, Stefan Sablatnög, Gerhard Kraetzschmar, and Günther Palm. Soccer-Robot Localization Using Sporadic Visual Features. In Enrico Pagello, Frans Groen, Tamio Arai, Rüdiger Dillmann, and Anthony Stentz, editors, *Intelligent Autonomous Systems 6 (IAS-6)*, pages 959–966, Amsterdam, The Netherlands, 2000. IOS Press.

[3] Stefan Enderle, Hans Utz, Stefan Sablatnög, Steffen Simon, Gerhard Kraetzschmar, and Günther Palm. Miró: Middleware for autonomous mobile robots. In *Telematics Applications in Automation and Robotics*, 2001.

[4] Gerhard K. Kraetzschmar, Pierre Bayerl, Alexander Neubeck, Peter Schaeffer, Marcus Rittter, Dominik Maschke, Stefan Sablatnög, Stefan Enderle, and Günther Palm. The Ulm Sparrows 2000. In Wiebe van der Hoek and Giovanni Adorni, editors, *Workshop for the RoboCup European Championships*, 2000. published as CD-ROM.

[5] Gerhard K. Kraetzschmar, Stefan Enderle, Stefan Sablatnög, Thomas Boß, Marcus Ritter, Hans Braxmayer, Heiko Folkerts, Gerd Mayer, Markus Müller, Heiner Seidl, Markus Klingler, Mark Dettinger, Robert Wörz, and Günther Palm. The Ulm Sparrows: Research into Sensorimotor Integration, Agency, Learning, and Multiagent Cooperation. In Minoru Asada, editor, *RoboCup-98*, volume 1604 of *Lecture Notes in Artificial Intelligence*. Springer, 1999.

[6] Object Management Group. *The Common Object Request Broker: Architecture and Specification*, 2.3 edition, June 1999.

[7] Stefan Sablatnög, Stefan Enderle, Mark Dettinger, Thomas Boß, Mohammed Livani, Michael Dietz, Jan Giebel, Urban Meis, Heiko Folkerts, Alexander Neubeck, Peter Schaeffer, Marcus Ritter, Hans Braxmeier, Dominik Maschke, Gerhard Kraetzschmar, Jörg Kaiser, and Günther Palm. The Ulm Sparrows 99. In *RoboCup-99: Robot Soccer World Cup III*, Lecture Notes in Computer Science, Berlin, Germany, 2000. Springer.

ASURA: Kyushu United Team in the Four Legged Robot League

Kentaro Oda[1], Takeshi Ohashi[2], Shuichi Kouno[1], Kunio Gohara[1], Toyohiro Hayashi[1], Takeshi Kato[1], Yuki Katsumi[1], and Toshiyuki Ishimura[1]

[1] Dept. of Artificial Intelligence, Kyushu Institute of Technology
ken@mickey.ai.kyutech.ac.jp
[2] Dept. of Biochemical Engineering and Science, Kyushu Institute of Technology
ohashi@bse.kyutech.ac.jp
Address: 680-4 Kawatsu, Iizuka 820-8502, Japan

Abstract. This paper presents our approach to the Sony Four Logged Robot League of RoboCup 2001. The components of our system consist of strategy, vision, localization, behaviors and walking module. We introduce several techniques for these modules: The image based cooperation, TSL color space for robust color classification, accurate localization through observing the nearest marker and enhanced parameterized walk. We also develop two powerful tools, the color classification tool which allows us to extract accurate thresholds fast, and the forth interpreting language environment which provides interactive access for the robot control.

1 Team Development

1.1 Team Members

[Kyushu Institute of Technology (KIT)]
Professors
Dr. Takaichi Yoshida, Dr. Takeshi Ohashi
Graduate Students
Kentaro Oda, Yuki Katsumi, Toshiyuki Ishimura, Yoichiro Furutani, Shuichi Kouno, Kunio Gohara, Toyohiro Hayashi, Takeshi Kato
Undergraduate Students
Yasuyuki Ogihara
[Fukuoka Institute of Technology (FIT)]
Professors
Dr. Takushi Tanaka, Dr. Masaru Ishii, Dr. Hiroaki Shirakawa
Undergraduate Students
Keiji Ikeda, Masato Shimokasa, Kenji Tetemichi, Daijirou Tsuru

1.2 Team introduction

We are the united team organized of Kyushu Institute of Technology(KIT) and Fukuoka Institute of Technology(FIT) highly involved members. KIT team is organized two

A. Birk, S. Coradeschi, and S. Tadokoro (Eds.): RoboCup 2001, LNAI 2377, pp. 681–684, 2002.

groups. One is involved in reinforcement learning, image processing and participated in the middle league in 1999 and 2000. The other is an expert in distributed systems field who also involved applying their techniques for robotics field.

For more information, please visit our web page:
http://www.mickey.ai.kyutech.ac.jp/asura/

Our strategy to participate the four legged league in this year 2001 is to catch up with the world competitive level because we are the fist participation. The last year 2000 winner, the UNSW team had great and successful results due to its technical advances in locomotion, vision, localization and the repertoire of behaviors [1]. According to its success, our basic system is based on their successful sophisticated system. Of course, it is not enough; we introduce the new techniques and tools discussed as follows.

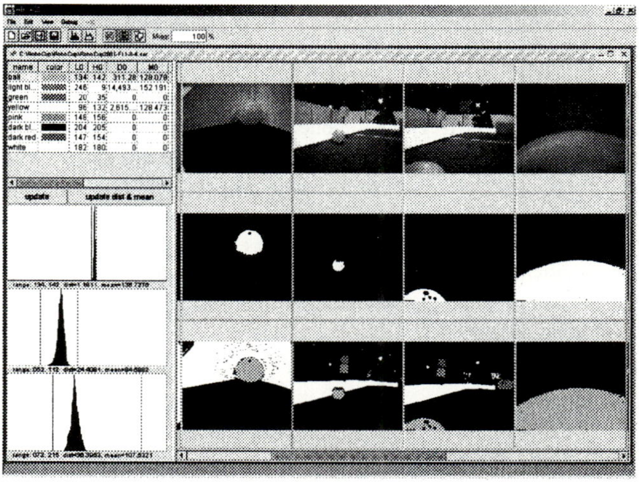

Fig. 1. The Color Classification Tool

2 Vision

The vision system must be robust and accurate since it is the main source of environmental information. Its main task, the color classification is to classify each pixel into the eight colors which appear in the field, for example, orange for the ball, blue, pink and green for the markers, yellow and blue for the goals and dark blue and dark red for the robots. As far as we know in the four legged league, the color classification was achieved by the built-in hardware color lookup table or software approaches like UNSW [1]. Both of them are based on YUV color space, U and V represent color and Y is luminance, which is native color space in the NTSC standard (YIQ for exact). Since historically reasons, YUV color space is not suitable for classify the colors; it was designed for keeping compatibility with monochrome televisions. Consequently, the luminance change seriously affects U and V components. To overcome this problem, we apply the TSL color space [2] for the color classification. T stands for tint, S for saturation and L for luminance. HSV or HSI spaces can be called similar color spaces in respect to T and H represent hue. TSL color space was originally designed to extract skin color from complex backgrounds. The advantage of TSL color model is that it can extract a given color robustly while minimizing luminance influence in compare with YUV, HSV or HSI color spaces. We can classify a given color reasonably accurate with only determining 6 parameters: minimum and maximum thresholds of T, S and L components (linear separation), whereas the UNSW team uses polygons to capture the colors on a given non-linear U-V plane (total up to 14 planes) with learning algorithm [1]. The dark blue color several teams reported difficult to separate, can be separated reasonably. Through minimizing the number of parameters, we can adapt the parameters easy and fast without complex tasks such as optimizing functions and learning algorithms. This advantage gives us the faster adaptation and manually tune-ups of the color table to a new lighting environment, for example, we will take about 15 minutes to whole setup including taking sample pictures. Figure 1 shows our color classification tool built with Java. Figure 2 demonstrates three-dimensional cumulative histograms of the soccer field pictures in RGB, YUV and TSL. In TSL color space, the colors are well clustered in the space. This tool will be available public, please visit http://www.mickey.ai.kyutech.ac.jp/asura/.

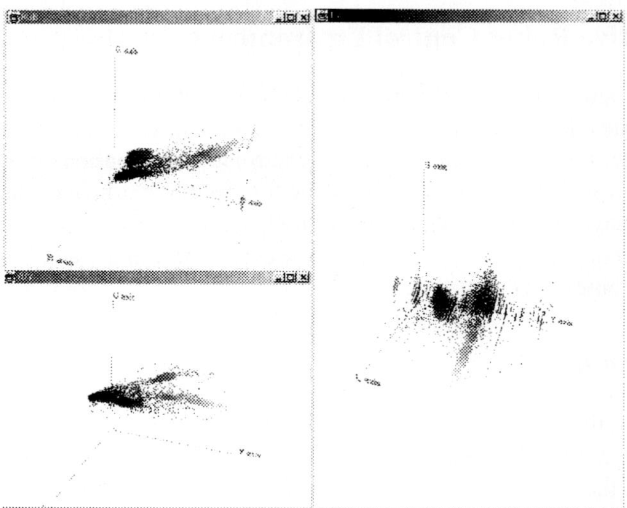

Fig. 2. Three-dimensional cumulative histograms of the soccer field pictures

3 Localization and Locomotion

Localization is the task to locate a robot itself in the field through observing 6 markers and goals. We employ stochastic gradient descent localisation [1], which is able to locate the robot with one marker observation. However, we experienced significant amount of the error occurred when the robot could not see near markers. Because a far marker occupies tiny area in the image, the estimated distance of the robot to far markers is unreliable. For example, 1pixel height error in the pixel area of the furthest marker could lead about 20 cm errors. Therefore, we introduced an observation strategy such that the robot trying to see the nearest marker if higher accuracy is needed. This strategy had been used in the localization challenge and the collaboration challenge, and brought successful result; the robot had visited the specified all positions.

Highly sophisticated locomotion module parameterized walk[1] invented by UNSW allows us to control the robot with three degree of freedom; forward, left and rotation parameters. Despite of its power, we have to find out a set of offset parameters which defines the posture during a walk. Since posture seriously affects stability and speed, it has to be determined carefully. UNSW chooses a set of offsets according to the context of game; aggressive mode takes faster but unstable offsets for approaching the ball around where seems to be no enemy. Our approach is that according to a given three control parameters (forward, left and rotation), enhanced parameterized walk module chooses the most suited a set of offset parameters automatically so as to gain speed and stability. This approach guarantees the offsets to be used as optimal. This enhanced parameterized walk was used for goalie but not for attacker because the offset transition sometimes causes the robot unstable. As a future work, a smooth transition of the offsets and automatic offset extraction (through like learning algorithms) are necessary.

4 Interactive Robot Control Environment

Expolation of new behaviors, gait, posture, etc is very time consuming task if it is done through compile and try iteration. We introduce an interative robot control environment for this purpose. The environment allows us control the robot online through commands typing. We choose forth interpreting language for this environment because of simplicity of its language semantics/syntax, small footprint and efficiency. Forth interpreting language is not designed for robot control, however its simplicity and flexibility is very suited in this context.

5 Conclusion

With these techniques, we got successful results, the best 8 position in the competition games, and the third and second position in the localization and collaboration challenge respectively in the RoboCup challenge. Even though we are the first participation of the league, we had very competing games against veteran teams. This cannot be achieved without the UNSW team highly sophisticated techniques and methods, we would like to express thanks for the UNSW team efforts.

References

[1] Bernhard Hengst, Darren Ibbotson, Son Bao Pham, John Dalgliesh, Mike Lawther, Phil Preston and Claude Sammut, The UNSW RoboCup 2000 Sony Legged League Team, in RoboCup 2000: Robot Soccer World Cup IV, Springer-Verlag, pp.64-75, 2000

[2] Jean-Christophe Terrillon and Shigeru Akamatsu, Comparative Performance of Different Chrominance Spaces for Color Segmentation Detection of Human Faces in Complex Scene Images, in Proc. of Vision Interface'99, Canada, pp.180-187, 1999

BabyTigers 2001: Osaka Legged Robot Team

Noriaki Mitsunaga, Yukie Nagai, Tomohiro Ishida, Taku Izumi, and Minoru Asada

Emergent Robotics Area, Dept. of Adaptive Machine Systems,
Graduate School of Engineering Osaka University, Suita, Osaka, 565-0871, Japan

1 Introduction

We have developed all kinds of motions such as walking, standing-up, and kicking towards Seattle. From a human designer's point of view, it is easier to write strategies based on the geometric, global position data but seems less attractive. Then, we are attacking learning issues such as action selection [1] [2], observation strategy without 3D-reconstruction.

2 Color Detection

In the RoboCup Legged Robot League field [3], seven colors (aqua-blue, yellow, orange, blue, red, pink, green) are used and robots need to detect and discriminate them. The SONY's legged robot has the color detection facility in hardware that can handle up to eight colors at frame rate. To detect these colors with this facility, we need to specify each color in terms of subspace in YUV color space. YUV subspace is expressed in a table called Color Detection Table(CDT). In this table, Y are equally quantized into 32 levels and at each Y level we specify one rectangle $(u_{\min i}, v_{\min i})$,$(u_{\max i}, v_{\max i})$ $(i = 1, ..., 32)$.

In order to make CDTs, we used the same procedure as in previous years [4] [5] [6], which consist of the followings;

1. take an image which includes the object to be detected,
2. specify appropriate pixels to be detected with GUI program and make a list of YUV components,
3. classify each pixel according to the Y level as they are classified in CDT and make a bounding box of UV in each level,
4. check if detection is OK, else return 1) .

We iterate these procedure for each color with several images.

We improved the GUI tool for making CDTs (Fig.1). The main point is to make the CDT checking by the operator smoother. We placed the file selector, original / filtered images, the color which mouse cursor is on, the list of the color specified, UV rectangle list (the CDT) and UV color box with a rectangle. We have found that the dynamic range of the on-board camera's color saturation was not so broad.

A. Birk, S. Coradeschi, and S. Tadokoro (Eds.): RoboCup 2001, LNAI 2377, pp. 685–688, 2002.

It seemed that the lightning conditions of this year (2001) were the best in RoboCup and discrimination of colors were much easier. However, there were situations where we specified all the shaded colors to be included in the CDT, the rectangle included the gray color which cause mis-detection of colors. We show a such an example in Fig.2. The figure shows the specified points by the human operator which consists of both shaded and non-shaded colors of the aqua-blue goal and landmark poles. We see that the bounding box of these points will includes the points (127, 128) in *UV* space, which means achromatic color. Also there were reflexions from the field (green) to the goal colors which may make this problem worse.

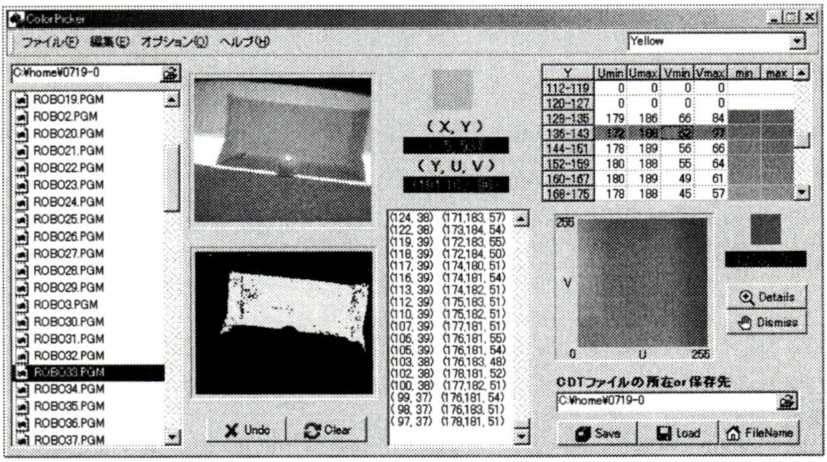

Fig. 1. Our new tool for constructing CDTs.

3 Behaviors

We show the state transition map of the goal keeper in Fig.3. The state changes when the size of the goal or the ball exceeds the thresholds (GOAL_S,M,L, BALL_S,M,L), or timeouts (*Counts). The basic behavior is; 1) to find and go to own goal, 2) to stay in front of the goal watching the opponent goal while try to find the ball, 3) if the ball size becomes high, try to clear the ball to the direction of the opponent goal.

One of our attacker is a modified version of the goal keeper. It does not wait for the ball to become large. It attempts to reach the ball if it finds. And it does not go inside the penalty area when it does not find the ball.

Our another attacker's basic behavior is, 1) to try to find the ball, 2) to go near the ball if the robot finds it, 3) to kick the ball according to the estimated

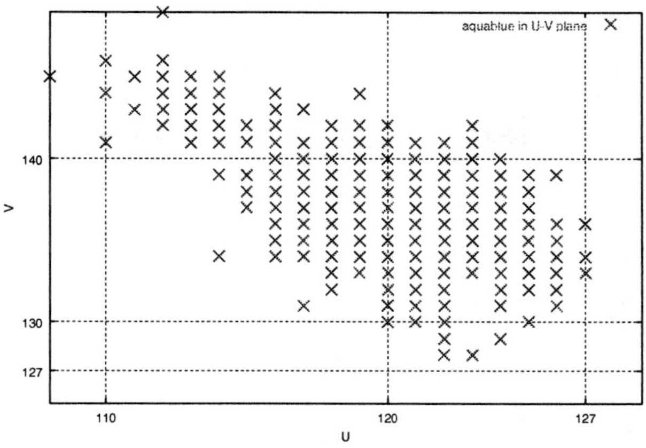

Fig. 2. Aquablue points in UV plane ($96 \leq Y \leq 133$) specified by the operator.

direction of the opponent goal. It estimates the direction by the dead-reckoning in the robot centered coordinate.

4 Actions and Walking

We developed walking based on trot, standing-up, kicking, and diving by ourselves. We adopted three kinds of methods for walking development. One was to design by human inspiration, the other two were to extract walking through robot-environment interaction. That is, we prepared basic walking motion with variable parameters and tuned the parameters with a real robot in the field. We used the steepest descent method for parameter tunings in RoboCup 2000 and hand-tuned in RoboCup 2001 based on that. And we also used genetic algorithm for tunings and we hand-tuned genetically acquired parameters. Other motions were developed by the human designer.

5 Conclusion

We implemented all strategies in the robot centered coordinates and our goal keeper showed good performance during the competition. Our keeper saved the goal well even from the champion team UNSW. Unfortunately the simple estimation of the goal direction did not seem to work well in the competitions. In addition to this problem, team works and learnings are our future issues.

References

[1] N. Mitsunaga and M. Asada. Observation strategy for decision making based on information criterion. In P. Stone, T. Balch, and G. K. Kraetzschmar, editors,

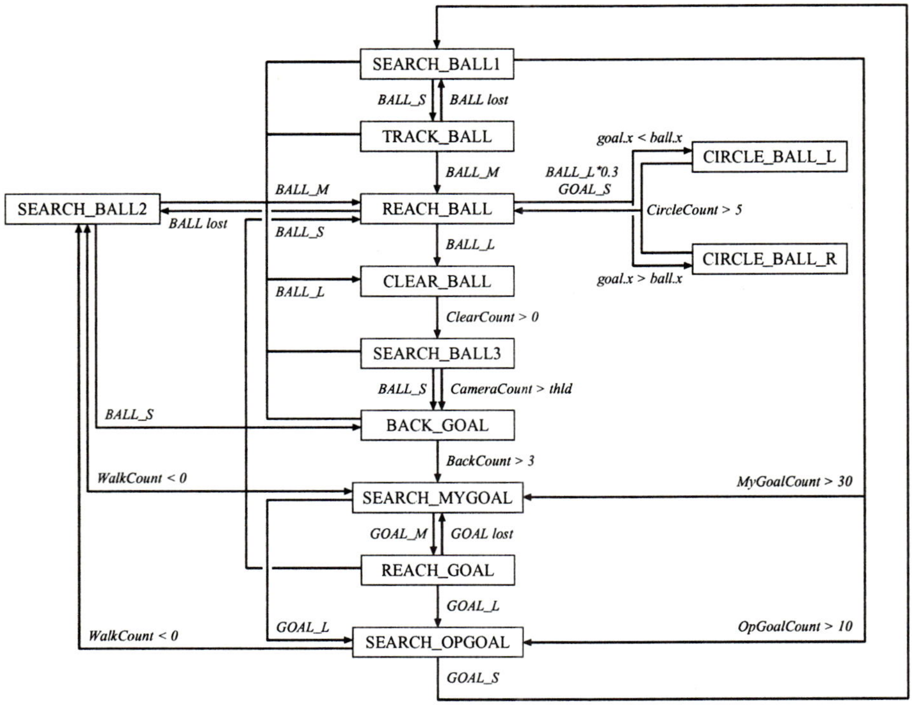

Fig. 3. State transition map of the goal keeper.

RoboCup 2000: Robot Soccer World Cup IV, pages 189–198. Springer, Lecture Note in Artificail Intelligence (2019), 2001.

[2] N. Mitsunaga and M. Asada. Visual attention control by sensor space segmentation for a small quadruped robot based on information criterion. In *The RoboCup 2001 International Symposium*, 2001.

[3] M. Veloso, W. Uther, M. Fujita, M. Asada, and H. Kitano. Playing soccer with legged robots. In *Proceedings of the 1998 IEEE/RSJ International Conference on Intelligent Robots and Systems*, volume 1, pages 437–442, 1998.

[4] N. Mitsunaga, M. Asada, and C. Mishima. BabyTigers-98: Osaka legged robot team. In M. Asada and H. Kitano, editors, *RoboCup-98: Robot Soccer World Cup II*, pages 498–506. Springer, Lecture Note in Artificail Intelligence (1604), 1999.

[5] N. Mitsunaga and M. Asada. BabyTigers-99: Osaka legged robot team. In M. Veloso, E. Pagello, and H. Kitano, editors, *RoboCup-99: Robot Soccer World Cup III*, pages 762–765. Springer, Lecture Note in Artificail Intelligence (1856), 2000.

[6] N. Mitsunaga, Y. Nagai, and M. Asada. BabyTigers: Osaka legged robot team. In P. Stone, T. Balch, and G. K. Kraetzschmar, editors, *RoboCup 2000: Robot Soccer World Cup IV*, pages 631–634. Springer, Lecture Note in Artificail Intelligence (2019), 2001.

Cerberus 2001 Team Description

H. Levent Akın[1], Andon Topalov[2], and Okyay Kaynak[3]

[1] Department of Computer Engineering, Boğaziçi University, 80815 İstanbul Turkey
akin@boun.edu.tr
[2] Department of Control Systems at the Technical University of Sofia Plovdiv Branch, Plovdiv, Bulgaria
[3] Department of Electrical and Electronics Engineering, Boğaziçi University, 80815 İstanbul Turkey
kaynak@boun.edu.tr

Abstract. Cerberus is one of the four new teams in the RoboCup Tournament, Legged Robot Category. The team is a joint effort of Boğaziçi University (Turkey) and Technical University of Sofia, Branch Plovdiv (Bulgaria). We have divided the robot software design into six major tasks, and worked on those tasks in parallel. The tasks were locomotion, localization, vision, behavior control, learning and communication. With limited time in our hands, our aims were a clear object oriented design, readable behavior architecture, and maximum code reuse.

1 Introduction

Cerberus is the first international team participating in Sony's Legged Robot League within RoboCup 2001. This has implied some difficulties in its organization: there where some difficulties to explain thoroughly the ideas each group had in mind because of the necessity to use a foreign language (English) and use predominantly e-mails to maintain communication between the team members. Making an international team has also some advantages: the possibility to use the know-how and the previous research developments in both countries.

The challenge of designing a robotic soccer team is best met with a lot of hands-on experience. The noisy real-time environment, the dynamic nature of the opponents, the different lighting conditions all necessitate robust algorithms for the robot and require arduous fine-tuning.

Our general strategy was to divide the robot software into six major tasks, and assign one team for each task. The tasks were locomotion, localization, vision, behavior control, learning and communication. Locomotion and vision were low-level tasks that involved interfacing the sensors and actuators of the AIBO robot via middleware. Localization relies on the information from vision and locomotion to determine robot's position. Learning was used whenever necessary. The behavior control involves the design of a conceptual finite state machine to guide the behavior of the robot. Cooperation and communication can be seen as integrated parts of the behavior.

A. Birk, S. Coradeschi, and S. Tadokoro (Eds.): RoboCup 2001, LNAI 2377, pp. 689–692, 2002.

2 Development Team

The team is a joint effort of Boğaziçi University (Turkey) and Technical University of Sofia, Branch Plovdiv (Bulgaria). The head of the Cerberus project is Prof. Okyay Kaynak, faculty member at the Electrical and Electronics Department of Boğaziçi University and is the UNESCO chair on Mechatronics.

2.1 Turkish Team

Assoc. Prof. H. Levent Akın, the head of the Artificial Intelligence Laboratory at the Department of Computer Engineering, leads the Turkish team. The Turkish team consists of the following students: Hatice Köse, Çiğdem Gündüz, Olcay Taner Yıldız, Ersin Başaran, Albert Ali Salah, Önder Tombuş, Onur Dikmen and Tamer Demir.

2.2 Bulgarian Team

The Bulgarian team is lead by Assoc. Prof. Andon Venelinov Topalov from the Department of Control Systems at the Technical University of Sofia Plovdiv Branch. The other members of the team are Christo Iliev Radev, Petya Emilova Pavlova, Nikola Georgiev Shakev, Ivan Kavalchev, Jordan Kirilov Tombakov and Anastasia Christova Radeva.

3 The Architecture

The Cerberus team has developed a behavior-based architecture for the robots. The components of the architecture are described briefly below.

3.1 Vision

The Turkish team worked on the vision system. Decision trees and Multi Layer Perceptrons were compared for object recognition. Decision trees were found to be faster and more robust for the successful classification of colors. After classification, we implemented the region finding codes.

3.2 Localization

Localization relies on the recognition of beacons and odometry. A global map is kept where the position of the robot and other locations of interest are recorded probabilistically. Each image frame processed by the vision module contributes in updating the relative probabilities.

3.3 Planner and Behaviors

There are two different planners for the goalkeeper and the attackers. The behavior module currently consists of some basic behaviors that can be switched by planner module in order to achieve more complex behaviors of the robotic players on the field and different team strategies during the game. The planner module is designed by the Turkish team. We use graph representations and algorithms to store the behavior finite state machine as shown in Figure 1.

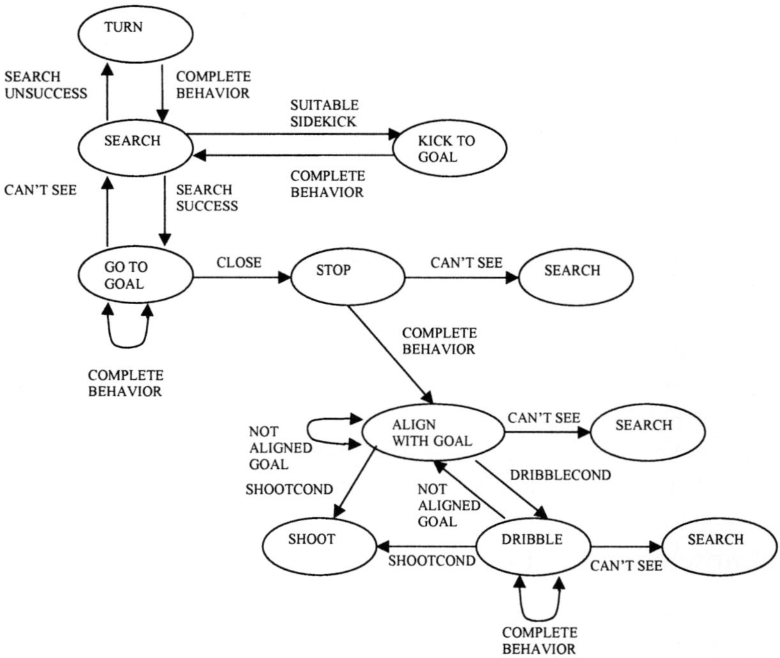

Fig. 1. Planner finite state diagram for the attacker

These behaviors are built by using the following low level behaviors: standing, lateral walking, turning around a specified rotation center, walking straight, dribbling, kicking, going to a specified point, goalkeeper standing, goalkeeper turning, goalkeeper walking forward and goalkeeper walking backward.

For reliable kicking and ball interception, the head control and locomotion are closely coupled with the vision module. For obstacle avoidance, we use an algorithm similar to the controller given in [1]. Some of the behaviors are directly implemented within the locomotion module (different kicking styles, walking straight, turning, etc.) and others like "going to a point" behavior, which is based on a genetically trained fuzzy-neural network [2],[3], are implemented in the behavior module.

While active most of the behaviors receive feedback information about the changes in robot's coordinates and orientation approximately every 1.2 seconds. After completing each behavior a notification message is sent to the planner module. Planner module can decide to interrupt at any time the implementation of the currently running behavior.

3.4 Locomotion

The Bulgarian group of the team developed the locomotion module. The basic structure of the module and its interface is very close to the one developed by the Swedish team for RoboCup 2000 competitions [4]. This structure has been adopted due to the limited time available and the command interface for passing the desired linear and angular velocity using robot's point model as data. This allows us to implement some behavior control approaches previously developed for wheeled mobile robots more easily.

3.5 Communication and Multi-agent Behaviors

The Bulgarian group of the team develops the communication modules. At this stage, the only way for the Sony's robotic dogs to communicate between them is by using their onboard speakers and stereo microphones.

The main goal of communications between the robotic dogs is to provide information exchange within the team to be used as a basis for multi-agent cooperation. There are currently two kinds of commands that are used in the communication: commands for cooperative play and localization. The commands contain the identification number of the sender and data.

4 Acknowledgements

Our work is supported by Boğaziçi University Foundation and Boğaziçi University Research Fund project 01A101.

References

1. Köse, H., Akın, H. L.: *Towards a Robust Cognitive Architecture for Small Autonomous Mobile Robots* ISCIS XV, The Fifteenth International Symposium on Computer and Information Sciences, October, 11-13, 2000, Istanbul, Turkey, pp.447–455, 2000.
2. Topalov, A.V., Tzafestas, S.G.: *Layered Multi-agent Reactive-behavior learning in a Robotic Soccer* Proc. of the 6th IFAC Symposium on Robot Control, SYROCO'00, Vienna, September, 2000, pp.325-330.,2000.
3. Topalov, A.V., Tzafestas, S.G.: *Fuzzy-neural-genetic Layered Multi-agent Reactive Control of a Robotic Soccer* Accepted article for publication in the book: Data Mining for Design and Manufacturing: Methods and Applications, D. Braha editor, Kluwer Academic Publishers,. 2001.
4. Safiotti, A., et al.: *Team Sweden* in P.Stone, T.Balch, G.Kraetzchmar (Eds.), Robocup 2000: Robot Soccer, World Cup IV, pp.643-646, Springer-Verlag, 2001.

CM-Pack'01: Fast Legged Robot Walking, Robust Localization, and Team Behaviors

William Uther, Scott Lenser, James Bruce, Martin Hock, and Manuela Veloso

School of Computer Science, Carnegie Mellon University, Pittsburgh, PA 15213, USA

CM-PACK'01 came in second place at RoboCup-2001 in the Sony Legged League. This is our fourth year competiting in this league. We used a simplified architecture this year. Our vision, localization, and behaviors ran synchronously with the camera input while our motions remained an asynchronous process. We focused on reducing latency and increasing timeliness and completeness of environment modelling for this year. We used more structured software engineering this year than in previous years and found this very useful[4].

1 Vision

We used a similar vision setup to our entry in last year's RoboCup. We used an updated version of our low level vision library CMVision [1] to do all of the low level vision processing of color segmentation and region identification. This new version uses a rather more expressive representation for defining colors. The new representation is a 3D table indexed by the high bits of Y,U,V pixels provided by the camera. The table is created using hand-labelled images and an offline learning algorithm. The algorithm counts examples that fall in each table entry and picks the most likely color from those above a confidence threshold. Examples are spread across the table using exponential decay to ensure good generalization ability while limiting the algorithm's runtime to under 2 minutes on a PII 366MHz.

High level vision used the colored regions identified by the low level vision to find objects of interest. Most of the high-level vision was the same as last year. The estimate of marker distance was changed to project rays through the center of the two colored regions of the marker and find the distance at which the rays are 20cm apart in the vertical direction. Robot detection was added for this year. The first stage of robot detection was finding regions that had the right color to be parts of a robot. After collecting regions of robot color, nearby regions of the same color were merged if the resulting region had a sufficiently high density of robot colored pixels in its bounding box. This usually resulted in regions that represented the major patches of color on the robot. Next, the vision system compared regions that had vertical overlap to see if they should be merged. Regions of the same color were merged if they were close enough (as compared to their boudning box) and seperated mostly by colors found on robots' bodies (black and white). The resulting merged regions were treated as robots and had a distance assigned based upon pixel size (assuming the robot was orientated to provide maximum colored area).

A. Birk, S. Coradeschi, and S. Tadokoro (Eds.): RoboCup 2001, LNAI 2377, pp. 693–696, 2002.

2 Localization

We developed a new simple probability-based method for localization in CM-PACK'01 . The theory is similar to previous Bayesian localization techniques used by our CMU legged robot teams [6, 7, 3]. We keep a probability distribution over possible current locations, P_{loc}. When the robot sees a marker, it updates P_{loc} using Bayes' rule: $P_{loc|sensor} \propto P_{loc} \times P_{sensor|loc}$. When the robot moves, it translates P_{loc}, and then it smooths it to account for the noise in the odometry.

This year, P_{loc} was decomposed into two gaussians. There is one gaussian representing X, Y location, and a second representing θ assuming that the robot is at the mean X, Y location.

When the robot sees a marker the vision returns a distance estimate for that marker and an egocentric angle to the marker. We make the approximation assumption that the robot is at its maximum liklihood X, Y location in P_{loc} and that the sensor reading distribution is gaussian. Once this approximation is made, Bayes' rule can be applied algebraically. The actual implementation follows closely the math in [5].

One can view our algorithm as similar to [2], but with the major refinement that the distance we move our position estimate depends upon our current confidence. This year's system was both fast and accurate.

3 The World Model

The world model tracked objects on the field even when they were out of view. Specifically, it tracked the ball, the four goalposts and all the robots. Initially the world model was very similar to the localization system. During the competition, the section tracking the robots was replaced with a monte-carlo system.

The ball was tracked as a single gaussian in ego-centric coordinates. These coordinates were used so that if our location estimate jumped around the ball estimate would still be accurate. The ball standard deviation was increased over time when the ball was not visible.

We tracked both edges of both goals separately. Tracking the edges of the goals allowed us to decide when we were accurately lined up with the goal for shooting. Sanity checking on the distance between the goal edges, and the corespondance with the localization were important.

During the competition we switched to a monte-carlo representation for robot locations. We kept 200 'samples' for each robot color. 100 of these were reseved for samples with IR PSD distance measurements, the other 100 for samples with image size based distance estimates. Each time a robot was seen, its location was recorded as a sample, displacing the oldest sample so far. Any time the robot moved the samples were moved (they were stored in ego-centric coordinates). Any time the robot should have been able to see a sample and couldn't the sample was aged faster than normal. Samples were timed out after 2 seconds. This value was decreased if the original vision confidence was low.

4 Behaviors

CM-PACK'01 used a very simple behavioral system this year. The main behavior system was a simple series of functions. The high level functions called lower level functions directly. There was none of the complex behavior arbitration used in previous years. This set of functions issued commands to the walking system. The behavior system was fast enough that it could be called by the vision system for each frame.

Active Localization: One behavior we found particularly useful this year was our active localization behavior. This behavior had the robot actively fixate on markers. The routine tried to fixate on each of the markers it thought it should be able to see given its current estimate of field location. Markers were fixated in order from right to left and then back from left to right.

This was significantly better than just waving the head about. Firstly, the pause to fixate on markers allows a better distance estimate. Secondly, it was able to know that it should look up for markers and hence was able to localize more effectively near the edges of the field. Finally, it was able to move its head faster between markers because it wasn't expecting to see anything useful at that point.

The one problem with the algorithm to this point is that if the robot is confused about its location then it will be focusing in the wrong places. To overcome this, we made two changes. Firstly, if the robot happens in passing to see a surprising marker (one that localization would say is over 2 standard deviations from where we think it should be) then we fixate on it. Secondly, if we are focussing where we think a marker should be and we cannot see it then we increment a counter. If we fail to see a few markers in a row then we fall back to waving our head side-to-side.

Attack: Our attack strategy was simple. If you are between the ball and the goal then walk to one side of the ball. If you are at 90° to the goal then use the head kick to kick the ball sideways. If you would kick the ball between the goalposts then kick. Otherwise, dribble the ball towards the goal. The dribble will tend to line you up with the goal so that you can then kick. It also tries to dribble around other robots, both allies and opponents.

We also implemented some very simple ally avoidance. Each of the attackers was assigned a side of the field. If you could see both the ball and your ally, and you were over on your allies side of the field, then back off slightly. This was enough to significantly reduce 'leg-lock'[1].

Defense: Our goalie tried to stay on the line between the ball and a point in the middle of the goal area. It tried to position itself at the goal mouth on that line. When the ball was too close it called the attacker functions to kick the ball away from our goal.

[1] The robots when ignoring each other would tend to get their legs entangled. This slowed both robots down and is a major problem.

5 Motion

The motion system is given requests by the behavior system for high level motions to perform, such as walking in a particular direction, looking for the ball with the head, or kicking using a particular type of kick. The system for CM-PACK'01 was based loosely on the design in used in the previous year [3], but features a completely rewritten, more flexible implementation.

The walking system implemented a generic walk engine that can encode crawl, trot, or pace gaits, with optional bouncing or swaying during the walk cycle to increase stability of a dynamic walk. The walk parameters were encoded as a 49 parameter structure. Each leg had 11 parameters; the neutral kinematic position (3D point), lifting and set down velocities (3D vectors), and a lift time and set down time during the walk cycle. The global parameters were the z-height of the body during the walk, the angle of the body (pitch), hop and sway amplitudes, and the walk period in milliseconds. In order to avoid compilation overhead during gait development, the walk parameters could be loaded from a file on bootup. Using this interface, we developed a high performance walk with a maximum walking speed of 200 mm/sec forward or backward, 170 mm/sec sideways, or 2.1 rad/sec turning.

Additional motions could be requested from a library of motion scripts stored in files. This is how we implemented getup routines and kicks, and the file interface allowed easy development since recompilation was not needed for modifications to an existing motion. Adding a motion still requires recompilation, but this is not seen as much of a limitation since code needs to be added to the behaviors to request that that motion be used.

References

[1] J. Bruce, T. Balch, and M. Veloso. CMVision (http://www.coral.cs.cmu.edu/cmvision/).

[2] Bernhard Hengst, Darren Ibbotson, Son Bao Pham, and Claude Sammut. The UNSW United 2000 Sony legged robot software system. Published on Sony Internal Web Site, 2000.

[3] S. Lenser, J. Bruce, and M. Veloso. CMPack: A complete software system for autonomous legged soccer robots. In *Autonomous Agents*, 2001.

[4] Steve McConnell. *Rapid Development: taming wild software schedules.* Microsoft Press, 1996.

[5] Ashley Stroupe, Martin C. Martin, and Tucker Balch. Merging gaussian distributions for object localization in multi-robot systems. In *Proc. of the Seventh International Symposium on Experimental Robotics (ISER '00)*. Springer-Verlag, December 2000.

[6] M. Veloso and W. Uther. The CMTrio-98 Sony legged robot team. In M. Asada and H. Kitano, editors, *RoboCup-98: Robot Soccer World Cup II*, pages 491–497. Springer Verlag, Berlin, 1999.

[7] M. Veloso, E. Winner, S. Lenser, J. Bruce, and T. Balch. Vision-servoed localization and behavior-based planning for an autonomous quadruped legged robot. In *Proceeding of AIPS-00*, 2000.

Essex Rovers 2001 Team Description

Huosheng Hu, Dongbing Gu, Dragos Golubovic, Bo Li, and Zhengyu Liu

Department of Computer Science, University of Essex
Wivenhoe Park, Colchester CO4 3SQ, UK
{hhu,dgu,dgolub,bli,zliu}@essex.ac.uk

Abstract. This article introduces our research efforts to build the Essex Rovers'01 robot soccer team participated in the RoboCup-2001 competition. A modular design for implementing a behavior-based hierarchy is introduced, which consists of three modules such as *Perception* module, *Cognition* module and *Action* module. This architecture is used for the team to achieve intelligent actions in real time. The implementation aspect of these three modules are briefly described.

1 Introduction

We are interesting in competing in the RoboCup Sony legged robot league since it is an excellent forum to promote the research in the many areas such as artificial intelligence, robotics, sensors, and agent technology. Currently, we have developed a behavior-based hierarchy for our Essex Rovers robot soccer team to achieve intelligent actions in real time [3]. This architecture is based on a modular design, which aims at easy design, flexible implementation and future extension. Moreover, neural networks based color detection algorithm and the fuzzy logic based controller are investigated, including an evolutionary approach to learning a fuzzy logic controller(FLC) employed for reactive behavior control of Sony legged robots. See [1] [2] for more details.

This article introduces our current research efforts to build a multi-agent system for cooperation and learning of multiple legged robots in the RoboCup domain. More specifically, a modular architecture designed for Essex Rovers is briefly introduced in section 2. Section 3 describes the method for color detection and object recognition. Both low-level behaviors for autonomy and high-level behaviors for cooperation are presented in section 4 . Section 5 addresses the motion control and walking problems in the locomotion of quadruped walking robots. Finally, brief conclusions and future work are summarized in section 6.

2 A Modular Architecture

The main objective for our Essex Rovers team is to build a firm research platform on which future work on multi-agent systems can be done. Currently, a multi-agent approach is adopted here to achieve real-time performance, and the modular architecture is adopted in overall agent implementation. More specifically:

A. Birk, S. Coradeschi, and S. Tadokoro (Eds.): RoboCup 2001, LNAI 2377, pp. 697–700, 2002.

- Perception – This module includes a multi-sensor system and a local map. The sensors being used are a color vision system, an infrared range sensor, five touch sensors, eighteen optical encoders, two microphones, and 3 gyros. Incoming visual, proximity, ranging and auditory information is processed by on-board computer. Neural networks based color detection algorithm is used to handle uncertain and changing lighting condition. A local map is then built and updated dynamically as long as new sensory data is available.
- Cognition – This module consists of both high-level behaviors for learning and team formation and low-level behaviors for safeguard and game playing. This is a typical hierarchical architecture that is to merge the advantages of traditional planning based approach and the behavior-based approach. Based on information from the perception module, it selects an action to perform and sends the result to the Actuators module for execution. This is the most complex module since it does the "thinking" and "reasoning" for each robot agent.
- Action – This module includes one speaker and 18 micro servomotors. Each leg has three joints driven by three servomotors. The synchronization of quadruped legs for each robot is extremely important for robot actions such as kicking the ball and moving towards the goal. A fuzzy logic controller is used here to deal with uncertainty in sensory data and imperfect actuators. The speaker can be used to communicate with teammates for team formation and cooperation.

3 Color Detection and Object Recognition

GCD is a threshold method presented by the researchers from the University of New South Wales (UNSW) in Australia. The motivation of this method is to avoid the rectangle threshold in YUV color space and achieve considerable high performance with software solution. The primitive method is to decide if a pixel belongs to a color or not by a huge table GCD.

We developed an interactive tool by using Microsoft Visual C++ to generate the GCD Table for image thresholding. It is a window-based tool and makes manual color separation easy by moving mouse only. Furthermore it also provides the ability to simulate the image processing with same C++ code employed in real robots. The most benefit of using the GCD method is the flexibility. By using it, the system can recognise up to 255 colors with a single table and achieve the high accuracy of color detection. With the non-rectangle threshold, there is nearly no confusion between two colors.

The process of object recognition in the *Perception* module is divided into three stages: image capture, image segmentation and image representation [3]. In the image segmentation stage, the threshold image may contain noise due to the luminance condition. We use morphology filters to de-noise the image since the detected object's shape is known a prior. For a binary image, there are two operators normally used in a morphology filter, namely dilation and erosion. Morphologically filtering an image by an opening or closing operation corresponds to the ideal non-realizable band-pass. In image representation, the similar adjacent pixels have to be grouped into the connected regions. This is typically expensive operation that severely impacts real time performance. We calculate the run length encoding (RLE) to represent the image in order to make our next image operation based on RLE not on individual pixels.

4 Robot Behaviors

One of important issues in the design of the *Cognition* module is to synthesis both low-level basic behaviors and high-level cooperative behaviors of multiple Sony legged robots. Low-level behaviors enable individual robots to play a role in a specified task or game. However, high-level behaviors enable a team of robots to accomplish missions that cannot easily be achieved with an individual mobile robot. Although many different behaviors can be synthesized for the co-ordination of multiple robots, several useful behaviors are identified in this application, which can be categorized into two levels as follows:

- Low-level behaviors in the Action module for mobility
  - Walking behavior - A number of simple behaviors such as *ForwardWalking, BackwardWalking, SideWalking* and *Rolling*, are designed by trial and error to make the robot move fast and more flexible.
  - Safeguard behavior - to achieve good stability and to protect robots from colliding with other objects.
  - Game-playing behaviors - to enable each robot to play a role in the competition, including *Approaching-the-ball, Kicking-the-ball, Passing-the-ball, Shooting-the-gaol.*
- High-level behaviors in the Cognition module for cooperation
  - Position behavior - to find out where each robot is and where is the goal in order to position itself at a good position at any moment. This is in fact a reactive planner to generate locally optimal strategies to direct low-level behaviors. It is very important for each robot to play an effective role in the competition.
  - Formation behavior - to enable multiple robots to form a team where each robot makes its own contribution towards a common goal.
  - Learning behavior - to learn from its own experience and from other mobile robots. Learning here includes both evolving fuzzy rules for low-level behaviors and searching optimal policy for high-level reactive planning.

Additional cooperative behaviors can be synthesized during the next stage of our research, for instance homing behavior and role switching behavior. This should be easy to implement in our modular design.

5 Action/Walking

The robot walking behavior plays key role in the RoboCup Sony Legged robot competition. Following successful results achieved by the UNSW team, we have developed a number of custom walking commands instead of using the normal walking commands provided by *MoNet* objects of *Aperios.*

A number of simple behaviors such as *ForwardWalking, BackwardWalking, SideWalking* and *Rolling*, are designed by trial and error to make the robot move fast and more flexible. Head motion for finding the ball is also designed to reduce time delay between the time when the robot finds the ball and the time when the command is executed. The head angles are recorded when the ball is founded and then head is moved back to that position after the current command is finished.

6 Conclusions and Future Work

This article presents our effect on training Sony legged robots for the competition in RoboCup-2001. The behavior construction and vision processing are two key aspects being focused by our team. The behaviors serve as basic building blocks for our modular software system. The structure of two level behaviors can decomposite the task into contextually meaningful units that couple perception and action tightly. Fuzzy logic implementation further enhances the abilities of the system in face of the uncertain situation.

Based on the current system, both the robot's learning and evolving abilities will be focused for the next competition, particularly on the vision system and cooperative behaviors. The main shortcomings for our team in RoboCup-2001 are inaccurate visual tracking and inefficient kicking, which will be addressed.

Acknowledgements

We would like to thank Masahiro Fujita and other staff members in the Digital Creatures Laboratory, Sony Corporation, Japan for their technical support. This research is financially supported by the Royal Society Research Grant 574006.G503 and the Essex University Research Promotion Fund DDPD40.

References

1. D. Gu and H. Hu, Evolving Fuzzy Logic Controllers for Sony Legged Robots, Proc. of the RoboCup 2001 International Symposium, Seattle, Washington, 4-10 August 2001.
2. D. Gu and H. Hu, Fussy Behaviour Learning for Sony Legged Robots, EUSFLAT 2001 - European Society for Fuzzy Logic and Technology Conference, De Montfort University, Leicester, UK, September 5-7, 2001.
3. H. Hu and D. Gu, Reactive Behaviours and Agent Architecture for Sony Legged Robots to Play Football, International Journal of Industrial Robot, Vol. 28, No. 1, ISSN 0143-991X, pages 45-53, January 2001.

French LRP Team's Description

Vincent Hugel, Olivier Stasse, Patrick Bonnin, and Pierre Blazevic

Laboratoire de Robotique de Paris, 10/12, avenue de l'Europe,
F-78140 Vélizy, France
hugel@robot.uvsq.fr

Abstract. This paper deals with the developments of the French team in RoboCup 2002 Sony legged league. For this event, Sony provided new quadruped robots that were able to compute and move faster, and to see further. The robots got improved skills in locomotion and vision. Compared to last year, soccer games were more lively with a lot of goals scored. Our team focused on designing a localization procedure while walking by taking into account significant and reliable information from the vision system.

1 Introduction

The French legged team took the 4th place in 2001. This is the worst result compared to our 1st and 2nd ranks in 1999 and 2000 respectively. But the number of participants increased from 12 to 16 and the competition becomes harder every year. In fact, UNSW and CMU were by far the best teams. None of the other teams could catch up with them. However, improvements have been significant this year. Robots are able to see much farther. Last year they could only see at a range of one half of the field, and now they are able to see over the whole length of the field. The camera is no more a CCD but a CMOS one. In addition, we improved our vision system. Confused pixels in the image are no more simply removed, but analysed together with its neighbours, and marked with one or the other colour. Regarding locomotion, parameters were tuned so that robots could walk faster using a standing trot. In the following of this paper, 3 sections are detailed, one about locomotion, the second about vision, and the third one about localization.

2 Tuning Locomotion Parameters

The new prototypes of quadruped provided by Sony are better than the previous ones. Motors are more powerful. Rear legs still have nails, but their design has far less influence on locomotion than the old one. As a matter of fact, rear leg nails of last year robots acted as support points for a great part of the stride. Therefore it was difficult to take those into account and to maintain locomotion balance. The new design of the rear leg nails allows to implement more various standing walking patterns. Our team tuned the walking parameters to make the

A. Birk, S. Coradeschi, and S. Tadokoro (Eds.): RoboCup 2001, LNAI 2377, pp. 701–704, 2002.

robot trot. Last year, the walk was based on a classic crawl gait with a duty factor β of 3/4 [2,3] - the duty factor is the fraction of the total time period T the leg is in support phase. Motion speed lay between 5 and 6 cm/s. Compared to crawl, trot features a duty factor of 0.5. To estimate the gain in speed from 3/4 to 1/2, we compute motion speed v as a function of the speed u of the leg in the swing phase. Knowing that the leg covers the same distance in stance and swing phases within the body reference frame, and neglecting the durations of the up-and-downward phases, we get: $\beta T v = (1 - \beta) \Rightarrow v = \frac{1-\beta}{\beta} u$. Setting u to a maximum for both values of β, the gain in motion speed is threefold.

To tune quadruped walk, the first parameter to set is the spacing between the fore and rear leg trajectories within the body reference frame, that we call longitudinal trajectory spacing. The other important parameter to set is the position of the leg trajectories as a whole with respect to the body centre (see Fig. 1). For the classic crawl gait with fixed leg height and stride length, the longitudinal trajectory spacing results from an optimisation that aims at minimizing the maximal torque of the hip joint over the stride. In addition the vertical of gravity centre G should pass through the middle of front and rear leg trajectories (I equals G in Fig. 1) to be sure that the projection of G remains inside or on the edges of the support polygon defined by the three legs on the ground. For the trot, longitudinal spacing should be set minimal. As a matter of fact, when the support diagonal legs change the projection of the gravity centre onto the ground must be as close as possible to the new support diagonal edge in order to reduce the backward overturning gravity momentum. This minimum depends on the bulkiness of the leg cover design, and should ensure that no collision occurs between a front and a rear legs. Moreover, leg trajectories must be shifted backwards as a whole (see Fig. 1). This also allows to bring the gravity centre nearer to the support diagonal legs that have just be placed on the ground. Other

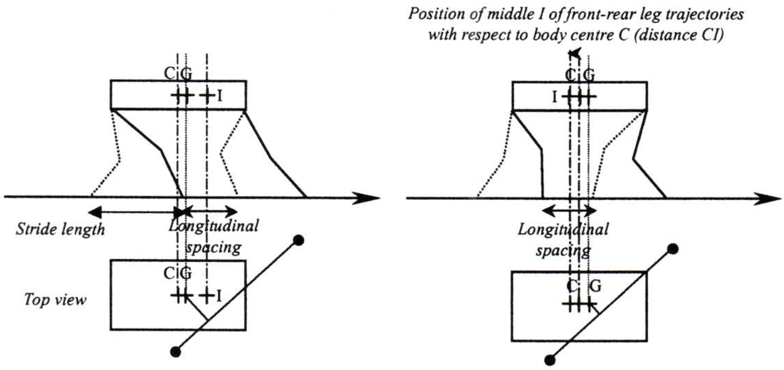

Fig. 1. Position of front-rear leg trajectories in trot. **Left:** bad configuration that can produce backward fall of the robot. **Right:** better configuration where trajectories are shifted backwards so that gravity centre is closer to the support diagonal that has just changed.

parameters that can be tuned are: stride length, sideways spacing, leg height, and amplitude and phase of sideways and vertical oscillations. Stride length and leg height result from a trade-off. Oscillation parameters are tuned to improve locomotion balance and support leg switching.

3 Improved Vision

In previous competitions, we tried to separate colours by avoiding regions of mixed colours [1]. For example, due to light reflections on the top, the orange ball could contain some areas that could be marked with orange or yellow, even white colour. Since these regions were removed, the size of the objects detected was reduced. In addition, it was not possible to remove all mixed colour areas, and it could result in misidentification of some objects that lead to unwanted behaviours. For RoboCup 2002, it was decided to allow regions of mixed colours to appear in the YUV space, and to treat them carefully. Therefore the vision system has been designed to take into account new mixed colours like *yellow-orange*, *orange-red*,... Looking at colour images, an object should be composed of regions of the right colour and other regions of mixed colour. A mixed colour is at the intersection of two colours in YUV space, the right colour of the object, and another colour that is close to the right colour in YUV space. There can be several regions of mixed colour related to the same right colour, each with a different overlapping colour (ex.: *yellow-orange*,*white-orange*). The idea is to merge connected components of the right colour with connected components of the related mixed colour regions. The decision to merge is made at low level. First, a dilatation procedure is performed on the *new* mixed connected components. Then the number of additional pixels generated by this dilatation is computed for each original colour (i.e. yellow and orange for a *yellow-orange* connected component). If the number of pixels of one of the original colours is twice the number of pixels of the other original colour, the mixed-colour connected component becomes a connected component of the first original colour. Fig. 2 shows the results of this procedure on the ball.

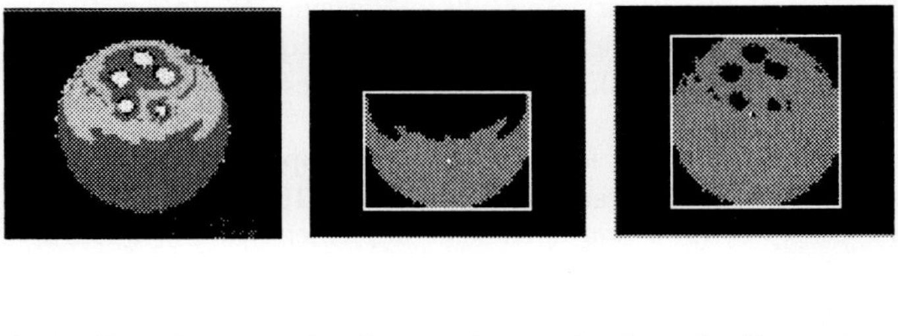

Fig. 2. Left: Ball with original, white and mixed colour (*yellow-orange*) regions. **Middle:** Bounding box extracted with last year vision system. **Right:** Ball original colour region after merging related connected components, and resulting bounding box extracted.

4 Localization

The localization technique used is inspired from the Monte Carlo method[5]. The technique is based on the evolution of a discrete set of sample robots. This population of *virtual* robots evolves when new vision data are generated, and moves as a whole while the robot walks. The surface covered by this population should shrink until becoming small enough to get an accurate estimation of the position/orientation of the real robot. Every sample robot has a probability of being at the right position. All sample robot probabilities are updated when a pole is detected in the image captured. Probabilities are computed using the distance and angle of the identified objects. The angle of the object detected is used to update the orientation of every sample robot. If its location probability is too low, the sample robot is removed and may be replaced by a new one that is a combination of the *updated* virtual robots. In addition odometry is used regularly to translate all sample robots as a whole in the same direction.

5 Conclusion

Our vision system was rather reliable except in the game for the 3rd place. We made a mistake by disabling the procedure that removed orange blobs between the two parts of the *yellow-pink* and *pink-yellow* landmarks. As a consequence of this, the robot could believe to see the ball in the middle of the pole. And these two poles could never be detected! Therefore it is necessary to design special visual debugging tools. It is not enough to check *static* images captured while the robots stays motionless. It would be worth checking sequences of images while the robot is walking. In addition localization could only be used in challenges - where robots have to achieve special tasks like goal defence, path tracking and ball passing -, because the robot had to scan the scene with the head to get reliable data about its position/orientation on the field. An effective strategy of localizing remains to be defined. Moreover we have to change our behaviors that are still too much reflex oriented [4].

References

1. Hugel, V., Bonnin, P., Bouramoué, J.C., Blazevic, P.: Integration of vision and walking skills together with high level strategies for quadruped robots to play soccer. Advanced Robotics, the Int. Journal of the Robotics Society of Japan, Special Issue on RoboCup.
2. Hugel, V.: On the control of legged locomotion applied to hexapod and quadruped prototypes. PhD thesis (in French), November 30, 1999.
3. Hugel, V., Blazevic, P.: Towards efficient implementation of quadruped gaits with duty factor of 0.75. Proceedings of ICRA 1999, Robotics and Automation, Detroit, May 10-15, 1999.
4. Hugel, V., Bonnin, P., Blazevic, P.: Using reactive and adaptive behaviors to play soccer. AI magazine. Vol. 21 – 3. Fall 2000.
5. Lenser, S., Veloso, M.: Sensor resetting localization for poorly modeled mobile robots. Proceedings of ICRA 2000, Robotics and Automation, San Francisco, 2000.

GermanTeam 2001

Ronnie Brunn[3], Uwe Düffert[1], Matthias Jüngel[1], Tim Laue[2], Martin Lötzsch[1], Sebastian Petters[3], Max Risler[3], Thomas Röfer[2], Kai Spiess[2], and Andreas Sztybryc[2]

[1] Institut für Informatik, LFG Künstliche Intelligenz, Humboldt-Universität zu Berlin, Rudower Chaussee 25, 12489 Berlin, Germany
[2] Bremer Institut für Sichere Systeme, TZI, FB 3 Mathematik/Informatik, Universität Bremen, Postfach 330440, 28334 Bremen, Germany
[3] Fachgebiet Simulation und Systemoptimierung, FB 20 Informatik, Technische Universität Darmstadt, Alexanderstr. 10, 64283 Darmstadt, Germany

1 Introduction

The GermanTeam is the successor of the Humboldt Heroes who already participated in the Sony Legged Robot League competitions in 1999 and 2000. Because of the strong interest of other German universities, in March 2001, the GermanTeam was founded. It consists of students and researchers of five universities: Humboldt-Universität zu Berlin, Universität Bremen, Technische Universität Darmstadt, Universität Dortmund, and Freie Universität Berlin. However, for the system presented in this document, the Humbold Heroes only had reinforcements from Bremen and Darmstadt. The two other universities will actively participate with the beginning of the winter semester.

2 Software Architecture

The basic concept of the GermanTeam is a shared memory architecture. In the shared memory, several data structures are stored for exchanging information between the different processes on the robots. The control software consists of several objects, which are running in parallel. The most important ones are:

HSensor reads data from the robot's sensors.
HMotion sends move commands to the robot.
HIntegra builds up the world model. Apart from image processing, the recognition of the ball, other robots, landmarks, and the self-localization are performed in this process.
HControl is responsible for the robot's behavior.

Due to the encapsulation of all robot specific functions in *HSensor* and *HMotion*, other processes may be also be run on a PC under the debugging tool *DogControl*. DogControl also offers many functions for visualization and communication, e.g., for showing the results of the image segmentation algorithm, or for controlling the robot from a PC.

A. Birk, S. Coradeschi, and S. Tadokoro (Eds.): RoboCup 2001, LNAI 2377, pp. 705–708, 2002.

3 Motion

In contrast to other RoboCup leagues, motion is still a research topic in the Sony Legged Robot League, because it means *walking*. The GermanTeam implemented a motion net that manages the transitions between different motions, and that ensures the *completeness* of these transitions. The motion net allows to combine all available motions, and it eases the development of new ones. As most simple motions (like kicking or standing up) can be defined by sequences of joint data vectors, a *motion description language* was developed, in which all motions are specified. For more complex motions, the description language provides commands to execute native C code functions. That way, all gaits are integrated.

4 Image Processing

The vision system processes YUV-images with a resolution of 176×144 pixels acquired by the robot's camera. The processing starts with a color segmentation. The color of every pixel is assigned to one of eight color classes, e.g. the orange of a ball, the yellow of a goal and some landmarks, or the green of the field and some landmarks. This assignment is performed by using a lookup table, the so-called color table. The color table assigns each color from a $64 \times 64 \times 64$ YUV color space to a color class by using the YUV-intensities as indices into the lookup table.

While segmenting an image, it is also run-length encoded to simplify the construction of the so-called *blobs*, and to compress the data. A blob is a data structure representing a region of connected runs of the same color class. For each region, eight characteristic points are determined that represent the maximum expansion of the blob in vertical, horizontal, and diagonal directions. The employed algorithm to calculate the blobs is based on a state machine that directly operates on the runs. The method follows the boundary of each eight-connected region, and was originally developed by Quek [3].

5 Object Recognition

The recognition of objects searches for areas or groups of areas with special shapes, colors, and positions. Before the object recognition takes place, the influences resulting from the three degrees of freedom of the robot's head are removed. Then, Allen's [1] interval calculus is employed to describe spatial relations between blobs. A goal is a yellow or sky blue rectangle lying above a green area, i.e. the carpet. A flag consists of a pink rectangle and a second green, yellow or sky blue rectangle above or below that has nearly the same size. The ball is an orange octagon with at least three of its corners lying on a circle. Its position is determined from intersecting the middle-perpendiculars, which allows to calculate the center of the ball even if most of it lies outside the image. The distances to objects are calculated from their sizes in the image. The directions

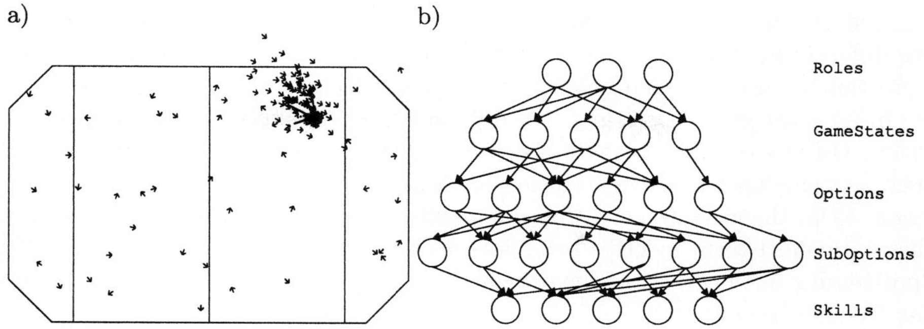

Fig. 1. a) Distribution of samples during Monte-Carlo localization. The large dotted arrow marks the real position of the robot, the solid arrow marks the estimated location. b) A five layer state machine for behavior. The circles represent behavior states.

to objects are determined from their positions in the image, and from the pose of the head of the robot, i.e. the field of view of the camera.

6 Self-Localization

The GermanTeam used a Markov-localization technique employing the so-called Monte-Carlo approach [2]. It is a probabilistic method, in which the current location of the robot is modeled as the density of a set of particles (cf. Fig. 1a). Each particle can be seen as the hypothesis of the robot being located at its position. Therefore, such particles mainly consist of a robot pose, i.e. a vector representing the robot's Cartesian coordinates and its rotation.

The localization approach works as follows: first, all particles are moved according to the *motion model* of the previous action of the robot. Then, the probabilities for all particles are determined on the basis of the *observation model* for the current sensor readings, i.e. bearings on landmarks calculated from the actual camera image. Based on these probabilities, the so-called *resampling* is performed, i.e. moving some of the particles to the location of the sample with the highest probability. Afterwards, the average of the probability distribution is determined, representing the best estimation of the current robot pose. Finally, the process repeats from the beginning.

7 Behavior

The GermanTeam used a *multi layer state machines* architecture to control the behavior of the robot. In this approach, each behavior is a state that is performed until the active state itself decides that another state should be executed. For instance, a state "kicking the ball to the goal" is executed until the ball is not seen anymore (the next state would be "search ball"), or the ball is to far away (the subsequent state would be "go to ball"). The underlying assumption is that the different behaviors themselves know best whether they should continue their execution, or if not, which behavior is a useful successor.

In a soccer game there are very different types of decisions that have to be made: On the one hand, there are very global and long-term behavior states such as "playing the ball", "waiting for a pass", "returning to the own goal", "waiting for the kick off", etc. that do not change very often. One the other hand, there are short-term reactive behaviors such as "turning left around the ball", "positioning in the center of the own goal", and "passing the ball sideward" that can frequently alternate.

To be able to make both long-term and short-term decisions, the different behavior states were distributed over five layers (cf. Fig 1b):

Roles define the general behavior of a robot during the whole game, e.g. "player1" or "goalie", but also "defensiveGoalie" or "offensiveGoalie".

Game-States are general states of the robot such as "waiting for a kickoff", "waiting for a pass", "searching the ball", "playing the ball", etc.

Options are long-term intentions, plans such as "passing the ball", "kicking the ball to the goal", "return to goal", etc.

Sub-Options are short-term intentions, plans such as "kicking the ball forward to the goal", "position in the left center of the own goal", etc.

Skills are basal capabilities, implemented in decision trees. They select gaits or kicks, and they are purely reactive, without internal states. They are parameterized from the sub-options layer.

In each layer, only one state is executed at the same time. The active state will determine the following state for the next lower layer. Only the skills initiate real motions of the robot.

8 Conclusion

Although the GermanTeam did not reach the quarter final, the fourth place in the challenge showed that many parts of the system are quite promising. For RoboCup 2002, the GermanTeam will replace the shared memory architecture by a more flexible communication scheme, and will revise all modules. Some parts of the functionality, i.e. the behavior architecture, will be formalized.

References

1. J. F. Allen. Maintaining knowledge about temporal intervals. *Communications of the ACM*, 26:832–843, 1983.
2. D. Fox, W. Burgard, F. Dellaert, and S. Thrun. Monte Carlo localization: Efficient position estimation for mobile robots. In *Proc. of the National Conference on Artificial Intelligence*, 1999.
3. F. K. H. Quek. An algorithm for the rapid computation of boundaries of run length encoded regions. *Pattern Recognition Journal*, 33:1637–1649, 2000.

McGill Reddogs
Robocup 2001 Final Report

Daniel Sud, Francois Cayouette, Gu Jin Hua, and Jeremy Cooperstock

McGill University, Montreal, Quebec, Canada

Abstract. For the Robocup 2001 competition, the McGill Reddogs team moved from a monolithic architecture to a distributed one and created a flexible motion module. The Monte Carlo localization and custom script interpreter for behaviors were also replaced by simpler algorithms.

1 Introduction

RoboCup 2001 was McGill University's third participation in the Sony four-legged league. However, it was the first competition for all but one of our team members and of course, our first experience with the ERS-210, Sony's second generation of the legged robot architecture. Our main objectives were to break up the previous years' monolithic architecture and implement a flexible walking module.

2 Architecture

The architecture used by our team in previous competitions consisted of a single object that interacted with Aperios, the robot's operating system. The main reason for this choice over a distributed architecture was to minimize the use of subject/observer communication channels, which were not well understood at the time. It also made the data easily accessible for all the different components of the robot. In this single-object architecture, all tasks the robot was programmed to perform would register themselves as "processes" to a kernel, which was responsible for starting their execution. However, as the kernel was non-preemptive, processes had to cede control of the processor voluntarily to ensure that the other processes had a chance to run. As a result, performance was unpredictable at best.

In order to get the new robots playing soccer as quickly as possible, we decided first to port our previous architecture and then to replace components individually. We faced many difficulties due to new restrictions, such as file name length and the limit on the number of virtual functions. Fortunately, the assistance provided by DCL support staff and Sony representatives in attendance at the Autonomous Agents conference (Montreal; May 28-June 1) were invaluable in resolving these problems. However, even after addressing all compilation problems, the robots would still not function correctly, eventually forcing us to

A. Birk, S. Coradeschi, and S. Tadokoro (Eds.): RoboCup 2001, LNAI 2377, pp. 709–712, 2002.

abandon the older architecture and begin working directly on different modules of the newer one.

The module responsible for high-level robot behavior, a script interpreter originally written in OCAML, proved to be problematic as none of the team members were familiar with this language. Its original purpose had been to speed up development cycles since behaviors could be written without having to recompile the entire system. This advantage was minimized with our new distributed architecture so we replaced the module by another, written directly in C++. Another module that disappeared in our new architecture was the Monte Carlo localization algorithm that had proved to be too CPU-hungry for real-time competition.

The current architecture consists of four objects: *Vision*, *Main*, *Zmove* and *Zled*. *Vision* accepts images from the CCD camera, processes these and sends relevant information to *Main*. *Main* receives information from the sensors and the *Vision* object and is responsible for localization and behaviors. It communicates its output to *Zmove* and *Zled*, which both act as effectors.

3 Motion

Last year's team identified the lack of custom walking as our weakest link so we made this our top priority. Since *Zmove* was planned as an independent module from the start, its development was not delayed by architectural issues. Initially, we sent test instructions to this object by writing a simple dummy *Main*, which was replaced with the real object once the rest of the architecture was in place.

Zmove is divided into two different sections, one for controlling the head and another for controlling the legs. It operates by receiving instructions from *Main*, computing all joint angles for 16 frames and sending the results to the virtual robot, which, in turn, is responsible for controlling the motors. A head instruction indicates the desired position of the head, reached by interpolating from the current position in a given amount of time. Alternatively, a single head instruction can tell the head to perform a predefined sequence of moves, which is useful for scanning for the ball or for localization posts.

Walking is based on UNSW's trot gait, introduced at the Robocup 2000 competition. Diagonally opposing legs lift simultaneously and their tips follow a rectangular trajectory. We also developed a statically stable crawl gait, where only a single leg is lifted at a time, but this turned out to be too slow. Walking instructions provide a center of rotation on a horizontal plane with the body of the robot at the origin and a direction. The individual leg trajectories are then rotated and resized, allowing the robot to turn. Walking forward, backward or sideways is performed by placing the center of rotation at infinity. The gait can be adjusted through many parameters. This has allowed for experimentation with different standing positions and leg trajectories. As evidenced by the results of RoboCup 2001, much more work needs to be done in this area, as many teams, including UNSW, demonstrated faster walking than ours, despite our use of a very similar algorithm.

To avoid the need for transitions to and from MoNet, we required a mechanism for restoring the robot to a standing position after a fall. Our strategy for this was to use the logging facilities developed last year to record joint angles of a dog getting up using MoNet. After plotting these values, we noted that the change in joint angles was linear and we could reproduce the same motion by interpolating between a few identifiable points. Once this mechanism was in place, we used it to explore some kicking behaviors. We developed a particularly effective kick in which the robot first raises its body to increase its reach, then drops on the ball. This kick turned out to be so strong that in a round-robin game against Rome, we scored on our own goal after the ball bounced against a post at the opposite end of the field!

4 Vision

Despite our hopes for improvement, the vision module remained an essentially untouched version of the same code used for the previous two competitions. Changes were made to take into account the new distributed architecture and the different camera settings of the new robots. In Seattle, this code was processing mid-resolution at a rate of approximately 8-12 Hz.

4.1 Color Space

As in previous years, we used a graphical color space trainer to create a preliminary color space along with class files for each color. We then used a command line tool to smear class files to increase the likelihood that a particular color will be selected. For example, smearing the orange color helps recognize balls that are far away but too much smearing causes the referee's hands and face to be recognized as false positives. Our current color space is stored as a simple 3-D array of length 64 for each dimension in Y, U and V. We explored alternative representations such as a spherical coordinate transform (SCT) color space, which has the advantage of reducing the color space to 2-D and being more tolerant to changes in lighting conditions. However, the SCT poses a serious disadvantage in that one of the colors, either red, blue or green, will be ill-behaved. We also tried to use a 24-bit color space where all 256 levels are stored for each color. While the results were more accurate, the processing overheard associated with compressed vectors, required to reduce storage requirements, resulted in an unacceptable level of performance.

4.2 Localization

Localization was previously done using a Monte Carlo algorithm, which turned out to be too slow. Instead we implemented a very simple single-point algorithm based on UNSW's steepest gradient descent localization. Every time a localization post is seen in the image, its distance and relative position to the robot are computed. The localization estimate is then moved from its previous position toward the newly computed one by a percentage of the difference between them. The percentage is chosen to reduce the effects of noise from the vision module, but has the adverse affect of increasing convergence time.

5 Competition Behavior

The Brain, a module that is part of the *Main* object, is now responsible for behaviors that used to be handled by Dogscript, a locally developed scripting language. This module makes decisions based on its current state and the input data available. Some of the functions it performs are *objectTracker*, *walkTowardBall* and *gotoPoint*. The Brain has internal states to indicate the length of time it has been performing a certain action or to tell which side the robot is defending. It is also responsible for generating the instructions for *Zmove* and *Zled* that result from the higher-level functions.

The normal behavior of a Brain cycle is to check whether the robot has fallen. Assuming this is not the case, it then checks whether the robot is paused. While in this mode, the robot's defending side can be selected by pressing the back switch, with the tail LED providing feedback. The robot can then be activated or paused using the head touch sensor. Selecting between the goalie and attacker behaviors is done at compilation time.

Having invested a great deal of time setting up the architecture, we had very little opportunity to work on behaviors, with much of this latter work done onsite during the competition in Seattle. The attackers' strategy to scoring was to reach the ball and kick it in the general direction of the opponent's goal. We considered this to be both a good defensive and offensive move, as it would keep the ball out of our half of the field as much as possible. The behavior provided very good results when tested on an empty field, but during actual games, there was often was an opponent between our attackers and the goal, sometimes immediately on the other side of the ball. Clearly, we need to improve our offensive behavior to avoid opponents while keeping control of the ball. The defender's task was simply to move sideways into its goal unless the ball enters the defense area. In this case, the defender leaves the goal, kicks the ball away and returns immediately to its defensive position.

6 Conclusions

RoboCup 2001 was the first year for all McGill University team members. During the late spring and summer months, we were able to modify our architecture and port it to the new robots, write a new walking module and set the foundations for complex behaviors. One of our objectives for next year is to process full-resolution images at twice the frame rate of this year. Another objective is to explore alternative object recognition techniques, such as edge detection, to make our robots less reliant on recognition by color segmentation. We look forward to demonstrating our progress at RoboCup 2002 in Japan.

RoboMutts++

Kate Clarke, Stephen Dempster, Ian Falcao, Bronwen Jones, Daniel Rudolph, Alan Blair, Chris McCarthy, Dariusz Walter, and Nick Barnes

Department of Computer Science and Software Engineering
The University of Melbourne
Parkville, Victoria, 3010
Australia
nmb@cs.mu.oz.au

1 Introduction

The University of Melbourne's RoboMutts++ team competed in the Sony Legged League at RoboCup for the second consecutive year in 2001. This paper describes the areas of research and development focused on in 2001. The software system developed evolved from the system implemented by the RoboMutts team in 2000. In order to direct development effort effectively, the 2000 system was evaluated and limitations were identified. The areas of development were: motion, system architecture, localisation and behaviours. For motion, a variable parameterised walk, and numerous kicks were developed. The system architecture was remodelled to remove multiple concurrent threads, and to facilitate easy integration of various behaviours. Localisation development included one beacon position determination. A number of attacking behaviours were developed for use in differing scenarios, as well as a defensive goalie behaviour.

2 Team Development

Team Supervisors: Dr. Nick Barnes, Dr. Alan Blair, Mr. Chris McCarthy, Mr. Dariusz Walter
Team Members: Ms. Kate Clarke, Mr. Stephen Dempster, Mr. Ian Falcao, Ms. Bronwen Jones, Mr. Daniel Rudolph
Web Page: http://www.cs.mu.oz.au/robocup/RoboMutts

3 Architecture and Behaviours

The architecture used in the 2001 system is shown in Fig. 1 and was evolved from the 2000 architecture. The *vision*, *head* and *field* modules combine to process visual information and sensor inputs. The *brain* module selects the appropriate behaviour for execution, while the *motion* module performs end-effector actuation.

The architecture implemented reactive behaviours. In order to facilitate rapid response to changes in perception, a world model was not employed. Behaviour

A. Birk, S. Coradeschi, and S. Tadokoro (Eds.): RoboCup 2001, LNAI 2377, pp. 713–716, 2002.

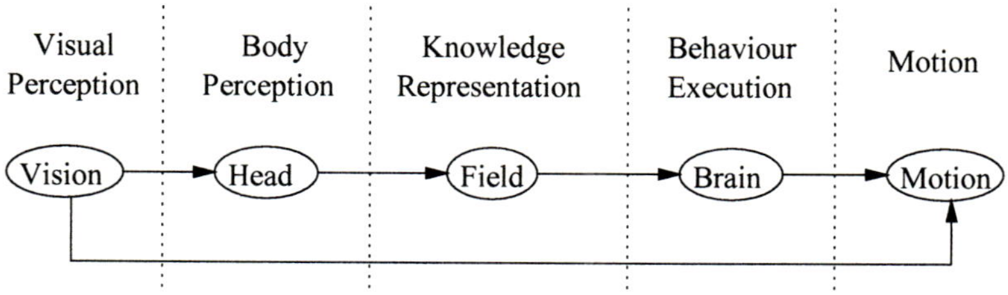

Fig. 1. RoboMutts++ 2001 system architecture

selection was based on current perceptions such as whether the ball was visible, and the location of landmarks in view. Decisions were made using a two-level state hierarchy: top level states (for example, "Active", "Fallen" and "Paused") described the robot's basic condition, while lower level states (including "Searching" and "Ball Visible") were used to switch between behaviours.

Behaviours were derived from a behaviour base class within the *brain* module. The behaviour base class contained methods for performing operations such as locomotion, kicking and LED illumination. By inheriting the methods to perform these operations, the development and integration of new behaviours was simplified, as the implementation details were abstracted.

4 Motion

The custom walk developed for the 2001 competition was a standout feature of the RoboMutts++ system. The walk was competitively fast and extremely stable (no RoboMutts++ robot fell over during a game). The development of the walk was focused on implementing a parameterised walk. The walk was easily varied, which meant experimentation to find an optimal walk was greatly simplified. The speed was considered the most important requirement of the walk. It was tested against the Sony MoNet walk as a measure of improvement, and was found to be approximately 2.5 times faster. The speed of the UNSW walk (running on Aibo ERS–110) was used as a target, as this walk was arguably the fastest in the 2000 competitions. The walk developed was marginally faster than the UNSW 2000 walk, and perhaps the fastest straight-line walk in the 2001 competition.

The parameterised walk uses a locus to define the cycle through which each end effector progresses. The locus is defined as a set of points in three dimensional space, with the origin located at the joint of each effector to the body of the robot. The walk was largely based on the walk implemented by UNSW United [1] in 2000, however it was more flexible as it allowed the locus to be defined by any number of points (see Fig. 2), as opposed to a quadrilateral. Using inverse kinematics, the angles required by the effectors to achieve each point in space were calculated. Different walks were generated by translating the locus around a pivot point.

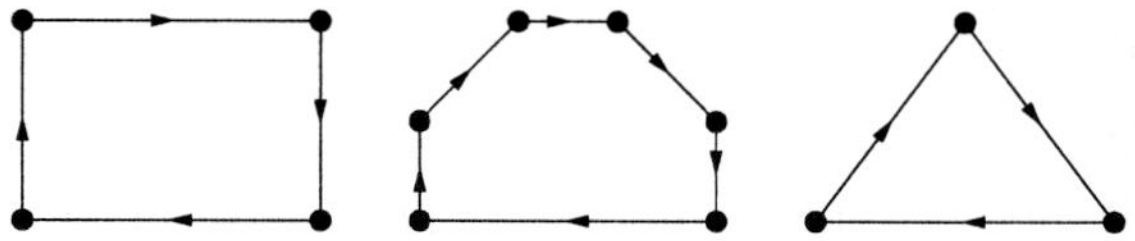

Fig. 2. Possible loci for the walk - each end effector progresses through the selected locus as shown

The walk was parameterised as follows:

- Length of the robot effectors (the lengths of the effectors were different in the ERS–210 and ERS–110 Aibo)
- Height and width of the stance
- Speed of the cycle
- Number of locus points
- Locus points
- Duty cycle
- Stride length

These parameters were read in from file during initialisation of the *motion* module. Reading parameters from file (as opposed to hard-coding them into the *motion* module) drastically reduced development time, as a change in walk did not require a recompilation and download of the program onto the robot.

While the walk code theoretically allowed for an infinite number of walks, twenty-three walks were generated for use in competition. It was found that calculating new walks on request resulted in a noticeable delay before the new walk action was carried out. To prevent this delay, walks were pre-calculated as the program loaded, and the angles required to achieve each point in the locus of any particular walk were stored in a look-up table. The following walks were created (each with varying speed): forwards walk, backwards walk, sideways walk, omni (diagonal) walk, spin, arc and an orbit.

The *motion* module also provided a variety of kicks to enhance the attacking capabilities of the robots. Kicks were developed for use in varying scenarios. The objectives of developing kicks were to gain control over the ball and to direct it in particular directions with maximum force. In total, nine kicks were developed: five forward kicks, and four sideways kicks. Kicks were implemented as the progression of the robot limbs through a set of positions or poses. The performance and suitability of each kick was measured by the distance achieved, and the directional accuracy.

Odometric information including the number of steps taken and the distance covered was provided to the *localisation* module. This was used to localise when visual localisation was not available. The odometric information was sent to the *localisation* module at the completion of a locus by either of the front legs of the robot.

5 Localisation

The system used a passive localisation scheme to determine position and orientation of the robot in the field environment. Passive localisation does not actively employ the motion of the camera to capture the required images. Instead, it uses the information present in images captured during the execution of various behaviours. Passive localisation is advantageous in the case of ball tracking. It does, however compromise the frequency and hence accuracy of the localisation being performed.

Only beacons were used in calculating localisation. Accurate angular information can be derived from beacons (as opposed to goals or line markings) as they remain largely unobscured during play. In 2000, two beacon localisation based on a triangulation algorithm [2] was used. Passive localisation reduces the frequency of observing two or more beacons in a single image, which results in rapid degradation of localisation. In the event that two beacons were not visible, odometric information was relied upon to update the localisation, however this information also quickly degrades.

To counter the degradation of the localisation, localisation based on the presence of one beacon in an image was implemented for 2001. Updating the robot's position based on one beacon requires the use of a triangulation algorithm (different to that used in two beacon localisation). The one beacon approach requires that one beacon be present in two consecutive images. Using the distance to the beacon (calculated from the size of the beacon in the image), and the angle of the head to the beacon for the two consecutive images, the position and orientation of the robot can be determined. A look-up table was used containing predetermined beacon size/distance to beacon tuples, which sped up the triangulation algorithm.

6 Conclusions

Research effort in 2001 was focused on the improvement of existing *behaviour* and *localisation* modules. The system architecture was revised to remove excessive threads, allow easy additions of behaviours, and to integrate the newly developed *motion* module. Performance issues arising from the use of the 2000 architecture prevented the improvement in individual modules from being fully realised.

Areas of future research include: redesigned architecture, modal vision and localisation, cooperative localisation, and communication and team play.

References

[1] B. Hengst, D. Ibbotson, S.B. Pham, and C. Sammut. Sony legged robot software system. In *Proceedings of the 2000 RoboCup Convention*, Melbourne, Australia, September 2000.

[2] A. Graham R. Sim, P. Russo and A. Blair. Robomutts. In *RoboCup 2000: Robot Soccer World Cup IV*, volume 2019 of *Lecture Notes in Artificial Intelligence*, pages 647–650. Springer, 2001.

S.P.Q.R. Legged Team

D. Nardi, V. Bonifaci, C. Castelpietra, U. Di Iorio, A. Guidotti, L. Iocchi, M. Salerno, and F. Zonfrilli

Dipartimento di Informatica e Sistemistica
Universitá di Roma "La Sapienza"
Via Salaria 113 - 00198 Roma, Italy
last_name@dis.uniroma1.it,
http://www.dis.uniroma1.it/~leggedro/

Abstract. The SPQR (Soccer Player Quadruped Robots, but also *Senatus PopolusQue Romanus*) team participated for the second time to Sony Legged League in RoboCup 2001. This work is a team description where it will be highlighted what the team development effort focused on: the realization of a motion module, the realization of a vision module and the improvement of the plans which characterize the different robot roles.

1 Introduction

S.P.Q.R. is the group of the Faculty of Engineering at University of Rome "La Sapienza" in Italy, that is currently involved in developing two RoboCup teams: the **S.P.Q.R. Legged** team [7] in the *Sony Legged League* and the **S.P.Q.R. Wheeled** team in the *Middle Size League.* Compared to the team which participated to RoboCup 2000, the team was substantially improved. In particular we have developed our own locomotion, thus abandoning the usage of the OMoNet module [4], which supervises the generation of movement commands, and our own object recognition module in substitution of the Sony's object recognition module. We also introduced localization for our robots. Moreover a great effort was involved in porting the software developed for the previous version of the Sony Aibo robotic platform to the new one which became available on April 2001.

2 Team Development

Team Leader: Daniele Nardi
Team Members: Luigia Carlucci Aiello, Luca Iocchi, Alessandro De Luca, Giuseppe Oriolo, Vincenzo Bonifaci, Claudio Castelpietra, Ugo Di Iorio, Alice Guidotti, Massimiliano Salerno, Fabio Zonfrilli
Sponsors: Facoltà di Ingegneria, University "La Sapienza" and NETikos Web Mobility.
Country: Italy

A. Birk, S. Coradeschi, and S. Tadokoro (Eds.): RoboCup 2001, LNAI 2377, pp. 717–720, 2002.

3 Architecture

The system relies on a two layer hybrid architecture [1] in order to integrate both deliberative and reactive capabilities [3]. The module named Perception

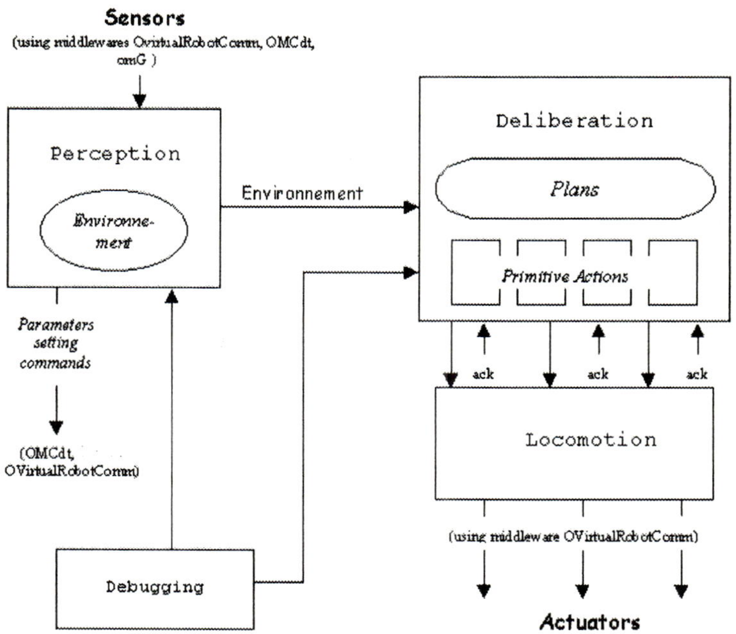

Fig. 1. Architecture

analyzes the sensor data (which are mainly obtained by the camera) and stores and updates the internal world model. The module Deliberation, consisting of a plan execution monitor and a set of primitive actions, receives the world model from Perception, controls the plan of actions to be executed, and generates a set of command for the module Locomotion. The latter provides an interface to the physical functionalities of robot by accepting abstract command from Deliberation and translating them into effective commands to the actuator. The two-level layered architecture is based on different representation of the information: an high-level symbolic one, and a low-level numerical one. The high-level plans are executing depending on the validity of abstract conditions [2], such as *KnownBallPosition* or *NearBall*, while low-level atomic actions allow for an immediate reaction to unexpected situations.

4 Vision

We have developed our own software for the blob formation and object recognition phases, while for image segmentation we have relied upon the robot hardware segmentation with the Sony's libraries. This allowed the recognition of objects to be much more stable and provided a solid basis for the design of a localization system. The blob formation routines are based on a RLE compression of the segmented image as described in [5]. This approach has the advantage of being fast and also allows the segmented image to be quickly transmitted through the serial cable to the personal computer for debugging purposes. The object recognition routine takes the blobs and classifies them on the basis of their color, while trying to recover from possibly wrong decisions through the use of some heuristics, depending on the Sony Legged League operational setting. These heuristics make the system less error-prone than before.

5 Localization

Building on the capabilities of the vision system, a localization procedure was written, based on trilateration (see [6]). This technique requires the distances and angles of the robot from two known points. Trilateration has the advantage that it does not necessarily require the robot to execute a particular procedure to recognize the markers, thus it allows the robot to extract the information needed without interrupting the task it is accomplishing, but simply by analyzing the acquired images. On the contrary, this technique needs two markers to be recently recognized and therefore the localization information is not always available, thus a reliability value, which is decreasing with time when the markers are not recognized, has been introduced to deal with the robot location.

6 Behaviors

The architecture of the deliberative module is based on the one described in [7]. Thus the development effort focused on the primitive action reorganization, in order to make them more parametric (with respect to speed, for instance). Since localization information has been made available, the higher level plans were rewritten in order to take advantage of the knowledge about the robot location. In particular, the goalkeeper can reposition itself on the goal line without having to turn back and look at it (which has proved to be a very risky action). Moreover, localization allowed for a safer and more appropriate choice of the kick to be activated by the robot when it is near the ball, for instance by choosing a right head kick when it knows the opponent goal being on its right, without the need to look at it.

7 Perception

The perception module was modified so that the new information needed to localize could be obtained. Moreover, a method, based on the robot kinematic model, was adopted to calculate the position of the robot camera with respect to the ground and thus obtain the ball position relative to the robot.

8 Locomotion

The Sony's OMoNet [4] module has been abandoned in favor of our own motion control. The new Locomotion allowed the robots to increase the motion speed and to have a larger selection of different movements to be used during the task execution. New walking styles and kicks were created through an off-line tool. The kicks developed include: a frontal head kick, lateral head kicks, a leg kick and a chest kick, the latter specific to the goalkeeper. Moreover many types of walking styles were developed so that both more stable and faster movements were obtained, and ad hoc movements for special robot behaviors were made available. The movement speed was increased by using a spline interpolation to smooth the joint control laws, thus avoiding discontinuities in the joint velocities.

9 Special Team Features and Conclusion

The main focus of the team work has been the realization of a team of cognitive soccer robots by following the design methodology presented in [2], whose main features are a hybrid architecture that is heterogeneous and asynchronous allowing for an effective integration of reasoning and reactive capabilities, a high level representation of the plans executed by the player, a formal account of the system for generating and verifying plans. The architecture proposed has been designed to consent the use of the high level representation and deliberation on both the SPQR Legged Team and the SPQR Wheeled Team. In the future the work will focus on the improvement of the motion, the vision and the localization modules, but mainly on communication based robot coordination, which has been proven to be a central point to obtain effective results.

References

1. Iocchi, L.: Design and Development of Cognitive Robots. PhD thesis. Dipartimento di Informatica e Sistemistica, Universitadi Roma "La Sapienza". (1999)
2. Castelpietra, C., Iocchi, L., Nardi, D., Rosati, R.: Design and Implementation of Cognitive Soccer Robots. In Procs. of The RoboCup 2001 International Symposium. (2001)
3. Saffiotti, A.:Some Notes on the Integration of Planning and Reactivity in Autonomous Mobile Robots. In Procs. of the AAAI Spring Symposium on Foundations of Automatic Planning. Stanford, CA. (1993) 122–126
4. Sony Corporation: OPEN-R Software Programmer's Guide. (2001)
5. Bruce, J., Balch, T., Veloso, M.: Fast and Cheap Color Image Segmentation for Interactive Robots. In Procs. of IROS-2000. Japan. (2000)
6. Dalgliesh, J., Lawther, M.: Playing soccer with quadruped robots. School of Computer Science and Engineering, University of New South Wales. (1999)
7. Nardi, D., Castelpietra, C., Guidotti, A., Salerno, M., Sanitati, C.: S.P.Q.R. in RoboCup-2000: Robot Soccer World Cup IV. Springer Verlag. (2000)

Team Description: UW Huskies-01

D. Azari, J.K. Burns, K. Deshmukh, D. Fox, D. Grimes, C.T. Kwok, R. Pitkanen, A.P. Shon, and P. Tressel

Department of Computer Science & Engineering
University of Washington, Seattle, WA
{azari, jkburns, kaustubh, fox, grimes, ctkwok, pitkanen, aaron, tressel}@cs.washington.edu

1 Introduction

This paper provides an overview of the robot control system of the UW Huskies, our entry in the RoboCup-2001 Sony AIBO legged robot league. Our development team consisted of one faculty member (D. Fox, team leader), one technical support staff (R. Pitkanen), one undergraduate student (J.K. Burns), and six graduate students. Since this was our first participation in RoboCup, we relied on the implementation of well-established techniques for robot control. These techniques include a layered control system, particle filters for robot localization, and efficient computer vision techniques for object recognition.

2 Architecture

Flexibility was the overriding design goal for our system. We were especially careful to make our system as modular as possible. This would permit us to adapt to unanticipated difficulties or novel tactics. Implementation efficiency and maintainability were important secondary goals. Figure 1 illustrates the control system. The different modules communicate using the Aperios inter-object message passing facility. When large data (e.g. world model, camera images) are to be shared between multiple modules, shared memory is used. The control system consists of the following five modules:

Hardware Interface: This module receives input from the raw sensors and passes it to the other modules. The most important information consists of time stamped camera images coupled with the corresponding position of the robot's head.

Vision: The vision module analyses images on a frame by frame basis. Efficient algorithms are applied to detect objects such as the ball, goals, and markers. Distances to these objects are extracted from their size. Information about the robot's head position is used to transform from camera-relative to robot-relative coordinates. The detected objects are passed to the state estimator module.

State estimator: This module integrates information over time and stores it in the world model. Most important information is the position of the robot on the soccer field and the position of the ball relative to the robot. The robot position is estimated using a particle filter approach [2]. The position of the ball is based on smoothed estimates of the vision system. From this information, it is rather straightforward to derive information such as the absolute position of the ball, or the relative position of the opponent's goal.

A. Birk, S. Coradeschi, and S. Tadokoro (Eds.): RoboCup 2001, LNAI 2377, pp. 721–724, 2002.

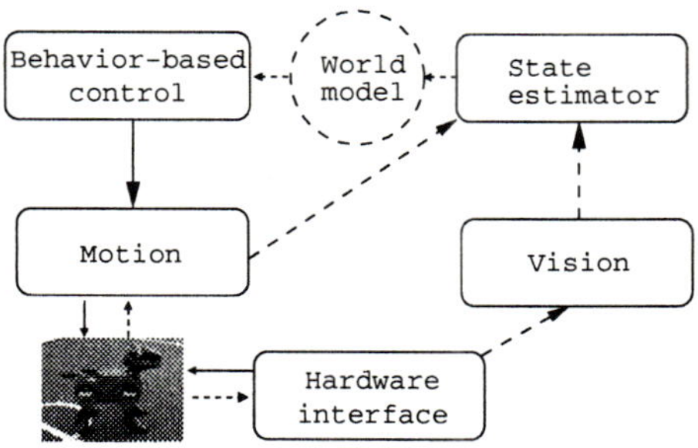

Fig. 1. System architecture. The arrows indicate the main flow of information (dashed) and commands (solid) between the different modules.

Behavior: Using the most recent information in the world model, this module decides which actions the robot should perform. Our control system follows a multi-layered approach (see also [3] and several chapters therein). Behaviors are modeled by hierarchical finite state machines (FSMs). Simple skills such as "kick" or "walk" form the basic skills of the hierarchy. At the top of the hierarchy a "goalie" or "forward" FSM determines the highest-level actions of the robot depending on a preassigned role.

Motion: We developed gaits and kicks offline by a combination of basic inverse kinematics and direct manipulation of the robot's joints. These gaits are saved on the memory stick as text files and read in at boot time. The motion module executes commands at the request of the behavior module. During the competition we relied on the gaits provided by Sony's OPEN-R robot control system.

3 Vision

Our basic vision package emphasizes efficiency and interface simplicity to maximize reuse of the code. The package possesses a memory footprint of less than 2MB and allows a frame rate of more than 100 frames per second on the robot itself (significantly faster than the camera update rate), including color clustering, blob identification, object recognition, and distance estimation. To achieve this high efficiency in combination with accurate object estimates, we follow a dual-resolution approach: Object detection is performed on the low-resolution images which are color-clustered by the robots vision hardware. Once the different objects are detected, we use the high-resolution image to get more accurate estimates for the positions and sizes of these objects.

The image processing pipeline begins with the hardware-clustered CDT image. The clusters are represented by rectangular boxes in YUV-space. These boxes are determined manually, using an offline tool that reads in images taken from the camera and allows a trainer to assign groups of pixels to specific color classes. During operation, the classified image pixels are grouped into blobs. We use a tree-based union find algorithm with path compression and 4-connectedness matching to locate blobs. The blobs of each color are sorted by size, and very small blobs (likely to be noise) are eliminated. Objects are detected by testing spatial relationships between the individual blobs (e.g. green blob above yellow blob for green/yellow marker).

All operations described so far are performed on the low-resolution CDT images. To get more accurate estimates for object sizes, we use the bounding boxes of the detected objects to focus on regions in a higher-resolution image. Color clustering of the higher-res images is performed in software using the same color bounding boxes as the CDT images. We found that the higher-resolution images tend to yield more accurate distance estimates, while the low-res pass allows reliable and efficient object detection[1]. Finally, before sending object information to the state estimator, the object positions are transformed from camera coordinates to positions relative to the robot.

4 State Estimation and World Model

The task of this module is to keep track of the dynamics of the environment. In our current system, each robot estimates its own position and the relative positions of the ball and the goals. The relative position of the goals is either determined by direct observation or extracted from the robot's location on the field. We estimate the robot position using particle filters, a highly efficient and robust approach to localization [2]. The key idea of this probabilistic approach is to represent the robot's position by sets of random samples. Position estimates are refined whenever the vision module detects a marker or a goal. Detections contain information about the distance and orientation of the objects relative to the robot. From these detections we can derive probabilities for the different robot positions on the field. Whenever the robot moves, the samples are shifted according to a model of robot motion. The noise parameters of this motion model were determined offline by measuring velocities for the different motion commands. Similar to [4], recovery from localization failures is achieved by adding random samples drawn according to the observation likelihoods.

Efficiency was one of the key goals of our implementation of particle filters. Using several approximations (such as pre-computing random samples and replacing Gaussian distributions by triangular distributions), one update cycle of our approach can be done in less than 0.05 seconds on average (400 samples).

5 Behaviors

This module receives world model updates from the state estimator, processes the information, and outputs commands to the motion module. The behavior module is implemented as a hierarchy of finite state machines. We have two top-level state machines, each corresponds to a different type of player, the Forward and the Goalie. A simplified Forward state machine is shown in Figure 5(a). Each state has a list of transitions and conditions associated with these transitions. Typical conditions test whether something holds true in the world model, for example, whether the ball is close to the robot. If true, the associated transition is made, during which motion commands may be issued. Each state in the state machine implements an abstraction called SKILL. Basic SKILLS

[1] To achieve more accurate estimates of the distance to the ball, we additionally use information from the IR sensor when the ball is already judged by the vision system to be relatively close and straight ahead.

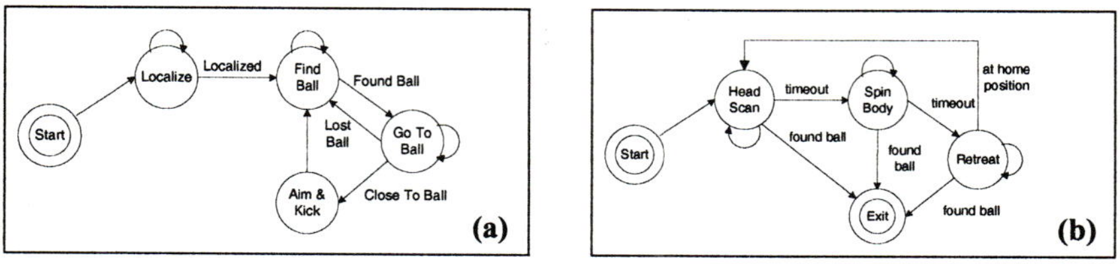

Fig. 2. Hierarchical finite state machines for robot control: a) Simplified version of the FSM implementing the Forward robot. Note shown are transitions from each state to the "Localize" state. The "Go to Ball" state is implemented by a docking motion field to guide the robot to the ball so that it can kick it toward the opponent's goal. b) The Find Ball skill of the Forward is implemented as a lower level FSM. When the exit state of this FSM is reached, control is transferred back to the main state machine.

are motion commands such as move head or walk. More complicated SKILLs are built out of state machines themselves, each state again corresponds to a SKILL.

Acknowledgments

This research is sponsored in part by NSF (contract number 0093406). The team wishes to thank Wolfram Burgard for his support and advice during the last weeks of preparation for RoboCup-2001.

6 Conclusions and Future Work

Our software system for RoboCup-2001 relied on well-established techniques from the robotics and vision community. Due to the lack of preparation time, our team was not competitive against the senior teams of our league. Our major problem was the fact that we were not able to develop fast and stable motion patterns within the given time. Since fast motion is of utmost importance for success in robot soccer, our next efforts will focus on the development of motion patterns. Another thread of our research will explore issues in multi-robot collaboration and the combination of information collected by different robot platforms (communication will be available in RoboCup-2002). Further research will involve reinforcement learning and efficient state estimation in the context of real-time robot control (see e.g. [1]).

References

1. D. Fox. KLD-sampling: Adaptive particle filters and mobile robot localization. In *Advances in Neural Information Processing Systems (NIPS)*, 2002. To appear.
2. D. Fox, W. Burgard, F. Dellaert, and S. Thrun. Monte Carlo Localization: Efficient position estimation for mobile robots. In *Proc. of the National Conference on Artificial Intelligence (AAAI)*, 1999.
3. D. Kortenkamp, R. P. Bonasso, and R. Murphy, editors. *Artificial Intelligence and Mobile Robots: Case Studies of Successful Robot Systems*. MIT/AAAI Press, Cambridge, MA, 1998.
4. S. Lenser and M. Veloso. Sensor resetting localization for poorly modelled mobile robots. In *Proc. of the IEEE International Conference on Robotics & Automation (ICRA)*, 2000.

Team Sweden

A. Saffiotti[1], A. Björklund[3], S. Johansson[2], and Z. Wasik[1]

[1] Center for Applied Autonomous Sensor Systems
Örebro University, S-701 82 Örebro, Sweden
[2] Department of Software Eng. and Computer Science
Blekinge Institute of Technology, S-372 25 Ronneby, Sweden
[3] Department of Computing Science
Umeå University, S-901 87 Umeå, Sweden

1 Introduction

"Team Sweden" is the Swedish national team that entered the Sony legged robot league at the RoboCup 2001 competition. The work was distributed over three universities in Sweden. This nationwide character has made the project organization particularly demanding, but has resulted in a rewarding cooperation experience, both scientific and human. The Team identity is as follows:

Team Leader: Alessandro Saffiotti (`asaffio@aass.oru.se`)
Team Members: include the authors, plus: R. Johansson (Örebro); P. Davidsson, D. Erman, and J. Kronqvist (Blekinge); J. Carstensen, P. Eklund, P. Larsson, and K. Prorok (Umeå).
Sponsors: Swedish KK foundation, Qualisys AB, Örebro University, Umeå University, and the Blekinge Institute of Technology.
Team home page: `http://www.aass.oru.se/Living/RoboCup/`.

We had two main requirements in mind when we started to work on RoboCup:

1. Our entry should effectively address the challenges of uncertainty in this domain, where perception and execution are affected by errors and imprecision;
2. it should illustrate our research in autonomous robotics, by incorporating general techniques that can be reused in different robots and environments.

While the first requirement could have been met by writing some *ad hoc* competition software, the second one led us to develop principled solutions that drew upon our current research in robotics, and that pushed it further ahead.

2 Architecture

Each robot is endowed with the layered architecture sketched in Fig. 1. This is a variant of the Thinking Cap, the autonomous robot architecture based on fuzzy logic in use at Örebro University.[1] The lower layer provides an abstract interface

[1] See `http://www.aass.oru.se/~asaffio/Software/TC/`.

A. Birk, S. Coradeschi, and S. Tadokoro (Eds.): RoboCup 2001, LNAI 2377, pp. 725–729, 2002.

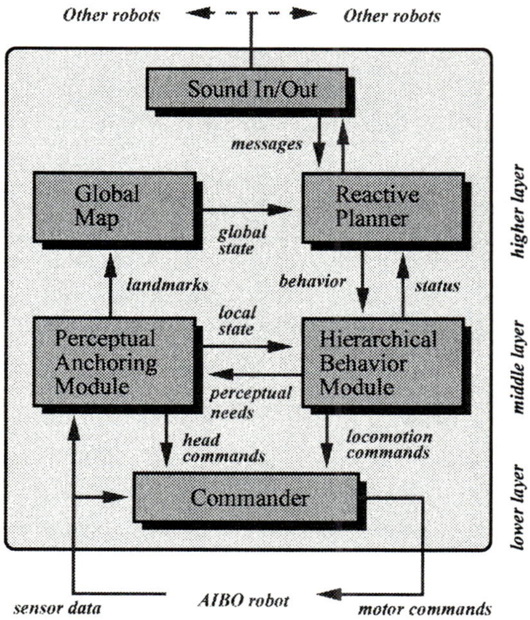

Fig. 1. The variant of the Thinking Cap architecture used by Team Sweden.

to the sensori-motoric functionalities of the robot. The middle layer maintains a consistent representation of the space around the robot (PAM), and implements a set of robust tactical behaviors (HBM). The higher layer maintains a global map of the field (GM) and makes real-time strategic decisions (RP). Sounds are used to exchange information with other robots for coordination purposes.

3 Motion Control

The Commander module accepts locomotion commands from the HBM in terms of linear and rotational velocities, and translates them to an appropriate walking style. This simplifies the writing of motion behaviors, that are easily portable between different (legged and/or wheeled) platforms. The Commander module also controls the head motion, and implements several types of kicks.

We have implemented a rich set of walking styles that realize several combinations of forward, lateral and rotational velocity. Most styles are dynamic, providing fast motion but limited controllability. A few semi-dynamic styles provide slower motion for precise maneuvers in the proximity of the ball.

4 Perception

The locus of perception is the PAM, which acts as a short term memory of the location of the objects around the robot. At every moment, the PAM contains the best available estimate of the position of these objects, based on a combination of current and past observations with self-motion information. The PAM also takes care of selective gaze control, by moving the camera according to the current perceptual needs, which are communicated by the HBM. (See [4] for details.)

Object recognition in the PAM relies on three mechanisms: (i) color segmentation, based on a new fast region-growing algorithm that takes as seeds the output of the color detection hardware in the Sony robot; (ii) model-based region fusion, to combine color blobs into features; and (iii) knowledge-based filters to eliminate false positives. For instance, a green blob over a pink one are fused into a landmark; this may rejected, however, if it is too low over the field.

5 Self-Localization

Self-localization in the Sony legged robot league is a challenging task: odometric information is extremely inaccurate; landmarks can only be observed sporadically since a single camera is needed for many tasks; and visual recognition is subject to unpredictable errors (e.g., mislabeling). To meet these challenges, we have developed a new self-localization technique based on fuzzy logic, reported in [1]. This technique only needs qualitative motion and sensor models, can accommodate sporadic observations during normal motion, can recover from arbitrarily large errors, and has a low computational cost.

6 Behaviors and Behavior Selection

The HBM implements a set of navigation and ball control behaviors realized using fuzzy logic techniques and organized in a hierarchical way [3]. As an illustration, the following set of fuzzy rules implement the "GoToPosition" behavior.

```
IF (AND(NOT(PositionHere), PositionLeft))    TURN (LEFT);
IF (AND(NOT(PositionHere), PositionRight))   TURN (RIGHT);
IF (OR(PositionHere, PositionAhead))         TURN (AHEAD);
IF (AND(NOT(PositionHere), PositionAhead))   GO (FAST);
IF (OR(PositionHere, NOT(PositionAhead)))    GO (STAY);
```

More complex behaviors are obtain by composing simpler ones using fuzzy meta-rules that activate concurrent sub-behaviors.

Game strategies for the players are dynamically generated by the RP. This implements a behavior selection scheme based on the artificial electric field approach (EFA) [2]. We attach sets of positive and negative electric *charges* to the nets and to each robot, and we estimate the heuristic value of a given field situation by measuring the electric potential at some *probe* position — for instance, the position of the ball. This heuristic value is used to select the behavior that would result in the best situation.

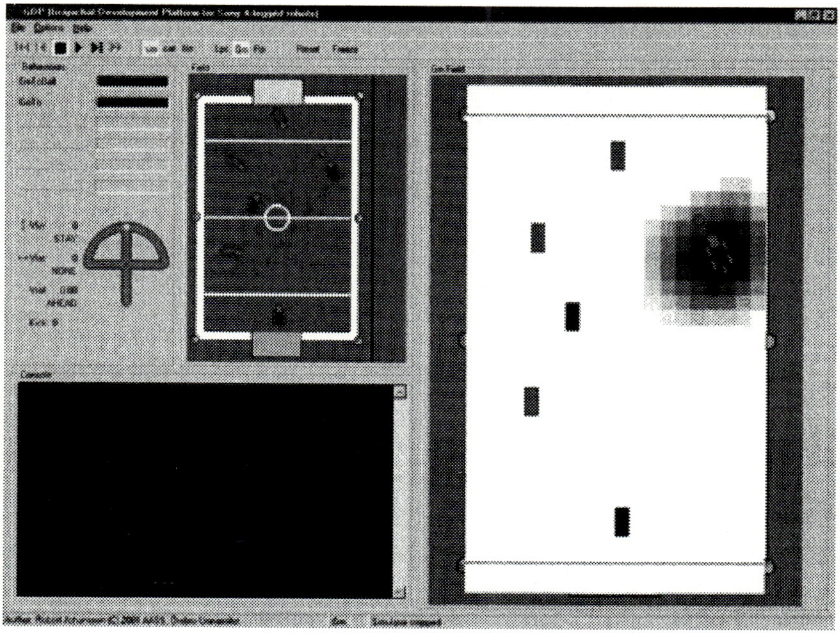

Fig. 2. A screen-dump of the GDP while inspecting the GM in a simulated run.

7 Coordination

The legged robot league does not allow radio communication. In order to perform some limited form of coordination, we have developed a simple sound communication protocol by which robots can communicate their intentions. In the Seattle games, we only used two signals, respectively to claim and release possession of the ball. The receiving robot uses this information to change its strategy from the normal one to a supportive one (wait for a pass) and back. Strategy switching is done by modifying the allocation of charges and probes in the RP.

8 Development Environment

We have built a Graphical Development Environment (GDP, see Fig.2) that allows real-time inspection and modification of the robot's behavior on a PC. The GDP can run some (or all) of the modules in Fig. 1 locally, while the others are run on-board the robot. The robot-PC communication is done via a serial cable. The GDP also includes a simple simulator for off-line development.

9 Conclusion

The general principles and techniques developed in our research could be successfully applied to the RoboCup domain. In particular, fuzzy logic was beneficial in writing robust behaviors and providing reliable self-localization, and the electric field approach proved to be a flexible way to encode high level strategies.

References

1. P. Buschka, A. Saffiotti, and Z. Wasik. Fuzzy landmark-based localization for a legged robot. *IEEE/RSJ Int. Conf. on Intell. Robots and Systems* (IROS), 2000.
2. S. Johansson and A. Saffiotti. Using the Electric Field Approach in the RoboCup Domain. Birk, Coradeschi, Tadokoro (eds) *RoboCup 2001*, Springer-Verlag, 2002.
3. A. Saffiotti. Using fuzzy logic in autonomous robot navigation. *Soft Computing*, 1(4):180–197, 1997. Online at http://www.aass.oru.se/Living/FLAR/.
4. A. Saffiotti and K. LeBlanc. Active perceptual anchoring of robot behavior in a dynamic environment. *IEEE Int. Conf. Robotics and Automation* (ICRA), 2000.

The Team Description of ARAIBO

Tamio Arai, Takeshi Fukase, Ryuichi Ueda, Yuichi Kobayashi, and Takanobu Kawabe

The Department of Precision Engineering, The University of Tokyo, Hongo, Bunkyo-ku, Tokyo, 113-8656, Japan
{arai, fukase, ueda, kobayasi, kawabe}@prince.pe.u-tokyo.ac.jp

Abstract. This paper describes a localization method and a navigation method for Sony Legged Robot League. They solve the problems caused by the robot's limited abilities (CPU power, sensor ability, control ability). These methods are implemented in the soccer program for the RoboCup 2001.

1 Introduction

On the soccer field, the robot must decide a proper motion quickly. This decision mainly depends on its location. In the Sony Legged League, it is prohibited to change the robot's hardware. So, following limitations of the robot's abilities make the localization and the decision making difficult.

- CPU Power: Impossibility of complex calculation.
- Sensor Ability: The low resolution and the narrow field of view of the camera.
- Control Ability: Discrete locomotion with a large error.

This year, we developed a real-time localization method which is a modification of Monte Carlo Localization(MCL)[1]. This method can be applied in the dynamic environment and achieves much lower computational power. We also developed a navigation method by utilizing this localization method. This method eliminates the unnecessary observations by judging the sufficiency of the information. Therefore the robot reaches the destination with the smallest observational cost.

2 Localization Method

We consider the followings when we design a self-localization method for the league.

- We should reduce the robot's behavior for the landmark observation.
- All image processings and calculations should finish within the robot's frame rate.
- The method should not be affected by the changes of lighting condition and the presence of other robots.

We design a new self-localization method so as to attain above things and name this method "Uniform Monte Carlo Localization (Uniform MCL)". This method branches from Monte Carlo Localization.

A. Birk, S. Coradeschi, and S. Tadokoro (Eds.): RoboCup 2001, LNAI 2377, pp. 730–733, 2002.

2.1 Uniform Monte Carlo Localization

Uniform MCL uses a set of N random samples, $S = \{s_i | i = 1 \ldots N\}$, to represent the position and the orientation of the robot stochasticly. Each s_i has the position and the orientation $l_i =< x_i, y_i, \theta_i >$. Uniform MCL considers the convex closure of all samples in the state space as the region where the robot exists(see Fig.1). These samples are operated when the robot moves or gets sensor readings.

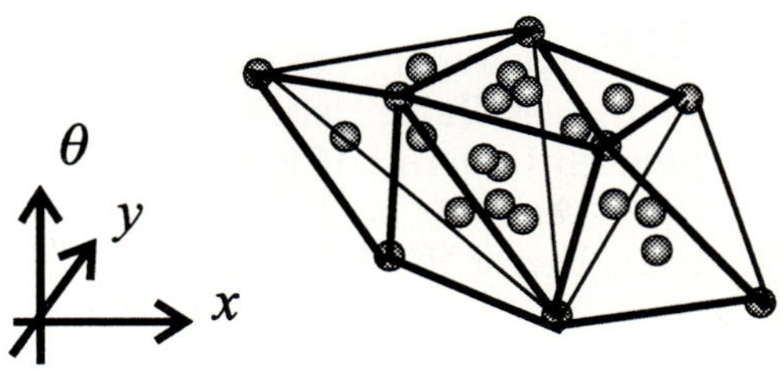

Fig. 1. The convex closure of all samples

When the robot moves: Uniform MCL firstly adds the average increment of the distance and the orientation to l_i $(i = 1, 2, \ldots, N)$. Uniform MCL nextly adds the random numbers to each l_i for the representation of the deadreckoning uncertainty.

When the robot gets sensor readings: Uniform MCL operates samples s_i $(i = 1 \ldots N)$ as follows:

- If the robot never gets the sensor readings at l_i, s_i is erased, and N is decremented.
- If the robot can get the sensor readings at l_i, s_i is not erased.

The samples decrease after sensor readings. Hence Uniform MCL increases samples as following way:

- If $N \geq 2$,
  1. Two samples are chosen at random from s_i $(i = 1 \ldots N)$.
  2. A new sample is put at the midpoint of the two samples.
  3. 1 and 2 are repeated $(N_{\max} - N)$ times.
  4. $N \leftarrow N_{\max}$.
- If $N = 1$, $(N_{\max} - 1)$ samples are put at l_1 and $N \leftarrow N_{\max}$.
- If $N = 0$, $N_{\max}$ samples are put at random in the state space and $N \leftarrow N_{\max}$.

Here $N_{\max}$ is the maximum of N.

2.2 Evaluation of the Performance

We implemented the method and evaluated by the experiments. The results of the experiments were compared with that of Sensor Resetting Localization (SRL)[2] which is another extension of MCL. The conditions of the experiments are as follows.

Precision

Conditions of experiment

- The robot walks between two points which are in front of each goal.
- The robot stops every 7 steps(total 154 steps). Then the maximum and minimum x, y, θ of samples are recorded.
- The robot rotates the camera horizontally through 180 degrees per 7 steps.

We also experimented the case that other two robots were located in front of the landmarks and goals as obstacles. Table.1 shows the result of the experiments. There is no difference between Uniform MCL and SRL though Uniform MCL.

Table 1. Average error (U-MCL means Uniform MCL)

	x[mm]	y[mm]	θ[deg]
SRL[2]	100	95	14.3
U-MCL	126	92	16.1
U-MCL with obstacles	199	146	15.8

This is because Uniform MCL conserves the past information. In fact, the state $N = 0$ happened only three times in this experiment.

Computational Cost We measured the speed of Uniform MCL with the robot. First we measured the speed of movement update. The result was 7msec with 1000 samples. Next we measured the speed of sensor update. The result was 7.5msec with 1000 samples includeding the image processing time. This means that Uniform MCL can handle the samples within the sampling time of camera image. It also indicates that the computational cost is much shorter than that of SRL.

3 Navigation Method

The robot decides the motion(walking command) according to the probability distribution of its position and distribution. When the distribution is too wide to decide the motion, the robot must stop walking and need to observe(rotate its camera). This observation is very time consuming and it is essential to eliminate the unnecessary ones. Therefore, the robot must judge whether it has adequate information for the decision making. For this judgment, we designed a State-Motion map by off-line calculation with Dynamic Programming. This map is a database of optimal walking commands which minimize the time to reach the destination at any robot's position and orientation. We call this map "Motion Planned Map(MPM)" (see Fig.2). The robot executes the walking command which corresponds to the region of probability distribution of robot's position and orientation. If the distribution crosses two or more regions, the robot judges

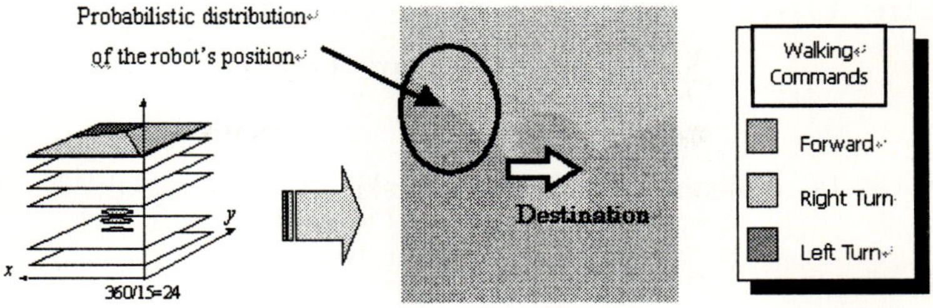

Fig. 2. Motion Planned Map. This is a State-Motion map designed offline.

Variables

- n : Number of the walking commands
- i : ID of the walking commands
- a_i : Walking command
- t_i : Cost of the walking command a_i
- $t_{observation}$: Cost of the observation
- E_0 : Expected time to the target region of samples Ω_0 before walking.
- E_i : Expected time to the target region of samples Ω_i after walking

Judging Method

$$d = \arg\min_{1 \le i \le n} (E_i - E_0 + t_i)$$

$$t_{min} = \min_{1 \le i \le n} (E_i - E_0 + t_i)$$

$$if\,(t_{min} \le t_{observation})\ do\ a_d$$

$$if\,(t_{min} > t_{observation})\ do\ \text{observation}$$

Fig. 3. Walk-Observe judging algorithm.

whether to walk or to observe according to the rule(Fig.3). This means "If the loss of time by executing the optimal walk command is larger than observational cost, the robot observes , else it walks without observation". As a result, the total observational cost to reach the destination is minimized.

4 Conclusion and Future Works

In this paper, we described a localization method and a navigation method for Sony Legged league. These were implemented in the soccer program for the RoboCup 2001, and some soccer behaviors such as approaching to the ball, returning to the home position were accomplished. The future works are the improvement of the precision of localization by increasing the information from one sensor input, and the advanced tasks with the ball and other robots.

References

1. Dieter Fox, Wolfram Burgard, Frank Dellaert and Sebastian Thrun, "Monte Carlo Localization: Efficient Position Estimation for Mobile Robots," *In Proc. of the Sixteenth National Conference on Artificial Intelligence (AAAI-99)*, 1999.
2. S. Lenser and M. Veloso: "Sensor Resetting Localization for Poorly Modeled Mobile Robot," *Proc. International Conference on Robotics and Automation*, pp.1225-1232, (2000)

The University of Pennsylvania RoboCup Legged Soccer Team

Sachin Chitta, William Sacks, Jim Ostrowski, Aveek Das, and P.K. Mishra

University of Pennsylvania

1 Introduction

This paper describes the University of Pennsylvania's RoboCup 2001 Legged League team, called the UPennalizers. This year's team was very successful in both the main competition, placing third out of sixteen teams, and the RoboCup Technical Challenges, where we placed second. Much of this success can be credited to our being able to build on a strong foundation, in both software architecture and low-level, fast locomotion, developed last year. We also were one of only two teams to utilize sound communication between robots, which gave a significant advantage in resolving conflicts between team members.

2 Team Development

Team Leader: James P. Ostrowski
- Department of Mechanical Engineering and Applied Mechanics, University of Pennsylvania, United States
- Assistant Professor, attended the competition

Team Members:

Sachin Chitta and Aveek K. Das
- Department of Mechanical Engineering and Applied Mechanics, University of Pennsylvania, United States
- Graduate Students, attended the competition

William Sacks
- Williams College
- Undergraduate junior, attended the competition

Caroline Klose, P.K. Mishra
- Computer and Information Science, University of Pennsylvania, U.S.
- Undergraduate Senior and Master's student, did not attend

Web page http://www.cis.upenn.edu/robocup/

3 Architecture

In [3] we presented our hierarchical structure for control of autonomous robots. This architecture is compact and can be applied across multiple platforms. Figure 1 shows the top level overview, along with a schematic of the main logical sensing and actuation blocks that are used. The major blocks are the ***planner***, the ***extended logical sensor module(XLSM)*** and the ***extended logical actuator module (XLAM)***.

A. Birk, S. Coradeschi, and S. Tadokoro (Eds.): RoboCup 2001, LNAI 2377, pp. 734–738, 2002.

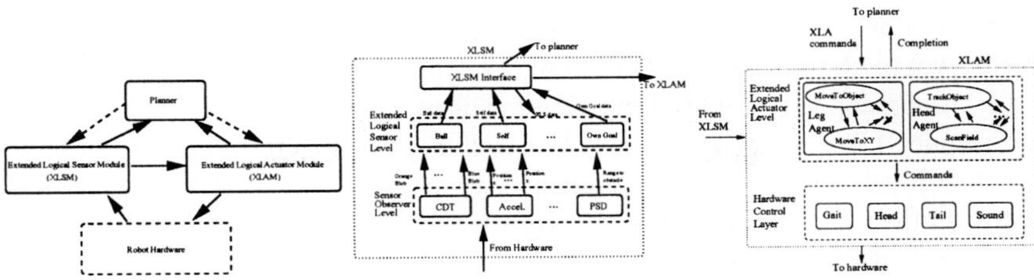

Fig. 1. (left) Top-level view of overall structure. Arrows indicate data flow; dashed arrows indicate commands. (center) XLSM (right) XLAM

Extended logical sensor module (XLSM) A ***logical sensor*** is defined as a software construct that converts raw sensory data into some ***application domain*** information. An ***extended logical sensor*** (XLS), then, is a sensory agent capable of modifying its operation (mode-switching) based on state information. An example would be an XLS for tracking a moving object in a noisy environment using vision. If the object can be seen, its position is recorded. If the object is occluded, the XLS can use Kalman filtering to give a best estimate of the object's position. If a second sensor were added to the system (radar for example), the XLS would be modified to use both sensors (or prediction) to detect the target. Decision making agents (planners) are only aware that they have access to a "sensor" that reads target position.

Extended logical actuator module (XLAM) A ***logical actuator*** is defined as the output counterpart to a logical sensor. It converts application domain commands into physical actuator outputs. An ***extended logical actuator*** (XLA) is an agent capable of performing some low-level "inner loop" control on physical actuators. XLA primitives for soccer might include moving to a position in the field, moving toward an object, kicking an object, etc. Planners need not be aware of operations and controllers that are involved in achieving these tasks. In fact, the operations are the same for a legged robot as for a wheeled robot–so the planner need not even know *how* the task is accomplished.

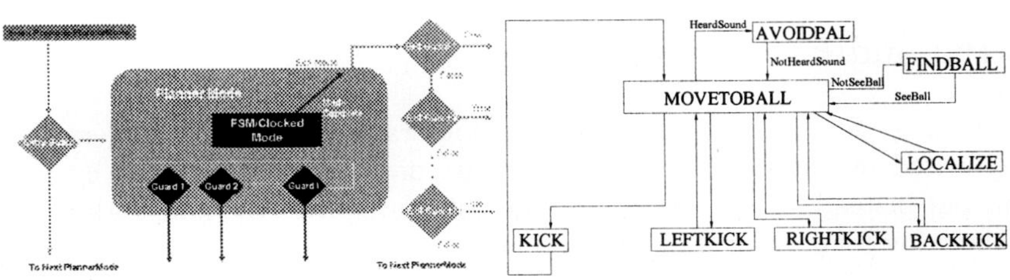

Fig. 2. Switching between Planner Modes (left) and a sample FSM used for generation of Plannertask table in example (right)

Planner The Planner is the decision making module of the robot and forms the highest level in the hierarchy. The main decision making structure of the planner is the Plannertask table, which represents the strategy for a robotic soccer player in a compact form. This formalism fits well within a hybrid systems approach, where low-level sensing and dynamics have been abstracted into higher-level, discrete modes. The left diagram in Figure 2 shows an example flow of a table, where a transition to a new mode results first in an entry guard being checked. If the guard condition is true, a new, mixed finite state machine and clocked, feedback loop is initiated (a "mode"). During the its execution, any number of guards may be checked to see if a mode switch is called for. If not, then at the completion of the mode a set of exit guards are used to determine the next mode.

A simple example is shown in Figure 2 of a PlannerTask table structure. The finite state machine executes the simple task of approaching and kicking a ball. The type of kick, e.g., RIGHTKICK, BACKKICK, (forward) KICK, is chosen depending on the player's and ball's location on the field. For example, if the robot is facing the wrong way it executes a back kick. As shown Figure 2, transitions between modes are selected by ***clocked guards***, such as SeeBall, and ***mode completions***, such as the transition from MoveToBall to a kick.

4 Omnidirectional Walking Capability

The robots have been programmed to have an omnidirectional walking capability. The XLA modes need to specify only the required translational and rotational components of velocity for the robot. These are then converted, using inverse kinematics, to required trajectories for the foot placement point of each foot. The robots use a trot walking gait which means that diagonally opposite pairs of legs move together. The foot placement point (which is the point of the foot that comes into contact with the ground) goes through a rectangular trajectory as it comes into and out of contact with the ground. The velocity of the robot is assumed to remain constant for each step. Thus, changes in velocity are initiated only at the end of a step. The duty factor for the gait is 0.5 which means that each foot of the robot spends as much time out of contact with the ground as it does in contact with the ground. This gait gives faster speeds and smoother transitions during changes in velocity. More details are in [1].

5 Pal and Foe Detection

The strategy for the game requires keeping track of all other players. This is complicated by the fact that the robots are not uniformly covered by a single color, but instead have multiple stickers distributed over the body. After sorting the blobs by team color and size, the largest blob is chosen to represent a particular player, and then other blobs are compared to see whether they belong to the same player. If the distance to the next largest blob is within a certain threshold (we have chosen this to be less than three times the length of the largest blob), the smaller blob is assigned to the same player as the larger blob.

This process is continued for all blobs representing the player types. The factor of three is conservative, sometimes merging two players that are near each other. This is preferable to a more aggressive approach where "ghost" players could be spuriously detected.

Keeping track of pals and foes The observer must next determine which player in memory corresponds to each player in the current view. Our algorithm for matching is similar to one presented by Veloso et al. [4]. For a player A in the current view and a player B in memory (with B on the same team as A), the likelihood that A and B are the same player is calculated as: likelihood$(A, B) =$ (B's uncertainty) $* \frac{1}{(d(A,B))^3} * \frac{1}{(\Delta\theta(A,B))}$. Uncertainty ranges from 0.5 (recently seen) to 100 (not seen in a long time). Thus, it is more likely that A and B are the same player if B has not been seen in a while, and also if A and B are close. Then, for each player A in the current view, the player B in memory that gives the highest likelihood(A,B) is replaced in memory by A.

6 Potential Field Approach

We have also utilized a potential field approach (see, e.g., Khatib [2]) to obstacle avoidance. In our situation, the goal exerts an attractive force on the robot, $F_{goal} = -k_1 * (x - x_{goal})$, where x and x_{goal} are the positions of the robot and goal, and k_1 is a positive constant; while obstacles exert repulsive forces

$$F_{obstacle} = k_2 * (\frac{1}{(x - x_{obstacle})} - \frac{1}{x_0}) * \frac{1}{(x - x_{obstacle})^2} * \nabla(x - x_{obstacle}), \quad (1)$$

where x and $x_{obstacle}$ are positions of the robot and obstacle, x_0 is the maximum distance at which an obstacle exerts a force, and k_2 is a positive constant.

Summing the forces on the robot gives the total force vector; the robot then moves in the direction of this vector. This effectively steers the robot toward the goal and away from the obstacles. A major advantage of this method is that it allows real-time planning, because the calculations at any point are not computationally expensive. Some important modifications had to be made before using this technique. When the obstacles were very far away they still exerted a repulsive force. As mentioned in [2], preventing forces from such obstacles leads to a more stable path. Hence, the force from obstacles further than a certain fixed distance was ignored. Similarly, obstacles behind the robot are ignored.

7 Conclusion

Contributions in RoboCup-2001 We felt our main contributions to RoboCup2001 were the full use of audio communication, a solid modular architecture (testing of many modules working together), and a very stable and fast dynamic trot gait based on inverse kinematics that is relatively easy to implement.

Future work for RoboCup-2001 We will enter a team in next year's competition and will focus on three areas: 1) wireless communication; 2) decentralized control and estimation strategies; and 3) high-resolution image processing.

References

[1] S. Chitta and J. P. Ostrowski. New insights into quasi-static and dynamic omnidirectional quadruped walking. To appear in IEEE Int. Conf. Intelligent Robots and Systems, IROS01, October 2001.

[2] O. Khatib. Real-time obstacle avoidance for manipulators and mobile robots. *International Journal of Robotic Research*, 5(1):90–98, 1986.

[3] K. A. McIsaac, A. K. Das, J. M. Esposito, and J. P. Ostrowski. A hierarchical structure for hybrid control applied to AIBO robotic soccer. In *Proc. IEEE/RSJ Intl. Conf. on Intelligent Robots and Systems (IROS)*, pages 1020–1025, Japan, November 2000.

[4] M. Veloso, M. Bowling, S. Achim, K. Han, and P. Stone. The CMUnited-98 champion small robot team. In M. Asada and H. Kitano, editors, *RoboCup-98: Robot Soccer World Cup II*, Berlin, 1999. Springer Verlag.

Wright Eagle 2001 - Sony Legged Robot Team

Xiang Li, Zefeng Zhang, Lei Jiang, and Xiaoping Chen

Computer Science Department, University of Science and Technology of China, Hefei, 230027, China

Abstract. This paper introduces team Wright Eagle - the only Chinese team in Sony legged robot league of RoboCup. The architecture and four main modules will be described.

1 Introduction

"Wright Eagle" is the only Sony legged robot soccer team in China.[1] The target of our participation in this league is to push our research in artificial intelligence, especially the multi-agent system.

As a new team in the Sony legged robot league, it is the first time that we use the AIBO robot and OPEN-R system.

2 Architecture

Each robot control system has four main modules — vision module, localization module, decision module and action module. There are also some functions to collect sensor data and other information.

The system has a serial controlling pattern. Vision module is the start point of each cycle. It receives image information taken by CCD sensor from AIBO robot and analyses it. Next, localization module uses these results and other information to make localization. And then, Decision module works to decide which action should be perform. Finally, action module generates actuator commands according to the decision.

Generally, the Wright Eagle has architecture as Fig 1.

3 Vision Module

The vision module is a very important part of the system. It is a main source of field information. The job of vision module contains several steps as blow.

(i) Distinguish colors — Instead using standard CDT, we draw polygons on YUV color plane to define colors.

(ii) Merge color blocks

[1] This work is supported by the National High-Tech Project of China and the national Natural Science Foundation of China (69875017)

A. Birk, S. Coradeschi, and S. Tadokoro (Eds.): RoboCup 2001, LNAI 2377, pp. 739–742, 2002.

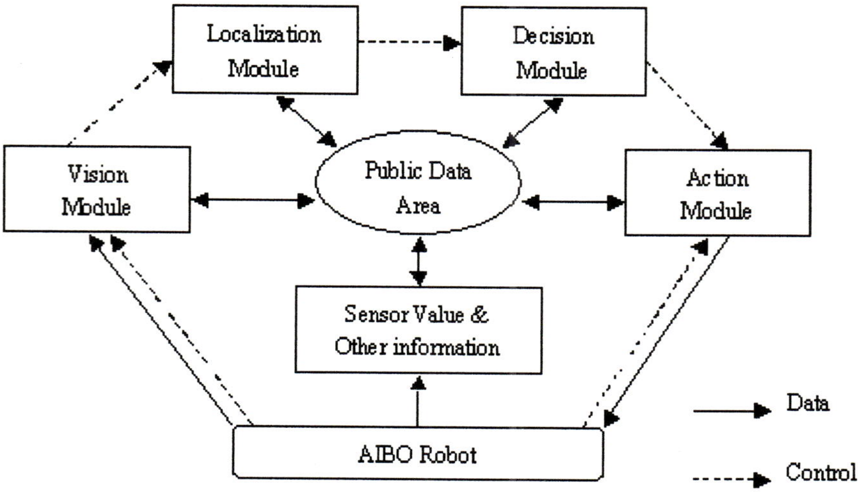

Fig. 1. Architecture of Wright Eagle 2001

(iii) Recognise objects — We use domain knowledge to analyze color blocks that we have gotten. If two blocks have a color relation pattern of a landmark, we will check them if they also have almost same area and very near distance. The blocks meet these conditions will be recognize as a landmark. Vision module also can recognize ball, goals and players.

(iv) Estimate distances and angles of objects — Vision module can estimate the distances and angles of ball, and landmarks. It also can provide the angles of goals and players.

Vision module also has a particular function called "Illegality Detection". We can use it to avoid recognition mistake. For instance, only the orange block that it is over green block can be recognize as a ball. This function is very successful to resist the interference from outside of field.

4 Localization Module

Localization of WrightEagle-2001 is quite simple. This module receives distance information and angle information of landmarks from vision module. When there is two landmarks' information in memory, localization illustrated in Fig2 will be performed. If only one landmark can be seen, information of it will be stored in memory and be used to correct player's angle.

In most situations of game, we do not need a very accurate localization result for making decisions. In this case, we partition the field into four areas and divide angles into four quadrants. For instance, player P illustrated in Fig3 is in rear right area and has a quadrant4 angle.

This module is right in most situations and acts well in games.

However, this localization is not very accurate. We are looking for a new approach to make better.

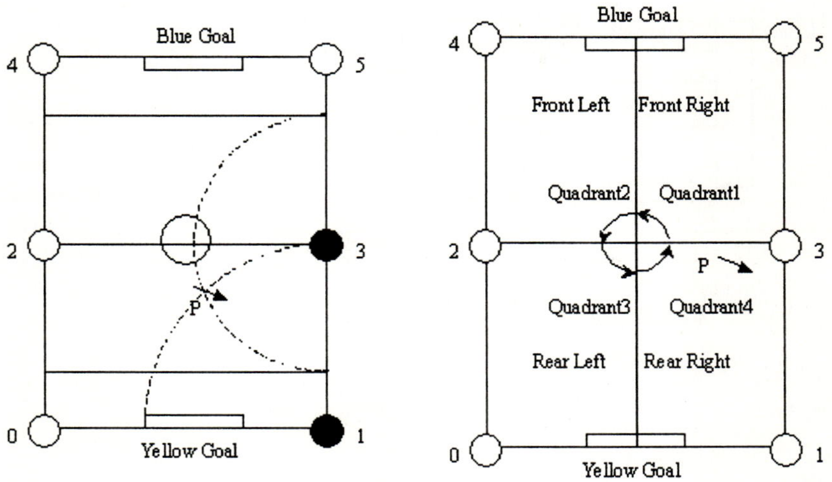

Fig. 2. Localization and Field Partition

5 Decision Module

The function of this module is making decisions that which kind of actions should be performed next step.

We define some situations for players. A set of actions will be performed when a player is under a given situation. Field players and goalie have difference strategies. Under most situations, goalie stays in penalty area and gaze the ball. When it finds the distance from itself to ball is short enough, goalie thinks the ball is in a danger location. It will walk out to clear ball away. Then, goalie backs to goal to keep it. The field player's strategy is as simple as goalie. First it turns around to search ball. After finding it, player walks to ball and try to adjust its direction to enemy's goal. Then, a kick or shoot action will be performed. These strategies work well in games. However, they also need promotion in future.

There is no communication between the players in games.

6 Action Module

We developed a set of custom actions, such as walking, turning and kicking. The advantage of these actions is stable. WrightEagle players seldom fall over in games. However, the main shortage of our actions is slow.

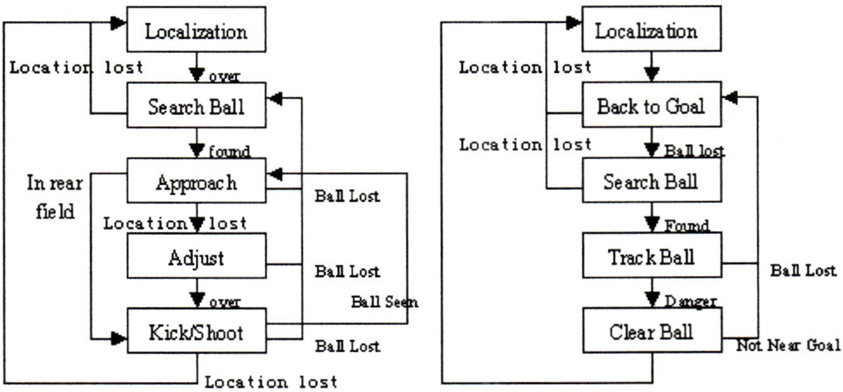

Fig. 3. Field Player Strategy and Goalie Strategy

We use OMoNet to achieve gaze and tracking ball.

7 Conclusion

It is the first time WrightEagle participate the RoboCup Sony legged robot games. We got a 5-8 rank. It shows that our approach to build a legged robot team is successful to some extent. However, compared with those very strong teams, our WrightEagle also has many parts to be refined. WrightEagle will be stronger next year.

References

[1]Manuela Veloso, William Uther, Masahiro Fujita, Minoru Asada, Hiroaki Kitano, Playing Soccer with Legged Robots, Proceedings of IROS-98, Intelligent Robots and Systems Conference.

[2]James Bruce, Tucker Balch,and Manuela Veloso, Fast and Cheap Color Image Segmentation for Interactive Robots, Workshop on Interactive Robotics and Entertainment, Pittsburgh, April, 2000.

[3]John Dalgliesh, Mike Lawther, Playing Soccer With Quadruped Robots, School of Computer Science and Engineering University of New South Wales

A Design of Agents for the Disaster Simulator on RoboCup-Rescue

Taku Sakushima, Tetsuya Esaki, Yoshiki Asai, Nobuhiro Ito, and Koichi Wada

Department of Electrical and Computer Engineering,
Nagoya Institute of Technology
Gokiso-cho, Showa-ku, Nagoya 466-8555, JAPAN
taku@phaser.elcom.nitech.ac.jp

Abstract. In a Multi-Agent System, it is important that an agent cooperate with the others appropriately. In our rescue agent team "NITRescue", each agent acts in cooperation with other kind of agent by using limited communication facility. Therefore they aim at coping with complicated disaster circumstances of Rescue simulator.

1 Introduction

In Robocup-Rescue simulation[1], agents act in order to minimize the damage of disaster. For example, fire fighting party(Fire Brigade) extinguish a burning building, and Ambulance Team saves a buried people by building collapse and injured people. However, as real disaster, there is limit to solve a situation by power of an individual agent. Therefore it is important an agent cooperates with other agent. In addition, it considers comparing with a Rescue simulation and a Soccer simulation[2]. In a Rescue simulation, the agent of two or more kinds exists in that environment. Therefore cooperation with a different kind agent is important same as cooperation with a same kinds agent.

In NITRescue, we implement cooperative activity between similar agents to fire brigade in order to extinguish a fire of a burning building more effectively. In addition, in order to cope with the circumstances that more than one element such as road confinement or collapse of a building was related with each other complicatedly, agent communicate with heterogeneous agent.

2 Cooperative Activity

2.1 Cooperative Activity of Homogeneous Agent

In Rescue Simulation, goal of an agent minimizes spread of a fire that occurred by a disaster, and it is to save a lot of human life as much as possible. However, power of single agent is limited. For example, we think about fire extinguishing activity of a fire brigade agent. It is difficult on a Rescue simulation for only one agent's power to extinguish a fire of a building. Also, this can say the same things also as fire-extinguishing activities in the real world.

A. Birk, S. Coradeschi, and S. Tadokoro (Eds.): RoboCup 2001, LNAI 2377, pp. 743–746, 2002.

Fire Brigade agent of NITRescue extinguishes a fire of a building in cooperation with a nearby fire brigade agent with using communication. We can use two different communication methods in rescue simulation. The first is "say" that communication range is limited (equivalent to communication of natural voice), and the other one is "tell" that communication range is unlimited (equivalent to communication method to use radio). When there is other fire brigade near it, fire brigade transmits information of the building which oneself extinguishes a fire with using "say" to that fire brigade. If the agent which received a message of the information does not extinguish a fire, they help extinguishing activity of the agent that sent the information. On the other hand, when that agent is extinguishing a fire, that agent judges which building can extinguish a fire effectively after having considered distance to a building and scale of a building.

Fire brigade extinguishes a fire of many buildings by extinguishing a fire of each building quickly.

2.2 Cooperative Activity of Heterogeneous Agent

In a RoboCup-Rescue simulation, in addition to the movable agent such as fire brigade, ambulance team and police force, there is headquarters agent having role to control an agent of each kind (Fire Station, Ambulance Center, Police Office). In a NITRescue team, these agents support action of a movable agent by undertaking communication of message between each kind of agent. In a rule of RoboCup-Rescue simulation[3], the number of message which each agent can receive in one simulation step is limited. Therefore it is impossible to receive every message from every agent.

In order to solve this problem, headquarters agent and a move agent of NITRescue limit the partner that each agent receives a message, like figure 1. By

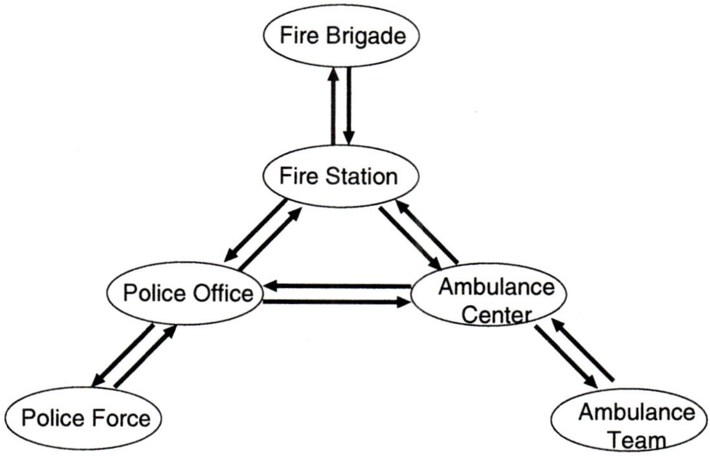

Fig. 1. Network Diagram of Rescue agent

using this method, an agent can do effective communication in limit of severe message.

As an example, we think message communication between fire brigade and police force. There is road confinement by disaster in Rescue simulator. Therefore there is a case that fire fighting party cannot arrive at a destination by road confinement and cannot act at all. In this case, fire brigade aims at a breakthrough of circumstances by sending in a relief request to the police force. Fire brigade can send in a direct relief request to the police party, but in such a case the police force needs to be prepared to always hear a message of fire brigade. In a fire brigade's message, the message which is not related to a police force at all is also contained. If the police force receives every message of fire brigade, it becomes difficult that the police force receives even a necessary message by limit of severe message. In order to solve this problem, based on network as figure 1, each agent transmits a message with the following procedure(figure 2).

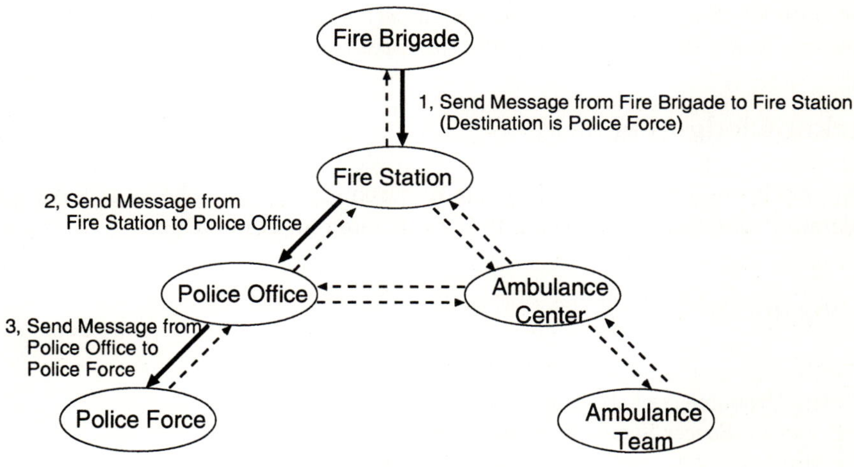

Fig. 2. Flow of Sending Message from Fire Brigade to Police Force

1. Fire brigade transmits the message whose transmission address is a police force to a fire station.
2. A fire station forwards the received message to a police station after having checked destination.
3. If the message which police station received from a fire station is addressed to the police force, a police office forwards the message to the police force.
4. If it is message from a police office, the police force checks contents.
5. The police force acts on the basis of contents of received message.

Based on such a procedure, fire brigade communicates with the police force. And an agent cooperates with heterogeneous agent in complicated disaster circumstances

And there is a limit in the capability of ambulance team which finds people who have been buried in a rescue simulation. In NITRescue, when police force and fire brigade found a buried person, they transmit the information with the same procedure to ambulance team. Therefore, in difficult situation, an agent aims at lessening damage of disaster by cooperating with other agents.

3 Conclusion

We designed the agent in the rescue simulation in which agent of more than one kind exists. In addition, in order to cope with restriction of the severe communication in the game rule of a rescue simulation, we developed the communication network which can communicate effectively among the agents of a different kind. However, it is mentioned that a headquarters agent's amount of communications increases and a bottleneck arises as a problem of the built network. Since the map of 1/10 scale is used, there is no problem in the present game. However,since the number of agents will increase if the scale of a map will become large from now on, it is a future work to solve this problem.

Acknowledgment

This work was partly funded by the research grants from the Hori Information Science Promotion Foundation and the Tatematsu Foundation.

References

1. RoboCup-Rescue Official Web Page, http://robomec.cs.kobe-u.ac.jp/robocup-rescue/index.html
2. RoboCup Soccer Server System Web Page, http://sserver.sourceforge.net/NEW/index.html
3. RoboCup-Rescue Simulation Tentative Rule and Evaluation, http://www.r.cs.kobe-u.ac.jp/robocup-rescue/tentrule2001.html

An Approach to Multi-agent Communication Used in RobocupRescue

Jafar Habibi, Ali Nouri, and Mazda Ahmadi

Department of Computer Engineering,
Sharif University of Technology, Tehran, Iran
arian@ce.sharif.edu, http://ce.sharif.edu/~arian

Abstract. This paper describes our strategies used in the RobocupRescue simulation contest in seattle 2001[3]. we first introduce our powerful communication manager for multi-agent systems which is specially designed to work in difficult ,let's say "Hard-zone", environment where number of communicated messages are drastically limited and also sensory noises and limited information about the environment are major problems. Also some hints are provided for developing multi-agent systems. Then we introduce our approach for solving some NP-Complete problems faced in the system.

1 Introduction

Arian team is the result of a research group in Robocup Field in Sharif University of Technology(SUT). We chose RobocupRescue as a research field for the sake of humanity.

Among all the other fields of Robocup, with 6 different types of agents and a very limited eyesight[2], rescue simulation has the heaviest network communication load . So it's crystal clear that use of a good communication manager in the system is inevitable[4].

This paper is organized as follows: In section 2, we introduce our communication manager used to set up a harmony between the agents. Then in section 3, we explain the methods used in solving time-consuming problems. Section 4 discusses future works and concludes this paper.

2 Communication System

Decision making in multi-agent systems is very complicated compared to single-agent ones. Since the knowledge of each agent from the world is very limited, a decision making algorithm based on self perception will result in actions far from the optimum ones. In order to provide more information about the world model, each agent can propagate the gained information to his partners. with this information modeling, agents will have the maximum available information from the world, while some of the faulty information generated by noises will be

A. Birk, S. Coradeschi, and S. Tadokoro (Eds.): RoboCup 2001, LNAI 2377, pp. 747–750, 2002.

recovered because of the redundant data. Also, agents will be able to synchronize their actions to provide the best results.

An overview of the whole communication system is shown in the figure below.

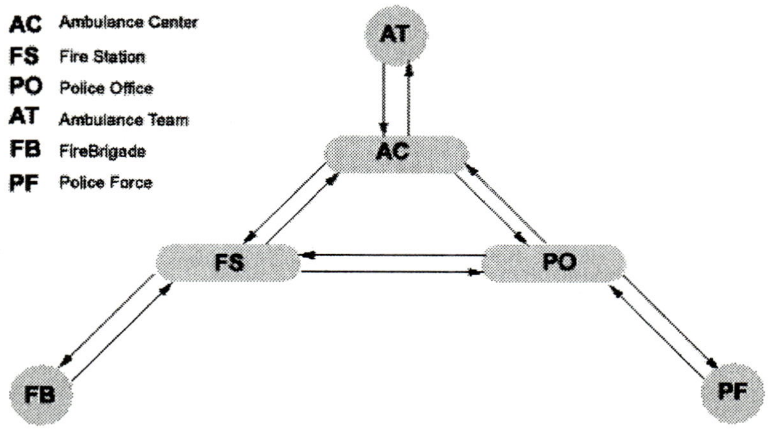

In the Rescue simulation scenario, there is a limit on the number of transmitted messages between the agents in each cycle, that is each agent can send or hear only 4 messages per cycle. Another problem in communication is that each agent can only send message to a certain set of other agents. For example a FireBrigade can only send message to another FireBrigade or FireStation. Because of these limitations, it is beneficial that we don't setup a bi-directional communication system (as it is in a TCP/IP connection) to overcome the message losses and delivery latency of the messages. The most common transmitted messages are shown below.

- FB/PF/AT to AT used when some agents are buried in the building
- FB/AT to PF used when some agents are trapped in a dead-end
- FB/PF to AT used to report buried civilians.
- AT to AT used to inform other partners about our decisions, etc.

The main problem for communication between agents here is that all the mentioned communications are multi-hopped. For example, there is only one-way from FB to AT in the above graph and that is: FB-FS-AC-AT. The hardness of this kind of communication in our system can be described as follows:

- Once a member of the path in the graph cannot hand in the message to the next one, the whole message will be lost.
- 90% of the messages is delivered very late. It takes more than 6 cycles for the messages to reach their destination.

To overcome these problems we have figured out some techniques to reduce the network traffic and to safe the whole system. Some of them are:

- All senders will keep track of sent and received messages in order to prevent overdoing.

- Some of the paths can be reduced as follows: A path of FB-FS-AC-AT can be reduced to FB-AT and a AT-AT piggybacked message, if there is an AT agent near FB agent so that the message can be sent via a "say" command instead of "tell" command.
- Fuzzy logic can greatly improve the system and prevent the system from too many messages. e.g. If the PF agent be, somehow, able to have a good fuzzy view of the world (about the crowded areas, fire places, etc.) Then it can dynamically change it's current algorithms for the new ones.
- Single messages can themselves be mixed. With use of a message splitter manager, we can embed more than one kind of message in a single one. And there are a lot more about communication manager in a multi agent system, which we took care of.

3 Reduction of Classic Algorithms

Once the information about the world's current state has been successfully passed between agents, they will enter the "Decision Making" state. In this state, we're faced with this question: "What is the best target?". To answer this question we have to take care of two significant issues about the problem:

- Picking the best target in our system is highly dependant on the other agents' actions. So the algorithms we choose for our agents should fit each other's.
- Our system is a little bit like "Real time systems", i.e. late results are worse than bad results. Because of the behavior of the kernel, the action of the agents should be transmitted back to kernel at the end of the current cycle and latecomers are discarded.

The best solutions, especially in graph related problems, are usually derived from classic algorithms. A Floyd algorithm can give the shortest path between every two nodes in a graph, course if we have enough time and information. But the problem here is that we lack them both. Most of the classic algorithms are highly time-consuming and cannot be easily used in our system. Also lack of information about the whole world's state prevents us from applying simple classic algorithms. For example a FireBrigade had to calculate the connected components of a huge graph with a lot of edges (near n^2). We tried some classic algorithms such as DFS[1] and Clustering but we soon found that the agents miss to send the information back to kernel on time and they loose some of the cycles. So we tried to find some new methods for the problem and we came up with a very tiny method to do the job and it really worked in more than 80% of the times. The idea was simply taken from some salutes usually done in the opening of a sport event, where lines of people, standing side by side, form a shape. Since the graph was based upon the location of the objects in the city, we applied the same technique for getting the connected components of the graph. There were several cases where we replaced heavy -duty classic algorithms by simpler ones.

4 Conclusion and Future Work

When dealing with a huge, multi-agent system, acquiring valid and valuable information is one of the most important jobs to do. For the sake of valuable information, having established a pioneer communication manager in the system, we're able to feed the agents with up-to-date, filtered and valuable information. The more the number of information resources are, the more reliable the gained information will be.

References

1. Thomes H Cormen, Charles E, leiserson, Ronald L. Rivest, Introduction to Algorithems, MIT Press, 1992.
2. he RoboCupRescue Technical committee, RoboCupRescue Simulator Manual.
3. The Robocup Federation, Robocup Regulations and Rules, http://www.robocup.org.
4. hristoph Aigne, Communication in a hybrid multi-agent system, DBAI TECHNICAL REPORT DBAI-TR-00-XX, APRIL 2000

Task Allocation in the RoboCup Rescue Simulation Domain: A Short Note

Ranjit Nair[1], Takayuki Ito[2], Milind Tambe[1], and Stacy Marsella[1]

[1] University of Southern California's Information Sciences Institute
4676 Admiralty Way, Suite 1001, Marina del Rey, CA 90292, USA
{nair,tambe,marsella}@ISI.edu
[2] Center for Knowledge Science
Japan Advanced Institute of Science and Technology (JAIST), Japan
itota@jaist.ac.jp
http://www.isi.edu/~nair/Rescue/index.html

1 Introduction

We consider the problem of disaster mitigation in the RoboCup Rescue Simulation Environment [3] to be a task allocation problem where the tasks arrive dynamically and can change in intensity. These tasks can be performed by ambulance teams, fire brigades and police forces with the help of an ambulance center, a fire station and a police office. However the agents don't get automatically notified of the tasks as soon as they arrive and hence it is necessary for the agents to explore the simulated world to discover new tasks and to notify other agents of these.

In this paper we focus on the problem of task allocation. We have developed two approaches, a centralized combinatorial auction mechanism demonstrated at Agents-2001 and a distributed method which helped our agents finish third in RoboCup-Rescue 2001. With regard to task discovery, we use a greedy search method to explore the world– agents count the number of times they have visited each node, and attempt to visit nodes that have been visited the least number of times.

2 Task Allocation

The problem of disaster mitigation in the RoboCup-Rescue Simulation Environment can be thought of as a task allocation problem where the tasks become known at different times and can change in intensity. The tasks can be thought of as civilians who need to be rescued, buildings which are on fire and roads that are blocked. The problem is then how to assign ambulance teams, fire brigades and police forces to these tasks. There are two categories of approaches to assigning agents to tasks: centralized approaches and distributed approaches. We will describe a centralized combinatorial auction mechanism demonstrated at Agents-2001 and a distributed method based on localized reasoning.

A. Birk, S. Coradeschi, and S. Tadokoro (Eds.): RoboCup 2001, LNAI 2377, pp. 751–754, 2002.

2.1 Combinatorial Auction Mechanism for Task Allocation

Combinatorial auctions have been used before for task allocation [2]. In our auction mechanism, the fire station, ambulance center and the police office take on the role of auctioneers, and the ambulances, fire brigades and police forces take on the roles of bidders . The items being bid for are the tasks. At the beginning of each cycle, each free agent makes several bids - each bid consists of a different combination of tasks and an estimate of the cost of performing sequentially the tasks in this combination. For example Fig. 1 shows fire brigades making bids to the fire station on different combination of buildings on fire that it wants to extinguish based on its proximity to these fires. We make an assumption here that each task can be performed by a single agent. This assumption can be removed by making agents bid on combination of roles in tasks.

The auctioneer receives all the bids and determines the winning bids – the set of bids that cover the most tasks and have the least cost . Most optimal winner determination algorithms that have been developed are branch-and-bound algorithms that fall into two broad-categories: those that branch on bids [6] and those that branch on items [1,5]. Our algorithm, while based on [6], branches on the bidders. This algorithm is based on one assumption:

- At most one bid per bidder wins. This assumption is a fair assumption to make in the rescue domain where the agent does not desire that more than one combination of tasks is assigned to it.

Our algorithm is a depth-first branch-and-bound tree search that branches on bidders. The branching factor of the search tree is $O(b)$, where b is the maximum number of bids that a single bidder makes. It is fairly trivial to see that winner determination can be performed in $O(b^k)$ time where k is the number of bidders. In domains like RoboCup-Rescue, the number of bidders is constant and thus we have a solution to the winner determination problem that is polynomial in the number of bids made by a single bidder.The other prominent algorithms for winner determination [1,5,6] all have complexity $O((n + m)^m)$ where n is the number of bids received and m is the number of items(tasks). Clearly, such solutions, though polynomial in domains where the auctioneer has control over the number of items, are exponential in a domain like RoboCup-Rescue where the auctioneer has no control over the number of items that are being bid on.

The key strength of this approach is that it considers all agents bids for multiple combinations of tasks and thus employs global reasoning. But there are several shortcomings of using combinatorial auctions for task allocation:

- **Exponential number of possible bids:** If the bidders bid on each and every possible combination of tasks then we are guaranteed that the solution obtained by winner determination is the optimal solution to the task allocation problem. However, this would lead to an exponential number of bids received by the auctioneer. Hence it was necessary to decide on what tasks a bidder should bid on. We used a heuristic based approach where the bidders bid on combinations of size $< l$. Therefore the number of bids received by

the auctioneer is $O(m^l * k)$ where m is the number of tasks and k the number of bidders. We also restricted the number of bids by not bidding on tasks that are far away.
- **Difficult to make cost estimate:** In highly dynamic and uncertain domains, it is difficult for bidders to make an accurate cost estimate for bids containing many tasks that need to be performed sequentially. This is another reason for restricting the size of bid to l tasks.
- **Domain-imposed communication constraints:** Constraints on communication could impact the use of combinatorial auctions for task allocation. In RoboCup-Rescue, there are two kinds of communication constraints. First, the number of messages heard by any agent is limited, in the 2001 competitions, this number was 4. This severely crippled the allocation mechanism as centers were responsible for receiving information on new tasks as well. In addition, all agents could not communicate to a centralized auctioneer and needed to go through an intermediate center, which slowed down the communication
- **Not globally optimal:** Even if we assume that this mechanism is optimal for each time step (which it would be if there were no restrictions on the bids made), the allocation is not optimal over all time steps.

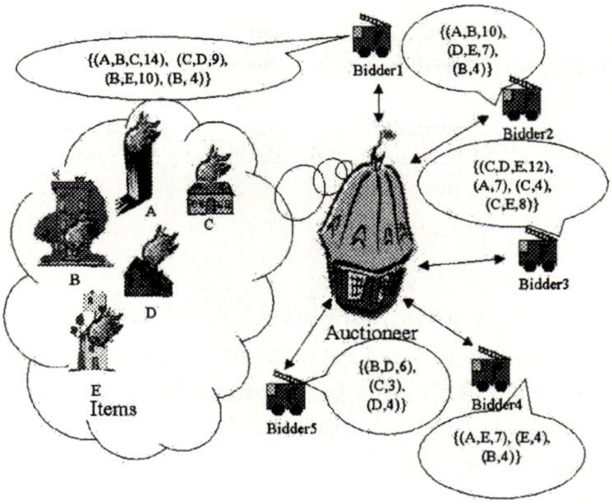

Fig. 1. Auctioning of fires by fire brigades.

2.2 Distributed Mechanism for Task Allocation

The distributed method, based on our agents described in [4], relies on each agent deciding for herself as to which task to perform. This localized reasoning allows agents to evaluate the seriousness of a task before committing to that task. Fire brigades select fires to put out based on intensity and proximity of the fire, ambulances select which civilian to save by evaluating the civilian's health, buriedness and proximity, while police agents choose which road to clear based on number of rescue agents trapped by this road.

The strength of this approach lies in the low number of messages that it requires. A major shortcoming of this approach is that the agents rely on local information and don't concern themselves much with what tasks the other agents were performing. Thus, this allocation scheme is clearly sub-optimal. In practice, by suitably adjusting the local reasoning, satisfactory allocations were found. Agents using this mechanism in RoboCup-Rescue 2001 finished in third place. Nevertheless, the sub-optimality of the distributed allocation is an issue for future work.

3 Conclusion

In this paper we described how search and rescue in the RoboCup Rescue Simulation Environment is a task allocation problem where the tasks arrive dynamically and change in intensity. We explained the need for exploration to discover new tasks and then discussed various approaches to solving the task allocation problem, viz. the centralized combinatorial auction approach and the distributed approach used by our agents in the RoboCup-Rescue 2001 competition where they finished third. In the future, we will be working on allocation methods that combine the benefits of both these approaches.

4 Acknowledgements

This research was made possible by a generous gift from Intel Corporation.

References

1. Fujishima, Y., Leyton-Brown, K., Shoham, Y.: Taming the Computational Complexity of Combinatorial Auctions: Optimal and Approximate Approaches. International Joint Conference on Artificial Intelligence 99(IJCAI-99), 1999.
2. Hunsberger, L., Grosz, B.: A Combinatorial Auction for Collaborative Planning. Proceedings of the Fourth International Conference on Multi-Agent Systems (ICMAS-2000), 2000.
3. Kitano, H. et al: RoboCup-Rescue: Search and Rescue for Large Scale Disasters as a Domain for Multi-Agent Research. Proceedings of IEEE Conference, **SMC-99**, October 1999.
4. Nair, R., Ito, T., Tambe, M., Marsella, S.: RoboCup-Rescue: A Proposal and Preliminary Experiences. Proceedings of RoboCup-Rescue Workshop, Fourth International Conference on Multi-Agent Systems(ICMAS-2000), July 2000.
5. Sandholm, T.: An Algorithm for Optimal Winner Determination in Combinatorial Auctions. International Joint Conference on Artificial Intelligence (IJCAI), 1999, pp. 542-547.
6. Sandholm, T., Suri, S.: Improved Algorithms for Optimal Winner Determination in Combinatorial Auctions and Generalizations. National Conference on Artificial Intelligence (AAAI), 2000, pp. 90-97.

Team Description for RMIT-on-Fire: Robocup Rescue Simulation Team 2001

Lin Padgham, John Thangarajah, David Poutakidis, and Chandaka Fernando

School of Computer Science and Information Technology, RMIT Melbourne, Australia
{linpa,johthan,davpout,mfernand}@rmit.cs.edu.au
http://www.cs.rmit.edu.au

Abstract. This team description paper outlines the reasons that RMIT Compter Science considers RoboCupRescue a valuable domain for ongoing research, and describes the preliminary architecture which has been developed as an infrastructure for ongoing work in the area of building BDI (Belief Desire Intention) intelligent agents and multi-agent cooperating teams.

1 Introduction

RMIT Computer Science has been involved with various aspects of RoboCup each year since its inception in 1996. The main reasons for this involvement is because RoboCup provides an international forum for discussing, comparing and evaluating implemented approaches for achieving multi-agent co-operation in a complex and changing world, an area we have been actively working in [CP98].

Of course it also provides many more challenges than just this one which was originally, and remains our central area of interest. One must manage uncertain information, develop high quality low -level behaviours, do localisation, and many other tasks.

With the start of RoboCupRescue we decided that this domain was better suited to our long term research interests than the soccer domain and we made a decision to focus this year on starting to set up the infrastructure for an ongoing participation in RoboCupRescue. Prior to the competition we had been unable to run on more than a single pentium PC, which somewhat hampered the infrastructure development. We are now in the process of setting up more adequate resources, in the form of 4-5 networked high end PC's.

2 Research Directions

RoboCupRescue offers an extremely broad research opportunity spanning social systems research, research in physical systems, both rescue robots and a variety of information gathering devices, disaster management research and research in multi-agent software systems. Our particular interest is in the latter area, although we welcome the opportunity to interact with people from the other

A. Birk, S. Coradeschi, and S. Tadokoro (Eds.): RoboCup 2001, LNAI 2377, pp. 755–757, 2002.

areas, particularly in order to better focus our research on addressing problems that lie within our area of expertise and are of importance to real and important applications.

The research questions which we are currently most interested in within this project, are those surrounding the issue of obtaining timely and effective co-operative behaviour, with sufficient flexibility for autonomous individual behaviour towards the agreed system goals of maximising rescue and minimising damage.

The fact that agents are inherently heterogenous, with different abilities requires at least implicit co-operation if the system as a whole is to be effective. There are however many questions as to how this co-operation can best be achieved. It is important that methods chosen are robust so that if some part of the system is rendered inactive, the rest of the system continues to operate in a reasonable if not optimal manner. Exploration of effective and robust means of co-operation is one of our foremost goals.

We are also interested in the process of software development and are consequently working on ways to reliably specify and program complex agent behaviour, including team behaviour. To this end we are using and evaluating JACK^{TM} Intelligent Agents development environment [HRHL01], which uses the wellestablished Belief, Desire, Intention, (BDI) agent paradigm [RG95]. We plan to explore the JACK^{TM} software support for building and programming agent teams [HRB00], which uses a different paradigm for buidling cooperative software teams than much other work in the area (e.g. [SGK00, Tam97]).

Another area of research interest for us is that of co-operation of software agents in open systems. How can we write our software so that it can sensibly interact with and co-operate with software written by others in the same domain, but without any planning for co-operation. We see this as an area which is a potential future focus for RoboCupRescue as a natural way to explore this would be to have agents written by different groups, come together at the competition where we could evaluate how well each managed to co-operate with previously unknown agents.

3 Implemented System

Our implemented system this year was quite minimal as most effort was towards setting up the infrastructure for future work. Two undergraduate students worked in their spare time on developing the framework for our java code to interact with the simulator. We used and modified the sample code supplied by ?? for processing of visual data and for obtaining routes to move.

We built an architecture where significant events are extracted from the visual data by each agent and are then passed to the agent reasoning mechanism for decisions as to how to react. Reactions are in terms of predefined plans which contain multiple steps, some of which may be further plans, or plan sets, where the appropriate plan is chosen from the set at run time, depending on circumstances.

At each time step the agent does the next action as determined by the plan which it has decided is the most appropriate to follow. Plans can be aborted if the situation changes, and if a plan fails to achieve its intended goal, a new plan can be tried.

The agent reasoning mecahnism uses the Belief Desire Intention (BDI) style of agent reasoning [RG95] which is implemented using the JACK^{TM} system.

This year we implemented only the fire engine agents and had only a skeletal reasoning structure with a minimal set of very primtive plans. One of the advantages of the BDI architecture is that once a system is in place it is reasonably straightforward to add more specialised plans which are effective in particular situations. We are hopeful that in the future we will have a much richer set of plans (as well as more reliable primitive behaviours) that will asllow our agents to react flexibly and effectively in the simulated disaster situation.

References

[CP98] Simon Ch'ng and Lin Padgham. From roles to teamwork: a framework and architecture. *Applied Artificial Intelligence*, 12:211–231, March 1998.

[HRB00] Andrew Hodgson, Ralph Rönnquist, and Paolo Busetta. Specification of coordinated agent behaviour (the simple team approach). Technical Report 5, www.agent-software.com.au, 2000.

[HRHL01] Nick Howden, Ralph Rönnquist, Andrew Hodgson, and Andrew Lucas. Jack intelligent agents – summary of an agent infrastructure. In *5th International Conference on Autonomous Agents*, 2001.

[RG95] Anand S. Rao and Michael P. Georgeff. Bdi agents: From theory to practice. In *Proceedings of the First International Conference on Multiagent Systems*, 1995.

[SGK00] D.J. Sullivan, B.J. Grosz, and S. Kraus. *Intention Teconciliation by Collaborative Agents.* IEEE Computer Society Press, 2000.

[Tam97] Milind Tambe. Towards flexible teamwork. *Journal of Artificial Intelligence research*, 7:83–124, 1997.

Author Index

Lecture Notes in Artificial Intelligence (LNAI)

Vol. 2226: K.P. Jantke, A. Shinohara (Eds.), Discovery Science. Proceedings, 2001. XII, 494 pages. 2001.

Vol. 2246: R. Falcone, M. Singh, Y.-H. Tan (Eds.), Trust in Cyber-societies. VIII, 195 pages. 2001.

Vol. 2250: R. Nieuwenhuis, A. Voronkov (Eds.), Logic for Programming, Artificial Intelligence, and Reasoning. Proceedings, 2001. XV, 738 pages. 2001.

Vol. 2253: T. Terano, T. Nishida, A. Namatame, S. Tsumoto, Y. Ohsawa, T. Washio (Eds.), New Frontiers in Artificial Intelligence. Proceedings, 2001. XXVII, 553 pages. 2001.

Vol. 2256: M. Stumptner, D. Corbett, M. Brooks (Eds.), AI 2001: Advances in Artificial Intelligence. Proceedings, 2001. XII, 666 pages. 2001.

Vol. 2258: P. Brazdil, A. Jorge (Eds.), Progress in Artificial Intelligence. Proceedings, 2001. XII, 418 pages. 2001.

Vol. 2275: N.R. Pal, M. Sugeno (Eds.), Advances in Soft Computing – AFSS 2002. Proceedings, 2002. XVI, 536 pages. 2002.

Vol. 2281: S. Arikawa, A. Shinohara (Eds.), Progress in Discovery Science. XIV, 684 pages. 2002.

Vol. 2293: J. Renz, Qualitative Spatial Reasoning with Topological Information. XVI, 207 pages. 2002.

Vol. 2296: B. Dunin-Keplicz, E. Nawarecki (Eds.), From Theory to Practice in Multi-Agent Systems. Proceedings, 2001. IX, 341 pages. 2002.

Vol. 2298: I. Wachsmuth, T. Sowa (Eds.), Gesture and Language in Human-Computer Interaction. Proceedings, 2001. XI, 323 pages.

Vol. 2302: C. Schulte, Programming Constraint Services. XII, 176 pages. 2002.

Vol. 2307: C. Zhang, S. Zhang, Association Rule Mining. XII, 238 pages. 2002.

Vol. 2308: I.P. Vlahavas, C.D. Spyropoulos (Eds.), Methods and Applications of Artificial Intelligence. Proceedings, 2002. XIV, 514 pages. 2002.

Vol. 2309: A. Armando (Ed.), Frontiers of Combining Systems. Proceedings, 2002. VIII, 255 pages. 2002.

Vol. 2313: C.A. Coello Coello, A. de Albornoz, L.E. Sucar, O.Cairó Battistutti (Eds.), MICAI 2002: Advances in Artificial Intelligence. Proceedings, 2002. XIII, 548 pages. 2002.

Vol. 2317: M. Hegarty, B. Meyer, N. Hari Narayanan (Eds.), Diagrammatic Representation and Inference. Proceedings, 2002. XIV, 362 pages. 2002.

Vol. 2321: P.L. Lanzi, W. Stolzmann, S.W. Wilson (Eds.), Advances in Learning Classifier Systems. Proceedings, 2002. VIII, 231 pages. 2002.

Vol. 2322: V. Mařík, O. Stěpánková, H. Krautwurmová, M. Luck (Eds.), Multi-Agent Systems and Applications II. Proceedings, 2001. XII, 377 pages. 2002.

Vol. 2333: J.-J.Ch. Meyer, M. Tambe (Eds.), Intelligent Agents VIII. Revised Papers, 2001. XI, 461 pages. 2001.

Vol. 2336: M.-S. Chen, P.S. Yu, B. Liu (Eds.), Advances in Knowledge Discovery and Data Mining. Proceedings, 2002. XIII, 568 pages. 2002.

Vol. 2338: R. Cohen, B. Spencer (Eds.), Advances in Artificial Intelligence. Proceedings, 2002. XII, 373 pages. 2002.

Vol. 2358: T. Hendtlass, M. Ali (Eds.), Developments in Applied Artificial Intelligence. Proceedings, 2002 XIII, 833 pages. 2002.

Vol. 2366: M.-S. Hacid, Z.W. Raś, D.A. Zighed, Y. Kodratoff (Eds.), Foundations of Intelligent Systems. Proceedings, 2002. XII, 614 pages. 2002.

Vol. 2371: S. Koenig, R. Holte (Eds.), Abstraction, Reformulation, and Approximation. Proceedings, 2002. XI, 349 pages. 2002.

Vol. 2375: J. Kivinen, R.H. Sloan (Eds.), Computational Learning Theory. Proceedings, 2002. XI, 397 pages. 2002.

Vol. 2377: A. Birk, S. Coradeschi, T. Satoshi (Eds.), RoboCup 2001: Robot Soccer World Cup V. XIX, 763 pages. 2002.

Vol. 2381: U. Egly, C.G. Fermüller (Eds.), Automated Reasoning with Analytic Tableaux and Related Methods. Proceedings, 2002. X, 341 pages. 2002 .

Vol. 2385: J. Calmet, B. Benhamou, O. Caprotti, L. Henocque, V. Sorge (Eds.), Artificial Intelligence, Automated Reasoning, and Symbolic Computation. Proceedings, 2002. XI, 343 pages. 2002.

Vol. 2389: E. Ranchhod, N.J. Mamede (Eds.), Advances in Natural Language Processing. Proceedings, 2002. XII, 275 pages. 2002.

Vol. 2392: A. Voronkov (Ed.), Automated Deduction – CADE-18. Proceedings, 2002. XII, 534 pages. 2002.

Vol. 2393: U. Priss, D. Corbett, G. Angelova (Eds.), Conceptual Structures: Integration and Interfaces. Proceedings, 2002. XI, 397 pages. 2002.

Vol. 2403: Mark d'Inverno, M. Luck, M. Fisher, C. Preist (Eds.), Foundations and Applications of Multi-Agent Systems. Proceedings, 1996-2000. X, 261 pages. 2002.

Vol. 2407: A.C. Kakas, F. Sadri (Eds.), Computational Logic: Logic Programming and Beyond. Part I. XII, 678 pages. 2002.

Vol. 2408: A.C. Kakas, F. Sadri (Eds.), Computational Logic: Logic Programming and Beyond. Part II. XII, 628 pages. 2002.

Lecture Notes in Computer Science

Vol. 2363: S.A. Cerri, G. Gouardères, F. Paraguaçu (Eds.), Intelligent Tutoring Systems. Proceedings, 2002. XXVIII, 1016 pages. 2002.

Vol. 2364: F. Roli, J. Kittler (Eds.), Multiple Classifier Systems. Proceedings, 2002. XI, 337 pages. 2002.

Vol. 2366: M.-S. Hacid, Z.W. Raś, D.A. Zighed, Y. Kodratoff (Eds.), Foundations of Intelligent Systems. Proceedings, 2002. XII, 614 pages. 2002. (Subseries LNAI).

Vol. 2367: J. Fagerholm, J. Haataja, J. Järvinen, M. Lyly. P. Råback, V. Savolainen (Eds.), Applied Parallel Computing. Proceedings, 2002. XIV, 612 pages. 2002.

Vol. 2368: M. Penttonen, E. Meineche Schmidt (Eds.), Algorithm Theory – SWAT 2002. Proceedings, 2002. XIV, 450 pages. 2002.

Vol. 2369: C. Fieker, D.R. Kohel (Eds.), Algebraic Number Theory. Proceedings, 2002. IX, 517 pages. 2002.

Vol. 2370: J. Bishop (Ed.), Component Deployment. Proceedings, 2002. XII, 269 pages. 2002.

Vol. 2371: S. Koenig, R. Holte (Eds.), Abstraction, Reformulation, and Approximation. Proceedings, 2002. XI, 349 pages. 2002. (Subseries LNAI).

Vol. 2372: A. Pettorossi (Ed.), Logic Based Program Synthesis and Transformation. Proceedings, 2001. VIII, 267 pages. 2002.

Vol. 2373: A. Apostolico, M. Takeda (Eds.), Combinatorial Pattern Matching. Proceedings, 2002. VIII, 289 pages. 2002.

Vol. 2374: B. Magnusson (Ed.), ECOOP 2002 – Object-Oriented Programming. XI, 637 pages. 2002.

Vol. 2375: J. Kivinen, R.H. Sloan (Eds.), Computational Learning Theory. Proceedings, 2002. XI, 397 pages. 2002. (Subseries LNAI).

Vol. 2377: A. Birk, S. Coradeschi, T. Satoshi (Eds.), RoboCup 2001: Robot Soccer World Cup V. XIX, 763 pages. 2002. (Subseries LNAI).

Vol. 2378: S. Tison (Ed.), Rewriting Techniques and Applications. Proceedings, 2002. XI, 387 pages. 2002.

Vol. 2379: G.J. Chastek (Ed.), Software Product Lines. Proceedings, 2002. X, 399 pages. 2002.

Vol. 2380: P. Widmayer, F. Triguero, R. Morales, M. Hennessy, S. Eidenbenz, R. Conejo (Eds.), Automata, Languages and Programming. Proceedings, 2002. XXI, 1069 pages. 2002.

Vol. 2381: U. Egly, C.G. Fermüller (Eds.), Automated Reasoning with Analytic Tableaux and Related Methods. Proceedings, 2002. X, 341 pages. 2002 .(Subseries LNAI).

Vol. 2382: A. Halevy, A. Gal (Eds.), Next Generation Information Technologies and Systems. Proceedings, 2002. VIII, 169 pages. 2002.

Vol. 2383: M.S. Lew, N. Sebe, J.P. Eakins (Eds.), Image and Video Retrieval. Proceedings, 2002. XII, 388 pages. 2002.

Vol. 2384: L. Batten, J. Seberry (Eds.), Information Security and Privacy. Proceedings, 2002. XII, 514 pages. 2002.

Vol. 2385: J. Calmet, B. Benhamou, O. Caprotti, L. Henocque, V. Sorge (Eds.), Artificial Intelligence, Automated Reasoning, and Symbolic Computation. Proceedings, 2002. XI, 343 pages. 2002. (Subseries LNAI).

Vol. 2386: E.A. Boiten, B. Möller (Eds.), Mathematics of Program Construction. Proceedings, 2002. X, 263 pages. 2002.

Vol. 2389: E. Ranchhod, N.J. Mamede (Eds.), Advances in Natural Language Processing. Proceedings, 2002. XII, 275 pages. 2002. (Subseries LNAI).

Vol. 2391: L.-H. Eriksson, P.A. Lindsay (Eds.), FME 2002: Formal Methods – Getting IT Right. Proceedings, 2002. XI, 625 pages. 2002.

Vol. 2392: A. Voronkov (Ed.), Automated Deduction – CADE-18. Proceedings, 2002. XII, 534 pages. 2002. (Subseries LNAI).

Vol. 2393: U. Priss, D. Corbett, G. Angelova (Eds.), Conceptual Structures: Integration and Interfaces. Proceedings, 2002. XI, 397 pages. 2002. (Subseries LNAI).

Vol. 2398: K. Miesenberger, J. Klaus, W. Zagler (Eds.), Computers Helping People with Special Needs. Proceedings, 2002. XXII, 794 pages. 2002.

Vol. 2399: H. Hermanns, R. Segala (Eds.), Process Algebra and Probabilistic Methods. Proceedings, 2002. X, 215 pages. 2002.

Vol. 2401: P.J. Stuckey (Ed.), Logic Programming. Proceedings, 2002. XI, 486 pages. 2002.

Vol. 2403: Mark d'Inverno, M. Luck, M. Fisher, C. Preist (Eds.), Foundations and Applications of Multi-Agent Systems. Proceedings, 1996-2000. X, 261 pages. 2002. (Subseries LNAI).

Vol. 2405: B. Eaglestone, S. North, A. Poulovassilis (Eds.), Advances in Databases. Proceedings, 2002. XII, 199 pages. 2002.

Vol. 2407: A.C. Kakas, F. Sadri (Eds.), Computational Logic: Logic Programming and Beyond. Part I. XII, 678 pages. 2002. (Subseries LNAI).

Vol. 2408: A.C. Kakas, F. Sadri (Eds.), Computational Logic: Logic Programming and Beyond. Part II. XII, 628 pages. 2002. (Subseries LNAI).

Vol. 2409: D. Mount, C. Stein (Eds.), Algorithm Engineering and Experiments. Proceedings, 2002. VIII, 207 pages. 2002.